Occupational Safety and Health Law

Occupational
Safety and Health
Law

Editors in Chief

Stephen A. Bokat   Horace A. Thompson III

Associate Editors

George Cohen   Willis Goldsmith
James English   Robert C. Gombar

Section of Labor and Employment Law
American Bar Association

The Bureau of National Affairs Inc., Washington, D.C.

Copyright © 1988
The Bureau of National Affairs Inc.

Library of Congress Cataloging-in-Publication Data

Occupational safety and health law / editors in chief, Stephen A.
  Bokat, Horace A. Thompson III ; associate editors, George Cohen
  . . . [et al.] ; Section of Labor and Employment Law, American Bar
  Association.
    p. cm.
  Includes index.
  ISBN 0-87179-527-2
    1. Industrial safety—Law and legislation—United
States.   2. Industrial hygiene—Law and legislation—United
States.   3. United States. Occupational Safety and Health
Administration.   I. Bokat, Stephen A.   II. Thompson, Horace
A.   III. American Bar Association. Section of Labor and
Employment Law.
KF3570.026   1988
344.73'0465—dc19   88-7252
[347.304465]   CIP

International Standard Book Number: 0-87179-527-2
Printed in the United States of America

Preface

Since 1945 when it began, the ABA Section of Labor and Employment Law has as its stated purposes (1) to study and report upon continuing developments in the law affecting labor relations, (2) to assist the professional growth and development of practitioners in the field of labor relations law, and (3) to promote justice, human welfare, industrial peace, and the recognition of the supremacy of law in labor-management relations.

Through the publication of the Second Editions of *The Developing Labor Law* and *Employment Discrimination Law*, and through annual and committee meeting programs designed to provide a forum for the exchange of ideas, the Section has pursued these stated goals. Slowly, the Section has built a library of comprehensive legal works intended for the use of the Section membership as well as the bar generally.

The Section of Labor and Employment Law is pleased to add this treatise on *Occupational Safety and Health Law* to its library of books published by BNA Books, a Division of The Bureau of National Affairs, Inc. Through the Occupational Safety and Health Law Committee of the Section, a distinguished panel of Editors was selected, and the combined efforts of many individual authors are reflected herein. The Editors have tried to maintain two primary objectives: (1) to be equally balanced and nonpartisan in their viewpoints, and (2) for the book to be of significant value to the practitioner and student alike, and the sophisticated nonlawyer as well. The Section wishes to express its appreciation to the Editors in Chief, Associate Editors, and the individual contributors for their outstanding work.

The views expressed herein do not represent the views of the American Bar Association, or its Section of Labor and Employment Law, or any other organization, but are simply the collective, but not necessarily the individual, views of the authors.

BERNARD T. KING
Chairman

WILLIAM L. KELLER
Chairman-Elect

Section of Labor
and Employment Law
American Bar Association

March 1988

Foreword

"To assure so far as possible every working man and woman in the Nation safe and healthful working conditions." On its face, this language in the first section of the Occupational Safety and Health Act is a simple and seemingly noncontroversial statement of the fundamental objective of the statute. Yet in the nearly two decades since its passage, very little, if anything, about the Act has been either simple or noncontroversial.

To be sure, a great deal of the controversy surrounding the Act and its enforcement has resulted from developments the framers of the legislation could not reasonably have foreseen. For example, while the legislators understood that much about occupational safety and health was yet unknown, they could not have anticipated the staggering pace—or volume—of learning in the field. Similarly, Congress knew that the Occupational Safety and Health Act, like most remedial social legislation, was likely to generate certain economic costs. But they had no reason to assume that the costs generated by the statute would be assessed in the context of a global economy—an economy in which those competing with American industry often operate in an environment where employee safety and health is not a major concern. The drafters of the Act undoubtedly knew that advocates on any and all sides of workplace safety and health issues would be creative and determined, but they could not have imagined the almost astonishing degree of creativity and determination that parties to proceedings under the Act have mustered. In short, in passing the Occupational Safety and Health Act, Congress may not have fully grasped the enormity, the complexity, or even the sen-

sitivity of the issues it addressed. This treatise was conceived
to identify and analyze those issues.

This volume explains the development of the law to date
under the Occupational Safety and Health Act. Predicting
future developments in any area of the law is almost always
an exercise in futility. Nevertheless, a reasonable prediction
can be made concerning the course of the law under the Act.
That is, the ever-expanding body of scientific knowledge about
occupational safety and health matters, the number of complex legal issues still open, and perhaps most importantly, the
increasing interest of our elected representatives in the area
virtually guarantee that the next two decades under the Act
will be even more challenging, and perhaps even more critical,
than the first two.

STEPHEN A. BOKAT
HORACE A. THOMPSON III

March 1988

Acknowledgments

This treatise was prepared by the Labor and Employment Law Section of the American Bar Association, through its Committee on Occupational Safety and Health Law, working closely with The Bureau of National Affairs, Inc. The project was the particular responsibility of the Editors.

Editors in Chief:

Stephen A. Bokat has served since 1983 as the Vice-President and General Counsel of the U.S. Chamber of Commerce and as Vice-President of the National Chamber Litigation Center. Prior to joining the Chamber of Commerce, Mr. Bokat was an attorney on the staff of a National Labor Relations Board member and then on the staffs of Occupational Safety and Health Review Commissioner James Van Namee and Chairman Frank Barnako. He also served as an appellate attorney, Office of Solicitor (OSHA) of the U.S. Department of Labor. He received his J.D. degree from George Washington University in 1972. Mr. Bokat served as the Management Co-Chair of the American Bar Association from 1983 to 1985.

Horace A. Thompson III is a founding partner in the New Orleans law firm of McCalla, Thompson, Pyburn & Ridley, specializing in representation of management in labor relations and employment law. He received his J.D. degree from Tulane University in 1968. He has served as Chairman of the Labor Relations Law Section of the Louisiana State Bar Association and of the Labor Law Committee of the Louisiana Association of Business and Industry. Mr. Thompson also served as the Management Co-Chairman of the OSHA Committee of the American Bar Association from 1981 to 1983.

Associate Editors:

George H. Cohen is a partner in Bredhoff & Kaiser in
Washington, D.C. He received his law degree from Cornell
University in 1960. Mr. Cohen is a member of the Advisory
Council for Environmental Health for the Johns Hopkins Uni-
versity School of Hygiene and Public Health. He also serves
as Counsel to the Industrial Relations Research Association.
Mr. Cohen served as Chairman of the Labor Law Committee
of the Federal Bar Association from 1971 to 1973. He also
served as Union Co-Chairman of the OSHA Committee of the
American Bar Association from 1976 to 1978.

James D. English, currently Associate General Counsel
of the United Steelworkers of America, has been associated
with the Union for nearly 20 years, handling matters in all
phases of labor-management relations. He received his law
degree from the Georgetown University Law Center in 1966
and his Masters in Labor Law from George Washington University in 1967. Mr. English served as the Union Co-Chairman
of the OSHA Committee of the American Bar Association from
1981 to 1983.

Willis J. Goldsmith is a partner in Jones, Day, Reavis &
Pogue, resident in the firm's office in Washington, D.C., where
he represents management in all areas of labor and employment law. He received his law degree from the New York
University School of Law in 1972. From 1972–1974, Mr. Goldsmith served as an attorney in the Office of the Solicitor of
the U.S. Department of Labor. He currently serves as an Adjunct Professor of Law at the Georgetown University Law
Center and as a Contributing Editor of the Employee Relations Law Journal.

Robert C. Gombar is a partner in Jones, Day, Reavis &
Pogue in Washington, D.C., where he represents and counsels
management in all aspects of occupational safety and health
law. He is a 1974 graduate of the Fordham University School
of Law. Mr. Gombar served as an attorney, General Counsel
and Chief Legal Officer of the U.S. Occupational Safety and
Health Review Commission. Mr. Gombar serves as an Adjunct
Professor of Law at the Georgetown University Law Center.
He also has served, since 1986, as the Management Co-Chairman of the OSHA Committee of the American Bar Association.

* * * * *

The effort began with a detailed outline of the treatise prepared by the Editorial Committee with the assistance of Russell J. Thomas, Jr., a partner in the law firm of Epstein, Becker & Green. Mr. Thomas' task was an essential one, performed extremely well.

To assist the Editorial Committee in bringing many contributions together into a single treatise, we acknowledge an extraordinary debt to Mary Green Miner, Publisher, BNA Books. Mary worked closely with the Editorial Committee throughout the evolution of this treatise and played a vital role in its development.

We also express our gratitude to Louise R. Goines, Senior Legal Editor, BNA Books, for her supervision, editorial skill, and advice. Her efforts are truly appreciated.

We also acknowledge the great contribution of Andrea Posner, who copyedited the manuscript. Since readers see only the final product, they may not fully appreciate where she started from and how much she improved the manuscript.

This treatise brings together the insights and experiences of contributors from all areas of the country who are outstanding occupational safety and health law specialists: attorneys responsible for writing, enforcing, and adjudicating the law; counsel to unions, management, and the government; and representatives from the academic community. In addition to the Editors, the following persons prepared chapters or portions of chapters:

Frank Barnako is Counsel to Reed, Smith, Shaw & McClay in Washington, D.C. He served as Chairman of the U.S. Occupational Safety and Health Review Commission and as the Chairman of the National Safety Council. Mr. Barnako is a 1936 graduate of the University of Michigan Law School.

John Billick is a partner in Buckingham, Doolittle & Burroughs in Akron, Ohio. He received his law degree from Georgetown University in 1973.

Frederick Braid is a partner in Rains & Pogrebin in Mineola, New York. He received his J.D. degree from St. John's University in 1971 and his LL.M. in Labor Law from New York University in 1979.

Charles M. Chadd is a partner of Pope, Ballard, Shepard & Fowle in Chicago. He received his law degree from Northwestern University Law School in 1967.

Jeremiah Collins is a partner in the law firm of Bredhoff & Kaiser in Washington, D.C. Mr. Collins also serves as an Adjunct Professor of Law at Georgetown University Law Center. He is a 1976 graduate of Stanford University Law School. Since 1985, Mr. Collins has served as Union Co-Chair of the OSHA Committee of the American Bar Association.

Donald R. Crowell, II, formerly a partner in the Washington, D.C., office of Pepper, Hamilton & Scheetz, now serves as Labor Counsel for Union Carbide Corporation in Danbury, Connecticut. He received his law degree from the Fordham University School of Law in 1973.

Raymond Donnelly is a Program Analyst Officer with the Occupational Safety and Health Administration's Office of Program Evaluation. He has a Bachelor of Arts degree in Political Science from Whittier College and a Master of Science degree in Environmental Health from the George Washington University. He also serves as an instructor at the University of Maryland.

Scott Dunham is a partner in the law firm of O'Melveny & Meyers in Los Angeles. He received his law degree from the University of Washington in 1975.

Harold Engel is the Deputy Director, Torts Branch, Civil Division of the Department of Justice. He formerly served as Counsel for Regional Litigation (OSHA) in the Solicitor's Office of the U.S. Department of Labor. Mr. Engel graduated from Howard University Law School in 1969.

Anthony E. Goldin was the Director of Policy of the Occupational Safety and Health Administration from 1981 until 1985. Mr. Goldin currently is with the Office of Management and Budget. He obtained his law degree from the Georgetown University Law Center in 1973.

Richard L. Gross is an associate with Sklarz and Early in New Haven, Connecticut. Mr. Gross formerly served as an attorney with the Office of the Solicitor, U.S. Department of Labor, and with the National Institute for Occupational Safety and Health. He obtained his law degree from Hofstra University in 1975.

Charles Hadden is an associate with Ross, Dixon & Masback in Washington, D.C. Mr. Hadden formerly served as an attorney with the Office of the Solicitor, U.S. Department of Labor. He is a 1975 graduate of the University of Michigan Law School.

John Hynan serves as Counsel for General Legal Advice in the Solicitor's Office of the U.S. Department of Labor. He received his law degree from the Georgetown University Law Center in 1958.

Thomas Holzman currently serves as Associate Counsel for Appellate Litigation, Employee Benefits Division, U.S. Department of Labor. He formerly served as a staff attorney in the Occupational Safety and Health Division of the Office of the Solicitor, U.S. Department of Labor. He is a 1977 graduate of the University of Pennsylvania Law School.

Bruce Justh is Counsel to the Occupational Safety and Health Review Commission. He is a 1972 graduate of the Northeastern University Law School.

Mark Kadzielski is a partner in Weissburg & Aronson, Inc., in Los Angeles. He graduated from the University of Pennsylvania Law School in 1976.

Peter Kilgore is a partner of Kirlin, Campbell & Keating in Washington, D.C. Mr. Kilgore previously served as Assistant Counsel, U.S. Occupational Safety and Health Review Commission. He received his J.D. degree from Valparaiso University in 1973 and his LL.M. degree from the Georgetown University Law Center in 1976.

Kenneth Kleinman is an associate of Kleinbard, Bell & Brecker in Philadelphia. He is a 1979 graduate of Harvard Law School.

Albert E. Lawson is presently an attorney with the International Brotherhood of Teamsters in Washington, D.C. He formerly served as a Commissioner of the Mine Safety and Health Review Commission. Mr. Lawson graduated from Harvard Law School in 1957.

Steven McCown is a partner in Jenkins & Gilchrist in Dallas, Texas. He received his J.D. degree from Southern Methodist University in 1975.

Benjamin Mintz is an Associate Professor of Law at the Columbus School of Law of The Catholic University of America in Washington, D.C. He previously served as Associate Solicitor for Occupational Safety and Health at the U.S. Department of Labor. Mr. Mintz received his law degree from Columbia University in 1952. We are particularly indebted to Professor Mintz for his substantial contributions to this treatise, particularly chapters 19 and 20. Mr. Mintz acknowledges the assistance of George Henschel of the Office of Solicitor, Department of Labor, Martha B. Kent of Meridian Research, Inc., and David C. Vladek of Public Citizen Litigation Group.

Dennis Morikawa is a partner in Morgan, Lewis & Bockius in Philadelphia. He is a 1974 graduate of Syracuse University College of Law.

Charles Newcom is a partner in Sherman and Howard in Denver, Colorado. He received his law degree in 1974 from Harvard Law School.

Mary Win O'Brien is the Assistant General Counsel of the United Steelworkers of America in Pittsburgh, Pennsylvania. She received her J.D. degree from the Georgetown University Law Center in 1972 and has a Masters in Public Health degree from the Harvard School of Public Health. From 1983 to 1985, Ms. O'Brien served as Union Co-Chair of the OSHA Committee of the American Bar Association.

Earl Ohman is the General Counsel of the Occupational Safety and Health Review Commission. He received his law degree from George Washington University in 1974.

Peter M. Panken is a partner in Parker, Chapin, Flattau & Klimpl in New York City. He is a 1962 graduate of Harvard Law School.

Andrew Peterson is a partner with Paul, Hastings, Janofsky & Walker in Los Angeles. He graduated from the University of Chicago Law School in 1971.

Lynn Pollan serves as General Counsel of Williamhouse-Regency, Inc., in New York City. She previously served as Assistant Counsel for Bunge Corporation. She received her law degree from the New York University School of Law in 1972.

W. Scott Railton is a partner of Reed, Smith, Shaw & McClay in Washington, D.C. Mr. Railton previously served as Senior Trial Attorney, Fair Labor Standards, and Assistant Counsel for Regional Litigation (OSHA) in the Solicitor's Office of the U.S. Department of Labor. He also served as Chief Counsel to Commissioner James Van Namee and Chairman Frank Barnako of the U.S. Occupational Safety and Health Review Commission. He is a 1965 graduate of George Washington University Law School.

Jonathan Ross is Senior Counsel to Chairman Stephens of the National Labor Relations Board in Washington, D.C. Mr. Ross previously served as Assistant General Counsel of the U.S. Occupational Safety and Health Review Commission. He received his law degree from the Georgetown University Law Center in 1976.

Arthur Sapper is a senior attorney with Jones, Day, Reavis & Pogue in Washington, D.C. Mr. Sapper previously served as the Deputy General Counsel of the U.S. Occupational Safety and Health Review Commission in Washington, D.C. He graduated from the Georgetown University Law Center in 1975.

Nina Stillman is a partner of Vedder, Price, Kaufman & Kammholz in Chicago. She is a 1973 graduate of Northwestern University Law School.

The Honorable Paul Tenney is Chief Judge of the Occupational Safety and Health Review Commission. Judge Tenney previously served as Chief Counsel to Chairman Timothy F. Cleary of the Occupational Safety and Health Review Commission. He received his LL.B. degree from Boston University in 1953 and his LL.M. degree from Georgetown University in 1959.

Patrick Tyson is a partner in Constangy, Brooks & Smith in Washington, D.C. Mr. Tyson previously served as Staff Attorney in the Office of the Solicitor of the U.S. Department of Labor; Director of Compliance, Occupational Safety and Health, U.S. Department of Labor; Deputy Assistant Secretary of Labor for Occupational Safety and Health; and as Acting Assistant Secretary of Labor for Occupational Safety and Health. He is a 1973 graduate of the University of Toledo Law School.

Richard Voigt is a partner with Siegel, O'Connor, Schiff, Zangari & Kainen in Hartford, Connecticut. Mr. Voigt was formerly Assistant Counsel for Regional Litigation (OSHA), Office of the Solicitor, U.S. Department of Labor. He received his law degree from the University of Virginia in 1974.

We also extend appreciation to those individuals who aided in research or otherwise assisted the named contributors, including Susan Brooks-Maher, Christopher P. Charlton, Joseph J. Costello, Cary A. DesRoches, Michael McDowell, Rory J. McEvoy, Maureen Moore, Carla Morrison, Sarah W. Payne, Christopher J. Perry, Donald Savelson, and Jeffrey Shobert. To those we may have excluded from this list, we are never-

theless grateful for your guidance and apologize for our inadvertent omission of your names.

    A. Renee Harris is Stephen Bokat's secretary in the Gen-
eral Counsel's Office of the United States Chamber of Commerce in Washington, D.C. Karen Lusco is Horace Thompson's
secretary at McCalla, Thompson, Pyburn & Ridley in New
Orleans. The Editors in Chief are extremely grateful to these
two individuals who demonstrated extraordinary skill—and
patience—in coordinating the preparation of the many drafts
of the treatise.

STEPHEN A. BOKAT
HORACE H. THOMPSON III

March 1988

Summary Contents

III. Enforcement of the Occupational Safety and Health Act

IV. Enforcement Proceedings Under the Occupational Safety and Health Act

Detailed Contents

II. Duties Imposed Upon Employers

III. Enforcement of the Occupational Safety and Health Act

IV. Enforcement Proceedings Under the Occupational Safety and Health Act

V. Development and Promulgation of Standards

VI. Other Issues Arising Under and Related to the Occupational Safety and Health Act

Part I

Overview and Historical Perspective

1

Safety and Health Law Prior to the Occupational Safety and Health Act of 1970

Because the law of master and servant as it existed in the eighteenth and nineteenth centuries seems irrelevant in today's industrial world, it is seldom discussed except as an occurrence in history. But to appreciate and understand our modern day laws, it is well to examine their roots in this earlier law.

In the late 1700s and early 1800s, the Industrial Revolution transformed a basically rural and agricultural society into an urban and industrial society. Manufacturing prior to that period had been done by workers at home and in rural areas. Their products had been sold either by themselves locally or by merchants who supplied the raw materials and distributed the products in markets which they developed. In that system employers and employees were in a close relationship and had a sense of responsibility to each.

With the onset of the Industrial Revolution, changes in manufacturing, the development of railroads, and the growth of large enterprises meant that employers could no longer deal with the workers personally, resulting in a lessening of their sense of responsibility for the workers, and, in turn, the workers' sense of responsibility for each other. Because of an abundance of available labor and competition among manufacturers, wages were held to a minimum. Workers were poor; and owing

3

to the migration of job seekers from rural to urban areas, housing was in short supply. As a result, overcrowding created unsanitary conditions.

The new manufacturing methods created hazards to health, either not understood or ignored, which resulted in frightful and vile working conditions. Great numbers of women and small children made up a large part of the work force, particularly in the textile mills. Employers paid little or no heed to health and safety. Dickens and other writers of his day did not exaggerate when they described the living and working conditions of the time.

Thus began the ongoing conflict between employers and employees over wages and working conditions.

The following review of the law of master and servant is designed to prepare the way toward understanding the development and impact of safety and health legislation in the United States.

I. Common Law Liability of the Master

A. Origin of the Master's Liability

Under the law of England's Henry I (1068–1135), "A master was held liable for the 'wergeld' of the servant if the latter lost his life in the service, and for payment of the appropriate amount (bot) if he was injured, insofar as the injury could not be imputed to some third person for whom the master (who had to answer for the misdeeds of his own people) was not responsible."[1] It was apparently this law which established, as early as the year 1200, the master's absolute liability for his own negligence as well as for the acts of his servants, with or without knowledge. The basis of this absolute liability was the fiction that though the master had not personally acted, he had either commanded or consented to the act of his servant.[2]

By the end of the 1200s the master could escape liability by pleading he had not commanded or consented to the act.

---

[1] Wigmore, *Responsibility for Tortious Acts: Its History*, 7 HARV. L. REV. 315, 335 (1894); for a discussion of wergeld and bot see *The New Encyclopedia Brittanica*, 15th ed., Vol. 12, p. 582.

[2] *Id.* at 317.

Although this theory relieved the master of a criminal charge, the idea of the master's civil liability continued to evolve.[3] In support of this theory, Larson states:

> "Beginning about 1700 the principle of vicarious liability of the master for torts of the servant was developed * * *. Since the statements of the respondeat superior rule at this time were in sweeping unqualified terms (the act of the servant is the act of his master), presumably a master would be liable to a servant injured by the negligence of a fellow servant."[4]

B. Limitations of the Master's Liability: England

Toward the end of the eighteenth century, when the Industrial Revolution was transforming England into a commercially dominant nation of teeming cities and mechanized factories, this economic upheaval had its impact on the courts, where judges concerned with protecting expanding industries developed restrictions on the servant's right of recovery for work injuries.

Technically, masters were still under an imposing set of common law duties that included the obligation to provide employees with a safe working environment; to employ servants of sufficient care and skill to make it probable that injury would not be caused to others; to provide sound and safe materials; and to avoid exposing servants to extraordinary risks they could not anticipate.[5] (But the master did not warrant the competency of a servant to the other servants, and was required to use only ordinary care in selecting tools and materials.)

Despite these duties, however, after the onset of the Industrial Revolution an employee could seldom recover from his employer for bodily injury or for diseases caused by workplace exposures.[6] Typically, claims based on an employer's

---

[3]*Id.*

[4]1 A. Larson, WORKMEN'S COMPENSATION LAW, §4.30, at 25 (1952).

[5]J. Sheerman & A. Redfield, A TREATISE ON THE LAW OF NEGLIGENCE 110ff. (2d ed. 1870).

[6]Although the common law cases most often involved on-the-job injuries, the English courts also addressed the problem of occupational disease. For example, in *Smith v. Baker*, 1891 A.C. 325, Lord Herschell observed that "one who has agreed to take part in an operation necessitating the production of fumes injurious to health would have no cause of action in respect of bodily suffering or inconvenience resulting therefrom . . . ."

negligence were now successfully rebutted on the grounds that
the servant had been injured by a fellow servant, had assumed
the risk of working under dangerous conditions, or had been
contributorily negligent (see *infra* Section I.B.1, Master's Defenses). In addition, as one scholar has pointed out,[7] an employee could have a good legal case and yet fail to carry the
burden of persuasion at trial. Apart from the defenses mentioned above, an additional handicap was the understandable
reluctance of fellow servants to testify against their mutual
employer.

1. Master's Defenses

The severe limitations that burdened the servant in re-
covering from his master evolved from *Priestly v. Fowler*,[8]
which first articulated a rationale for limiting the liability of
the master.

(a) Fellow servant rule and theory of assumption of risk.
In *Priestley* a servant, injured when another servant of the
same master overloaded a carriage which overturned, sued
the master. A distinguished authority on English law states:
"It is certain that no case in which an action was brought
against an employer for injury caused by one of his servants
to another was known to have occurred until [this case]."[9]
Given the principle, well-established in England as noted above,
that the master was responsible for the acts of his servant,
the court might have been expected to rule in favor of the
plaintiff.[10] But the court instead carved out a sizable exception
to the rule of respondeat superior, stating:

> "[T]o allow this sort of action to prevail would be an encour-
> agement to the servant to omit that diligence and caution which
> he is bound to exercise on behalf of his master to protect him
> against the negligence and misconduct of others who serve him
> and which that diligence and caution while they protect the
> master are a much better security against any injury the servant
> may sustain by the negligence of others engaged under the same
> master than any recourse the master could possibly offer."[11]

---

[7]W. Dodd, ADMINISTRATION OF WORKMEN'S COMPENSATION 8–9 (1936).
[8]150 Eng. Rep. 1030, 1032 (1838).
[9]7 W. Holdsworthy, A HISTORY OF ENGLISH LAW 480 (1925).
[10]*Id.* n.4.
[11]*Priestley, supra* note 8, at 1033.

The judge was even more concerned with the possible consequence of allowing such a suit to be brought, suggesting that a master's liability for the negligence of his "inferior agents" could be extended to a suit by his coachman for a defect in the harness caused by the carriage maker; for want of skill; for drunkenness of his coachman in injuring another servant; and even for negligence of the cook in serving employees tainted food.[12] Further, the employee could not fairly complain since he was free "to decline any service in which he reasonably apprehends injury to himself."[13]

Some years later Pollock, C.B., in a case involving the same issue, affirmed the rule, stating that "[f]ew rules of law are of greater practical importance * * *. We ought not to allow so important a decision to be frittered away by minute distinctions or the ingenuity of advocates."[14]

From *Priestley*, then, developed both the fellow servant rule—the principle that employees could not recover against masters for the negligence of fellow servants—and the parallel theory of assumption of risk—that an employee who works under dangerous conditions and subsequently suffers an injury cannot expect compensation if he entered into the job with full knowledge of those risks. Underpinning these theories in *Priestley* was the principle that the master and servant were free to contract for the services which they had mutually understood they owed and desired. Later English cases echoed this theme. Thus, in *Hutchison v. York, Newcastle & Berwick Railway*,[15] a case which solidified the fellow servant rule in England, plaintiff's claim against the railroad for negligence of other employees was rejected, Alderson, B., stating that the deceased

> "knew when he was engaged in the service that he was exposed to the risk of injury, not only from his own want of skill or care, but on the part of his fellow servant also, and he must be supposed to have contracted on the terms that as between himself and master he would run that risk."[16]

Needless to say, these rationales reflected assumptions about the bargaining relationship between workmen and their

---

[12]*Id.* at 1032.
[13]*Id.* at 1033.
[14]*Vose v. Lancashire & Y. Ry.*, 157 Eng. Rep. 300, 303 (1858).
[15]155 Eng. Rep. 150 (1850).
[16]*Id.* at 154.

employers that were grossly unrealistic and particularly at odds with the facts of the new industrial age. No doubt the desire to sweep the path of industrial progress clear of as many obstacles as possible helped distort judicial perceptions.

(b) Contributory negligence. Even if a worker entered into employment ignorant of some unusual risk that the job entailed, he still would be unable to win a judgment if he were found guilty of contributory negligence. The doctrine of contributory negligence found its way into English law in a case that did not involve a master-servant dispute, *Butterfield v. Forrester.*[17] In *Butterfield,* the court held that the plaintiff, who ran his horse into a barrier in a roadway, could not recover for an injury, stating that his misfortune could be attributed as much to his own recklessness in not seeing the barrier as to the negligence of the defendant in placing it. The concept that an individual should not profit from his own wrong is the centerpiece of the law of contributory negligence. In the master-servant context, the common law courts warmly embraced this point of view.[18] The doctrine persisted in England for a surprisingly long period of time, and finally receded only with the passage of the Law Reform (Contributory Negligence) Act of 1945.[19]

(c) Common employment. In order to establish the defense based on the fellow servant rule, the master need only prove that the injured and the negligent servants were engaged in "common employment." This prerequisite was consistent with the fiction that an employee, in contracting with an employer, assumed the risk of negligent acts by his fellow servant. The English courts took an extremely broad view of what constituted common employment, holding to the position that simply because two employees performed different tasks of the same master did not mean they were not engaged in common employment. In *Morgan v. Vale of North Railway,*[20] the court ruled that a carpenter and porters who worked for a railroad

---

[17]103 Eng. Rep. 926 (1809).

[18]*Hannah Senior v. Ward,* 120 Eng. Rep. 954 (1859).

[19]8 & 9 Geo. 6, ch. 28. The Act introduced to English law the concept of comparative negligence. While under the previous law a negligent plaintiff was completely barred from recovery, the new system reduced damages only in proportion to the injured party's degree of fault.

[20][1865] 1 L.R.-Q.B. 149.

company were in common employment. Fearing a "flood of litigation" if "the employees in every large establishment" are split up "into different departments of service," Pollock, C.B., said in this case: "[W]e must not over refine, but look at the common objective and not at the common immediate objective."[21]

(d) Agent versus fellow workman. Just as the English courts did not distinguish between employees who performed different duties, they were unwilling to make distinctions based on rank. In America, there evolved the "vice-principal rule," *viz.*, that a superior officer was the alter ego of the master for whose acts the master was liable when such act caused injury to another in the same employ.[22] In England, however, this doctrine was soundly rejected in *Wilson v. Merry.*[23] Lord Cranworth concluded summarily that "[w]orkmen do not cease to be fellow workmen because they are not all equal in point of station or authority."[24]

2. *Plaintiff's Burden*

Because of the master's defenses, a plaintiff could expect to recover, assuming he could prove a breach of duty by the master, only by showing (a) complete ignorance of some extraordinary risk involved in his work, and (b) the absence of contributory negligence. Thus, the duties technically imposed on masters by the English courts,[25] significant as they may appear taken in isolation, had little discernible impact on workplace safety.[26]

C. American Adoption and Modification of English Common Law

At the outset courts in the United States followed the English common law conception of the master-servant rela-

---

[21]*Id.* at 155.

[22]*Chicago, M. & St. P. Ry. v. Ross*, 112 U.S. 377, 319 (1884).

[23][1868] 1 L.R.-S. & D. App. 326.

[24]*Id.* at 334.

[25]See text at note 5, *supra.*

[26]See *infra* text at notes 138 and 139 for discussion of subsequent English legislation aimed at reducing obstacles to recovery by injured or health-impaired employees.

tionship. The fellow servant rule first appeared in America four years after *Priestley*, in *Murray v. South Carolina Railroad*,[27] a South Carolina decision which adopted the principle of *Priestley* without citing the case. One year later, *Priestley* itself crossed the Atlantic and was cited, as was *Murray*, in Chief Justice Shaw's landmark opinion in *Farwell v. Boston & Worcester Railroad*.[28]

1. Rule of Priestley *Followed*

The English cases, in developing the master-servant doctrine, were clearly bringing public policy considerations to the forefront. Both *Murray* and *Farwell* repeat this emphasis. The *Murray* court found that the respondeat superior doctrine was required by "the public security" but that it would not promote safety in the workplace since employees would not be compelled "to look to each other for protection and safety."[29] In *Farwell*, Chief Justice Shaw noted the fact that "it is competent for courts of justice to regard considerations of policy and general convenience, and to draw from them such rules as will * * * best promote the safety and security of all concerned."[30]

The free market concept that a worker assumed all the ordinary risks of employment by entering into a contractual relationship with his employer was also repeated by Shaw in *Farwell* and succeeding cases.

The chief justice in *Farwell* also based his decision on freedom of contract, and thus brought the harsh rule of *Priestley* to America. Consistent with that principle, Shaw in the same term decided *Commonwealth v. Hunt*,[31] holding that it was not a conspiracy for employees to band together and strike for a closed shop, stating that persons "are free to work or not to work as they prefer."

American courts, like their British counterparts, believed that their view of employer liability smoothed the path of industrial progress. An Indiana case presented an especially blunt articulation of that belief: "[A]ny departure from it [the

---

[27]26 S.C.L. (1 McMul.) 166 (1841).
[28]45 Mass. (4 Met.) 49 (1842).
[29]*Supra* note 27, at 171.
[30]*Supra* note 28, at 58.
[31]45 Mass. (4 Met.) 111 (1842).

fellow-servant rule] is dangerous to the prosperity and perpetuity of the enterprises of manufacturing, mining, railroading, and those industries requiring the services of many servants."[32]

2. Common Employment Redefined

The early U.S. decisions also did not differ markedly from the English model in specifying when workers were engaged in common employment. In *Farwell*, Shaw summarily disposed of the argument that employees in different departments unable to observe each other's conduct were therefore not in common employment by saying that "this is founded on a supposed distinction in which it would be extremely difficult to establish a practical rule."[33]

More than 50 years later, the U.S. Supreme Court held that the fellow servant rule would apply unless the possibility of employees coming into contact with each other and thus causing injury "would not be said to be within contemplation of the person injured."[34] Even when courts acknowledged the internal inconsistencies of this doctrine, they were often unwilling to adopt a more flexible stance. A Texas decision observed that realistically, the fellow servant rule should only include those workers who could observe each other's actions. Yet, the court held, the settled view was "too firmly established to be changed, except by action of the legislature."[35]

Soon after the creation of the fellow servant and assumption of risk rules and with the coming of the railroads, the courts in England and America recognized the impact of the railroads in changing the relationship between master and servant. Duties once performed by the master personally were now exercised by a large number of servants acting for, on behalf of, or at the direction of the master. However, the English courts considered the fellow servant rule to be protection to the master, as exemplified by Pollock's statement in *Priestley*,[36] and continued to apply the strict rule of *Priestley*. United

---

[32]*New Pittsburgh Coal & Coke Co. v. Peterson,* 136 Ind. 398, 405, 35 N.E. 7, 8 (1893).

[33]*Supra* note 28, at 60.

[34]*Northern Pac. R.R. v. Hambly,* 154 U.S. 349 (1893).

[35]*St. Louis, A. & T. Ry. v. Welch,* 72 Tex. 298, 302, 10 S.W. 529, 531 (1888).

[36]See text at note 11, *supra.*

States courts gradually viewed it otherwise, and "shortly after the introduction of the railroad entered upon a slow but marked transition on the subject of fellow employees."[37] This transition was shaped by the rules of vice-principal, agency, and consociation.

(a) Vice-principal rule. One important change "on the subject of fellow employees" was the holding that while the servant assumed the risks of the employment, including the negligence of a fellow servant, the master's duties were nondelegable duties and the person to whom the master assigned them was held to be a vice-principal or superior officer, for whose negligence the master was liable.[38] Thus, the exemption accorded the master did not apply where the master was negligent—either personally or through a superior officer or vice-principal—in furnishing a safe workplace, tools, and appliances, or in the selection or retention of fellow servants.[39]

Although England and, in the United States, Massachusetts (to name one state) did not follow this vice-principal rule, many courts did. One court commented, "In Massachusetts we think the courts have a tendency to narrow the remedies for negligence by technical and unsound decisions and especially to favor corporations and employers at the expense of servants."[40]

(b) Agency. Other courts went further, holding that when a corporation assigned an employee to carry out a responsibility it owed to its servants, such as repair of tools, equipment, or machinery, the employee charged with that duty was acting as an agent for whose negligence the master was liable regardless of the grade of service.[41]

(c) Consociation. Finally, a few courts, acknowledging the rule that as a matter of public policy a servant accepted the ordinary risks of employment as part of his employment contract, including the risk of injury by a fellow servant, limited the master's exemption to those employment situations where

---

[37]*Union Pac. R.R. v. Erickson*, 41 Neb. 1, 59 N.W. 347 (1894).
[38]*Chicago, M. & St. P. Ry. v. Ross*, 112 U.S. 377 (1884).
[39]*Baltimore & O.R.R. v. Baugh*, 149 U.S. 368 (1893).
[40]*Quoted in* Sheerman & Redfield, *supra* note 5, at 127.
[41]E.g., *Northern Pac. R.R. v. Herbert*, 116 U.S. 642, 652 (1886).

there was "consociation"[42] in the same department or line of employment.[43] Agreeing "that the fellow servant exemption rule was simple and easy to apply," these courts refused to follow that rule absent consociation for the reason that "[w]hen the law of fellow servant was first announced business enterprises were comparatively small and simple. The servants of one master were not numerous. They were all engaged in the pursuit of a simple and common undertaking. Now things have changed."[44]

By enunciating the rules of vice-principal, agency, and consociation, the courts slowly diluted the exemption of the master from liability and recognized that the rule was not consistent with the facts of industrial employment.

Coincidentally, employers were made aware of their responsibility to employ skilled and safe employees. As one court said, "[W]hen it is thoroughly understood that it is not profitable to employ careless and inefficient agents or reckless and indolent servants, better men will take their places and not before."[45]

II. Government Regulation in the Workplace

A. Beginning of Social Concern

1. England

Charles Beard designates 1760 as the eve of the Industrial Revolution in England.[46] Unfortunately, the rapid succession of industrial inventions and the rise of the factory system of production would long precede general recognition of the impact of the Industrial Revolution on the health and safety of the worker.

Still standing in the Lancashire district as late as the 1920s were many of the first factories, witness to poor working conditions with their low ceilings, small windows, and absence

---

[42]Consociation: Association, fellowship, partnership—RANDOM HOUSE DICTIONARY OF THE ENGLISH LANGUAGE, unabridged (1966).

[43]E.g., *Union Pac. R.R. v. Erickson, supra* note 37.

[44]*Louisville & N.R.R. v. Dillard*, 114 Tenn. 240, 86 S.W. 313, 350 (1905).

[45]*Goddard v. Grand Trunk Ry.*, 57 Me. 202 (1869).

[46]C. Beard, THE INDUSTRIAL REVOLUTION 22 (1927).

of sanitary arrangements. Such factories allowed the worker little space and little light.[47] Among the causes of the most common work-related illness and injuries, a parliamentary report of 1833 listed confinement in a heated and cramped atmosphere, admission of foreign matter into the lungs, long hours of work in a constantly upright position, dampness and sudden transitions of temperature, hurried eating, accidents involving machinery, and general uncleanliness.[48] Mortality from fevers of various kinds caused by the overcrowding of ill-housed and ill-nourished people reached epidemic proportions. At the same time, the employment of little children was a well-established practice, dating back as early as 1671. The two evils—the general unhealthiness of town life and the societal toleration of child labor—had been spreading since the early 1700s and were now to be worsened by the incipient Industrial Revolution, which would further increase both urban density and the demand for child labor.[49]

Labor was a plentiful commodity to be bought and sold in the market. Conditions of life were secondary to the production of wealth. The worker of this era has been described as being far worse off than an American slave, who at least was valued and cared for as property. Parliament was under the control of the Crown, the peers, and the wealthy. The industrial areas with their masses either were not represented in Parliament or were outnumbered; Manchester, the largest industrial city at the time, had not a single member. Business and manufacture were unhindered by legislation or public opinion.

The workers expressed their rage through riots and strikes, which were illegal as conspiracy. At the same time Parliament and the landed gentry feared a result similar to the French Revolution. In August 1819, these conflicts erupted in the Peterloo (Manchester) Massacre, when the local magistrates called on the local cavalry to arrest the leaders of the 60,000 workers and families assembled on St. Peter's Field in Manchester calling for Parliamentary reform. The result was a clash leaving 16 dead and more than 400 injured.[50] However, not until 1832 did reform and change occur.

---

[47]*Id.* at 58.

[48]*Id.* at 60.

[49]*Id.* at 9–10, 12.

[50]J. Marlow, THE PETERLOO MASSACRE (1969).

2. America

Before the early 1800s, large-scale manufacturing in the United States lagged behind that of England. This was so largely because before the American Revolution, England had discouraged industrial development in the colonies so that they would continue to serve England as a source of raw materials and a captive market for English manufactured products. After the revolution, U.S. industrialization could proceed. Although it took time to amass sufficient investment capital, to develop an adequate system of transportation, and to acquire by immigration a labor force, by the early 1800s the United States began to replicate British inventions like the power loom and the spinning jenny. America also invented its own manufacturing products such as the cotton gin.[51] As in England, even obvious requirements for worker safety and health were not provided for, partially because so little was known about occupational diseases. In 1837, the medical society of New York offered a prize for the best essay on the subject "The Influence of Trades, Professions and Occupations in the United States on the Production of Diseases." The winner, Benjamin McCready, had to base his essay on foreign data because, according to McCready, American workers then paid "little attention to the causes which [affected] their health, and their views were often warped by prejudice."[52]

B. Government Intervention

1. Early Measures in England

Addressing the widespread social discontent evident within the English working class in the early and middle 1800s, Beard has said that "when it became a physical impossibility to endure this terrible state of affairs any longer, the agitation for factory legislation began."[53]

Reports and pamphlets on conditions in the English factories seem to have attracted little attention from the general

---

[51]Altman, *Growing Pains: A Portrait of Developing Occupational Safety and Health in America,* 4 JOB SAFETY & HEALTH 24, 24–26 (August 1976).

[52]*Id.* at 26.

[53]Beard, *supra* note 46, at 69–70.

public until epidemics broke out in the factory districts. The rising agitation for factory legislation was counteracted, however, by the unwillingness of policy makers to take a position on the issue, which would be seen as trampling upon "the sacred rights of freedom of contract." At the height of the English elite's support for the philosophy of laissez faire, "[human] life was nothing when measured against the holy obligation of precedent," especially when that precedent supported laissez faire.[54]

Nonetheless, even before the worst incidents of rioting and property destruction, an early step toward legislative control of factory conditions was taken. In 1802, the Health and Morals of Apprentices Act, passed through the efforts of Sir Robert Peel, reduced the hours of labor of children employed and housed in the cotton mills to 72 hours per week. Children who lived at home were not covered by the Act because they were presumably under the care of their parents. The children who were covered were to be provided with some schooling and a suit of clothes each year. The factories were to be whitewashed annually and adequately ventilated and the sleeping quarters of the sexes were to be separated. This measure was enforced by a penalty of not more than £5.[55]

In 1815, Robert Owen began to make public pleas for more comprehensive factory regulation. His efforts led to the passage of a second act in 1819 which forbade the employment of children under 9 years of age in the cotton mills and limited the hours of work to 12 per day for children between 9 and 16. The first act only made the state a guardian for the children it covered, but the second was clearly a legislative interference between the free laborer and his employer. It was the first legislative expression of the doctrine that the state should protect the interests of its weaker citizens, and it led to more progressive legislation later on.

In 1825, the hours of child labor in the cotton mills were further restricted. In 1831, the administration of the above acts was given over to the justices of the peace. Unfortunately, the justices tended to be sympathetic to the pleas of local industrialists against the acts. Some effort to remedy this problem was made in 1831 when mill owners and their re-

---

[54]*Id.*
[55]42 Geo. 3, ch. 73.

lations were excluded from hearings in cases under the acts. Note that all of the above legislation applied only to employees in cotton mills.[56]

Throughout the history of the Factory Acts, as they came to be called, employers resisted new legislation with claims that it would ruin industry. Proponents made appeals to humanitarian concerns. Neither side ever appealed to the results of direct experiment; perhaps the proponents should have, as each legislative step taken was later found to increase productivity.[57]

In 1833, another child labor act was passed which applied to all textile workers. Children 9 to 13 years old could work only 48 hours per week. Children 13 to 18 years old could work only 69 hours per week. All night work by children was forbidden. Government-paid factory inspectors could be and were appointed under the Act, and 1 of every 11 English mill owners was convicted for its violation.[58] In 1842, the Mines and Collieries Act was passed, prohibiting the employment of women and children in underground work. Finally, after a long period of controversy regarding hours of labor, what became known as the Ten Hours Act was passed in 1847. The Act and its supplements restricted the labor of all women and children to 10 hours a day. Also, since sending the women and children home generally meant closing the mills, the Act effectively applied to male mill workers as well and later became the unofficial rule for all factory work.[59]

All these statutes regulated textiles or mines and collieries, and it was not until 1864 that legislation was extended to manufacturing other than textile. After that, safety and health legislation both specific and general proliferated. This is well-documented in a report presented to Parliament by the Secretary of State for Employment by Command of Her Majesty.[60]

---

[56]*Id.* at 73–76; E. Collis & M. Greenwood, THE HEALTH OF THE INDUSTRIAL WORKER 26–27 (L. Stein ed. 1977).

[57]Collis & Greenwood, *supra* note 56, at 25.

[58]*Id.* at 28–29; Beard, *supra* note 46, at 79.

[59]Beard, *supra* note 46, at 79–80; Collis & Greenwood, *supra* note 56, at 29–31.

[60]SAFETY AND HEALTH AT WORK: REPORT OF THE COMMITTEE APPOINTED BY THE SECRETARY OF STATE FOR EMPLOYMENT AND PRODUCTIVITY, at 181 (Her Majesty's Stationery Office, London, 1970).

2. Regulation in America

In America, as in England, the regulation of labor conditions to protect the health and safety of the workers was established in spite of the protests of employers. Such regulation was more easily implemented at the outset in England, however, where government institutions had long dealt with economics and trade. In the United States, regulatory bodies had to be built from nothing. Much of American safety legislation mirrored or built on the English statutes.[61]

For the most part, the nineteenth century was well into its second half before the states and the federal government began to regulate workplace conditions. But while it was slower than England to legislate safety and health requirements, the United States, once embarked, made rapid progress. A complete history of this progress, which is still continuing today, would require a separate volume. What immediately follows is a rapid review of the main outlines and highlights (a) at the state level, (b) in the private sector, and (c) at the federal level. Following this review is a more detailed recapitulation of the train of federal legislation and regulation that began in 1935 and led to the comprehensive Occupational Safety and Health Act of 1970 (see Section II.C. below).

(a) State regulation. In 1836, Massachusetts passed the nation's first child labor law. In 1869 it established the first bureau of labor statistics, to be followed shortly by eight other states. In 1877, it passed the nation's first state factory inspection law, which provided for the appointment of government-paid factory inspectors. It required the guarding of belts, shafts, gears, and drums of any machines determined by the inspectors to be hazardous; various fire prevention efforts; and general cleanliness. In 1887, the Wisconsin legislature passed a similar safety act, which further required that belt shifters and other contrivances used to pull belts or pulleys on or off be automatic. In rapid succession, many other states followed suit, passing similar safety acts. Some of these acts, however, were conditioned on the employee being engaged in "ordinary duties." In 1869 Pennsylvania passed the first coal mine inspection act. In 1890 and 1891, various states passed similar

[61]Pausner & Clack, *OSHA's Ancestors: Previous Laws of the Workplace*, 2 JOB SAFETY & HEALTH 16, 20 (April 1974).

mining legislation, providing that coal mines be inspected for unsafe conditions. Also during the late 1800s, many states passed legislation dealing with other specific occupational hazards such as elevators, scaffolds, and boilers.[62]

By 1890, 21 states had passed occupational safety laws, which included requirements pertaining to ventilation, heating, lighting, and fire safety. In the late 1890s states began regulating the conditions of tenements in which work was being done (i.e., sweatshops). In 1900, Ohio passed the first American statute to place a duty of general safety upon the employer, requiring that the employer "make suitable provision to prevent injury to persons who may come in contact with" certain specified machinery.[63]

The peak of state legislative concern over safety and health was reached during the period from 1890 to 1910. For example, in the early 1900s legislation was passed mandating meal breaks for workers. Massachusetts passed the first of such laws in 1902. It required a half-hour meal break for any woman or young person working 16 hours or more. Other states followed, requiring meal breaks of 45 minutes to an hour for all workers. Following England's lead, five states passed legislation regulating the place where workers took their meal breaks, requiring an area separated from noxious fumes or poisonous substances.[64]

The most consistently regulated industry during this period was the railroad industry. The first statutes passed concerned the threat of fire from locomotives. Thus, in 1893, New York enacted a statute regulating the height of covered bridges. Many states adopted laws requiring that a reasonable amount of space be left on either side of all railroad tracks. In the first decade of the twentieth century, most states passed legislation requiring that all frogs, switches, and guardrails be guarded. In 1912, South Carolina empowered its railroad commission to compel the installation of safety devices which in its opinion contributed materially to the safety and welfare of the workers.

In 1906 the Russell Sage Foundation financed a study commonly known as the Pittsburgh Survey, which docu-

---

[62]E.g., KAN. STAT. ANN. §4676 (1909); MASS. GEN. LAWS ANN. ch. 1041, §43 (West 1909).

[63]Pausner & Clack, *supra* note 61.

[64]5 C. Labatt, MASTER AND SERVANT 5726 (1913).

mented the tragic extent of industrial accidents. The investigators found that Allegheny County (Pennsylvania) annually sent out of its mills, factories, railroad yards, and mines 45 one-legged men; 100 hopeless cripples with crutches or canes; 45 men with a twisted, useless arm; 30 men with an empty sleeve; 20 men with only one hand; 60 with half a hand gone; 70 with one eye; and so on—more than 500 disabled humans in all. An additional 500 men died each year in accidents. The survey aroused the public's awareness of the extent of industrial injuries and deaths in one county.[65]

At the same time, a general awareness of industrial hazards was growing, as manifested by Maryland's 1906 law requiring shirt factories to sprinkle floors with water to keep cotton dust down. In 1911, Wisconsin legislated what was to become known as the Wisconsin Plan and passed comprehensive occupational safety laws, creating an industrial safety board which had broad authority to set occupational safety standards.[66] New York and four other major industrial states followed Wisconsin's lead by passing their own comprehensive occupational health and safety laws. In 1911, the 10-story Asch Building in New York City caught fire and 146 Triangle Shirtwaist Company employees either burned, choked, or jumped to their deaths. This tragedy served to crystallize public sentiment and spur legislators and social workers to renewed concern and action.

After 1920, states began using safety standards developed by the American Standards Association and other private entities (see discussion below). However, because of wide variation among the states respecting promulgation of standards, enforcement, and funding, progress in safety was far from uniform and in some states was minimal.

(b) Early private sector contributions. The labor movement, particularly after 1850, steadily advocated and used its influence to secure state legislation promoting safety and health in the workplace. Much of the legislation enacted directly resulted from the efforts of unions, especially in the mining and railroad industries. And as early as 1878 the Knights of Labor demanded federal legislation on occupational safety and health.

---

[65]National Safety News, May 1963, p.38.
[66]Pausner & Clack, *supra* note 61.

Insurance companies worked for fire safety in factories and formed the National Board of Fire Underwriters in 1866 and the National Fire Protection Association (NFPA) in 1896. Later, the voluntary safety movement directed attention to the need for on-the-job safety programs and education, and 1911 saw the beginning of what became the National Safety Council in 1913 and the American Society of Safety Engineers.

In 1920 the American Standards Association (ASA), originally formed in 1918 as the American Engineering Standards Committee, reorganized at the request of the National Bureau of Standards and the National Safety Council to develop safety standards. These standards, in addition to those of other organizations such as NFPA and ASME, became available to industry and the states as safe practice guides.[67]

(c) Federal regulation. The federal government also became active in worker health and safety regulation. In 1840 an executive order restricted naval shipyard workers to a 10-hour day. In 1888 all government contracts required an eight-hour day. Also in 1888 the first Federal Bureau of Labor was established; in 1903 it became part of the Department of Commerce and Labor. In 1893 the Safety Appliance Act was passed to regulate the safety of railroad employees.

In 1900 the National Bureau of Standards was established as part of the Commerce Department to conduct research into technological matters that affect workers, and 1902 saw the establishment of the U.S. Public Health Service. In 1907 railroad workers were restricted to a 16-hour day; this was shortened further to eight hours in 1916. In 1910, the Federal Bureau of Mines was created.

In 1912—late by British standards—the first important American occupational health legislation, the Esch Act, was passed. The Esch Act put a prohibitive tax on the domestic sale of white phosphorus matches but it came too late for the many workers in match factories who had already contracted a painful necrosis of the jaw called "Phossy-jaw" caused by exposure to phosphorus.[68]

In 1913 the Department of Labor was established as a separate department. In 1915, the Public Health Service added

---

[67]*Id.*
[68]*Id.*

a Division of Industrial Hygiene. In 1920, the Department of Labor established the Women's Bureau. In 1934 the Bureau of Labor Standards and the Federal Interdepartmental Safety Council were organized.

In 1936 Congress passed the Walsh-Healey Act[69] to use the purchasing power of the federal government to regulate not only wages and hours but safety and health (see discussion below). Beginning in 1952 a steady progression of federal legislation, starting with the Federal Coal Mine Safety and Health Act[70] (see discussion below), brought safety and health further into the federal regulatory and enforcement ambit. In 1958 the Longshoremen's and Harbor Workers' Compensation Act was amended to authorize the Secretary of Labor to promulgate and enforce safety and health regulations for employment covered by that Act.[71]

By 1962 Congress enacted the Contract Work Hours and Safety Standards Act covering employment on public works of the United States.[72] It was followed in 1965 by the McNamara-O'Hara Service Contract Act, under which safety and health regulations were promulgated for employees of federal maintenance and service contractors;[73] in the same year this coverage was extended to actors and other workers providing services under grants administered under the National Foundation on the Arts and Humanities.[74]

Congress turned to mineral mines in 1966 with the passage of the Metal and Nonmetallic Mine Safety Act[75] (see discussion below), and in 1969 broadly amended and strengthened the Federal Coal Mine Safety and Health Act of 1952[76] (see discussion below). In 1970 it added the Occupational Safety and Health (OSH) Act[77] (see Chapter 2).

Although at the time, the OSH Act seemed to burst upon the scene from nowhere, in retrospect it appears as inevitable.

---

[69] 41 U.S.C. §35 (1982).

[70] 30 U.S.C. §471 (1982).

[71] 33 U.S.C. §901 (1982).

[72] 40 U.S.C. §327 (1982).

[73] 41 U.S.C. §351.

[74] 20 U.S.C. §951.

[75] 30 U.S.C. §§721 *et seq.* (repealed 1977).

[76] 30 U.S.C. §§801 *et seq.* (1976).

[77] 29 U.S.C. §651.

C. Historical Review of Federal Safety and Health Regulation, 1935–1977

For those not already subject to the Walsh-Healey Act, the OSH Act was a new and perhaps traumatic event. But businesses subject to the Walsh-Healey Act and similar subsequent legislation had long been familiar with government safety and health regulations, although inspection and enforcement had not been as pervasive as the OSH Act was to make it.

For years organized labor had lobbied for a federal safety and health law, but the right of Congress to delegate similar functions remained in doubt until *Jones & Laughlin Steel Corp. v. NLRB*.[78]

1. Walsh-Healey Public Contracts Act[79]

In 1935 Senator Walsh at the request of the Roosevelt administration introduced a bill in the Senate designed to use the purchasing power of the government to regulate hours and wages on government contracts for supplies, materials, and services exceeding $10,000. This was a substitute for a general minimum wage law, which was then thought to be unconstitutional. The bill, which did not mention safety and health, was passed by the Senate in 1935 and sent to the House, which did not act upon it, however, out of concern with its constitutionality. In 1936 Representative Healey introduced a similar bill. As passed by the House, it provided, in addition to the minimum wage and hour sections, that no part of such contracts were to be performed under working conditions "which are unsanitary, or hazardous or dangerous to the health and safety" of employees so engaged. The bill was sent to the Senate, where Senator Walsh, without holding a meeting or conference, moved that the Senate concur, which it did.[80]

In 1938, the Bureau of Labor Standards issued an inspection manual covering all aspects of the Act, including general guidelines for safety and health.[81] For the most part

---

[78]301 U.S. 1 (1937).

[79]41 U.S.C. §35 (1982).

[80]Walsh-Healey Act Analysis, Labor Legislation Committee—Radio-Electronics-Television Manufacturers Ass'n 1954 (available in Labor Dep't library).

[81]U.S. Dep't of Labor Bulletin No. 20.

these guidelines had slight impact until World War II, when in accordance with the authority conferred by the Walsh-Healey Act, the Department of Labor entered into cooperative agreements with selected states to inspect war production plants under the Act.[82] In 1942 the Wage and Hours Division and the Public Contracts Division of the Department were merged and became the Department's primary enforcement arm. The 1938 regulations, though revised and extended periodically, remained vague until 1952, at which time what came to be known as the Green Book, designed to be used by Walsh-Healey inspectors, was issued; it was last revised in 1956.[83]

In 1960, deciding that the Green Book had become outmoded, Labor Secretary James Mitchell promulgated new safety and health standards for federal supply contracts.[84] At that time 23 states were under cooperative agreements that continued the type of agreements begun during World War II. These cooperative agreements were informal and by authority only of the wartime administrative order.[85] In 1962 the Department finally issued a regulation specifically authorizing the Secretary to enter into agreements with the states for inspection and enforcement.[86] As early as 1950, however—since the Walsh-Healey Act provided that compliance with state safety and health laws "shall be prima facie evidence of compliance" with the federal law—the Department had begun measuring the adequacy of state codes against such national voluntary consensus standards as those of ASA, NFPA, and ASME.

Long dissatisfied with state safety codes and enforcement, organized labor now found a tool—Walsh-Healey—with which to bring pressure on that segment of American industry engaged in interstate commerce and accepting federal contracts subject to the Walsh-Healey Act. Almost simultaneously, however, and apparently in response to complaints by labor organizations that the states were not adequately performing their agreements, the Department began canceling the agreements and expanding its own inspection and enforcement activities.

---

[82]U.S. Dep't of Labor Administrative Order No. 103.
[83]The Green Book has been discontinued.
[84]41 C.F.R. §50.204 (1968).
[85]Note 82, *supra*.
[86]41 C.F.R. §50.205.

In October 1963, it was once again decided that the existing (1960) regulations needed to be updated. The revised standards that were proposed provoked a storm of protest from the business community. Public hearings followed in March 1964; 1,374 pages of testimony were taken.

The standards were not issued, largely because of the vigorous opposition. Instead the Secretary appointed an Ad Hoc Safety Advisory Committee representing all interested segments, which met twice in 1966 and was chaired by the Under Secretary of Labor. The committee recommended that rather than follow the "cut and paste" system of developing standards, as was done with the 1960 and 1963 proposals, the Department should simply adopt the national standards of ASA, NFPA, and others. This recommendation was not acted upon, although in 1964 radiation standards based largely on AEC standards had been adopted, raising little protest. Nonetheless, in 1967 the Department did follow the Ad Hoc Committee's recommendation, not for Walsh-Healey, but for the McNamara-O'Hara Service Contract Act.[87] A public hearing brought little criticism, and the standards became effective in February 1968. (Meantime, in July 1966, Labor Secretary Willard Wirtz transferred the administration of the Walsh-Healey Act to the Bureau of Labor Standards, which already had responsibility for the Service Contract Act and others.[88])

Finally, on the basis of the experience with the Radiation and Service Contract Act standards, the Department proposed new Walsh-Healey standards in September 1968 using consensus standards and adding rules for hazards for which there were no standards, such as noise, gases, fumes, vapor, and coal dust. The 1968 proposals were in part a response to a study prepared for the Department that pointed out the duplication and overlapping of existing state and federal standards, and the superiority of many state standards to federal standards.[89]

The new proposals virtually eliminated the "prima facie" test for compliance, since the Secretary of Labor could elect to inspect and enforce state, federal, or local standards, or

---

[87]See note 73 and accompanying text, *supra*.

[88]Secretary's Order No. 12-66, July 19, 1966.

[89]An Evaluation of Federal-State Relationships in the Administration of Labor Standards Laws. Kirschner Assocs., June 1968.

standards set by collective bargaining agreements, whichever were the most stringent. Despite vehement objection, particularly to the "in house" standards for such things as noise and dust the Secretary issued the new regulations on January 17, 1969 (the last day of the outgoing administration), with an effective date of 30 days hence.[90]

On February 17, 1969, the new Secretary of Labor, George Shultz, postponed the effective date in order to allow a review.[91] He also appointed a National Safety Advisory Committee to recommend standards on noise and dust and to advise him with respect to pending safety and health federal legislation. Following submission of the Committee's recommendations, the standards were finally made effective in May 1969.

The Secretary's policy of using state standards under Walsh-Healey did not succeed in substantially improving workplace safety and health conditions or safety programs of all employers. While the large employers generally recognized and met their responsibilities, the smaller employers as a group lagged far behind. Consequently, labor organizations continued to lobby for a federal safety and health law.

2. Coal Mine Safety and Health

A very important segment of American business is the mining industry. From its inception, the extremely hazardous nature of mining was well known to miners, their families, and the communities in which the miners worked. Mine safety, like industrial safety, was a province of the states. Even prior to the state industrial legislation discussed in Section II.B.2.a, above, state legislatures had enacted laws to regulate safety and health in coal mines in varying degrees.[92] The first federal legislative concern occurred in 1865 when a bill to create a Federal Mining Bureau was introduced in Congress but was not passed.[93] Only in 1910, after a decade of coal mine disasters, did Congress finally establish the Bureau of Mines in the Department of the Interior. While the Bureau was charged to

---

[90] 41 C.F.R. §50.204 (1968).
[91] 34 C.F.R. §2207 (1969).
[92] See Labatt, *supra* note 64.
[93] S. REP. No. 411, 91st Cong., 1st Sess. (1970).

make "diligent investigations" of methods of mining, the treatment of ores, and the safety of miners, it had no authority to inspect mines or to develop or enforce any safety rules or regulations.[94]

After further coal mine disasters, demand for further action by Congress resulted in the passage of the Coal Mine Inspection and Investigation Act in 1941.[95] This law created a federal authority to inspect mines for safety and health hazards, and accordingly an Inspection Bureau was formed in the Bureau of Mines. Tentative standards were issued as a "guide" for the inspectors, but there was still no federal authority to establish mandatory standards or enforce compliance. That was left to the states, despite the evidence that states were lax in their laws and enforcement.[96]

A coal mine strike in 1946 over a new labor contract caused President Truman to seize the coal mines in the interest of national security and the economy.[97] That order authorized the Secretary of the Interior to take possession of and operate all bituminous coal mines not operating owing to the strike. On May 21, 1946, the Secretary of the Interior established the Office of Coal Mines Administrator. On May 29, 1946, the Secretary of the Interior J.A. Krug, acting as Coal Mines Administrator, and President John L. Lewis of the United Mine Workers signed a contract known as the Krug-Lewis Agreement. In addition to covering terms and conditions of employment, the agreement also directed the Bureau of Mines to issue a reasonable code of standards on safety conditions and practices, subsequently promulgated as a Federal Mine Safety Code for Bituminous and Lignite Mines.[98] The code was enforced by the Administrator until the mines were returned to the owners in 1947.

In 1947, during the period of the seizure, and following another mine disaster at Centralia, Illinois, Congress directed the Secretary of the Interior to report all violations of the Federal Mine Safety Code to the state agencies having jurisdiction and required a report of corrective actions taken.[99] The

---

[94]30 U.S.C. §1 (1982).
[95]30 U.S.C. §451 (1982).
[96]Investigations in Coal Mines Report No. 168, 77th Cong., 1st Sess. (1941).
[97]Exec. Order No. 9728 (1946).
[98]32 C.F.R. §304 (1946).
[99]30 U.S.C. §48 (1982).

Senate Committee on Public Lands, in reporting the enabling legislation favorably, limited it to one year, refusing to "clothe the Federal authorities with police powers to enforce the code in the States in all its details."[100] It stated that until Congress had an opportunity to study the matter, the states had the burden of making the mines safe. It made clear that "if the States do not guard the safety of the miners, the Congress will act further."

Four years later, on December 21, 1951, a Franklin Coal Company mine in Illinois exploded, killing 119 miners. In short order, on July 16, 1952, Congress enacted the Federal Coal Mine Safety and Health Act.[101] While it was a major step forward, it focused only on conditions that might lead to major disasters, leaving the principal responsibility for normal operating hazards to the states; and the Act did not apply to mines employing 14 or fewer miners underground. In 1953, in another step forward by the federal government, the Federal Mine Safety Code was revised and made applicable to anthracite mines.

Despite these measures, however, the coal mining safety record following 1952 was not much improved on the prior record. In 1963 President Kennedy, disturbed by the continuing poor record, requested the Secretary of the Interior to review existing legislation and make proposals for new legislation. As a result of a task force report carrying out the presidential directive, Congress amended the existing law, and eliminated the small mines exemption.[102] The intent of the 1952 Act had been to eliminate fatalities caused by major disasters (such as explosions and roof falls). However, 90 percent of the causes of death or injury (from normal operating hazards, particularly at the face where the coal was cut and loaded into cars) remained under the jurisdiction of state laws and the Federal Mine Safety Code.[103]

On November 20, 1968, Consol Mine No. 9 at Farmington, West Virginia, exploded, leaving 78 miners dead. Whatever remaining opposition there was to an all-encompassing federal coal mine safety law evaporated. On December 30, 1969, Pres-

---

[100]S. Rep. No. 431, 80th Cong., 1st Sess., *reprinted in* 1947 U.S. Code & Cong. Serv. 1549.

[101]30 U.S.C. §471 (1982).

[102]30 U.S.C. §476.

[103]H. Rep. No. 563, 91st Cong., 1st Sess. (1969).

ident Nixon signed the Federal Coal Mine Health and Safety Act of 1969,[104] which established mandatory safety rules and, for the first time, health standards. The Act also directed and authorized the Secretaries of the Interior and of Health, Education, and Welfare to promulgate mandatory safety and health standards to be enforced by the Secretary of the Interior.[105] In 1977 this Act was substantially amended (see Section II.C.4, below).

3. Metal and Nonmetallic Mine Safety and Health

Less well known but equally serious was the problem of safety and health in the mining of minerals other than coal, in so-called metal and nonmetallic mines. Indeed, on an hours-of-exposure basis, the mineral mines experience was almost as poor as that of the coal mines. Although as early as 1891 Congress passed an act "for the protection of the lives of miners in the Territories," and later, as we have seen, enacted safety and health laws for various segments of industry, it was not until 1966 that the mining of minerals other than coal received attention.[106] Here, also, safety and health legislation and enforcement was the province of the states.

Although there had been explosions, fire, and inundations in mineral mines, none was of a scale comparable to that in coal mines in the sense of lives lost. But mounting evidence of a high rate of serious injuries and deaths in such mines led to a federal legislative proposal in 1961, which proved to be another in a series of many unsuccessful attempts to pass a mineral mines safety and health law. As a compromise, however, Congress passed a bill authorizing the Secretary of the Interior to investigate the causes and prevention of injuries and health hazards in metal and nonmetallic mines.[107] The results of the investigation were submitted in January 1963 by the Mine Safety Board of the Interior Department. As a

---

[104]30 U.S.C. §§801 *et seq.* (1976).

[105]30 U.S.C. §803.

[106]Subcommittee on Labor of the Senate Comm. on Human Resources, LEGISLATIVE HISTORY OF THE FEDERAL MINE SAFETY AND HEALTH ACT OF 1977, 95th Cong., 2d Sess. 501 (1978).

[107]5 U.S.C. §485 (1982).

result of those findings Congress enacted the Federal Metal and Nonmetallic Mine Safety Act on August 31, 1966.[108]

Under the Act the Secretary of the Interior was required to promulgate safety and health standards with the assistance of advisory committees, and to designate as mandatory those standards "dealing with conditions or practices which could reasonably be expected to cause death or physical harm." States were authorized to continue existing jurisdiction except "where to the extent that such law is in conflict with the Act or Orders issued pursuant thereto." Not until three years later were standards issued, of which half were advisory.

Then on May 2, 1972, a disastrous fire at Sunshine Silver Mine in Kellogg, Idaho, took the lives of 91 miners. Subsequent oversight hearings on that disaster and its causes dramatically revealed the deep dissatisfaction of labor with the administration of the Act. Specifically, witnesses alledged delay in promulgating standards, failure to revise original standards or promulgate additional mandatory or voluntary standards, and conflict of interest between the Bureau's responsibility for promoting safety and health and its simultaneous aim of encouraging production through developing efficient mining methods.[109] Several witnesses urged transfer of the administration of both mining acts to the Department of Labor, which as a result of the OSH Act of 1970 had "developed competency for safety of workers." The pressure for more stringent legislation and for a single safety authority continued to grow.

A few months prior to the Sunshine Mine disaster, in February 1972, a coal mining impoundment dam in Buffalo Creek, West Virginia, burst, causing widespread destruction in the valley and the loss of 125 lives. There appeared to be no single federal authority over such dams at mine locations, and even the Bureau of Mines questioned its own authority to regulate impoundment dams. In July 1972 at Blacksville, West Virginia, nine miners were trapped by a machinery fire, apparently because of lack of training in emergency procedures. These events, coupled with congressional pressure to remove mining safety and health issues from the Department

---

[108]30 U.S.C. §§721 *et seq.* (repealed 1977).

[109]*Oversight Hearings on the Federal Metal and Nonmetallic Mine Safety Act Before the Select Subcomm. on Labor of the House Comm. on Education and Labor*, 92 Cong., 2d Sess. (1972).

of the Interior, led the Department to separate the research from the safety activities and to establish the Mining Enforcement and Safety Administration (MESA), responsible for administering both safety and health acts. Mining research and development remained in the Bureau of Mines. However, both the Bureau of Mines and MESA continued reporting to the same Assistant Secretary as before, which tended to annul the desired separation of the Department's simultaneous goals of increased mining production and increased safety and health protection for miners.[110] Additionally, criticism of the Act continued for its failure to provide mandatory penalties for a violation of the standards; its small or minimal penalties when they were assessed at all; and its failure to resolve a conflict between the Occupational Safety and Health Administration (OSHA) and MESA as to jurisdiction over the milling operations at mining locations.

The Subcommittee on Labor of the Senate Committee on Human Resources had already held oversight hearings on the administration of both acts and investigated the mine disasters mentioned above.[111] The Subcommittee concluded that stricter legislation was needed, particularly for mineral mining. A series of bills to correct the deficiencies was introduced in the 93rd Congress (1973–1974) and in the 94th Congress (1975–1976), but they were not considered or not brought to a vote.

4. Federal Mine Safety and Health Act of 1977

On February 11, 1977, Senator Harrison Williams introduced a bill to amend the Federal Coal Mine Health and Safety Act of 1969. Designed to merge and transfer the administration of the mining laws and substantially strengthen the law applicable to mineral mining, the bill received extended hearings and stirred bitter debate in the Senate. A major force behind the bill was the conclusions of an investigative report made by the Senate Subcommittee on Labor regarding two explosions within a few days of each other at the Scotia Coal Mine near Whitesburg, Kentucky, in March

---

[110]*U.S. Bureau of Mines.* Prepared by the Senior Specialist Division, Congressional Research Service, Library of Congress, at the request of Henry M. Jackson, Chrmn, Senate Comm. on Interior and Insular Affairs, 94th Cong., 2d Sess. (1976).

[111]*Hearings on H.R. 4287 Before the Subcomm. on Compensation, Health, and Safety of the House Comm. on Education and Labor,* 95th Cong., 1st Sess. (1977).

1976. Twenty-six deaths resulted, including those of three federal mine inspectors who were in the mine at the time of the second explosion, investigating the cause of the first explosion.

The investigative report included the statement that "the Company's safety education and training program was a sham."[112] Congress passed Senator Williams' bill (S. 717) and President Nixon signed it on November 9, 1977.[113] It repealed the Metal and Nonmetallic Mine Safety Act and several others. It revised the rulemaking procedures, adopted the existing metal and nonmetal standards, and directed the Secretary to revise the standards and make them mandatory. The Act also revised the inspection, enforcement, and penalty procedures and mandated inspections of mines on a scheduled basis. It also transferred the functions of the Secretary of the Interior under both existing acts to the Secretary of Labor; established in the Labor Department a Mine Safety and Health Administration (MSHA) with its own Assistant Secretary of Labor; and established a Review Commission with five members.

5. *Occupational Safety and Health Act of 1970*

As we have seen, starting in 1936 Congress little by little extended its presence in the safety and health field. Using the Walsh-Healey Act as a wedge, organized labor urged Congress to enact a generally applicable federal safety and health law. But opposition from business to such legislation was too strong. Following the war years, and in large part owing to the emphasis on safety brought about by the need to conserve manpower during that time, the record on occupational injury and death had been steadily improving. However, in the decade of the 1960s the record worsened. Manufacturing injury frequency rates increased from 11.4 per million hours worked (the frequency measurement prior to the OSH Act) in 1958 to 15.2 in 1970.[114] The injury frequency experience of the companies reporting to the National Safety Council showed a surprising change, increasing from 6.04 in 1960 to 9.37.[115]

---

[112]*Supra* note 106.

[113]30 U.S.C. §801.

[114]Bureau of Labor Statistics, U.S. Dep't of Labor, HANDBOOK OF LABOR STATISTICS, BULLETIN No. 1735, Table 163 (1972).

[115]National Safety Council, ACCIDENT FACTS (1972).

Congress had already shown its concern for safety and health in the legislative initiatives discussed above. In 1965, a proposed Occupational Safety and Health Act was actually introduced, but no action was taken. This background, combined with the accident trend, led to the introduction in 1968 of another bill to enact a federal safety and health law, which was supported by President Johnson. Although hearings were held, Congress took no further action. As noted previously, new Walsh-Healey safety and health standards were promulgated in January 1969.[116]

In early 1969, momentum continued to gather as bills proposing a federal occupational safety and health law were again introduced, in both Houses, including an administration bill supported by President Nixon. By that time a consensus had been reached in support of a comprehensive federal safety and health law, but there was disagreement as to the precise method. Then the Federal Coal Mine Health and Safety Act of 1969 was passed and signed.[117] With that in mind, and the emotions engendered by the November 1968 Farmington mine disaster still forceful, Congress enacted the Occupational Safety and Health Act of 1970. (See Chapter 2, Legislative History of the Occupational Safety and Health Act of 1970.)

III. Recovery for Workplace Injuries Based on Safety and Health Legislation

A. Under Specific Statutes

The earliest legislation at the state level to protect workers from the hazards of machinery and equipment took the form of requiring (a) guarding of dangerous machinery, (b) guarding where some public official deemed it necessary, and (c) guarding of machinery without qualification.[118] The enactment of these statutes was closely followed by court decisions deriving civil liabilities for workplace injury from the statutes.

---

[116]See note 90 and accompanying text, *supra.*
[117]See note 104 and accompanying text, *supra.*
[118]Labatt, *supra* note 64, at 5933ff.

Tort actions for injuries resulting from violations of these statutes were sustained,[119] particularly when the statute was remedial and for worker protection.[120] But despite the intent that such statutes benefit employees, many courts held that the employee assumed the risk of the employer's violation,[121] although others disagreed.[122]

Since some courts had previously allowed damages at common law where the injury was the result of employer failure to provide the protection now required by statute, employers argued that the common law remedy was superseded and the statute provided the exclusive remedy. This argument was generally rejected.[123] Additionally, some statutes banned the defense of assumption of risk.[124]

Some legislatures also enacted general statutes intended to protect the public, such as requiring the guarding of railway crossings or the equipping of trains with whistles or bells to warn the public. Suits alleging that employees also came under the protection of these statutes generally were dismissed,[125] although a few courts agreed.[126]

1. Strict Construction of Safety Legislation

While employees seemingly received added protection under these statutes, they were strictly construed with respect to the person, corporation, or business intended to be regulated, or as to those responsible for complying with the statute, such as owner, lessee, or contractor.[127] Although these statutes improved chances of recovery, the courts continued to apply tort principles as the test of liability. Plaintiffs were required to prove the master's knowledge of the condition,[128] failure to comply with the statute as the proximate cause of injury,[129]

---

[119]E.g., *Consolidation Coal Co. v. Bokamp*, 181 Ill. 954, 54 N.E. 567 (1899).

[120]*Hane v. Midcontinent Petroleum Co.*, 43 F.2d 406 (N.D. Okla. 1930).

[121]E.g., *St. Louis Cordage Co. v. Miller*, 126 F. 495 (8th Cir. 1903).

[122]E.g., *Fitzwater v. Warren*, 208 N.Y. 355, 99 N.E. 1042 (1912).

[123]Labatt, *supra* note 64.

[124]See *Chicago & N.W. Ry. v. Booten*, 57 F.2d 786 (8th Cir. 1932).

[125]E.g., *Union Pac. Ry. v. Elliot*, 54 Neb. 199, 74 N.W. 627 (1898).

[126]E.g., *Illinois Cent. Ry. v. Gilbert*, 157 Ill. 354, 41 N.E. 724 (1895).

[127]*Hane, supra* note 120.

[128]*Akron & N.Y.R.R. v. Whittaker*, 31 Ohio App. 507, 166 N.E. 694 (1929).

[129]*Davis v. Pennsylvania Coal Co.*, 209 Pa. 153, 58 A. 271 (1904).

a causal connection between the hazard and the injury,[130] and that the accident was a natural and probable consequence of the condition, with no independent intervening cause.[131]

2. *Statutory Duty and Negligence*

The English courts held that breach of a statutory duty created an absolute right to sue with no need to prove negligence.[132] This was not followed in the United States. Here the majority view was that a violation of the statute was negligence *per se*.[133] Other courts found that a breach established a presumption of negligence,[134] or was a question for the jury; but all courts agreed that such statutes did not make the employer an insurer.[135]

B. General Statutes and Defenses

As noted above, the early statutes usually applied to specified industries such as railroads or construction; or to named hazards in elevators, shafts, and chutes; or to machinery in particular locations, such as at exits or where flying particles could injure those passing in the area. Very few covered all workplaces, or all manufacturers. But virtually paralleling and coincidental with the enactment of the specific statutes, which were in effect preventive, were other statutes of a remedial nature.

As early as 1855 Georgia put in place a law, applicable only to railroads, which abolished the rules of fellow servant and assumption of risk.[136] But it was during the period from 1885 to 1910 that most states enacted some type of law either declaratory of the existing common law or a substantial modification of it. Originally, railroads were the principal subject of such legislation;[137] but before long general statutes were

---

[130]*Piepho v. Gesse*, 106 Ind. App. 450, 18 N.E.2d 868 (1939).
[131]*Coal v. Seaboard Airline Ry.*, 199 N.C. 389, 154 S.E. 682 (1930).
[132]Labatt, *supra* note 64, at 5942.
[133]*American Car & Foundry Co. v. Armentrack*, 214 Ill. 509, 73 N.E. 766 (1905).
[134]E.g., *Mosgrave v. Zimbleman Coal Co.*, 110 Iowa 169, 81 N.W. 227 (1899).
[135]Labatt, *supra* note 64.
[136]W. Dodd, ADMINISTRATION OF WORKMEN'S COMPENSATION LAWS 13 (1936).
[137]Labatt, *supra* note 64, at 5411.

enacted or the existing railroad statutes were amended to that effect.

The English forerunner and model for many of these general statutes was the Employer's Liability Act of 1880. Although it was designed to reduce the obstacles to recovery in master-servant suits and resolve disputes, the prediction that it was "so framed as to provoke rather than minimize litigation" proved to be accurate.[138] The Act, while revolutionary, fell short of removing the impediments to employee suits against employers. Under the Act, the employer was liable for defects in the ways, works, machinery, or plants used in the employer's business; and for acts of those in charge of designated equipment. The Act also adopted the vice-principal rule.[139] But it did not abolish the fellow servant rule; negligence of the employer continued as the basis for liability.

In the United States similar statutes generally did no more than codify what was judge-made law, but usually by making statutory the earlier court modifications of the fellow servant rule.[140] Liability of the employer was predicated on negligence of the employer, and existing common law rules were the guide for determining recovery. The defenses of assumption of risk (including express or implied terms of the employment contract), fellow servant, and contributory negligence were intact.

In fact, new issues served to confound the litigants and courts, such as to what "works," "ways," "machinery," "plant," "tools," and "defects" a statute applied; who was a vice-principal, superintendent, or superior; and, surprisingly, whether a statute was intended to apply to laborers or only operating employees.

In 1908 Congress passed the Federal Employer's Liability Act,[141] a statute applicable to interstate railroads and one of the most advanced of its time. It abolished the fellow servant rule, but did not abolish the defense of assumption of risk until it was amended in 1939. Contributory negligence was not abolished as defense but was to be considered under the question of damages. However, the Jones Act, which gave to

---

[138]*Id.* at 3412.
[139]See notes 22 and 38–39 and accompanying text, *supra.*
[140]Malone et al., WORKMEN'S COMPENSATION AND EMPLOYMENT RIGHTS 11 (1980).
[141]45 U.S.C. §51.

seamen a right of action for injuries or death in the service of a vessel, provided that "seamen having command shall not be held to be fellow servants with those under their authority."[142]

Indeed, despite an intent by statute to make recovery for work injuries less difficult, in practice this did not come about. Settlements and verdicts were minimal or nonexistent. As before, proof of the master's negligence was difficult because witnesses were unwilling to testify out of fear of reprisal from employers. Disabled workers or families left without a provider, needing money and faced with the cost and delays of litigation, were pressured to settle for small sums, although they were the very people the statutes were designed to benefit. Additionally, since there was a large pool of available labor to replace the injured and deceased, employers had little incentive to reduce injuries or to settle lawsuits quickly.[143] As before, the cost of injuries was borne by workers, then estimated at from 70 to 94 percent,[144] leaving them no better off. Furthermore, employers no longer dealt directly with injured employees; they turned that task over to insurance companies, whose concerns were the employers rather than the employees, the cost of premiums, and a profit—a situation not conducive to largesse on the part of the carriers.[145]

The continuing search for solutions to the problem of adequate compensation for the injured worker and assignment of financial resopnsibility for this compensation is discussed in Section II of Chapter 25 (Impact of the OSH Act on Private Litigation and Workers' Compensation Laws).

*   *   *   *   *

As we have seen, the occupational safety and health reforms embodied in modern legislation are rooted literally in centuries of legal and social evolution. The early common law doctrines grounded in public policy considerations favoring the preservation and growth of industry, at the expense of workplace safety, were gradually supplanted by a patchwork of state and federal laws codifying only aspects of our occupational and safety health policy. The lack of a uniform and comprehensive statutory scheme impeded the realization of

---

[142]46 U.S.C. §688.
[143]W. Prosser, THE LAW OF TORTS 530 (4th ed. 1971).
[144]Dodd, *supra* note 136.
[145]*Id.*

workplace safety and health. In 1970 Congress acted to remedy this situation by enacting the Occupational Safety and Health Act, the centerpiece of our nation's occupational safety and health policy. This book will focus on the various aspects of that law—its administration, enforcement, interpretation, and application in conjunction with other occupational safety and health laws.

2

Legislative History of the Occupational Safety and Health Act of 1970

While there were sharply conflicting political and economic interests at stake in the debate over safety and health legislation in Congress in 1970, the difference in the debate centered less on what needed to be done than on how it was to be done and by whom.

I. Passage of the Senate and House Measures[1]

By November 1970, the Senate had before it two safety and health bills. One was S. 2193, the "Williams bill" reported by the Senate Labor and Public Welfare Committee;[2] the other was S. 4404,[3] the bill supported by the Nixon Administration. S. 4404, introduced by Senator Peter Dominick (R.-Colo.), was

---

[1]See also discussion of legislative history in Chapter 4 (The General Duty Clause), Section II.

[2]S. 2193, 91st Cong., 2d Sess. (1970), *reprinted in* Subcommittee on Labor of the Senate Comm. on Labor and Public Welfare, 92d Cong., 1st Sess., LEGISLATIVE HISTORY OF THE OCCUPATIONAL SAFETY AND HEALTH ACT OF 1970, at 204–295 (Comm. Print 1971) (hereinafter cited as LEGIS. HIST.).

[3]S. 4404, 91st Cong., 2d Sess. (1970), *reprinted in* LEGIS. HIST. at 73–140.

the Senate companion bill to the House "substitute bill," H.R.
19200, introduced by Congressman William Steiger (R.-Wis.)
on the floor of the House.[4] The Steiger substitute bill had been
offered in opposition to the Daniels bill, H.R. 16785,[5] which
the House Committee on Education and Labor had reported
out over the marked objections of most of the Republican members of the Committee.[6]

The version of S. 2193 that was brought before the full
Senate in November 1970 was not the original version introduced by Senator Williams in 1969,[7] but an amended version
that reflected attempts within the Senate Committee to resolve differences between the Administration and the legislative majority in their respective approaches to a
comprehensive bill on occupational safety and health.[8] The
resulting compromise was not entirely satisfactory, as evidenced by the attempt of Senator Dominick, a member of the
Committee, to substitute S. 4404 on the Senate floor. The
Senate defeated Senator Dominick's move,[9] however, and passed
the Williams bill (S. 2193) with a number of amendments.[10]

A week later, the House took up its two bills, with an
opposite result: the House accepted the Administration-backed
substitute (H.R. 19200, identical to S. 4404) in place of the
Committee bill.[11]

Congress was now faced with the task of resolving the
differences between the Administration-backed substitute bill
passed by the House and the Senate Committee (Williams)
bill as amended by the full Senate. Since the Senate's amendments to the Committee-reported bill were the outcome of still
further compromise between the Senate majority and the Administration forces (see below), the more telling comparison
for historical purposes is not between the House- and Senate-

---

[4]H.R. 19200, 91st Cong., 2d Sess. (1970), *reprinted in* Legis. Hist. at 763–830.

[5]H.R. 16785, 91st Cong., 2d Sess. (1970), *reprinted in* Legis. Hist. at 721–762.

[6]H.R. Rep. No. 1291, 91st Cong., 2d Sess. 47–60 (1970), *reprinted in* Legis. Hist. at 877–890 (Minority Views on H.R. 16785).

[7]S. 2193, 91st Cong., 1st Sess. (1969), *reprinted in* Legis. Hist. at 1–30.

[8]See, e.g., S. Rep. No. 1282, 91st Cong., 2d Sess. 54 (1970), *reprinted in* Legis. Hist. at 193 (Individual Views of Mr. Javits).

[9]See Senate debate on OSH Act of 1970 (Nov. 16, 1970), in Legis. Hist. at 448–450.

[10]See Senate debate on OSH Act of 1970 (Nov. 17, 1970), in Legis. Hist. at 451–528; and S. 2193 as passed by the Senate, *reprinted in* Legis. Hist. at 529–597.

[11]House debate on OSH Act of 1970 (Nov. 24, 1970), in Legis. Hist. at 1091–1117.

passed versions but between the final House (substitute) bill[12] and the Senate Committee (Williams) bill.

II. Comparison of the Senate and House Measures

Despite differences between the Williams bill and the substitute bill, the two measures shared common goals and contained many similar provisions. Both bills sought to attack the problem of workplace injuries by enacting a comprehensive federal statute that would apply to virtually every workplace in the United States. Each measure contained a three-pronged approach to reducing workplace hazards through regulation, research, and education. Both bills would have implemented this approach by (a) authorizing the adoption of safety and health standards; (b) providing the means of enforcing the standards; (c) enhancing research and statistical efforts; (d) providing an active role for the states; and (e) enhancing education and participation by employers and employees and others with occupational safety and health responsibilities.[13]

The major difference between the Williams bill and the substitute bill lay in the methods each proposed for administering and enforcing the legislation.

A. Locus of Rulemaking, Enforcement, and Adjudication Authority

In its essential provisions, the Williams bill, which was supported by organized labor, resembled S. 2864 and H.R. 14816, the bills introduced in the previous Congress as a Johnson Administration measure. It would have housed the three key functions, namely, rulemaking, enforcement, and adjudication, in the Labor Department. The Secretary of Labor would have developed the safety and health standards and issued them according to informal rulemaking procedures including a hearing, if requested. Labor Department inspectors

---

[12]As passed, the substitute included two amendments adopted during floor proceedings; see LEGIS. HIST. at 1060–1061 and at 1067–1068.

[13]S. 2193, *supra* note 2, at §2; H.R. 19200, *supra* note 4, at §2.

would have conducted investigations to find violations; hearings for adjudicating the penalties and abatement orders proposed by the Labor Department inspectors would have been conducted in the Labor Department before administrative law judges.[14]

In contrast, the Administration-supported substitute bill would have set up a separate agency, a board, to issue the safety and health standards, and a second separate agency, a commission, to adjudicate the enforcement cases. The standards-setting board was to have five members and the commission, three, all appointed by the President.[15] Like the Williams bill, the substitute authorized the Secretary of Labor to conduct inspections and investigations to enforce the legislation.[16] Both bills provided review by the U.S. courts of appeals.[17]

Senator Jacob Javits of New York, the ranking Republican member of the Senate Labor and Public Welfare Committee, summed up the two sides in the controversy as follows: "The general attitude of labor is to have the Secretary of Labor administer the bill, including the promulgation of standards and the adjudication of complaints of violations. The position of the administration is that it wants a board to handle the standards and a commission to handle the violations."[18]

The division described by Senator Javits affected most of the other key provisions of the legislation, and was particularly crucial in the context of standards-setting. This division created a crisis that threatened the survival of the legislation, with the then Secretary of Labor, James Hodgson, stating publicly that he would recommend a veto if a bill that placed rulemaking and adjudicatory responsibilities in the Labor Department reached the President for signature.

The Administration proposal for an independent standards-setting board ultimately failed, but not before it served as a critically important bargaining chip: key Republicans

---

[14]S. 2193 §§6, 7, 8, 10. See Senate debate (Nov. 16, 1970) (statement of Sen. Williams) in LEGIS. HIST. at 414.

[15]H.R. 19200 §§6, 8, 10, 11. See House debate (Nov. 23, 1970) (statement of Rep. Steiger) in LEGIS. HIST. at 989, 991.

[16]H.R. 19200 §9.

[17]S. 2193 §10(d); H.R. 19200 §13.

[18]Senate debate on the OSH Act of 1970 (Nov. 16, 1970) (remarks of Sen. Javits) in LEGIS. HIST. at 417. See generally MacLaury, *The Job Safety Law of 1970: Its Passage Was Perilous*, MONTHLY LAB. REV. (March 1981).

agreed to place authority for issuing and enforcing the standards with the Secretary instead of a board, provided that an independent Review Commission would adjudicate enforcement cases brought before it by the Secretary.[19] This trade-off cleared the way to enactment of the Occupational Safety and Health Act.

In addition to the question of who would have responsibility for standards and adjudication, however, various other points required extended debate.

B. General Duty Clause[20]

The Act's general duty clause was one of its most controversial provisions. Because Congress considered it unlikely that specific, detailed standards to cover all hazardous working conditions in some 5 million workplaces could be developed, it sought to include in the legislation a provision requiring employers, in very general terms, to provide safe and healthful employment. Agreement on the principle was relatively easy. However, only after protracted debate was Congress able to agree on language to implement the principle.[21]

The Williams bill had a general requirement that employers furnish employment "which is free from recognized hazards so as to provide safe and healthful working conditions." The supporters of the substitute bill, fearing possible enforcement abuse under such a broadly worded provision, narrowed the requirement in the substitute to "[employment] free from readily apparent hazards which are causing or are likely to cause death or serious physical harm."[22] Because of the narrower version in the substitute, that bill provided for penalties upon first discovery of a violation of its general duty clause. The supporters of the Senate version defended the broad language, maintaining that the fears of abuse expressed by the opponents were groundless because the Williams bill, un-

---

[19]See B. Mintz, OSHA: HISTORY, LAW, AND POLICY 21–26 (1984).

[20]For a more detailed review see Chapter 4 (The General Duty Clause) at Section II.

[21]The final language is "(a) Each employer—(1) shall furnish to each of his employees employment and a place of employment which are free from recognized hazards that are causing or are likely to cause death or serious physical harm to his employees[.]" 29 U.S.C. §654 (1985).

[22]See House debate on OSH Act of 1970 (Nov. 23, 1970) (statement of Rep. Steiger) in LEGIS. HIST. at 991, 992.

like the substitute bill, did not authorize imposing any civil
penalties upon first discovery of a violation of the general duty
clause.[23] The Senate bill authorized civil penalties only for
failure to correct a condition found in violation of that clause.[24]

The final version of the general duty clause reflected a
compromise of the two versions. The substitute version, authorizing penalties upon first discovery of a violation, was
adopted, as was the restricted language of the substitute bill
with one change: the words "readily apparent" were dropped
and in their place Congress wrote in the Senate's "recognized."[25]

C. Standards

A number of debates involved the types of mandatory
standards and the procedures used for issuing the standards.[26]

1. "Early Standards"

Both the Senate bill and the substitute bill recognized the
need for so-called "early standards." Proponents of early standards were expressing their frustration with the length of time
it took the executive branch to issue regulations. To avoid the
problem of lengthy delays, both bills permitted the quick adoption of national consensus standards which organizations in
the private sector had issued under a consensus arrangement.[27] Both bills also permitted the quick adoption of existing
standards that the Labor Department had already issued under other statutes it enforced. While there was considerable
discussion about these standards in the legislative history,

---

[23]See Senate debate (Nov. 16, 1970) (statement of Sen. Williams) in Legis. Hist.
at 415–416.

[24]*Id.* at 416.

[25]H.R. Rep. No. 1765 (Conference Report), 91st Cong., 2d Sess. 33 (1970), *reprinted in* Legis. Hist. at 1186.

[26]See House debate, *supra* note 22, at 994–996.

[27]S. Rep. No. 1282, 91st Cong., 2d Sess. (1970), *reprinted in* Legis. Hist. at 141–
203, noted that there were two major sources of consensus standards: the American
National Standards Institute and the National Fire Protection Association. The report further noted that because these consensus standards had been adopted under
procedures that gave diverse views an opportunity to be considered, and that indicated
that interested and affected persons had reached substantial agreement on their
promulgation, the Secretary could adopt the standards without regard for the provisions of the Administrative Procedure Act. *Id.* at 6, Legis. Hist. at 146.

the early standards concept did not give rise to serious differences. Much of the discussion centered on the relatively minor differences between the bills' provisions as to the procedures to be followed in adopting these standards.

2. *Emergency Temporary Standards*

Both bills called for emergency temporary standards to be adopted and to be immediately effective in order to combat grave dangers involving toxic substances and new hazards. Here, again, although there was agreement on a basic concept, that of allowing the speedy adoption of temporary standards, there were differences in the precise language of the bills.

3. *Procedures for Setting New Standards*

There was one essential difference between the substitute bill and the Williams bill with respect to the type of procedures to be used in setting new standards. The substitute called for using the formal rulemaking procedures of the Administrative Procedure Act (APA),[28] while the Senate bill provided for only informal procedures with a requirement for a hearing if any interested person requested one.[29] This difference generated considerable discussion and floor debate. The Administration's proponents, emphasizing procedural fairness, argued that stricter, more exacting formal procedures provided the better means of insuring a wider and more thorough public scrutiny of a proposal. Those favoring the Senate-reported bill argued that the formal APA procedures were designed essentially for adjudication and were inappropriate in promulgating rules of general application like those contemplated under the safety and health bills.

The version of the bill that became law included the Senate provisions, i.e., informal, legislative-type rulemaking procedures with no hearing unless requested. However, under Section 6(f) of the Act, those adversely affected by a recently published standard have 60 days after publication of the standard in its final form in which to challenge the validity of the standard in a federal court of appeals.[30] Section 6(f) states that

[28]5 U.S.C. §553 (1977).
[29]S. REP. NO. 1282, *supra* note 27, at 6, LEGIS. HIST. at 146.
[30]29 U.S.C. §655(f).

"[t]he determinations of the Secretary shall be conclusive if supported by substantial evidence in the record considered as a whole." In a sense, then, there was a compromise: The Senate's informal rulemaking procedures won the day, except as to the test the Secretary had to meet when the standards were challenged in the courts; for the "substantial evidence" test is derived from the formal rulemaking procedures of the APA.[31]

D. Imminent Danger

The enforcement procedures for handling situations of imminent danger provoked extensive debate. There was a consensus in Congress that the legislation should provide a rapid response to eliminate imminent dangers that could not be abated under the usual procedures for abating hazards where urgency was not a factor. The controversy, however, was over the form of this response.

The response the substitute bill provided for in imminently dangerous situations was injunctive relief in the federal district courts upon petition of the Secretary of Labor. The Senate version, on the other hand, would have permitted Labor Department inspectors to shut down any imminently dangerous operation if there was insufficient time to seek an injunction in the district courts. The proponents of the substitute argued that giving a Labor Department inspector the power to shut down an operation, or indeed an entire plant, was excessive and needlessly ignored the elements of due process; they maintained that federal district courts can respond quickly with temporary restraining orders that are far more likely to be obeyed than a safety inspector's on-the-spot shutdown order.[32] Although opponents of the substitute measure simply did not agree that a federal injunction proceeding would always be fast enough to meet an imminently dangerous situation, the substitute version prevailed.[33]

As a matter of actual experience, the Labor Department has sought an injunction to combat an imminent danger only a few times in a dozen years.

---

[31]5 U.S.C. §702(2)(E). For a detailed discussion, see Chapter 20 (Judicial Review of Standards).

[32]See House debate, *supra* note 22, at 1021 (statement of Rep. Erlenborn).

[33]29 U.S.C. §§662(a), (b).

E. Relationship to Other Federal Laws

While Congress intended the Occupational Safety and Health (OSH) Act to be a comprehensive statute covering virtually all employers and employees in the nation, it recognized that there were other federal laws that affected worker safety and health. To achieve the comprehensive protection of workers under federal laws, Congress included Section 4(b)(1) in the OSH Act, which section provided that if workplace hazards were dealt with under the regulations of the other federal agencies, the Labor Department could not apply the OSH Act in those circumstances.[34] But where the other agencies had no applicable regulation, Section 4(b)(1) permitted the so-called residual jurisdiction of the OSH Act to provide the basis of protection.

The legislative history of Section 4(b)(1) is important because of two critical language changes. In earlier versions of the legislation, the section sought to make the Act inapplicable to employees "with respect to whom" other federal agencies had laws affecting occupational safety and health. Under such language the Act would not have applied to large groups of employees because those employees are covered by other federal laws. The final version of Section 4(b)(1), however, speaks not of "employees with respect to whom" but of "working conditions with respect to which." Thus, the OSH Act became inapplicable only to certain working conditions. The other change was that the other federal agency had to "exercise" its statutory authority over the working condition by issuing "standards or regulations affecting occupational safety and health."[35]

F. Penalties

Both the Senate bill and the substitute bill relied essentially on civil monetary penalties as the primary sanction to ensure compliance. There were differences between the bills

---

[34]§653(b)(1): "Nothing in this act shall apply to working conditions of employees with respect to which other Federal agencies, and State agencies acting under section 2021 of Title 42, exercise statutory authority to prescribe or enforce standards or regulations affecting occupational safety and health."

[35]*Southern Pac. Transp. Co. v. Usery*, 539 F.2d 385, 4 OSHC 1693 (5th Cir. 1976), *cert. denied*, 434 U.S. 874 (1977).

as to amounts and circumstances under which the civil penalties would be imposed, but these differences were relatively minor.

The substitute bill had one criminal sanction, which made assaulting an inspector a federal felony. The Senate bill, on the other hand, had, in addition to an assault-felony provision somewhat like the one in the Steiger bill, a provision making it a misdemeanor to willfully violate the standards and certain other requirements. This additional criminal penalty generated no real controversy and was included very late in the legislative process, with the added requirement that the willful violation result in an employee death.[36]

The Senate bill had another criminal provision, however, that did give rise to considerable controversy. This provision made it a misdemeanor for any person to give advance notice of any impending inspection. The Administration forces objected strongly to this criminal provision, arguing that the only people in a position to give such advance notice would be governmental employees, and that therefore the Secretary of Labor or state commissioners of labor should handle the problem administratively just as they would in the case of any inappropriate conduct on the part of a civil servant.

The Senate version prevailed, however, and it is a misdemeanor for any person to give advance notice of an inspection.[37]

G. Employee Complaints

The Senate bill had a provision requiring the Secretary to conduct an inspection in response to an employee's complaint where there were reasonable grounds for believing that the danger of which the employee complained existed. The Administration-backed measure had no such provision, Congressman Steiger having argued that it was already the accepted practice for the Secretary to respond to written employee complaints, resources permitting; and that in view of the Secretary's limited resources for enforcing a law as comprehensive as the OSH Act, it would be imprudent to require the

---

[36]Conference Report, *supra* note 25, at 41, Legis. Hist. at 1194.
[37]§666(f).

Secretary to respond to every valid employee complaint. This position did not prevail, however, and a provision essentially like the Senate version became law.[38]

III. Resolution of Differences Between the Senate and House Measures[39]

A. Statutory Definitions

The meaning of "national consensus standard" presented the only issue at the conference stage. The House amendment would have qualified a standard as a national consensus standard if, in the private sector procedures that produced the standard, interested parties' views were considered. According to the Senate version, qualification as a national consensus standard required that interested parties reach substantial agreement on its adoption after diverse views had been considered. The Senate version was adopted.[40]

B. Geographical Application

The Senate bill did not cover Guam and the House amendment did; the House coverage of Guam prevailed.[41]

C. Duties of Employers and Employees

The House version of the general duty provision was adopted, but Steiger's particular words holding employers responsible for "readily apparent" hazards were rejected in favor of the Senate language, "recognized hazard."[42] Also, the Senate measure had a section imposing duties on employees, which

---

[38]§657(f)(1).

[39]This section summarizes how the House and Senate conferees resolved differences between their respective versions of the Act, which was then signed into law by President Nixon on Dec. 29, 1970. The summary is based on the Statement of the Managers on the Part of the House, in Conference Report, *supra* note 25, 32–45, LEGIS. HIST. at 1185–1198.

[40]§652(9).

[41]§653(a).

[42]§654(a)(1).

the House measure lacked. The Senate provision was adopted,[43] but the legislative history made it clear that the obligation on employees in no way relieved employers of their obligations under the Act.[44]

D. "Secretary Versus Board"

On the much-debated issue of whether the Secretary of Labor or a board would issue standards, the Senate version, which conferred the authority upon the Secretary of Labor instead of a board, prevailed.[45] This was perhaps the single most crucial issue of this legislation. (See Section II.A above.)

E. Early Standards

Both the Senate and House measures allowed the Secretary of Labor to adopt the private sector's national consensus standards, as well as already existing federal standards, with a bare minimum of procedural steps.[46] While the House bill allowed a three-year period for this process, the Senate restriction to two years prevailed.[47] A Senate provision to allow the Secretary to issue existing proprietary standards under shortened procedures during the two-year period was rejected.

F. Procedures for Setting New Standards

The Senate measure prevailed, calling only for informal APA rulemaking procedures but with the added provision that a hearing be held if any interested person objected to the published proposal.[48]

G. Effective Dates of Standards

Both the Senate and House measures permitted a delay in the effective date of a new standard to give employers and

---

[43]§654(b).
[44]S. Rep. No. 1282, *supra* note 27, at 10–11, Legis. Hist. at 150–151.
[45]§655.
[46]§655(a). See note 27 and accompanying text, *supra*.
[47]§655(a).
[48]§655(b)(3). See text accompanying note 29, *supra*.

employees time to familiarize themselves with its terms. The conferees rejected the open-ended familiarization period of the Senate bill, choosing instead the House 90-day familiarization period.[49]

H. Setting Standards for Toxic Substances

The Senate bill required the Secretary of Labor to set a standard that assures that no employee will suffer material impairment of health or functional capacity even if the employee is exposed for all his working life. The Senate bill also required that the standards be based on research, demonstration, experiments, and other relevant information; that they take into account the latest scientific data, feasibility, and results of experience gained under the present and other health and safety laws; and that, whenever practicable, they be expressed in terms of objective criteria and desired performance.

The House amendment had no comparable provisions. The Senate version was adopted,[50] and was to become the center of much controversy in later years. (See Chapter 3, Specific Occupational Safety and Health Standards.)

I. Labels and Warnings

Both the Senate bill and the House amendment called for the use of labels and warnings as the means of informing employees of existing hazards. There were minor differences between the two measures, but the Senate version was approved.[51]

J. Emergency Temporary Standards

Both the Senate bill and the House amendment provided for the issuance of such standards with shortened rulemaking procedures. The House amendment limited such standards to toxic substances or new hazards "resulting from the intro-

---

[49]§655(b)(4).
[50]§655(b)(5).
[51]§655(b)(7).

duction of new processes." The conferees rejected the House version, adopting instead the Senate version, which allowed the issuance of emergency standards where "employees are exposed to grave danger from exposure to substances or agents determined to be toxic or physically harmful or from new hazards."[52]

K. Court Review of Standards

The Senate bill permitted review by any U.S. court of appeals of any safety or health standard, provided such review is sought within 60 days following issuance of the standard in final form. The House measure limited the time period to 30 days and made the U.S. Court of Appeals for the District of Columbia Circuit the exclusive forum for such review. The Senate version prevailed.[53]

The House amendment provided as the test for upholding the standard on review that the agency action (the standard) be based on the "substantial evidence" of record. The Senate bill relied on the "arbitrary and capricious" test. The House amendment prevailed.[54]

L. Delegation of Inspection Authority

The conferees rejected the House amendment authorizing the Secretary of Labor to delegate his inspection authority to other agencies of the federal government or to the state agencies. The Senate bill had no such provision.

M. Inspections and Subpoena Power

The statute's words "without delay" in regard to permitting a compliance officer to enter a workplace to conduct an

---

[52]§655(c)(1)(A).

[53]§655(f).

[54]*Id.* See generally *Industrial Union Dep't v. American Petroleum Inst.*, 448 U.S. 607, 8 OSHC 1586 (1980), in which the Supreme Court held that before promulgating a new standard, the Secretary must find a significant risk of harm to employees.

inspection[55] came from the House amendment; the Senate bill did not contain these words. The Senate bill let the compliance officers "privately" question employees and employers during the inspection; the House measure did not have the word "privately" and the conferees adopted the Senate choice.[56] As to the provisions authorizing compulsory process for the production of evidence and witnesses, the conferees chose the House version.[57]

N. Recordkeeping and Self-Inspections

The statute's dual intended use of recordkeeping both as an enforcement aid and as a research tool[58] comes from the Senate measure; the House version had only the compliance-related purpose. The Senate bill allowed the Secretary of Labor to require employers to conduct self-inspections and certify the results. The House amendment had no comparable provision, and the conferees compromised by retaining the Senate language but deleting the certification language.[59] The conferees also adopted Senate language authorizing the Secretary to require employers to post notices as a means of supplying employees with information as to their protection under the statute.[60]

O. Employer Reports

The conferees adopted the Senate provision authorizing the Secretary to require employers to maintain records and make reports as to all work-related deaths, injuries, and illnesses; they included an amendment excluding injuries and illnesses of a minor nature.[61]

---

[55]§657(a)(1). See *Marshall v. Barlow's, Inc.*, 436 U.S. 307, 6 OSHC 1571 (1978), in which the Supreme Court ruled that §8a of the Act (29 U.S.C. §657(a)), authorizing workplace inspections, was unconstitutional under the Fourth Amendment to the extent that it purported to allow nonconsensual warrantless inspections.

[56]§657(a)(2).

[57]§657(b).

[58]§657(c)(1).

[59]*Id.*

[60]*Id.*

[61]§657(c)(2).

P. Employee Exposure Records

Senate language, adopted by the conferees, required the recording of employee exposure to potentially toxic materials or harmful physical substances that are required to be monitored or measured.[62] With one word change, the conferees also adopted the Senate provision affording employees the right to be notified when they have been exposed to toxic substances.[63]

Q. Walkaround Provision

The Senate version, providing that both a management and an employee representative be given an opportunity to accompany the inspector during an inspection, was adopted.[64]

R. Employee Complaints

In general, the Senate version on the issue was adopted in conference; that is, the Secretary is required to conduct an inspection in response to a reasonable employee complaint.[65]

S. Citations and Enforcement

The House and Senate bills essentially agreed in that citations were to be issued to compel abatement under sanction of civil monetary penalties.[66] There were, however, many technical differences on relatively minor connected points; the Senate version of these details prevailed in most instances.[67]

T. The OSH Review Commission

Both the Senate and House measures called for a commission to adjudicate enforcement cases. This was a major

---

[62]§657(c)(3).

[63]*Id.*

[64]§657(e).

[65]§657(f)(1). See text at note 38.

[66]§§658, 666.

[67]§659. For a more complete discussion of details, see LEGIS. HIST. at 1145–1152.

issue in the legislative history of the OSH Act, and it is discussed in detail in Chapter 16 (The Occupational Safety and Health Review Commission).

U. Judicial Review of Review Commission Decisions

The House bill limited the right of appeal to the employer and the Secretary of Labor. The conferees rejected this limitation and adopted the Senate version giving the right of appeal to any person adversely affected or aggrieved by a Commission action.[68] The House amendment provided for judicial review in the court of appeals for the circuit in which the violation occurred or the employer had his principal place of business. The conferees added the Senate's additional words, "or in the Court of Appeals for the District of Columbia Circuit."[69] Under the Senate bill all Commission orders became final 15 days after issuance unless stayed by a court of appeals. The conferees adopted this but changed the time period to 30 days.[70]

V. Discrimination Against Employees

The Senate bill authorized administrative action to obtain relief for an employee discriminated against for asserting rights under the statute, including reinstatement with back pay. The House measure, however, called for criminal and civil penalties against employers who discriminated against employees in such circumstances. The conferees compromised, requiring that the Secretary seek relief (reinstatement with back pay) but that this be done in the district courts, not through administrative process.[71]

---

[68]§660(a).

[69]*Id.*

[70]§659(c).

[71]§660(c). See *Whirlpool Corp. v. Marshall*, 445 U.S. 1, 8 OSHC 1001 (1980), the case in which the Supreme Court unanimously upheld the Secretary of Labor's regulation (at 29 C.F.R. §1977.12(b)(2) (1985)) which forbids employers from retaliating against employees who, under certain limited circumstances, refuse to perform tasks which they reasonably believe would subject them to serious injury or death. This decision contains a detailed discussion of the legislative history of this section and of the basic purpose of the statute. The legislative history was of crucial importance in this decision. The Court held that the Secretary's regulation advances the purpose of the Act, complements the legislative scheme, and is not inconsistent with the legislative history.

W. Imminently Dangerous Situations

Essentially, the House view of the much-debated issue of imminent dangers—that the only way to compel abatement is by seeking injunctive relief in the federal district courts—prevailed.[72]

X. Penalties

In general, differences with respect to penalties were of relatively minor importance except as to criminal penalties in the case of advance notice of an impending inspection.[73]

The conferees adopted the House amendment authorizing a district court suit to collect civil monetary penalties.[74] The Senate bill had no counterpart provision.

Y. State Plans

Section 18 of the OSH Act,[75] dealing with state plans, is of major importance in the administration and enforcement scheme of the Act, providing a unique statutory arrangement for state participation under a comprehensive federal statute. There is, however, relatively little legislative history on this section, and very few changes were made in its provisions during the legislative development of the Act. The most important issue for the conferees was the test to be applied by the courts in reviewing state plan withdrawal or rejection by the Secretary of Labor. The Senate bill's "substantial evidence" test won out over the House's "arbitrary and capricious" test.[76]

---

[72]§§662(a), (b). See text at notes 32 and 33, *supra*.
[73]§666(f). See text at note 37, *supra*.
[74]§666(*l*).
[75]§667.
[76]§667(g).

Part II

Duties Imposed Upon Employers

3

Specific Occupational Safety and Health Standards

I. Statutory Elements and Sources

The explicit purpose of the OSH Act, as stated by Congress in its findings and declaration of purpose and policy, is to "assure as far as possible every working man and woman in the Nation safe and healthful working conditions."[1] Congress chose to implement this goal by establishing a mechanism for the promulgation of standards by the Secretary of Labor to govern health and safety conditions in the workplace. Section 3(8) of the Act defines "occupational safety and health standard" as a "standard which requires conditions, or the adoption or use of one or more practices, means, methods, operations, or processes, reasonably necessary or appropriate to provide safe or healthful employment or places of employment."[2] In addition to the general duty imposed on employers to maintain hazard-free workplaces,[3] Section 5(a)(2) of the Act[4] requires that employers comply with standards promulgated under the Act.

---

[1] 29 U.S.C. §651(b) (1982).
[2] §652(8).
[3] See Chapter 4 (The General Duty Clause).
[4] §654(a)(2).

The Act does not define the difference between a "safety" and a "health" standard. The only guidelines given to the Secretary by the Act are contained in Sections 6(a)[5] and 6(b).[6] Under the Act, the Secretary was directed to promulgate immediate safety and health standards pursuant to Section 6(a), utilizing existing national consensus and federal standards.[7] Ultimately, these standards were to be refined over time, and new standards promulgated as necessary, pursuant to the rulemaking and notice and comment procedures contemplated by Section 6(b).[8] In practice, virtually all of the safety standards currently in effect were issued shortly after the effective date of the Act, pursuant to Section 6(a).[9] Conversely, although maximum exposure limits were accorded to approximately 400 substances under Section 6(a) authority to protect employee health,[10] most of the rulemaking activity under Section 6(b) has occurred with respect to health standards.[11] A discussion of the legislative background of these two sections is set forth in Chapter 2.

A. Section 6(a) Standards

To begin implementing the Act as soon as possible after passage, Congress provided in Section 6(a) that during the first two years after its effective date, i.e., from April 28, 1971, to April 28, 1973,[12] the Secretary was to promulgate as a 6(a) standard any national consensus standard[13] or established

---

[5]§655(a).

[6]§655(b).

[7]§655(a).

[8]§655(b).

[9]OSHA has rarely utilized the procedures set forth in §6(b) of the Act to promulgate safety standards. Recent examples of such standards adopted in that manner include fire protection, 29 C.F.R. §1910.155–.165, and service of multi-piece and single-piece rim wheels, 29 C.F.R. §1910.177. A complete discussion of the adoption of safety and health standards is set forth in Chapter 19 (The Development of Occupational Safety and Health Standards). See also Chapter 2, Sections II.C.1 and II.C.3.

[10]29 C.F.R. §1910.1000.

[11]See discussion of health standards promulgated under §6(b), in Section IV.B below.

[12]Under the express terms of §6(a), the Secretary's authority to promulgate regulations under that section expired two years after the effective date of the Act. §655(a).

[13]A national consensus standard is defined in §3(9) of the Act as a standard "which (1), has been adopted and promulgated by a nationally recognized standards-producing organization under procedures whereby it can be determined by the Sec-

federal standard.[14] In the same Section (6(a)), the Secretary was directed to implement two principles: first, the standards were to be adopted unless the Secretary determined that the promulgation of a standard "would not result in improved safety or health for specifically designated employees"; and second, where possible standards conflicted, the Secretary was to promulgate the one that assured "the greatest protection of the safety or health of the affected employees."

Linked with the congressional directive as to what standards to adopt was the provision that the standards were to be promulgated without the traditional notice and comment protections associated with administrative rulemaking.[15] National consensus standards and established federal standards for promulgation by the Secretary were to become final upon publication.[16]

Based on the definition of national consensus standard, the Secretary determined that the American National Standards Institute (ANSI) and the National Fire Protection Association (NFPA) were consensus organizations, thus rendering their standards eligible for Section 6(a) status.[17] Many of these standards have been incorporated by reference rather than produced *in toto* in the regulations, and may be obtained from the issuing organization.[18] The established federal standards were chiefly utilized for general industry safety standards, such as personal protective equipment[19] and general machine guarding,[20] which, in turn, were primarily derived from

---

retary that persons interested and affected by the scope or provisions of the standard have reached substantial agreement on its adoption, (2) was formulated in a manner which afforded an opportunity for diverse views to be considered and (3) has been designated as such a standard by the Secretary after consultation with other appropriate Federal agencies." 29 U.S.C. §652(9).

[14]An established federal standard is defined under §3(10) of the Act as "any operative occupational safety and health standard established by any agency of the United States and presently in effect, or contained in any Act of Congress, in force on [Dec. 29, 1970]." 29 U.S.C. §652(10).

[15]"Sec. 6. (a) Without regard to chapter 5 of title 5, United States Code, or to the other subsections of this section, the Secretary shall *** by rule promulgate *** standard[s] ***." 29 U.S.C. §655(a).

[16]36 FED. REG. 10,466 (1971).

[17]In addition, the American Society for Testing and Materials was designated as a consensus organization in 1972.

[18]One of the most common examples is the National Electrical Code, which is issued by ANSI, and incorporated within several regulations, including powered platforms, 29 C.F.R. §1910.66, and powered industrial trucks, §1910.178.

[19]29 C.F.R. §1910.132.

[20]§1910.212.

the Walsh-Healey Public Contracts Act.[21] Although the majority of these Section 6(a) standards relate to safety rather than health, the exposure limits for air contaminants listed in 29 C.F.R. §1910.1000 were derived from both ANSI national consensus standards and established federal standards adopted under the Walsh-Healey Act.[22] Pursuant to Section 6(a), the Secretary adopted the initial package of health and safety standards on May 29, 1971.[23]

In response to criticism that many of the Section 6(a) standards were unnecessary or inappropriate,[24] OSHA deleted approximately 600 safety standards in 1978 on the grounds that the standards were either obsolete, directed to comfort rather than safety, directed to the public rather than to employees, enforced by other agencies, contingent on manufacturer approval, too detailed, or covered by other standards.[25] In addition, numerous Review Commission decisions held that any standard written in the advisory "should" rather than the mandatory "shall" language was unenforceable.[26] Consequently, OSHA formally removed the 153 provisions of the general industry standards containing advisory language adopted from ANSI, effective February 1984.[27] Rather than replacing these provisions with standards adopted pursuant to Section 6(b), OSHA has relied on other mandatory language general industry standards, or upon the general duty clause, to enforce the Act in areas previously covered by the deleted standards.[28]

---

[21]41 U.S.C. §§35–45. Established federal standards were also derived from the Longshoremen's and Harbor Workers' Compensation Act, 33 U.S.C. §§901–950, and the Contract Work Hours and Safety Standards Act, 40 U.S.C. §§327–333.

[22]Table Z-2 of 29 C.F.R. §1910.1000 was derived from ANSI Z-37, while Tables Z-1 and Z-3 were derived from the Walsh-Healey established federal standards, 41 C.F.R. §50-204.50, which were based on recommendations of the American Conference of Governmental Industrial Hygienists.

[23]36 FED. REG. 10,466 (1971).

[24]See, e.g., *Oversight on the Administration of the Occupational Safety and Health Act, 1980: Hearing Before the Senate Comm. on Labor and Human Resources*, 96th Cong., 2d Sess. 730–731 (1980); *Occupational Safety and Health Act Review, 1974: Hearings Before the Subcomm. on Labor of the Senate Comm. on Labor and Human Resources*, 93d Cong., 2d Sess. 52-56 (1974).

[25]43 FED. REG. 49,726–727 (1978).

[26]See, e.g., *Edward Hines Lumber Co.*, 4 OSHC 1735 (Rev. Comm'n 1976); *Pan Am. World Airways*, 4 OSHC 1203 (Rev. Comm'n 1976). Also, OSHA may not modify advisory standards to include mandatory language without formal rulemaking procedures. *Marshall v. Union Oil Co.*, 616 F.2d 1113, 8 OSHC 1169 (9th Cir. 1980); *Usery v. Kennecott Copper Corp.*, 577 F.2d 1113, 1117 (10th Cir. 1977); *Cargill, Inc.*, 5 OSHC 1832 (Rev. Comm'n 1976).

[27]49 FED. REG. 5318 (1984).

[28]47 FED. REG. 23,477–478 (1982).

B. Section 6(b) Standards

While Section 6(a) standards provide a baseline of protection, it was understood by Congress that the preexisting standards in most instances represented the lowest common denominator[29] and thus would need to be modified to provide meaningful protection. Section 6(b) contains the detailed procedures for accomplishing that objective, i.e., for promulgating new "health or safety" standards and for modifying or revoking any standard, including the existing 6(a) ("start-up" or "early") standards.[30] These procedures have proven to be extremely time-consuming, and have frequently resulted in lengthy legal challenges.[31] Consequently, there have been extremely few modifications or additions to the safety standards since 1971.[32]

Virtually all the standards promulgated under Section 6(b) have been health standards, primarily because a prominent factor triggering the passage of the Act had been the growing awareness of occupational illness and death resulting from exposure to toxic substances in the workplace.[33] Accordingly, the Secretary was directed, pursuant to Section 6(b)(5), to promulgate "standards dealing with toxic materials or harmful physical agents" so that "no employee will suffer material impairment of health or functional capacity."[34] In issuing such standards, the Secretary was instructed to rely on research, demonstrations, experiments, and appropriate in-

---

[29]"These standards may not be as effective or as up-to-date as is desirable, but they will be useful for immediately providing a nationwide minimum level of health and safety." S. REP. NO. 1282, 91st Cong., 2d Sess. 6 (1970), *reprinted in* Subcommittee on Labor of the Senate Comm. on Labor and Public Welfare, 92d Cong., 1st Sess., LEGISLATIVE HISTORY OF THE OCCUPATIONAL SAFETY AND HEALTH ACT OF 1970, at 146 (Comm. Print 1971).

[30]29 U.S.C. §655(b).

[31]For example, OSHA issued a benzene standard in February 1978, 43 FED. REG. 5918, which was ultimately vacated by the courts in 1980. See *Industrial Union Dep't v. American Petroleum Inst.*, 448 U.S. 607 (1980). A final rule lowering the PEL from 10 ppm to 1 ppm was promulgated on Sept. 1, 1987. It was immediately challenged by the United Steelworkers. For another example, OSHA issued a new lead standard in November 1978, 43 FED. REG. 52,952, which was upheld in large part by the District of Columbia Circuit in 1980. (D.C. Cir. 1980, *United Steelworkers v. Marshall*, 647 F.2d 1189 (D.C. Cir. 1980), *cert. den.*, 453 U.S. 913 (1981). OSHA was required to submit additional documentation supporting feasibility of compliance findings for certain industries, which remains under consideration by the court of appeals to date.

[32]See note 9 above. OSHA has also issued final revisions to the national consensus standard for electrical hazards. 46 FED. REG. 4034 (1981), codified at 29 C.F.R. §1910.301 *et seq.*

[33]H. REP. NO. 1291, 91st Cong., 2d Sess. 15 (1970).

[34]29 C.F.R. §655(b)(5).

formation and to consider "the latest available scientific data in the field, the feasibility of the standards, and experience gained under this and other health and safety laws." Finally, the standards were to be written in terms "of objective criteria and the performance desired," whenever that was practicable.[35]

Section 6(b)(7) and Section 8(c) of the Act list a series of substantive provisions that are to be incorporated into the 6(b) standards where appropriate.[36] These include *labels*[37] to provide hazard information, *protective equipment*[38] and *controls* or other technological procedures, *monitoring*[39] and *measuring employee exposure*,[40] and *medical examinations* (Section 6(b)(7)).[41] Sections 8(c)(1) and (3) emphasize the employee information elements of a standard through provisions for posting of notices, observation of exposure monitoring and access to the results, and notice to workers of the corrective action being taken to reduce excessive exposure.[42] Implementation of these requirements in the context of standards governing toxic substances is discussed in Section IV below. These detailed requirements of employee information are also covered by general provisions, e.g., the Access to Medical and Exposure Records Standard[43] and the Hazard Communication Standard,[44] both of which are discussed in Section II below. The remainder of this chapter is divided between discussions of general health and safety standards (Section II), specific safety standards (Section III), and specific health standards (Section IV).[45]

---

[35]*Id.*

[36]See appendix to Chapter 5 (Employer Obligations to Obtain, Maintain, and Disseminate Information) for examples of §6(b)(7) provisions in health-related standards.

[37]See Section IV.B.2.e, below.

[38]See Section IV.B.2.d, below.

[39]See Section IV.B.2.b, below.

[40]See Section IV.B.2.a, below.

[41]See Section IV.B.2.h, below.

[42]29 U.S.C. §657(c)(1) and (3). See generally Chapter 5.

[43]29 C.F.R. §1910.20 *et seq.*

[44]§1910.1200 *et seq.*

[45]This chapter does not deal with a third category of standards, Emergency Temporary Standards (ETSs), promulgated under §6(c) of the Act, 29 U.S.C. §655(c). Under this section, if the Secretary determines that employees are "exposed to grave danger from exposure to substances or agents determined to be toxic or physically harmful or from new hazards," an ETS may be issued, to be effective for 6 months. Thereafter, the Secretary must promulgate a permanent standard under §6(b). For a discussion of the ETS procedure and examples of such standards, see Chapter 19 (The Development of Occupational Safety and Health Standards).

II. General Industry Safety and Health Standards

Safety and health standards have been categorized in various ways, depending upon the extent of coverage, the specificity of detail provided, and the authority under which the standard is promulgated (i.e., 6(a) or 6(b)).[46] As a general rule, safety standards have been based on objective and easily definable criteria, designed primarily to protect workers from their own or others' negligence. Unlike health standards, which involve medically and scientifically determined relationships between exposure and long-term potential for adverse health consequences, safety standards involve protection from hazards that could cause immediate physical injury, such as unguarded machines and exposed fall hazards.

In light of the detailed requirements for safeguards within the employees' physical environment, safety standards are generally "specification standards"; that is, the particular means of protection are specifically set forth in the OSHA regulations. For example, a minimum width is prescribed for the space between siderails of ladders,[47] and machine guards for blades less than 7 feet above the floor may not have openings greater than ½ inch.[48] By contrast, OSHA has utilized "performance standards" for employee protection against many toxic materials and harmful physical agents; under these standards OSHA sets forth requirements in terms of end results, and the employer may select the method by which these results may be achieved.[49]

Safety and health standards have also been categorized as "vertical," i.e., applicable to a specific industry, or "horizontal," i.e., applicable to general industry.[50] This chapter focuses on horizontal standards; the vertical standards applicable to the construction, maritime, agriculture, diving, and telecommunications industries are discussed elsewhere.[51] Fi-

---

[46]See Section I above for a description of the procedures underlying the promulgation of standards under §§6(a) and 6(b).

[47]29 C.F.R. §1910.25(c)(2)(i)(C).

[48]§1910.212(a)(5).

[49]See, e.g., the hearing conservation amendment to the OSHA Standard for Occupational Exposure to Noise, 29 C.F.R. §1910.95(c). See Section IV.A.2 below, and Tables Z-1, Z-2, and Z-3 of 29 C.F.R. §1910.1000, establishing threshold limit values for certain air contaminants.

[50]See Chapter 6 (Special Obligations of Certain Employers).

[51]See Chapter 6.

nally, there is a frequently utilized distinction between "general" and "specific" standards that relates to whether the standard applies to a variety of workplace settings or to particular equipment or operations. Thus, for example, there are general provisions regarding respirator use when employees may be exposed to toxic substances,[52] as well as respiratory protection requirements contained in the standards covering specific toxic substances.[53] This section focuses on general safety and health standards applicable to general industry.

A. Coverage of General Industry Safety and Health Standards

General Industry and Health Standards,[54] which are contained in Part 1910 of Title 29 of the *Code of Federal Regulations*, apply to nearly all private employers engaged in a business affecting commerce in the United States and its territories.[55] In addition, each federal agency must establish safety and health programs including these standards for coverage of federal employees,[56] and states may establish such programs for state and municipal employees.[57]

1. *Employment Conditions Regulated by Other Federal Agencies*

The General Industry Standards do not apply in some cases to employers such as airlines, mines, government con-

---

[52]29 C.F.R. §1910.134.

[53]See, e.g., the lead standard, 29 C.F.R. §1910.1025(f).

[54]This section discusses those standards that have both safety and health aspects, such as first-aid requirements. Specific safety standards are discussed in Section III below, while specific health standards are discussed in Section IV.

[55]29 U.S.C. §§652(5) and 653. See, e.g., *Austin Road v. OSHRC*, 683 F.2d 905, 10 OSHC 1943 (5th Cir. 1982); *Val-Pak, Inc.*, 11 OSHC 2094 (Rev. Comm'n 1984). Agricultural employers are not subject to the general industry standards, with minor exceptions, for they are covered by agricultural standards set forth in 29 C.F.R. §§1928.1–.57. Churches that do not employ persons in secular activities and individuals who employ housekeepers for their residences are not covered. 29 U.S.C. §§1975.4(c) and 1975.6.

[56]29 U.S.C. §668.

[57]In order to assume responsibility for occupational safety and health, a state must obtain OSHA approval for a state plan. 29 U.S.C. §667; see Chapter 23 (Role of States in Occupational Safety and Health). State plans must include provisions for coverage of state and municipal employees. 29 U.S.C. §667(c)(6).

tractors, motor carriers, and railroads whose working conditions are governed by regulations promulgated by other federal agencies.[58] This exception is applied according to specific working conditions rather than on an industrywide or companywide basis. Thus, for example, a railroad is covered by OSHA regulations with respect to certain nonoperating personnel who do not work near the tracks, but is governed exclusively by the Department of Transportation with respect to specific over-the-road operations.[59] In addition, OSHA has reached jurisdictional agreements with the Federal Mine Safety and Health Administration and the Food and Drug Administration, which define health and safety matters covered by each agency.

2. Precedence of Industry-Specific Standards

Certain industries, such as construction, longshoring, and ship repair, shipbuilding, and shipbreaking are subject to separate, industry-specific safety standards, which deal with particular safety problems inherent in those industries.[60] Although the general industry regulations also apply to these industries, the specific standards must take precedence.[61] To the extent that the specific standards do not cover a particular safety issue, the general standards will apply.[62] This principle has been applied, for example, to a shipbuilding and drydock firm where an administrative law judge found that general industry standards were applicable to a scaffold used for paint-

---

[58] 29 U.S.C. §653(b)(1).

[59] *Velasquez v. Southern Pac. Transp. Co.*, 734 F.2d 216, 11 OSHC 2060 (5th Cir. 1984); *Southern Ry. v. OSHRC*, 539 F.2d 335, 3 OSHC 1940 (4th Cir.), *cert. denied,* 429 U.S. 999, 4 OSHC 1936 (1976). See also *Donovan v. Red Star Marine Servs.*, 739 F.2d 774, 11 OSHC 2049 (2d Cir. 1984); *Illinois Cent. Gulf R.R.*, [1977–1978] OSHD ¶22,148 (Rev. Comm'n 1977). A more complete discussion of this exemption is set forth in Chapter 15 (Defenses), Section II.B.

[60] Specific construction standards are set forth in 29 C.F.R. §1926.1 *et seq.*; longshoring in §1981.1 *et seq.*; ship repair, shipbuilding, and shipbreaking standards in §1951.1 *et seq.*; pulp, paper, and paperboard mills in §1910.261; textiles in §1910.262; bakeries in §1910.263; laundries in §1910.264; sawmills in §1910.265; pulpwood logging in §1910.266; and telecommunications in §1910.268. Moreover, as noted in note 55, *supra*, agricultural employers are covered by agricultural standards, to the exclusion of general industry standards, under 29 C.F.R. pt. 1928. All of these standards are discussed in detail in Chapter 6 (Special Obligations of Certain Employers).

[61] 29 C.F.R. §1910.5(c)(1).

[62] §1910.5(c)(2).

ing a sign, despite broad coverage of the employer by maritime standards.[63]

3. Qualification on Employer Responsibilities

Although an employer may be covered by general industry safety standards, not all individuals working on the employer's behalf will be considered "employees" subject to the general safety requirements. Responsibility for compliance with OSHA standards depends on the extent to which the employer exercises control over working conditions and activities. Regardless of contractual arrangements between a general contractor and a subcontractor, for example, employees loaned to the subcontractor will remain the responsibility of the general contractor.[64] Where the employees are exposed to a hazard created by a second company, the employer may be cited for failure to take steps to protect its employees, and the second company may be cited for creation of, and failure to abate, the hazard.[65] (A more complete discussion of employee coverage is set forth in Chapter 15 (Defenses)).

B. General Safety and Health Standards Requiring Protection of Employees

1. Personal Protective Equipment

The standard for personal protective equipment is set forth in Subpart I to Part 1910 of C.F.R. Title 29.

(a) The general requirements provision. The general requirements provision directs the employer to ensure that protective equipment such as helmets, protective clothing, respirators, safety shoes, gloves, and safety belts is used "wherever it is necessary by reasons of hazards of processes or environment."[66] The five federal circuit courts of appeals

---

[63]*Maryland Shipbuilding & Drydock Co.*, 75 OSAHRC 85/E9 (Rev. Comm'n J. 1974), *vacated on other grounds*, 3 OSHC 1985 (Rev. Comm'n 1976).

[64]*Scotty Smith Constr. Co.*, 2 OSHC 1329 (Rev. Comm'n 1974); *Grossman Steel & Aluminum Corp.*, 2 OSHC 3200 (Rev. Comm'n 1974).

[65]*Anning-Johnson Co.*, 4 OSHC 1193 (Rev. Comm'n Nos. 3694 & 4409, 1976); *Grossman Steel & Aluminum Corp.*, 4 OSHC 1185 (Rev. Comm'n 1975).

[66]29 C.F.R. §1910.132. Safety belts have only recently been found by the Review Commission to fall within this standard. *Bethlehem Steel Corp.*, 10 OSHC 1607 (Rev. Comm'n 1982).

that have considered the "wherever it is necessary" language of 29 C.F.R. §1910.132 have ruled that the standard is not unenforceably vague.[67] Like the general duty clause, this section is a "catchall" provision, designed to provide protection for employees in situations not covered by more specific standards.

The application of Section 1910.132 to specific situations involves evaluating whether a "reasonable man" would require the use of protective equipment, considering common understanding, industry practice, the employer's safety record, and the particular circumstances of the job in question.[68] Though industry custom and practices are important factors in determining whether employer procedures are reasonable, an employer must make an independent judgment in deciding the hazards for which to seek protection.[69] Reliance on industry practice in the face of a readily apparent hazard and experienced injuries is not a defense to a citation under this section.[70]

The employer is not required to pay for the purchase of protective equipment unless explicitly required to do so by statute or by regulation.[71] Otherwise, such payment is a matter of negotiation between the parties.[72] The employer must nonetheless provide appropriate protective equipment and, in addition, must enforce its use pursuant to a consistently applied and adequately communicated program.[73]

(b) Specific standards. The specific personal protective equipment safety standards of Subpart I are national consen-

---

[67]*Arkansas-Best Freight Sys. v. OSHRC*, 529 F.2d 649, 3 OSHC 1910 (8th Cir. 1976); *Brennan v. Smoke-Craft, Inc.*, 530 F.2d 843, 3 OSHC 2000 (9th Cir. 1976); *McLean Trucking Co. v. OSHRC*, 503 F.2d 8, 2 OSHC 1165 (4th Cir. 1974); *Ryder Truck Lines v. Brennan*, 497 F.2d 230, 2 OSHC 1075 (5th Cir. 1974); *Cape & Vineyard Div., New Bedford Gas & Edison Light Co. v. OSHRC*, 512 F.2d 1148, 3 OSHC 1401 (1st Cir. 1975). See also *General Dynamics Corp.*, 10 OSHC 1188 (Rev. Comm'n 1981).

[68]*Marshall v. Haysite Div. of Synthane Taylor Corp.*, 9 OSHC 1443 (3d Cir. 1980); *McLean Trucking Co. v. OSHRC, supra* note 67; *Ryder Truck Lines v. Brennan, supra* note 67; *Cape & Vineyard Div., New Bedford Gas & Edison Light Co. v. OSHRC, supra* note 67; *Coca-Cola Bottling Co.*, 11 OSHC 1728 (Rev. Comm'n 1983).

[69]*Cotter & Co.*, 5 OSHC 2046 (Rev. Comm'n 1977).

[70]*Farrar Corp.*, 7 OSHC 1193 (Rev. Comm'n Nos. 78-190 & 78-503, 1978).

[71]*Budd Co. v. OSHRC*, 513 F.2d 201, 2 OSHC 1698 (3d Cir. 1975).

[72]*Arkansas-Best Freight Sys. v. OSHRC, supra* note 67; *Budd Co. v. OSHRC, supra* note 71. Certain health standards, such as lead and arsenic, specifically require employees to pay for protective equipment mandated by those standards. See 29 C.F.R. §1910.1018(h)(2)(i), (j)(1) (inorganic arsenic); §1910.1025(f)(1), (g)(1) (lead).

[73]*Miller Well Serv.*, 10 OSHC 1966 (Rev. Comm'n 1982); *Mack Trucks*, 1 OSHC 3169 (Rev. Comm'n 1972).

sus standards, derived from ANSI.[74] Protective eye and face equipment is specifically required where "machines or operations present the hazard of flying objects, glare, injurious radiation, or a combination of these hazards."[75] Such equipment must meet ANSI standards and fit comfortably, snugly, and in a manner adequate to provide protection.[76] In general, only those protective measures the industry would deem appropriate under the circumstances must be provided, except in instances where the employer has acknowledged that a hazard requires extra protection.[77] Before a citation will be affirmed, the Secretary of Labor must show that a hazard exists which could be prevented or protected against by such personal equipment.[78]

As with the general requirements section, the standard implicitly mandates that employers ensure use of the equipment by employees.[79] Whether employees actually desire to utilize the protective equipment is not relevant; the employer must implement the policy embodied in the standard.[80] Safety responsibility is not delegable to employees.[81] Even if the employer shows good faith in establishing a safety program, it may be held responsible under the standard if the program is inadequately enforced,[82] or if the employer could have prevented the non-use of protective equipment through specific steps shown by the Secretary.[83]

Additional requirements contained in Subpart I incorporate ANSI standards for head and foot protection and for electrical protective devices.[84] These standards do not define

---

[74]See 29 C.F.R. §1910.139.

[75]29 C.F.R. §1910.133(a)(1).

[76]§1910.133(a)(2)–(6).

[77]*Florida Mach. & Foundry v. OSHRC*, 693 F.2d 119, 10 OSHC 1049 (11th Cir. 1982).

[78]*Kit Mfg. Co.*, 2 OSHC 1672 (Rev. Comm'n 1975).

[79]*General Elec. Co. v. OSHRC*, 540 F.2d 67 (2d Cir. 1976).

[80]*Florida Mach. & Foundry*, 10 OSHC 1049 (Rev. Comm'n 1981), *aff'd*, 693 F.2d 119, 10 OSHC 1049 (11th Cir. 1982).

[81]*Utica Steam Engine & Boiler Works*, 10 OSHC 1011 (Rev. Comm'n 1981).

[82]*Mako Constr. Co.*, 10 OSHC 1135 (Rev. Comm'n 1981).

[83]*General Elec. Co. v. OSHRC*, 540 F.2d 67, 4 OSHC 1512 (2d Cir. 1976).

[84]29 C.F.R. §§1910.135–137. OSHA is currently in the process of revising existing standards for eyes, face, head, and foot protection, and anticipates final action by June 1988. 51 FED. REG. 38,560 (1986). OSHA has also proposed adding to Subpart I standards governing "personal fall protection systems" aimed at enhancing employee protection from injury and death due to falls from high elevations. Final action is anticipated by August 1988. 51 FED. REG. 38,560 (1986).

any specific situations in which protection is necessary. Instead, they simply mandate that hard hats, safety-toe footwear, and insulating equipment that are required by other OSHA standards must meet ANSI standards. Failure to ensure use of such protection may be the subject of a citation under a specific standard, the general protective equipment standard, or the general duty clause. The "reasonable man" test has been applied to the use of hard hats in warehousing operations.[85] Familiarity with industry conditions will also be considered in evaluating a head or foot protection violation.[86]

(c) Respiratory protection. Respiratory protection must be provided as set forth in the general respirator provision of Subpart I[87] and in several specific (Section 6(b)) health standards.[88] OSHA has deemed respirators necessary to protect employee health from the adverse effects of air contaminants such as dust, gases, and vapors. The standard is derived from a 1969 ANSI standard.[89] Since its adoption in 1971, there has been only one set of amendments, which were finalized on February 10, 1984. On that occasion, the Secretary amended several provisions and revoked one subsection to delete references to advisory language that had been held unenforceable when contained in other standards.[90]

---

[85]*Safeway Stores*, 9 OSHC 1880 (Rev. Comm'n 1981).

[86]See, e.g., *Pratt & Whitney Aircraft v. Secretary of Labor*, 649 F.2d 96, 9 OSHC 1554 (2d Cir. 1981).

[87]29 C.F.R. §1910.134.4.

[88]Virtually every specific health standard contains a requirement for the use of respiratory protection, and many incorporate the provisions of §1910.134. See, e.g., 29 C.F.R. §1910.1003(c)(4)(iv) (4-nitrobiphenyl); §1910.1025(f)(4)(i) (lead).

[89]OSHA has proposed revising the respiratory protection standards to reflect recent technological advances and to remove or modify redundancies and advisory provisions. Final action is anticipated by January 1988. 51 FED. REG. 38,560 (1986).

[90]49 FED. REG. 5318 (1984). The main impediment to enforcement results from the use of the advisory term "should" rather than the mandatory "shall." As noted in Section I.A above, the courts have held that OSHA cannot change "should" to "shall" without rulemaking. *Usery v. Kennecott Copper Corp.*, 577 F.2d 1113, 1117 (10th Cir. 1977). Also, OSHA cannot enforce a standard written in advisory terms. *Marshall v. Pittsburgh Des Moines Steel Co.*, 584 F.2d 638, 643–644 (3d Cir. 1978). Therefore, OSHA deleted any advisory references that set out a preference for individual workers to be assigned their own respirators. However, when OSHA revised 29 C.F.R. §1910.134, some of the advisory language was retained (e.g., §1910.134(b)(10) dealing with physical examinations) because OSHA thought that such guidance would be helpful (49 FED. REG. 5399–5420 (1984)).By doing so, however, OSHA may have restricted its ability to issue general duty citations for failure to provide a medical evaluation of a respirator user since the Review Commission has held that as long as the "should" standard remains on the books, a general duty citation is not appropriate. *A. Prokosch & Sons Sheet Metal*, 8 OSHC 2077 (Rev. Comm'n 1980).

As presently constituted, the respiratory protection standard contains provisions that emphasize the primacy of engineering controls[91] but recognizes that in certain limited circumstances—where such controls are either not feasible or are still being implemented—appropriate respirator usage can be required as an interim compliance technique. In such a case, the basic requirement of the respirator standard mandates the preparation of written standard operating procedures for the selection and use of the respirators.

Unlike the Section 6(b) standards, which generally contain a respirator selection table that enables an employer to determine the differing types of respirators required to protect employees at various exposure levels, the general respirator standard refers to the source ANSI standard[92] and requires that respirators be selected "on the basis of *hazards to which the worker is exposed.*"[93] For example, a gas respirator would not be sufficient to protect an employee against a dust hazard.

The standard also mandates procedures for the use of respirators.[94] A typical respirator program includes inspection by a qualified individual "to assure that respirators are properly selected, used, cleaned and maintained."[95] Moreover, the users and supervisors must be instructed by competent persons in the proper selection, use, and maintenance of the respirators. Such training includes fitting the respirator, testing its facepiece-to-face seal, wearing it in a normal atmosphere, and wearing it in a test atmosphere that simulates workplace conditions.[96]

---

[91]Such controls include enclosure or confinement of the operation, general or local ventilation, and substitution of less toxic materials. 29 C.F.R. §1910.134(a)(1).

[92]§1910.134(c).

[93]§1910.134(b)(2).

[94]§1910.134(e).

[95]§1910.134(e)(4).

[96]§1910.134(e)(5). Fit testing can be done by either qualitative or quantitative means. Qualitative fit testing involves a test subject's responding (either voluntarily or involuntarily) to a chemical substance such as isoamyl acetate (banana oil) or irritant smoke that is released around the outside of the respirator seal. If the respirator does not seal properly, the wearer will either smell the banana oil or react to the irritant smoke by coughing or sneezing. Quantitative fit testing involves exposing the respirator wearer to a test containing an easily detectable, relatively nontoxic aerosol, vapor, or gas as the test agent and then measuring the penetration of the test agent into the respirator. OSHA Industrial Hygiene Technical Manual, March 30, 1984, p. V-9. Either method meets the test atmosphere requirement of the standard. The facepiece seal can be spot-checked by the individual for most types of negative pressure respirators by either a negative or positive pressure fit test. The latter two tests are frequently described in the manufacturer's instructions.

Because the purpose of the respirator is to filter out the contaminated air, it is important that nothing interfere with the fit of the respirator to the face. Therefore, the general respirator standard requires that respirators shall not be worn when conditions prevent a proper face seal, for example, beards, sideburns, frames for glasses, or missing dentures.[97] In retrospect, the greatest impact of this provision has been to limit beards in areas where respirators are required.[98]

The standard specifies four basic elements as part of a required maintenance program: inspection for defects, cleaning and disinfecting, repair, and storage.[99] Each of these elements is further described in the standard. For example, respirators are to be stored so as to protect them from dust, sunlight, heat, extreme cold, excessive moisture, or damaging chemicals. Emergency respirators are to be located in clearly marked accessible compartments.[100]

Many companies have their own centralized program of inspecting, cleaning, and repairing respirators: the respirators are turned in after each shift and returned or replaced the next day in good condition with the proper protective elements. However, the standard does not specify who is to maintain the respirators. While the employer has the ultimate responsibility under the Act to provide a safe workplace, the standard by its silence allows employers to delegate to employees the day-to-day job of inspecting, cleaning, and storing respirators. In such circumstances, the employer still must provide training and the appropriate facilities for the employees to do this. Furthermore, if an OSHA inspector finds unclean or broken respirators in use, the employer is subject to citation.

Most of the preceding description applies to negative pressure respirators and those respirators powered by a battery pack, i.e., positive pressure respirators. There are, however, respirators that depend on a source of supplied air other than that filtered directly from the work environment. These include supplied-air respirators or self-contained breathing apparatus. The standard also discusses these devices in terms

---

[97]§1910.134(e)(5)(i).

[98]See, e.g., *American Smelting & Refining Co.*, 69 Lab. Arb. 824 (1977); *Phillips Petroleum Co.*, 74 Lab. Arb. 400 (1980).

[99]§1910.134(f)(1).

[100]§1910.134(f)(5)(i).

of the requisite degree of air quality and procedures for safe usage.[101]

2. Safety Training

In connection with the implementation of many safety standards, OSHA has required various safety training programs to be established by employers. In preparing such a program, employers must provide information that is required under specific safety and health standards.[102] Safety programs must also include training in other recognized hazards in the workplace which are not covered by specific standards, based on the general duty clause. Thus, employees must be provided with information as to all safety considerations associated with various components of their jobs. An established safety training program may be helpful in defending against a citation; for if the program is normally adequate to prevent violations of a particular standard, a violation may be found not to have been preventable.[103] Moreover, the program may be evidence of an employer's good faith with respect to the assessment of a penalty.[104]

OSHA has enacted "voluntary training guidelines" to assist employers in establishing health and safety training programs.[105] Although the guidelines are not mandatory, OSHA intended to aid employers in determining whether a health and safety program would be helpful in their particular workplace. The voluntary guidelines are not, of course, designed to override the particular training mandated by many safety standards. OSHA also offers employers numerous pamphlets and audiovisual presentations pertaining to training programs.[106]

---

[101]In 29 C.F.R. §1910.134(d), the air quality for airline respirators that are fed from compressors is described in terms intended to avoid the influx of contaminated air. In addition, the need for standby personnel and safety lines is spelled out when these self-contained or airline respirators are used in emergencies or in atmospheres immediately dangerous to life or health. 29 C.F.R. §1910.134(e)(3).

[102]See, e.g., 29 C.F.R. § 1910.134(a) (standards for respirator training).

[103]*Howard P. Foley Co.*, 5 OSHC 1501 (Rev. Comm'n 1977).

[104]*United Parcel Serv.*, 4 OSHC 1421 (Rev. Comm'n 1976).

[105]49 FED. REG. 30,290 (1984).

[106]Contact either the National Audiovisual Center, Washington, D.C. 20409, or the Superintendent of Documents, Government Printing Office, Washington, D.C. 20402.

3. Medical and First Aid

Under this standard,[107] which was adopted as an established federal standard,[108] employers must provide first-aid equipment and personnel where an "infirmary, clinic or hospital" is not in "near proximity" to the workplace. In *Brennan v. Santa Fe Trail Transport Co.*,[109] the Tenth Circuit upheld the "near proximity" language in the face of a challenge that it was unconstitutionally vague. In that case, the court found that medical facilities located 4 to 10 minutes from the workplace were not in near proximity since serious harm or death could occur within several minutes in the absence of prompt first-aid training. Therefore, first-aid training was required.[110]

The requirements of the section will be met where the employer has a commercial first-aid kit or first-aid supplies, and an employee available with Red Cross or appropriate military training.[111] Medical personnel have been held to be "readily available," within the language of the section, where doctors and nurses lived nearby and guards and supervisors at the workplace were trained in first aid.[112]

4. Employee Access to Information

The need for employee access to records is addressed in Section 6(b)(7) and Section 8(c)(3) of the Act. Initially, such access was provided on a case-by-case basis in each standard; but in 1980 OSHA promulgated a regulation governing access to and maintenance of employee exposure, monitoring, and medical records.[113] That regulation governs access not only to all records that the employer voluntarily maintains but also to the records required to be compiled and maintained by

---

[107]29 C.F.R. §1910.151.

[108]§1910.153.

[110]Similar disputes have arisen under 29 C.F.R. §1910.151(c) dealing with "suitable facilities" to provide water to remove injurious corrosive materials from a worker's body or eyes. See, e.g., *Lee Way Motor Freight*, [1973–1974] OSHD ¶17,693, *aff'd on other grounds, Lee Way Motor Freight v. Secretary of Labor*, 511 F.2d 864 (10th Cir. 1975).

[111]*Lee Way Motor Freight*, 1 OSHC 3197 (Rev. Comm'n 1973).

[112]*Toms River Chem. Corp.*, 6 OSHC 2192 (Rev. Comm'n Nos. 76-5197 & 76-5282, 1978).

[113]29 C.F.R. §1910.20.

OSHA, including those under the specific health standards promulgated under Section 6(b).[114]

Employees may also secure information regarding health conditions in their workplace under the generic Hazard Communication Standard, promulgated by OSHA in November 1983.[115] That standard requires labels and material safety data sheets for all chemicals determined by chemical manufacturers to be hazardous, including those for which standards have been promulgated under Section 6(b) and those which are listed in 29 C.F.R. §1910.1000. In addition, all employers in the manufacturing sector (Standard Industries Classification Codes 20-39) have the obligation to prepare a written Hazard Communication Program, which must include provisions for employee training in workplace hazard identification and protection, and maintenance or preparation of labels and material safety data sheets for all hazardous substances before they are handled by employees or sent to downstream users. In mid-1987, this requirement was extended to all employers.

III. Specific Safety Standards Requiring Protection in the Workplace

A. Structural Protection

1. Subpart D: Walking-Working Surfaces

Subpart D is designed to guard against falling and tripping hazards.[116] The initial two sections set forth specific def-

---

[114]This regulation was upheld in its entirety by the Fifth Circuit in *Louisiana Chem. Ass'n v. Bingham*, 550 F. Supp. 1136 (W.D. La. 1982), *aff'd per curiam*, 731 F.2d 280 (5th Cir. 1984).

[115]29 C.F.R. §1910.1200. This standard was upheld by the Third Circuit in *United Steelworkers v. Auchter*, 763 F.2d 728 (3d Cir. 1985). In addition, the court ordered OSHA to extend the standard to nonmanufacturing employers or to justify why it should only apply to manufacturing employers. The United Steelworkers petitioned the court to hold the Assistant Secretary of Labor in contempt for failure to comply with the court's order. *United Steelworkers v. Pendergrass*, No. 83-3554 (3d Cir. filed Jan. 21, 1987). In August 1987, OSHA published a final rule extending the standard to all workers. 52 FED. REG. 31,852 (Aug. 24, 1987).

[116]29 C.F.R. §1910.21–.31.

initions for terms used in Subpart D and establish general requirements for walking and working surfaces, respectively.[117] These general requirements include a housekeeping provision requiring that all places of employment be "clean" and "orderly,"[118] which has been held not to be unenforceably vague.[119] So long as an employer can show reasonable efforts to maintain cleanliness, no violation of this section will be found.[120] On the other hand, where conditions are so disorderly as to create an unsafe environment, compliance with the standard will be enforced.[121]

Under this subpart, aisles and passageways must be provided with sufficient "safe clearance" where mechanical equipment (e.g., forklifts) is utilized.[122] Sufficient clearance has been defined by OSHA as that which will prevent an employee from being caught between an obstruction and a moving object.[123] This is a performance standard, for the regulations do not mandate any particular method for ensuring clearance. Likewise, employees must be protected by use of "covers and/or guardrails" against the dangers of open pits, tanks, vats, and ditches.[124] The existence of a hazard will be assumed with respect to this provision, and need not be proven by the Secretary.[125] The final general requirements provision mandates "floor loading" protection, under which a building owner must prepare plates setting forth the maximum weights approved by a "building official" for the building.[126]

---

[117]§1910.21–.22.

[118]§1910.22(a)(1).

[119]*Plessy, Inc.*, 2 OSHC 1302 (Rev. Comm'n 1974).

[120]*Con Agra, Inc.*, 12 OSHC 1016 (Rev. Comm'n 1984); *Hughes Tool Co.*, 7 OSHC 1666 (Rev. Comm'n Nos. 78-1490 & 78-1932, 1979); *Central Meat Co.*, 5 OSHC 1432 (Rev. Comm'n 1977).

[121]*Farmers Coop. Grain & Supply Co.*, 10 OSHC 2086 (Rev. Comm'n 1982); *Myron Nickman Co.*, 1 OSHC 3215 (Rev. Comm'n 1973).

[122]29 C.F.R. §1910.22(b)(1). See, e.g., *Winfrey Structural Concrete Co.*, 10 OSHC 1270 (Rev. Comm'n J. 1981).

[123]U.S. Dep't of Labor, OSHA Form 2095, Q's and A's to Part 1910 (December 1973).

[124]29 C.F.R. §1910.22(c). See, e.g., *Consolidated Rail Corp.*, 9 OSHC 1245 (Rev. Comm'n 1980).

[125]*Greyhound Line-West v. Marshall*, 575 F.2d 759, 6 OSHC 1636 (9th Cir. 1978); *Lee Way Motor Freight v. Secretary of Labor*, 511 F.2d 864, 3 OSHC 1843 (10th Cir. 1975).

[126]29 C.F.R. §1910.22(d). Citations will not be upheld under the section against employers who rent their workplaces, or in instances where a building official responsible for the design, construction permits, or consultation with respect to a building could not be located. *Cole Div. of Litton Sys.*, 7 OSHC 2145 (Rev. Comm'n Nos.

Under the more detailed requirements of Subpart D, floor and wall openings, platforms, runways, and stairways must be guarded by use of standard railings, guards, toeboards, covers, and screens.[127] This section does not apply to roof openings, even where a roof is used as a walking or working surface.[128] Platforms and open-sided floors 4 or more feet above ground must generally be guarded by a standard railing or other feasible method of protection, such as guardrails or safety belts.[129] OSHA has issued an instruction defining a platform as "any elevated surface designated or used primarily as a walking or working surface, and any other elevated surface worked or walked on in a regular performance of assigned tasks."[130] Where a work plank is found to be a "scaffold" rather than a platform, depending on the permanence of the device, guardrails are only required if the plank is higher than 10 feet.[131] Unlike the platform requirements, which apply only to surfaces used on a regular basis,[132] stairways must be fitted with guardrails even if they are temporary.[133] An exception from guarding "special purpose" runways is available, however, where operating conditions necessitate.[134]

OSHA has also promulgated detailed safety requirements for industrial stairs, portable wood and metal ladders, and

---

77-3432 & 78-2939, 1979); *Lee Way Motor Freight*, 4 OSHC 1968 (Rev. Comm'n 1977).

[127]29 C.F.R. §1910.23. See *Weatherby Eng'g Co.*, 9 OSHC 1292 (Rev. Comm'n 1981); *Central Soya of Puerto Rico*, 8 OSHC 2074 (Rev. Comm'n 1980).

[128]*Arkansas Rice Growers Coop. Ass'n*, 10 OSHC 1616 (Rev. Comm'n 1982).

[129]29 C.F.R. §1910.23(c)(1).

[130]OSHA Instruction STD 1.13 (April 16, 1984), OSHR [Reference File 21:8114]. OSHA Instructions are directives which remain in effect until specifically canceled or revised. The directives are issued identifying numbers by OSHA's Directives Officer. Those instructions relating to standards are labeled "STD" and those relating to compliance "CPL." On October 30, 1978, OSHA issued a number of directives pertaining to 29 C.F.R. §1910.23. See OSHA Instructions STD 1-1.4, OSHR [Reference File 21:8102] (markings for aisles and passageways); STD 1-1.5, OSHR [Reference File 21:8102–8103] (loading rack platforms); STD 1-1.6, OSHR [Reference File 21:8103] (clearance requirements between handrails and railings); STD 1-1.7, OSHR [Reference File 21:8103–8104] (guarding floor and wall openings); STD 1-1.8, OSHR [Reference File 21:8104] (open-sided metal pouring platforms).

[131]*Fleetwood Homes*, 8 OSHC 2125 (Rev. Comm'n 1980).

[132]*Donovan v. Anheuser-Busch*, 666 F.2d 315, 10 OSHC 1193 (8th Cir. 1981); *General Elec. Co. v. OSHRC*, 583 F.2d 61, 6 OSHC 1868 (2d Cir. 1978).

[133]*Morris Enter.*, 5 OSHC 1248 (Rev. Comm'n 1977); *Chromalloy Am. Corp.*, 8 OSHC 1188 (Rev. Comm'n 1979).

[134]29 C.F.R. §1910.23(c)(2). It is not sufficient to show merely that railings on both sides would render the work more difficult. *Agrico Chem. Co.*, 4 OSHC 1727 (Rev. Comm'n 1976).

fixed ladders.[135] These standards set forth inspection requirements, measurement specifications, materials to be used, and wear/corrosion rules for all stairs and ladders. In general, these regulations are explicit specification standards that must be reviewed carefully for compliance purposes.

Scaffolds, towers, and other working surfaces, such as dockboards and forging machine areas, are also addressed in Subpart D.[136] These regulations require employers to furnish scaffolds for work that cannot be performed safely from the ground or from solid construction. One court has held that this standard is not enforceably vague.[137] The Review Commission has refused to enforce part of the scaffolding standard, however, to the extent it was derived from an advisory ANSI standard.[138]

2. Subpart E: Means of Egress

OSHA standards in Subpart E are designed to establish safe methods of escape for employees in case of emergency.[139] "Egress" is divided into three distinct parts: the way of exit access, the exit, and the exit discharge.[140] Under the general requirements, exits must be "sufficient to permit prompt escape of occupants,"[141] and a reasonable man standard will be used in evaluating the efficacy of egress.[142] Mobile structures, such as vehicles and vessels, are exempt from these requirements.[143] The subpart also addresses the marking of exits

---

[135]29 C.F.R. §§1910.24–.27. See, e.g., *Nicor Drilling Co.*, 10 OSHC 1215 (Rev. Comm'n 1981); *Miller Well Serv.*, 10 OSHC 1966 (Rev. Comm'n 1982). OSHA has proposed revising the existing standard for ladders and similar climbing devices so as not to restrict "technological innovation." Final action is anticipated by August 1988. 51 FED. REG. 38,560 (1986).

[136]29 C.F.R. §§1910.28–.30.

[137]*Allis-Chalmers Corp. v. OSHRC*, 542 F.2d 27, 4 OSHC 1633 (7th Cir. 1976).

[138]*Weatherby Eng'g Co.*, 9 OSHC 1292 (Rev. Comm'n 1981); *Kennecott Copper Corp.*, 4 OSHC 1400 (Rev. Comm'n 1976), *aff'd*, 577 F.2d 1113, 6 OSHC 1197 (10th Cir. 1977). OSHA has proposed revising its Subpart D Scaffolding regulations and anticipates that the new standards will be enacted by August 1988. 51 FED. REG. 38,560 (1986).

[139]29 C.F.R. §§1910.35–.40. See, e.g., *Wolf Auto Sales*, 9 OSHC 1947 (Rev. Comm'n 1981).

[140]29 C.F.R. §1910.35(a).

[141]§1910.36(b)(1).

[142]*United Aircraft Corp., Pratt & Whitney Aircraft Div.*, 2 OSHC 1713 (Rev. Comm'n 1975).

[143]29 C.F.R. §1910.36(a). See *Fairbanks Well Serv.*, 5 OSHC 1873 (Rev. Comm'n 1977).

and locked exit doors, and defines the circumstances under which exits may be locked for security reasons.[144] These regulations are designed to avoid reliance for safety on a single exit,[145] and to allow for well-marked, unobstructed, and safely constructed exits.[146] OSHA may use advisory NFPA standards as guidelines for interpretation of the standard.[147]

3. *Subpart F: Powered Platforms, Manlifts, and Vehicle-Mounted Work Platforms*

Powered platforms for exterior building maintenance, manlifts, and vehicle-mounted elevating and rotating work platforms are addressed by Subpart F.[148] These standards do not mandate the use of such equipment, but rather set forth requirements that must be met if permanent powered platforms or similar devices are utilized in the workplace. "Basket" requirements for vehicle-mounted work platforms apply only to equipment designed as a personnel carrier.[149] OSHA stated in 1983 that some revisions may be forthcoming in Section 1910.66(b)(3), relating to stabilization systems for powered platforms. Proposed revisions that would permit use of "intermittent" rather than "continuous" stabilization systems, owing to new technology, were expected to become final by March 1987.[150]

4. *Subpart S: Electrical*

In 1981, a general industry standard regarding design safety requirements for electrical systems was adopted and designated new Subpart S.[151] The new regulations replace the detailed provisions of the National Electrical Code (NEC), which had been previously incorporated by reference, with

---

[144]*Spot-Bilt Inc.*, 11 OSHC 1998 (Rev. Comm'n 1984); 29 C.F.R. §§1910.36(d)(4), 1910.37(g)(i).

[145]*Fred Wilson Drilling Co. v. Marshall*, 624 F.2d 38, 8 OSHC 1921 (5th Cir. 1980); *United Aircraft Corp., Pratt & Whitney Aircraft Div., supra* note 142.

[146]29 C.F.R. §1910.36(b)(4), (d)(1); 29 C.F.R. § 1910.37(k)(2), (q)(1).

[147]*Gold Kist, Inc.*, 7 OSHC 1855 (Rev. Comm'n 1979).

[148]29 C.F.R. §§1910.66–.70.

[149]§1910.67. See Q's and A's to Part 1910, *supra* note 123.

[150]51 FED. REG. 38,560 (1986).

[151]46 FED. REG. 4034 (1981). See 29 C.F.R. §§1910.301–.39.

more general performance requirements. The subpart is designed to cover "all electrical equipment and installations used
to provide electric power and light to employees."[152] General
requirements for the safe maintenance of electrical systems,[153]
wiring design and wiring methods,[154] and special circumstances and hazardous locations are all set forth in the regulation rather than by the NEC.[155] Ground fault circuit
interrupters are required to be used by employers on construction sites.[156] Much of Subpart S has been reserved for future
promulgation of standards, covering such areas as design safety
standards for electrical systems, safety-related work practices,
safety-related maintenance requirements, and safety requirements for special equipment.[157] A performance-oriented safety-
related work practices standard has been proposed by OSHA
and final action is anticipated by April 1988.[158]

B. Machine Protection

1. Subpart O: Machinery and Machine Guarding

(a) General standards. Subpart O of the General Industry
Standards addresses the problem of machine guarding in the
workplace.[159] The first two sections in this subpart establish
definitions and general requirements for all machines in order
to "protect the operator and other employees in the machine
area from hazards such as those created by point of operation,
ingoing nip points, rotating parts, flying chips and sparks."[160]
The standard is performance-oriented, but does offer three
examples of a guarding system: barrier guards, two-hand tripping devices, and electronic safety devices.[161] The regulation
primarily deals with point of operation guarding, although
general requirements for revolving drums, fan blades, and

---

[152]§1910.301(a).
[153]§1910.303.
[154]§§1910.304–.305.
[155]§§1910.306–.307.
[156]§1910.304(b).
[157]§§1910.309–.398. See, e.g., *General Motors Corp. Chevrolet Motors Div.,* 10
OSHC 1293 (Rev. Comm'n 1982).
[158]51 FED. REG. 38,560 (1986).
[159]29 C.F.R. §§1910.211–.222.
[160]§§1910.211–.212.
[161]§1910.212(a)(1).

anchoring machinery are also included.[162] The Commission has held that machines must be guarded only where the Secretary can prove that a hazard exists.[163] Nonetheless, the lack of past injuries or the low probability of harm with proper machine operation will bear only on the gravity of the violation, not the existence of a hazard, where an employee may be exposed to injury.[164]

Where a hazard does exist, employers may not rely on employee compliance with safety rules as a substitute for machine guards.[165] On the other hand, because the physical guarding methods required do not depend on correct employee behavior, a machine may be found to be adequately guarded despite the occurrence of a serious injury to an employee who improperly reached into the machine.[166] The burden of proving impossibility of compliance lies with the employer. To meet this standard, the employer must show that a guard would preclude performance of the work or would pose a greater hazard and that alternative means of protection were unavailable.[167] Neither the failure of the manufacturer of the machine to supply a guard nor the increased cost of production justifies an employer's noncompliance with the standard.[168]

(b) Specific standards. The general requirements for machine guarding do not apply, and a citation based on those sections will be vacated, where a specific machine guarding standard is applicable, such as for woodworking equipment, abrasive wheel machinery, mills and calenders, mechanical power presses, hot forging machines, and power transmission belts.[169] Woodworking equipment such as saws, boring machines, sanding machines, and lathes are covered by a regu-

---

[162]§1910.212.

[163]*Collator Corp.*, 3 OSHC 2041 (Rev. Comm'n 1975).

[164]*Mayhew Steel Prods.*, 8 OSHC 1919 (Rev. Comm'n 1980); *Bethlehem Steel Corp.*, 7 OSHC 1607 (Rev. Comm'n 1979).

[165]*Akron Brick & Block Co.*, 3 OSHC 1876 (Rev. Comm'n 1976).

[166]*Westdale, Inc.*, 11 OSHC 2150 (Rev. Comm'n 1984).

[167]*Firman L. Carswell Mfg. Co.*, 11 OSHC 1871 (Rev. Comm'n 1984); *F.H. Lawson Co.*, 8 OSHC 1063 (Rev. Comm'n 1980); OSHA Instruction STD 1-12.9 (Oct. 30, 1978), OSHR [Reference File 21:2503]. For a more detailed discussion of employer defenses, see Chapter 15 (Defenses).

[168]*Mercury Metal Prods.*, 11 OSHC 1704 (Rev. Comm'n 1983); *K&T Steel Corp.*, 3 OSHC 2026 (Rev. Comm'n 1975).

[169]*Irvington Moore, Div. of U.S. Nat'l Resources v. OSHRC*, 556 F.2d 431, 5 OSHC 1585 (9th Cir. 1977); *Queen City Sheet Metal & Roofing*, 75 OSAHRC 2/A2 (Rev. Comm'n J. 1975), *aff'd*, 3 OSHC 1696 (Rev. Comm'n 1975).

lation derived from an ANSI standard, which combines both performance and specification requirements for woodworking equipment.[170] Safety standards for saws are relatively stringent. If a saw is available for use, it must be guarded; the Secretary need not show that the saw was actually used.[171] The infrequency of use may be considered, however, as a mitigating factor.[172] The method of guarding may be selected by the employer, so long as the employee is adequately protected.[173] Employers may also use the defenses of impossibility of compliance and greater hazard,[174] noted in the preceding paragraph, and discussed in detail in Chapter 15 (Defenses).

The standard for abrasive wheel machinery establishes requirements for cup wheels, guard exposure angles, bench and floor standards, cylindrical grinders, surface grinders and cutting-off machines, swing frame grinders, automatic snagging machines, top grinding, exposure adjustment, material requirements and minimum dimensions, band type guards, and guard design specifications.[175] The standard generally requires safety guards and flanges to be installed on all machinery within the purview of the section.[176] Work rests must be installed on all offhand grinding machines, although a violation will be considered *de minimis* if a side guard offers sufficient protection owing to the size of the machine.[177]

Safety requirements are also mandated for mills and calenders in the rubber and plastic industries. Mill safety controls include safety trip controls to operate on contact and pressure-sensitive body bars, which become operable by pressure of the mill operator's body.[178] Safety cables and safety tripping devices must be installed on calender machines.[179]

---

[170]29 C.F.R. §1910.213.

[171]*Huber, Hunt & Nichols Inc. & Blount Bros.*, 4 OSHC 1406 (Rev. Comm'n 1976).

[172]*James L. Price*, 4 OSHC 1024 (Rev. Comm'n 1976).

[173]OSHA Instruction STD 1-12.17 (Oct. 30, 1978), OSHR [Reference File 21:8505]; OSHA Instruction STD 1-12.18 (Oct. 30, 1978), OSHR [Reference File 21:2505–2506.

[174]*House Wood Prods.*, 3 OSHC 1993 (Rev. Comm'n 1976); *Shelvie Summerlin, A&S Millworks & Rentals*, 6 OSHC 1212 (Rev. Comm'n 1977).

[175]29 C.F.R. §1910.215.

[176]§1910.215(a).

[177]OSHA Instruction STD 1-12.8 (Oct. 30, 1978), OSHR [Reference File 21:8502–8503].

[178]29 C.F.R. §1910.216(b).

[179]§1910.216(c).

This section also provides for stopping-distance limitations on both mills and calenders.[180]

Mechanical power presses are covered in Subpart O by Section 1910.217, which is primarily derived from an ANSI standard. [181] The section establishes requirements pertaining to guarding and construction; setting and feeding of dies; inspection, maintenance, and modification of presses; operation of presses; and reports of press operator injuries. Certain types of presses (primarily hydraulic and pneumatic power presses), hammers, and riveting machines are specifically excluded from this section,[182] and therefore the general machine guarding requirements apply. The standards set forth rules for single-stroke mechanisms and anti-repeat features. The single-stroke mechanism is a mechanical arrangement limiting the motion of the slide to one stroke at each engagement of the rod, even if the engaging means is hand-operated.[183] The anti-repeat feature is a powered part of the control system, which requires the release of all tripping mechanisms before another stroke can be initiated.[184] Various requirements for safeguarding the point of operation for the power press include the construction of adjustable barrier guards and other devices to protect the operator from injury.[185] Owing to the incidence of injuries with respect to power press operations, OSHA requires that inspection and recordkeeping be completed on a regular basis.[186]

Unlike the detailed specification requirements for mechanical power presses, OSHA has established general performance standards for forging machines.[187] This standard applies only to the forging of hot metal.[188]

The final section of the machine guarding regulations is addressed to all types and shapes of mechanical power-transmission apparatus used in the workplace.[189] This includes pulleys, belts, ropes and chaindrives, gears, sprockets and chains,

---

[180]§1910.216(f).
[181]§1910.217.
[182]§1910.217(a)(5).
[183]Q's and A's to Part 1910, *supra* note 123.
[184]*Id.*
[185]29 C.F.R. §1910.217(c).
[186]§1910.217(e).
[187]§1910.218.
[188]Q's and A's, *supra* note 123.
[189]29 C.F.R. §1910.219.

and friction drives. The section is designed to ensure that such belts are "fully enclosed" to provide for safe operation.[190] Where the drive of the belt is unguarded or not enclosed, a violation will be found, unless the belts are close to walls or other objects, or employees are otherwise not exposed to a hazard.[191] The standard does not require guards that would prohibit maintenance operations.[192]

2. Subpart P: Hand and Portable Power Tools

Subpart P of the General Industry Standards focuses on portable power tools and other hand-held equipment.[193] The standards in this section contain design and work practice requirements for all hand and portable power tools and other hand-held equipment including portable powered saws, drills, sanders, abrasive wheels, grinders, and mowers. This section also deals with explosive actuated fastening tools, jacks, and cleaning with compressed air. The general section requires that each employer shall be responsible for the safe condition of tools and equipment used by employees.[194] This is not a strict liability standard, however, and the employer will not be held responsible if it could not have known of the defect even with the exercise of due diligence.[195] Specific requirements are also set forth for the guarding of portable powered tools.[196] For example, a tool retainer must be installed on each piece of pneumatic powered equipment to prevent the ejection of the tool.[197] The section also contains requirements designed to ensure the safety of those workers using compressed air for cleaning.[198]

---

[190]§1910.219(c)(1), (3).

[191]*United Resins, Inc.*, 12 OSHC 1004 (Rev. Comm'n 1984).

[192]*Schundler Co.*, 6 OSHC 1343 (Rev. Comm'n 1978); *Grayson Lumber Co.*, 1 OSHC 1234 (Rev. Comm'n 1973).

[193]29 C.F.R. §§1910.241–.247.

[194]§1901.242(a).

[195]*Mountain States Tel. & Tel. Co.*, 1 OSHC 1077 (Rev. Comm'n 1973).

[196]29 C.F.R. §1910.243. See, e.g., *Waupaca Foundry*, 11 OSHC 1094 (Rev. Comm'n Nos. 81-1543, 81-1749, & 81-1750, 1982).

[197]29 C.F.R. §1910.243(b)(1).

[198]29 C.F.R. §1910.242(b).

3. Subpart Q: Welding, Cutting, and Brazing

The welding, cutting, and brazing standards[199] establish requirements for the installation and operation of oxygen-fuel gas systems, and arc welding, cutting, and resistance welding equipment.[200] Specific employee protection is also provided by standards governing fire prevention, personal protective equipment, and ventilation.[201] Only approved apparatus such as torches, regulators or pressure-reducing valves, acetylene generators, and manifolds may be used,[202] although replacement tips may be installed.[203] Helmets or handshields must be worn during arc welding and cutting, and goggles are also recommended during operations.[204]

4. Subpart N: Materials Handling and Storage

Subpart N imposes requirements for the use and storage of various pieces of materials handling equipment.[205] The general requirements section covers clearances for mechanical handling equipment in aisles, at loading docks, and through doorways and turns or passages.[206] Aisles and passageways must be kept clear and in good repair, and free of obstructions. Permanent aisles and passageways must be appropriately marked, and storage of materials should not create a hazard. This standard has been applied to exterior passageways, where molten metal was transported over rutted roads.[207]

Operating procedures for multi-piece and single-piece rim wheel servicing on vehicles were added to the regulations in 1980 and 1984, respectively, and employers must train employees who service rim wheels in appropriate safety procedures.[208] Additional requirements are designed to ensure the

---

[199] §§1910.251–.254.

[200] §1910.252(a)–(c). OSHA has proposed revising Subpart Q and anticipates that revisions will be enacted by December 1988. 51 FED. REG. 38,560 (1986).

[201] 29 C.F.R. §1910.252(d)–(f).

[202] §1910.252.

[203] OSHA Instruction STD 1-14.1 (Oct. 30, 1978), OSHR [Reference File 21:8511].

[204] 29 C.F.R. §1910.252(c)(2).

[205] §§1910.176–.189.

[206] §19100.176.

[207] *Titanium Metals Corp. of Am.*, 5 OSHC 1164 (Rev. Comm'n 1977).

[208] 29 C.F.R. §1910.177.

safe operation of fork trucks, tractors, platform lift-trucks, motorized hand-trucks, and other specialized industrial trucks powered by electric motors or internal combustion engines.[209] This regulation addresses designations of different trucks and the locations in which such trucks may be used.[210] In general, the OSHA requirements for the use of trucks in differing locations are drawn from ANSI standards. The section also sets forth requirements for fuel handling and storage, changing and charging storage batteries, and other safety precautions.[211] Several OSHA standards governing loading trucks, railroad cars, and truck operations have been held to be preempted by Department of Transportation regulations.[212]

Overhead cranes and derricks also fall within the materials handling subpart.[213] Detailed requirements for wind indicators and rail clamps, rated load markings, clearances from obstruction, designated personnel, and cab location are included.[214] The section also includes various safety requirements for overhead and gantry cranes which must comply with ANSI specifications.[215] Crawler and locomotive cranes, wheel-mounted cranes of both truck and self-propelled wheel type, and other equipment powered by internal combustion engines or electric motors and using drums and ropes must also comply with the OSHA regulations.[216] Cranes designed for railway and automobile wreck clearances are not included and the section is applicable only to machines used as lifting cranes.[217] While no specific requirements regarding licensing of crane operators are in effect, OSHA has stated that only trained and authorized personnel should be permitted to operate the units.[218] Once again, this section mandates regular inspection and test-

---

[209]§1910.178.

[210]*Id.* See, e.g., *Celotex Corp.*, 11 OSHC 1127 (Rev. Comm'n Nos. 81-212 & 81-2195, 1983).

[211]29 C.F.R. §1910.178(f), (g), (i).

[212]*Mushroom Transp. Co.*, 1 OSHC 1390 (Rev. Comm'n 1973); OSHA Instruction STD 1-11.5 (Oct. 30, 1978), OSHR [Reference File 21:8403].

[213]29 C.F.R. §§1910.179–.182.

[214]*Id.*

[215]§1910.179. This section only applies to cranes installed after Aug. 31, 1971, because of the advisory nature of the ANSI standard. *U.S. Steel Corp.*, 5 OSHC 1289 (Rev. Comm'n Nos. 10825 & 10849, 1977).

[216]29 C.F.R. §1910.180. See, e.g., *Farthing & Weidman, Inc.*, 11 OSHC 1069 (Rev. Comm'n 1982).

[217]29 C.F.R. §1910.180(b)(1).

[218]Q's and A's to Part 1910, *supra* note 123.

ing for the crane equipment, including maintenance procedures and inspection of ropes.[219]

Various safety procedures must also be followed with respect to permanent and temporary derricks, including load ratings, inspections, testing, and rope inspection.[220] The regulation also applies to any modification of a derrick which retains its fundamental features, except for floating derricks.[221]

The two other substantive standards in Subpart N concern helicopters and slings. Helicopter cranes must comply with any applicable rules of the Federal Aviation Administration.[222] The standard also covers various other safety considerations relating to helicopters and operating and maintenance personnel. OSHA has provided a poster of helicopter hand signals to accompany the standards, in Figure N-1 (immediately following the text of the standard, at 29 C.F.R. §1910.183). Signs used in conjunction with other materials handling equipment to move objects by hoisting must conform to OSHA regulations as well.[223] Specific design requirements and safe operating practices and inspections are included for slings made from alloy steel chain, wire rope, metal mesh, natural or synthetic fiber rope, and synthetic web.[224]

C. Protection Against Hazardous Materials

1. *Subpart H: Hazardous Materials*

Subpart H is directed at the safe storage and handling of compressed gases and various flammable and combustible materials.[225] General safety requirements for compressed gases direct employers to conduct visual and other inspections as prescribed in the Hazardous Materials Regulations of the De-

---

[219] 29 C.F.R. §1910.180(d), (e), (f), (g).

[220] §1910.181(c), (d), (e), (g).

[221] §1910.181(b).

[222] §1910.183(a).

[223] §1910.184.

[224] *Id.*

[225] §§1910.101–.116. OSHA has proposed revising Subpart H and anticipates that new standards will be enacted by December 1988. 51 FED. REG. 38,560 (1986).

partment of Transportation.[226] Specific safety requirements also relate to acetylene, hydrogen, oxygen, and nitrous oxide.[227] These requirements specify methods for the safe storage and maintenance of the gases, where the gases are utilized in systems on industrial and institutional consumer premises.[228]

The remainder of Subpart H applies to the handling, storage, and use of other hazardous gases and liquids. These regulations cover flammable and combustible liquids with flash points below 100 degrees Fahrenheit.[229] The design, installation, support, and testing of tanks used to store flammable or combustible liquids are the subject of detailed specification requirements.[230] Storage of such liquids in containers and portable tanks is also governed by these regulations, which establish various safety rules for such containers.[231] Lastly, the use of flammable and combustible liquids in industrial plants, service stations, and processing plants, but not refineries, must comply with OSHA requirements.[232]

Flammable and combustible liquids are also subject to specific requirements when used in conjunction with spray finishing and dip tanks.[233] With respect to spray finishing, employers must comply with the regulations governing spray booth design, ignition sources, ventilation, liquids storage and handling, sprinkler systems, spray operations and maintenance, electrostatic equipment, drying, curing, or fusion apparatus, vehicle undercoating in garages, powder coating, organic peroxides, and dual component coatings.[234] The section does not apply to outdoor spray application.[235] Spray booths must have independent exhaust duct systems and a fan-rotating element.[236] Although the Secretary must establish a

---

[226]§1910.101(a).

[227]§§1910.102–.105.

[228]See, e.g., §1910.104(a).

[229]§1910.106(a)(18), (19).

[230]§1910.106(b).

[231]§1910.106(d).

[232]*Texaco, Inc.*, 11 OSHC 1713 (Rev. Comm'n 1983); 29 C.F.R. §1910.106(d), (e), (g), (h).

[233]§§1910.107–.108.

[234]§1910.107.

[235]§1910.107(n).

[236]§1910.107(c).

lack of "adequate ventilation" to prove a violation,[237] the Commission has held that the existence of a hazard will be presumed wherever employees may work in the direct path of spray.[238]

Dip tanks containing flammable or combustible liquids must meet detailed requirements for ignition sources, operations and maintenance, and extinguishers.[239] Employers must equip dip tanks containing over 500 gallons of flammable or combustible liquid with bottom drains, unless specific combinations of alternate extinguishment systems are used.[240]

General safety provisions govern the storage, handling, transportation, and use of explosives and blasting agents.[241] The regulations also contain specific coverage of explosives at piers, railway stations, and cars or vessels, the use of water gel (slurry) explosives and blasting agents, the storage of ammonium nitrate, and the use of small arms ammunition primers and propellants.[242] OSHA will not penalize employers that comply with Department of Transportation or Bureau of Alcohol, Tobacco, and Firearms regulations in this area.[243] OSHA has established additional requirements for the storage and handling of liquified petroleum gases and anhydrous ammonia.[244]

2. Subpart M: Compressed Gas, Compressed Air Equipment

At the present time, the sole substantive section remaining in Subpart M[245] pertains to air receivers.[246] This section sets safety requirements for air receivers and other equipment used for cleaning, drilling, hoisting, and chipping. The regu-

---

[237]*Keystone Body Works*, 4 OSHC 1523 (Rev. Comm'n 1976).

[238]*Fusibles Westinghouse de Puerto Rico v. OSHRC*, 658 F.2d 21, 9 OSHC 2176 (1st Cir. 1981); *Highway Motor Co.*, 8 OSHC 1796 (Rev. Comm'n 1980).

[239]29 C.F.R. §1910.108.

[240]§1910.108(c)(3); OSHA Instruction STD 1-5.6A (Oct. 30, 1978), OSHR [Reference File 21:8217].

[241]29 C.F.R. §1910.109.

[242]*Id.*

[243]OSHA Instruction STD 1-5.12 (Oct. 30, 1978), OSHR [Reference File 21:8215–8216].

[244]29 C.F.R. §§1910.110–.111.

[245]§1910.166–.176.

[246]§1910.169. The remaining provisions in this subpart, §§1910.166–.168, were revoked by OSHA in February 1984. 49 FED. REG. 5318 (1984).

lation governs the installation of air receivers, and establishes requirements for the gauges and valves on such receivers.[247]

D. Subpart L: Fire Protection

Subpart L, as revised effective December 1980, covers fire brigades, portable fire extinguishers, standpipe and hose systems, automatic sprinkler systems, extinguishing systems, fire detection systems, and employee alarm systems.[248] Subpart L generally consists of performance-oriented provisions which permit employers discretion to choose compliance programs for fire safety.[249] For example, employers have three alternatives at the worksite with respect to the use of portable fire extinguishers. No portable fire extinguishers are required where an employer has a fire prevention and emergency action plan which calls for the evacuation of employees, or where an employer uses standpipe systems or hose stations.[250] Likewise, although the employer is not required to do so, he may establish a fire brigade.[251] Once a brigade is organized, however, the employer must prepare a written statement or policy describing its functions, and must provide training and education for the members of the brigade.[252]

Despite the overall performance-oriented approach of the standard, specific requirements have been established for fixed extinguishing systems and for fire detection systems.[253] Though these standards impose no obligations on employers to install standpipes, hoses, or automatic sprinkler systems, if such systems are mandated by other OSHA standards they meet the requirements of Subpart L.[254]

*   *   *   *   *

This section has only highlighted the numerous areas of safety standards for which OSHA has issued regulations; it is intended to apprise employers and employees of the need

---

[247]§1910.169(b)(1), (3).
[248]§1910.155–.165.
[249]*Id.*
[250]§1910.157(d).
[251]§1910.156(a)(1).
[252]§1910.156.
[253]§1910.158–.163.
[254]§1910.158(a)(1), .159(a)(1).

to review carefully the standards applicable to a given work-place. In many cases, the safety standards are so detailed and specific that employers have had little discretion in the manner in which the required protection has been provided. Accordingly, employers must comply with these standards unless they apply for a variance (see Chapter 21 (Other Issues Related to Standards)), or can prove one of the defenses to compliance, such as the greater hazard defense (see Chapter 15 (Defenses)).

IV. Specific Health Standards

The health standards promulgated by OSHA can be categorized as either air contaminant standards or general occupational health and environmental control standards. As described above,[255] in 1971, pursuant to Section 6(a), OSHA issued a list of air contaminants for which permissible exposure limits or threshold limit values had been established by ANSI or pursuant to the Walsh-Healey Act. In addition, OSHA has promulgated 26 toxic and hazardous substances standards pursuant to Section 6(b)(5) which contain detailed requirements for monitoring, medical surveillance, engineering controls, and training.[256] The current general occupational health and environmental control standards, concerning such areas as ventilation, radiation, sanitation, and noise, were also issued under Section 6(a) in 1971, with the sole exception of the Hearing Conservation Amendment to the Occupational Exposure to Noise Standard,[257] which became a final rule in 1983.[258] This section will briefly review the elements of these standards.

A. Subparts G and J: Occupational Health and Environmental Control

1. Ventilation

Subpart G contains a ventilation section which establishes ventilation requirements with respect to abrasive blast-

---

[255]See Section I.A., above.
[256]See Section IV.B.2, below.
[257]29 C.F.R. §1910.95(c).
[258]48 FED. REG. 9738 (1983). See Section IV.A.2, below.

ing work; grinding, polishing, and buffing operations; spray-finishing operations; and open-surface tanks.[259] The abrasive blasting ventilation regulation applies to all operations in which an abrasive is forcibly applied to a surface by pneumatic or hydraulic pressure or by centrifugal force.[260] The standard requires the prevention of dust hazards and the implementation of blast cleaning enclosures, exhaust ventilation systems, and personal protective equipment, as well as clean air for abrasive-blasting respirators.[261] With respect to grinding, polishing, and buffing operations, a local exhaust ventilation system must be provided and used to maintain employee exposure within prescribed permissible exposure limits.[262] Ventilation for spray-finishing operations must be supplied within spray booths or spray rooms which are used to enclose or confine all such operations.[263] A physical barrier is not required, however, so long as the operation is maintained separate from other activities and no hazard exists.[264] Finally, ventilation requirements have been established for open-surface tanks where materials are immersed in or removed from liquids, or vapors of such liquids, for cleaning, altering, finishing, or changing such materials, such as in electroplating and tanning operations.[265] The regulation applies only, however, where airborne concentrations of toxic substances at or near the tank exceed specified limits or where a fire or explosion hazard exists.[266] In any event, a hazard will not be presumed under this section, and the Secretary bears the burden of proving such a hazard exists.[267] Personal protective equipment, such as goggles and face shields, must be used in connection with hazards covered by this section.[268]

---

[259]29 C.F.R. §1910.94.

[260]§1910.94(a)(1)(xii).

[261]§1910.94(a).

[262]§1910.94(b)(2).

[263]§1910.94(c).

[264]*Westinghouse Elec. Corp. v. OSHRC*, 617 F.2d 497 (7th Cir. 1980), *on remand*, 9 OSHC 1183 (Rev. Comm'n 1981); *Concord Collision Corp.*, 5 OSHC 1607 (Rev. Comm'n 1977).

[265]29 C.F.R. §1910.94(d)(1)(i).

[266]*Republic Steel Corp.*, 5 OSHC 1238 (Rev. Comm'n 1977).

[267]*Pratt & Whitney Aircraft, Div. of United Technologies Corp. v. Donovan*, 715 F.2d 57, 11 OSHC 1641 (2d Cir. 1983).

[268]29 C.F.R. §1910.94(d)(9).

2. Noise

One of the more important and controversial provisions within Subpart G requires employers to provide hearing protection and engineering controls against occupational exposure to noise.[269] The source of the permissible exposure limit of 90 decibels as an 8-hour time-weighted average and the requirement for engineering controls to reduce noise exposure, as set forth in 29 C.F.R. §1910.95(a) and (b), was the Walsh-Healey Act.[270]

The basic structure of the noise standard, like the general air contaminant standards, establishes a priority among the methods to be used to control worker exposure to noise. Where it is determined that employee exposure exceeds the specified levels of noise, the employer first has the obligation to implement feasible engineering or administrative controls. Only if those controls do not reduce exposure to the specified levels can personal protective equipment such as hearing protectors be relied upon to achieve compliance with the exposure levels.[271] The noise standard has a sliding scale of exposure levels so that if the level is above a certain amount, a correspondingly shorter exposure time is permitted. For example, while an employee can be exposed to 90 decibels time-weighted average (dBA) for 8 hours, at 97 dBA the permissible exposure time is only 3 hours before the requirement for engineering or administrative controls is triggered.

The most frequently litigated issue regarding these portions of the noise standard involved whether an employer could utilize a cost-benefit analysis to determine whether engineering controls are "feasible," as required by the standard.[272] At the present time, the Review Commission takes the posi-

---

[269] §1910.95.

[270] See 36 FED. REG. 10,466 (1971); 41 C.F.R. §50-204.10.

[271] At a minimum, an employer must use protective equipment to reduce employee exposure. To fail to do so subjects an employer not only to citations under the applicable engineering control standards but also to violation of the protective equipment standards, 29 C.F.R. §1910.132 (the general requirements provisions; see notes 66–86 and accompanying text, *supra*) or §1910.134 and §1910.95(c).

[272] Compare *Continental Can Co.*, 4 OSHC 1541 (Rev. Comm'n 1976) (cost-benefit analysis appropriate), with *Sun Ship, Inc.*, 11 OSHC 1028 (Rev. Comm'n 1982) (cost-benefit analysis not appropriate).

tion that a cost-benefit analysis is not material to a determination of feasibility, which generally is limited to technological feasibility.[273]

On January 16, 1981, OSHA promulgated a Hearing Conservation Amendment to the Noise Standard, which imposed
a comprehensive program for audiometric testing, exposure
monitoring, and training for employees who are exposed to 85
dBA and who have experienced a "significant threshold shift,"
or loss, of hearing capacity.[274] This regulation was subsequently stayed under President Reagan's administration,[275]
and ultimately was replaced by a new, simplified, more performance-oriented Hearing Conservation Amendment.[276] The
current amendment has been upheld by the Fourth Circuit.[277]

Under the new amendment, employers may select their
own methods of complying with the hearing conservation requirements.[278] The employer must establish a baseline audiometric measurement from which to compare subsequent
annual measurements to determine whether a shift in hearing
capability has occurred.[279] The employer must also provide
hearing protection to those employees who are exposed to 85
dBA and who have experienced a shift in hearing capability.[280]
Employers may use area monitoring of noise, however, rather
than personal monitoring,[281] and many of the detailed specifications for hearing protection methods and determination of
hearing capability changes were deleted from the previous
version.[282]

---

[273]*Sun Ship, Inc., supra* note 272. But see *Donovan v. Castle & Cooke Foods,* 692
F.2d 641 (9th Cir. 1982) (cost-benefit analysis defense available for noise standard).

[274]46 FED. REG. 4078 (1981). Although an undefined hearing conservation program had been required under 29 C.F.R. §1910.95(c) since 1971, this provision was
rarely enforced and was arguably unenforceably vague. See *B. W. Harrison Lumber
Co.,* 4 OSHC 1091 (Rev. Comm'n 1976).

[275]46 FED. REG. 42,622 (1981).

[276]48 FED. REG. 9738 (1983).

[277]*Forging Indus. Ass'n v. Donovan,* 773 F.2d 1436, 12 OSHC 1472 (4th Cir. 1985)
(*en banc*).

[278]48 FED. REG. 9738 (1983).

[279]29 C.F.R. §1910.95(g).

[280]§1910.95(i).

[281]§1910.95(d).

[282]Compare the standard set forth in 46 FED. REG. 42,622 (1981) with 48 FED.
REG. 9738 (1983).

3. Radiation

Subpart G also includes various standards pertaining to ionizing and nonionizing radiation.[283] The standard on ionizing radiation covers employee exposure to alpha rays, beta rays, gamma rays, X-rays, neutrons, high-speed electrons, high-speed protons, and other atomic particles.[284] To a large extent, however, OSHA coverage appears to be limited to machinery such as X-ray machines and particle accelerators.[285] Jurisdiction as to other sources of radiation is preempted under Section 4(b)(1) of the Act[286] by other federal agencies which regulate source material, special nuclear material, and by-product material.[287] In addition, the nonionizing radiation levels have been considered advisory, and thus nonenforceable, by the Review Commission.[288]

4. Sanitation

Sanitation standards contained in Subpart J specify requirements for drinking and washing water, toilet facilities, washing facilities, change rooms, clothes drying facilities, worksite consumption of food and beverages, and food handling at all permanent places of employment.[289] The standards prohibit eating or drinking in a toilet room or in any area exposed to toxic material.[290] A lunchroom need not be provided, however, if employees are not permitted to eat on the premises.[291] The standards also set forth the minimum number of toilets that must be provided for a given number and

---

[283]29 C.F.R. §§1910.96–.97.

[284]§1910.96(a)(1).

[285]§653(b)(1).

[286]Such agencies include the Energy Research and Development Administration, the Atomic Energy Commission, the Coast Guard, the Food and Drug Administration, the Department of Transportation, and the Federal Aviation Administration.

[287]OSHA Instruction STD 1-4.1 (Oct. 30, 1978), OSHR [Reference File 21:8211–8212.

[288]*Swimline Corp.*, [1977–1978] OSHD ¶21,656 (Rev. Comm'n 1977).

[289]29 C.F.R. §1910.141. OSHA's sanitation standards currently do not apply to agricultural workers. However, the District of Columbia Circuit has recently ordered OSHA to issue a field sanitation rule giving agricultural workers access to drinking water and toilets. The court characterized OSHA's delay in issuing such a rule "a disgraceful chapter of legal neglect." *Farmworker Justice Fund v. Brock*, 811 F.2d 613, 13 OSHC 1049 (D.C. Cir. 1987), *vac. as moot*, 13 OSHC 1288 (May 7, 1987).

[290]29 C.F.R. §1910.141(g)(2).

[291]Q's and A's to Part 1910, *supra* note 123.

sex of employees,[292] and explicit sanitary standards must be
followed in establishing toilet facilities.[293] The sanitation regulation also includes standards for housekeeping, vermin control, waste disposal, water supply, and showers. The shower
specifications are only applicable, however, whenever showers
must be provided on the basis of another specific OSHA standard.[294]

Site and shelter specifications have been established under Subpart J for temporary labor camps, which must be
equipped with adequate water supply, toilet facilities, sewage
disposal facilities, laundry, bathing facilities, lighting, refuse
disposal, kitchens, vermin control, and first aid.[295] Housing
constructed prior to April 3, 1980, however, will not be found
in violation of the standard, so long as it complies with prior
regulations of the Employment Training Administration (ETA)
of the Department of Labor.[296] Camp inspections will be conducted on the basis of coordinated enforcement activities by
OSHA, ETA, and the Employment Standards Administration.[297] Employers that do not require employees to reside in
company houses as a condition of employment are not covered
by this regulation.[298]

OSHA has also proposed regulations governing entry into
confined spaces and the release of stored energy.[299] Final action on both proposals is anticipated sometime in 1988.[300]

B. Toxic and Hazardous Substances

1. Air Contaminants

The Section 6(a) air contaminant standards apply to a
variety of substances such as carbon monoxide, silica, and

---

[292]No female toilet facilities need be provided if an employer has no women
employees. *Southern Ry.*, 2 OSHC 1396 (Rev. Comm'n 1974), *aff'd on other grounds*,
539 F.2d 335, 3 OSHC 1940 (4th Cir.), *cert. denied*, 429 U.S. 999 (1976).
[293]29 C.F.R. §1910.141(c).
[294]§1910.141(d)(3).
[295]§1910.142.
[296]45 FED. REG. 14,180 (1980).
[297]45 FED. REG. 39,486 (1980).
[298]*Spencer Farms, Inc.*, 5 OSHC 1891 (Rev. Comm'n 1977).
[299]To be codified at 29 C.F.R. §§1910.147 and 1910.146, respectively.
[300]51 FED. REG. 38,560 (1986).

hydrogen sulfide, and consist of tables, labeled Z-1, Z-2, and Z-3, of those substances with accompanying exposure levels and descriptions of how exposures are to be calculated. The standards require that the levels of permissible exposure be met through the use of engineering controls supplemented, where necessary, by respirators. The exposure level for the substance is usually based upon an 8-hour time-weighted average (TWA) measured either by weight in a set volume as in milligrams per meter cubed (mg/m^3) or by parts per million (ppm). Other substances such as manganese or chlorine have ceiling levels. A ceiling level means that the exposure is never to exceed that level regardless of the duration.[301] Table Z-2 lists several substances that have maximum peak concentrations in addition to TWAs and ceiling concentrations. The peak concentration defines a concentration over the ceiling value that is not to be exceeded in an 8-hour shift beyond the period of time set in the standard. For example, no person is to be exposed to carbon disulfide at over 100 ppm for more than 30 minutes during an 8-hour shift.[302] Finally, like the noise standard, the general air contaminant standards permit the use of personal protective equipment as a method of abatement of the hazard only after it can be shown that no feasible engineering or administrative controls are available to sufficiently reduce the exposure below the permissible limit.[303]

2. Section 6(b)(5) Standards for Toxic and Hazardous Substances

Since its inception in 1971, OSHA has promulgated the following toxic and hazardous substances standards pursuant to the rulemaking procedures of Section 6(b) and the requirements of Section 6(b)(5) of the Act:[304]

---

[301]*Chlorine Inst. v. OSHA*, 613 F.2d 120, 121 (5th Cir.), *cert. denied*, 449 U.S. 826 (1980).

[302]OSHA has proposed adding 1,3-butadiene and methylene chloride to Table Z-1 and expects to issue a notice of proposed rulemaking sometime in 1987. 51 FED. REG. 38,560 (1986).

[303]29 C.F.R. §1910.1000(e).

[304]OSHA is also developing standards for 4,4'-methylenedianiline and is in the final rulemaking stages with respect to ethylene dibromide and formaldehyde. 51 FED. REG. 38,560 (1986). In addition, OSHA is examining whether special standards should be developed for laboratory workers who are exposed to a multitude of toxic substances under frequently changing or unpredictable conditions. *Id.*

1. Asbestos, 29 C.F.R. §1910.1001[305]
2. Coal tar pitch volatiles, 29 C.F.R. §1910.1002[306]
3. 4-Nitrobiphenyl, 29 C.F.R. §1910.1003[307]
4. alpha-Naphthylamine, 29 C.F.R. §1910.1004
5. 4,4'-Methylene bis (2-chloroaniline), 29 C.F.R. §1910.1005
6. Methyl chloromethyl ether, 29 C.F.R. §1910.1006
7. 3,3'-Dichlorobenzidine (and its salts), 29 C.F.R. §1910.1007
8. bis-Chloromethyl ether, 29 C.F.R. §1910.1008
9. beta-Naphthylamine, 29 C.F.R. §1910.1009
10. Benzidine, 29 C.F.R. §1910.1010
11. 4-Aminodiphenyl, 29 C.F.R. §1910.1011
12. Ethyleneimine, 29 C.F.R. §1910.1012
13. beta-Propiolactone, 29 C.F.R. §1910.1013
14. 2-Acetylaminofluorene, 29 C.F.R. §1910.1014
15. 4-Dimethylaminoazobenzene, 29 C.F.R. §1910.1015
16. N-Nitrosodimethylamine, 29 C.F.R. §1910.1016
17. Vinyl chloride, 29 C.F.R. §1910.1017[308]
18. Inorganic arsenic, 29 C.F.R. §1910.1018[309]

---

[305]On June 20, 1986, OSHA published a revised final standard governing occupational exposure to asbestos, tremolite, anthophyllite, and actinolite. 29 C.F.R. §1910.1001. The standard was immediately challenged in three federal circuit courts by various industry and labor groups. On July 11, 1986, the District of Columbia Circuit ordered that all challenges to the standard be transferred to the District of Columbia Circuit. *Building & Construction Trades Dep't v. Brock*, Nos. 86-1359 & 86-1360 (D.C. Cir. July 11, 1986) (order issued). OSHA subsequently temporarily stayed the effective date of the standard as it applied to nonasbestiform tremolite, anthophyllite, and actinolite. In the interim, OSHA promulgated a temporary standard governing occupational exposure to these substances during the pendency of the stay. 29 C.F.R. §1910.1101. OSHA anticipates that a final decision will be reached with respect to these substances sometime in 1987. 51 FED. REG. 38,560 (1986).

[306]This standard was considered an interpretative definition of the substances covered by the existing exposure level in Table Z-1 and thus was promulgated without full rulemaking. It was amended in 1983, 48 FED. REG. 2764, to delete the reference to asphalt.

[307]The standards listed above as Nos. 3–16 govern exposure to 14 carcinogens and were, for the most part, sustained on review in *Synthetic Organic Chem. Mfrs. Ass'n v. Brennan*, 506 F.2d 385 (3d Cir. 1974), *cert. denied*, 420 U.S. 973, *reh. denied*, 423 U.S. 886 (1975). However, 4,4'-methylene bis (2-chloroaniline) was removed from the list of standards in 1976 (41 FED. REG. 35,184) based on the court's decision that OSHA did not comply with the requirements of §6(b) in promulgating the standard for that substance.

[308]This standard was sustained on review in *Society of Plastics Indus. v. OSHA*, 509 F.2d 1301 (2d Cir. 1975).

[309]The Ninth Circuit initially remanded the inorganic arsenic standard to OSHA for a determination of whether the substance posed a "significant risk" to worker's health, as required by the Supreme Court's decision in *Industrial Union Dep't v. American Petroleum Inst.*, 448 U.S. 607 (1980). *ASARCO v. OSHA*, 647 F.2d 1 (9th Cir. 1981). OSHA made that determination and again promulgated the standard,

19. Lead, 29 C.F.R. §1910.1025[310]
20. Benzene, 29 C.F.R. §1910.1028[311]
21. Coke oven emissions, 29 C.F.R. §1910.1029[312]
22. Cotton dust, 29 C.F.R. §1910.1043[313]
23. 1,2-Dibromo-3-chloropropane, 29 C.F.R. §1910.1044[314]
24. Acrylonitrile, 29 C.F.R. §1910.1045[315]
25. Ethylene oxide, 29 C.F.R. §1910.1047[316]
26. Hearing Conservation Amendment to Noise Standard, 29 C.F.R. §1910.95(c)

All of these standards contain substantive provisions concerning the same subjects. Those subjects include definitions, permissible exposure limit, monitoring, methods of compli-

---

which was upheld on review by the Ninth Circuit. *ASARCO v. OSHA*, 746 F.2d 483 (9th Cir. 1984).

[310]The lead standard was substantially upheld by the District of Columbia Circuit in *United Steelworkers v. Marshall*, 647 F.2d 1189 (D.C. Cir. 1980), *cert. denied*, 453 U.S. 913 (1981). However, the court stayed the application of the standard to miscellaneous lead ("other") industries, remanding to OSHA the determination of the feasibility of compliance with the standard. In 1981, OSHA concluded that "other industries" could comply with the 50 μg/m³ standard within 2½ years of the stay being lifted, which it assumed would fall on December 19, 1983. For no apparent reason, the stay was never lifted. OSHA finally realized this and in July 1984, issued Notice CPL 52-2 establishing the following compliance dates with the 50 μg/m³ standard:

Primary lead production—June 29, 1991

Secondary lead production—June 29, 1986

Lead acid battery manufacturing—June 29, 1986

Automobile mfg./solder grinding—June 29, 1988

Electronics, gray iron, foundries, ink mfg., paints and coatings mfg., can mfg., and printing—June 29, 1982

Lead pigment mfg., nonferrous foundries, leaded steel mfg., shipbuilding and ship repair battery breaking in the collection and processing of scrap (excluding collection and processing of scrap which is part of secondary smelting operation), secondary lead smelting of copper and lead casting—feasibility has yet to be determined.

All other industries—2½ years from lifting of stay.

Notice CPL 52-2 is set forth in its entirety in [1983–1984 Developments] EMPL. SAFETY & HEALTH GUIDE ¶8248.

[311]The benzene standard was held to be invalid by the Supreme Court in *Industrial Union Dep't v. American Petroleum Inst., supra* note 309. Proceedings for its modification and repromulgation were concluded and OSHA published a final rule in 1987. But see *supra* note 31.

[312]This standard was sustained in review of *American Iron & Steel Inst. v. OSHA*, 577 F.2d 825 (3d Cir. 1978), except for the provision requiring employers to research or develop additional controls.

[313]The Supreme Court upheld this standard in *American Textile Mfrs. Inst. v. Donovan*, 452 U.S. 490 (1981).

[314]No court review sought.

[315]No court review sought.

[316]The District of Columbia Circuit upheld most of the final standard for ethylene oxide. *Public Citizen Health Research Group v. Tyson*, 796 F.2d 1479 (D.C. Cir. 1986).

ance, personal protective equipment (including respirators), signs and labels, hygiene facilities, employee training, medical surveillance, and recordkeeping and access to records.

The following discussion illustrates how these various subjects have been handled in the standards that have been promulgated to date.

(a) Permissible exposure limit. All of the standards except 3 through 16 listed above (29 C.F.R. §§1910.1003–.1016) contain an 8-hour TWA exposure level that limits employee exposure to no more than that level. The excepted standards, which deal with certain carcinogens,[317] do not contain a numerical limit. Rather, they apply to any area in which the specified substance is "manufactured, processed, repackaged, released, handled or stored."[318] They require that these materials be used in either isolated or closed systems and that transfer operations be controlled through local exhaust ventilation.[319]

In addition, the asbestos, acrylonitrile, and vinyl chloride standards have ceilings that place an upward limit on any one-time exposure to that substance.[320]

Unlike the air contaminant standards issued under Section 6(a), many of the toxic and hazardous substance standards issued under Section 6(b) use the concept of an "action level" to trigger several requirements.[321] The action level is set below the permissible exposure level so as to build in a margin of safety. Exposure at or above the action level is used to trigger such provisions as medical surveillance,[322] training,[323] or monitoring frequency.[324] In addition, for some substances, the

---

[317]This regulation was adopted by OSHA in 1980 to set out procedures and policy determinations to govern standard setting for carcinogens. 29 C.F.R. §1990. It has never been implemented and it is still under court challenge in the Fifth Circuit. *American Petroleum Inst. v. OSHA*, No. 80-3040 et al.

[318]See, e.g., 29 C.F.R. §1910.1003(a)(1). These standards, however, do not apply to the substance as it is shipped in sealed containers or to solid or liquid mixtures that contain less than a specified volume or weight. See, e.g., §1910.1008(a)(1) and (2).

[319]See, e.g., benzidine, §1910.1010(c)(2), (4).

[320]§§1910.1001(b)(3), 1910.45(c)(1)(ii), and 1910.1017(c)(2), respectively.

[321]These standards include inorganic arsenic (§1910.1018(b)), lead (§1910.1025(b)), and acrylonitrile (§1910.1045(b)).

[322]E.g., see vinyl chloride, §1910.1017(k).

[323]E.g., see inorganic arsenic, §1910.1018(o).

[324]E.g., see acrylonitrile, §1910.1045(b)(3).

standards require protection for skin or eye contact regardless of the airborne exposure level.[325]

(b) Monitoring. All of the standards except 3 through 16 listed above[326] require periodic determination of employee exposure to the regulated substance to determine if the standard is applicable to that work area. Since the excepted standards do not set numerical exposure limits but rather operate based on the presence of the substance in a given area, no monitoring is required. For the remainder of the standards, the monitoring occurs in several stages—an initial determination that may or may not require actual sampling, and periodic monitoring at different intervals depending on what concentration levels are found. To illustrate, the inorganic arsenic standard[327] requires initial monitoring by taking air samples.[328] If the levels are below the action level, no further monitoring is required. If the results are in excess of the action level but below the permissible exposure level, then further monitoring is required at 6-month intervals. Quarterly monitoring is required for those exposed over the permissible exposure level. Finally, in order to deal with process changes in the plant's operations that may affect exposure, monitoring is required regardless of previous levels if there is reason to believe a change in exposure could have occurred because of the process change.

Where monitoring is required, all the standards require the employer to notify employees in writing of the results regardless of whether the level is above or below the exposure limit;[329] and where exposure is excessive, the notice must describe the corrective action being taken.[330]

(c) Methods of compliance. It is the general philosophy of OSHA not to specify in detail the methods employers must

---

[325]See acrylonitrile, §1910.1045(c)(2), and vinyl chloride, §1910.1017(c)(3).

[326]§§1910.1003–.1016.

[327]§1910.1018(e).

[328]Other standards do not rely solely upon new air samples but permit the use of other accurate information to be used to supplement the air monitoring. Lead standard, §1910.1025(d).

[329]The Act (§8(c)(3)) requires that employees or their representatives have the right to observe the monitoring and to have access to the results. Therefore, the standards include such a provision.

[330]The sole exception is the hearing conservation standard, §1910.95(3), which only requires notice, including oral notice, of overexposure. The reason for this departure from past OSHA policy was not explained in the preamble. 48 FED. REG. 9748 (1983).

use to comply with a given health standard ("specification standard"). The standards set out a hierarchy of preferred methods of control and then the employer is free to choose the specific technology consistent with that priority system ("performance standard").[331]

The priority system derives from the scientific discipline of industrial hygiene. This discipline establishes the control of the substance at the source as the first priority,[332] followed by work practices and administrative controls,[333] with personal protective equipment[334] as the alternative of last resort.

Under such a system, the employer is required to determine what controls are feasible and, depending on the standard, to prepare a written compliance program, the required degree of specificity of which varies with the standard.

(d) Personal protective equipment. Since the largest number of 6(b) standards deals with air contaminants, the most common protective equipment dealt with is the respirator.[335] In many respects, protective equipment requirements in the toxic and hazardous substances standards are similar to those general standards adopted under Section 6(a) since the requirements in 29 C.F.R. §1910.134 are incorporated into the 6(b) standards.[336]

---

[331]The coke oven standard, §1910.1029, is the primary exception. That standard deals with one homogeneous industry and one basic industrial process, namely, metallurgical or foundry cokemaking. The coke oven standard specifies a set of minimum engineering controls and work practices that must be implemented for all existing batteries. Additional controls and controls for new and rehabilitated batteries are handled in a manner similar to that for other standards, namely, the implementation of all feasible controls without further specificity.

[332]Such controls include substitution of materials, as well as engineering controls.

[333]Administrative controls, primarily rotation of workers, have not been permitted in carcinogen standards since the absence of any known "safe level of exposure" militates against reducing exposure to all employees by exposing more employees to carcinogens even for shorter time periods.

[334]OSHA has recently proposed deviating from this long-standing determination by issuing an advance notice of proposed rulemaking (48 FED. REG. 7473 (1983)) seeking information on a change to 29 C.F.R. §1910.1000(e) and §1910.134(a) that would make all methods equal. In the case of asbestos, such a proposal was in fact made for exposures below the current standard of 2 fibers. 49 FED. REG. 14,116 (1984).

[335]A number of the standards also require specified clothing and change frequencies to avoid skin contamination or ingestion hazards resulting from dust or chemicals adhering to dirty clothes. Employers are obliged to provide the clothes. See, e.g., vinyl chloride, 29 C.F.R. §1910.1017(h)(1)(ii) and (2), and acrylonitrile, §1910.1045(j)(1) and (2). In addition, the noise standard spells out the need for hearing protection and the procedures for selecting the proper hearing protection for the specific noise levels in the workplace. §1910.95(i)–(j). See Section IV.A.2 above.

[336]See, e.g., §§1910.1003–.1016, where the carcinogen standards merely refer to §1910.134, with the sole addition being the specification of the minimum type of acceptable respirators for each substance.

With respect to certain toxic substances, such as lead, specific types of respirators are required based on exposure levels. More precise procedures for fitting respirators through qualitative fit tests to determine whether the respirator selected provides the requisite protection are also required.[337]

(e) Signs and labels. All of the standards covering the toxic and hazardous substances described above require signs and/or labels to advise employees of the areas where the regulated substance is present and to provide some basic information on the nature of the hazard. In addition, several standards require labeling of containers which hold the regulated substance, including those that leave the workplace.[338]

(f) Hygiene facilities. To reduce employee exposure to a regulated substance, hygiene facilities are required by the toxic and hazardous substances standards. Various hygiene facilities such as showers, change rooms, washing facilities, and lunchrooms are dealt with in these standards. Where such facilities are required, the employer also must assure that employees use them. Further, to avoid the occupational health hazard of exposure through ingestion, the standards also prohibit eating, smoking, or applying cosmetics in the regulated areas.[339]

Where protective clothing is required, the standards normally specify how frequently such clothing must be changed and how it is to be handled at home, and may even prohibit its removal from the workplace to avoid contamination of the home environment. Those employed by the employer to launder the contaminated clothes must be notified of the hazards associated with the contaminated clothing.

(g) Employee training. All of the standards require the employer to train employees on an annual basis concerning

---

[337]The lead standard, §1910.1025(f), is the most comprehensive in terms of defining a minimally acceptable respirator program.

[338]For examples of the types of signs and labels required, see §1910.1006(e)(1)–(4), which governs employee exposure to methyl chloromethyl ether and covers such details as the size of the letters; and see §1910.1045(p), which governs employee exposure to acrylonitrile and mandates the actual wording of the sign. See also the Hazard Communication Standard, §1910.1200.

[339]The cotton dust standard, §1910.1043, does not have such a requirement. The asbestos standard, §1910.1001, allows eating, drinking, and smoking, but requires that the employer ensure that employees wash their hands and faces before engaging in these activities. §1910.1001(i)(3)(iii).

the hazards posed by the regulated substance and the provisions of the relevant standard such as medical surveillance, respiratory protection, the compliance program through engineering and work practices, and a general review of the standard.[340]

(h) Medical surveillance. In addition to reducing exposures through controls and protective equipment and advising employees concerning hazards, the Act authorizes the Secretary to require, where appropriate, that the employer make available, at no cost to employees,[341] medical examinations to detect the effects, if any, of exposure to the regulated substance.[342]

The basic elements of the medical surveillance program require that certain information about the standard, the job duties of the employee, and the employee's levels of exposure be provided to the physician. In turn, the physician is to conduct a series of tests, develop work and personal histories, and on the basis of the foregoing provide a written opinion to the employer as to whether the employee has any medical condition which would place him at increased risk of material impairment due to exposure to the regulated substance. The employer must give the employee a copy of the physician's opinion letter. In addition, the physician can offer the employer his recommendations about limitations on exposure or use of protective equipment.[343]

The Act does not empower OSHA to require employees to take medical examinations.[344] Since employee participation is essential to the medical surveillance program, OSHA has been faced with the problem of how to obtain employee cooperation given the realistic employee fear that a detected

---

[340]See also the Hazard Communication Standard, §1910.1200.

[341]The phrase "at no cost to employees" has been interpreted to include time and transportation if the medical examination is given outside of normal working hours. *Phelps Dodge v. OSHRC*, 725 F.2d 1237 (9th Cir. 1984).

[342]The noise standard does not require any medical examination but does provide for audiograms (§1910.95(g)), with possible referral to a physician in certain circumstances.

[343]See, e.g., DBCP, §1910.1044(m). The physician is prohibited from revealing to the employer findings or diagnoses unrelated to the occupational exposure.

[344]§6(b)(7). Several arbitrators have held that *employers* can require employees to take medical examinations as a condition of employment in order to fulfill the employer's responsibility to maintain a safe and healthful workplace. See, e.g., *Lever Bros.*, 66 Lab. Arb. 211 (1976), and *Copperweld Steel Co.*, Steelworkers Arbitration Awards, Vol. XXII at 16,558 (1981).

medical condition could result in loss of employment or transfer to another job with a loss in pay.

Two regulatory initiatives have been pursued. The first deals with situations where a doctor certifies that an employee is physically unable to wear a respirator. Both the asbestos[345] and cotton dust[346] standards provide that such employees cannot be allowed to remain in jobs where exposure exceeds the PEL and require the employer to transfer them to other available jobs without economic loss as defined in each standard.[347]

In the lead standard, OSHA went a major step further, by mandating a medical removal protection (MRP) program. Under that program, workers with an elevated blood lead level or those who receive a medical determination that continued exposure will likely lead to medical impairment must be removed by their employer from jobs where the PEL is exceeded. If alternative jobs are available where the workers will not be exposed to excessive lead levels, the employer is to assign the workers to those jobs; if none are available, the employee is sent home. In either case, under the MRP program the worker suffers no loss of earnings, seniority, or other employment benefits for a maximum period of 18 months. Given the fact that lead is not a carcinogen and excess blood lead levels over time in many cases can be reduced to normal, the standard contemplates that earnings protection will not be needed for extended periods.[348]

(i) Recordkeeping and access to records. Exposure records and medical records are the two broad categories of information that employers are required to retain under the typical occupational health standards. The length of retention time varies from 2 years for the noise exposure measurements[349] to 40 years or the duration of employment plus 20 years for

[345]29 C.F.R. §1910.1001(g)(3)(iv).

[346]§1910.1043(f)(2)(iv).

[347]The asbestos standard provision was upheld by the District of Columbia Circuit in *Industrial Union Dep't v. Hodgson*, 499 F.2d 467, 485 (D.C. Cir. 1974). Cf. *American Textile Mfrs. Inst. v. Donovan*, 452 U.S. 490, 536–540 (1981).

[348]29 C.F.R. §1910.1025(k). This provision, in its entirety, was upheld by the District of Columbia Circuit in *United Steelworkers v. Marshall*, 647 F.2d 1189, 1228–1238 (D.C. Cir. 1980), *cert. denied*, 453 U.S. 913 (1981). See also Ch. 20, text at nn. 265–270.

[349]§1910.95(m)(3)(i).

medical records in standards such as coke oven emissions[350] and lead.[351]

In recognition of the realities of employment relationships, the Secretary also has required that records are to be transferred to successor employers or to NIOSH during the above-mentioned maintenance period. Even after that period has elapsed, and before the records are destroyed, NIOSH has the option of obtaining the records upon being notified by the employer.[352]

---

[350]§1910.1029(m((2)(iii).

[351]§1910.1025(n)(2)(iv).

[352]See, e.g., the lead standard at §1910.1025(n)(5.) and see Access to Employee Medical and Exposure Records, §1910.20. See also Ch. 20, Sec. V.B.

4

The General Duty Clause

I. Overview

One of the most frequently litigated provisions of the OSH Act is the general duty clause. This provision imposes on each employer the broad obligation to

> "furnish to each of his employees employment and a place of employment which are free from recognized hazards that are causing or are likely to cause death or serious physical harm to his employees."[1]

This requirement has withstood due process challenges premised upon its alleged vagueness. In addressing this issue, courts have found that since employers were obligated to deal only with hazards that were "recognized," the constitutional requirement that fair warning be given of the prohibited or required conduct was satisfied.[2]

Despite the breadth of the general duty clause, it does not impose absolute liability on employers or render the employer a guarantor of employee safety and health:[3]

---

[1]§5(a)(1), 29 U.S.C. §654(a)(1) (1979).

[2]*Donovan v. Royal Logging Co.*, 645 F.2d 822, 9 OSHC 1755 (9th Cir. 1981); *Bethlehem Steel Corp. v. OSHRC*, 607 F.2d 871, 875, 7 OSHC 1802, 1805 (3d Cir. 1979) ("We also reject Bethlehem's argument that the general duty clause is unconstitutionally vague. The 'recognized hazard' standard, in our view, gives industrial employers fair notice of the conduct they must avoid").

[3]*National Realty & Constr. Co. v. OSHRC*, 489 F.2d 1257, 1 OSHC 1422 (D.C. Cir. 1973). See also *Whirlpool Corp. v. OSHRC*, 645 F.2d 1096, 9 OSHC 1362 (D.C.

"An employer's duty under Section 5[(a)](1) is not an absolute one. It is the [House] Committee's intent that an employer exercise care to furnish a safe and healthful place to work and to provide safe tools and equipment. This is not a vague duty, but is protection of the worker from preventable dangers."[4]

The legislative history of the OSH Act makes it clear that the purpose of the general duty clause is to fill any gaps in the protection afforded employees by the standards promulgated under the Act,[5] and not to act as a substitute for standards setting pursuant to Section 6 of the Act.[6]

Section 5(a)(1) applies to recognized hazards that are "causing or are likely to cause death or serious physical harm." Section 17(k) of the Act defines a serious violation as one for which "there is substantial probability that death or serious physical injury could result."[7] Therefore, a general duty clause violation must be of a serious nature.[8] Because general duty citations are issued as alleged serious violations, the Act's mandatory initial penalty requirement is applicable. This requirement is stated at Section 17(b), which provides that an employer who receives a citation for a serious violation *"shall*

Cir. 1981); *Babcock & Wilcox Co. v. OSHRC*, 622 F.2d 1160, 8 OSHC 1317 (3d Cir. 1980); *Champlin Petroleum Co. v. OSHRC*, 593 F.2d 637, 7 OSHC 1241 (5th Cir. 1979); *General Dynamics Corp. v. OSHRC*, 599 F.2d 543, 7 OSHC 1373 (1st Cir. 1979); *Central of Ga. R.R. v. OSHRC*, 576 F.2d 620, 6 OSHC 1784 (5th Cir. 1978); *Dunlop v. Rockwell Int'l*, 540 F.2d 1283, 4 OSHC 1606 (6th Cir. 1976); *Getty Oil Co. v. OSHRC*, 530 F.2d 1143, 4 OSHC 1121 (5th Cir. 1976); *Brennan v. Butler Lime & Cement Co.*, 520 F.2d 1011, 3 OSHC 1461 (7th Cir. 1975); *Brennan v. OSHRC (Hanovia Lamp Div., Canrad Precision Indus.)*, 502 F.2d 946, 2 OSHC 1137 (3d Cir. 1974); *REA Express v. Brennan*, 495 F.2d 822, 1 OSHC 1651 (2d Cir. 1974).

[4]H.R. REP. NO. 1291, 91st Cong., 2d Sess. 21 (1970), *reprinted in* Senate Comm. on Labor and Public Welfare, 92d Cong., 1st Sess., LEGISLATIVE HISTORY OF THE OCCUPATIONAL SAFETY AND HEALTH ACT OF 1970, at 851 (Comm. Print 1971) (hereinafter cited as LEGIS. HIST.).

[5]S. REP. NO. 1282, 91st Cong., 2d Sess. 10 (1970), *reprinted in* LEGIS. HIST. at 150 ("The general duty clause in this bill would not be a general substitute for reliance on standards, but would simply enable the Secretary to insure the protection of employees who are working under special circumstances for which no standard has yet been adopted").

[6]See, e.g., House debate on conference report on S. 2193 (Dec. 17, 1970) (statement of Rep. Steiger), *reprinted in* LEGIS. HIST. at 1217 ("It is also clear that the general duty requirement should not be used to set ad hoc standards. The bill already provides procedures for establishing temporary emergency standards. It is expected that the general duty requirement will be relied upon infrequently and that primary reliance will be placed on specific standards which will be promulgated under the act").

[7]29 U.S.C. §666(k).

[8]In analyzing §§5(a)(1) and 17(k), the court of appeals in *Pratt & Whitney Aircraft Div. v. Secretary of Labor*, 649 F.2d 96, 98, 9 OSHC 1554, 1555 (2d Cir. 1981), reasoned:
"[A]ny condition at a place of employment that violates the general duty clause would have to be deemed serious, since an element of a general duty violation is that the condition is 'likely to cause death or serious physical harm.' There seems to be little distinction between 'substantial probability' as employed in §17(k) and 'likely' as employed in §5(a)(1)."

be assessed a civil penalty of up to $1,000 for each violation [emphasis added]."[9]

II. Legislative History[10]

In numerous cases, the Review Commission and the courts have been called upon to address the scope of coverage of the general duty clause. Unfortunately, the legislative history provides little assistance beyond the fact that Section 5(a)(1) is not intended to be a substitute for setting standards.[11] The absence of legislative history explaining the meaning of the phrase "free from recognized hazards" is perhaps the central problem that has arisen in interpreting the general duty clause.[12]

The occupational safety and health bills introduced in 1969 in the House by Congressman O'Hara[13] and in the Senate by Senators Williams, Kennedy, Mondale, and Yarborough[14]

---

[9]29 U.S.C. §666(b). Early in the administration of the Act, the Solicitor of Labor issued an opinion holding that a penalty *must* be assessed for any serious violation, because of the language of this section. That position continues to be followed by OSHA and the Review Commission. See, e.g., *Logan County Farm Enters.*, 7 OSHC 1275 (Rev. Comm'n 1979); *Continental Steel Corp.*, 3 OSHC 1410 (Rev. Comm'n 1975), *remanded on other grounds sub nom. Penn-Dixie Steel Corp. v. OSHRC*, 553 F.2d 1078, 5 OSHC 1315 (7th Cir. 1977).

[10]For the history of the entire Act, see Chapter 2 (Legislative History of the Occupational Safety and Health Act of 1970).

[11]See notes 5 and 6 and accompanying text, *supra*.

[12]The confusion in the legislative history over the general duty clause is due in part to the highly political nature of the proceedings leading up to passage of the Act. See Chapter 2 (Legislative History of the Occupational Safety and Health Act of 1970). As explained by an early commentator:

"[T]he legislative history of the Act, although extensive, is not dispositive of many of the questions arising under the general duty clause. The basic issue debated in the Congress was whether to include a general duty requirement in the statute at all. Perhaps the time and attention centered on this issue explain why many particular questions relating to the meaning of the requirement were never considered. To complicate matters, the coverage of the general duty provision was narrowed somewhat as the clause passed through several versions during its consideration; on the other hand, the original bills with the 'broader' general duty requirements provided no penalties for violation. Thus, a number of the explanations which do appear in the legislative record with regard to the precise scope of the clause, in addition to reflecting the usual degree of inconsistency among themselves, refer to earlier versions of the provision which were quite different in concept from the final version."

Morey, *The General Duty Clause of the Occupational Safety and Health Act of 1970*, 86 HARV. L. REV. 988, 990–1991 (1973) (footnotes omitted). See also Gombar, *OSHA's General Duty Clause: Not a License to Change Industry Practice*, BARRISTER (Winter 1982). See note 54, *infra*, for further discussion of the meaning of the phrase "recognized hazard."

[13]H.R. 3809, 91st Cong., 1st Sess. (1969), *reprinted in* LEGIS. HIST. at 629–658.

[14]S. 2193, 91st Cong., 1st Sess. (1969), *reprinted in* LEGIS. HIST. at 1–30.

contained no general duty clause. Later that year, a bill was proposed in both houses of Congress by the Nixon administration;[15] Section 4(a) of this bill later contained a requirement that employers "furnish employment and a place of employment which are as safe and healthful as prescribed by occupational safety and health standards" promulgated under that Act.

During the House and Senate hearings on these bills, held intermittently during 1969 and 1970, the president of the National Safety Council recommended that a general duty provision be included "to provide a means for requiring correction of hazardous situations which happened not to be covered by a specific standard."[16] After those hearings, the House Select Subcommittee on Labor reported a new bill,[17] introduced in the House by Congressman Daniels, which contained the open-ended requirement that each employer "furnish to each of his employees employment and a place of employment which is safe and healthful."[18]

The Daniels bill was approved by the full House Committee on Education and Labor over the opposition of the Administration and most of the Republican members of that committee.[19] Congressman Steiger then introduced a bill backed by the Administration[20] which narrowed the Daniels bill's open-ended safety and health obligation by requiring each employer to furnish to each of his employees "employment and a place of employment which are free from *any hazards which are readily apparent* and are causing or are likely to cause death or serious physical harm to his employees [em-

---

[15]H.R. 13373, 91st Cong., 1st Sess. (1969), *reprinted in* LEGIS. HIST. at 679–720; and S. 2788, *reprinted in* LEGIS. HIST. at 31–72.

[16]Governor Howard Pyle, president of the National Safety Council, testified: "If national policy finally declares that all employees are entitled to safe and healthful working conditions, then all employers would be obligated to provide a safe and healthful workplace rather than only complying with a set of promulgated standards. The absence of such a 'general obligation' provision would mean the absence of authority to cope with a hazardous condition which is obvious and admitted by all concerned for which no standard has been promulgated." *Quoted in* S. REP. NO. 1282, *supra* note 5, at 10, *reprinted in* LEGIS. HIST. at 150; *also quoted in* H.R. REP. NO. 1291, *supra* note 4, at 21, *reprinted in* LEGIS. HIST. at 851.

[17]H.R. 16785, 91st Cong., 2d Sess. (1970), *reprinted in* LEGIS. HIST. at 721–762.

[18]*Id.* §5(1).

[19]H.R. REP. NO. 1291, *supra* note 4, at 47–60, *reprinted in* LEGIS. HIST. at 877–890.

[20]H.R. 19200, 91st Cong., 2d Sess. (1970), *reprinted in* LEGIS. HIST. at 763–830.

phasis added]."[21] The Steiger bill, which also imposed penalties for first-instance citations,[22] was substituted for the Daniels bill and passed by the House.[23]

Meanwhile, the Senate Committee on Labor and Public Welfare reported out S. 2193,[24] an amended version of Senator Williams' 1969 proposal.[25] S. 2193 closely paralleled the Daniels bill; and, like Congressman Steiger in the House, Senator Dominick, a member of the Committee, sought to substitute an Administration-backed bill (S. 4404)[26] on the floor of the Senate. S. 4404 was identical to the Steiger substitute, but the Senate rejected it[27] in favor of passing S. 2193 with, however, a number of amendments.[28]

The Senate-passed bill,[29] unlike the Steiger bill, did not provide penalties for first-instance violations of the general duty clause but required each employer to "furnish to each of his employees employment and a place of employment free from *recognized hazards* so as to provide safe and healthful working conditions [emphasis added]."[30]

The differences in the language of the House and Senate provisions—together with the overriding issue of where to place responsibility for establishment of standards, enforcement, and adjudication of violations—resulted in the appointment of a conference committee to attempt to reach a compromise. The version of the statute that was enacted resulted from the conference committee's (a) adopting the House general duty provision but substituting the Senate's "recognized hazards" for the House's "readily apparent hazards," while (b) retaining the mandatory initial penalty provision of the House bill.[31]

There was little debate when the conference report was taken up on the floor of each chamber and therefore there is

[21]*Id.* §5(a).
[22]*Id.* §17(b).
[23]Legis. Hist. at 1091–1117.
[24]S. 2193, 91st Cong., 2d Sess. (1970), *reprinted in* Legis. Hist. at 204–295.
[25]See note 14 and accompanying text, *supra.*
[26]S. 4404, 91st Cong., 2d Sess. (1970), *reprinted in* Legis. Hist. at 73–140.
[27]See Legis. Hist. at 448–450.
[28]See *id.* at 451–528.
[29]S. 2193, as passed by the Senate, *reprinted in* Legis. Hist. at 529–597.
[30]*Id.* §5(a)(1).
[31]§666(b) of Title 29. See note 9 and accompanying text, *supra.*

not much guidance for determining why Congress enacted the "recognized hazards" version of the general duty clause. The conference report itself contains no explanation for the adoption of the Senate's "recognized hazards" terminology. Nor are any reasons provided for adoption of the mandatory initial penalty from the House bill.[32] What the conference report does reveal in its single-sentence explanation of the general duty clause is the nature of the compromise that was reached:

> "The Senate bill required workplaces to be free from 'recognized hazards.' The House amendment required such places to be free from 'any hazards which are readily apparent and are causing or are likely to cause death or serious bodily harm.' The House provision was adopted with the Senate's 'recognized hazard' term replacing the House's 'readily apparent hazard.' "[33]

With little assistance from the legislative history, then, the Review Commission and the courts have had to resolve questions raised by Section 5(a)(1). As more fully described below, the Commission and the courts have developed a series of principles of general applicability for determining when a general duty clause violation has occurred.[34] However, significant issues remain concerning, in particular, the proper interpretation of the term "recognized hazards" and whether or not a particular hazard is "likely" to cause death or serious physical injury.

---

[32]Indeed, the conference report explains very little about any of the Act's provisions.

[33]H. REP. NO. 1765, Conference Report, 91st Cong., 2d Sess. at 33 (1970), *reprinted in* LEGIS. HIST. at 1186. The most detailed explanation of the general duty clause in the legislative history of the Act is given by Congressman Steiger during the consideration of the conference report by the House of Representatives:

"The conference bill takes the approach of this House to the general duty requirement that an employer maintain a safe and healthful working environment. The conference-reported bill recognizes the need for such a provision where there is no existing specific standard applicable to a given situation. However, this requirement is made realistic by its application only to situations where there are 'recognized hazards' which are likely to cause or are causing serious injury or death. Such hazards are the type that can readily be detected on the basis of the basic human senses. Hazards which require technical or testing devices to detect them are not intended to be within the scope of the general duty requirements."

LEGIS. HIST. at 1217. This interpretation was rejected early in the Act's history by the first court of appeals to address the applicability of the general duty clause to a nonobvious hazard. *American Smelting & Ref. Co. v. OSHRC*, 501 F.2d 504, 2 OSHC 1041 (8th Cir. 1974).

[34]In addition, early in the Reagan administration, OSHA established uniform guidelines for use of §5(a)(1) in an attempt to clarify appropriate circumstances in which to use it as a basis for a violation. OSHA Instruction CPL 2-2.50 (Mar. 17, 1982).

III. Elements of a General Duty Clause Violation

The elements of a general duty clause violation identified by the first court of appeals that addressed the interpretation of Section 5(a)(1) have been quite uniformly used by both the Review Commission and the courts in subsequent cases. In *National Realty & Construction Co. v. OSHRC*, the court listed three elements that OSHA was required to prove in order to establish a general duty clause violation: "(1) that the employer failed to render its workplace 'free' of a hazard which was (2) 'recognized' and (3) 'causing or likely to cause death or serious physical harm.' "[35] Based on language in *National Realty*, the Review Commission has added a fourth element, requiring OSHA to specify the steps the employer should have taken to avoid citation and demonstrate the feasibility and likely utility of those steps. As will be illustrated in the next section, this fourth element is really a reformulation of the employer's duty to "free" the workplace of hazards. It is, however, a useful reformulation, for it assures that the employer is placed on notice of the precise steps OSHA believes the employer must take to comply with the general duty clause. Also, by imposing under Section 5(a)(1) a feasibility requirement corresponding to the feasibility finding required for the promulgation of a specific standard,[36] it upholds the congres-

---

[35]489 F.2d 1257, 1265, 1 OSHC 1422, 1426 (D.C. Cir. 1973). The decision has been criticized for failing to take into account the fact that the legislative history passages it relied upon in interpreting an employer's obligations under §5(a)(1) were addressed to earlier legislative proposals for a general duty clause that contained no mandatory initial penalty requirement. Additionally, it has been criticized for misquoting passages of the legislative history upon which it purports to rely. See Gombar, *supra* note 12. However, to date, neither the Review Commission nor the courts have embraced these criticisms, and the *National Realty* decision continues to be routinely cited as a landmark decision. See, e.g., *Ensign-Bickford Co. v. OSHRC*, 717 F.2d 1419, 11 OSHC 1657 (D.C. Cir. 1983); *Donovan v. Missouri Farmers Ass'n*, 674 F.2d 690, 10 OSHC 1460 (8th Cir. 1982); *S&H Riggers & Erectors v. OSHRC*, 659 F.2d 1273, 10 OSHC 1057 (5th Cir. 1981); *Pratt & Whitney Aircraft Div. v. Secretary of Labor*, 649 F.2d 96, 9 OSHC 1554 (2d Cir. 1981); *Magma Copper Co. v. Marshall*, 608 F.2d 373, 7 OSHC 1893 (9th Cir. 1979); *Bethlehem Steel Corp. v. OSHRC*, 607 F.2d 871, 7 OSHC 1802 (3d Cir. 1979); *R.L. Sanders Roofing Co. v. OSHRC*, 620 F.2d 97, 8 OSHC 1559 (5th Cir. 1980); *Brennan v. OSHRC (Reilly Tar & Chem. Corp., Republic Creosoting Div.)*, 501 F.2d 1196, 2 OSHC 1109 (7th Cir. 1974).

[36]See Chapter 3 (Specific Occupational Safety and Health Standards). In addition, there is a factor distinguishing general duty citations from those for violation of specific standards: an employer's obligation under §5(a)(1) runs only to his own employees, whereas the obligations imposed by specific standards runs to employees of any employer. See *Ronsco Constr. Co.*, 10 OSHC 1576, 1577 n.3 (Rev. Comm'n 1982).

sional intent that the general duty clause be a supplement to, and not a substitute for, the adoption of standards.[37]

A. The "Free" From Recognized Hazards Element

The requirement that an employer provide a workplace "free" from recognized hazards appears to make the employer responsible for all on-the-job hazards to which its employees are exposed. In *National Realty*, however, the court rejected this interpretation. The court explained, "Congress quite clearly did not intend the general duty clause to impose strict liability: The duty was to be an achievable one. * * * Congress intended to require elimination only of preventable hazards."[38]

Thus, in order to prove that a workplace was not "free" of a hazard under *National Realty*, OSHA must prove that the employer could have prevented the hazard. This aspect of OSHA's burden is illustrated by the facts of *National Realty*. The citation resulted from a supervisory employee riding the running board of a front-end loader, which tipped over and killed him. The court recognized that "a generic form of hazardous conduct, such as equipment riding, may be 'recognized' "; but the court reasoned further that "unpreventable instances of it are not" and therefore concluded that "the possibility of their occurrence at a workplace is not inconsistent with the workplace being 'free' of recognized hazards."[39]

The court therefore tied employer liability to more than a simple showing that hazardous conduct by employees had occurred.[40] Rather, it required a showing that the hazardous conduct was preventable, i.e., "that demonstrably feasible measures would have materially reduced the likelihood that such misconduct would have occurred."[41] The court recognized that "[a] demented, suicidal or willfully reckless employee

---

[37]Cf. *Kastalon, Inc.*, 12 OSHC 1928, 1932 (Rev. Comm'n 1986) ("Congress intended that the general duty clause would provide protection when no standard had yet been adopted, not provide protection that would go beyond what standards could permissibly provide").

[38]*Supra* note 35, at 1265–1266, 1 OSHC at 1426–1427 (footnote omitted).

[39]*Id.* at 1266, 1 OSHC at 1427.

[40]It should be emphasized that the occurrence of an accident is not required for a violation to be found. It is only necessary that the existence of a hazardous condition be proved by OSHA.

[41]*National Realty, supra* note 35, at 1267, 1 OSHC at 1428. See Section D, *infra*, for a more detailed discussion of the "feasibility of abatement" element.

may on occasion circumvent the best conceived and most vigorously enforced safety regime."[42] And it recognized that hazardous conduct by employees

> "is not preventable if it is so idiosyncratic and implausible in motive or means that conscientious experts, familiar with the industry, would not take it into account in prescribing a safety program. Nor is misconduct preventable if its elimination would require methods of hiring, training, monitoring, or sanctioning workers which are either so untested or so expensive that safety experts would substantially concur in thinking the methods infeasible."[43]

Thus, the court in *National Realty* viewed the effectiveness of the employer's safety program as the key to whether the employer had freed the workplace of the hazard. The court vacated the citation because it found no evidence in the record that indicated what OSHA believed the company should have done to prevent the accident. The court stated, "the hearing record is barren of evidence describing, and demonstrating the feasibility and likely utility of the particular measures which National Realty should have taken to improve its safety policy. Having the burden of proof, the Secretary must be charged with these evidentiary deficiencies."[44]

As this discussion illustrates, the court did not intend that the requirement that OSHA prove the feasibility and likely utility of abatement be a separate element of a general duty clause violation; the requirement merely was intended to clarify what OSHA had to show in order to prove that a workplace was not "free" of a recognized hazard. Nevertheless, in cases following *National Realty*, the Review Commission listed the feasibility and likely utility of abatement as a fourth element distinct from the "free from recognized hazards" element,[45] and applied this element to cases that involved the feasibility of physical means of abatement as well as those that involved the effectiveness of an employer's safety program. Although this fourth element was added with no rationale beyond citation to *National Realty*, it has been accepted by the courts

---

[42]*National Realty, supra* note 35, at 1266, 1 OSHC at 1427.

[43]*Id.* (footnote omitted).

[44]*Id.* at 1267, 1 OSHC at 1428.

[45]E.g., *Jones & Laughlin Steel Corp.*, 10 OSHC 1778, 1781 (Rev. Comm'n 1982); *Beaird-Poulan*, 7 OSHC 1225, 1228 (Rev. Comm'n 1979).

of appeals and has not been challenged by OSHA.[46] This is probably because the ultimate holding of *National Realty*— that Section 5(a)(1) requires employers to eliminate only preventable hazards—is an accurate reflection of the intent of Congress; and requiring OSHA to specify and prove what the employer should have done is a reasonable way of implementing that intent.

Many Section 5(a)(1) citations, like that in *National Realty*, involve situations where OSHA argues that the employer's safety program was not adequate to free the workplace of recognized hazards. Because the Review Commission views the employer's safety program as being relevant to whether OSHA proved the feasibility and likely utility of a means of abatement, those cases will be discussed under that topic. (See Section D below.)

B. The "Recognized Hazard" Element

Perhaps the most frequently litigated aspect of general duty clause citations is whether the cited condition constitutes a "recognized hazard" under Section 5(a)(1). As discussed previously, the initial impetus for a general duty provision was concern that OSHA would lack authority to require correction of "a hazardous condition which is obvious and admitted by all concerned"[47] if the condition was not yet covered by a specific standard. As also discussed previously, however, the final version of the general duty clause emerged only as a compromise between the House and Senate versions,[48] lacking any legislative history that unequivocally established the intent of Congress in adopting the "recognized hazards" terminology and providing the Review Commission and the courts with little guidance.

1. Industry Knowledge

The earliest Review Commission cases contained no significant analysis of what constituted a recognized hazard.

---

[46]*Baroid Div. of NL Indus. v. OSHRC*, 660 F.2d 439, 446–447, 10 OSHC 1001, 1005 (10th Cir. 1981); *St. Joe Minerals Corp. v. OSHRC*, 647 F.2d 840, 844, 9 OSHC 1646, 1648 (8th Cir. 1981); *Babcock & Wilcox Co. v. OSHRC*, 622 F.2d 1160, 1164, 8 OSHC 1317, 1319 (3d Cir. 1980).

[47]See note 16 and accompanying text, *supra*.

[48]See note 33 and accompanying text, *supra*.

Rather, the hazards involved, e.g., operation of a crane near a power line,[49] riding the running board of a front-end loader,[50] and staying clear of a loading area while a crane was in operation,[51] were described as "obvious" or "generally recognized within the industry." In later cases involving situations where no particular industry or employer knowledge of a hazard could be demonstrated but where the nature of the hazard was obvious, the recognition involved would be referred to as "common sense."[52]

From the outset, however, OSHA generally viewed recognized hazards as based on the standard of knowledge in the particular industry. In its 1972 Compliance Operations Manual, OSHA described a recognized hazard as

"a condition that is [a] of common knowledge or general recognition in the particular industry in which it occurs, and (b) detectable (1) by means of the senses (sight, smell, touch and hearing), or (2) is of such wide, general recognition as a hazard in the industry that even if it is not detectable by means of the senses, there are generally known and accepted tests for its existence which should make its presence known to the employer."[53]

Thus, the "recognition" element of a general duty clause violation has been considered by OSHA from its earliest days to be focused upon constructive employer knowledge of a hazard based upon the employer's knowledge of the industry. Such constructive knowledge based on industry knowledge of a haz-

[49]*Dale M. Madden Constr.*, 1 OSHC 1031 (Rev. Comm'n 1972).

[50]*National Realty & Constr. Co.*, 1 OSHC 1049 (Rev. Comm'n 1972), *rev'd*, 489 F.2d 1257, 1 OSHC 1422 (D.C. Cir. 1973).

[51]*Hansen Bros. Logging*, 1 OSHC 1060 (Rev. Comm'n 1972).

[52]Such cases include *Usery v. Marquette Cement Mfg. Co.*, 568 F.2d 902, 5 OSHC 1793 (2d Cir. 1977), in which the dumping of bricks and debris into an unguarded alley without warning was found to be a general duty clause violation although no applicable standard or industry practice was identified to establish employer recognition of a hazard. Rather, the court pointed out,
"[i]t scarcely required expertise in the industry to recognize that it is hazardous to dump bricks from an unenclosed chute into an unbarricaded alleyway, twenty-six feet below, between buildings in which unwarned employees worked."
568 F.2d at 910, 5 OSHC at 1798. See also *Litton Sys.*, 10 OSHC 1179, 1182 (Rev. Comm'n 1979) (30-foot blind spot in front of 30-ton straddle carrier is obvious hazard); *Eddy's Bakeries Co.*, 9 OSHC 2147, 2150 (Rev. Comm'n 1981) (danger of gasoline vapors near source of ignition is matter of common knowledge); *Richmond Block, Inc.*, 1 OSHC 1505 (Rev. Comm'n 1974) (cleaning inside of cement mixer is recognized hazard).
The most recent guidelines issued by OSHA also refer to the common-sense basis for recognized hazards. CPL 2.50, OSHA Industrial Hygiene Field Operations Manual, ch. IV, §A.2.

[53]OSHA Compliance Operations Manual, ch. VIII, §A.2.b(1).

ard was contemplated by Congress as providing the minimum notice consistent with due process considerations.[54]

The District of Columbia Circuit's *National Realty* decision provided the first comprehensive analysis of the recognized hazard element in a footnote[55] now routinely cited in general duty clause cases.[56] In *National Realty*, the Review Commission had found a general duty clause violation based on its determination that equipment riding was a recognized hazard in the construction industry as well as being an obvious hazard. On appeal, the court analyzed the issue of whether the cited employer must have actual knowledge of the hazard in order for a general duty violation to be found, or if industry knowledge could be attributed to the cited employer.

Relying upon the legislative history, the court concluded that actual knowledge by the cited employer was not necessary for a violation to be sustained. It was sufficient for constructive knowledge of the hazard to be imputed to the cited employer on the basis of knowledge of the hazard by its industry. The court explained its reasoning as follows:

> "An activity may be a 'recognized hazard' even if the defendant employer is ignorant of the activity's existence or its potential for harm. The term received a concise definition in a floor speech by Representative Daniels when he proposed an amendment which became the present version of the general duty clause:
>
> > " 'A recognized hazard is a condition that is known to be hazardous, and is known not necessarily by each and every individual employer but is known taking into account the standard of knowledge in the industry. In other words, whether or not a hazard is "recognized" is a matter for objective determination; it does not depend on whether the particular employer is aware of it.'
>
> "116 CONG. REC. (Part 28) 38377 (1970). The standard would be the common knowledge of safety experts who are familiar

---

[54]"A recognized hazard is a condition that is known to be hazardous, and *is known not necessarily by each and every individual employer but is known taking into account the standard of knowledge in the industry.* In other words, whether or not a hazard is 'recognized' is a matter for objective determination; *it does not depend on whether the particular employer is aware of it.*" House debate on OSH Act (Sept. 22, 1970) (statement of Rep. Daniels), *reprinted in* LEGIS. HIST. at 1007 (emphasis added).

[55]In text accompanying note 57, *infra*.

[56]See, e.g., *Continental Oil Co. v. OSHRC*, 630 F.2d 446, 449, 8 OSHC 1980, 1982 (6th Cir. 1980), *cert. denied*, 450 U.S. 965 (1981); *General Dynamics Corp. v. OSHRC*, 599 F.2d 453, 464, 7 OSHC 1373, 1375 (1st Cir. 1979); *Horne Plumbing & Heating Co. v. OSHRC*, 528 F.2d 564, 569, 3 OSHC 2060, 2062 (5th Cir. 1976); *Brennan v. OSHRC (Hanovia Lamp Div., Canrad Precision Indus.)*, 502 F.2d 946, 951–952, 2 OSHC 1137, 1141–1142 (3d Cir. 1974).

with the circumstances of the industry or activity in question. The evidence below showed that both National Realty and the Army Corps of Engineers took equipment riding seriously enough to prohibit it as a matter of policy. Absent contrary indications, this is at least substantial evidence that equipment riding is a 'recognized hazard.' "[57]

2. Actual Knowledge

A few months after deciding *National Realty*, the Review Commission was faced with the issue of whether industry knowledge of a hazard was an absolute prerequisite to finding a general duty clause violation or if an employer's actual knowledge of a hazard could be relied upon. The issue was presented in *Vy Lactos Laboratories*,[58] in which three employees died and two others were injured after entering a basement to perform clean-up work when a fish solubles slurry overflowed a holding pit. The employees entered the basement and were overcome by hydrogen sulfide gas, a by-product of the decomposition of the slurry. The employer's chemist knew that hydrogen sulfide was produced in this process and that in sufficient concentration the gas posed a danger of death or serious injury. The government's experts testified that the gas could result from the decomposition process and that it was dangerous, but the Commission held that these facts did not establish that "it is recognized to be [a hazard] by Respondent's industry or by the public in general."[59] The Commission majority found irrelevant whether the employer's own chemist knew of the hazard because there was no showing that others in the industry possessed such knowledge.[60] Thus, the majority held that the recognition element was not proven by a showing that an employer had actual knowledge of a hazard;

---

[57]*National Realty, supra* note 50, 489 F.2d at 1265 n.32, 1 OSHC at 1426, rev'g 1 OSHC 1049 (Rev. Comm'n 1972). See also *Pratt & Whitney Aircraft Div. v. Secretary of Labor*, 649 F.2d 96, 100, 9 OSHC 1554, 1557 (2d Cir. 1981); *Usery v. Marquette Cement Mfg. Co., supra* note 52, 568 F.2d at 910, 5 OSHC at 1798; *H-30, Inc. v. Marshall*, 597 F.2d 234, 7 OSHC 1253 (10th Cir. 1979); *Brennan v. Smoke-Craft, Inc.*, 530 F.2d 843, 845, 3 OSHC 2000, 2001 (9th Cir. 1974); *Cape & Vineyard Div., New Bedford Gas & Edison Light Co. v. OSHRC*, 512 F.2d 1148, 1152, 2 OSHC 1628, 1631 (1st Cir. 1975).

[58]1 OSHC 1141 (Rev. Comm'n 1973), *rev'd and remanded sub nom. Brennan v. OSHRC and Vy Lactos Laboratories*, 494 F.2d 460, 1 OSHC 1623 (8th Cir. 1974).

[59]*Vy Lactos Laboratories, supra* note 58, at 1143.

[60]*Id.*

it was necessary to show that employers in the industry generally recognized the hazard.

In a vigorous dissent, Commissioner Burch concluded that Congress had not intended to exclude an employer's actual knowledge of a hazard from the concept of a recognized hazard, arguing that no logical reason existed for ignoring an employer's actual knowledge of a hazardous condition in his own workplace. Even absent industry recognition, a cited employer with actual knowledge of a hazardous condition clearly would have had the requisite notice to fulfill due process requirements.

On appeal, the Eighth Circuit refused to accept the Review Commission majority's narrow interpretation of "recognized hazard" and concluded that actual knowledge of a hazard by an employer was intended by Congress to be sufficient to establish a violation, stating: "Even a cursory examination of the Act's legislative history clearly indicates that the term recognized was chosen by Congress not to exclude actual knowledge, but rather to reach beyond an employer's actual knowledge to include the general recognized knowledge of the industry as well."[61] Subsequent decisions of the courts and the Review Commission have adhered to this view,[62] accepting the rationale that where "an employer is shown to have actual knowledge that a practice is hazardous, the problem of fair notice does not exist."[63]

3. Other Sources of Constructive Knowledge

As the use of industry knowledge to demonstrate employer recognition of hazards gained routine acceptance before the Review Commission and the courts, OSHA attempted to expand the concept of constructive knowledge. The first such attempts occurred in cases where recognition of a hazard in another industry was sought to be used as proof of recognition

---

[61]*Brennan, supra* note 58, 494 F.2d at 464, 1 OSHC at 1625.

[62]See, e.g., *Pratt & Whitney Aircraft Div. v. Secretary of Labor,* 649 F.2d 96, 9 OSHC 1554 (2d Cir. 1981); *St. Joe Minerals Corp. v. OSHRC,* 647 F.2d 840, 9 OSHC 1646 (8th Cir. 1981); *Continental Oil Co. v. OSHRC, supra* note 56; *Magma Copper Co. v. Marshall,* 608 F.2d 373, 7 OSHC 1893 (9th Cir. 1979); *General Elec. Co.,* 10 OSHC 2034 (Rev. Comm'n 1982); *Copperweld Steel Co.,* 2 OSHC 1602 (Rev. Comm'n 1975)); *Vy Lactos Laboratories, supra* note 58.

[63]*Cape & Vineyard Div., New Bedford Gas & Edison Light Co. v. OSHRC, supra* note 57, at 1152, 2 OSHC at 1631.

in a cited employer's industry. The Review Commission and the courts have for the most part rejected these attempts.[64] However, where the hazard is obvious or is generally known within a profession, it is not necessary for OSHA to show that the cited employer's industry specifically recognizes the hazard.[65]

Other vehicles for efforts to broaden the concept of constructive knowledge, which have had mixed success, include advisory standards,[66] National Institute for Occupational Safety and Health (NIOSH) documents,[67] manufacturers' warnings

---

[64]See, e.g., *R.L. Sanders Roofing Co. v. OSHRC*, 620 F.2d 97, 8 OSHC 1559 (5th Cir. 1980) (construction industry recognition of hazard of unguarded edges of platforms inapplicable to unguarded edge of roof involving roofing industry); *Magma Copper Co. v. Marshall, supra* note 62 (expert testimony about explosion hazard due to lack of noninterchangeable hose fittings in oxygen system in hospitals and welding operations does not demonstrate recognition of hazard in smelting or refining industry); *H-30, Inc. v. Marshall*, 597 F.2d 234, 7 OSHC 1253 (10th Cir. 1979) (standards applicable to truck cranes, railroad and crawler cranes, and derricks or cranes used in construction did not establish recognized hazard in oil-well drilling industry); *Brennan v. OSHRC (Reilly Tar & Chem. Corp., Republic Creosoting Co.)*, 501 F.2d 1196, 2 OSHC 1109 (7th Cir. 1974) (that failure to barricade or band piles of green ties is recognized hazard in material hauling industry does not prove recognition of hazard in wood treatment industry); *P&D Trucking*, 7 OSHC 2139 (Rev. Comm'n J. 1979) (adoption by construction and maritime industries of ANSI Hand Signal Standards which recommend training for crane hookers does not establish recognition of hazard in trucking industry).

[65]See note 52, *supra*. See also *Kelly Springfield Tire Co.*, 10 OSHC 1970, 1973 (Rev. Comm'n 1982), *aff'd*, 729 F.2d 317, 11 OSHC 1889 (5th Cir. 1984) (explosion hazard from combination of combustible dust, oxygen, and ignition source in enclosed space is not confined to one industry but is "known to all chemical engineers").

[66]*St. Joe Minerals Corp. v. OSHRC, supra* note 62, at 845 n.8, 9 OSHC at 1649 (ANSI standards for elevators); *Kansas City Power & Light Co.*, 10 OSHC 1417 (Rev. Comm'n 1982) (NFPA standards for coal pulverizers); *Cargill, Inc.*, 10 OSHC 1398 (Rev. Comm'n 1982) (NFPA standard for grain elevators); *Betten Processing Corp.*, 2 OSHC 1724 (Rev. Comm'n 1975) (ANSI standard for cranes); *Hamilton Erection*, 8 OSHC 2174 (Rev. Comm'n J. No. 79-2379, 1980) ((ANSI standard for cranes); *Whaley Eng'g Corp.*, 8 OSHC 1644 (Rev. Comm'n J. 1980) (ANSI standard for cranes); *Valley Center Farmers Elevator*, 8 OSHC 1061 (Rev. Comm'n J. 1979) (NFPA standards for grain elevators); *Consolidated Edison Co.*, 8 OSHC 1224 (Rev. Comm'n J. 1979) (ANSI standard for lift platforms); *Madison Foods, Inc.*, 7 OSHC 2072 (Rev. Comm'n J. 1979) (ANSI standard for cranes); *L.H.C., Inc.*, 7 OSHC 1672 (Rev. Comm'n J. 1979) (Manual on Uniform Traffic Control Devices); *Morrow County Grain Growers*, 7 OSHC 1661 (Rev. Comm'n J. 1979) (NFPA standards for grain storage facilities); *Mon-Clair Grain Co.*, 7 OSHC 1658 (Rev. Comm'n J. 1979) (NFPA standard for grain storage facilities); *Pierce Packing Co.*, 6 OSHC 1849 (Rev. Comm'n J. 1978) (American Standard Safety Code for elevator operation); *Pittston Stevedoring Corp.*, 5 OSHC 1808 (Rev. Comm'n J. 1977) (ANSI standards for elevators).

[67]*Carlton & Wilkerson*, 8 OSHC 1721 (Rev. Comm'n J. 1980) (NIOSH Safety Guide for Auto and Home Supply Stores); *Toms River Chem. Corp.*, 6 OSHC 2192 (Rev. Comm'n J. Nos. 76-5197 & 76-5282, 1978) (NIOSH criteria document for phosgene gas).

or instructions,[68] industry publications,[69] and state[70] and local[71] laws.

The concept of constructive knowledge found its most expansive interpretation by OSHA in the area of health hazards. In addition to the use of advisory standards and standards recommended by NIOSH but not yet adopted, in 1979 OSHA issued a directive that in general duty clause cases a health hazard was to be considered "recognized" if a substance was deemed hazardous by nationally recognized voluntary standards-setting associations[72] or other government agencies[73] or was so identified in "health literature."[74] The problem created by such a broad definition of constructive knowledge is the same problem faced by Congress when it adopted the recognized hazards terminology: the due process concern of notice to the employer. In most cases, the recommended standards of NIOSH and other government agencies were rejected as evidence.[75]

---

[68]*Young Sales Corp.*, 7 OSHC 1297 (Rev. Comm'n 1979); *Great S. Oil & Gas Co.*, 10 OSHC 1996 (Rev. Comm'n J. 1982); *Georgia Steel, Inc.*, 7 OSHC 1408 (Rev. Comm'n J. 1979); *Bethlehem Steel Corp.*, 7 OSHC 1027 (Rev. Comm'n J. 1978).

[69]*Georgia Elec. Co.*, 5 OSHC 1112 (Rev. Comm'n 1977) (industry safety manual); *Atlantic Sugar Ass'n*, 4 OSHC 1355 (Rev. Comm'n 1976); *Bomac Drilling*, 9 OSHC 1681 (Rev. Comm'n Nos. 76-450 & 76-2131, 1981) (American Petroleum Institute publication); *Parker Drilling Co.*, 8 OSHC 1717 (Rev. Comm'n J. 1980) (International Association of Drilling Contractors guidelines).

[70]*Bomac Drilling, supra* note 69; *Ford Motor Co.*, 5 OSHC 1765 (Rev. Comm'n 1977); *Sugar Cane Growers Corp.*, 4 OSHC 1320 (Rev. Comm'n 1976); *M.A. Swatek & Co.*, 1 OSHC 1191 (Rev. Comm'n 1973); *Imbus Roofing Co.*, 8 OSHC 2166 (Rev. Comm'n J. 1980); *Spring Valley Water Co.*, 8 OSHC 1680 (Rev. Comm'n J. 1980); *Parker Drilling Co., supra* note 69.

[71]*Williams Enters.*, 4 OSHC 1663 (Rev. Comm'n 1976).

[72]OSHA Instruction CPL 2.20, OSHA Industrial Hygiene Field Operations Manual (hereinafter IHFOM), ch. II, §A.12.b (1979) ("Exposure to a substance that has been added to the ACGIH [American Conference of Governmental Industrial Hygienists] TLV [threshold limit value] list since 1968, or so identified as a serious hazard by federal agencies such as NIOSH, EPA or the National Cancer Institute, should be considered for a possible 5(a)(1) citation where an OSHA PEL [permissible exposure limit] does not exist"). Note: The 1968 ACGIH TLV list was adopted by OSHA at 29 C.F.R. §1910.1000, which sets the permissible exposure limit for some 400 substances. TLV lists from later years have not been adopted although they include many more such substances that are not the subject of OSHA health standards.

[73]See, e.g., *Toms River Chem. Corp., supra* note 67 (NIOSH criteria document on phosgene gas exposure limits not evidence of violation).

[74]IHFOM ch. IV §A.12.a.

[75]See, e.g., *Missouri Farmers Ass'n*, 9 OSHC 2101 (Rev. Comm'n J. 1981) (NIOSH recommendations not evidence of recognized hazard); *Archer Daniels Midland Co.*,

In 1982, OSHA issued new guidelines for general duty clause citations which retreated from the very broad 1979 guidelines.[76] Rather than imputing knowledge to employers or industry on the basis of governmental recommendations or "health literature," the revised guidelines require more critical analysis of a cited employer's knowledge of such sources or relegate them to the role of corroborative evidence.[77] Similarly, advisory standards of private standards-setting organizations are to be used only where the cited employer's industry participated in their development.[78] However, while the revised guidelines specify that recognition in another industry does not establish recognition in the cited employer's industry, this restriction is relaxed by a provision permitting the cited employer's industry to be broadly defined.[79]

4. Types of Hazards Covered

Until recently there were few issues raised concerning the types of hazards encompassed by the general duty clause. Despite some legislative history supporting the proposition that "nonobvious" hazards, such as those not detectable by the basic human senses,[80] were intended by Congress to be outside the ambit of recognized hazards, the early rejection of this position by the Review Commission quickly foreclosed such arguments.

---

8 OSHC 2051 (Rev. Comm'n J. 1981) (advisory government publication on hazard of grain dust explosions not evidence of industry recognition). Cf. *GAF Corp.*, 3 OSHC 1686, 1691 (Rev. Comm'n 1975) *aff'd*, 561 F.2d 913, 5 OSHC 1555 (D.C. Cir. 1977) (NIOSH recommendations not binding on OSHA when it promulgates standard). But see *Carlton & Wilkerson, supra* note 67 (NIOSH safety guide evidence of industry recognition of hazard where employer had actual knowledge of problem through employee safety complaints).

[76]OSHA Instruction CPL 2.50, IHFOM ch. IV, §A.2 (1983).

[77]IHFOM ch. IV, §A.2.b(2)(a)(8).

[78]IHFOM ch. IV, §A.2.b(2)(a)(7) (industry recognition may be established by "[s]tandards issued by the American National Standards Institute (ANSI), the National Fire Protection Agency (NFPA), and other private standards-setting organizations, if the relevant industry participated on the committee drafting the standards. Otherwise, such private standards normally shall be used only as corroborating evidence of recognition").

[79]*Id.* §A.2.b(2)(a). The exact meaning of the qualification allowing a broad definition of the cited employer's industry conceivably applies where there are recognized divisions within an industry and it is reasonable to presume familiarity on the part of the cited employer with the practices of the other division. One possible example would be imputing knowledge of construction industry practices to an employer in the roofing industry.

[80]See statement of Rep. Steiger, *supra* note 33.

In *American Smelting & Refining Co.*,[81] a Section 5(a)(1) citation was issued for exposing employees to airborne concentrations of inorganic lead in excess of the generally recognized safe levels. The company argued that the general duty clause was applicable only to recognized hazards detectable by the human senses and not to hazards, such as airborne lead concentrations, that were detectable only by the use of testing instruments. The company's argument was derived from a statement made by Congressman Steiger during consideration by the House of Representatives of the conference report on the Act.[82]

The Review Commission rejected the Steiger statement as an inaccurate reflection of congressional intent, relying instead upon the explanation of Senator Javits, who was the author of the amendment that substituted the term "recognized hazards" for the Administration bill's "readily apparent hazards":

> "This is a significant improvement over the Administration Bill which requires employers to maintain the workplace free from readily apparent hazards. *That approach would not cover nonobvious hazards discovered in the course of the inspection.*"[83]

From this statement, the Review Commission concluded that nonobvious hazards were intended by Congress to be covered and rejected the employer's argument to the contrary. The court of appeals affirmed and the issue has never again been seriously questioned.

Recently, however, another issue concerning the scope of coverage of the term "hazard" was raised in *American Cyanamid Co.*[84] The employer instituted a policy that excluded women of age 16 to 50 from production jobs in its lead pigment department unless they presented proof of surgical sterilization. The stated purpose of the policy was to protect the fetuses of women employees exposed to lead, especially during the earliest stages of pregnancy when the employee may not have

---

[81]1 OSHC 1256 (Rev. Comm'n 1973), *aff'd*, 501 F.2d 504, 2 OSHC 1041 (8th Cir. 1974).

[82]The statement is reproduced at note 33, *supra*.

[83]1 OSHC at 1257 (quoting S. REP. NO. 1282 (*supra* note 5, at 58, *reprinted in* LEGIS. HIST. at 197), citing 3 U.S. CODE CONG. & ADMIN. NEWS at 5222 (1970)) (emphasis added by Review Commission).

[84]9 OSHC 1596 (Rev. Comm'n 1981), *aff'd sub nom. Oil, Chem. & Atomic Workers v. American Cyanamid Co.*, 741 F.2d 444, 11 OSHC 2193 (D.C. Cir. 1984).

known that she was pregnant and when the fetus would be extremely susceptible to adverse effects from lead contained in the mother's blood. The employees were notified that there were only a few jobs to which some of them could be transferred with no loss of pay and that the rest would be transferred to any available lower-paying positions or would be fired.

The union filed a complaint with OSHA, which subsequently issued a citation alleging a general duty violation by the adoption and implementation of

> "a policy which required women employees to be sterilized in order to be eligible to work in those areas of the plant where they would be exposed to certain toxic substances."[85]

The employer contended, among other things, that the fetal protection policy was not a "hazard" within the meaning of the general duty clause. OSHA, on the other hand, argued that "any condition of employment which can ultimately result in reduced functional capacity is a hazard within the meaning of the general duty clause."[86]

Although the Review Commission agreed that the policy was a condition of employment, it concluded that Congress did not intend every aspect of employer-employee relations to be subject to the Act. Rather, the Commission concluded that Congress intended the Act to apply to "occupational hazards in terms of processes and materials which cause injury or disease by operating directly upon employees as they engage in work or work-related activities."[87]

The decision then went on to explain:

> "The fetus protection policy is of a different character altogether. It is neither a work process nor a work material, and it manifestly cannot alter the physical integrity of employees while they are engaged in work or work-related activities. An employee's decision to undergo sterilization in order to gain or retain employment grows out of economic and social factors which operate primarily outside the workplace.

> "The employer neither controls nor creates these factors as he creates or controls work processes and materials. For these reasons we conclude that the policy is not a hazard within the meaning of the general duty clause.

---

[85]*Supra* note 84, at 1597.
[86]*Id.* at 1598.
[87]*Id.* at 1600 (footnote omitted).

"It is true, as the Secretary points out, that the Act is broad in scope and may be fairly described as intended to protect employees from reduced functional capacity as a result of the work experience. However, it does not follow that the general duty clause applies to an employment policy whose physical impact on employees is indirect and derives not from work processes and materials but from social and economic factors outside the work places."[88]

Accordingly, the Review Commission concluded that the citation failed to allege a violation because the policy was not a hazard within the scope of the general duty clause.

On appeal, the District of Columbia Circuit affirmed, relying in part upon the definition of working conditions and hazards set forth by the Supreme Court in an Equal Pay Act case:

"[T]he element of working conditions encompasses two subfactors: 'surroundings' and 'hazards.' 'Surroundings' measures the elements, such as toxic chemicals or fumes, regularly encountered by a worker, their intensity, and their frequency. *'Hazard' takes into account the physical hazards regularly encountered, their frequency, and the severity of injury they can cause.* This definition of 'working conditions' is * * * well accepted across a wide range of American industry."[89]

Noting that the definition had previously been applied to a different section of the OSH Act by the Fourth Circuit in *Southern Railway v. OSHRC,*[90] the court concluded there was no reason why the definition should not also "influence the concept of 'hazards' in the general duty clause."[91] Presuming that Congress legislated about industrial relations with the "language of industrial relations" in mind, the court held that

---

[88]*Id.* (footnote omitted).

[89]*Oil, Chem. & Atomic Workers v. American Cyanamid Co., supra* note 84, 741 F.2d at 448, 11 OSHC at 2196 (footnote omitted) (citing *Corning Glass Works v. Brennan,* 417 U.S. 188, 202) (emphasis added by court).

[90]539 F.2d 335, 339, 3 OSHC 1940, 1943 (4th Cir.), *cert. denied,* 429 U.S. 999 (1976). The District of Columbia Circuit quoted the Fourth Circuit's statement that
" '[w]e think this aggregate of "surroundings" and "hazards" contemplates an area broader in its contours than * * * "particular discrete hazards" * * * but something less than the employment relationship in its entirety[.] * * * [W]e are *of the opinion that the term "working conditions"* as used in Section 4(b)(1) *means the environmental area in which an employee customarily goes about his daily tasks.*' "
741 F.2d at 448, 11 OSHC at 2196 (emphasis added by District of Columbia Circuit).

[91]*Oil, Chem. & Atomic Workers v. American Cyanamid Co., supra* note 84, 741 F.2d at 448, 11 OSHC at 2196.

"the general duty clause does not apply to a policy as contrasted with a physical condition of the workplace."[92]

C. The "Causing or Likely To Cause" Element

Once the existence of a recognized hazard has been demonstrated, OSHA has the burden of proving that the hazard is one which is, in the language of the general duty clause, "causing or likely to cause death or serious physical harm to [the employer's] employees."[93] The issue raised by the "likely to cause" terminology is whether, as a requisite for finding a violation, (a) the occurrence of an accident must be probable, or (b) it is sufficient to prove that the possibility of "death or serious physical harm" exists should an accident occur (the "possibility" test). A comparison with the burdens of proof placed upon OSHA where a *specific* standard is cited under Section 5(a)(2)[94] is enlightening in this regard.

When a serious violation of a specific standard is alleged, OSHA's burden of proof is determined by the definition in Section 17(k) of a serious violation, viz., that there exists a "substantial probability that death or serious physical harm

---

[92]*Id.* The court also rejected the union's argument that the sterilization policy was analogous to a situation where employees were exposed to a sterilizing chemical and faced the same choice of being sterilized or quitting. The union argued that since "the Commission would not suggest that the presence of such factors would remove the sterilizing *chemical* from the ambit of the Act, * * * there is no logical basis for suggesting that the presence of such factors should remove the sterilization *policy* from the ambit of the Act." *Id.*

The court rejected the argument as being an overly broad construction of the general duty clause, stating that the hazards in the two cases were not identical "unless it can be said that anything, no matter what its nature or how it operates, is a 'hazard' within the meaning of the general duty clause if it has a harmful effect." *Id.* at 449, 11 OSHC at 2196. It then concluded that to accept the union's interpretation would be

"to adopt a broad principle of unforeseeable scope: any employer policy which, because of employee economic incentives, left open an option exercised outside the workplace that might be harmful would constitute a 'hazard' that made the employer liable under the general duty clause. It might be possible to legislate limitations upon such a principle but that is a task for Congress rather than courts. As it now stands, the Act should not be read to make an employer liable for every employee reaction to the employer's policies. There must be some limit to the statute's reach and we think that limit surpassed by petitioners' contentions. The kind of 'hazard' complained of here is not, as the Commission said, sufficiently comparable to the hazards Congress had in mind in passing this law."

Id.

[93]29 U.S.C. §654(a)(1). See *supra* note 36 concerning an employer's safety and health obligation being limited to his own employees.

[94]29 U.S.C. §654(a)(2) (1979).

could result from a [hazardous] condition [emphasis added]."[95] Application of the possibility test is therefore consistent with this burden of proof: where a specific standard is cited, OSHA only need demonstrate that a proscribed hazardous condition exists.

From the earliest of such cases brought before the Review Commission, the decisions have held that the test to be applied is whether there is a substantial probability that death or serious physical harm could result if an accident occurs, not whether there is a substantial probability that an accident could occur.[96] Deference will generally be accorded the Review Commission's determination of whether such an accident would result in death or serious physical harm. As the *National Realty* court explained in a footnote:

> "If evidence is presented that a practice could eventuate in serious physical harm upon other than a freakish or utterly implausible concurrence of circumstances, the Commission's expert determination of likelihood should be accorded considerable deference by the courts."[97]

The general duty clause, on the other hand, specifies not simply that severe results "could" occur *if* there were an accident, but that a recognized hazard must be "causing or likely to cause death or serious physical harm."

1. R.L. Sanders Roofing Co.: *The Possibility Test Enunciated*

In *R.L. Sanders Roofing Co.*,[98] the Review Commission made explicit what previously could only be inferred from its decisions concerning the proper interpretation of the phrase "likely to cause":

---

[95]29 U.S.C. §666(k) (1979).

[96]See, e.g., *California Stevedore & Ballast Co.*, 1 OSHC 1305, 1307 (Rev. Comm'n 1973) ("[W]e do not consider the evidence introduced to establish that there was little likelihood the beam would dislodge. That evidence is a relevant consideration in determining the appropriate penalty"), *aff'd*, 517 F.2d 986, 988, 3 OSHC 1174, 1175 (9th Cir. 1975) (under §17(k), OSHA must only demonstrate that "any accident which should result from violation of a regulation would have a substantial probability of resulting in death or serious physical harm").

[97]*National Realty & Constr. Co. v. OSHRC*, 489 F.2d 1257, 1265 n.33, 1 OSHC 1422, 1426 (D.C. Cir. 1973).

[98]7 OSHC 1566 (Rev. Comm'n 1979), *rev'd on other grounds*, 620 F.2d 97, 8 OSHC 1559 (5th Cir. 1980).

"Although the Commission has previously reviewed 5(a)(1) citations in which the likelihood was great that the consequences would be serious if an accident did occur, our decisions have not always discussed this distinction. To the extent that prior Commission decisions have not clearly defined the phrase 'likely to cause' in section 5(a)(1), we do so now. We hold that the proper question is not whether an accident is likely to occur, but whether, if an accident does occur, the result is likely to be death or serious physical harm."[99]

Earlier appellate decisions had implicitly adopted substantially the same position, i.e., that the likelihood of an accident need not be considered in evaluating whether a recognized hazard existed.[100]

2. Bomac Drilling: *The Reasonable Foreseeability Test*

The Review Commission adhered to the position announced in *R.L. Sanders Roofing Co.* for only a short time, however. Slightly less than two years later, in *Bomac Drilling*,[101] the Commission modified its position to require OSHA to show that an "incident is 'reasonably foreseeable' and that the likely consequence in the event of an incident would be death or serious physical harm to employees."[102] It continued to adhere to the position that OSHA was under no obligation to prove the "probability that exposure to a recognized hazard will result in a hazardous incident."[103]

3. *Tests Rejected by Court of Appeals*

The *Bomac Drilling* interpretation was to endure for only a year when both it and the *R.L. Sanders Roofing Co.* defi-

---

[99]*Supra* note 98, 7 OSHC at 1569.

[100]See, e.g., *Usery v. Marquette Cement Mfg. Co.*, 568 F.2d 902, 910, 5 OSHC 1793, 1798 (2d Cir. 1977) ("Here the serious nature of the harm that could result if [an accident occurred] warranted precautions against even the slightest possibility of its occurrence"); *Titanium Metals Corp. v. Usery*, 579 F.2d 536, 543, 6 OSHC 1873, 1878 (9th Cir. 1978) ("In applying the 'likely to cause' element of the general duty clause, it is improper to apply mathematical tests relating to the probability of a serious mishap occurring * * * given the Act's prophylactic purpose to prevent employee injuries" (citations omitted)).

[101]9 OSHC 1681 (Rev. Comm'n 1981).

[102]*Id.* at 1696. See also *Brown & Root, Inc.*, 8 OSHC 2140 (Rev. Comm'n 1980); *Armstrong Cork Co.*, 8 OSHC 1070 (Rev. Comm'n 1980), *petition dismissed*, 636 F.2d 1207, 9 OSHC 1416 (3d Cir. 1980).

[103]*Bomac Drilling, supra* note 101, at 1696.

nition of "likely to cause" were rejected by the Second Circuit in *Pratt & Whitney Aircraft v. Secretary of Labor*.[104] There the company had been cited for a general duty violation based upon the storage of large quantities of acids and cyanides in a common, indoor bulk storage facility with a common drain. It was conceded that the accidental mixing of the two would result in the production of hydrogen cyanide, a lethal gas. But the company argued that the possibility of hydrogen cyanide being formed by the two chemicals being spilled into the drain was too remote to be deemed "likely" to occur since such an occurrence depended upon "a series of disconnected events."[105] The Review Commission rejected the company's argument and held that the occurrence of an incident need not be likely but only reasonably foreseeable in order for a violation to be proved.

On appeal, the Second Circuit first rejected the possibility test of *R.L. Sanders Roofing Co.* as inconsistent with the concept of a recognized hazard, since

> "an employer could be held liable notwithstanding that neither he nor any responsible member of his industry knew or could have known of the potential danger of a condition or activity, so long as there existed a plausible, theoretical possibility that the condition or activity could cause an employee to incur serious injury. Under such a standard, it would seem that whenever the occurrence of an incident capable of causing an employee serious injury is not impossible, an employer is in violation of §5(a)(1)."[106]

Similarly, the *Bomac Drilling* reasonable foreseeability test was rejected on the basis that "there is a vast difference between what is known or recognized and what is reasonably foreseeable."[107] Instead, the court relied upon its earlier holding in *Usery v. Marquette Cement Manufacturing Co.*[108] that to be a recognized hazard "the dangerous potential of the condition or activity being scrutinized either must be known by the employer or known generally in the industry."[109] According to the court, "[a]pplying the appropriate standard avoids

---

[104]649 F.2d 96, 9 OSHC 1554 (2d Cir. 1981).

[105]*Id.* at 99, 9 OSHC at 1556.

[106]*Id.* at 101, 9 OSHC at 1557.

[107]*Id.* (footnote omitted).

[108]568 F.2d 902, 6 OSHC 1052 (2d Cir. 1977).

[109]*Pratt & Whitney Aircraft v. Secretary of Labor, supra* note 104, 649 F.2d at 101, 9 OSHC at 1557.

completely the task of speculation about probabilities of hazardous incidents occurring."[110] Because the record evidence included testimony that only two of the 300 to 400 plants visited by OSHA's expert had a common drain for acids and cyanides, and that those two plants had corrected the condition when it was brought to their attention, the court concluded that there was industry knowledge of the hazard and upheld the citation.

4. Review Commission Response

After the court's *Pratt & Whitney* decision, the Review Commission explicitly overruled the *Bomac Drilling* reasonable foreseeability test "to the extent it held that the reasonable foreseeability of an incident is a distinct element of a Section 5(a)(1) violation."[111] It did not, however, discuss the court's rejection of the *R.L. Sanders Roofing Co.* possibility test. Thus, the Review Commission's position on the "likely to cause" element of a general duty clause violation appears to have reverted to the possibility test. Whether that criterion will be applied remains in question since the other courts of appeals have not focused upon this issue.[112]

D. The Feasible Abatement Method Element

As the final element of its proof of a general duty violation, OSHA must specify the steps the employer should have taken to avoid citation and demonstrate the feasibility and likely utility of those steps. As noted earlier, this element was born in *National Realty* as a way of analyzing whether the employer rendered the workplace free of the hazard. However, it later came to be recognized as a distinct element, and in many cases is the key to OSHA's case.

In general, cases in which the feasibility of abatement is an issue fall into two categories: those in which OSHA seeks

---

[110]*Id.*

[111]*United States Steel Corp.*, 10 OSHC 1752, 1757 (Rev. Comm'n 1982).

[112]See, e.g., *Baroid Div. of N.L. Indus. v. OSHRC*, 660 F.2d 439, 1 OSHC 1001 (10th Cir. 1981); *St. Joe Minerals Corp. v. OSHRC*, 647 F.2d 840, 9 OSHC 1646 (9th Cir. 1981); *Whirlpool Corp. v. OSHRC*, 645 F.2d 1096, 9 OSHC 1362 (D.C. Cir. 1981); *Empire Detroit Steel Div., Detroit Steel Corp.*, 579 F.2d 378, 6 OSHC 1693 (6th Cir. 1978); *Getty Oil Co. v. OSHRC*, 530 F.2d 1143, 4 OSHC 1121 (5th Cir. 1976); *National Realty & Constr. Co. v. OSHRC*, 489 F.2d 1257, 1 OSHC 1422 (D.C. Cir. 1973).

to have the employer install physical means of protection, and those in which OSHA wants the employer to upgrade its safety training program to prevent hazardous conduct by employees.

1. Physical Means of Abatement

In some cases, OSHA seeks to have the employer install certain equipment or use other physical means of abatement to eliminate a recognized hazard. Issues that arise include whether it is physically possible to install the equipment and whether it would accomplish the desired result. In *Cargill, Inc., Nutrena Feed Division*,[113] OSHA wanted the employer to install a rooftop ventilation system in a grain elevator to reduce the danger of fire and explosion from the buildup of grain dust. The employer presented evidence that installing the ventilation system would require piercing structural steel beams that supported the roof, seriously weakening the roof. Based on this evidence, the Commission found the proposed abatement method infeasible.

In *Phillips Petroleum Co.*,[114] the citation concerned the possibility that leakage of a flammable liquid could result in an excessive accumulation of flammable vapors in a building. OSHA wanted the employer to install a device that would monitor flammable vapors and, upon detecting excessive vapors, would automatically shut off the flow of flammable liquid, ventilate the building to remove the flammable vapors, shut off the electricity in the building, and sound a warning alarm. There was, however, no evidence that such a system had been used successfully elsewhere, and there was evidence that the device would be inaccurate, unreliable, and difficult to maintain. Upon weighing the evidence, the Commission concluded that feasibility was not proven.

Cases in which the Commission has found that OSHA proved it was feasible for employers to install certain equipment to reduce or eliminate a recognized hazard include *Kelly Springfield Tire Co.*[115] There, the hazard was the possibility of fire or explosion from the accumulation of rubber dust in a

---

[113]10 OSHC 1398 (Rev. Comm'n 1982).

[114]11 OSHC 1776 (Rev. Comm'n 1984), *aff'd without published opinion*, No. 84-1425 (10th Cir. Sept. 10, 1985).

[115]10 OSHC 1970 (Rev. Comm'n 1982), *aff'd*, 729 F.2d 317, 11 OSHC 1889 (5th Cir. 1984).

ventilation system used to carry the dust away from a grinding operation. Protection against the hazard could be achieved by assuring that the vacuum in the ventilation system was maintained. This in turn depended on the water level in the system being kept at the proper level. The Commission found that the water level could feasibly be controlled by the installation of a "bounced-air" system. The Commission noted that such a system was being used successfully at other plants operated by the employer. The Commission also found that the hazard could be reduced by adhering to the manufacturer's recommendation to clean the ventilation system every week.[116]

A recurring problem in steel mills is the accumulation of water on the floor combined with the possibility of a molten metal spill. If the molten metal covers the water so that the water is entrapped by the metal, the sudden generation of steam as the water is vaporized by the metal can result in an explosion. In affirming a 5(a)(1) citation because a steelmaker permitted water to accumulate where it could have become entrapped by a spill of molten metal, the Commission found that it would have been feasible for the company to put absorbent material on the water, to drain the water away from the area, and to build dams or dikes to keep the water away from areas of potential molten metal spills.[117]

A case in which the proposed means of abatement was found to be partially feasible involved a steel mill that transported molten iron in railroad cars. The company used wheel chocks to keep the cars from moving inadvertently, but OSHA alleged that the use of chocks did not free the workplace of the hazard and insisted that the company set the brakes on the cars to keep them stationary. The Commission found that the use of brakes was generally feasible, but noted that there were certain locations in the plant where the brakes could not be set safely and therefore limited its affirmance of the citation to those locations where the brakes could safely be set.[118]

The previous case illustrates that a means of abatement sought by OSHA may itself create hazards that must be bal-

---

[116]*Supra* note 115, at 1974–1975.

[117]*United States Steel Corp.*, 12 OSHC 1692, 1700 (Rev. Comm'n 1986). See also *Babcock & Wilcox Co. v. OSHRC*, 662 F.2d 1160 (3d Cir. 1980).

[118]*Wheeling-Pittsburgh Steel Corp.*, 10 OSHC 1242 (Rev. Comm'n 1981), *aff'd without published opinion*, 688 F.2d 828 (3d Cir. 1982), *cert. denied*, 459 U.S. 1203 (1983).

anced against the hazard sought to be eliminated. As discussed in Chapter 15, the Commission permits an employer to defend against a violation of a standard on the basis that compliance would be more hazardous than noncompliance—the "greater hazard" defense. In order to establish that defense, the employer must show not only that compliance would result in a greater hazard but that alternative means of protection are unavailable and that a variance application would be inappropriate.[119] Under Section 5(a)(1), evidence of hazards that the means of abatement would create tends to rebut OSHA's showing of feasibility and is not a matter of affirmative defense.[120] The practical result of this distinction is that an employer who is able to show in a Section 5(a)(1) case that the Secretary's proposed abatement method would be more dangerous than existing practices does not also have to show that alternative means of protection are unavailable.[121]

The abatement method advanced by OSHA as feasible will be sustained even if it exceeds the standard of protection used by the employer or the employer's industry,[122] provided, of course, that the technological and economic feasibility of the proposed abatement method is demonstrated: "The question is whether a precaution is recognized by safety experts as feasible, not whether the precaution's use has become customary."[123] Of course, widespread use in an industry of a certain means of protection is strong evidence of the precaution's feasibility.[124]

---

[119]E.g., *M.J. Lee Constr. Co.*, 7 OSHC 1140 (Rev. Comm'n 1979).

[120]*Royal Logging Co.*, 7 OSHC 1744, 1751 (Rev. Comm'n 1979), *aff'd*, 645 F.2d 822, 9 OSHC 1755 (9th Cir. 1981).

[121]*Id.*

[122]See, e.g., *Magma Copper Co. v. Marshall*, 608 F.2d 373, 7 OSHC 1893 (9th Cir. 1979); *General Dynamics Corp., Quincy Shipbuilding Div. v. OSHRC*, 599 F.2d 453, 7 OSHC 1373 (1st Cir. 1979); *Empire Detroit Steel Div., Detroit Steel Corp. v. OSHRC*, 579 F.2d 378, 6 OSHC 1693 (6th Cir. 1978); *Brennan v. OSHRC (Hanovia Lamp Div., Canrad Precision Indus.)*, 502 F.2d 946, 2 OSHC 1137 (3rd Cir. 1974); *Wheeling-Pittsburgh Steel Corp., supra* note 118.

[123]*National Realty & Constr. Co. v. OSHRC*, 489 F.2d 1257, 1266 n.37, 1 OSHC 1422, 1427 (D.C. Cir. 1973). Accord *Magma Copper Co. v. Marshall, supra* note 122; *General Dynamics Corp. v. OSHRC, supra* note 122; *Donovan v. Royal Logging Co.*, 645 F.2d 822, 9 OSHC 1755 (9th Cir. 1981).

[124]E.g., *Pratt & Whitney Aircraft*, 8 OSHC 1329, (Rev. Comm'n 1980), *aff'd in pertinent part*, 649 F.2d 96, 9 OSHC 1554 (2d Cir. 1981) (practice followed in more than 300 plants of storing acids completely separate from cyanides shows feasibility of precaution); *Litton Sys.*, 10 OSHC 1179, 1182 (Rev. Comm'n 1981) (evidence that front-mounted mirrors were standard equipment on new straddle carriers shows that such mirrors are feasible).

2. Abatement by Safety Training and Supervision

In many workplace situations, avoidance of hazards depends on proper employee conduct. Many citations have been issued under the general duty clause either because actions by employees created hazards[125] or because employees did not take precautions necessary to avoid hazards.[126] In such cases, the action OSHA seeks to have the employer take is to upgrade its safety program to avoid hazardous conduct by employees. The improvements to employer safety programs that OSHA proposes fall into two general categories: proper instruction of employees so that they are aware of potential hazards, and enforcement of work rules designed to prevent hazardous conduct.

(a) Instruction. General Dynamics Corp.[127] is a case in which the employer was found to have failed to properly instruct employees how to avoid a recognized hazard. Employees were welding heavy vertical steel plates to a horizontal bulkhead. In preparation for welding, the plates were supported by devices called monuments to keep them from falling. After an employee prematurely removed the monuments supporting one of the plates, the plate collapsed and killed another employee. At the hearing, a number of the employer's welders testified that they had not been instructed in when it was safe to remove the monuments. Absent such instructions, the Commission concluded that the employer had not freed the workplace of the hazard.[128] The Commission went on to say:

> "Moreover, even if such instructions were generally given, the fact that none of the employees who testified knew the proper procedure establishes that the instructions were not effectively communicated. An employer must do more than simply give safety instructions; it must also take reasonable steps to see that the instructions are understood and are carried out."[129]

---

[125]*National Realty & Constr. Co. v. OSHRC, supra* note 123 (employee rode on running board of front-end loader).

[126]*Alabama Power Co.*, 13 OSHC 1240 (Rev. Comm'n 1987) (employee failed to remain safe distance from dump truck delivering coal).

[127]*General Dynamics Corp., Quincy Shipbuilding Div.*, 6 OSHC 1753 (Rev. Comm'n 1978), *aff'd*, 599 F.2d 453, 7 OSHC 1373 (1st Cir. 1979).

[128]*Id.* at 6 OSHC 1757.

[129]*Id.* at 1758. See *United States Steel Corp.*, 9 OSHC 1641 (Rev. Comm'n 1981) (employee's action in opening gate that isolated high voltage line contrary to effectively communicated work rule).

Instructions are not inadequate simply because they require employees to exercise a degree of judgment.[130] But the instructions must be more than general admonitions to avoid a hazard or to act in a safe manner.[131] When OSHA alleges that employees were inadequately instructed, it is not sufficient for the agency to show that the employer failed to give instructions that might have prevented the hazard. Because OSHA bears the burden of proving the feasibility and likely utility of abatement, OSHA must present sufficient evidence to establish that the additional instructions would have materially reduced the chance that hazardous conduct would occur.[132]

(b) Enforcement. An employer's duty to provide a workplace free of recognized hazards requires it to not only instruct employees in how to avoid hazards but to take steps to assure that those instructions are followed. In Section 5(a)(1) cases in which hazardous conduct has occurred even though the employer has issued appropriate instructions or established work rules designed to prevent the hazard, OSHA may allege that the feasible means of abatement is for the employer to enforce its instructions and rules.[133] Inadequate enforcement has been found where violations of a work rule were common[134] or where the employer took essentially no steps to enforce a rule.[135]

In order to show that enforcement of work rules is a feasible abatement method, OSHA must prove that the employer knew or should have known that the work rule would be violated. In *Jones & Laughlin Steel Corp.*,[136] a highly experienced employee was killed when he attempted to board an overhead crane in a manner contrary to the employer's safety

---

[130]*Alabama Power Co., supra* note 126, 13 OSHC at 1244.

[131]*Brown & Root, Inc.*, 8 OSHC 2140 (Rev. Comm'n 1980).

[132]*Champlin Petroleum Co. v. OSHRC*, 593 F.2d 637, 640–642, 7 OSHC 1241, 1243–1244 (5th Cir. 1979) (instructing employees to follow procedure they already know is not feasible means of abatement); *Inland Steel Co.*, 12 OSHC 1968, 1977–1979 (Rev. Comm'n 1986) (OSHA did not show safety rules inadequate or inadequately communicated); *Pelron Corp.*, 12 OSHC 1833, 1836–1838 (Rev. Comm'n 1986) (same).

[133]*Western Mass. Elec. Co.*, 9 OSHC 1940, 1945 (Rev. Comm'n 1981).

[134]*Id.*

[135]*Babcock & Wilcox Co. v. OSHRC*, 622 F.2d 1160, 1164, 8 OSHC 1317, 1319 (3d Cir. 1980).

[136]10 OSHC 1778 (Rev. Comm'n 1982).

rules. OSHA argued that the employer should have taken further steps by way of additional supervision to determine if employees were violating the rules. The Commission, however, concluded that it was not reasonable for the company to more closely monitor the experienced employees involved in the particular work activity.[137] Similarly, in *Cerro Metal Products Division*,[138] another case in which an experienced employee violated a well-established safety rule, the Commission found no evidence that the employer either knew of violations of its rule or that it inadequately supervised its experienced employees.

As discussed earlier, the leading case under the general duty clause, *National Realty & Construction Co. v. OSHRC*,[139] involved a situation in which a foreman violated his company's work rule by riding on the running board of a front-end loader. Although noting that the violation of a safety rule by a foreman was "strong evidence that implementation of the policy was lax,"[140] the court emphasized that the burden was on OSHA to show what the employer should have done to improve its safety policy and that general principles could not substitute for specific evidence on the point.[141]

---

[137]*Id.* at 1783.

[138]12 OSHC 1821 (Rev. Comm'n 1986). See also *Inland Steel Co., supra* note 132, 12 OSHC at 1979–1983; *Alabama Power Co.*, 13 OSHC 1240, 1245 (Rev. Comm'n 1987).

[139]489 F.2d 1257, 1 OSHC 1422 (D.C. Cir. 1973).

[140]*Id.* at 1267 n.38, 1 OSHC at 1428.

[141]*Id.* at 1267–1268 n.40, 1 OSHC at 1428.

5

Employer Obligations to Obtain, Maintain, and Disseminate Information

I. Scope

The OSH Act requires employers to maintain and post records and make reports concerning occupational illnesses and injuries and citations. The Act empowers the Secretary of Labor to promulgate regulations detailing the obligations expressed in the Act. This chapter examines an employer's obligations under the Act and the regulations.[1]

II. Statutory Authority

Section 8(c)(1) of the Act[2] grants the Secretary the authority to prescribe recordkeeping requirements to employers "by regulation as necessary and appropriate for the enforcement of this Act or for developing information regarding the causes and prevention of occupational accidents and illnesses." That section also requires employers to post notices informing employees of their rights and obligations under the Act. Section 8(c)(2) requires the creation of records and reports on

---

[1] The regulation concerning access to medical and exposure records is addressed in Chapter 20 (Judicial Review of Standards).

[2] 29 U.S.C. §657(c)(1) (1976 & Supp. IV 1980).

work-related deaths, injuries, and illnesses.[3] Section 9[4] further requires employers to post copies of citations issued pursuant to an inspection.

The regulations promulgated under the Act are not safety and health standards in that they do not prescribe means or methods of protecting workers within the meaning of Section 3(8) of the Act. However, the regulations are substantive rules governing employer conduct.[5] The particulars of the regulations are discussed below as they relate to the different employer obligations.

III. Logs and Summaries of Occupational Illnesses and Injuries

The regulations in Part 1904 of Title 29 of the *Code of Federal Regulations* require employers with 11 or more employees[6] to "maintain in each establishment a log and summary of all recordable occupational injuries and illnesses."[7] Recordable injuries are those that result in fatality or lost workdays; require transfer to another job, termination of employment, or medical attention (more than first aid); or result in loss of consciousness or restriction of work or motion.[8] Such injuries must be recorded no later than 6 working days after the employer learns of the injury.[9] OSHA Form 200 can be used as the log, or an equivalent form supplying the same information may be used.[10] The log must be kept on a calendar year basis.[11]

Employers must also maintain supplemental records of each illness or injury on OSHA Form 101.[12] An alternative

---

[3]§657(c)(2).

[4]§658(b).

[5]The recordkeeping regulations are contained in 29 C.F.R. pts. 1903 and 1904 (1985). There are, however, recordkeeping requirements within some standards, a summary of which is set forth in the appendix at the end of this chapter. The Secretary's standards, including safety standards, appear in 1 CCH Employment Safety & Health Guide ¶3043.

[6]29 C.F.R. §1904.15 (1985).

[7]29 C.F.R. §1094.2(a)(1).

[8]§1904.12(c).

[9]§1904.2(a)(2).

[10]*Id.*

[11]§1904.2(b)(2).

[12]§1904.4.

form supplying the same information may be used as a substitute.[13] Between February 1 and March 1 of each year, employers must post, for each establishment, an annual summary of injuries and illnesses.[14]

In late 1986, the Bureau of Labor Statistics published new guidelines concerning the recordkeeping requirements of the Act. These guidelines are likely to be the focal point for future government enforcement activity in the recordkeeping area.

The following sections illustrate how the administrative law judges and the Review Commission have interpreted the regulation.

A. Who Is an Employer

As noted above, 29 C.F.R. Part 1904 requires that certain forms be maintained by an employer. In *Dayton Tire & Rubber Co.*,[15] the Review Commission considered whether a warehouser or a personnel referral service should maintain OSHA records. The Commission held that the policies behind the recordkeeping requirements would be served only by requiring the warehouser to maintain the records. The referral service was not able to conduct inspections or correct hazards, and it depended upon the warehouser to report injuries.[16] Therefore, the warehouser had a "non-delegable duty"[17] to maintain the records.

The Commission also noted that the warehouser trained, equipped, supervised, and disciplined employees, as well as set working hours and vacation times.[18] The decision focused on which entity actually controlled the employees, rather than who was labeled as the employer.

Employers may be exempted from the recordkeeping requirements by Section 4(b)(1) of the Act[19] if their activities are governed by other federal agencies that have prescribed

---

[13]*Id.*

[14]§1904.5.

[15]2 OSHC 1528 (Rev. Comm'n 1975).

[16]*Id.* at 1529. The Commission indicated that the referral service was "nothing more than agent to which [the warehouser] ha[d] contracted the undertaking of its personal activities." *Id.*

[17]*Id.* at 1530.

[18]*Id.* at 1529.

[19]29 U.S.C. §653(b)(1).

or enforced regulations affecting occupational safety and health.[20] This issue, however, is still unsettled.

In 1974, in *Southern Pacific Transportation Co.,*[21] the Commission held the railroad industry exempt from the Act's accident reporting requirements. In 1982, the Commission overruled *Southern Pacific* in *Consolidated Rail Corp.,*[22] holding that even though OSHA did not exercise substantive jurisdiction over the railroad industry, it "continues to have an interest in information about occupational illnesses and injuries occurring in the railroad industry."[23]

In *Puget Sound Tug & Barge,*[24] the Commission refused to recognize a Section 4(b)(1) exemption based on Coast Guard authority. Since the Coast Guard had not exercised its authority to promulgate recordkeeping requirements, there was nothing to preempt OSHA's regulations. Accordingly, the citations were affirmed.[25]

B. What Constitutes an "Establishment"

The log and summary, supplemental record, and annual summary must be kept at each "establishment" where business is conducted.[26] If the employer's activities are dispersed, as in the construction industry, the records must be kept at a location where employees report daily.[27] If employees do not report regularly to a single location, records must be kept where employees are paid or where they base their activities.[28]

---

[20]The issue of §4(b)(1) exemptions is treated more extensively in Chapter 15 (Defenses).

[21]2 OSHC 1313 (Rev. Comm'n 1974), *aff'd*, 539 F.2d 386, 4 OSHC 1693 (5th Cir. 1976), *cert. denied*, 434 U.S. 874 (1977).

[22]10 OSHC 1706 (Rev. Comm'n 1982).

[23]*Id.* at 1708. See also *Southern Pac. Transp. Co.*, 11 OSHC 1621 (Rev. Comm'n 1983) (vacating citation for recordkeeping violations owing to employer's reliance on Commission precedent).

[24]9 OSHC 1764 (Rev. Comm'n Nos. 76-4905, 75-5155, & 78-617, 1981).

[25]9 OSHC at 1777–1778. See also *Donovan v. Red Star Marine Servs.*, 739 F.2d 774, 11 OSHC 2049 (2d Cir. 1984), *cert. denied*, 470 U.S. 1003 (1985) (OSHA can regulate conditions on "uninspected vessels when the Coast Guard has failed to exercise jurisdiction"); *Bettendorf Terminal Co. & LaClair Quarries*, 1 OSHC 1695 (Rev. Comm'n 1974).

[26]29 C.F.R. §§1904.2, -.4, -.5, -.12(g)(1), -.14.

[27]§1904.12(g)(1).

[28]§1904.14.

In *Abdo S. Allen*,[29] a citation for failing to maintain the log at the construction site was vacated since the log was kept at a location where employees reported daily.[30] In *Alder & Neilson Co.*,[31] the Review Commission held that an employer need only compile an annual summary *for* each establishment and it need not be maintained *at* each establishment. Nonetheless, the summary must be posted at each establishment during the month of February.[32]

C. Injuries and Illnesses

Despite the definitions of recordable injuries and illnesses contained in the regulations,[33] whether an injury or illness is recordable may be unclear. In *Brilliant Electric Signs*,[34] a citation for failing to report a death was vacated because the Secretary did not prove that the cause of death was occupational, even though the employee died while working and the cause of death was listed as muscular injury, occlusion of the coronary vessels, effects of certain drugs, or electrical stimuli.[35] Similarly, in *Ohio Edison Co.*,[36] the citation for failure to report skin rashes was vacated because the conditions were undiagnosed, even though the administrative law judge inferred that the conditions resulted from exposure to creosote.[37]

The Commission, however, encourages employers to resolve doubts in favor of reporting. In *General Motors Corp., Inland Division*,[38] three employees were hospitalized with respiratory problems after working in areas containing toluene diisocyanate. Even though none of the treating physicians cited the chemical as the sole cause of the respiratory problems

---

[29]2 OSHC 1460 (Rev. Comm'n 1974).

[30]See also *Truland Corp.*, 6 OSHC 1896 (Rev. Comm'n 1978) (upholding citation for failure to maintain log at construction site that constituted establishment); *Parnon Constr.*, 5 OSHC 1232 (Rev. Comm'n 1977) (vacating citation because employees worked in dispersed locations and log maintained in employer's main office); *Otis Elevator Co.*, 3 OSHC 1736 (Rev. Comm'n J. 1975).

[31]5 OSHC 1130 (Rev. Comm'n 1977).

[32]See 29 C.F.R. §1904.5.

[33]See text accompanying notes 6–9, *supra*.

[34]1 OSHC 3222 (Rev. Comm. 1973).

[35]*Id.*

[36]9 OSHC 1450 (Rev. Comm'n J. 1981).

[37]See also *Congoleum Indus.*, 5 OSHC 1871 (Rev. Comm'n 1977) (vacating citation for failure to report skin rashes because Secretary could not establish rashes resulted from working with chemicals).

[38]8 OSHC 2036 (Rev. Comm'n 1980).

and a workers' compensation board had ruled one employee's illness not work-related, the Commission found the illnesses to be sufficiently connected to the employees' work to be recordable.[39] An explanation of the legislative history of the Act indicated that a primary purpose of the Act's recording obligations is to develop information for future scientific use.[40] The Commission rejected the argument that only illnesses directly caused by the occupational environment are recordable. Otherwise, "[u]nknown medical correlations between diseases and the work place would be obscured * * *."[41]

Distinguishing between reportable "medical treatment" and unreportable "first aid" may also be more complex than the regulations indicate. In *La Biche's, Inc.*,[42] the Secretary cited the employer for failure to report a bruise treated once by a physician. Since the Secretary produced no evidence that the treatment was more extensive than first aid or that the bruise resulted in loss of work or motion, the citation was vacated.

In *Chrysler Corp.*,[43] however, the citation was affirmed when the employer failed to record a hematoma suffered by an employee hit by a powered industrial truck. The injury was recordable because the employer's physician recommended transfer to a lighter duty job and the employee was still experiencing pain upon returning to his former job.

In *General Motors Corp., Warehousing & Distribution Division*,[44] the employer failed to record seven injuries[45] pursuant to its physician's analysis that only first aid was involved, or that the injuries were not sufficiently work-related. All of the employees, however, missed work as a result of the inju-

---

[39]*Id.* at 2038.

[40]*Id.* at 2039. The Commission also noted that the Senate committee report indicated "[t]he distinction between occupational and non-occupational illnesses is growing increasingly difficult to define." *Id.*, citing S. REP. NO. 1282, 91st Cong., 2d Sess. 2 (1970), *reprinted in* Subcommittee on Labor of the Senate Comm. on Labor and Public Welfare, 92d Cong., 1st Sess., LEGISLATIVE HISTORY OF THE OCCUPATIONAL SAFETY AND HEALTH ACT OF 1970 at 142 (Comm. Print 1971).

[41]*General Motors Corp.*, *supra* note 38, at 2040.

[42]2 OSHC 3110 (Rev. Comm'n J. 1974).

[43]7 OSHC 1578 (Rev. Comm'n J. 1979).

[44]10 OSHC 1844 (Rev. Comm'n J. 1982).

[45]*Id.* at 1845. The injuries were as follows: (1) an employee suffered a strained hip and pelvis; (2) an employee crushed the middle finger on her right hand; (3) an employee suffered a contusion of his right elbow; (4) an employee pulled his back; (5) an employee suffered back strain; (6) an employee hurt her back; and (7) an employee twisted her back. *Id.*

ries. The administrative law judge not only found violations as "there was sufficient objective evidence of the relation between the work accident and the subsequent diagnosis,"[46] but also found that "[t]he employer's decision not to record the incidents as work-related injuries was a conscious, intentional, deliberate decision, and therefore willful."[47]

The logs need only be maintained if a recordable injury or illness has occurred. Citations for failure to maintain records may not be issued where no injury or illness has occurred.[48]

In *Joseph Bucheit & Sons*,[49] the general contractor was held to have the duty to record the fatality of a subcontractor's employee. The employee was "on loan" to the general contractor as a bulldozer operator. This relationship made him an employee of the general contractor for purposes of recordkeeping.[50]

D. Forms

As long as a single recordable injury occurs, an employer must maintain the proper forms. Failure to do so may result in a citation.[51] Although certain information is necessary, it

---

[46]*Id.*

[47]*Id.* The administrative law judge also held that the employer was aware of the recording requirements because of the decision in *General Motors Corp., Inland Div.*, *supra* note 38. 10 OSHC at 1845.

[48]See, e.g., *Atlas Steel Erectors Co.*, 3 OSHC 1998 (Rev. Comm'n J. 1976); *Taysom Constr. Co.*, 2 OSHC 1606 (Rev. Comm'n 1975); *La Biche's, Inc.*, 2 OSHC 3110 (Rev. Comm'n 1974); *Central Tire Co.*, 1 OSHC 3315 (Rev. Comm'n Nos. 720 & 737, 1974). But cf. 29 C.F.R. §1904.5(a) (requiring forms to be posted with zeros in the totals columns); *Thermal Reduction Corp.*, 12 OSHC 1264 (Rev. Comm'n 1985) (holding burden of proving affirmative defense of no injuries or illnesses is on employer); *Cash-Way Bldg. Supply*, 2 OSHC 3314 (Rev. Comm'n Nos. 5487 & 5488, 1975) (requiring employer to post form even though no recordable incidents for 47 years).

[49]2 OSHC 1001 (Rev. Comm'n Nos. 2684 & 2716, 1974).

[50]See also *Dayton Tire & Rubber Co.*, 2 OSHC 1528 (Rev. Comm'n 1975), discussed *supra* at text accompanying notes 15-18. See also Chapter 15 (Defenses) for further discussion on the definition of "employee."

[51]See, e.g., *Atlantis Mfg. Co.*, 11 OSHC 1123 (Rev. Comm'n J. 1983) (upholding citation, even though employer contended it kept log, because none was produced at either inspection or hearing); *Howard P. Foley Co.*, 5 OSHC 1161 (Rev. Comm'n 1977) (upholding citation for maintaining log with no entries); *R.T. Pugh Motor Transp.*, 3 OSHC 1592 (Rev. Comm'n J. 1975) (upholding citation for failure to maintain and post OSHA Form 102, the annual summary of occupational illnesses and injuries). But cf. *A. Perlman Iron Works*, 5 OSHC 1875, 1877 (Rev. Comm'n J. 1977) ("[t]he burden was upon the Secretary to prove that the summary was neither mailed to

is not a violation to handwrite, rather than type, the forms.[52]
It is not sufficient, however, to attempt to excuse noncompli-
ance by asserting that the proper forms were not supplied by
OSHA,[53] or that the employer's supply of forms was depleted.[54]

Employers are allowed to use alternate forms provided
the same information is included.[55] Whether the alternate
form is sufficient is generally determined on a case-by-case
basis. Thus, employers have been permitted to utilize insurance reports,[56] state forms,[57] and workers' compensation reports.[58]

An employer risks OSHA's rejection of substitute forms,
however, unless it successfully petitions the Bureau of Labor
Statistics' Regional Commissioner for relief.[59] Substitutes for
the annual summary are not permitted without the Regional
Commissioner's approval.[60]

E. Access to Records[61]

Under the recordkeeping regulations, employers, upon request, are to provide access to the log, the supplementary
record, and the annual summary "for inspection and copying

---

employees nor posted in a conspicuous place at one of its work places. The employer's
inability to testify that the report had been compiled was suspicious but did not
constitute a prima facie case of violation. The secretary failed to meet his burden of
proof, and the item was vacated").

[52]See *Simmons, Inc.*, 6 OSHC 1157 (Rev. Comm'n J. 1977).

[53]See *Aluminum Coil Anodizing Corp.*, 5 OSHC 1381 (Rev. Comm'n J. 1977).
Cf. text accompanying note 87, *infra*.

[54]See *Kent Nowlin Constr. Co.*, 5 OSHC 1051 (Rev. Comm'n Nos. 9483, 9485 &
9522, 1977).

[55]29 C.F.R. §§1904.2(a)(2), 1904.4.

[56]See, e.g., *Maxine Power & Equip. Co.*, 5 OSHC 1915 (Rev. Comm'n J. 1977)
(accepting reports for insurance carrier). But see *Automotive Prods. Corp.*, 1 OSHC
1772 (Rev. Comm'n 1974) (rejecting use of insurance reports).

[57]*B.C. Crocker Cedar Prods.*, 4 OSHC 1775 (Rev. Comm'n 1976) (accepting use
of Idaho state forms). But see *Cam Indus.*, 1 OSHC 1564 (Rev. Comm'n J. 1974)
(rejecting Washington state forms).

[58]*Thompson Mfg. Co.*, 11 OSHC 1682 (Rev. Comm'n J. 1983) (accepting Texas
workers' compensation report); *Williams & Davis Boilers*, 8 OSHC 2148 (Rev. Comm'n
J. 1980) (same). But see *Quality Stamping Prods.*, 10 OSHC 1010 (Rev. Comm'n J.
1981) (rejecting use of workers' compensation form); *New Hampshire Provision Co.*,
1 OSHC 3071 (Rev. Comm'n 1974) (same).

[59]See 29 C.F.R. §1904.13. See also *Automotive Prods.*, 1 OSHC 1772 (Rev. Comm'n
J. 1974).

[60]See *Puterbaugh Enters.*, 2 OSHA, 1030 (Rev. Comm'n 1974).

[61]Medical access provisions are discussed in Chapter 20 (Judicial Review of Standards).

by any representative of the Secretary of Labor for the purpose of carrying out the provisions of the act."[62] Employers also, upon request, must make available the log and summary "to any employee, former employee, and to their representatives for examination and copying in a reasonable manner and at reasonable times."[63]

Although the records maintained pursuant to the record-keeping regulations are "required," that is, they are mandated by law, the Review Commission, in *Taft Broadcasting Co., Kings Island Div.*,[64] has concluded that employers nevertheless have a reasonable expectation of privacy in such records.[65] Accordingly, the Review Commission has held that the regulation at 29 C.F.R. §1904.7(a) "violates the Fourth Amendment to the extent that it purports to authorize an inspection of required records without a warrant or its 'equivalent', *e.g.*, the employer's consent or an administrative subpoena under section 8(b) of the Act."[66]

Access, if it is provided, must be provided promptly. In *RSR Corp.*,[67] a violation was affirmed because the employer refused to furnish the log to an employee representative while a personal injury lawsuit involving the representative and other employees was pending. According to this decision, the regulation cannot be satisfied by eventual production of the log.

F. Criminal Penalties

Section 17(g) of the Act[68] permits criminal sanctions of not more than $10,000 and/or imprisonment for not more than six months for knowingly making a false statement, representation, or certification in any record or report required to be maintained under the Act. Even though omissions are not mentioned in Section 17(g), the provision can be invoked against employers who knowingly omit information from the log and

---

[62]29 C.F.R. §1904.7.

[63]*Id.*

[64]13 OSHC 1137 (Rev. Comm'n 1987).

[65]*Id.* at 1143.

[66]*Id.* at 1146.

[67][1981] OSHD ¶25,207 (Rev. Comm'n J. Nos. 79-3813 & 80-1602, 1981) *rev'd in part*, 11 OSHC 1163 (1983), *aff'd*, 764 F.2d 355, 12 OSHC 1413 (5th Cir. 1985).

[68]29 U.S.C. §666(g).

summary because the regulations require the employer to certify that the annual summary "is true and complete."[69] The certification will be false if relevant information has been omitted deliberately.

G. Enforcement of the Recordkeeping Regulations

In 1986, OSHA began a series of enforcement actions involving recordkeeping requirements which resulted in the issuance of citations carrying the largest proposed penalties in the history of the Act. In relatively quick succession, penalties for alleged recordkeeping violations were proposed against Union Carbide Corporation ($1.28 million), Chrysler Corporation ($910,000), Fina Oil & Chemical Co. ($184,000), USX ($130,000), and Shell Oil ($244,000), among others. While the case involving Chrysler Corporation was settled with that company agreeing to accept 182 willful violations and a penalty in the amount of $284,030, no case has yet been litigated. Therefore, as of this writing it is too early to draw any conclusions, or even to reasonably speculate about, the ultimate direction and result of OSHA's efforts in this area.

IV. Reports of Fatalities or Multiple Hospitalizations

Section 1904.8 of Title 29 of the *Code of Federal Regulations* requires an employer to report, within 48 hours, accidents that result in a fatality or hospitalization of 5 or more employees.[70] The purpose of reporting these incidents is to

---

[69]29 C.F.R.§1904.5(c).

[70]*Id.* §1904.8. That section provides:
"Within 48 hours after the occurrence of an employment accident which is fatal to one or more employees or which results in hospitalization of five or more employees, the employer of any employees so injured or killed shall report the accident either orally or in writing to the nearest office of the Area Director of the Occupational Safety and Health Administration, U.S. Department of Labor. The reporting may be by telephone or telegraph. The report shall relate the circumstances of the accident, the number of fatalities, and the extent of any injuries. The Area Director may require such additional reports, in writing or otherwise, as he deems necessary concerning the accident."
Id.

allow OSHA to investigate the accident while the evidence is still fresh and to take action to avoid further injuries if hazardous conditions continue to exist.[71]

OSHA Instruction CPL 2.43 clarifies the definition of hospitalization as meaning "to be sent to; to go to; or to be admitted to a hospital or an equivalent medical facility."[72] If an employment accident results in 5 or more employees going to the hospital, it "must be reported whether or not treatment was provided and without regard to the length of stay in the hospital."[73]

A. Duty to Report

A citation may be issued for failure timely to report a fatality or multiple hospitalization.[74] The employer's responsibility to report accidents resulting in such incidents is nondelegable. In *Hampton Pugh Co.*,[75] the citation was upheld even though the employer notified the insurance agent "and requested [he] notify everyone who was supposed to know."[76] Notifying the insurance company was not an adequate substitute for notifying the OSHA Area Director.[77]

In *Knapp Brothers*,[78] the administrative law judge held that a report to the insurance company was sufficient because the decedent was the employer's brother and the insurance

---

[71]See *Yelvington Welding Serv.*, 6 OSHC 2013 (Rev. Comm'n, 1978).

[72]OSHR [1 Reference File 21:8190] (1979).

[73]*Id.* However, it has been held that "hospitalization" refers only to in-patient care. See, e.g., *Simplex Time Recorder Co. v. Secretary of Labor*, 766 F.2d 575, 12 OSHC 1401 (D.C. Cir. 1985); *Western Airlines*, 12 OSHC 1084 (Rev. Comm'n J. 1984); *General Motors Corp., Truck & Bus Group*, 11 OSHC 2013 (Rev. Comm'n J. 1984).

[74]*L.R. Brown, Jr., Painting Contractor*, 3 OSHC 1318 (Rev. Comm'n J. 1975) (upholding citation for failure to report fatality local compliance officer learned of through newspaper); *Anderson Co.*, 2 OSHC 3024 (Rev. Comm'n J. 1974) (upholding citation for reporting fatality 41 days after it occurred). But cf. *Forth Worth Enters.*, 2 OSHC 1103 (Rev. Comm'n 1974) (vacating citation where Area Director learned of fatality on day it happened and compliance officer was at worksite following day and employer reported details).

[75]5 OSHC 1063 (Rev. Comm'n J. 1977), *aff'd without review* (Rev. Comm'n Final Order).

[76]*Id.* at 1064.

[77]See also *Structural Steel*, 10 OSHC 1601 (Rev. Comm'n J. 1982) (upholding citation where fatality was reported by warehouse representative because employer did not present evidence that it requested representative to report it); *Abild Builders*, 9 OSHC 1988 (Rev. Comm'n J. 1981) (upholding citation where fatality was reported only to insurance agent). Cf. *Anderson Co., supra* note 74 (vacating penalty because employer had requested its project manager to make timely report).

[78]3 OSHC 1344 (Rev. Comm'n J. 1975).

agent had agreed, although failed, to notify all necessary authorities. In justifying this delegation, account was taken of the unusual circumstances of the case, including problems of family bereavement, the employer brother's good faith, and the lack of experience with such procedures. In *F.F. Green Construction Co.*,[79] a report to the state agency was held sufficient because the OSHA office had opened only 2 days prior to the fatality. The state agency immediately contacted the new federal office. The Commission stated that "[b]y providing notification in the manner known to it at the time, Respondent has demonstrated its intention to comply with the Act."[80]

The 48 hours allotted for reporting a fatal accident runs from the time of the accident, not from the time of the employee's death. In *William Edwin Reitz*,[81] the employee died 7 days after a tree fell on him. The citation for failing to report within 48 hours was affirmed, even though the employer did not know within the 48 hours death would result.[82] In addition, 48 hours are consecutive, not work hours.[83]

B. Statute of Limitations

In a failure-to-report case, the 6-month statute of limitations for citing an employer does not begin to run until OSHA learns of the accident. In *Yelvington Welding Service*,[84] OSHA learned of a fatal accident from a state agency 9 months after its occurrence. The Review Commission rejected the employer's contention that the 6 months runs from the end of the 48 hours:

> "Enforcement of the reporting regulation is especially important because reports reveal particularly hazardous working conditions that might otherwise continue unchecked, such as the conditions alleged in this case. We cannot read the reporting regulation to require the filing of a report within 48 hours after a fatality and, at the same time, deny the Secretary the ability

---

[79]1 OSHC 1494 (Rev. Comm'n 1973).

[80]*Id.* at 1495.

[81][1974–1975] OSHD ¶19,373 (Rev. Comm'n J. 1975).

[82]See also *Yelvington Welding Serv.*, *supra* note 71, at 2014 n.9 ("[u]nder respondent's interpretation, an employer would not be obligated to report an accident if the fatality it causes does not occur within 48 hours. Such an interpretation is unacceptable").

[83]*Fort Worth Enters.*, *supra* note 74.

[84]*Yelvington*, *supra* note 71.

to enforce by requiring him to discover a failure to report within six months."[85]

V. Employer's Posting Obligations

Section 1903.2 of Title 29 of the *Code of Federal Regulations* requires employers to post a notice, furnished by OSHA, that informs employees of their protections and obligations under the Act and directs them to their employer or the nearest Department of Labor office for assistance and information, including copies of the Act and specific standards.[86] Part of the Secretary's burden of proof in establishing a violation of this section is to show that OSHA has furnished a copy of the poster to the employer.[87]

A. Posting the Notice

Employers must post the notice at each establishment.[88] A separate notice must be posted at the location of each activity to the extent that OSHA has furnished copies.[89] The issue of what constitutes an establishment for employers who maintain multiple worksites and for employees who have no fixed workplace has led to diverse opinions. In *Western Waterproofing*,[90] the Review Commission vacated a citation for failure to post the informational notice at an Albuquerque construction site at which employees had worked continuously for 6 weeks. The decisive facts in that case were that the employees had been assigned from, and regularly

---

[85]*Id.* at 2015.

[86]29 C.F.R. §1903.2.

[87]See, e.g., *Larene Elec. Co.*, 8 OSHC 1771 (Rev. Comm'n J. 1980); *Dawson Co. Mfg.*, 3 OSHC 1534 (Rev. Comm'n J. Nos. 9404 & 9406, 1975); *Mangus Firearms*, 3 OSHC 1214 (Rev. Comm'n J. 1975); *WKRG-TV*, 2 OSHC 3053 (Rev. Comm'n J. 1974); *Woerfel Corp.*, 1 OSHC 3299 (Rev. Comm'n J. 1974); *Oak Lane Diner*, 1 OSHC 1248 (Rev. Comm'n 1973).

[88]See text accompanying notes 26–28, *supra*. See also 29 C.F.R. §1903.2(b).

[89]29 C.F.R. §1903.2(b). See also *Ira Holliday Logging Co.*, [1971–1973] OSHD ¶15,167A (Rev. Comm'n J. 1972) *rev'd in part on other grounds*, 1 OSHC 1200 (Rev. Comm'n 1973). Contra *Charles H. Thompkins Co.*, [1980] OSHD ¶24,245 (Rev. Comm'n J. 1980) (Secretary meets obligation by furnishing original notice; employer can make copies or request additional ones if he has to post notice in several locations).

[90]7 OSHC 1499 (Rev. Comm'n 1979).

reported to, the Dallas headquarters where the notice was displayed, and that the employees had a full opportunity to read it.[91]

In *Well Tech, Inc.*,[92] the employer's posting of the notice at its main office where the employees reported daily for work, rather than at the drilling rig where employees worked, was held to be a *de minimis* violation. In *Cyclone Drilling & Workover*,[93] however, posting at the employer's headquarters was insufficient because "all hiring, except for supervisory personnel, took place at the drilling rig and * * * employees made only occasional, haphazard visits to the headquarters."[94]

B. Defenses Asserted for Failure to Post

In *Lake Butler Apparel Co. v. Secretary of Labor*,[95] the Fifth Circuit held that the employer's freedom of speech under the first amendment was not infringed by the posting requirement.[96] Another defense that has met with little success is that the Secretary failed to prove the poster was not displayed. Where a compliance officer was unable to find the notice during his inspection and neither the foreman nor the general superintendent could tell him where it was located, the administrative law judge found that the Secretary had met his burden of proof.[97] It is risky to depend on the Secretary's inability to prove a negative fact; if the notice was posted, the employer should introduce evidence to that effect during the hearing. It should also be noted that ignorance of the requirement will not excuse a failure to post the notice.[98]

---

[91]See also *Haring Contractors*, 9 OSHC 1301 (Rev. Comm'n J. 1981) (posting at main office sufficient where employees worked at transitory locations); *Bi-Co Paver, Inc.*, 2 OSHC 3142 (Rev. Comm'n J. 1974) (posting at main office sufficient). Contra *Para Constr. Co.*, 4 OSHC 1779 (Rev. Comm'n 1976).

[92]10 OSHC 1189 (Rev. Comm'n J. 1981).

[93]9 OSHC 1439 (Rev. Comm'n J. Nos. 80-1636 & 80-1721, 1981).

[94]*Id.* at 1440.

[95]519 F.2d 84, 3 OSHC 1522 (5th Cir. 1975).

[96]See also text accompanying notes 121–122, *infra*.

[97]*Southeast Contractors v. Dunlop*, 1 OSHC 1713 (Rev. Comm'n 1974), *rev'd on other grounds*, 512 F.2d 675 (5th Cir. 1975). See also *Thunderbolt Drilling*, 11 OSHC 1047 (Rev. Comm'n J. 1982).

[98]See, e.g., *Clarence M. Jones*, 11 OSHC 1529 (Rev. Comm'n 1983); *Eckel Mfg. Co.*, 9 OSHC 2145 (Rev. Comm'n J. 1981); *Campbell Constr. Co.*, 2 OSHC 3217 (Rev. Comm'n J. 1974).

The poster is also available in Spanish. The assertion that employees would probably not understand it will not serve as a defense. In *Belau Transfer & Terminal Co.*,[99] the poster's contents had been explained to employees upon its receipt and at monthly safety meetings, but it was not posted because only half the employees could read English. The citation was affirmed as a technical violation without penalty.

The poster must be conspicuously displayed; mere posting at an inaccessible location is not sufficient. In *Kinney Steel*,[100] the employer unsuccessfully contended that he satisfied the posting requirement, even though the poster was kept in his briefcase, because he occasionally showed it to employees. In *Crushed Toast Co.*,[101] the citation for failure to post was upheld because the poster was on the back of the men's room door and was partially blocked by clothes hanging over it.[102] In *Western Steel Erectors*,[103] however, the citation against the subcontractor was dismissed because the employer posted the notices on the inside of toolbox lids mounted in his pickup truck.

VI. Posting of Citations and Other Notices of Proceedings

Section 9(b) of the Act[104] requires employers to prominently post citations "at or near each place a violation referred to in the citation occurred."[105] The regulations require that the citation be posted immediately upon receipt.[106] If it is not practical to post the citation at or near each place where a violation allegedly occurred, it must be posted "in a prominent place where it will be readily observable by all affected employees."[107] The employer must take steps to ensure that the

---

[99] 6 OSHC 1592 (Rev. Comm'n J. 1978).

[100] 3 OSHC 1453 (Rev. Comm'n J. 1975).

[101] 5 OSHC 1360 (Rev. Comm'n J. 1977).

[102] See also *San Juan Constr. Co.*, 3 OSHC 1445 (Rev. Comm'n J. Nos. 9662 & 9823, 1975) (posting in shack 300 meters from worksite insufficient).

[103] 3 OSHC 1466 (Rev. Comm'n J. 1975).

[104] 29 U.S.C. §658(b).

[105] *Id.*

[106] 29 C.F.R. §1903.16.

[107] §1903.16(a).

notice is not altered, defaced, or covered by other material.
Notices of *de minimis* violations, as well as the portion of the
citation containing the proposed penalties need not be posted.
The citation must remain posted for 3 days or until the violation is abated, whichever is longer.[108] If the citation is contested, it must remain posted until it has been vacated by a
final order or abatement is achieved.[109]

A. Location and Duration of Posting

In *San Juan Construction Co.*,[110] the employer was successfully cited for not prominently posting citations where
they were posted in an unused latrine. In *Ashton Industries*,[111]
however, the citation was vacated because the compliance officer did not check the time clock where the employer claimed
to have posted the citation.[112] In *Mallory Electric Co.*,[113] posting in the workplace was sufficient as it was impractical to
post at or near the alleged violations and employees were able
to observe the citation.[114] It is not sufficient to pass a citation
around to employees rather than post it.[115]

The Secretary has the burden of proving the citation was
not posted. Where compliance officers have not looked in the
locations where employers assert the citations were posted
and those locations are obvious and prominent, citations for
failure to post earlier citations have been vacated.[116]

An employer may be cited if it fails to post the citation
for the proper length of time. It has been stated that to permit
the employer to decide for himself the duration of citation

---

[108]§1903.16(b).

[109]*Id.*

[110]3 OSHC 1445 (Rev. Comm'n J. Nos. 9662 & 9823, 1975).

[111]2 OSHC 3097 (Rev. Comm'n J. 1974).

[112]See also *Bay State Smelting Co.*, [1982] OSHD ¶25,940 (Rev. Comm'n J. Nos. 80-344 & 80-1968, 1982) (posting citations next to time clock on advice of counsel sufficiently conspicuous, even though compliance officer did not see them).

[113]2 OSHC 3155 (Rev. Comm'n J. Nos. 6349 & 6396, 1974).

[114]Cf. *Kesler & Sons Constr. Co.*, 2 OSHC 1096 (Rev. Comm'n 1974) (upholding violation where citation posted in main office rather than with other government notices).

[115]*Caloric Corp.*, 1 OSHC 3008 (Rev. Comm'n J. 1973).

[116]See, e.g., *Easterbrooke Textile Co.*, [1974–1975] OSHD ¶19,385 (Rev. Comm'n J. 1975). Cf. *Chobee Steel Erectors*, 8 OSHC 1094 (Rev. Comm'n J.1979) (upholding violation where compliance officer could not find posted citation and employees indicated none had been posted).

posting would eliminate both the employer's and employees' awareness of the uncorrected violations.[117]

B. Penalties

Section 17(i) of the Act[118] requires the assessment of monetary penalty. The penalty is mandatory and an administrative law judge does not have discretion to dismiss the notice of contest.[119] Failure to post a notice of contest or other notice required to be posted by the Review Commission's rules of procedure may, however, result in dismissal of the notice of contest.

C. Constitutional Challenges

As in the notice-posting cases,[120] constitutional challenges to the requirement of posting citations have not been successful. The Tenth Circuit, in *Stockwell Manufacturing Co. v. Usery*,[121] rejected the employer's argument that posting resulted in forced "self-vilification." In *Cullen Industries*,[122] the employer's claim that the posting requirement violated the Fifth and Ninth Amendments was also rejected.

D. Posting of Other Documents

Settlement agreements must also be posted at the workplace or otherwise served on affected employees.[123] The employees must be notified before the settlement becomes effective

[117]*Osborn Mfg. Co.*, 2 OSHC 3221 (Rev. Comm'n J. Nos. 6901 & 7124, 1974). Cf. *Amity Drum Serv.*, 3 OSHC 1328 (Rev. Comm'n J. 1975) (finding sufficient evidence to show improper duration, but no violation because Secretary only alleged failure to post).

[118]29 U.S.C. §666(i).

[119]*James L. Brussa, Masonry*, 1 OSHC 1185 (Rev. Comm'n 1973). See also *Acme Metal*, 3 OSHC 1932 (Rev. Comm'n Nos. 1811 & 1931, 1976). But cf. *Caribtow Corp.*, 1 OSHC 1503 (Rev. Comm'n J. 1973) (dismissing notice of contest for failure to post citation); *Hamilton Metal Prods.*, 1 OSHC 1025 (Rev. Comm'n 1973) (same).

[120]See, e.g., text accompanying note 95, *supra*.

[121]536 F.2d 1306, 4 OSHC 1332 (10th Cir. 1976).

[122]6 OSHC 2177 (Rev. Comm'n J. Nos. 77-4267 & 77-4268, 1978).

[123]29 C.F.R. §2200.100(c). See also *Farmers Export Co.*, 11 OSHC 1289 (Rev. Comm'n Nos. 78-910 & 78-2809, 1983); *Canton Elevator & Mfg. Co.*, 3 OSHC 1206 (Rev. Comm'n 1975).

to allow employees or their representatives the opportunity to file objections.[124] Employers must similarly notify employees or their representatives when a petition for modification of abatement date is filed.[125] The Secretary has also promulgated and enforced other posting requirements for temporary orders granting a variance[126] and a petition for an exception to the Bureau of Labor Statistics' recordkeeping requirements.[127]

VII. Recordkeeping and Information Dissemination Requirements of the Review Commission and of the National Institute for Occupational Safety and Health

A. Review Commission

The Review Commission has adopted Rules of Procedure, pursuant to the authority granted it by Sections 10(c)[128] and 12(g)[129] of the Act, to ensure the orderly transaction of its proceedings and to provide employees and their representatives an opportunity to participate.[130] Every pleading in a Review Commission proceeding must be served upon every other party (which may include employees or their representatives) and intervenors.[131] If affected employees are not represented, employers must maintain copies of all pleadings and make them available for employee inspection and copying. (Since most workplaces have at least one management employee who is "affected," even in unionized plants employers should keep a file of pleadings available for employee inspection.)

---

[124]*Brockway Glass Co.*, 6 OSHC 2089 (Rev. Comm'n 1978).

[125]29 C.F.R. §1904.14(a)(1), §2200.34(c)(1). See also *Auto Bolt & Nut Co.*, 7 OSHC 1203 (Rev. Comm'n 1979).

[126]See 29 U.S.C. §655(d).

[127]29 C.F.R. §1904.13(c)(5).

[128]29 U.S.C. §659(c).

[129]*Id.* §661(g).

[130]For a full discussion, see Chapter 16 (The Occupational Safety and Health Review Commission).

[131]29 C.F.R. §2200.7.

Immediately upon receipt of the Review Commission's notice that the employer's notice of contest or petition for modification of the abatement period has been docketed, an employer with "unrepresented" employees must post the notice of contest and the Review Commission's notice. The Commission's notice informs affected employees of their right to elect party status and of the availability of all pleadings for their inspection.[132] The employer must serve both the notice of contest and the Review Commission's notice upon the authorized employee representative.[133]

Upon receiving notice of the hearing date, the employer must post it near the posted citation for affected employees who are not represented and must serve it upon the employee representative, if any.[134] Notices posted may not be removed until the hearing commences or the case is disposed of, whichever occurs first.

B. National Institute for Occupational Safety and Health

The Director of NIOSH has promulgated regulations for workplace investigations pursuant to Section 8(g)(2) of the Act,[135] permitting his agents to "review, abstract or duplicate employment records, medical records, records required by the act and regulations, and other related records."[136] NIOSH can perform workplace investigations, take exposure measurements during an investigation, and conduct medical examinations and other tests.[137] A report of the investigation is prepared and made available first to the employer for comment and trade secret review and then to the public.[138] Specific findings from employee medical examinations and tests are released to others only with the employee's written authorization.[139] Not even the company physician is entitled to obtain the agency's specific findings without such authorization.

---

[132]§2200.7(g).

[133]§2200.7(h).

[134]§§2200.7(i), (j).

[135]29 U.S.C. §657(g)(2).

[136]42 C.F.R. §85a.3(a).

[137]§85a.5. For a full discussion, see Chapter 24 (National Institute for Occupational Safety and Health).

[138]§85a.8(a).

[139]§85a.8(b).

The Director also has promulgated regulations for conducting health hazard evaluations at the request of an employer or its employees.[140] These regulations contain record access provisions similar to the investigation rules. Copies of NIOSH's determination regarding the health hazard will be mailed to the employer and to the authorized employee representative.[141] The employer must post the determination at or near the workplace of affected employees for 30 calendar days or provide NIOSH with the names and addresses of all affected employees so that they can each be mailed a copy of the determination.[142]

C. Secretary of Health and Human Services

The Secretary of Health and Human Services has also been delegated broad rulemaking authority in order to obtain information on exposure to substances and physical agents that may endanger employee safety or health.[143] Rather than promulgate its own rules, the Department of Health and Human Services has opted, for the most part, to collaborate with the Department of Labor so that OSHA's standards and regulations afford Health and Human Services agents access to the information they may need.[144]

---

[140]§§85.1 *et seq.*

[141]§85.11(b).

[142]§85.11(c).

[143]29 U.S.C. §§657(g)(2), 669(a). See also Chapter 24 (National Institute for Occupational Safety and Health), Section I.

[144]A notable exception is the records access rule, which does not accord direct access rights to NIOSH.

Appendix to Chapter 5

The following health-related standards contain provisions authorized by Section 6(b)(7) of the Act for the creation and retention of records and the dissemination of information:

(a) Occupational noise, 29 C.F.R. §1910.95 (1985). The standard prescribes monitoring for decibel exposure; audiometric testing; warning signs; a training program for affected employees; and access to the standard, to training materials, and to all records the standard requires for present and former employees, their designated representatives, and OSHA. The employer or its successor must retain records of employee exposure measurements for 2 years and of audiometric tests for the duration of the affected employee's employment. Employees must be notified of monitoring results if they are exposed at or above an 8-hour time-weighted average of 85 decibels, and of any standard threshold shift in their audiograms (in writing and within 21 days of the determination). Affected employees must be allowed to observe noise measurements.

(b) Ionizing radiation, 29 C.F.R. §1910.96 The standard requires the creation and retention (for an unspecified length of time) of radiation exposure records; the use of caution signs, labels, and warning signals; the training of affected employees in, and the posting of, the standard; immediate notification of OSHA regarding certain incidents (with longer periods in which to notify OSHA of other incidents); written reports to OSHA about incidents; at least annual notification to employees of

their exposure; access to their exposure reports for former employees; and record retention.

(c) Commercial diving, 29 C.F.R. §1910.401–.441. The standard prescribes the creation of the following records and access thereto by current and former employees, their representatives, OSHA, and NIOSH: medical records on dive team members (to be kept for 5 years, then sent to NIOSH); a record of each dive (to be kept for 1 year or, if decompression illness occurs, for 5 years, then sent to NIOSH); tagging or logging records for work done on equipment; a record of any injury or illness requiring the hospitalization of a dive team member for 24 hours or more (to be kept for 5 years, then sent to NIOSH); depth-time profiles; and a safe practices manual.

(d) Asbestos, 29 C.F.R. §1910.1001. Under this standard, records must be created and retained for 20 years of the results of personal and environmental monitoring and of employee medical examinations. Access to all records, which the standard mandates be created, must be afforded to OSHA, NIOSH, employees, and their designated representatives. Caution signs and labels must be used and employees must be notified within 5 days of any excessive exposure to asbestos and of the steps taken to reduce their exposure.

(e) 4-Nitrobiphenyl, 29 C.F.R. §1910.1003. The standard requires signs, labels, and posting of emergency procedures; notification to OSHA's Area Director of emergency incidents within 24 hours; a written report to him of such incidents within 15 days; a written report to him of the first use, or a change in use, of the substance within 15 days; a medical surveillance program, including the creation and retention of certain medical records for the duration of an employee's employment and access to the medical records by NIOSH, an employee's new employer, or his designated physician, and by the employee, his representative, and OSHA pursuant to the records access rule, 29 C.F.R. §1910.20 (1981).

(f) alpha-Naphthylamine, 29 C.F.R. §1910.1004. This standard's record creation, keeping, and access requirements are virtually identical to those of *4-Nitrobiphenyl, supra,* as are those of:

(g) Methyl chloromethyl ether, 29 C.F.R. §1910.1006.
(h) 3,3'-Dichlorobenzidine (and its salts), §1910.1007.

(i) *bis-Chloromethyl ether, §1910.1008.*
(j) *beta-Naphthylamine, §1910.1009.*
(k) *Benzidine, §1910.1010.*
(l) *4-Aminodiphenyl, §1910.1011.*
(m) *Ethyleneimine, §1910.1012.*
(n) *beta-Propiolactone, §1910.1013.*
(o) *2-Acetylaminofluorene, §1910.1014.*
(p) *4-Dimethylaminoazobenzene, §1910.1015.*
(q) *N-Nitrosodimethylamine, §1910.1016.*
(r) *Vinyl chloride, 29 C.F.R. §1910.1017.*

Employers must conduct exposure monitoring, employee training (including review of the standard), and a medical surveillance program. They must create and retain personal and environmental monitoring and measuring records for 30 years and employee medical records for 30 years or the length of employment plus 20 years, whichever is longer. Access to such records is available to NIOSH. Access to these records shall be provided to employees, their designated representatives, and OSHA pursuant to the records access rule. Also required are signs and labels, reports to OSHA's Area Director of emergencies within 24 hours, and notification to employees within 10 days of any excessive exposure to the substance, of the results of measurements, and of the steps taken to reduce exposure.

(s) *Inorganic Arsenic, 29 C.F.R. §1910.1018.* The standard calls for exposure monitoring, a medical surveillance program, and the creation and retention of records of both for 40 years or the length of employment plus 20 years, whichever is longer. Employees must receive written notification of monitoring results and, if excessive exposure to the substance occurs, of the corrective action to be taken. A report of excessive exposure must be made to OSHA within 60 days. Also required are written plans for compliance with the standard, caution labels and signs, and access to records for NIOSH. Access to these records shall be provided to employees, their designated representatives, and OSHA pursuant to the records access rule. Successor employers must undertake to retain such records or the records must be transmitted to NIOSH.

(t) *Lead, 29 C.F.R. §1910.1025.* This standard prescribes exposure monitoring and creation of records of the results; written notification to employees of such results; and, if there has been excessive exposure to the substance, a statement to that effect and a description of the corrective action to be taken. Employers must create and maintain a written compliance program available upon request to OSHA, NIOSH, affected employees, and their authorized representatives. The standard requires a medical surveillance program with written records thereof and written notification to employees who have excessive blood lead levels informing them of their blood lead levels and of the standard's requirement that they be removed temporarily, with benefits, from the area of lead exposure. Access to medical and exposure records is required pursuant to the records access rule (with employee access to the standard and other materials); or employers must create records of each employee removed from lead exposure and retain the records for the length of his employment. Exposure monitoring records and employee medical records must be kept for 40 years or the length of employment plus 20 years, whichever is longer.

(u) Coke oven emissions, 29 C.F.R. §1910.1029. The standard prescribes exposure monitoring with written notification to employees of their exposure measurements; and, if the permissible exposure limit was exceeded, a statement to that effect and a description of the corrective action to be taken. Also required are written procedures for work practices; a written compliance program which contains monitoring data; access to such written programs for OSHA, NIOSH, and the authorized employee representative (but seemingly not the employees); a written respiratory protection program; a medical surveillance program which includes furnishing employees with a physician's written opinion regarding their examinations; an employee training program; warning signs and labels; the creation and retention of exposure measurement records and of employee medical surveillance records for 40 years or the length of employment plus 20 years, whichever is longer; and access to such records pursuant to the records access rule (except for NIOSH).

(v) Cotton dust, 29 C.F.R. §1910.1043. This standard requires exposure monitoring with written notification to employees within 5 days of their exposure measurements; and, if the permissible exposure limit was exceeded, a statement to that effect and a description of the corrective action to be

taken. Also required are a written compliance program, including monitoring data, with access thereto for OSHA, NIOSH, employees, and their designated representatives; a written respiratory protection program; a written program of work practices; a medical surveillance program which includes furnishing employees with a physician's written opinion regarding their examinations; an employee training program which includes employee access to the standard; warning signs; the creation and retention of exposure measurement records and of medical surveillance records for 20 years, and access to such records pursuant to the records access rule (except for NIOSH).

(w) 1,2-Dibromo-3-chloropropane, 29 C.F.R. §1910.1044. This standard prescribes exposure monitoring with written notification to employees within 5 days of their exposure measurement, and, if the permissible exposure limit was exceeded, a statement to that effect and a description of the corrective action to be taken; a written compliance program with access thereto for OSHA, NIOSH, employees, and their designated representatives; a report to OSHA within 10 days of the first use of the substance; a written respiratory protection program; a written emergency plan; a medical surveillance program which includes furnishing employees with a physician's written opinion regarding their examinations; an employee training program; warning signs and labels; creation and retention of all exposure monitoring records and of employee medical surveillance records for 40 years or the length of employment plus 20 years, whichever is longer; and access to such records pursuant to the records access rule (except for NIOSH).

(x) Acrylonitrile, 29 C.F.R. §1910.1045. This standard prescribes reports to OSHA of the first use of the substance within 30 days and of emergencies within 72 hours; exposure monitoring with written notification to employees within 5 days of their exposure measurements and, if the permissible exposure limit was exceeded, a statement to that effect and a description of the corrective action to be taken; a written compliance program with access thereto for OSHA, NIOSH, employees, or their representatives; a written respiratory protection program; a written emergency program; a medical surveillance program which includes furnishing employees with a physician's written opinion regarding their examinations; an employee training program which includes employee access

to the standard; warning signs and labels; recordkeeping for data that exempt the employer from the standard; the creation and retention of all exposure monitoring records and of employee medical surveillance records for 40 years or the length of employment plus 20 years, whichever is longer, and access to such records pursuant to the records access rule (except for NIOSH).

(y) Maritime standard for carbon monoxide, 29 C.F.R. §1918.93. The standard requires that tests be made when internal combustion engines exhaust into the hold or a deck of a vessel and that records of same be kept for 30 days.

(z) Temporary labor camps, 29 C.F.R. §1910.142. Employers must report immediately to local health authorities the occurrence of communicable diseases and certain other illnesses.

6

Special Obligations of Certain Employers

I. Introduction

To address particular safety problems inherent in certain types of employment, the Secretary of Labor has imposed special obligations on certain industries or categories of employers. Most of the special obligations were incorporated into the original standards promulgated when the OSH Act became effective, although some have been defined recently in an effort to make the obligations of particular types of employers more explicit.

When the Act was originally passed, Congress sought to consolidate national safety and health enforcement policy by authorizing the Secretary of Labor under Section 6(a) of the Act to incorporate preexisting federal safety and health standards[1] and voluntary consensus standards. Pursuant to Section 6(a), standards were adopted from previous statutes which

---

[1]The Secretary of Labor was given the authority to adopt these standards under 29 U.S.C. §655(a) (1971). This section provided for adoption of established federal standards and national consensus standards for an initial 2-year period during which public participation was not required. The two kinds of standard are defined in §§3(9) and 3(10) of the Act, respectively. See Chapter 3 (Specific Occupational Safety and Health Standards).

had been directed toward specific industries, including construction standards originally promulgated under the Walsh-
Healey Public Contracts Act[2] and the Construction Safety
Act[3]; the standards of the Service Contract Act[4]; and maritime[5] and longshoring[6] standards developed under the Longshoremen's and Harbor Workers' Compensation Act.[7] Also
pursuant to Section 6(a), the Secretary incorporated an as-
sortment of national consensus standards, some of which imposed specific requirements on particular types of operations.
Some were adopted as part of the Maritime Standards[8] or the
Construction Standards,[9] and others became part of the General Industry Standards of Part 1910.[10]

Although the General Industry Standards of Part 1910
apply to all covered employers, certain subparts are specifically directed toward a certain type of business. Industries
addressed within subparts of the General Industry Standards
include commercial diving activities covered under Subpart
T and a wide range of diverse activities covered under Subpart
R, including pulp, paper, and paperboard mills, textile industries, bakery equipment, laundry machinery, sawmills, pulpwood logging, and telecommunications.[11]

---

[2] 41 U.S.C. §§35–45 (1976 & Supp. V 1981).

[3] Standards promulgated under the Construction Safety Act of 1969, 40 U.S.C.
§§327–333 (1976), and published in 29 C.F.R. pt. 1926 (1983), provided the majority
of the established federal standards adopted as OSHA construction standards. Section
1910.12(c) of the OSHA regulations provides:

   "(c) *Construction Safety Act Distinguished.* This section adopts as occupational
   safety and health standards * * * the standards which are prescribed in Part
   1926 of this chapter. Thus, the standards (substantive rules) published in Subpart
   C and the following subparts of Part 1926 of this chapter are applied. This section
   does not incorporate Subparts A and B of Part 1926 * * *. * * *"

29 C.F.R. §1910.12(c) (1981). See generally Section III below.

[4] 41 U.S.C. §§351–358 (1978).

[5] 29 C.F.R. pts. 1915 and 1917.

[6] 29 C.F.R. pt. 1918.

[7] 33 U.S.C. §§901–950 (1976 & Supp. V 1981).

[8] The national consensus standards adopted as Maritime Standards pursuant to
§655(a) were generally taken from a consensus safety code covering dockside operations developed in 1969 by the American National Standards Institute (ANSI).

[9] The bulk of the national consensus standards which became part of the Construction Industry Standards were taken from ANSI and the National Fire Protection
Association.

[10] 29 C.F.R. pt. 1910 (1983).

[11] Many of the special industry provisions specifically list the General Industry
Standards incorporated by reference. See, e.g., 29 C.F.R. §1910.262(a)(2) (1983) (covering textile industries; see Section VI. B. 1, below), incorporating all general industry
standards.

Following Section II immediately below, the remainder of this chapter discusses the special obligations imposed on employers in the particular industries and operations mentioned above. Section III discusses standards addressing the construction industry; Section IV discusses standards addressing maritime industries, including standards for shipyards, marine terminals, longshoring, and gear certification; Section V discusses agricultural operations; Section VI discusses standards applying to commercial diving, the various activities covered under Subpart R, and helicopters and helicopter cranes. There is also a final section that discusses exemptions of small employers.

II. Hierarchy of Standards

As standards were added and revised after the promulgation of the original standards, the Secretary declared an intent to codify standards pertinent to particular industries or sets of employers within groups of "vertical" standards applying exclusively to certain industries, as differentiated from the "horizontal" General Industry Standards of Part 1910 applicable to many employers in various industries. The concept of vertical versus horizontal standards must be distinguished from the concepts of "specific" versus "general" standards and "specification" versus "performance" oriented standards. Vertical or horizontal standards may be either general, such as housekeeping or toilet-seat requirements which are applicable to a variety of workplace settings, or specific, such as machine guarding or guardrail requirements which are applicable to particular equipment or operations. Moreover, either horizontal or vertical standards may be specification oriented—indicating a precise method for abating a hazard—or performance oriented—indicating only the hazard to be abated, making the employer responsible for determining the precise method of abatement.

As a general rule, vertical and specific standards apply with hierarchical precedence over horizontal and general standards. Any vertical standard applicable to a specific hazard will preempt application of a horizontal standard to the same hazard.[12] A specific horizontal standard will also take precedence over a more general horizontal standard.[13] For example, Section 1910.213 regulating machine guards for woodworking saws applies to such saws rather than Section 1910.212, the more general machine-guarding requirements for power saws.[14] And of course there can be citation under Section 5(a)(1) of the OSH Act, the general duty clause, if a specific standard does not apply.[15]

The first vertical standard to be promulgated as a new standard applicable exclusively to a specific industry was the Marine Terminals standard.[16] The Secretary stated:

"This final rule is a vertical standard, i.e., one that applies to this industry exclusively and is designed specifically to address the hazards associated with marine cargo handling ashore. However, certain sections and subparts of Part 1910, the General Industry Standards, will continue to apply to marine terminals and have been referenced in this vertical standard."[17]

OSHA has stated its intent to continue to draft vertical standards until all special industry standards are compiled in sets of comprehensive vertical standards.[18] The construction, maritime, agricultural, diving, and telecommunications industries are the only industries currently subject to special comprehensive vertical standards.

---

[12]*Western Waterproofing*, 7 OSHC 1499, 1502 (Rev. Comm'n 1979) ("Pre-emption occurs only if a construction standard is addressed to the particular hazard arising from the cited conditions."). See also *Lloyd C. Lockrem, Inc.*, 3 OSHC 2045 (Rev. Comm'n 1976); *Diebold, Inc.*, 3 OSHC 1897 (Rev. Comm'n 1976), *rev'd in part on other grounds*, 585 F.2d 1327 (6th Cir. 1978); *Advance Specialty Co.*, 3 OSHC 2072, 2075 (Rev. Comm'n 1976).

[13]29 C.F.R. §1910.5(c)(1) states: "If a particular standard is specifically applicable to a condition, practice, means, method, operation or process it shall prevail over any different general standard which might otherwise be applicable to the same condition, practice, means, method, operation or process."

[14]See *General Elec. Co.*, 5 OSHC 1448 (Rev. Comm'n 1977); *Saturn Constr. Co.*, 5 OSHC 1686 (Rev. Comm'n 1977); *Matson Terminals*, 3 OSHC 2071 (Rev. Comm'n 1976); *Bristol Steel & Iron Works v. OSHRC*, 601 F.2d 717, 7 OSHC 1462 (4th Cir. 1979); *Davis/McKee, Inc.*, 2 OSHC 3046 (Rev. Comm'n 1974); *Trojan Steel Co.*, 3 OSHC 1384 (Rev. Comm'n 1975); *Roxbury Constr. Corp.*, 2 OSHC 3253 (Rev. Comm'n 1974); *Rocky Mountain Prestress*, 5 OSHC 1888 (Rev. Comm'n 1977).

[15]29 C.F.R. 1910.5(f); Field Operations Manual ch. VIII. But see *United Automobile, Aerospace & Agricultural Implement Workers v. General Dynamics Land Sys. Div.*, No. 85-1760, slip op. (D.C. Cir. April 14, 1987). For a detailed discussion of the requirements for citation pursuant to the general duty clause, see Chapter 4.

[16]Codified at 29 C.F.R. pt. 1917.

[17]48 FED. REG. 30,886 (1983).

[18]The Secretary has also compiled for the construction industry a listing of General Industry Standards it deems applicable to that particular industry. OSHA, Dep't of Labor, Pub. No. 2207, Construction Industry (1983). See Field Operations Manual, OSHR [Reference File 77:4301].

III. Construction Industry

A. Origin

Occupational safety and health standards for the construction industry originated in 1969, when the Secretary of Labor promulgated standards applicable to federal jobs and federal contractors pursuant to the Construction Safety Act. In 1971, OSHA adopted the standards of the Construction Safety Act, which were published in Part 1926 of Title 29 of the *Code of Federal Regulations*.[19]

B. Applicability

The standards promulgated in Subparts A and B of Part 1926 remain applicable only to government contractors.[20] All construction contractors, however, are covered by the standards encompassed by Sections 1926.20–.1051.[21] Of these, Subparts C, D, E, F, and G provide general guidelines concerning such matters as safety training, first aid, fire prevention, housekeeping, illumination, sanitation, personal protective equipment, and exposure to noise, radiation, and harmful fumes. The remainder of the subparts cover dozens of specific construction activities, including concrete placement, materials handling, use of hand and power-operated tools, welding and cutting, installation of electrical equipment, use of ladders and scaffolding, guarding of floor and wall openings, use of cranes and conveyors, use of motor vehicles, excavating, trenching and shoring, steel erection, work in tunnels, and demolition and blasting.

C. Coverage

The Construction Industry Standards of Part 1926 apply only to the working conditions of those employees "engaged

---

[19]See note 3, *supra*.

[20]See note 3, *supra*.

[21]29 C.F.R. §1910.12 (1981) provides in pertinent part:
"(a) *Standards*. The Standards prescribed in Part 1926 * * * shall apply * * * to every employment and place of employment of every employee engaged in construction work. Each employer shall protect the employment and places of employment of each of his employees engaged in construction work by complying with the appropriate standards prescribed in this paragraph."

in construction work,"[22] defined broadly as "work for construction, alterations, and/or repair, including painting and decorating."[23] Most work on electric transmission and distribution lines and equipment is specifically mentioned as covered types of construction work.[24] The Secretary bears the burden of proving the applicability of the Construction Industry Standards to a cited workplace,[25] and in several cases the Review Commission has been required to define the types of "construction work" covered under Part 1926.[26]

The term "alteration and/or repair" has been construed to require more than normal maintenance or upkeep of equipment, looking to whether the work caused a defined improvement over what existed prior to the maintenance work.[27] Thus breakdown maintenance, such as the replacement of existing defective parts,[28] or renewal maintenance, such as the reconstruction of an existing system,[29] have been considered covered construction, but preventative maintenance such as the trimming of trees along right-of-way corridors for power lines[30] has not qualified as covered construction.

A contractor may not avoid application of the Part 1926 construction standards to the contractor's operations by subcontracting an essential part of the construction process.[31] But an exemption from the construction standards is provided for construction activity that is merely incidental to the function of a nonconstruction operation: "Activities that could be re-

---

[22]*Id.*

[23]§1910.12(b).

[24]§1910.12(d).

[25]See *Snyder Well Servicing*, 10 OSHC 1371, 1373 (Rev. Comm'n 1982) (citing *United Geophysical Corp.*, 9 OSHC 2117, 2121 (Rev. Comm'n 1981)).

[26]See *Heede Int'l*, 2 OSHC 1466 (Rev. Comm'n 1975), and *Skidmore, Owings & Merrill*, 5 OSHC 1762 (Rev. Comm'n 1977).

[27]In *Consumers Power Corp.*, 5 OSHC 1423 (Rev. Comm'n 1977), the Commission noted that maintenance work was explicitly excluded from the coverage of Subpart V of 29 C.F.R. pt. 1926, as indicated by the remarks on applicability published with the standards in 37 FED. REG. 24,882 (1972).

[28]See *Claude Neon Fed. Co.*, 5 OSHC 1546, 1547–1548 (Rev. Comm'n 1977) (replacement of light bulbs in neon sign treated as "repair" and thus construction work rather than maintenance).

[29]See *United Tel. Co.*, 4 OSHC 1644, 1645–1646 (Rev. Comm'n 1976) (erecting and removing telephone poles is "alteration" qualifying as construction work and thus not "field work" as employer contended).

[30]*Consumers Power Co.*, 5 OSHC 1423 (Rev. Comm'n 1977).

[31]*A. A. Will Sand & Gravel Corp.*, 4 OSHC 1442 (Rev. Comm'n 1976) (mere delivery not construction work, but unloading combined with other activities on worksite may be).

garded as construction work should not be so regarded when they are performed as part of a non-construction operation."[32]

Conversely, the Review Commission has noted that employers not directly engaged in construction may be subject to construction standards if "their operations are an integral part of or intimately involved with the performance of construction work."[33] Thus, in *Bechtel Power Corp. v. Secretary of Labor*,[34] the Eighth Circuit affirmed the Commission's determination[35] that a construction management firm was "engaged in construction work" when it performed no construction but exercised responsibility for the administration and coordination, including the site safety program, of the construction project.[36] But an architectural engineering firm which supervised construction did not engage in substantial control sufficient to warrant coverage by the Construction Industry Standards.[37] Only employers performing "actual construction" work,[38] or exercising "substantial control" over the construction work being performed,[39] are coverd by Part 1926 standards. Factors considered as indicative of an employer's "substantial control" over construction activities include (a) contractual language allocating risks to particular employers; (b) ability to control, and hold apart, personnel on the job; (c) ability to control aspects of safety on the job site; (d) similarity of functions to those of a typical general contractor; and (e)

---

[32]*B.J. Hughes, Inc.*, 10 OSHC 1545, 1547 (Rev. Comm'n 1982) (cement servicing operations performed as part of oil drilling operation not subject to construction standards because oil drilling operations constitute mining, not construction). See also *Royal Logging Co.*, 7 OSHC 1744 (Rev. Comm'n 1979), *aff'd*, 645 F.2d 822 9 OSHC 1755 (9th Cir. 1981) (logging company that built roads for delivering timber not subject to construction standards because road building ancillary to nonconstruction function of cutting and delivering logs); *Aluminum Coil Anodizing Corp.*, 5 OSHC 1381, 1384 (Rev. Comm'n J. 1977) (moving machinery out of plant in course of dismantling plant not construction work). Cf. *Heede Int'l*, 2 OSHC 1445, 1467 (Rev. Comm'n 1975) (dismantling crane within building under construction is construction work); *Perini Corp.*, 5 OSHC 1596 (Rev. Comm'n 1977) (handling of construction materials was construction, not longshoring, work since materials were installed, not handled as cargo).

[33]*Royal Logging Co.*, 7 OSHC 1744 (Rev. Comm'n), *aff'd*, 645 F.2d 822, 9 OSHC 1755 (9th Cir. 1979).

[34]548 F.2d 248, 4 OSHC 1963 (8th Cir. 1977).

[35]*Bechtel Power Corp.*, 4 OSHC 1005 (Rev. Comm'n 1976).

[36]548 F.2d 248, 4 OSHC 1963 (8th Cir. 1977).

[37]*Skidmore, Owings & Merrill*, 5 OSHC 1762 (Rev. Comm'n 1977) (standards did not apply to architects and engineers who had more limited functions and authority over construction work).

[38]*Id.* at 1764.

[39]See *Bechtel Power Corp., supra* note 35.

performance of activities within the realities of the construction industry.[40]

The Commission has found "construction activity" in replacement of a utility pole,[41] running telephone cable in new construction,[42] dismantling a crane,[43] and refitting two towers and installing a furnace at an oil refinery.[44]

The Commission found no construction activity in building logging trails,[45] and repairing cement on an offshore rig.[46]

IV. Maritime Industries

A. Part 1915: Shipyard Employment Standards

1. Scope of Application

The standards applicable for shipyard employment were initially issued under the Longshoremen's and Harbor Workers' Compensation (LHWC) Act. In 1971, they were incorporated as separate OSHA standards and recodified as 29 C.F.R. Parts 1915, 1916, and 1917.

The OSHA standards for shipyard employment apply to all ship repairing, shipbuilding, and shipbreaking employment. Shipbuilding is defined as the construction of a vessel; ship breaking is the tearing down of a vessel's structure to scrap the vessel. A "vessel" is defined in the standards to include all watercraft used or capable of use as a means of transportation on water, including special purpose floating structures not primarily designed for transportation. The Review Commission has expanded the definition "vessel" to cover a land-based superstructure intended to become part of a drill-

[40]*Skidmore, Owings & Merrill, supra* note 37, at 1764–1765 (citing *Anning-Johnson Co. v. OSHRC,* 516 F.2d 1081, 3 OSHC 1166 (7th Cir. 1975), and *Grossman Steel & Aluminum Corp.,* 4 OSHC 1185 (Rev. Comm'n 1975)).

[41]*Mississippi Power & Light Co.,* 7 OSHC 2036 (Rev. Comm'n 1979).

[42]*Southwestern Bell Tel. Co.,* 7 OSHC 1058 (Rev. Comm'n 1979).

[43]*Heede Int'l,* 2 OSHC 1466 (Rev. Comm'n 1975).

[44]*Gulf Oil Co.,* 6 OSHC 1240 (Rev. Comm'n 1977).

[45]*Royal Logging Co.,* 7 OSHC 1744 (1979), *aff'd,* 645 F.2d 822, 9 OSHC 1755 (9th Cir. 1981).

[46]*B.J. Hughes, Inc.,* 10 OSHC 1545 (Rev. Comm'n 1982).

ing rig. The term "vessel" also has been applied to a burned-out hulk of a ferry boat[47] and to an uncompleted hull.[48]

Employers covered by Parts 1915–1918 are those engaged in operations on the navigable waters of the United States and on dry docks, graving docks, and marine railways.[49] According to the Review Commission, the scope of the shipyard employment standards is not restricted to navigable waters, but applies to every employee and place of employment of every employee engaged in shipbuilding or related employment.[50] The Third Circuit, however, has held that the standards did not apply to a structural shop located on an island where vessels were assembled and outfitted. In so holding, the court looked to the LHWC Act. That Act, the court noted, limits coverage to maritime work on navigable waters and does not include structural fabrication shops.[51]

2. Substantive Regulations Governing Shipyard Employment

Sections 1915.32–.36 govern surface preparation and preservation and the use of toxic or flammable substances. Mechanical paint removers are covered in Section 1915.34, which requires ventilation, personal protective equipment, and safe equipment design for the mechanical removal of paint. A choice of method is permissible, but use of equipment within a range of listed types is mandatory.[52] Employees using paints mixed with toxic solvents must also be protected by respiratory equipment under Section 1915.35.[53]

Standards governing welding, cutting, and heating are covered by Sections 1915.51–.57. Ventilation of areas in which welding, cutting, or heating is performed, or protection of employees by means of personal respiratory equipment, is required. Failure to provide ventilation is not excusable on the basis of the employer's lack of knowledge or isolated employee

---

[47]*Cable Car Advertisers*, 1 OSHC 1446, 1448 (Rev. Comm'n 1973).

[48]*Bethlehem Steel Corp.*, 7 OSHC 1053 (Rev. Comm'n J. 1976).

[49]29 C.F.R. §1915.4(c).

[50]*Dravo Corp.*, 10 OSHC 1651, 1654 (Rev. Comm'n 1982).

[51]*Dravo Corp. v. OSHRC*, 613 F.2d 1227, 7 OSHC 2089, 2093 (3d Cir. 1980) (OSHA coverage with respect to shipbuilding regulations may not be "broader than that of the LHWCA under which [those] regulations were originally promulgated").

[52]*National Steel & Shipbuilding Co.*, 5 OSHC 1039, 1040 (Rev. Comm'n J. 1977).

[53]*Dravo Corp.*, 3 OSHC 1085 (Rev. Comm'n 1975) (Cleary, Comm'r dissenting).

misconduct when this misconduct stems from the employer's failure to provide respiratory equipment.[54] Fire prevention precautions for welding, cutting, or heating operations include a requirement of additional trained personnel to guard against fire.[55] Stringent procedures must be implemented to eliminate the substantial hazard of employee noncompliance with fire safety instructions.[56]

Shipyard employment standards governing scaffolds, ladders, and other working surfaces are contained in 1915.71–.77. General safety standards require the use of certain types of planking and the provision of life vests if backrails are not available.[57] Deck openings and edges must be guarded unless the nature of the work prohibits the use of guardrails.[58]

Standards governing general working conditions are covered in Sections 1915.91–.98. Housekeeping requirements mandate maintenance of adequate aisles and passageways and keeping working areas free of obstructions, debris, and slippery conditions.[59] Employers must always eliminate hazards or slippery conditions if doing so would not interfere with the use of the equipment.[60] Employers are also prohibited under Section 1915.97 from using chemical products, structural materials, or process materials until the potential hazards of these materials are ascertained. Standards for moving gear and rigging equipment are covered in Sections 1915.111–.136. In general, these provisions require employers not to exceed safe working loads when instructing employees to use such equipment.

The regulations governing personal protective equipment in the ship repairing, shipbuilding, and shipbreaking industries are covered in Sections 1915.151–.154. Under regulations governing the design, condition, and hygiene of eye

[54]*National Steel & Shipbuilding Co., supra* note 52, at 1040.

[55]29 C.F.R. §1915.32(e) (recodified effective May 20, 1982, at §1915.52(b)(3)); *Bethlehem Steel Corp.*, 3 OSHC 1652, 1653 (Rev. Comm'n 1975).

[56]29 C.F.R. §1915.52(4); see, e.g., *Tacoma Boatbuilding Co.*, 1 OSHC 1309, 1310 (Rev. Comm'n J. 1972).

[57]*Marine Repairs*, 1 OSHC 3142, 3143 (Rev. Comm'n J. 1972).

[58]*National Steel & Shipbuilding Co.*, 8 OSHC 1807, 1808 (Rev. Comm'n J. 1980).

[59]*Bethlehem Steel Corp.*, 7 OSHC 1053, 1055 (Rev. Comm'n J. 1976) ("excessive accumulation" of debris allowed to exist for "unreasonable" length of time); *Bethlehem Steel Corp.*, [1976–1977] OSHD ¶21,147 (Rev. Comm'n J. 1976) (housekeeping citation vacated if alternate means of access to work areas).

[60]*Dravo Corp.*, 578 F.2d 1373, 6 OSHC 1933, 1934 (3d Cir. 1978) (if slippery conditions caused by bad weather, no violation since employer compliance virtually impossible).

protection equipment, employers must show that a rule requiring goggles was enforced at the worksite, and that an employee's failure to wear goggles was a deviation from normal procedures followed there.[61] Requirements concerning respiratory, head, foot, and body protection and lifesaving equipment are addressed in Sections 1915.152–.154.

B. Part 1917: Marine Terminal Standards

On October 3, 1983, Part 1917 took effect. This sets forth safety and health requirements for work within marine terminals. Marine terminals had previously been subject to the longshoring standards of Part 1918. "Marine terminals" subject to the new standards are defined as wharves, piers, docks, and other berthing locations, adjacent storage or contiguous areas, and structures devoted to receiving, holding, loading, or delivery of waterborne shipments.[62]

The Marine Terminal Standards apply only to the foot of the gangway; from that point on, the shipboard longshore regulations apply.[63] Application of the Marine Terminal Standards does not extend to construction activities at a marine terminal, to facilities used solely for the storage and handling of combustible liquids, or to fully automated coal-handling facilities adjacent to generating plants.[64]

C. Part 1918: Longshoring Standards

The standards for longshoring encompass employment involving the loading, unloading, moving, or handling of cargo, ship's stores, and gear in, on, or out of any vessel on the navigable waters of the United States. As in the shipbuilding areas, these regulations do not apply where there is the Coast Guard regulation of seamen.[65] Of course, Part 1918 covers employers only; shipowners are not covered unless they are

---

[61]*Murphy Pac. Marine Salvage Co.*, 2 OSHC 1464, 1465–1466 (Rev. Comm'n 1975) (Moran, Chrmn, concurring).

[62]29 C.F.R. §1917.2(u).

[63]48 FED. REG. 30,886 (1983).

[64]29 C.F.R. §1917.1(a)(1).

[65]See, e.g., *Clary v. Ocean Drilling & Exploration Co.*, 609 F.2d 1120, 7 OSHC 2209 (5th Cir. 1980).

also employers.[66] According to OSHA, employers must ensure that their employees do not use ship facilities that do not meet the applicable standards even if the facilities are not owned or controlled by the employer. Under such circumstances, employers are not required to correct the violations themselves.[67]

Sections 1918.11–.25 establish requirements for gangways and other means of access, and the requirements for inspection and certification of a vessel's cargo handling gear and shore-based material handling devices such as cranes and derricks. Pursuant to Section 1918.12, before using any vessel's cargo handling gear, longshoring employers are required to ascertain whether the vessel has a current and valid cargo gear register and a certificate that indicates that the gear has been tested and examined. The employer must check the vessel's cargo gear register; reliance on others, e.g., a Coast Guard inspection, is not a valid defense.[68]

Sections 1918.31–.43 describe requirements for the protection of employees working near or on potentially hazardous conditions. Safe passage is required over and around deck loads and open weather deck hatches for employees permitted to pass fore and aft. Temporary guardrails or cable barriers must be put up if necessary.[69] However, if it is not necessary for employees to be in a dangerous area, and if the employer has ordered them to keep out of the area, the employer will not be held in violation.[70]

Sections 1918.91–.99 cover general working conditions such as housecleaning, lighting, ventilations, and sanitary conditions. Gear and equipment must be removed from work areas if they present a hazard.[71] Ventilation is required for the protection of workers exposed to hazardous atmospheres. Testing and records are required to ensure the ventilation system works.[72]

---

[66]See, e.g., *Brown v. Mitsubishi Shintaku Ginko*, 550 F.2d 331 (5th Cir. 1977).

[67]OSHA Instruction STD 2-1.8, Oct. 30, 1978.

[68]*Oswego Stevedore & Warehousing Corp.*, 1 OSHC 3037, 3038 (Rev. Comm'n J. 1973) (violation despite register having been checked by Coast Guard); *Belau Transfer & Terminal Co.*, 6 OSHC 1592 (Rev. Comm'n J. 1978) (testing gear by lifting small loads still a violation when no check for valid cargo gear register certificate).

[69]See *Seattle Stevedore Co.*, 4 OSHC 2049 (Rev. Comm'n 1977).

[70]*Morelli Overseas Export Serv.*, 1 OSHC 3305 (Rev. Comm'n J. 1974).

[71]*Concrete Technology Corp.*, 5 OSHC 1751, 1752 (Rev. Comm'n 1977).

[72]*Metropolitan Stevedore Co.*, 6 OSHC 1329 (Rev. Comm'n J. 1978).

Personal protective equipment for longshoring is covered by Sections 1918.101–.106. At times the vertical, or industry-specific, standard for longshoring is less protective than the general standard would be. For example, the general industry standard requires employers to require and assure the use of safety shoes by their employees, whereas longshoring employers need only make the shoes available and encourage their use.[73]

Hardhat requirements have led to significant disagreement between the industry and OSHA over the extent of employer action necessary to avert a violation. Thus, the Third Circuit has held that longshoring employers must take all "demonstrably feasible measures" to ensure compliance with this requirement, even at the risk of work stoppage.[74] In a narrower holding, the First Circuit has ruled that erecting signs urging compliance and verbally announcing the requirement may not constitute sufficient effort by the employers.[75] The Review Commission has taken the position that employers must terminate or suspend noncomplying employees.[76] In addition, employers' arguments that hardhats are required only when their employees are exposed to hazards have been rejected by the Commission.[77]

Longshoring standards governing a ship's cargo handling gear require that the safe working load not be exceeded and that splicing be performed properly.[78] All gear and equipment provided by the employer must be inspected before use and at regular intervals; test procedures are described in the standard.[79] The proper methods for handling various types of cargo are set forth at Sections 1918.81–.86. Precautions are to be taken to protect employees from shifting or falling cargo.[80] Employers must ascertain whether cargo to be handled is haz-

[73]*Matson Terminals*, 3 OSHC 2071 (Rev. Comm'n 1976).

[74]*Atlantic & Gulf Stevedores v. OSHRC*, 534 F.2d 541, 4 OSHC 1061, 1065 (3d Cir. 1976) (threat of wildcat strikes over hardhat use not enough to avert liability).

[75]*I.T.O. Corp. v. OSHRC*, 540 F.2d 543, 4 OSHC 1574 (1st Cir. 1976) (since likelihood of work stoppage if employees disciplined "remote," company has not taken all "demonstrably feasible" steps to comply).

[76]*John T. Clark & Son*, 4 OSHC 1913 (Rev. Comm'n 1976); *Seattle Stevedore Co.*, 5 OSHC 2041, 2042 (Rev. Comm'n 1977).

[77]*Seattle Stevedore Co.*, 4 OSHC 1119, 1120 (Rev. Comm'n J. 1976).

[78]29 C.F.R. §§1918.51–55.

[79]§§1918.61–76.

[80]*Texports Stevedore Co.*, 484 F.2d 465, 1 OSHC 1322, 1323 (5th Cir. 1973) (dislodging of bales not unexpected or untoward incident).

ardous and, if so, take precautions in its handling and warn employees of the hazards involved.[81]

D. Part 1919: Gear Certification

Part 1919 provides regulations for accreditation of persons to perform gear certification required under Sections 1918.12(c) and (d)(3) and Section 1918.13 and specifies certification tests. These standards specify the duties of certificating personnel, which include recordkeeping and regularly checking both the ship's cargo handling gear and the shore-based material handling equipment. The regulations do not apply to certification performed for vessels certified by the U.S. Coast Guard for foreign vessels certified by foreign nations; nor do they apply to certification of shore-based material handling devices under standards set by states and local governments if the local standards meet the requirements of Section 1918.13(b)(2).[82] The gear certification standards are only applicable to gear that is a "permanent" part of the vessel's equipment used for handling cargo.[83]

If the cargo handling gear is considered permanent and covered by the standards of Part 1918, a citation will issue for failure to ascertain whether the vessel had a current, valid cargo gear register even if the employees performed perfunctory tests to determine whether the gear was functioning properly.[84]

Agency policies and procedures on the enforcement of gear certification standards in the shipyard and longshoring area are described in an OSHA instruction.[85] The instruction notes that the accredited certifying agency is not an "employer" for the purpose of certification and should not be cited by the inspector for the manner in which certification is performed. If the compliance officers determine, however, that employees of an accredited agency are exposed to hazards unrelated to

---

[81]*Pittston Stevedoring Corp.*, 8 OSHC 1744, 1745 (Rev. Comm'n J. 1980) (employer had duty to specifically warn employees about poison; labels describing hazardous nature not enough).

[82]29 C.F.R. §1919.1(b)(1) and (2).

[83]§1918.13(a)(1).

[84]*Belau Transfer & Terminal Co.*, 6 OSHC 1592 (Rev. Comm'n J. 1978).

[85]OSHA Instruction CPL 2-1.38, June 14, 1982.

their assigned duties, the accredited agency will be cited under the appropriate standards.

V. Agricultural Operations

A. Applicability of the Standards

The OSHA standards applicable to agricultural operations are set forth at 29 C.F.R. pt. 1928. The term "agricultural operations" is not defined within 29 C.F.R. pt. 1928. Indeed, disputes over the definition of the term have provided one of the few grounds for litigation involving the standard. The Review Commission has defined "agricultural operations" broadly to include any activity that is "integrally related" to an employer's business of agriculture.[86] For example, the Review Commission found agricultural operations to encompass the maintenance and repair of a crop irrigation system and for this reason declined to subject a citrus grower to the personal protective equipment standards for general industry or for the construction industry.[87] The Review Commission also applied the integrally related test to find that agricultural operations included the delivery of grain to feed bins where the employer was engaged in the business of poultry farming but where the bins were not located on the employer's premises.[88] However, the employer was not subject to the equipment standards of 29 C.F.R. pt. 1910 for the condition of equipment located on other farmers' premises.

B. Substantive Provisions Under the Agriculture Standards

The substantive provisions of 29 C.F.R. pt. 1928 are found in Subparts B, C, and D of the standards. These subparts are comprehensive and well illustrated and the terms of the stan-

---

[86]See e.g., *Darragh Co.*, 9 OSHC 1205, 1208 (Rev. Comm'n Nos. 77-2555, 77-3074, 77-3075, 1980); *Chapman & Stephens Co.*, 5 OSHC 1395, 1396 (Rev. Comm'n 1977).

[87]*Chapman & Stephens Co., supra* note 86, at 1396, 1397.

[88]*Darragh Co., supra* note 86, at 1207, 1208.

dards are quite clearly defined. This probably accounts for the fact, as noted above, that there has been practically no litigation related to these provisions.

Subpart B, for example, not only specifies the provisions under 29 C.F.R. pt. 1910 that apply to Part 1928, but also clearly identifies those standards under Part 1910 that do not apply to agricultural operations under Part 1928.[89] Subpart C sets forth the standards for roll-over protective structures (ROPS) for the following categories of vehicles and structures: (1) tractors used in agricultural operations; (2) "protective frames" for wheel-type agricultural tractors, as well as the test procedures and performance requirements for such frames; and (3) "protective enclosures" for wheel-type tractors and the corresponding test procedures and performance criteria for the enclosures.[90]

Although the standards' provisions do not define "protective enclosure" or "protective frame," the illustrated appendix to Subpart C suggests that a protective frame is a structure similar to a roll-over bar, whereas a protective enclosure is a cabin-type structure in which a vehicle operator sits.[91] Subpart C ROPS standards also require ancillary measures to protect employees from injury in the event of a tractor rollover, such as injury caused by the spillage of vehicle fluids or by contact with sharp surfaces on such vehicles.[92]

The ROPS standards for tractors used in agricultural operations apply only to vehicles manufactured after October 25, 1977. Furthermore, it should be noted that at least two states, Washington and Oregon, have received OSHA approval to use revised standards for farm vehicles.[93]

Finally, Subpart D provides the standards for safety features and procedures that are required for farm field equipment, farmstead equipment, and cotton gin machinery used in agricultural operations.[94] Generally, the standards set re-

---

[89]29 C.F.R. §1928.21.

[90]§§1928.51–53.

[91]See §1928, Subpart C, app. A.

[92]§1928.51(b)(3), (b)(4).

[93]See, e.g., "Tractor Roll-Over Protection Rule Approved With Track-Type Exemption," 13 O.S.H. Rep. No. 13, at 318 (OSHA found "significant evidence on record" indicating that relatively flat terrain of state of Oregon would permit safe operation of track-type tractors without ROPS); "State Standard on ROPS Approved by OSHA," 13 O.S.H. Rep. No. 11, at 239.

[94]§1928.57.

quirements for features such as guard covers or railings which must be designed to protect employees from the hazards of working near equipment with exposed, moving machinery parts.[95] With regard to farm-field and farmstead equipment manufactured after October 25, 1977, the standards require guard covers as well as other safety devices or procedures to shield employees for all power take-off shafts and other power transmission components, including the mesh points of all power-driven gears, belts, chains, and sprockets.[96] As for cotton ginning equipment, the standards require that the main drive and miscellaneous drives of gin stands be completely enclosed, and that all drives between gin stands be guarded to prevent access to the area between the machinery.[97]

VI. Industries Subject to Limited Special Standards

A. Special Standards Applicable to Commercial Diving

1. Definition and Coverage

The standards that cover commercial diving operations are set out at 29 C.F.R. Part 1910, Subpart T.[98] Commercial diving is not expressly defined in any provision of Subpart T. However, the term "diver" is defined as "[a]n employee working in water using underwater apparatus which supplies compressed breathing gas at the ambient pressure."[99] The term "underwater apparatus" is not defined within Subpart T. Any employer who has doubts as to the applicability of Subpart T

---

[95]*Id.*

[96]*Id.*

[97]*Id.*

[98]The standards governing commercial diving operations apply to all diving and connected operations conducted in relation to all types of employment, including general industry, construction, ship repairing, ship building, ship breaking, and longshoring. The exceptions to coverage are enumerated at §1910.401(a)(2). The standards offer guidelines as to personnel requirements, general operations procedures, specific operations procedures, equipment procedures, and recordkeeping. 29 C.F.R. pt. 1910, Subpart T.

[99]§1910.402.

to its operation should examine activities that are specifically excluded from coverage.[100] Essentially, there are seven areas exempted from coverage: (1) scuba diving operations performed solely for instructional purposes;[101] (2) search and rescue operations performed for public safety by or under the control of a governmental operation;[102] (3) diving operations specifically governed by 45 C.F.R. Part 46;[103] (4) scientific diving under the direction and control of a diving program;[104] (5) emergency circumstances that require an employer to deviate from the standards in order to avert a fatality, serious physical harm, or major environmental damage;[105] (6) pursuant to Section 3(5) of the OSH Act, employees of the U.S. government;[106] and (7) individuals engaged in recreational or sport diving not connected with their respective employments.[107]

---

[100]42 FED. REG. 37, 654 (1977) states:
"1. *Scope and application (§ 1910.401).* The standard applies wherever OSHA has statutory jurisdiction. Consequently, unless specifically excluded from the standard, diving in any natural or artificial inland body of water, as well as diving along the coasts of the United States and possessions listed in Section 4(a) of the Act, 29 U.S.C. 655, or within the Outer Continental Shelf surrounding them, is covered. Diving outside of the Outer Continental Shelf is not covered by this standard."

[101]"[U]sing open-circuit, compressed-air SCUBA [self-contained underwater breathing apparatus] and conducted within the no-decompression limits." §1910.401(a)(2)(i). See also 42 FED. REG. 37,654 (1977). But scuba diving for commercial purposes is covered by the standard. 21 FED. REG. 37,655.

[102]§1910.401(a)(2)(ii). See also 42 FED. REG. 37,655 (1977), where the "by or under control of" language is explained.

[103]§1910.401(a)(2)(iii). 45 C.F.R. pt. 46 is entitled "Protection of Human Subjects, U.S. Department of Health, Education and Welfare." The exemption also extends to activities governed by "equivalent rules or regulations established by another federal agency, which regulates research, development or related purposes involving human subjects."

[104]§1910.401(a)(2)(iv) sets forth the minimal elements constituting a "diving program." Prior to the addition of this section in 1982, OSHA had considered scientific diving comparable to commercial diving, so that it was covered by the standards. However, a substantial lobbying effort by the scientific and education community forced the change. The main contentions of these groups was that scientific diving has always had self-regulating standards and could minimize exposure to dangerous situations.

[105]§1910.401(b). In such cases the employer must notify the OSHA Area Director "within 48 hours of the onset of the emergency situation indicating the nature of the emergency and extent of the deviation from the prescribed regulations" and submit such information in writing if requested to. §1910.401(b)(1), (2). The exemption does not apply to an emergency purely economic or property damage. 42 FED. REG. 37,655 (1977).

[106]42 FED. REG. 37,654 (1977). Such employees are protected instead under §19 of the Act. *Id.*

[107]42 FED. REG. 37,655 (1977).

2. *Employer Obligations and Judicial Interpretations*

The Commercial Diving Standards have not received concrete substantive interpretation since their promulgation in October 1977.[108] One factor contributing to this may be their comprehensive nature, compelled by the high risk involved in this type of employment, and exemplified by the provisions for general diving procedures[109] and specific diving operations[110] as well as the qualifications for dive team personnel[111] and medical requirements[112] imposed upon employers for their employees.

Outside of the medical requirements provision of Section 1910.411 (see discussion in Section VI.A.3 below), the courts have very broadly interpreted the Commercial Diving Standards in order to comply with the Act's purpose of providing safe and healthy working conditions for a very dangerous occupation.[113] Under the general diving section of Subpart T,[114] an employer has a substantial obligation of employee safety. It must anticipate any and all contingencies of the dive[115] by engaging in a "planning" and "assessment" procedure to consider surface and water hazards before a dive is made. The Review Commission in assessing liability will determine whether prior experience showed the need to anticipate the hazard or occurrence.[116] Once a diving procedure has begun, an employer continues to have a very broad general duty imposed on it by the requirements of Sections 1910.410(c)(1) and (2) to have a designated person in charge.[117] The Review Commission interprets this requirement of employer contact at the dive location very narrowly.[118]

---

[108] 29 C.F.R. §1910.441 (1977).

[109] §§1910.420–.423. See notes 114–116 and accompanying text, *infra.*

[110] §§1910.424–.427. See notes 119–120 and accompanying text, *infra.*

[111] §1910.410. See notes 117–118 and accompanying text, *infra.*

[112] §1910.411. But see notes 122–132 and accompanying text, *infra.*

[113] *Jones/Schiavone, A Joint Venture,* 11 OSHC 1928, 1929 (Rev. Comm'n 1984).

[114] §1910.420, Safe Practices Manual; §1910.421, pre-dive procedures; §1910.422, procedures during dive; and §1910.423, post-dive procedures.

[115] *Jones/Schiavone, A Joint Venture, supra* note 113. In this case, the employer was cited under three sections of §1910.421, pre-dive procedures. The citations were (1) failure to plan the diving operation adequately, (2) failure to brief divers of any unusual hazards or conditions, and (3) failure to inform divers of modifications to the operating procedures necessitated by the operation).

[116] *Id.*

[117] *Bultema Dock & Dredge Co.,* 8 OSHC 1937, 1938 (Rev. Comm'n 1980).

[118] *Id.* at 1938 (court rejected employer's argument that radio contact by supervisor with diver constituted requisite contact "at the dive location").

The employer also must be aware of its obligations under the more specific provisions that precede the general procedural diving provisions.[119] The employer's obligations are more exact as to the specific operation procedures[120] and equipment requirements. The employer also has special recordkeeping requirements to meet.[121]

3. Taylor Diving: *Medical Requirements Provision Vacated*

In *Taylor Diving & Salvage Co. v. United States Department of Labor*,[122] the Fifth Circuit struck down the medical requirements provision of Section 1910.411. Prior to this decision, the procedure followed under this section required the employer to provide a medical examination of all employees exposed to, or likely to be exposed to, hyperbaric conditions at no cost to the employee.[123] The purpose of the examination was to determine whether dive team members exposed to hyperbaric conditions were "medically fit to perform assigned tasks in a safe and healthful manner."[124] The procedure was structured so that if the employee failed the first examination, he had the right to get a cost-free exam from a second doctor of the employee's choosing. If the second opinion disagreed with the first, the procedure required that a third physician, whose selection was agreed upon by the first two, render a binding decision, also free of cost to the employee, as to his medical fitness.[125] In *Taylor Diving*, the employer objected to the examination procedure and to the cost allocation system that required the employer to pay for all the examinations.[126]

The court concluded that in promulgating the medical requirements provision, the Secretary of Labor had exceeded

---

[119]Specific Operations Procedures are outlined in §§1910.424–.427. Equipment Procedures and Requirements are set forth in §1910.430.

[120]*R.C. Diving Co.*, 7 OSHC 2108, 2109 (Rev. Comm'n 1979). In *R.C. Diving*, an employer was cited for violations of the commercial diving operations standards, including the failure of its divers to wear three specific safety devices. In affirming the citation, the Review Commission effectively spelled out in detail the requirements for complying with the commercial diving standards.

[121]See appendix to Chapter 5 (Employer Obligations to Obtain, Maintain, and Disseminate Information).

[122]599 F.2d 622, 7 OSHC 1507 (5th Cir. 1979).

[123]§1910.411(a)(3).

[124]§1910.411(a)(1).

[125]Final Standard for Diving Operations, 42 FED. REG. 37,650, 37,658 (1977).

[126]*Taylor Diving, supra* note 122, at 623, 7 OSHC at 1507.

the scope of his authority under the OSH Act.[127] In explaining its decision to vacate the provision, the court commended OSHA "for trying to avoid doing harm with regulations."[128] But, citing this intention of the agency to avoid creating job restrictions through regulations and to "be cognizant of the employees' countervailing rights to be protected in their choice of occupation,"[129] the court concluded that the provision imposing upon employers "a mandatory job security provision" was invalid because "OSHA is simply not authorized to regulate job security as it has done here."[130] The court also concluded that the provision imposed a maximum standard on the medical requirements an employer could set for his employees,[131] while the OSH Act is intended to set minimum, not maximum, standards of safety and health.[132]

B. Special Industry Standards Arising From Subpart R

1. *Definition and Coverage*

Subpart R, Special Industry Standards,[133] covers diverse industries from pulp, paper, and paperboard mills to telecommunications.[134]

---

[127]*Taylor Diving* did not, however, prohibit multiphysician review for other purposes. See, e.g., *United Steelworkers v. Marshall*, 647 F.2d 1189, 8 OSHC 1810 (D.C. Cir. 1980) (multiphysician review upheld in regulation requiring medical examinations of employees exposed to lead).

[128]*Id.* at 625, 7 OSHC at 1509. In explaining the medical requirements provision, OSHA had said it "must endeavor not to create, through a health and safety standard, a situation which restricts entry into a profession or allows employees to be dismissed for a cause which is less than substantial." Final Standard for Diving Operations, *supra* note 125, at 37,658.

[129]Final Standard for Diving Operations, *supra* note 125, at 37,658.

[130]*Taylor Diving, supra* note 122, at 625, 7 OSHC at 1509; " 'When adopting [the OSH Act], Congress deliberately sought to achieve job safety while maintaining proper employee-employer relations.' " *Id.* (quoting *Marshall v. Daniel Constr. Co.*, 563 F.2d 707, 716 (5th Cir. 1977), *cert. denied*, 439 U.S. 880 (1978).

[131]*Id.* ("the employer has no control over the third doctor's fitness standards, so that the employer is prevented from setting higher health standards * * * than the examining doctors choose to set").

[132]*Id.*

[133]29 C.F.R. §§1910.261–.275 (1983).

[134]The following are the current substantive provisions covered under Subpart R Special Instructions: §1910.261, pulp, paper, and paperboard mills; §1910.262, textile industries; §1910.263, bakery equipment; §1910.264, laundry machinery and operations; §1910.265, sawmills; §1910.266, pulpwood logging; and §1910.268, telecommunications. (§1910.267, agricultural operations, have been transferred to 29 C.F.R. §1928.21.)

(a) Pulp, paper, and paperboard mills. Section 1910.261 "applies to establishments where pulp, paper, and paperboard are manufactured and converted." However, this standard does not define a pulp, paper, or paperboard mill. It contains only a negative definition and provides that the standard does "not apply to logging and the transportation of logs to pulp, paper, and paperboard mills."[135] This section also expressly incorporates certain other standards by reference[136] and provides for the application of General Industry Standards to certain defined provisions of Section 1910.261.[137]

(b) Textile industries. Section 1910.262 applies to all phases of industries "engaged in the manufacture and processing of textiles."[138] However, those engaged exclusively in processes concerned with the manufacture of synthetic fibers are exempt.[139] There is no definition of textiles or synthetics within the provisions of this section, so that judicial interpretations of these provisions will be applied in contested situations. The General Industry Standards are incorporated into all provisions of Section 1910.262.[140]

(c) Bakery equipment. The bakery standards, Section 1910.263, have been severely curtailed in scope.[141] The standards apply "to the design, installation, operation and maintenance of machinery and equipment used within a bakery."[142] The term "bakery" is not defined and therefore would become a substantive issue for the courts.

(d) Laundry machinery. Section 1910.264 applies to laundry machinery and operations. As with the bakery standards,

---

[135]§1910.(261)(a)(1).

[136]§§1910.261(a)(3) and (4).

[137]§1910.261(a)(2).

[138]"The requirements of this subpart for textile safety apply to the design, installation, processes, operation, and maintenance of textile machinery, equipment, and other plant facilities in all plants engaged in the manufacture and processing of textiles, except those processes used exclusively in the manufacture of synthetic fibers." §1910.262(a)(1).

[139]*Id.*

[140]§1910.262(a)(2).

[141]43 FED. REG. 49,751 (1978). Two hundred sixty-three provisions under Section 1910.263 were eliminated in response to a request by Congress to simplify detailed provisions in an industry with a low rate of injury. The cost of compliance could not be justified in light of the protection provided by the General Industry Standards, and all provisions that were not cost effective were revoked.

[142]§1910.263(a).

a large number of the substantive provisions have been re-voked.[143] The remaining standards apply to "moving parts of equipment used in laundries and conditions particular to this industry."[144]

(e) Sawmills. The sawmill standards at Section 1910.265 are defined as "including, but not limited to" all products, handling, and equipment aspects of sawmill operations but "excluding the manufacture of plywood, cooperage, and veneer."[145] Whether an operation is covered by the sawmill standards or the pulpwood logging standards (see next paragraph) is determined by the classification of the majority of the finished product. The General Industry Standards incorporated by reference are specifically set out in the sawmill provisions.[146]

(f) Pulpwood logging. Section 1910.266 applies to pulpwood logging, which is defined as including, but not limited to, all operations "associated with the preparation and movement of pulpwood from the stump to the point of delivery, but excludes any "logging operations relating to saw logs, veneer, bolts, poles, piling and other forest products."[147] This standard does not apply to subsequent manufacturing processes which are covered under Section 1910.261 (see paragraph (a), *supra*). The General Industry Standards are specifically included by reference in the applicable provisions of this standard and their use is governed by 29 C.F.R. §1910.5.[148]

(g) Telecommunications. The applicability of this standard has been the most difficult issue of any arising from Subpart R for the Review Commission and the courts to resolve. Section 1910.268, which covers the telecommunications industry, was promulgated subsequent to the majority of the Subpart R standards.[149] The standards were meant to cover

---

[143] 43 FED. REG. 49,751 (1978). Sixty-four provisions under Section 1910.264 were revoked. See note 141, *supra*.

[144] §1910.264(b)(1).

[145] §1910.265(a)(1).

[146] §1910.265(a)(2).

[147] §1910.266(a)(1).

[148] §1910.266(a)(2).

[149] §1910.268 was initially published at 40 FED. REG. 13,441 (1975), and later amended, 43 FED. REG. 49,751 (1978), and 47 FED. REG. 14,706 (1982). The rest of the Subpart R provisions were published in the *Federal Register* in 1974.

the unique working conditions of the telecommunications industry to the extent that existing general industry or construction standards did not provide the necessary safety protection.[150] There has been considerable litigation involving confusion between the Construction Industry Standards and the Telecommunications Standards. The provisions apply to activities "specifically associated with" the telecommunications industry,[151] but they allow Construction Industry Standards to take precedence[152] if the principal object of the work relates to construction. Work performed for electric utilities is not governed by the Telecommunications Standards,[153] and General Industry Standards specifically apply wherever the Telecommunications Standards do not.[154]

2. *Substantive Rulings*

In Subpart R cases, as in other OSHA cases, the Review Commission must determine, first, whether the employer was in compliance with the standard; second, whether the employer had knowledge of the violative condition; and finally, whether there are any valid employer defenses such as employee misconduct.[155]

(a) Compliance. The first line of inquiry concerning compliance with the cited standard involves whether the specific Subpart R standard or a more general provision is applicable.

---

[150]40 FED. REG. 13,437 (1975).

[151]§1910.268 provides in relevant part:
"(a) *Application.* (1) This section sets forth safety and health standards that apply to the work conditions, practices, means, methods, operations, installations and processes performed at telecommunications centers and at telecommunications field installations which are located outdoors or in building spaces suited for such field installations * * *. 'Center' work includes the installation, operation, maintenance, rearrangement, and removal of communications equipment and other associated equipment in telecommunications switching centers. 'Field' work includes the installation, operation, maintenance, rearrangement, and removal of conductors and other equipment used for signal or communication service, and of their supporting or containing structures, overhead or underground, on public or private rights of way, including buildings or other structures."

[152]§1910.268(a)(2).

[153]29 C.F.R. §1910.268(a)(2)(ii).

[154]§1910.268(a) provides in relevant part:
"(3) Operations or conditions not specifically covered by this section are subject to all the applicable standards contained in this Part 1910. *See* §1910.5(c). Operations which involve construction work, as defined in §1910.12, are subject to all the applicable standards contained in Part 1926 of this chapter."

[155]*New England Tel. & Tel. Co.,* 11 OSHC 1501, 1509 (Rev. Comm'n 1983).

This inquiry often involves applicability of a general provision such as a General Construction Industry Safety and Health standard versus that of the more specific Special Industry standard. The telecommunications field has provided a substantial amount of the litigation on this particular issue (see paragraph (g) above). The test used to determine which standard applies has been whether "the task being performed was * * * integral to the business of construction."[156] If the Review Commission determines the task was not integral to the construction job, then the more specific Telecommunications Standard applies.[157]

Another frequently occurring issue is which special industry standard is applicable to the type of activity engaged in by the employer. The Review Commission will resolve this issue by ascertaining what constitutes the "major portion" of the employer's final product and then apply the correct standard.[158]

If the employer has been cited under the correct section, the extent of coverage under the section must be determined. The telecommunications field provides an example of the Review Commission's struggle to determine the extent of coverage of the applicable standards. Under that section, the issue centers on whether a "functional" or a "geographic" approach should be used.[159] The early decisions applied the geographic approach, according to which the Review Commission determined, for example, that the "employer and employees were working in an apartment complex, which was neither a 'switching center' nor a 'field installation,' "[160] and that therefore the Telecommunications Standards were inappropriate. The Review Commission eventually overturned the geographic limitation in *New England Telephone & Telegraph Co.,*[161] where the Commission stated that "the telecommuni-

---

[156]*South Cent. Bell Tel. Co.*, 5 OSHC 1614, 1614 (Rev. Comm'n 1977).

[157]*Id.* at 1614 (citation under §1926.150(c)(1)(iv) vacated when applicable work performed was only wire splicing at existing open manhole).

[158]*Antonio Levesque & Sons*, 11 OSHC 1093, 1094 (Rev. Comm'n 1982) (Pulpwood Logging Standards applicable because major portion of employer's tree-harvesting operation was to produce pulpwood and not saw logs).

[159]*New England Tel. & Tel. Co.*, 8 OSHC 1478, 1488–1489 (Rev. Comm'n 1980).

[160]*Southwestern Bell Tel. Co.*, 5 OSHC 1746, 1746 (Rev. Comm'n 1977). The Review Commission ultimately held that the "general construction standards" should be applied in this case.

[161]8 OSHC 1478, 1489 (Rev. Comm'n 1980).

cations standards * * * were intended to hinge primarily on the function and the nature of the work operation being performed, rather than on the location where the work was being performed."[162] The Review Commission's rationale was based on the idea that telecommunications employees are expected to be trained to work safely near highly energized lines. Therefore, the Review Commission found the General Industry Standard for safe distance from energized lines was not applicable.[163]

(b) Defenses.[164] An employer cited under the correct section often contends that the particular standard does not apply to its operations. A common situation in which an employer raises this defense is where the citation refers to the necessary safety features for a specific piece of machinery and the employer is able to contend successfully that this machinery is different from that covered by the cited standard.[165] The Review Commission generally examines the actual definition contained in the standard and uses expert testimony in reaching its decision.[166]

Another defense to a citation under Subpart R is that the cited section is not a substantive provision but rather a "connecting link provision" applying to regulations found in other parts of the *Code of Federal Regulations*.[167] This defense is raised most often where an employer is cited under a specific subpart standard that incorporates outside standards by ref-

---

[162]*Id.* at 1489.

[163]*Id.* The Commission found the intent precisely set out in the preamble of §1910.268:

"'The reason for this [ten foot] requirement is to protect employees who lack familiarity with and training in hazards associated with working near electric power lines. * * * But telecommunications workers * * * are familiar with the hazards and techniques associated with working on or near energized lines * * *. Therefore, the final rule provides an exception from the ten foot clearance requirement * * * for telecommunications operations * * * to clearly reflect the intent of the proposal * * *.'"

New England Tel. & Tel. Co., supra note 161, at 1489 (quoting from 40 FED. REG. at 13,436 (1975)).

[164]See also Chapter 15 (Defenses).

[165]*J. W. Black Lumber Co.*, 3 OSHC 1678, 1682 (Rev. Comm'n 1975) (citation under §1910.265(c)(21)(i) vacated because employer's machine not a whole-log chipper but a "hog").

[166]*Ricardo's Mexican Enters.*, 4 OSHC 1081, 1081–1082 (Rev. Comm'n 1976) (citation under §1910.263(e)(1) vacated because "masa feeder' not a "mixer," according to expert testimony and definition contained in standard).

[167]*Kellog Transfer, Inc.*, 1 OSHC 3046 (Rev. Comm'n 1973).

erence.[168] The validity of a citation hinges on whether the citation states with particularity the substantive standards violated and not merely the nonsubstantive "incorporation by reference" standard.[169] Thus, if the citation states which provision of the standards incorporated by reference has been violated[170] or declares that all of the incorporated standards have been violated,[171] the Review Commission generally will hold that the citation has enough specificity to withstand the challenge.[172]

Another common defense under Subpart R is vagueness of the cited standard.[173] The Review Commission decides whether the standard is unenforceably vague by determining if a reasonable employer in a similar industry[174] would be aware of the precautions that are required.[175]

(c) Knowledge. With Subpart R, the Secretary must prove the employer had knowledge of the violative condition.[176] The issue generally turns on whether "[the employer] knew, or with the exercise of reasonable diligence, could have known of the violative conditions."[177] The frequently used defensive strategy concerning reasonable diligence is that the employer acknowledges the cited condition existed but contends that it did not violate the Act because the Secretary could not show potential employee harm.[178]

---

[168]*Id.* at 3046.

[169]*Id.* at 3046 (§1910.265(c)(22) not a substantive provision but a connecting link provision making applicable all standards contained in §1910.219).

[170]*Idaho Forest Indus.*, 2 OSHC 3147 (Rev. Comm'n 1974) (citation under §1910.265(c)(22) specifically stated §§1910.219(c)(2), (c)(4), (f)(3), and (h)(1) had been violated).

[171]*Weyerhaeuser Co.*, 1 OSHC 3164 (Rev. Comm'n 1973) (upheld citation under §1910.265(c)(22) alleging all 30 items of incorporated statute §1910.219 had been violated).

[172]*Id.* at 3164 (found violations of 14 different items).

[173]*Georgia-Pac. Corp.*, 1 OSHC 1282 (Rev. Comm'n 1973).

[174]*Id.* at 1282: "Under these circumstances, we concluded that persons of common intelligence in Respondent's industry are apprised of the conduct required by the standard and need not guess as to its meaning."

[175]*Idaho Forest Indus., supra* note 170, at 3148 (incorporated standard not vague in failing to give fair warning of requirement to guard specific parts of power transmission apparatus).

[176]*New England Tel. & Tel. Co.*, 8 OSHC 1478 (Rev. Comm'n 1980).

[177]*New England Tel. & Tel. Co.*, 11 OSHC 1501, 1509 (Rev. Comm'n 1983).

[178]*J.W. Black Lumber Co.*, 3 OSHC 1678, 1681–1682 (Rev. Comm'n 1975) (although employees failed to wear gloves as required under §1910. 265(c)(17)(ii), Secretary could prove no past or potential harm from exposure to toxic materials).

Another defense is to assert unpreventable employee misconduct. To claim this defense, the employer must show that the action of its employee was a departure from a safety rule which the employer had communicated to its employee and had uniformly enforced.[179] This determination bears a close connection to whether the employer had knowledge of the violative condition, because the employer must show that its employees received adequate instruction to prevent harm from known hazardous conditions.[180] The employer must do more than show that the particular employee was experienced or skilled,[181] or that the employer was coerced by the employee into permitting the violation to exist.[182] However, the employer may avoid liability by showing it provided to its employees "specific safety instructions and work rules concerning particular hazards that may be encountered on the job."[183] Also, if an employer provides necessary safety equipment on the job site but employees, after receiving instruction on how to use it, failed to do so, the employer is not liable for a violation unless the employer knew or should have known of the employees' actions.[184]

C. Helicopter Cranes

Safety requirements for the use of helicopter cranes are covered by Section 1910.183 Helicopters[185] and Section 1926.551 Helicopters.[186] The standards were promulgated under the

[179]*New England Tel. & Tel. Co., supra* note 176, at 1490.

[180]*Moser Lumber Co.*, 1 OSHC 3108, 3109 (Rev. Comm'n 1973) (citation for failure of driver to secure unloading lines on truck before releasing binders on truck vacated when employer contended employee had received proper instructions prior to accident). See also *Del-Cook Lumber Co.*, 6 OSHC 1363, 1365 (Rev. Comm'n 1978) (citing *Enfield's Tree Serv.*, 5 OSHC 1142 (Rev. Comm'n 1977)).

[181]*Antonio Levesque & Sons*, 11 OSHC 1093, 1093 (Rev. Comm'n 1982) (employee skill no defense to citation for failure to instruct employees to plan retreat path and clear path, because employer failed in responsibility to instruct).

[182]*Weyerhaeuser Co.*, 3 OSHC 1107 (Rev. Comm'n 1975) (citation upheld even though employer contended employees would strike if forced to wear life jackets, because employer has duty to enforce compliance with standards).

[183]*Del-Cook Lumber Co.*, 6 OSHC 1362, 1365 (Rev. Comm'n 1978).

[184]*Northern Cheyenne Forest Prods.*, 8 OSHC 1421, 1421 (Rev. Comm'n 1980).

[185]These standards are contained within 29 C.F.R. pt. 1910, Subpart N, Materials Handling and Storage. See note 222 and accompanying text in Chapter 3 (Specific Occupational Safety and Health Standards).

[186]These standards are contained within 29 C.F.R. pt. 1926, Subpart N, Cranes, Derricks, Hoists, Elevators and Conveyors.

Construction Industry Standards of Part 1926 and later adopted, with only one slight modification, in Part 1910.[187] Therefore, an employer with any operation involving the use of a helicopter should initially consult one of the above-referenced provisions.

At present, there are no Commission or court decisions under the substantive provisions of either Section 1910.183 or Section 1926.551. This lack of litigation may be due in part to the clarity of the safety requirements for helicopter cranes. While helicopter cranes are not specifically defined, the standard specifically applies to the use of any helicopter to move materials or equipment attached underneath the helicopter by means of slings and tag lines[188] and cargo hooks[189] to and from job sites.

The majority of the remaining helicopter provisions deal with the necessary conduct of the ground operation crew[190] and the pilot.[191] Standards require a daily briefing of all involved personnel[192] and "reliable communications" during all phases of loading or unloading.[193] The standards also regulate the clothing and safety equipment of the ground personnel to maximize employee safety.[194]

VII. Exemptions of Small Employers

Special exemptions from such requirements as record-keeping and inspections have been allowed to small businesses in order to reduce the potential for substantial adverse economic impact.

---

[187]See 40 FED. REG. 13,436 (1975). The standards were meant to duplicate the existing construction standards and were enacted to provide similar protection to nonconstruction employees. The only differences were to be minor and to help make more explicit the employer's obligation.

[188]§1926.551(c).

[189]§1925.551(d).

[190]§1926.551(p). (All employees shall remain in full view and in crouched position during all approach and departure maneuvers.)

[191]§1926.551(h). (Operator of the helicopter shall bear responsibility for correct load limits.)

[192]*Del-Cook Lumber Co.*, 6 OSHC 1362, 1365 (Rev. Comm'n 1978).

[193]§1926.551(r). Section 1926.551(m) provides a uniform system of air-to-ground signals.

[194]§1926.551(e).

A. Small Business Exemptions From Certain Recordkeeping

Normal recordkeeping requirements are not applicable to an employer of 10 or fewer employees during the previous calendar year, counting all full-time, part-time, or seasonal employees of the employer at as many locations as the employer may operate.[195] The employer can lose its exemption if chosen to participate in the Annual Survey of Occupational Injury and Illness.[196] Even if covered by this exemption, the employer should also ascertain recordkeeping requirements under state law. Even employers otherwise exempt from reporting requirements are required to report any fatality or the hospitalization of five or more employees.[197]

B. Small Business Exemptions From Certain Inspections

A small business exemption from certain inspection provisions prohibits OSHA from conducting inspections of employers who qualify on the date of the inspection, or during the preceding 12-month period, under the recordkeeping test of 10 or fewer employees and who are in an industry that has a lost workday rate lower than the national average rate.[198] This exemption does not prohibit inspections that may occur in response to employee complaints; but, unless a fatality or the hospitalization of five or more employees results, penalties may be imposed only for willful violations or for failure to correct hazardous conditions. The exemption does not limit imminent danger or health hazard inspections.[199]

---

[195]29 C.F.R. §1904.15.

[196]29 C.F.R. §1904.15(b).

[197]29 C.F.R. §1904.15(a).

[198]Fiscal Year 1980 Appropriations Act, Pub. L. No. 96-86 §101(j), 93 Stat. 656, 658 (1979).

[199]*Id.*

Part III

Enforcement of the Occupational Safety and Health Act

7

Inspections

Section 8 of the OSH Act, 29 U.S.C. §657, establishes the Secretary of Labor's authority to conduct inspections of worksites. The present chapter describes the types of inspections, the order of priority in which these inspections are conducted, and the manner in which a particular establishment is selected for an inspection.

Also addressed in this chapter are certain employee rights established by Section 8, OSHA's consultation program, and voluntary protection programs. Finally, this chapter discusses the inspection itself and the warrant requirement in OSHA inspections.

I. The Right to Inspect—A General Overview

Section 8(a) of the Act provides for a right of entry and inspection for the Secretary's representative:

"In order to carry out the purposes of this Act, the Secretary, upon presenting appropriate credentials to the owner, operator, or agent in charge, is authorized—
"(1) to enter without delay at reasonable times any factory, plant, establishment, construction site, or other area, workplace or environment where work is performed by an employee of an employer; and
"(2) to inspect and investigate during regular working hours and at other reasonable times, and within reasonable limits and in a reasonable manner, any such place of employment and all pertinent conditions, structures, machines, apparatus, devices,

197

equipment, and materials therein, and to question privately any such employer, owner, operator, agent or employee."[1]

OSHA's implementing regulations for Section 8(a)[2] incorporate the Supreme Court's mandate in the *Barlow's* case,[3] and provide for obtaining warrants either in the case of refusal of entry[4] or in advance of an inspection[5] when the employer's previous conduct has put the Secretary on notice that a warrantless inspection will not be allowed. The regulations also provide for securing a warrant in advance of an inspection when obtaining a warrant after a visit to the worksite would require significant time and resources in extra travel between distant worksites and the Area Office, or when the inspection requires the presence of special equipment or personnel and having a warrant would ease coordination with the availability of the expert or the equipment.[6]

Preinspection warrants, however, are not common. OSHA's Field Operations Manual notes that "agency policy is generally not to seek warrants without evidence the employer is likely to refuse entry."[7]

Instructions for compliance officers who are refused entry for inspection are spelled out at CPL 2.45A III.D.1(d). The instruction acknowledges the right of an employer to request a warrant, whether the request occurs at the beginning of or during the inspection. The instruction also sets forth the procedures to be followed when the compliance officer is refused entry or is not allowed to look at records to decide whether to perform a comprehensive inspection. (See discussion in Section IV below.)

---

[1]29 U.S.C. §657(a) (1985).

[2]29 C.F.R. §1903.3 (1985).

[3]*Marshall v. Barlow's, Inc.*, 436 U.S. 307 (1978) (Section 8(a) of the Act violates constitutional prohibition against unreasonable searches insofar as it purports to authorize warrantless, nonconsensual searches of commercial workplaces; absent consent, OSHA inspections must be conducted pursuant to valid inspection or search warrant).

[4]§1903.4(a).

[5]§1903.4(b).

[6]*Id.*

[7]OSHA Instruction CPL 2.45AIII.B.3a OSHR [Reference File 77:2502] ("[T]he Regional Administrator may, on a case-by-case basis, authorize the Area Director to seek compulsory process in advance of an attempt to inspect or investigate whenever circumstances indicate the desirability of such warrants.").

II. OSHA Inspection Policies

A. Types of Inspections

OSHA's Field Operations Manual describes four categories of inspections: unprogrammed, unprogrammed related, programmed, programmed related.[8]

1. Unprogrammed

Inspections in which alleged hazardous working conditions have been identified at a specific worksite are "unprogrammed." This type of inspection includes responses to complaints of imminent dangers, investigations of fatalities or catastrophes, responses to complaints, follow-up investigations, and investigations of referrals.

2. Unprogrammed Related

Unprogrammed related inspections are inspections of employers on multiemployer worksites whose operations are not directly related to the subject of the unprogrammed activity, such as a complaint, a referral, or an accident.

3. Programmed

"Programmed" inspections are inspections done according to a regular inspection schedule. Not all worksites are inspected in this schedule. The worksites are selected through use of the planning guide for safety and health, or special or local emphasis programs.

4. Programmed Related

These include inspections of worksites not included in the programmed inspections, for example, a low-injury-rate em-

---

[8]CPL 2.45A II.B, OSHR [Reference File 77:2301].

ployer at a worksite where programmed inspections are being conducted for all high-injury-rate employers.

There are different degrees of thoroughness of inspections. Comprehensive inspections check the entire establishment except clearly low-hazard areas such as offices. Partial inspections are restricted to certain operations, areas, or conditions. Records only inspections are confined to the establishment's injury and/or illness record and an evaluation of compliance with the hazard communication standard.

B. Inspection Priorities

Imminent danger inspections have the highest priority, followed by fatality investigations, investigations of complaints or referrals, and, finally, programmed inspections.[9] Unprogrammed inspections are usually scheduled and conducted before programmed inspections. Under certain circumstances, however, programmed inspections may receive a higher priority. For example, a programmed inspection may be conducted during the response period allowed for a formal, other-than-serious employee complaint.

The scheduling of both programmed safety inspections and programmed health inspections is a multiple-step process. The initial selection of a category of employment (e.g., construction, maritime, high-rate general industry) is made in accordance with the annual Field Operations Program Plan Projection. Within a category, establishments are selected for inspection from the Statewide Industry Rank Report or Lost Time Claims Rate List for that category and placed in an inspection cycle. Within an inspection cycle, establishments may be inspected in any order that makes efficient use of resources, but generally all establishments in a cycle must be inspected before a new cycle is begun.[10]

C. Records Request

Section 8 of the Act requires each employer to keep and to make available records concerning its activities relating to occupational health and safety. In addition, the employer is

---

[9]CPL 2.45A II.D, OSHR [Reference File 77:2302].

[10]CPL 2.45A II.E.2, OSHR [Reference File 77:2304]; but see FOM ch. II.E.2.b(1)(e), OSHR [Reference File 77:2307].

required to keep records on work-related deaths, injuries, and illnesses, and records of employee exposure to potentially toxic materials.[11]

The compliance officer may request any or all of the records related to the inspection. In manufacturing establishments scheduled for a safety inspection, the injury records will be requested along with annual employment data for the appropriate number of years. These figures are required for the purpose of calculating Lost Workday Injury (LWDI) rates for the establishment. If the employer refuses to produce these records, or any other required records, voluntarily, the Area Director may request the issuance of an administrative subpoena by the Secretary under authority of Section 8(b) of the Act.[12]

The administrative subpoena will require the employer to produce the records needed to complete the inspection or the records review. If the records are produced, the compliance officer will calculate an LWDI rate for the establishment in accordance with the procedures outlined in OSHA Instruction CPL 2.45 III.D.4(a). This procedure recognizes the employer's right to request that OSHA issue an administrative subpoena before voluntarily producing requested records or data.

If the employer refuses to honor the administrative subpoena, the Area Director will proceed as usual for refusals of entry and the matter will be turned over to the Solicitor of Labor to proceed with the warrant procedure.

D. Employee Complaints

Section 8(f) of the Act grants to employees and authorized representatives of employees the right to request an inspection where there is a reasonable belief that a violation of a safety or health standard exists that threatens physical harm or presents an imminent danger.[13]

For complaints made to OSHA about hazardous conditions, an "employee" is defined as a current employee of the

---

[11]See §8(c)(1)–(3) of the Act, reproduced at the end of this volume. Also see Chapter 5 (Employer Obligation to Obtain, Maintain, and Disseminate Information).

[12]Procedures for obtaining an administrative subpoena are described in CPL 2.45A III.D.1.d(4), OSHR [Reference File 77:2506].

[13]See §8(f)(1) of the Act. Chapter IX of CPL 2.45A, OSHR [Reference File 77:3701], implements §8(f)(1); also cf. Chapter 17 (Employee and Union Participants in Litigation Under the Act).

employer at an establishment or an employee of some other employer who also is exposed to the hazards of the workplace during his work hours.[14] Former employees are *not* considered to be employees for purposes of filing a formal complaint.

A formal complaint may be filed by an employee's authorized bargaining representative, his attorney, or any bona fide representative, such as a family member.[15]

A person who submits a formal complaint must do so in writing. He may use a standard OSHA form, but this is not required. Further, the complaint must allege that an imminent danger or violation threatening physical harm exists in the workplace. This allegation must describe with "reasonable particularity" the nature of the hazard. This does not mean that the complaint must specify a particular standard that has been violated; it need only specify a condition or practice that is believed to be hazardous. Finally, in order to meet the formality requirements, the complaint must be signed.[16]

If the Area Director decides that a signed complaint will not be investigated because it fails to meet all of the formality requirements, a letter is sent to the complainant explaining that decision and the reasons for it. The complainant also is informed that he has a right to appeal the decision to the Regional Administrator for an informal review.

Not all complaints that reach OSHA are considered "formal." Nonformal complaints include oral complaints and written complaints that have not been signed or descriptions of conditions that appear not to threaten physical harm. OSHA responds to them without an immediate inspection. Imminent danger complaints, however, are inspected whether formal or nonformal.

In the case of a nonformal complaint not alleging an imminent danger, the employer will be sent a letter notifying it that such a complaint has been received and informing it of the standards allegedly violated. In such cases, the letter includes copies of the relevant section of OSHA standards and describes ways in which corrective action might be taken. Employers who receive such a letter are requested to post a copy of the letter along with a copy of their response to OSHA.

---

[14]CPL 2.45A IX.A.2(a), OSHR [Reference File 77:3701].
[15]CPL 2.45A IX.A.2(b), OSHR [Reference File 77:3701].
[16]CPL 2.45A IX.A.2(c), OSHR [Reference File 77:3701].

OSHA also notifies the complainant, if it is possible to do so, that a letter has been sent to the employer and requests that the complainant notify the Area Director if no corrective action is taken within 30 calendar days.

E. Consultation Visits

OSHA conducts a consultation program through which health and safety personnel visit workplaces at the request of the employer for the purpose of providing advice regarding safety and health hazards, preventive measures, and workplace safety and health programs. This service is not part of the inspection and enforcement process. It has its origin in other sections of the Act and regulations. (Section 2(b)(1) of the Act declares the intention of Congress to assure occupational safety and health "by encouraging employers and employees in their efforts to reduce the number of occupational safety and health hazards at their places of employment, and to stimulate employers and employees to institute new and to perfect existing programs for providing safe and healthful working conditions.")

OSHA has implemented its consultation program through a series of cooperative agreements with state governments for provision of consultation services to employers. These services are offered free of charge. Historically, these consultation services have focused on specific workplace hazards and have provided assistance in the elimination of those hazards. Consultation services are independent of the OSHA enforcement system, although employers remain under a statutory obligation to protect their employees and are required to take necessary protective action.

In 1985, OSHA promulgated new regulations governing its consultation programs[17] and expanded the scope of the regulations to increase the program's emphasis on the prevention of illness and injuries through development of employer safety and health programs. Briefly, this shift in emphasis encompasses a concern for the effectiveness of the employer's total management system for promoting a safe and healthful workplace, and it provides an exemption from general sched-

---

[17]29 C.F.R. pt.1908 (1985).

ule OSHA inspections for employers that meet specified conditions. This regulation intends to further OSHA's goal of achieving a balanced mix of program activities by strengthening, encouraging, and assisting voluntary employer and employee safety and health efforts.

1. Consultation Activities

Because consultants have expertise similar to OSHA inspectors, and because they observe hazardous conditions in the workplace, the line between consultation activity and enforcement activity must be carefully drawn in order to ensure, on the one hand, that hazardous conditions are eliminated so that employees are protected, and, on the other hand, that the cooperative relationship with the employer can be firmly established and maintained so as to provide long-term protection in the workplace.

Thus, the consultant must advise the employer of both hazards addressed in the employer's request for assistance and any other safety and health hazards observed in the workplace during the course of the on-site consultative visit. This advice includes possible solutions to hazardous conditions and describes the general form of the remedy. The consultant may give the employer material on ways to eliminate or control the kinds of hazards observed. When a hazard is identified in the workplace, the consultant must indicate to the employer whether the situation would be classified as a "serious" or "other-than-serious" hazard.

If the consultant determines that a serious hazard exists, he must help the employer to develop a plan to correct the hazard. This plan affords the employer sufficient time to complete the necessary action. The employer is encouraged to advise affected employees of the identified hazard and to notify them of the correction of the hazard or hazards.[18]

An employer must take immediate action to eliminate exposure to a hazard that presents an imminent danger to employees. If the employer fails to take necessary action, the consultant must immediately notify the affected employees and the appropriate OSHA enforcement authority.[19]

---

[18]§1908.6(e).
[19]§1908.6(f).

In order to demonstrate that the necessary action is being taken as the result of a consultation visit, the employer may be required to submit periodic reports, to permit a follow-up visit, or to take similar action. If the employer fails to take the action necessary to correct a serious hazard within the established time, the consultation program manager must notify the appropriate OSHA enforcement authority and provide the relevant information.[20]

This does not mean that OSHA consultants are "spying" on the business that invites them for an inspection. OSHA regulations provide for independent consultation and enforcement activities, including separate managerial staffs.[21] The identity of employers requesting consultations and the file of the consultant's visit may not be provided to OSHA for use in compliance inspections or inspection scheduling, with the exceptions noted above.

Historically, cases that resulted in referrals for unabated serious hazards or imminent dangers have been extremely rare. In general, the independence of the consultation program has been preserved almost without exception.

2. *Inspection Exemption Program*

One other link between consultation and enforcement is described in the regulations.[22] This has to do with the inspection exemption program. To qualify for an exemption from general schedule (programmed) inspections, employers must:

(1) Obtain a consultation visit at the establishment, covering all conditions and operations related to occupational safety and health;
(2) Correct all hazards identified during the consultative visits within the established schedule;
(3) Post notice of correction of hazards when such correction is completed;
(4) Demonstrate to the consultant that certain elements of an effective safety and health program will be implemented within a reasonable time;

---

[20]*Id.*
[21]§1908.7(a).
[22]§1908.7(b)(4).

(5) Agree to request a consultative visit if major changes in working conditions or processes occur which may introduce new hazards.

The employer also must post a notice of participation at the time that it elects to participate in the inspection exemption program described above.

F. Voluntary Protection Programs

Like the consultation program, OSHA's voluntary protection programs (VPP) operate as an adjunct to enforcement. OSHA announced implementation of its three voluntary protection programs in the *Federal Register* of Friday, July 2, 1982.[23] The *Federal Register* notice contains a full description of these programs.

The purpose of the voluntary protection program is to seek out and to recognize exemplary workplace safety and health programs as a means of expanding worker protection. The agency locates companies, general contractors, and small business organizations that go beyond OSHA standards in providing safe and healthful workplaces for their employees. In return, OSHA removes participants from general schedule inspection lists and gives priority attention to any participant that requests a variance.[24]

As noted above, the consultation and voluntary protection programs are kept largely separate from the enforcement programs, although establishments in the VPP are removed from the list of programmed inspections. For example, different OSHA employees work in the different programs, and information gained from consultation or VPP inspections is not shared with the enforcement personnel.[25] Where hazards are found, they are expected to be discussed with management and resolved. In such cases, a memorandum to the files will describe the abatement agreement.[26]

---

[23]47 FED. REG. 29,025 (1982).
[24]See Chapter 21 (Other Issues Related to Standards), Section I.
[25]CPL 5.1 II.C.1(c).
[26]CPL 5.1 IV.K.

III. Conduct of the Inspection

A. Role of the Compliance Officer

In brief, the compliance safety and health officer's role in the inspection is to represent OSHA to the public and to carry out the policies and procedures of the agency.[27] In this regard, the primary responsibility of the OSHA compliance officer is "to build cooperative relationships in the interest of workplace safety and health."

OSHA policy is to remain neutral in dealing with management and labor; the compliance officer is an agent of neither side, but rather of OSHA, and is, therefore, charged with the ensuring of a safe and healthful workplace. Bias or even the appearance of partiality toward one side or the other is seen as an impediment to carrying out the agency's objectives.

B. Advance Notice of Inspection

Section 17(f) of the Act and the regulation at 29 C.F.R. §1903.6 generally prohibit the giving of advance notice of inspections, except as authorized by the Secretary. Such "surprise" inspections allow the compliance officer to observe working conditions without "alteration and disguise."[28] Advance notice includes cases where the Area Director sets up a specific date or time with the employer for OSHA to begin inspection, under the limited circumstances prescribed in the instruction. It does not include general indications of potential future inspections. This instruction reflects the understanding that there may be occasions when advance notice is necessary to conduct an effective investigation within the framework of the Act, such as the following:

(1) In cases of apparent imminent danger, to enable the employer to correct the danger as quickly as possible;

(2) When the inspection can most effectively be conducted after regular business hours or when special preparations are necessary;

---

[27]CPL 2.45A III.B, OSHR [Reference File 77:2501].
[28]CPL 2.45A III.C, OSHR [Reference File 77:2503].

(3) To ensure the presence of employer and employee representatives or the appropriate personnel who are needed to aid in inspection; and,

(4) When the Area Director determines that giving advance notice would enhance the probability of an effective and thorough inspection, for example, in complex fatality investigations.

Inspections are normally made during the regular working hours, and, typically, during the daytime working hours. At the beginning of the inspection, the compliance officer locates the owner, operator, or agent in charge at the workplace, and presents his credentials. Special instructions consistent with this intention are provided for multiemployer worksites.[29] The inspection is not postponed or unreasonably delayed because of the unavailability of the person in charge; another manager may be identified for presentation of the credentials. Similarly, the inspection is not to be delayed for more than one hour to await the arrival of the employee representative.

C. Opening Conference

The purpose of the worksite visit may be to investigate an alleged imminent danger situation, to investigate a fatality or catastrophe, to conduct an investigation based upon a formal complaint, or to conduct a programmed health or safety inspection. At the opening conference, the compliance officer is instructed to inform the employer of the purpose of the inspection and to request the employer's permission to include participation of an employee representative, when it is appropriate to do so (see discussions of employer and employee rights during an inspection in Sections III.F and III.G below).[30]

In the opening conference for a health inspection, the compliance officer may request certain information, such as plant layouts, relevant to the inspection. If these are not available, a sketch of the plant layout may be made to facilitate the inspection process.

---

[29]CPL 2.45A III.D.1(c), OSHR [Reference File 77:2505].
[30]CPL 2.45A III.D.3.

In the opening conference for a programmed safety inspection in the manufacturing sector (but not in the construction sector), a records review may be performed to determine whether a comprehensive workplace inspection will be conducted. In such cases, the opening conference will be abbreviated and completed only if a comprehensive inspection is found to be necessary.[31]

The opening conference is normally not more than one hour in length. During the conference, both the employer and employee representatives are told of the opportunity to participate in the physical inspection of the workplace.[32] A joint opening conference for employer and employee representatives may be held, at the option of the employer. Where it is determined that separate conferences will unacceptably delay the observation or evaluation of the workplace conditions, each conference will be brief, and if appropriate, reconvened after the inspection of alleged hazards.[33]

During the conference, the compliance officer outlines in general terms the scope of the inspection, including private employee interviews, physical inspection of the workplace and records, possible referrals, and other matters. The compliance officer is instructed to provide employer and employee representatives with copies of applicable law, standards, regulations, and informational handouts and materials.[34]

During the opening conference, the compliance officer also ascertains whether or not the employer is covered by any of the exemptions or limitations that might apply. Any safety and health hazards that may be encountered by the walka-round party should be reviewed at this time (so that appropriate personal protective equipment may be used). Also, if the employer representative wishes to identify areas in the establishment that might reveal trade secrets, or if he has objections to the taking of photographs, he discusses this with the compliance officer during the opening conference.[35]

As already noted, a records review is conducted whenever a manufacturing establishment is scheduled for a safety inspection. The purpose of the records review is to calculate a

---

[31]CPL 2.45A III.D.3(a)(5).

[32]CPL 2.45A III.D.3.

[33]CPL 2.45A III.D.3(c).

[34]CPL 2.45A III.D.3(d) and (e).

[35]29 C.F.R. §1903.7(b) (1985); CPL 2.45A III.D.3(j), OSHR [Reference File 77:2513].

Lost Workday Injury (LWDI) case rate for the establishment.[36] If the calculated LWDI rate is below the rate for the manufacturing sector, as calculated by the Bureau of Labor Statistics, a comprehensive safety inspection is not conducted. The results of the records review, including the calculated LWDI rate, are given to the employee representative; or if there is no employee representative, the employer is asked to post a form letter for the employees' information.

For all unprogrammed inspections, and in certain special circumstances for programmed safety inspections in the manufacturing sector, a partial inspection may be conducted even when the establishment's LWDI rate is below the rate for the manufacturing sector. Such circumstances might include the presence of an imminent danger; the discovey during the records review of an unusual number, pattern, or type of injuries; or the filing of a safety or health complaint during the opening conference by an employee representative.[37]

D. Trade Secrets

Effective enforcement of the Act requires that the compliance officer and all OSHA personnel preserve the confidentiality of all information and investigations that might reveal a trade secret (see discussion of employer rights during an inspection, below). OSHA regulations specify the penalties for unauthorized disclosure of trade secret information. This fact, and the corresponding need for full information and understanding about the processes and materials being used in the workplace, are reflected in Section 15 of the Act:

> "All information reported to or otherwise obtained by the Secretary or his representatives in connection with any inspection or proceeding under this Act which contains or which might reveal a trade secret * * * shall be considered confidential * * * except that such information may be disclosed to other officers or employees concerned with carrying out this Act. In any such proceeding, the Secretary, the Commission, or the court shall issue such orders as may be appropriate to protect the confidentiality of trade secrets."

---

[36]CPL 2.45A III.D.4, OSHR [Reference File 77:2514].
[37]CPL 2.45A III.D.4(b)(2), OSHR [Reference File 77:2516].

OSHA regulations elaborate on this provision, as follows:

"(c) At the commencement of an inspection, the employer may identify areas in the establishment which contain or which might reveal a trade secret. If the Compliance Safety and Health Officer has no clear reason to question such identification, information obtained in such areas, including all negatives and prints of photographs, and environment samples shall be labeled 'confidential—trade secret' and shall not be disclosed except in accordance with the provisions of section 15 of the Act.

"(d) Upon request of an employer, any authorized representative of employees under §1903.8 in an area containing trade secrets shall be an employee in the area or an employee authorized by the employer to enter that area. Where there is no such representative or employee, the Compliance Safety and Health Officer shall consult with a reasonable number of employees who work in that area concerning matters of safety and health."[38]

E. Closing Conference

At the conclusion of an inspection, the compliance officer conducts a closing conference with the employer and the employee representative.[39] At multiemployer worksites, separate closing conferences may be held with each employer representative. As in the case of the opening conference, it is the employer's option whether to hold a *joint* conference with the employee representative. Where the employer chooses to have a separate conference, or when it is not practical to hold a joint closing conference, separate closing conferences are held.

At the conference, the compliance officer describes apparent hazards found during the inspection and indicates the sections of the OSHA standards that may have been violated. Copies of the standards are given both to the employer and to the employee representatives if such materials were not provided earlier. If the results of collected industrial hygiene or other samples are not available prior to the initial closing conference, a second conference (in person or by telephone) is held to inform the employer as to whether the establishment is in compliance.

If citations are issued, copies will be sent to the employer representative at the inspected establishment and, in the case of a nonfixed establishment, to the employer's headquarters.

---

[38] §1903.9.
[39] §1903.7(e).

If there is an authorized employee representative, copies of the citation also are sent to the representative after they are received by the employer. The citation or a copy of it must be posted at or near the place where each violation occurred to inform the employees of hazards to which they may be exposed. If, because of the nature of the employer's operation, it is not practical to post the citation at or near the places where the alleged violation occurred, the citation must be posted where it can be observed by all affected employees. The citation must remain posted for 3 working days or until the violation is corrected, whichever is longer.

The compliance officer also explains at the closing conference that if the employer receives a citation, a follow-up inspection may be conducted to verify that the citation has been posted as required, that the violations have been corrected as required, and that during the abatement period employees are adequately protected.

Other topics covered in the closing conference include failure-to-abate penalties, prohibitions against reprisals against employees, the availability of variances, consultative services, and other services such as the voluntary protection program, employer abatement assistance programs, and employee training programs.[40]

F. Rights of the Employer During an Inspection

In addition to the rights regarding the issuance of warrants discussed below, OSHA regulations provide certain rights to employers during the course of the inspection process. First, a representative of the employer may accompany the compliance officer during the inspection.[41] Second, trade secrets are protected from disclosure.[42] At the commencement of an inspection, the employer may identify areas in the establishment that might reveal a trade secret. If there is no reason to question such information, all samples and photographs and other information obtained in such areas will be labeled "confidential" and protected from disclosure.[43]

[40]CPL 2.45A III.D.9, OSHR [Reference File 77:2525].
[41]§1903.8(a).
[42]§1903.9. See also discussion at Section IV.D, *supra*.
[43]§1903.9(e).

Third, the employer is entitled to a copy of the employee/complainant's statement of position requesting an informal review of the Area Director's decision not to investigate. The employer is not entitled to a copy of the Area Director's letter to the employee/complainant, but the employer is entitled to a copy of the employee's complaint.[44] Such copies will not contain the names of the employee/complainants. The employer also is entitled to a copy of the OSHA Area Director's notification to employee/complainants when it is determined that no inspection is warranted.[45]

Fourth, an employer may petition for modification of an abatement date, if a good faith effort has been made to abate the cited hazard but such abatement has not been completed because of factors beyond the employer's control.[46]

G. Rights of Employees During an Inspection

Employee rights pertaining to the conduct of an inspection have to do generally with participation in the opening and closing conferences, and in the walkaround.[47] An employee representative may accompany the compliance officer during the physical inspection of the workplace.[48] The compliance officer may decide who is the authorized representative by asking the employer. If there is not an authorized representative, the compliance officer consults with a "reasonable number of employees concerning matters of safety and health in the workplace."[49]

The term "employee representative," for these purposes, refers to a representative of the certified or recognized bargaining agent, or if none, an employee member of a safety and health committee, or an individual employee who has been selected as the walkaround representative by employees of the establishment.[50]

---

[44]§1903.11.
[45]§1903.12.
[46]§1903.14a.
[47]§1903.8(a).
[48]§1903.8(a).
[49]§1903.8(b).
[50]CPL 2.45A III.D.2, OSHR [Reference File 77:2509].

The right of the employee representative to accompany the compliance officer is subject to agency rules designed to protect trade secrets.

Employee representatives may attend the opening and closing conferences either separately or jointly with the employer representatives, at the option of the employer.[51]

As part of the closing conference, employees are informed of their rights in connection with the citation and adjudication process. Employee representatives have the right to participate in any subsequent conferences, meetings, and discussions concerning the inspection, citation, or abatement.[52] Employees are also notified that they have a right to elect party status before the Review Commission if the employer contests citations. Employees also are notified of any contests.[53]

Finally, the employees or their representatives also have the right to contest any or all of the abatement dates set for a violation if they believe the date(s) to be unreasonable.[54]

IV. The Warrant Requirement in OSHA Inspections

Since the Supreme Court's May 1978 decision in *Marshall v. Barlow's Inc.*,[55] OSHA has been required to perform nonconsensual worksite inspections pursuant to an administrative search warrant. *Barlow's* has spawned many lower court decisions on a variety of issues relating to the issuance of such warrants.

A. Need for a Warrant and Exceptions

1. Inspection Authority

To achieve the purposes of the OSH Act, Congress invested the Secretary with authority to enter places of em-

---

[51]CPL 2.45A III.D.3(c), OSHR [Reference File 77:2511], III.D.9, OSHR [Reference File 77:2525].

[52]CPL 2.45A III.D.9(a), OSHR [Reference File 77:2525].

[53]CPL 2.45A III.D.9(b)(4), OSHR [Reference File 77:2526].

[54]CPL 2.45A III.D.9(b), OSHR [Reference File 77:2526].

[55]*Marshall v. Barlow's, Inc.*, 436 U.S. 307 (1978).

ployment and conduct safety and health inspections.[56] The statute provides that inspections must take place at reasonable times, within reasonable limits, and in a reasonable manner; they may extend to all pertinent conditions, structures, machines, apparatus, devices, equipment, and materials in the workplace; they may include the private questioning of employers, owners, operators, agents, and employees.[57] By regulation, inspections may also extend to "records required by the Act and regulations * * *, and other records which are directly related to the purpose of the inspection."[58] Entry is authorized upon presentation of appropriate credentials, at reasonable times and "without delay."[59] No search warrant or other process is expressly required under the Act.

2. *Imposition of a Warrant Requirement*

In *Marshall v. Barlow's, Inc.*,[60] the Supreme Court examined the Secretary's inspection authority. The Secretary had sought to inspect the Barlow's workplace, but was denied entry by Barlow, the company's president and general manager, who contended that an inspection without a warrant violated the Fourth Amendment.[61] The Secretary petitioned

---

[56]§8(a) of the Act; 29 U.S.C. §657(a) (1982).

[57]29 U.S.C. §657(a)(2).

[58]29 C.F.R. §1903.3(a) (1985). But cf. §1913.10(d)(1) (inspection of personally identifiable employee medical records must generally be made pursuant to a written access order approved by the Assistant Secretary of Labor for Occupational Safety and Health upon recommendation of the OSHA Medical Records Officer).

[59]29 U.S.C. §657(a)(1).

[60]*Marshall v. Barlow's, Inc.*, supra note 55.

[61]*Id.* at 309–310. The Fourth Amendment provides:
"The right of the people to be secure in their persons, houses, papers, and effects against unreasonable searches and seizures, shall not be violated, and no Warrants shall issue, but upon probable cause, supported by Oath or affirmation, and particularly describing the place to be searched, and the persons or things to be seized."
The Amendment's protections were, at one time, not applied to civil, administrative searches. See *Frank v. Maryland*, 359 U.S. 360 (1959). However, *Frank* was subsequently overruled in *Camara v. Municipal Court*, 387 U.S. 523 (1967), and *See v. City of Seattle*, 387 U.S. 541 (1967).
Prior to *Barlow's*, the lower courts took three approaches with respect to OSHA inspections. In *Brennan v. Buckeye Indus.*, 374 F. Supp. 1350 (S.D. Ga. 1974), the court held such inspections to be reasonable and, therefore, not violative of the Fourth Amendment. In *Brennan v. Gibson's Prods.*, 407 F. Supp. 154 (E.D. Tex. 1976), *vacated on other grounds*, 584 F.2d 668 (5th Cir. 1978), a three-judge court construed the Act constitutionally to require a warrant in order to inspect over an employer's objection. And, in *Barlow's Inc. v. Usery*, 424 F. Supp. 437 (D. Idaho 1976), a three-judge court held warrantless inspections pursuant to §8(a) unconstitutional and enjoined the Secretary from conducting them.

the U.S. District Court for the District of Idaho for an order requiring Barlow to permit the inspection.[62] The court issued the requested order, which the Secretary then presented to the employer.[63] Barlow again refused entry, and this time sought injunctive relief against the warrantless searches authorized by the Act.[64] A three-judge court ruled in the employer's favor, concluding that the Fourth Amendment required a warrant for the search at issue, and that Section 8(a) of the Act was unconstitutional because it purported to allow warrantless inspections.[65] Accordingly, the court enjoined inspections pursuant to that provision of the Act.[66]

On appeal, the Supreme Court affirmed the decision of the three-judge panel. In so doing, the Court relied on its decisions in *Camara v. Municipal Court*[67] and *See v. City of Seattle*.[68] In *Camara* the Court held that the Fourth Amendment prohibits warrantless nonconsensual administrative inspections because they are "unreasonable,"[69] while in *See* the Court extended this rule's application to commercial premises as well as to homes.[70] The Court did not agree[71] with the Secretary that OSHA searches fell within the exception it had created to the general rule for "pervasively regulated businesses" in *United States v. Biswell*[72] and for "closely regu-

---

[62]*Marshall v. Barlow's, Inc., supra* note 55, at 310.
[63]*Id.*
[64]*Id.*
[65]*Barlow's, Inc. v. Usery, supra* note 61, at 441–442.
[66]*Marshall v. Barlow's, Inc.*, 436 U.S. 307, 310 (1978).
[67]387 U.S. 523 (1967).
[68]387 U.S. 541 (1967).
[69]*Camara v. Municipal Court, supra* note 67, at 540.
[70]*See v. City of Seattle, supra* note 68, at 545–546.
[71]*Marshall v. Barlow's, Inc., supra* note 66, at 313. Cf. *Donovan v. Dewey*, 452 U.S. 594, 598–602 (1981). In *Dewey*, the Court held that warrantless inspections of stone quarries, like similar inspections of other mines covered by the Federal Mine Safety & Health Act of 1977, 30 U.S.C. §801 *et seq.* (1982), are constitutionally permissible. It noted the "substantial federal interest in improving the health and safety conditions in the Nation's underground and surface mines," and that the mining industry's poor health and safety record "has significant deleterious effects on interstate commerce," and concluded that Congress could reasonably have determined a system of warrantless inspections was necessary for effective enforcement. 452 U.S. at 602. It deferred to this legislative determination because "the statute's inspection program, in terms of the certainty and regularity of its application, provides a constitutionally adequate substitute for a warrant." *Id.* at 603. The Court contrasted the Mine Act with the OSH Act, which it characterized as not applying to "industrial activity with a notorious history of serious accidents and unhealthful working conditions," *id.*, and as not requiring "the periodic inspection of business covered by [the Mine Act]." *Id.* at 604 n.9.
[72]406 U.S. 311, 316 (1972).

lated" industries "long subject to close supervision and inspection" in *Colonnade Catering Corp. v. United States.*[73]

The Court did not consider the "balance between the necessities of OSHA inspections and the incremental protection of privacy of business owners"[74] to be so weighted in favor of inspections as to make warrantless searches permissible.[75] On the one hand, it viewed a warrant requirement as not imposing a serious burden on OSHA or the courts: It expected the great majority of businessmen to consent to warrantless inspections;[76] it considered obtaining a warrant no more harmful to surprise and prompt inspections than obtaining compulsory process under 29 C.F.R. §1903.4,[77] the Secretary's "Objection to Inspection" regulation;[78] and the Court doubted that "consumption of enforcement energies in the obtaining of such warrants will exceed manageable proportions."[79] On the other hand, the Court regarded a warrant as providing important protections against the "almost unbridled discretion" that "[t]he authority to make warrantless searches devolves * * * upon executive and administrative officers, particularly those in the field, as to when to search and whom to search."[80]

---

[73]397 U.S. 72, 74–77 (1970).

[74]*Marshall v. Barlow's, Inc., supra* note 66, at 316.

[75]*Id.* at 316–324.

[76]*Id.* at 316.

[77]At the time *Barlow's* was decided, §1903.4 provided:
"Upon a refusal to permit a Compliance Safety and Health Officer, in the exercise of his official duties, to enter without delay and at reasonable times any place of employment or any place therein, to inspect, to review records, or to question any employer, owner, operator, agent, or employee, in accordance with §1903.3 or to permit a representative of employees to accompany the Compliance Safety and Health Officer during the physical inspection of any workplace in accordance with §1903.8, the Compliance Safety and Health Officer shall terminate the inspection or confine the inspection to other areas, conditions, structures, machines, apparatus, devices, equipment, materials, records, or interviews concerning which no objection is raised. The Compliance Safety and Health Officer shall endeavor to ascertain the reason for such refusal, and he shall immediately report the refusal and the reason therefor to the Area Director. The Area Director shall immediately consult with the Assistant Regional Director and the Regional Solicitor, who shall promptly take appropriate action, including compulsory process, if necessary."
29 C.F.R. §1903.4 (1977). The Secretary subsequently amended the regulation. 45 FED. REG. 65,916 *et seq.* (1980); 43 FED. REG. 59,839 (1978). See *infra* notes 84–91 and accompanying text.

[78]*Marshall v. Barlow's, Inc.*, 436 U.S. 307, 316–320 (1978).

[79]*Id.* at 320–321.

[80]*Id.* at 323. The Court also rejected the Secretary's argument that requiring a warrant for OSHA inspections means that warrantless-search provisions in other regulatory statutes are also constitutionally infirm. *Id.* at 321. Such a determination "will depend upon the specific enforcement needs and privacy guarantees of each statute." *Id.*

Unlike the lower court, however, the Court did not enjoin all inspections performed pursuant to Section 8. Instead, it construed the Act constitutionally to permit nonconsensual inspections under authority of a warrant or its equivalent.[81] The Court explained that to obtain this authorization the Secretary need only satisfy the relaxed probable cause standards set forth in *Camara*:

> "Whether the Secretary proceeds to secure a warrant or other process, with or without prior notice, his entitlement to inspect will not depend on his demonstrating probable cause to believe that conditions in violation of OSHA exist on the premises. Probable cause in the criminal law sense is not required. For purposes of an administrative search such as this, probable cause justifying the issuance of a warrant may be based not only on specific evidence of an existing violation but also on a showing that 'reasonable legislative or administrative standards for conducting an * * * inspection are satisfied with respect to a particular [establishment].' *Camara v. Municipal Court*, 387 U.S. at 538. A warrant showing that a specific business has been chosen for an OSHA search on the basis of a general administrative plan for the enforcement of the Act derived from neutral sources such as, for example, dispersion of employees in various types of industries across a given area, and the desired frequency of searches in any of the lesser divisions of the area, would protect an employer's Fourth Amendment rights."[82]

Because the district court had not considered whether the order to inspect the Barlow's workplace was the functional equivalent of a warrant, and because the Secretary had declined to rely on that order, the Court did not decide whether the *Camara* probable cause standards had been satisfied.[83] It simply affirmed the judgment below.[84]

---

[81] *Id.* at 325 n.23.

[82] *Id.* at 320–321 (footnote omitted).

[83] *Id.* at 325 n.23.

[84] *Id.* at 325. Justice Stevens wrote a dissenting opinion in which Justices Blackmun and Rehnquist joined. The dissent first noted that the OSHA inspection program falls within the category of searches that are reasonable within the meaning of the first clause of the Fourth Amendment even though the probable cause requirement of the Warrant Clause cannot be satisfied. *Id.* at 326. It chided the majority's substitution of its judgment for Congress's on the necessity for warrantless inspections, noting that the rate of refusals of entry is likely to increase if covered businesses have a right to deny warrantless searches, and that the Secretary had promulgated 29 C.F.R. §1903.4 in the belief that employers could not lawfully object to inspection. *Id.* at 329–331. The dissent further noted that the "new fangled inspection warrant" required by the Court provides very little protection beyond that already provided by the Act, and that the warrant requirement was, therefore, an unnecessary burden. *Id.* at 331–334. Finally, the dissent read the Court's other administrative warrant cases as requiring deference to Congress's determination of the need for warrantless inspections. *Id.* at 334–339.

3. *Exceptions to the Warrant Requirement*

There are exceptions to the requirement that the Secretary obtain a warrant to perform an OSHA inspection. For example, a warrantless search is proper if the employer consents to the inspection[85] or if an authorized third party gives consent.[86] It is only "absent consent" that a warrant is necessary.[87] The Supreme Court has also recognized the importance of "prompt inspections, even without a warrant * * * in emergency situations," if there is a compelling need for official action and no time to secure a warrant.[88] Similarly, no warrant is necessary if the conditions seen by the Secretary's compliance officers are observable by the public,[89] or if those conditions are in "plain, obvious view" of the inspectors while they are lawfully on the employer's premises.[90]

B. **Obtaining the Warrant**

As noted above, if an employer refuses to permit an OSHA inspector to enter its worksite, *Barlow's* requires that entrance

---

[85]As the *Barlow's* Court explained, "[t]he critical fact in this case is that entry over Mr. Barlow's objection is being sought by the government agent." *Id.* at 314 (footnote omitted). The Court distinguished Mr. Barlow from "the great majority of businessmen [who] can be expected in normal course to consent to inspection without warrant." *Id.* at 316.

[86]In *Donovan v. A.A. Beiro Constr. Co.*, 746 F.2d 894, 12 OSHC 1017 (D.C. Cir. 1984), all the prime contractors (except for Berio Construction) working at a construction site owned by the District of Columbia consented to an inspection. The Court upheld the inspection as valid, concurring with the administrative law judge's determination that the inspection was made "pursuant to consent properly sought and obtained from appropriate and authorized representatives of the D.C. Government." Although Berio had not given its consent, the D.C. Circuit found that, because the other contractors on the worksite consented and the District of Columbia had access to and use of the "common areas of the *** site," Beiro could not claim a reasonable expectation of privacy. Third-party consent was thus sufficient to validate the inspection.

[87]*Donovan v. Dewey*, 452 U.S. 594, 600 (1981). Accord *Kropp Forge Co. v. Secretary of Labor*, 657 F.2d 119, 121–122 (7th Cir. 1981); *Stephenson Enters. v. Marshall*, 578 F.2d 1021, 1023, 1025 (5th Cir. 1978); *Daniel Int'l Corp.*, 9 OSHC 1980, 1985 n.13 (Rev. Comm'm 1981), *rev'd on other grounds*, 683 F.2d 361 (11th Cir. 1982).

[88]*Camara v. Municipal Court*, 387 U.S. 523, 539 (1967). Accord *Michigan v. Tyler*, 436 U.S. 499, 509–510 (1978).

[89]*Marshall v. Barlow's, Inc.*, 436 U.S. 307, 315 (1978). See *Marshall v. Western Waterproofing Co.*, 560 F.2d 947, 951 (8th Cir. 1978). Cf. *Air Pollution Variance Bd. v. Western Alfalfa Corp.*, 416 U.S. 861, 865 (1974).

[90]*Stephenson Enters. v. Marshall*, supra note 87, at 1024 n.2; *Lake Butler Apparel Co. v. Secretary of Labor*, 519 F.2d 84, 88 (5th Cir. 1975). See *Michigan v. Tyler*, supra note 88, at 509.

be obtained pursuant to a search warrant.[91] Even prior to *Barlow's*, OSHA had adopted a regulation directing OSHA officials, upon refusal of entry, "to take appropriate action, including compulsory process, if necessary."[92] This regulation authorizes OSHA personnel to obtain warrants. Challenges to the procedure used in obtaining the warrant have tended to focus on three major issues: (1) whether the warrant may be obtained *ex parte*; (2) whether the district courts have subject matter jurisdiction to issue warrants; (3) whether magistrates have authority to issue them.

1. *The* Ex Parte *Nature of the Warrant*

In October 1980, after notice and comment rulemaking, OSHA amended 29 C.F.R. §1903.4 to provide that "[e]x parte inspection warrants shall be the preferred form of compulsory process."[93] Prior to that amendment, the courts and the Commission had divided on the question of whether the original version of the regulation or the amended version promulgated as an interpretative and procedural rule in December 1978[94] authorized OSHA to seek *ex parte* warrants.[95] Those courts that denied OSHA the authority relied primarily on dictum in *Barlow's* interpreting 29 C.F.R. §1903.4: "Indeed, the kind of process sought in this case and apparently anticipated by the regulation provides notice to the business operator."[96] They viewed the Supreme Court as having decided the meaning of the regulation in a manner that precluded OSHA from amend-

---

[91]*Marshall v. Barlow's, Inc., supra* note 89, at 325.

[92]29 C.F.R. §1903.4. See 36 FED. REG. 17,851 (Sept. 4, 1971). The warrant is one form of "compulsory process." *Cerro Metal Prods. v. Marshall*, 620 F.2d 964, 983 (3d Cir. 1980) (Seitz, C.J., dissenting).

[93]29 C.F.R. §1903.4(d). See *Ingersoll Rand Co. v. Donovan*, 540 F. Supp. 222, 226 (M.D. Pa. 1982); *Erie Bottling Corp. v. Donovan*, 539 F. Supp. 600, 606 (M.D. Pa. 1982). See also 45 FED. REG. 65,916–924 (Oct. 3, 1980) (promulgation of amendments to 29 C.F.R. §1903.4). Section 1903.4 was also amended to provide that warrants may be obtained in advance of a refusal to permit entry, §1903.4(b), and that the OSHA Area Director or his designee may obtain them, §1903.4(c). Cf. *Marshall v. Milwaukee Boiler Mfg. Co.*, 626 F.2d 1339, 1346 (7th Cir. 1980) (reaching the same result based on 29 C.F.R. §1903.21(e)).

[94]43 FED. REG. 59,838–839 (Dec. 22, 1978). This amendment specifically defined "compulsory process" to include *ex parte* warrants.

[95]See *Rockford Drop Forge Co. v. Donovan*, 672 F.2d 626, 630 (7th Cir. 1982) (citing the conflicting decisions). The Commission upheld the Secretary's authority to seek *ex parte* warrants in *Davis Metal Stamping Co.*, 10 OSHC 1741 (Rev. Comm'n No. 78-5775, 1982).

[96]*Marshall v. Barlow's Inc., supra* note 89, at 318.

ing it without resort to notice and comment rulemaking procedures.[97] In any event, the October 1980 amendment resolved the issue concerning OSHA's authority to seek *ex parte* warrants after that date.[98]

A related issue which courts have addressed is whether, despite the fact that OSHA is empowered to seek warrants *ex parte*, a magistrate nonetheless may give an employer notice of filing of a warrant application and an opportunity to be heard in opposition.[99] The courts generally favor denying the employer notice and a hearing, primarily relying on Section 8(a) of the Act, which provides that OSHA inspectors shall "enter [a workplace] without delay"[100] and Section 17(f) which creates criminal liability for persons who give an employer advance notice that an inspection will take place.[101] The courts have recognized that Congress designed both provisions to ensure surprise and prompt inspection.[102]

In any case, an employer has several methods at its disposal for challenging a warrant even if it cannot be present

---

[97]*Cerro Metal Prods. v. Marshall, supra* note 92, at 976–983; *Marshall v. Huffines Steel Co.*, 488 F. Supp. 995, 997–1001 (N.D. Tex. 1979), *aff'd without opinion*, 645 F.2d 288 (5th Cir. 1981). The *Cerro* court relied primarily on its perception that the Secretary's litigation strategy and representations in *Barlow's* had produced the adverse dictum because it had been designed to bolster the constitutional argument. *Cerro Metal Prods. v. Marshall, supra* note 92, at 976–978. The *Huffines* court simply believed that the Supreme Court had decided the issue and that the Secretary was powerless to seek *ex parte* warrants under the regulation without a notice and comment amendment. *Marshall v. Huffines Steel Co.*, 488 F. Supp. at 999–1001.

[98]See *West Point-Pepperell, Inc. v. Donovan*, 689 F.2d 950, 955 n.6 (5th Cir. 1982); *Rockford Drop Forge Co. v. Donovan, supra* note 95, at 629.

[99]Compare *Rockford Drop Forge Co. v. Donovan, supra* note 95, at 630–631 (employer may not receive notice and appear); *Marshall v. Seaward Int'l*, 510 F. Supp. 314, 316–317 (W.D. Va. 1980) (same), *aff'd mem.*, 644 F.2d 880 (4th Cir. 1981); *Donovan v. Red Star Marine Servs.*, 739 F.2d 774, 783 (2d Cir. 1984); and *In re Worksite Inspection of S.D. Warren, Div. of Scott Paper*, 481 F. Supp. 491, 494–495 (D. Me. 1979) (same); with *In re Establishment Inspection of C.F. & I. Steel Corp.*, 489 F. Supp. 1302, 1305–1308 (D. Colo. 1980) (employer may be entitled to notice); and *Donovan v. Red Star Marine Servs.*, 739 F.2d 774, 783 (2d Cir. 1984).

[100]29 U.S.C. §657(a).

[101]§666(f).

[102]See *Marshall v. Barlow's Inc.*, 436 U.S. 307, 317 (1978); *Rockford Drop Forge Co. v. Donovan, supra* note 95, at 630–631; *Marshall v. Shellcast Corp.*, 592 F.2d 1369, 1371–1372 (5th Cir. 1979); *Marshall v. Gibson's Prods.*, 584 F.2d 668, 672–673 (5th Cir. 1978); *In re Worksite Inspection of S.D. Warren, Div. of Scott Paper, supra* note 99, at 495; Subcommittee on Labor of the Senate Comm. on Labor and Public Welfare, 92d Cong., 1st Sess., LEGISLATIVE HISTORY OF THE OCCUPATIONAL SAFETY AND HEALTH ACT OF 1970, 1076 (Comm. Print 1971) (hereinafter cited as LEGIS. HIST.) (remarks of Reps. Galifianakis and Steiger; words "without delay" inserted in Section 8(a) to prevent certain employer stalling tactics which could delay inspections); LEGIS. HIST. at 856–857 (remarks of Rep. Daniels; major flaw in previous federal occupational health and safety legislation was lack of advance notice prohibition).

at a hearing on the warrant's issuance. These include moving to quash the warrant, refusing to comply with the warrant and litigating its validity in civil contempt proceedings, and moving for suppression of evidence after the warrant's execution.[103] For these reasons as well, courts have held that OSHA can obtain *ex parte* warrants notwithstanding an employer's request for notice and a hearing on the warrant application.[104]

2. Subject Matter Jurisdiction of the District Courts to Issue Warrants

Section 8(a)[105] authorizes the Secretary to enter workplaces to perform inspections. However, it does not, by its terms, empower him to seek a warrant in district court if an employer refuses entry. The courts that have directly addressed the absence of language in the OSH Act permitting the Secretary to seek such a warrant have nevertheless concluded that subject matter jurisdiction exists for issuance of a warrant.[106]

Two general grants of jurisdiction furnish a basis for a court to grant a warrant application. 28 U.S.C. §1337 grants jurisdiction over "any civil action or proceeding arising under any Act of Congress regulating commerce." 28 U.S.C. §1345 grants district courts jurisdiction over "civil actions, suits or proceedings commended by the United States, or by any agency or officer thereof expressly authorized to sue by Act of Con-

---

[103]*Rockford Drop Forge Co. v. Donovan, supra* note 95, at 631; *In re Worksite Inspection of S.D. Warren, Div. of Scott Paper, supra* note 99, at 495.

[104]*Rockford Drop Forge Co. v. Donovan, supra* note 95, at 630–631; *In re Worksite Inspection of S.D. Warren, Div. of Scott Paper, supra* note 99, at 494–495. See *Marshall v. Seaward Int'l, supra* note 99, at 316. But can a magistrate or judge, on his motion, order that the employer be given an opportunity to contest the warrant application? One judge has suggested that such opportunity may be appropriate in cases in which OSHA asks for sweeping inspection authority to inspect conditions which are not alleged to create an emergency. See *In re Establishment Inspection of C.F. & I. Steel Corp., supra* note 99, at 1306–1308.

[105]29 U.S.C. §657(a).

[106]*In re Establishment Inspection of Gilbert & Bennett Mfg. Co.*, 589 F.2d 1335, 1344 (7th Cir.), *cert. denied*, 444 U.S. 884 (1979); *Marshall v. Seaward Int'l, supra* note 99, at 316; *Marshall v. Huffines Steel Co.*, 478 F. Supp. 986, 988–989 (N.D. Tex.) (denying motion to dismiss), *complaint dismissed on other grounds*, 488 F. Supp. 995 (N.D. Tex. 1979), *aff'd without opinion*, 645 F.2d 288 (5th Cir. 1981); *United States v. Cleveland Elec. Illuminating Co.*, No. M80-2128, slip op. at 3 (N.D. Ohio Feb. 27, 1981), *aff'd in pertinent part and rev'd in part*, 689 F.2d 66, 68 n.2 (6th Cir. 1982) (alternative holding). See *Marshall v. Shellcast Corp., supra* note 102, at 1370–1371 n.3 (dictum suggesting federal courts have jurisdiction to issue warrants).

gress." In addition, 29 U.S.C. §663 authorizes the Solicitor of Labor under the Attorney General's supervision to appear in court on behalf of the Secretary "in any civil litigation" brought under the Act.[107] Some courts have also found the *Barlow's* decision to contain an implicit holding that jurisdiction exists.[108]

3. *Authority of Magistrates to Issue Warrants*

Once the district court has jurisdiction to issue a warrant, a judge may grant a warrant application.[109] A number of cases have held that magistrates are also empowered to do so.[110] In many cases, local court rules specifically authorize magistrates to issue warrants.[111] In addition, federal magistrates are authorized to exercise "all powers and duties conferred or imposed upon United States Commissioners [the magistrates' predecessors] by law"[112] and "such additional duties as are not inconsistent with the Constitution and laws of the United States."[113] The only limitation of this broad authority is found in 28 U.S.C. §636(b)(1), which prohibits a judge from desig-

---

[107]But see *Marshall v. Gibson's Prods.*, *supra* note 102, at 676–677, which holds that this provision does not grant jurisdiction for OSHA suits which are not based on specific authority to sue contained in the Act itself.

[108]*In re Establishment Inspection of Gilbert & Bennett Mfg. Co.*, *supra* note 106, at 1344; *id.* at 679–681 (Tuttle, J., dissenting). The Fifth Circuit has held that the district courts lack jurisdiction to issue any type of process which is not *ex parte*. *Id.* at 672–678. See *Marshall v. Shellcast Corp.*, *supra* note 102, at 1372. In reaching this result, the court held that since §8(a) of the Act, 29 U.S.C. §657(a), did not authorize such suits and since Congress desired surprise and undelayed inspections, the Act not only failed expressly to authorize such jurisdiction but implicitly precluded the district courts from assuming jurisdiction under the more general grants contained in 28 U.S.C. §§1337 and 1345. However, the court has suggested in dictum that it would reach a different result if called upon to determine whether courts could issue *ex parte* warrants. See *Marshall v. Shellcast Corp.*, *supra* note 102, at 1370–1371 n.3, 1372; *Marshall v. Gibson Prods.*, *supra* note 102, at 673 n.6. The district court in *Marshall v. Huffines Steel Co.* has extensively discussed its view that the *Gibson's Products* case does not preclude the conclusion that district courts may issue *ex parte* warrants, *supra* note 106, at 988–989.

[109]See *Marshall v. Horn Seed Co.*, 647 F.2d 96, 98 (10th Cir. 1981); *Babcock & Wilcox Co. v. Marshall*, 610 F.2d 1128, 1135 n.17 (3rd Cir. 1979).

[110]E.g., *In re Worksite Inspection of Quality Prods.*, 592 F.2d 611, 613 n.2 (1st Cir. 1979); *In re Establishment Inspection of Gilbert & Bennett Mfg. Co.*, *supra* note 106, at 1340–1341; *Pelton Casteel, Inc. v. Marshall*, 588 F.2d 1182, 1186 (7th Cir. 1978); *Marshall v. Seaward Int'l*, 510 F. Supp. 314, 316 (W.D. Va. 1980) (same), *aff'd mem.*, 644 F.2d 890 (4th Cir. 1981); *United States v. Cleveland Elec. Illuminating Co.*, *supra* note 106, at 3–4.

[111]See, e.g., *Marshall v. Seaward Int'l*, *supra* note 110, at 316.

[112]28 U.S.C. §636(a)(1).

[113]§636(a)(3).

nating a magistrate to determine certain matters not relevant to the issue at hand.[114] Since Section 636(b)(1) does not prohibit a magistrate from issuing a warrant, the courts have upheld[115] the magistrate's authority in the absence of an express designation as either a power conferred previously on U.S. Commissioners[116] or as a duty consistent with the Constitution and laws of the United States.[117]

C. Probable Cause Showing Required by *Barlow's*

The *Barlow's* decision imposed upon the Secretary a bifurcated probable cause standard. The Secretary may base his warrant application "not only on specific evidence of an existing violation but also on a showing that 'reasonable legislative or administrative standards for conducting an * * * inspection are satisfied with respect to a particular [establishment].' "[118] The issue of the precise showing that satisfies these criteria has spawned much litigation. In the "specific evidence" cases, employers have attacked evidence proffered by the Secretary based on its source, specificity, lack of corroboration, and staleness. In the case of legislative and administrative standards warrants, employers have challenged the nature and neutrality of the standards and the method used to choose a particular company for inspection pursuant to the standards. Neither the *Barlow's* decision nor *Michigan v. Tyler*,[119] a decision issued almost contemporaneously and also addressing Fourth Amendment doctrines, provides much concrete guidance concerning these questions. Accordingly, the lower courts have fashioned their own, sometimes conflicting, answers to these challenges.

---

[114]Section 636(b)(1) primarily precludes a judge from authorizing a magistrate to decide motions that request relief in the form of a final judgment in a case (e.g., dismissal, summary judgment).

[115]See note 110, *supra.*

[116]§636(a)(1).

[117]§636(b)(3).

[118]*Marshall v. Barlow's, Inc.*, 436 U.S. 307, 320 (1978) (quoting *Camara v. Municipal Court*, 387 U.S. 523, 538 (1967)).

[119]436 U.S. 499 (1978).

1. Specific Evidence for Probable Cause

Courts have agreed that specific evidence for probable cause may come from a variety of sources. OSHA most frequently receives such evidence in the form of an employee complaint of unsafe conditions at a particular worksite. The courts and the Review Commission have held that the Secretary may base his probable cause showing on an employee complaint.[120] In *Marshall v. Barlow's, Inc.*, the Supreme Court itself suggested that a written, signed complaint of unsafe conditions received from an employee under Section 8(f)(1) of the Act could satisfy the requirement of specific evidence.[121] However, the courts have rejected the argument that OSHA may inspect only on the basis of a written complaint meeting the criteria established by Section 8(f)(1) and have adopted the agency's view that oral complaints may be proffered in support of a warrant application.[122]

Courts also regard sources of information other than employee complaints as providing the specific evidence required by *Barlow's*. For example, the need for a follow-up inspection to determine whether a hazardous condition has been cor-

---

[120]E.g., *West Point-Pepperell, Inc. v. Donovan*, 689 F.2d 950 (5th Cir. 1982); *Marshall v. Horn Seed Co.*, 647 F.2d 96 (10th Cir. 1981); *Burkart Randall Div. of Textron v. Marshall*, 625 F.2d 1313 (7th Cir. 1980); *Weyerhaeuser Co. v. Marshall*, 592 F.2d 373 (7th Cir. 1979); *B.P. Oil, Inc. v. Marshall*, 509 F. Supp. 802 (E.D. Pa.), *aff'd mem.*, 673 F.2d 1298 (3d Cir. 1981); *Sarasota Concrete Co.*, 9 OSHC 1608 (Rev. Comm'n 1981), *aff'd*, 693 F.2d 1061 (11th Cir. 1982).

[121]*Marshall's v. Barlow's, Inc.*, *supra* note 118, at 320 n.16. Section 8(f)(1), 29 U.S.C. §657(f)(1), requires the Secretary to perform a "special inspection" in response to a written, signed complaint from an employee or employee representative if that complaint provides reasonable grounds to believe that an imminent danger or serious hazard exists in a workplace.

[122]*Marshall v. Horn Seed Co.*, *supra* note 120, at 100 n.3; *Burkart Randall Div. of Textron v. Marshall*, *supra* note 120, at 1321–1322; *Donovan v. Metal Bank of Am.*, 516 F. Supp. 674, 679 (E.D. Pa.), *purge order entered*, 521 F. Supp. 1024 (E.D. Pa. 1981), *vacated in part on other grounds, appeal dismissed in part on other grounds*, 700 F.2d 910 (3d Cir. 1981); *Sarasota Concrete Co.*, 9 OSHC 1608, 1616 (Rev. Comm'n 1981). See *Rockford Drop Forge Co. v. Donovan*, 672 F.2d 626, 631 (7th Cir. 1982). The courts have diverged on the related issue of whether written complaints submitted by striking employees are "employee" complaints for purposes of §8(f)(1). The Seventh Circuit held that they are, *Rockford Drop Forge Co. v. Donovan*, 672 F.2d at 632–633 and n.4, while one district court has implied that such complaints may not furnish the basis for an inspection, *Amoco Oil Co. v. Marshall*, 496 F. Supp. 1234, 1241 (S.D. Tex. 1980), *vacated as moot*, No. 80-2235 (5th Cir. Mar. 27, 1981). One district court has stated in dictum that complaints from employees on strike do not fall within §8(f)(1), *B.P. Oil, Inc. v. Marshall*, *supra* note 120, at 808. Under the National Labor Relations Act, 29 U.S.C. §§151 *et seq.*, both unfair labor practice and economic strikers are considered employees. See *Mastro Plastics Corp. v. NLRB*, 350 U.S. 270, 284–287 (1956) (unfair labor practice strikers); *Laidlaw Corp. v. NLRB*, 414 F.2d 99, 105 (7th Cir. 1969) (economic strikers), *cert. denied*, 397 U.S. 920 (1970).

rected will furnish probable cause.[123] A referral[124] or a newspaper article[125] may form the basis of a probable cause showing. OSHA may also derive evidence of possible violations from the observations of hazardous conditions by one of its compliance officers,[126] admissions concerning use of toxic substances made by company officers,[127] and accident reports.[128] In general, courts have accepted information of hazardous conditions from almost any source, so long as certain conditions, discussed *infra*, are met when that evidence is presented to a magistrate in a warrant application.

Much of the litigation in the specific evidence area has focused on the nature and quality of the information presented to the magistrate in terms of specificity; the demonstrated likelihood of an OSHA violation; corroboration by other sources; and how "stale" it is. The somewhat divergent views of the Seventh[129] and Tenth[130] Circuits dominate these areas analytically; and their views, in turn, appear to depend on the applicability of *Michigan v. Tyler*[131] to OSHA "specific evidence" warrant applications.[132] Both circuits agree, based on the broadest holding of *Tyler*, that administrative probable cause standards apply to warrant applications based on specific evidence.[133] However, they differ with respect to what

---

[123]*Pelton Casteel, Inc. v. Marshall*, 588 F.2d 1182, 1188 (7th Cir. 1978); *Donovan v. Metal Bank of Am.*, *supra* note 122, at 678.

[124]See *Donovan v. Metal Bank of Am.*, *supra* note 122, at 678–679.

[125]See *Donovan v. Federal Clearing Die Casting Co.*, 655 F.2d 793 (7th Cir. 1981).

[126]*Plum Creek Lumber Co. v. Hutton*, 452 F. Supp. 575 (D. Mont. 1978), *aff'd*, 608 F.2d 1283, 1287 (9th Cir. 1979).

[127]See *Marshall v. Seaward Int'l*, 510 F. Supp. 314, 317 (W.D. Va. 1980) (same), *aff'd mem.*, 644 F.2d 880 (4th Cir. 1981); *Quality Prods.*, 6 OSHC 1663, 1665 (D.R.I. 1978), *vacated on other grounds*, 592 F.2d 611 (1st Cir. 1979).

[128]*Plum Creek Lumber Co. v. Hutton*, *supra* note 126, at 1287.

[129]*Burkart Randall Div. of Textron v. Marshall*, *supra* note 120.

[130]*Marshall v. Horn Seed Co.*, *supra* note 120.

[131]*Michigan v. Tyler*, 436 U.S. 499 (1980). *Tyler* involved a search for the cause of a fire and presented, in that context, a "specific evidence" basis for probable cause. *Id.* at 507–508.

[132]Compare *Burkart Randall Div. of Textron v. Marshall*, 625 F.2d 1313, 1318 n.5 (7th Cir. 1980) (*Tyler* limited to fire searches) with *Marshall v. Horn Seed Co.*, 647 F.2d 96, 101–103 (10th Cir. 1981) (relying heavily on *Tyler* with regard to information OSHA must present to magistrate and standard for reviewing warrant application).

[133]*Marshall v. Horn Seed Co.*, *supra* note 132, at 99 n.2; *Burkart Randall Div. of Textron v. Marshall*, *supra* note 132, at 1317–1319 n.4. Employers contended in those cases that the language in *Barlow's* was to some extent ambiguous on this point. The broad holding of *Michigan v. Tyler* makes clear that administrative probable cause standards apply to any warrant application until the government investigators have probable cause to believe a crime has been committed and require

mandate, if any, the narrow holding of *Tyler* imposes in the OSHA specific evidence context.[134]

(a) Specificity of the information. In terms of the detail and specificity required of the information in the application, all courts require at a minimum that the nature of the hazardous condition be disclosed.[135] This information permits the magistrate to determine whether the conditions alleged to exist constitute a violation of the Act.[136] Apparently, even a conclusory description of the conditions will likely pass muster.[137]

---

further evidence. *Supra* note 131, 436 U.S. at 511–512. At that point, criminal probable cause standards become controlling. 436 U.S. at 511–512. Other courts have reached the same conclusion as the Tenth and Seventh Circuits based on those cases or on the *Barlow's* language. E.g., *West Point-Pepperell, Inc. v. Donovan*, 689 F.2d 950, 957 (5th Cir. 1982); *Plum Creek Lumber Co. v. Hutton*, *supra* note 126, at 1287; *Marshall v. W & W Steel Co.*, 604 F.2d 1322, 1326 (10th Cir. 1979); *Donovan v. Metal Bank of Am.*, 516 F. Supp. 674 (E.D. Pa.), *purge order entered*, 521 F. Supp. 1024 (E.D. Pa. 1981), *vacated in part on other grounds, appeal dismissed in part on other grounds*, 700 F.2d 910 (3d Cir. 1981). See *Blackie's House of Beef v. Castillo*, 659 F.2d 1211, 1224–1225 (D.C. Cir. 1981), *cert. denied*, 455 U.S. 940 (1982). Cf. *United States v. Consolidation Coal Co.*, 560 F.2d 214, 218 (6th Cir. 1977), *reinstated after remand*, 579 F.2d 1011, 1012 (6th Cir. 1978), *cert. denied*, 439 U.S. 1069 (1979).

[134]The Tenth Circuit believes the *Tyler* holding renders "specific evidence" cases under *Barlow's* more analogous to criminal searches than routine administrative ones. In *Tyler*, the state argued that no warrant should be needed to investigate the cause of a fire because the magistrate could "do little more than rubberstamp a [warrant] application to search fire-damaged premises for the cause of the blaze," given the "simple" and "obvious" justification for the search. *Supra* note 131, at 507. The Supreme Court rejected this argument. Distinguishing fire searches from routine building inspections, it stated:

"In the context of investigatory fire searches, which are not programmatic but are responsive to individual events, a more particularized inquiry may be necessary. The number of prior entries, the scope of the search, the time of day when it is proposed to be made, the lapse of time since the fire, the continued use of the building, and the owner's efforts to secure it against intruders might all be relevant factors [to the magistrate's probable cause determination]."

Id. at 507. This holding probably responds to a particular aspect of fire searches of which the *Tyler* case furnishes a classic example, i.e., the potential in the fire search context of repeated, intrusive searches based on the same "specific evidence" and not limited by any legislative or administrative standards. This problem is one that occurs only rarely under the OSH Act. See *Marshall v. Seaward Int'l*, *supra* note 127, at 319 (Part H).

[135]*Marshall v. Horn Seed Co.*, *supra* note 132, at 103; *Burkart Randall Div. of Textron v. Marshall*, *supra* note 132, at 1319. Accord *Weyerhaeuser Co. v. Marshall*, 592 F.2d 373, 378 (7th Cir. 1979); *Donovan v. Metal Bank of Am.*, *supra* note 133, at 678.

[136]*Marshall v. Horn Seed Co.*, *supra* note 132, at 103; *Burkart Randall Div. of Textron v. Marshall*, *supra* note 132, at 1319.

[137]The operative language in the *Horn Seed* (*supra* note 132, at 98) and *Burkart Randall* (*supra* note 132, at 1315 n.1), warrant applications is set forth in those decisions. Neither court appears to have been overtly troubled by the description of workplace hazards contained in those applications. 625 F.2d at 1319, 1320–1321; see 647 F.2d at 103, 104. A fuller, more detailed statement of the hazardous conditions is contained in the warrant application upheld in *In re Establishment Inspection of Gilbert & Bennett Mfg. Co.*, 589 F.2d 1335, 1339 (7th Cir.), *cert. denied*, 444 U.S. 884 (1979).

(b) Likelihood of a violation. The Tenth Circuit's statement of the required showing of likelihood that a violation will be found comports with the concept that administrative probable cause standards are less stringent than criminal standards. Thus, the application need not show that it is more probable than not that a violation of the Act will be found:[138] "There must [only] be some plausible basis for believing that a violation is *likely* to be found. The facts offered must be sufficient to warrant further investigation or testing."[139] In addition to a description of the alleged hazards at the worksite which shows some likelihood of a violation, the Seventh Circuit's *Burkart Randall* decision calls only for a sworn recitation that OSHA received the specific evidence from a particular source, and that OSHA believed that the information furnished reasonable grounds for believing that violations of the Act existed.[140] The Tenth Circuit would require not only that information but other facts including, in appropriate cases, those the Supreme Court identified as relevant in the *Michigan v. Tyler* holding.[141]

(c) Corroboration by other sources. Another area where the Seventh and Tenth Circuits appear to diverge is the extent to which they require corroboration of information proffered in support of the warrant application.[142] This divergence prob-

---

[138]*Marshall v. Horn Seed Co., supra* note 132, at 102.

[139]*Id.* (emphasis in original). The *West Point-Pepperell* decision employs a similar formulation, *supra* note 133, at 958. The Fifth Circuit has articulated the standard as follows: "The agency need show only that the facts are susceptible of being true and supported." *Service Foundry Co. v. Donovan,* 721 F.2d 492, 498 (5th Cir. 1982). The *Horn Seed* decision correctly recognizes that in the OSHA context and particularly with respect to toxic substances, both the government and its informants face greater difficulty than criminal investigators in evaluating the likelihood that the facts in their possession indicate the existence of a violation. *Marshall v. Horn Seed Co., supra* note 132, at 102. Consistent with this view, two district courts have suggested that the mere use of toxic substances at a worksite supplies probable cause for an inspection for violations of the Act. See *Marshall v. Seaward Int'l,* 510 F. Supp. 314, 317 (W.D. Va. 1980) (same), *aff'd mem.,* 644 F.2d 880 (4th Cir. 1981); *Quality Prods.,* 6 OSHC 1663, 1663 n.4 (D.R.I 1978).

[140]*Burkart Randall Div. of Textron v. Marshall, supra* note 120, at 1319.

[141]*Marshall v. Horn Seed Co., supra* note 132, at 103. See note 134 *supra.* A problem the Tenth Circuit's opinion presents is the large amount of dictum it contains. Courts may have difficulty judging in an individual case which particular pieces of information the warrant application must furnish.

[142]Most courts now accept the position that a written employee complaint need not be corroborated. *In re Establishment Inspection of Gilbert & Bennett Mfg. Co., supra* note 137, at 1339; *B.P. Oil, Inc. v. Marshall,* 509 F. Supp. 802, 806–807 (E.D. Pa.); *Marshall v. Horn Seed Co., supra* note 132, at 103 ("[a] signed written employee complaint containing detailed information demonstrating first hand knowledge may be so compelling that further verification is unnecessary"). See also *United States v.*

ably derives to some extent from the silence of *Michigan v. Tyler* in this area. The Supreme Court's decision in *Illinois v. Gates*,[143] overruling previous criminal probable cause cases requiring corroboration of informant's tips and adopting a "totality of the circumstances" test of probable cause, may require courts to reconsider the issue in the administrative probable cause context. An absolute requirement of corroboration in any particular case likely would be precluded under *Illinois v. Gates*. A sliding scale, such as that adopted in *Horn Seed*, making the need for corroboration depend upon the extent of apparent firsthand knowledge of the affiant or informant (where a signed, written employee complaint detailing firsthand knowledge would require little or no corroboration but a simple allegation by a competitor or unknown caller would require more) might pass muster under *Gates*.[144]

(d) Staleness. A final challenge that employers frequently raise is whether probable cause has dissipated because the evidence of a violation is out of date or "stale." The concept of staleness is applied strictly in cases involving criminal search warrants.[145] Consistent with the notion that criminal probable cause standards are inapplicable, the courts have resorted to a more lenient test in the OSHA context.[146] If "the conditions alleged * * * are not of a type that will likely disappear through mere passage of time [and without affirmative corrective measures]," the magistrate may conclude that they still exist at the worksite.[147] Under this test courts have up-

---

Consolidation Coal Co., supra note 133, at 222 (complaint of ex-employee need not be corroborated). The Seventh Circuit has extended that holding to oral complaints. *Burkart Randall Div. of Textron v. Marshall*, 625 F.2d 1313, 1320 (7th Cir. 1980). On the other hand, the Seventh Circuit has held that an uncorroborated newspaper article does not furnish probable cause. *Donovan v. Federal Clearing Die Casting Co.*, 655 F.2d 793, 797 (7th Cir. 1981). The Tenth Circuit has suggested that some form of corroboration may be necessary for information provided by a business competitor or unknown caller. *Marshall v. Horn Seed Co., supra* note 132, at 103. Cf. *Donovan v. Metal Bank of Am., supra* note 133, at 679 (magistrate entitled to consider earlier referrals and citations as corroboration of more recent anonymous complaints from employee and doctor.)

[143]462 U.S. 213 (1983).

[144]Compare *Marshall v. Horn Seed Co.*, 647 F.2d 96, 103–104 (10th Cir. 1981), with *Illinois v. Gates, supra* note 143, at 230–239.

[145]E.g., *Sero v. United States*, 287 U.S. 206, 211–212 (1932); *Schoeneman v. United States*, 317 F.2d 173, 177 (D.C. Cir. 1963).

[146]*Hern Iron Works v. Donovan*, 670 F.2d 838, 840 (9th Cir.) cert. denied, 459 U.S. 830 (1982). See *B.P. Oil, Inc. v. Marshall, supra* note 142, at 806.

[147]*Burkart Randall Div. of Textron v. Marshall*, 625 F.2d 1313, 1322 (7th Cir. 1980).

held warrant applications based on information furnished as much as 15 months previously.[148]

2. Reasonable Legislative or Administrative Standard for Probable Cause

In *Barlow's*, the Supreme Court explained that an inspection based on "reasonable legislative or administrative standards" could be supported by a "showing that a specific business has been chosen for an OSHA search on the basis of a general administrative plan for the enforcement of the Act derived from neutral sources.[149] The Court suggested that a warrant application must (1) describe the plan and (2) present facts indicating why the inspection of a particular worksite fits within the plan.[150] Prior to *Barlow's*, and in response to the decision, OSHA developed administrative programs for choosing worksites. Typically, they are based on illness-injury data for various industries compiled by the government and have as their priority objective the selection of high-hazard workplaces for inspection. For the most part, warrants based on these selection mechanisms have been upheld in the absence of some unexplained deviation from the plan or inadequate explanation of the plan's basis or operation.[151] The Tenth and First Circuits have pointed out that the magistrate reviewing such a warrant has a more limited role than in a specific evidence case: he merely verifies that the proposed search conforms to the proffered plan.[152]

---

[148]*Hern Iron Works v. Donovan, supra* note 146, at 840 (15 months); *Burkart Randall Div. of Textron v. Marshall, supra* note 147, at 1322 (6 months); *In re Inspection of Central Mine Equip. Co.*, 7 OSHC 1185, 1189 (E.D. Mo.) (8 months), *vacated on other grounds*, 608 F.2d 719 (8th Cir. 1979). See *B.P. Oil, Inc. v. Marshall, supra* note 142, at 806. In the context of a warrant based on an administrative plan as opposed to specific evidence, an even longer delay has been upheld. See *Federal Castings Div., Chromalloy American Corp. v. Donovan*, 684 F.2d 504, 511 (7th Cir. 1982).

[149]*Marshall v. Barlow's, Inc.*, 436 U.S. 307, 320–321 (1978) (quoting *Camara v. Municipal Court*, 387 U.S. 523, 538 (1967)).

[150]*Barlow's, supra* note 149, at 323 n.20.

[151]*Brock v. Gretna Mach. & Ironworks*, 769 F.2d 1110, 1112–1113 (5th Cir. 1985) (failure to describe manner in which company was selected for inspection); *In re Northwest Airlines*, 587 F.2d 12, 13 (7th Cir. 1978) (failure to explain how company fit within program); *In re Establishment Inspection of Urick Foundry*, 472 F. Supp. 1193, 1195 (W.D. Pa. 1979) (failure to justify choice of one foundry for inspection within 17-county area); *Marshall v. Weyerhaeuser Co.*, 456 F. Supp. 474, 482–484 (D.N.J. 1978) (failure to follow "worst-first" program).

[152]*Marshall v. Horn Seed Co.*, 647 F.2d 96, 100–101 (10th Cir. 1981). Accord *Donovan v. Wollaston Alloys, Inc.*, 695 F.2d 1, 5–6 (1st Cir. 1982). Cf. *Michigan v.*

The plans OSHA has used rely on government-collected statistics which produce a ranking of industries based on the number of employees who suffer work-related illnesses and injuries.[153] Courts have approved the use of such statistics to derive a neutral administrative scheme[154] and have encouraged magistrates to rely on the Secretary's expertise in constructing the ranking that classifies the workplace.[155] These plans meet the *Barlow's* criteria as they focus on high-hazard industries,[156] whether they inspect those industries on some "worst-first" basis[157] or in another manner.[158]

In reviewing a warrant application based on a plan, the magistrate is entitled to information concerning the nature of the plan and the manner in which the plan operated to select the particular employer.[159] Perhaps the most perfunctory showing that any court has upheld as sufficient is the following language from the *Barlow's* application in the *Gilbert & Bennett* case: "[Entry was sought] pursuant to 'a Na-

---

Tyler, 436 U.S. 499, 507 (1978). The magistrate does not replicate the structuring of the plan or operation of its selection procedure with the data the Secretary used in reviewing the warrant application. *Donovan v. Wollaston Alloys, Inc., supra*, 695 F.2d at 5–6. See *In re Establishment Inspection of Gilbert & Bennett Mfg. Co.*, 589 F.2d 1335, 1342 (7th Cir.) *cert. denied*, 444 U.S. 884 (1979); *Erie Bottling Corp. v. Donovan*, 539 F. Supp. 600, 605–606 (M.D. Pa. 1982).

[153]See, e.g., *Donovan v. Wollaston Alloys, Inc., supra* note 152, at 2–3 n.15; *Stoddard Lumber Co. v. Marshall*, 627 F.2d 984, 988 (9th Cir. 1980); *Marshall v. Milwaukee Boiler Mfg. Co.*, 626 F.2d 1339, 1341 (7th Cir. 1980); *In re Establishment Inspection of Gilbert & Bennett Mfg. Co., supra* note 152, at 1343; *Reynolds Metals Co. v. Secretary of Labor*, 442 F. Supp. 195, 197 (W.D. Va. 1977).

[154]*Ingersoll-Rand Co. v. Donovan*, 540 F. Supp. 222, 224 (M.D. Pa. 1982); *Erie Bottling Corp. v. Donovan, supra* note 152, at 605. See also *Donovan v. Wollaston Alloys, Inc., supra* note 153, at 5; also *Marshall v. Milwaukee Boiler Mfg. Co., supra* note 153, at 1346; *Reynolds Metals Co. v. Secretary of Labor, supra* note 153, at 200–201. Cf. *In re Establishment Inspection of Gilbert & Bennett Mfg. Co., supra* note 153, at 1343 (magistrate entitled to rely on Secretary's ability to use statistics to "form a reasoned opinion" of what constitutes high incidence of injuries).

[155]*In re Establishment Inspection of Gilbert & Bennett Mfg. Co., supra* note 152, at 1343; *Erie Bottling Corp. v. Donovan, supra* note 152, at 605–606. But cf. *Marshall v. Milwaukee Boiler Mfg. Co., supra* note 154, at 1346 (inadequate explanation of source of statistics for Milwaukee).

[156]See *Marshall v. Milwaukee Boiler Mfg. Co., supra* note 153, at 1342–1343; *In re Establishment Inspection of Gilbert & Bennett Mfg. Co., supra* note 152, at 1343; *Erie Bottling Corp. v. Donovan, supra* note 152, at 605; *Reynolds Metals Co. v. Secretary of Labor, supra* note 153, at 200–201. This is not to say that some other program with a different focus would not withstand scrutiny. See *Reynolds Metals Co. v. Secretary of Labor, supra* note 153, at 200.

[157]See *Marshall v. Milwaukee Boiler Mfg. Co., supra* note 153, at 1346; *Reynolds Metals Co. v. Secretary of Labor, supra* note 153, at 200–201.

[158]*Ingersoll-Rand Co. v. Donovan, supra* note 154, at 225; *Erie Bottling Corp. v. Donovan, supra* note 154, at 605; *In re Peterson Builders, Inc.*, 525 F. Supp. 642 (E.D. Wis. 1981).

[159]E.g., *Marshall v. Barlow's, Inc.*, 436 U.S 307, 323 n.20 (1978).

tional-Local plan' designed to achieve significant reduction in the high incidence of occupational injuries and illnesses found in the metal-working and foundry industry, * * *."[160]

More recent warrant applications have provided a detailed description of the plan's criteria and method of operation which satisfied the reviewing court.[161] In most cases, courts have not required any specific statistical or other showing with regard to an individual company as long as the agency demonstrates that the company is in a high-hazard industry upon which the plan is focusing.[162]

D. What the Warrant May Authorize

An OSHA inspection warrant supported by an adequate showing of probable cause will authorize the agency to search

---

[160]*In re Establishment Inspection of Gilbert & Bennett Mfg. Co., supra* note 152, at 1341. This case involved a warrant issued prior to the *Barlow's* decision; whether the warrant would survive scrutiny given further development of the law in this area is problematic.

[161]E.g., *Donovan v. Hackney, Inc.,* 769 F.2d 650, 652 (10th Cir. 1985); *Stoddard Lumber Co. v. Marshall, supra* note 153, at 985–986 n.1, 988; *Marshall v. Milwaukee Boiler Mfg. Co., supra* note 153, at 1341 n.1, 1342–1343; *Erie Bottling Corp. v. Donovan, supra* note 154, at 603, 604, 606; *In re Peterson Builders, Inc., supra* note 158, at 643, 645. OSHA's current program, Instruction CPL 2.25B CH-1, is typical of the type of detailed scheduling mechanism used since the *Barlow's* decision. CPL 2.25B CH-1 is reproduced at Empl. Safety & Health Guide (Developments 1981–82), ¶ 12,537. It requires establishment of a list based on the Standard Industrial Classification (SIC) of high-hazard industries ranked in descending order according to their lost workday rate ("worst-first") and a list of worksites in each classification. The OSHA Area Officers select businesses for a local inspection register based on a worst-first ranking of the industries to which they belong until the office inspection quota is reached. Each establishment is inspected (in a random order) until the register is exhausted. Firms are exempt from a full inspection if their lost workday rate is below the national average. The operation of CPL 2.25B, the very similar predecessor to CPL 2.25B CH-1, is described in the *Ingersoll-Rand* and *Erie Bottling* decisions. Both cases upheld its use as a selection mechanism. *Ingersoll-Rand Co. v. Donovan, supra* note 154, at 224–225; *Erie Bottling Corp. v. Donovan, supra* note 154, at 605–606.

[162]E.g., *Stoddard Lumber Co. v. Marshall, supra* note 153, at 988–989. *In re Establishment Inspection of Gilbert & Bennett Mfg. Co., supra* note 152, at 1343; *Ingersoll-Rand Co. v. Donovan, supra* note 154, at 225. Several courts have suggested one exception to this rule: a situation where a plan calls for inspection of one worksite among a number of possible ones which are similarly situated in terms of the neutral criteria the plan employs. *Ingersoll-Rand Co. v. Donovan, supra* note 154, at 222; *Erie Bottling Corp. v. Donovan, supra* note 154, at 605; *In re Establishment Inspection of Urick Foundry,* 472 F. Supp. 1193, 1195 (W.D. Pa. 1979). One issue employers raise in challenging warrants based on an administrative plan is whether the plan is valid absent promulgation through notice and comment rulemaking under the Administrative Procedure Act, 5 U.S.C. §553. The Ninth Circuit has held that rulemaking is not required. *Stoddard Lumber Co. v. Marshall, supra* note 153, at 986–988. Another issue which one court has explored is whether reasonable legislative or administrative standards may be a substitute for a warrant. See *United States v. Mississippi Power & Light Co.,* 638 F.2d 899, 907 (5th Cir. 1981).

an employer's workplace for violations. However, this does not necessarily end the inquiry concerning the warrant's validity because an employer can challenge the warrant on two additional grounds. It may allege that OSHA seeks to inspect places or things beyond the scope of what has been justified by the probable cause showing or the warrant. It may also challenge the methods the warrant specifies for performing the inspection. Examples of the first type of challenge are that the warrant is overbroad in its scope or that the Act does not permit records to be inspected pursuant to a warrant. Inspection methods that have been challenged have focused on the use of private employee interviews at the worksite during the inspection and the attachment of sampling devices to employees.

1. *Scope of a Warrant*

Section 8(a)(1) of the Act authorizes the OSHA inspector to enter the workplace and inspect "all pertinent conditions, structures, machines, apparatus, devices, equipment and material therein."[163] The scope of the warrant issue arises in circumstances in which an employer challenges a warrant's authorization of a search which is overly broad.[164]

In cases involving probable cause based on an administrative plan, courts have uniformly rejected the argument that a full-scope warrant is overbroad.[165] The rationale for the full-scope inspection lies in the courts' conclusion that no meaningful limitation can be devised which does not defeat the purpose of a general inspection.[166]

---

[163]29 U.S.C. §657(a)(1).

[164]E.g., *In re Establishment Inspection of Gilbert & Bennett Mfg. Co.*, 5 OSHC 1375, 1376 (N.D. Ill. 1977), *aff'd*, 589 F.2d 1335, 1343 (7th Cir.), *cert. denied*, 444 U.S. 884 (1979). A warrant for the inspection of an entire worksite is referred to hereafter as a "full-scope" warrant.

[165]*In re Establishment Inspection of Gilbert & Bennett Mfg. Co.*, *supra* note 164, 589 F.2d at 1343–1344; *Ingersoll-Rand Co. v. Donovan*, 540 F. Supp. 222, 225 (M.D. Pa. 1982); *Erie Bottling Corp. v. Donovan*, 539 F. Supp. 600, 606 (M.D. Pa. 1981); *In re Peterson Builders, Inc.*, *supra* note 158, at 645. See *Marshall v. Central Mine Equip. Co.*, 608 F.2d 719, 720 n.1 (8th Cir. 1979) (dictum). Cf. *Marshall v. North Am. Car Co.*, 626 F.2d 320, 323 (3d Cir. 1980) (inspections pursuant to administrative plan distinguished from §8(f) complaint inspections).

[166]*In re Establishment Inspection of Gilbert & Bennett Mfg. Co.*, *supra* note 164, at 1343–1344. See *Erie Bottling Corp. v. Donovan*, *supra* note 165, at 606. CPL 2.25B only requires inspection of an entire worksite if the employer's lost workday rate derived from injury-illness statistics exceeds the national average. *Erie Bottling Corp.*, *supra* note 165, at 605.

On the other hand, inspections exceeding the scope of an employee complaint have been the subject of a great deal of controversy.[167] The courts have sharply divided on the issue, with some upholding full-scope inspections based on employee complaints[168] and some holding that the scope of such warrants must be limited to the complaint area.[169] In attempting to decide the issue, courts have focused on two possible rationales for limiting the scope of a complaint inspection: (1) that Section 8(f)(1) of the Act requires such a constraint;[170] and (2) that the Fourth Amendment imposes it.[171]

The holding of the *North American Car* decision, that Section 8(f)(1) limits a complaint inspection to an area bearing an "appropriate relationship" to the complaint,[172] treats Section 8(f)(1) as a separate source of inspection authority from

---

[167]Some courts have recognized that a complaint or complaints may involve conditions of such a pervasive nature that a full-scope warrant is justified. *Burkart Randall Div. of Textron v. Marshall*, 625 F.2d 1313, 1326 (7th Cir. 1980) (Fairchild, C.J., concurring). See *Marshall v. North Am. Car Co., supra* note 165, at 324 (dictum). Analytically, these holdings are implicitly based on the rationale of the administrative plan cases.

[168]A full-scope, wall-to-wall search was upheld in *In re Establishment Inspection of Cerro Copper Prods. Co.*, 752 F.2d 280, 283 (7th Cir. 1985), where (1) there was no evidence of harassment in the filing of the complaint, (2) the evidence showed the employer to be a high-hazard workplace in a high-hazard industry, (3) there had not been such an inspection within the previous fiscal year, and (4) OSHA's limited resources would be conserved by allowing a full-scope inspection (citing *In re Inspection of Workplace Carondelet Coke Corp.*, 741 F.2d 172 (8th Cir. 1984)). *Donovan v. Burlington N., Inc.*, 694 F.2d 1213, 1216 (9th Cir. 1982), *cert. denied*, 463 U.S. 1207 (1983); *Hern Iron Works v. Donovan*, 670 F.2d 838, 841 (9th Cir.), *cert. denied*, 459 U.S. 830 (1982); *Marshall v. Seaward Int'l*, 644 F.2d 880 (4th Cir. 1981) (unpublished order), *aff'g mem.* 510 F. Supp. 314, 317–318 (W.D. Va. 1980); *J.R. Simplot Co. v. OSHA*, 640 F.2d 1134, 1138 (9th Cir. 1981), *cert. denied*, 455 U.S. 939 (1982); *Burkart Randall Div. of Textron v. Marshall, supra* note 167, at 1322–1326 (Sprecher, J.), at 1326 (Fairchild, C.J.); *In re Establishment Inspection of Gilbert & Bennett Mfg. Co., supra* note 164.

[169]*Donovan v. Fall River Foundry*, 712 F.2d 1003, 1111 (7th Cir. 1983); *Donovan v. Sarasota Concrete Co.*, 693 F.2d 1061, 1068–1070 (11th Cir. 1982); *Marshall v. North Am. Car Co., supra* note 165, at 324; *Burkart Randall Div. of Textron v. Marshall, supra* note 167, at 1326–1328 (Wood, J., dissenting); *Marshall v. Central Mine Equip. Co., supra* note 165, at 720–721 n.1 (dictum); *Sarasota Concrete Co.*, 9 OSHC 1608, 1616–1617 (Rev. Comm'n 1981).

[170]Compare *Burkart Randall Div. of Textron v. Marshall, supra* note 167, at 1326 (Sprecher, J.) (§8(f)(1) does not limit scope), with *Marshall v. North Am. Car Co., supra* note 165, at 324 (§8(f)(1) imposes scope limitation).

[171]Compare *Burkart Randall Div. of Textron v. Marshall, supra* note 167, at 1322–1326 (Sprecher, J.) (Fourth Amendment imposes no scope limit except in extraordinary circumstances), and *Marshall v. Seaward Int'l, supra* note 168, at 317–318 (same), with *Donovan v. Sarasota Concrete Co., supra* note 169, at 1068–1070 (Fourth Amendment limits scope of complaint inspection); *Burkart Randall Div. of Textron v. Marshall, supra* note 167, at 1326–1328 (Wood, J., dissenting) (same); and *Marshall v. Central Mine Equip. Co., supra* note 165, at 720–721 n.1 (same) (dictum).

[172]*Marshall v. North Am. Car Co., supra* note 165, at 324.

Section 8(a). Other courts have attributed to Section 8(f)(1) a role in Section 8 that is more limited: that the section provides not a separate source of inspection authority, but rather a mechanism by which employees who file a signed written complaint of unsafe conditions with OSHA may require that an inspection be performed "as soon as practicable" to see if those conditions exist.[173] Thus, the provision is viewed as an employee rights provision within the inspection scheme rather than as a provision purported to limit the inspection's scope.

With regard to the constitutional issue, those courts favoring limitations on the scope of a warrant rely either on the traditional Fourth Amendment notion of limiting the scope to the underlying probable cause[174] or on the *Tyler* language.[175] The courts that do not choose to limit the scope of a complaint point to considerations such as advancing the Act's remedial purposes, avoiding the problem of employers who present "sanitized areas" to OSHA inspectors "while concealing real violations,"[176] enhancing administrative efficiency, and avoiding the disruption to employers' operations from repetitive limited inspections.[177]

---

[173]*Marshall v. Horn Seed Co.*, 647 F.2d 96, 100 n.3 (10th Cir. 1981); *Burkart Randall Div. of Textron v. Marshall*, *supra* note 167, at 1321–1326 (Sprecher, J.); *Donovan v. Metal Bank of Am.* 516 F. Supp. 674, 678 (E.D. Pa.), *purge order entered*, 521 F. Supp. 1024 (E.D. Pa. 1981), *vacated in part on other grounds*, *appeal dismissed in part on other grounds*, 700 F.2d 910 (3d Cir. 1981). In fact, §8(f)(1) incorporates the other provisions of §8 by reference. *Burkart Randall Div. of Textron v. Marshall*, *supra* note 167, at 1326 (Sprecher, J.). But see *Marshall v. North Am. Car Co.*, *supra* note 165, at 323.

[174]See *Donovan v. Sarasota Concrete Co.*, *supra* note 169, at 1068–1070; *Burkart Randall Div. of Textron v. Marshall*, *supra* note 167, at 1328 (Wood, J., dissenting). Cf. *Donovan v. Fall River Foundry Co.*, *supra* note 169, at 1106–1109.

[175]See note 134, *supra*. See *Marshall v. Central Mine Equip. Co.*, 608 F.2d 719, 720–721 n.1 (8th Cir. 1979) (dictum).

[176]*In re Establishment Inspection of Gilbert & Bennett Mfg. Co.*, 589 F.2d 1335, 1343 (7th Cir.), *cert. denied*, 444 U.S. 884 (1979); *Burkart Randall Div. of Textron v. Marshall*, *supra* note 167, at 1324–1325 (Sprecher, J.). See *Hern Iron Works v. Donovan*, *supra* note 168, at 841. *Gilbert & Bennett Mfg. Co.* is the first appellate case in which the argument was adopted. The concept of "sanitized areas" has never been fully explained by the Seventh Circuit. Probably it derives from the notion that most employers know which parts of their plant are likely to contain the most violations. When the OSHA inspector arrives, the employer may instruct safety personnel to clean up these areas as quickly as possible during the opening conference at which the OSHA official explains the nature of his visit. After the opening conference, the inspection may be structured by management to concentrate first in less hazardous areas of the plant before going to the areas being cleaned up. Or, if the cleanup has or is about to be terminated, the inspection can proceed to those areas. Either scenario would plausibly explain the Seventh Circuit's reasoning.

[177]*Burkart Randall Div. of Textron v. Marshall*, *supra* note 167, at 1324 (Sprecher, J.); *Marshall v. Seaward Int'l*, *supra* note 168, at 318. But see *Burkart Randall Div.*, *supra* note 167, at 1328 (Wood, J., dissenting).

Courts have also commented on the anomaly that results from limiting complaint inspections: A programmed inspection may cover the entire worksite even though OSHA has no knowledge of any violations in that worksite; yet if OSHA has specific evidence of actual violations in that worksite, it may only conduct a limited inspection.[178] In response, the Eighth Circuit reversed the magistrate's decision in *In re Inspection Workplace Carondelet Coke Corp.*,[179] which had limited the scope of the warrant to work areas described in the employee complaint. Taking what it described as a "middle or moderate approach," the court held that in the instant case, a wall-to-wall search was not unreasonable even though it was conducted on the basis of a specific employee complaint. In doing so, it relied upon the holding of the Eleventh Circuit in *Donovan v. Sarasota Concrete Co.*[180] that "[w]hen *nothing more* is offered than a specific complaint relating to a localized condition, probable cause exists to determine only whether the complaint is valid," but that in instances where other circumstances exist, such as a "specific violation plus a past pattern of violations" or an allegation of a violation "which permeates the workplace," a full-scope inspection may be reasonable.[181] The warrant procedure, in any case, is viewed as adequately protecting employers from fishing expeditions or harassment. Accordingly, these courts have concluded that they are entitled to strike a balance which permits an inspection of greater scope than in the criminal context.

Finally, it should be noted that the Supreme Court decisions provide little guidance on this issue. The *Barlow's* majority contemplated that an OSHA inspection warrant would "advise the owner [of an establishment] of the scope and objects of the search beyond which limits the inspector is not expected to proceed[,]" but was silent concerning any limits on scope which it viewed the Fourth Amendment as requiring.[182] *Michigan v. Tyler*, involving administrative

---

[178]*Burkart Randall Div. of Textron v. Marshall*, 625 F.2d 1313, 1324 (7th Cir. 1980) (Sprecher, J.). See *Marshall v. Seaward Int'l*, 510 F. Supp. 314 (W.D. Va. 1980), aff'd mem., 644 F.2d 880 (4th Cir. 1981).

[179]741 F.2d 172 (8th Cir. 1984).

[180]693 F.2d (1061) (11th Cir. 1982).

[181]*Id.* at 1068.

[182]*Marshall v. Barlow's, Inc.*, 436 U.S. 307, 323 (1978) (footnote omitted). The dissenters believed that the majority did not intend, in fact, to impose scope limitations. *Id.* at 334 (Stevens, J., dissenting). It has been argued that since *Barlow's*

probable cause based on specific evidence, also offers little assistance.[183]

2. Access to Records Pursuant to a Warrant

OSHA inspectors routinely seek to examine employer records during a worksite inspection because of the information such records may contain regarding possible violations. Although Section 8(a)(2) of the Act does not explicitly authorize inspection of records,[184] Section 8(c)(1) requires an employer to "make available" to OSHA the records it is obligated by law to keep.[185] In addition, OSHA has promulgated a regulation which permits a compliance officer during his inspection "to review records required by the Act and regulations published in this chapter, and other records which are directly related to the purpose of the inspection."[186] OSHA may also subpoena employer records.[187]

Most courts have permitted a records inspection pursuant to a warrant,[188] and the *Barlow's* decision appears implicitly

---

involved a plan rather than a complaint inspection, the Court did not fail to focus on the scope issue.

In *Camara v. Municipal Court*, 387 U.S. 523 (1967), and *See v. City of Seattle*, 387 U.S. 541 (1967), the Court held that an administrative search warrant should be "suitably restricted," although it relies on cases involving the scope of a subpoena which suggest that a full-scope inspection under the Act could be based on an employee complaint. See *See v. City of Seattle*, 387 U.S. at 544–545; *Camara v. Municipal Court* 387 U.S. at 539. The Court's citation to *Oklahoma Press Publishing Co. v. Walling*, 327 U.S. 186 (1946), and *United States v. Morton Salt Co.*, 338 U.S. 632 (1950), suggests that an administrative search warrant, like a subpoena, is suitably restricted under the Fourth Amendment if it is of no broader scope than the relevant statutory investigation authority. See *Walling*, 327 U.S. at 206–210, 215–216; *Morton Salt Co.*, 338 U.S. at 652–653. Thus, under *Camara* and *See*, a full-scope warrant would appear to be proper since §8(a) delineates the relevant investigatory power.

[183]436 U.S. 499 (1978). As discussed *supra* note 134, the decision's narrow holding does require the magistrate granting a fire search warrant to consider the scope proposed. *Id.* at 507. At least one court has considered the particular context of fire searches—where government officials may offer the same specific evidence to justify repeated searches—to be *sui generis*, and it is one that occurs very rarely in OSHA cases. See *Burkart Randall Div. of Textron v. Marshall*, *supra* note 167, at 1318 n.5 (holding *Tyler* is limited to fire searches). Even assuming that *Tyler* is directly applicable in the OSHA context, the Court only requires the magistrate to consider the scope as a "relevant factor" in deciding whether probable cause exists to issue a warrant. *Tyler*, 436 U.S. at 507.

[184]29 U.S.C. §657(a)(2). The provision does permit inspection of "materials," a term that arguably includes records.

[185]§657(c)(1).

[186]29 C.F.R. §1903.3(a).

[187]29 U.S.C. §657(b).

[188]See *Hern Iron Works v. Donovan*, 670 F.2d 838, 841 (9th Cir.), *cert. denied*, 459 U.S. 830 (1982); *Marshall v. W & W Steel Co.*, 604 F.2d 1322, 1326–1327 (10th Cir. 1979); *Blocksom & Co. v. Marshall*, 582 F.2d 1122, 1125 (7th Cir. 1978); *Donovan v. Metal Bank of Am.*, 516 F. Supp. 674, 681 (E.D. Pa.), *purge order entered*, 521 F. Supp. 1024 (E.D. Pa. 1981), *vacated in part on other grounds, appeal dismissed on*

to authorize this course.[189] Several courts, however, have held that documents may only be inspected through use of a subpoena.[190] This latter conclusion is based on two rationales: First, the lack of an explicit authorization in Section 8(a)(2) to examine records as part of a physical inspection coupled with the subpoena power in Section 8(b) gives rise to application of the maxim *expressio unius est exclusio alterius* or "the expression of one thing is the exclusion of another."[191] Second, Congress is viewed as implicity deciding to require a subpoena because of the superior protection it affords the employer.[192]

Courts have declined to require a subpoena for records in one circumstance. Section 8(c)(1) mandates that the employer "make available" to OSHA records it is required by law to keep.[193]

3. Private Employee Interviews

Some employers have objected to the authorization in OSHA warrants of private employee interviews on the employer's premises. All of the courts that have considered the issue have rejected this challenge, although one court placed certain limitations on OSHA's right to do so.[194]

---

other grounds, 700 F.2d 910 (3d Cir. 1981). Cf. *Marshall v. Able Contractors*, 573 F.2d 1055, 1056 (9th Cir.) (use of subpoena authority discretionary), *cert. denied*, 439 U.S. 826 (1978). Of these cases, only *Metal Bank* and *Able Contractors* address in any manner the issue of whether a warrant may authorize a records inspection.

[189]The Supreme Court noted that particular care is required in delineating the scope of a documents search and rejected the Secretary's argument that an inspection of documents of broad scope "may be effected *without a warrant.*" *Marshall v. Barlow's, Inc.*, 436 U.S. 307, 324 n.22 (1978) (emphasis added).

[190]*In re Establishment Inspection of Kulp Foundry*, 691 F.2d 1125, 1132–1133 (3d Cir. 1982); *In re Establishment Inspection of Inland Steel Co.*, 492 F. Supp. 1310, 1313–1316 (N.D. Ind. 1980). One court has concluded that OSHA may inspect records the employer is required to keep without resort to a subpoena, but not other records. *Erie Bottling Corp. v. Donovan*, 539 F. Supp. 600, 606 (M.D. Pa. 1982).

[191]BLACK'S LAW DICTIONARY 521 (5th Ed. 1979). *In re Establishment Inspection of Kulp Foundry*, *supra* note 190, at 1132; *In re Establishment Inspection of Inland Steel Co.*, *supra* note 190, at 1314–1315.

[192]*In re Establishment Inspection of Kulp Foundry*, *supra* note 190, at 1131–1132; *In re Establishment Inspection of Inland Steel Co.*, *supra* note 190, at 1315.

[193]29 U.S.C. §657(c)(1). See *Donovan v. Wollaston Alloys, Inc.*, 695 F.2d 1, 8 (1st Cir. 1982); *Donovan v. Metal Bank of Am.*, *supra* note 188, at 681. In any event, no Fourth Amendment protection attaches to required records. *United States v. Snyder*, 668 F.2d 686, 690 (2d Cir. 1982).

[194]*Donovan v. Wollaston Alloys, Inc.*, *supra* note 193, at 9; *In re Establishment Inspection of Keokuk Steel Castings, Div. of Kast Metals Corp.*, 638 F.2d 42, 46 (8th Cir. 1981); *Donovan v. Metal Bank of Am.*, *supra* note 188, at 681; *Marshall v. Rochester Shoe Tree Co.*, Misc. No. 306, slip op. at 10 (N.D.N.Y. Apr. 4, 1981); *In re*

The Act itself specifically authorizes the inspector during his inspection "to question privately any * * * employee."[195] In addition, two regulations grant the inspector that same authority.[196] However, a question has been raised as to the potential conflict between private employee interviews and the employer's right under Section 8(e) "to accompany the Secretary or his authorized representative during the physical inspection of [the] workplace * * * for the purpose of aiding such inspection."[197] The legislative history discloses Congress's awareness of this issue and explains Congress's intention to favor private employee interviews: "[A]n employer should be entitled to accompany an inspector on his physical inspection, although the inspector should have an opportunity to question employees in private so that they will not be hesitant to point out hazardous conditions which they might otherwise be reluctant to discuss."[198] In addition, Congress specifically subjected the walkaround right to "regulations issued by the Secretary."[199]

However, one court has imposed limitations on private interviews to avoid significant interruption of an employer's operations.[200] In *Urick Foundry*,[201] the employer moved to have an inspection warrant authorizing a wall-to-wall inspection quashed on the ground that the warrant allowed OSHA

Establishment Inspection of Inland Steel Co., *supra* note 190, at 1313. Cf. *Marshall v. Wollaston Alloys, Inc.*, 479 F. Supp. 1102, 1103 (D. Mass. 1979) (OSHA could be empowered to conduct private interviews if warrant contained appropriate language). One court has upheld OSHA's right to conduct such interviews but placed some limitations on that right to avoid substantial disruption of employer operations. *In re Establishment Inspection of Urick Foundry*, 3 Empl. Safety & Health Guide [1982 OSHD] ¶26,194 at 33,054 (W.D. Pa. 1982); *Erie Bottling Corp. v. Donovan*, *supra* note 190, at 608. It should be noted that the *Urick Foundry* decision modifies *Erie Bottling* to some extent concerning those limitations.

[195]29 U.S.C. §657(a)(2).

[196]29 C.F.R. §§1903.3(a), 1903.7(b). See §1903.10.

[197]29 U.S.C. §657(e). This provision is known as the employer's "walkaround" right.

[198]S. Rep. No. 1282, 91st Cong., 2d Sess. 11 (1970), *reprinted in* Legis. Hist. at 151. The court in *Donovan v. Metal Bank of Am.*, explained that the two provisions of the Act did not really conflict; employers could participate in the physical inspection but not the private interviews accompanying the inspection, 516 F. Supp. 674, 681 (E.D. Pa. 1981). Accord *Donovan v. Wollaston Alloys, Inc.*, *supra* note 193, at 9. But see *Erie Bottling Corp. v. Donovan*, *supra* note 190, at 608.

[199]29 U.S.C. §657(e). Even in the absence of the explicit legislative history, the regulations allowing private employee interviews during the inspection could arguably preempt the walkaround right to that minor extent. See text accompanying note 197, *supra*.

[200]*In re Establishment Inspection of Urick Foundry*, *supra* note 194.

[201]*Id.*

unreasonable latitude in conducting worksite interviews. The district court disagreed, holding that OSHA may conduct private interviews on the worksite during work breaks and lunch periods. Its decision provided further that interviews may be conducted during production time so long as they do not disturb an employee who is actually at work and thereby "create a risk of injury or unduly disrupt production."[202]

4. Attachment of Personal Samplers

An issue on which the courts have differed is whether a warrant may authorize the Secretary to attach samplers to employees for the purpose of monitoring exposure to toxic substances.[203] According to the applicable regulation, use of such devices is necessary to conduct "effective and efficient health inspections."[204] OSHA's authority to attach them to employees derives from its regulation empowering inspectors to take "environmental samples" and employ other reasonable investigative techniques.[205] OSHA has interpreted this regulation to permit attachment of samplers as evidenced by references in various compliance field operations manuals to permit samplers as a method for measuring exposure to contaminants.[206] The courts have sustained the government position, favoring the attachment of sampling devices to employees.[207]

---

[202]*Id.*

[203]Compare *Service Foundry Co. v. Donovan*, 721 F.2d 492, 498 (5th Cir. 1982); *United States v. Cleveland Elec. Illuminating Co.*, 689 F.2d 66, 68 n.2 (6th Cir. 1982) (alternative holding) (attachment permitted), *aff'g in pertinent part and rev'g in part* No. M80-2128, slip op. at 5–7 (N.D. Ohio Feb. 27, 1981), and *In re Establishment Inspection of Keokuk Steel Castings, supra* note 194, at 46 (same), with *In re Establishment Inspection of Metro-East Mfg. Co.*, 655 F.2d 805, 811–812 (7th Cir. 1981) (attachment not permitted), and *Plum Creek Lumber Co. v. Hutton*, 452 F. Supp. 575 (D. Mont. 1978), *aff'd*, 608 F.2d 1283, 1290 (9th Cir. 1979) (same).

[204]47 FED. REG. 6530, 6532 (1982).

[205]29 C.F.R. §1903.7(b). This regulation has been amended to specify that attachment of personal samplers is one meaning of the phrase "employ other reasonable investigative techniques." 47 FED. REG. 6530–33 (1982) (interpretative and procedural rule); 47 FED. REG. 55,478–481 (1982) (legislative rule).

[206]See *Service Foundry Co. v. Donovan, supra* note 203, at 495; *In re Establishment Inspection of Metro-East Mfg. Co., supra* note 203, at 807 n.4. Two OSHA cases, including one of the earliest, have noted OSHA's use of personal samplers. See *Deering Milliken, Inc., Unity Plant v. OSHRC*, 630 F.2d 1094, 1102 (5th Cir. 1980); *American Smelting & Ref. Co. v. OSHRC*, 510 F.2d 504, 507, 513, 514 (8th Cir. 1974).

[207]*Service Foundry Co. v. Donovan, supra* note 203, at 496–498; *United States v. Cleveland Elec. Illuminating Co., supra* note 203, at 68 (alternative holding); *In re Establishment Inspection of Metro-East Mfg. Co., supra* note 203 at 812 (Swygert, J., dissenting); *In re Establishment Inspection of Keokuk Steel Castings*, 638 F.2d 42, 46

Even those courts that have disagreed have found use of personal samplers by OSHA to be reasonable.[208] None of the latter courts have held that OSHA lacks authority to attach samplers to cooperating employees in the absence of a contrary employer policy.

Both the *Plum Creek* and *Metro-East* cases involved employers whose policies prohibited employees from wearing samplers.[209] Each district court held, and each appeals court affirmed, that it was powerless to order rescission of the company policy.[210] In *Plum Creek*, the court of appeals held that, absent a state law or regulation authorizing attachment of samplers, it could not order the employer to rescind the policy.[211] In *Metro-East*, the court found that the regulation was not explicit enough to provide employers fair notice of its requirements.[212] OSHA has subsequently promulgated both an interpretative and procedural amendment and a legislative rule which specifically authorizes attachment of sampling devices to employees.[213]

E. Challenging the Warrant—Procedural and Substantive Issues

An employer that decides to challenge a warrant may take one or more of several steps to do so.

(8th Cir. 1981); *Ingersoll-Rand Co. v. Donovan*, 540 F. Supp. 222, 225–226 (M.D. Pa. 1982) and cases cited therein; *Erie Bottling Corp. v. Donovan*, 539 F. Supp. 600, 608–609 (M.D. Pa. 1982); *Marshall v. Miller Tube Corp. of Am.*, 6 OSHC 2042, 2044–2245 (E.D.N.Y. 1978).

[208]See *In re Establishment Inspection of Metro-East Mfg. Co.*, *supra* note 203, at 809; *Plum Creek Lumber Co. v. Hutton*, *supra* note 203, at 1290.

[209]*Plum Creek v. Hutton*, *supra* note 203; *In re Establishment Inspection of Metro-East Mfg. Co.*, *supra*, note 203.

[210]See *In re Establishment Inspection of Metro-East Mfg. Co.*, *supra* note 203, at 811–812; *Plum Creek Lumber Co. v. Hutton*, *supra* note 203, at 1289–1290.

[211]*Plum Creek Lumber Co. v. Hutton*, *supra* note 203, at 1289–1290. The court concluded that it lacked any authority to issue such an order under either the All Writs Act, 28 U.S.C. §1651 (1982), or any inherent power. *Id.*

[212]*In re Establishment Inspection of Metro-East Mfg. Co.*, *supra* note 203.

[213]See *supra* note 205. The original version of 29 C.F.R. §1903.7(b) might have provided the regulation on which the district court in *Plum Creek* sought initially to base an order rescinding the company's policy prohibiting sampler wearing—an order which the district court later refused to enforce in favor of a principle that "in the absence of law the OSHA inspectors have no power to make Plum Creek do something simply because the inspectors think it reasonable." 452 F. Supp. 575, 577 (D. Mont. 1978), 6 OSHC 1839, 1840, *aff'd*, 608 F.2d 1283 (9th Cir. 1979).

1. *Procedural Aspects*

An employer may turn away the OSHA inspector and not permit execution of the warrant. At that point, OSHA will petition the federal district court in which the warrant was issued to hold the employer in civil contempt.[214] The employer may file a motion or, if OSHA has filed for contempt, a cross-motion to quash the warrant.[215] These steps will permit the employer to obtain a preexecution hearing on the warrant's validity in the district court. However, the employer must be prepared to suffer any contempt sanctions which the court imposes.[216]

The employer may also permit the inspection and challenge the warrant in administrative proceedings before the Review Commission and, if available, the court of appeals on review of the Commission's final order.[217] The risk presented

---

[214]See, e.g., *Rockford Drop Forge Co. v. Donovan*, 672 F.2d 626, 631 (7th Cir. 1982); *Babcock & Wilcox Co. v. Marshall*, 610 F.2d 1128, 1136 (3d Cir. 1979); *Blocksom & Co. v. Marshall*, 582 F.2d 1122, 1124 (7th Cir. 1978); *In re Worksite Inspection of S.D. Warren, Div. of Scott Paper*, 481 F. Supp. 491, 495 (D. Me. 1979). OSHA may also petition for criminal contempt. See *In re Ohio New & Rebuilt Parts*, No. C-79-241 (N.D. Ohio May 1, 1979). Cf. *Blocksom & Co. v. Marshall, supra*, 582 F.2d at 1124. OSHA very rarely invokes this power although a court may *sua sponte* convert the contempt petition from civil to criminal. See *In re Establishment Inspection of Consolidated Rail Corp.*, 631 F.2d 1122, 1125 (3d Cir. 1980). A criminal contempt petition presents a serious problem for the employer because in such cases, in contrast to civil contempt, the "validity of the underlying court order which the employer has disobeyed is not a defense." *Blocksom & Co. v. Marshall, supra*, 582 F.2d at 1124.

[215]See, e.g., *Rockford Drop Forge Co. v. Donovan, supra* note 214, at 631; *In re Worksite Inspection of S.D. Warren, Div. of Scott Paper, supra* note 214, at 495.

[216]See *Rockford Drop Forge Co. v. Donovan, supra* note 214, at 631; *Babcock & Wilcox Co. v. Marshall, supra* note 214, at 1136; *Blocksom & Co. v. Marshall, supra* note 214, at 1124. As discussed *supra*, a criminal contempt petition by OSHA would deprive the employer of its day in court on the warrant's validity. In addition, OSHA could accomplish the same end by executing warrants using force. Legally, use of force is permissible. *Marshall v. Shellcast Corp.*, 592 F.2d 1369, 1372 (5th Cir. 1979); *See v. City of Seattle*, 387 U.S. 541, 545 (1967). OSHA's internal guidelines and regulations, however, prohibit such self-help. OSHA Field Operations Manual ch. V D.a.c (9). Cf. 29 C.F.R. §1903.4(a) (OSHA will obtain "compulsory process" upon refusal to permit inspection). Some isolated allegations of OSHA's use of force have appeared notwithstanding the policy. See *Baldwin Metals Co. v. Donovan*, 642 F.2d 768, 771 n.6 (5th Cir.), *cert. denied*, 454 U.S. 893 (1981). It should be noted that courts consistently refuse to adjudicate, in the context of a contempt or motion to quash hearing, issues of agency jurisdiction over a workplace. E.g., *Marshall v. Burlington N., Inc.*, 595 F.2d 511, 513 (9th Cir. 1979); *Marshall v. Northwest Orient Airlines*, 574 F.2d 119, 122 (2d Cir. 1978); *Marshall v. Able Contractors*, 573 F.2d 1055, 1057 (9th Cir.), *cert. denied*, 439 U.S. 826 (1978); *In re Restland Memorial Park*, 540 F.2d 626, 628 (3d Cir. 1976); *In re Establishment Inspection of Sauget Indus. Research & Waste Treatment Ass'n*, 477 F. Supp. 88, 90 (S.D. Ill. 1979). Exhaustion of administrative remedies before the Review Commission is required as to these challenges. *Id.*

[217]E.g., *Rockford Drop Forge Co. v. Donovan, supra* note 214, at 631; *In re Worksite Inspection of S.D. Warren, Div. of Scott Paper, supra* note 214, at 495, and cases cited

by choosing this course of action is that OSHA's inspection may not result in the issuance of citations that trigger Commission proceedings. The issue may then be moot[218] or the court may find itself without jurisdiction.[219]

Another issue which has been raised is whether the Commission itself has the authority to adjudicate the validity of a warrant. The Commission believes it may do so and two courts have supported its position.[220] Other courts have concluded that the Commission's authority is unimportant since a court of appeals can review the warrant on petition from a final Commission order.[221]

2. Scope of the Challenge

In a contempt or motion-to-quash proceeding, the scope of the challenge to a warrant is much more limited as an evidentiary matter than would be the case in a normal civil suit which involves discovery. The probable cause showing made to the magistrate or judge who issued a warrant is the focus of subsequent challenges to the warrant's validity by the employer.[222] Evidence which OSHA introduces at a sub-

---

therein. The Seventh Circuit takes the position that the district court may hear postexecution challenges to a warrant. *Federal Castings Div., Chromalloy Am. Corp. v. Donovan*, 684 F.2d 504, 507–508 (7th Cir. 1982); *Weyerhaeuser Co. v. Marshall*, 592 F.2d 373, 376–377 (7th Cir. 1979). All other courts have taken the position that such challenges must be exhausted before the Review Commission prior to federal court review. *Bell Enters. v. Donovan*, 710 F.2d 673, 675 (10th Cir. 1983), *cert. denied*, 464 U.S. 1041 (1984), and cases cited therein; *Baldwin Metals Co. v. Donovan, supra* note 216, 771–777. The *Baldwin Metals* decision presents the most complete discussion of the majority view.

[218]*In re Establishment Inspection of Kulp Foundry*, 691 F.2d 1125, 1128–1130 (3d Cir. 1982); *United States v. Cleveland Elec. Illuminating Co.*, No. M80-2128, slip op. at 3 (N.D. Ohio Feb. 27, 1981), *aff'd in pertinent part and rev'd in part*, 689 F.2d 66, 67–68 (6th Cir. 1982); *Donovan v. Holbrook Drop Forge*, No. 81-1634 (1st Cir. Oct. 26, 1981). Cf. *Marshall v. Milwaukee Boiler Mfg. Co.*, 626 F.2d 1339, 1342 (7th Cir. 1980) (issue of suppression of fruits of search provides independent case or controversy). Employers could avoid mootness by demanding money damages as part of the relief sought under *Bivens v. Six Unknown Named Agents*, 403 U.S. 388 (1971). However, government officials enjoy a good faith immunity defense to requests for such relief. *Bivens v. Six Unknown Named Agents*, 456 F.2d 1229 (2d Cir. 1972).

[219]See *Baldwin Metals Co. v. Donovan, supra* note 216, at 777; *Marshall v. Central Mine Equip. Co.*, 608 F.2d 719, 722 (8th Cir. 1979).

[220]*Sarasota Concrete Co.*, 9 OSHC 1608, 1611–1612 (Rev. Comm'n 1981). See *Donovan v. Sarasota Concrete Co.*, 693 F.2d 1061, 1066–1067 (11th Cir. 1981); *Babcock & Wilcox Co. v. Marshall, supra* note 214, at 1138–1139.

[221.]E.g., *Baldwin Metals Co. v. Donovan, supra* note 216, at 773–774 n.11; *Marshall v. Central Mine Equip. Co., supra* note 219, at 721–722; *In re Worksite Inspection of Quality Prods.*, 592 F.2d 611, 615–616 nn. 7–9 (1st Cir. 1979).

[222]*West Point-Pepperell, Inc. v. Donovan*, 689 F.2d 950, 959 (5th Cir. 1982); *Marshall v. Horn Seed Co.*, 647 F.2d 96, 104 (10th Cir. 1981); *J.R. Simplot Co. v. OSHA*,

sequent hearing on the warrant's validity may not be considered by the court.[223] The magistrate who initially decided to issue the warrant based the decision on the information provided by or in conjunction with the original application, and judicial review of the validity of the warrant must be based on the same evidence.

Under limited circumstances the employer is entitled to go behind the warrant application and adduce additional facts. In *Franks v. Delaware*, the Supreme Court explained that

> "[t]here must be allegations of deliberate falsehood or of reckless disregard for the truth, and those allegations must be accompanied by an offer of proof. They should point out specifically the portion of the warrant affidavit that is claimed to be false; and they should be accompanied by a statement of supporting reasons. Affidavits or sworn or otherwise reliable statements of witnesses should be furnished, or their absence satisfactorily explained. Allegations of negligence or innocent mistake are insufficient."[224]

In addition, the challenged position of the affidavit must be essential to the probable cause finding.[225] If these two conditions are met, the party challenging the warrant is entitled to an evidentiary hearing at which the party may adduce facts outside of the face of the warrant application.[226]

A related challenge involves an allegation of abuse of process in obtaining the warrant.[227] As with the *Franks* type of challenge, however, abuse of process allegations do not result in discovery or a hearing unless the employer is able to make a fairly strong preliminary showing of government bad faith. A two-part test is used: (1) whether the agency's informant had an improper motive; and (2) whether the agency

---

640 F.2d at 1138; *In re Establishment Inspection of Gilbert & Bennett Mfg. Co.*, [1977–1978] OSHD ¶21,798, 26,235 (N.D. Ill. 1977), *aff'd*, 589 F.2d 1335, 1342 (7th Cir.), *cert. denied*, 444 U.S. 884 (1979); *Sarasota Concrete Co.*, *supra* note 220, at 1611–1612.

[223]*In re Establishment Inpsection of Gilbert & Bennett Mfg. Co.*, *supra* note 222, at 1342 n.7.

[224]438 U.S. 154, 171 (1978). Accord *Donovan v. Hackney*, 769 F.2d 650, 653 (10th Cir. 1985); *West Point-Pepperell, Inc. v. Donovan*, *supra* note 222, at 959–960; *Marshall v. Milwaukee Boiler Mfg. Co.*, *supra* note 218, at 1346; *In re Worksite Inspection of Quality Prods.*, *supra* note 221, at 616–617; *Pelton Casteel, Inc. v. Marshall*, 588 F.2d 1182, 1188 n.12 (7th Cir. 1978).

[225]*Franks v. Delaware*, *supra* note 224, at 171.

[226]*Id.* Cf. *In re Establishment Inspection of Gilbert & Bennett Mfg. Co.*, *supra* note 220, at 1340 (no discovery permitted where warrant application on its face supplies adequate probable cause).

[227]See *B.P. Oil, Inc. v. Marshall*, 509 F. Supp. 802, 810 (E.D. Pa. 1981).

either adopted the improper motive of its informant or displayed its own bad faith in causing its investigation to be pursued.[228]

In the context of challenges based on allegations that OSHA is relying on invalid regulations either in seeking the warrant[229] or requesting authorization to employ certain investigative techniques,[230] one court has discussed the preliminary showing which an employer must make to obtain discovery beyond the public record on which those regulations were based.[231] That showing is analogous to the requirements in the *Franks* case and in the abuse of process context.[232]

3. Remedies for Invalidity of the Warrant

Once the warrant or any aspect of it fails to survive court scrutiny, the issue of the appropriate remedy arises. In the context of a contempt proceeding or motion to quash, the court must decide whether it will quash the entire warrant or only the tainted portion.[233] If the warrant has been executed, the court must consider whether evidence obtained should be suppressed.[234]

No court has explicitly considered whether a partially invalid warrant must be quashed in its entirety, although one has suggested that it need not be.[235] However, courts frequently modify warrants in order to remove the tainted aspects.[236] This approach is consistent with that taken by courts in enforcing subpoenas; a court may limit a subpoena which is overbroad or unduly burdensome and enforce it as modified.[237] This position also comports with the "reduction" doc-

---

[228]See, e.g., *SEC v. Wheeling-Pittsburgh Steel Corp.*, 648 F.2d 118, 123–128 (3d Cir. 1981) (en banc).

[229]See text accompanying notes 85–91, *supra*.

[230]See text accompanying notes 175–206, *supra*.

[231]*Texas Steel Co. v. Donovan*, 93 F.R.D. 619 (N.D. Tex. 1982).

[232]*Id.* at 621.

[233]See *Donovan v. Wollaston Alloys, Inc.*, 695 F.2d 1, 8 (1st Cir. 1982); *West Point-Pepperell, Inc. v. Donovan, supra* note 224, at 960.

[234]See *Marshall v. Milwaukee Boiler Mfg. Co.*, 626 F.2d 1339, 1342 (7th Cir. 1980).

[235]Cf. *Donovan v. Wollaston Alloys, Inc., supra* note 233, at 8.

[236]See, e.g., *Rockford Drop Forge Co. v. Donovan*, 672 F.2d 626, 629 n.3 (7th Cir. 1982); *B.P. Oil, Inc. v. Marshall, supra* note 227, at 807.

[237]See, e.g., *FTC v. Shaffner*, 626 F.2d 32, 38 (7th Cir. 1980); *FTC v. Texaco, Inc.*, 555 F.2d 862, 881–882 (D.C. Cir.) (en banc), *cert. denied*, 431 U.S. 974 (1977); *SEC v. Wall Street Transcript Corp.*, 422 F.2d 1371, 1381 (2d. Cir. 1970); *Adams v.*

trine in the criminal law warrant area which permits courts to limit application of the exclusionary rule to evidence seized during execution of invalid parts of the warrant.[238]

The more difficult question is the applicability of the exclusionary rule in cases involving executed warrants. The courts and the Commission all agree, on the basis of Supreme Court precedent, that the exclusionary rule does not apply retroactively, i.e., to pre-*Barlow's* inspections.[239] Although a recent Supreme Court decision suggests that these cases may have been incorrectly decided,[240] the issue is significant only if there are pre-*Barlow's* cases involving Fourth Amendment challenges still in litigation.

With respect to prospective application of the exclusionary rule to post-*Barlow's* searches, some courts have suggested,[241] and the Commission and Eleventh Circuit have held,[242] that courts may exclude illegally seized evidence. Other courts have held or suggested the contrary.[243]

The Supreme Court has given guidance on this point in the criminal context,[244] but has never applied the exclusionary rule in a civil proceeding[245] and it is therefore presumably

---

FTC, 296 F.2d 861, 866–867 (8th Cir. 1961), *cert. denied*, 369 U.S. 864 (1962). The use of the subpoena procedure would be appropriate because of the Supreme Court's reliance on the analogy between subpoenas and administrative warrants in imposing the warrant requirement. See *See v. City of Seattle*, 387 U.S. 541, 544–545 (1967).

[238]See *United States v. Christine*, 687 F.2d 749, 754–760 (3d Cir. 1982); *United States v. Cardwell*, 680 F.2d 75, 78–79 (9th Cir. 1982); *United States v. Cook*, 657 F.2d 730, 734–736 (5th Cir. 1981).

[239]*Robberson Steel Co. v. OSHRC*, 645 F.2d 22 (10th Cir. 1980); *Savina Home Indus. v. Secretary of Labor*, 597 F.2d 1358, 1364 (10th Cir. 1979); *Todd Shipyards Corp. v. Secretary of Labor*, 586 F.2d 683, 690–691 (9th Cir. 1978).

[240]See *United States v. Johnson*, 457 U.S. 537, 557–565 (1982). *Johnson* appears to cut back the reach of the precedent upon which *Savina*, *Todd*, and *Meadows Industries* are based.

[241]See *Savina Home Indus. v. Secretary of Labor, supra* note 185, at 1363 (dictum); *Weyerhaeuser Co. v. Marshall*, 592 F.2d 373, 375, 378 (7th Cir. 1979) (affirming district court order suppressing evidence). The *Weyerhaeuser* case is of doubtful value as precedent since it involved a pre-*Barlow's* warrant and the government did not raise or litigate the issue of the applicability of the district court's exclusion of evidence. See *Marshall v. Milwaukee Boiler Mfg. Co., supra* note 234, at 1347.

[242]*Donovan v. Sarasota Concrete Co.*, 693 F.2d 1061, 1070–1072 (11th Cir. 1982); *Sarasota Concrete Co.*, 9 OSHC 1608, 1612–1615 (Rev. Comm'n 1981).

[243]*Donovan v. Federal Clearing Die Casting Co. & OSHRC*, 695 F.2d 1020, 1022–1025 (7th Cir. 1982) (good faith exception). See *Robberson Steel Co. v. OSHRC, supra* note 239, at 22 (good faith exception); *Todd Shipyards Corp. v. Secretary of Labor, supra* note 239, at 689 (dictum) (inapplicable to civil proceedings).

[244]See *United States v. Leon*, 468 U.S. 897 (1984); *Massachusetts v. Sheppard*, 468 U.S. 981 (1984).

[245]*United States v. Janis*, 428 U.S. 433, 447 (1976). In *Janis* the Court also made clear that the term "quasi-criminal" proceeding, in which the rule applied, encom-

reluctant to extend the rule beyond criminal trials.[246] The Seventh and the Tenth Circuits have adopted an approach of refusing to suppress illegally seized evidence where the inspectors acted in good faith and pursuant to a reasonable though mistaken belief.[247]

The Commission has held that the exclusionary rule is applicable to its proceedings and has declined to recognize the good faith exception.[248] A related question is the extent to which the Commission, pursuant to its supervisory power, may exclude evidence which the courts would admit under the Fourth Amendment. The Eleventh Circuit has held, and the Third Circuit has suggested, that it may exclude such evidence.[249] This holding appears inconsistent with the subsequent Supreme Court decision in *United States v. Payner*.[250] The Commission might nonetheless be able to continue to apply the exclusionary rule if it can articulate a rationale for doing so that does not depend on the Fourth Amendment or the deterrence concept.[251]

---

passed the narrow category of forfeiture case involving articles used in the commission of a crime. *Id.* at 447 n.17.

[246]*Franks v. Delaware*, 438 U.S. 154, 171 (1978). The test it has adopted for deciding whether to apply the rule to such areas requires a balancing of the government's need for the evidence against the potential benefits of the sanction's deterrent effect. *United States v. Calandra*, 414 U.S. 338, 349 (1974).

[247]*Donovan v. Federal Clearing Die Casting Co. & OSHRC*, 695 F.2d 1020 (7th Cir. 1982); *Robberson Steel Co. v. OSHRC*, 645 F.2d 22 (10th Cir. 1980).

[248]*Sarasota Concrete Co.*, 9 OSHC 1608, 1612–1615 (Rev. Comm'n 1981).

[249]See *Babcock & Wilcox Co. v. Marshall*, 610 F.2d 1128, 1139 (3d Cir. 1979); *Donovan v. Sarasota Concrete Co.*, 693 F.2d 1061, 1072 (11th Cir. 1982).

[250]447 U.S. 727, 735 (1980). In *Payner*, the Supreme Court held that the federal courts, based on their supervisory power, could not exclude as illegally obtained evidence which was nonetheless admissible against a particular defendant under the Fourth Amendment. The Seventh Circuit has held that *Payner* precludes the Commission from excluding evidence which would be admissible under the Fourth Amendment. See *Donovan v. Federal Clearing Die Casting Co. & OSHRC*, *supra* note 247, at 1023 n.6.

[251]Cf. *Donovan v. Federal Clearing Die Casting Co. & OSHRC*, *supra* note 247, at 1023 n.6.

8

Citations

If the Area Director believes from his review of a compliance officer's report that a violation has occurred, Section 9(a) of the Act requires that he issue a written citation to the employer with reasonable promptness. This citation must describe the violation with particularity; identify the violated rule, standard, or regulation; and set a reasonable date for its correction or abatement.[1] Section 9(b) requires that the citation be posted if possible at or near each place where a violation has occurred.[2] Section 9(c) prohibits any citation from being issued more than 6 months after violations have occurred.[3]

I. Particularity Requirement

The purpose of the particularity requirement for citations contained in Section 9(a) of the Act is to provide employers

---

[1]Section 9(a) of the Act, 29 U.S.C. §658(a) (1985). See also 29 C.F.R. §§1903.14(a) and (b) (1985).

[2]29 U.S.C. §658(b). See Chapter 9 (Types and Degrees of Violations).

[3]§658(c); 29 C.F.R. §1903.14(a). This 6-month time period is a maximum time period and is not the same as the "reasonable promptness" requirement of §9(a). *Dravo Corp.*, 3 OSHC 1085 (Rev. Comm'n 1975). Violations of these §9 requirements may be raised as a procedural defense warranting the dismissal of a citation. See Chapter 15 (Defenses).

with adequate notice of the violation.[4] For example, the precise identification of hazardous equipment, its location, and a description of the nature of the hazard observed will afford sufficient information for the employer to make a decision whether and how to abate the violation or to contest the citation. To comport with Section 9(a), such identification need not be made with minute detail as long as the employer has received fair notice.[5] It need not identify or specify how to abate a hazardous condition.[6]

A typographical error in a citation's reference to a cited standard did not mean that the citation lacked particularity when the citation taken as a whole made the violation clear.[7] Likewise, when the employer was aware of the particulars of the cited violation through knowledge of its own business, there was no lack of particularity because the date of inspection was omitted or the exact location or the extent of the hazard was left out of a citation.[8]

The Fifth Circuit affirmed the Review Commission's vacation of a citation in *Marshall v. B.W. Harrison Lumber Co.* for lack of particularity. The Commission had held that:

> "The test of particularity is whether the citation provided fair notice of the alleged violation. In determining whether fair notice has been afforded, consideration may be given to factors external to the citation, such as the nature of the alleged violation, the circumstances of the inspection, and the employer's knowledge of his own business."[9]

Any deficiencies in the specific details of the violation may well be cured by reference to the factual context in which the citation was issued. The Commission has made this clear by

---

[4]See, e.g., *Brabham-Parker Lumber Co.*, 11 OSHC 1201, 1202 (Rev. Comm'n Nos. 78-6060 & 78-6061, 1983).

[5]*Meadows Indus.*, 7 OSHC 1709 (Rev. Comm'n 1979).

[6]*Pabst Brewing Co.*, 11 OSHC 1203 (Rev. Comm'n 1983); *Del Monte Corp.*, 4 OSHC 2035 (Rev. Comm'n 1977).

[7]*H.S. Holtze Constr. Co. v. Marshall*, 627 F.2d 149, 8 OSHC 1785 (8th Cir. 1980), *aff'g in part, rev'g in part* 7 OSHC 1753 (Rev. Comm'n 1979). See also *Savina Home Indus. v. Secretary of Labor*, 594 F.2d 1358, 7 OSHC 1154 (10th Cir. 1979).

[8]*Ringland-Johnson, Inc. v. Dunlop*, 551 F.2d 1117, 6 OSHC 1137 (8th Cir. 1977) (failure to state inspection date); *General Motors Corp.*, 8 OSHC 1735 (Rev. Comm'n 1980) (failure to specify hazardous areas); *Meadows Indus., supra* note 5 (failure to specify hazardous areas); *B.F. Goodrich Textile Prods.*, 5 OSHC 1458 (Rev. Comm'n 1977) (failure to identify hazardous equipment).

[9]*Marshall v. B.W. Harrison Lumber Co.*, 4 OSHC 1091 (Rev. Comm'n 1976), *aff'd*, 569 F.2d 1303, 6 OSHC 1446 (5th Cir. 1978).

holding that an apparently knowledgeable employer who failed to inquire about the precise locations of violations or to discover them prior to the hearing could not raise lack of particularity as a defense.[10]

Lack of particularity is an affirmative defense that must be pleaded and timely raised by the employer or it will be considered waived.[11] To successfully sustain this defense, the employer must show it was prejudiced in preparing its defense.[12]

II. Reasonable Promptness Requirement

Section 9(a) of the Act requires that a citation must be issued to an employer with "reasonable promptness" after an inspection or investigation.[13] Initially OSHA attempted to comply with this requirement by issuing citations within 72 hours of a violation's discovery. However, it was held that failure to issue a citation did not constitute prejudice to the employer since that time period is not required by either the Act or its legislative history.[14] The test for reasonable promptness has thus become a determination of whether the employer has actually been prejudiced by the delay in issuing the citation.[15]

---

[10]*General Motors Corp., supra* note 8.

[11]Commission Rule 36(b)(1), 29 C.F.R. §2200.36(b)(1) (1987); *Wheeling Pittsburgh Steel Corp.*, 7 OSHC 1581 (Rev. Comm'n 1979).

[12]*Ringland-Johnson v. Dunlop, supra* note 8, at 1118.

[13]29 U.S.C. 658(a).

[14]*Brennan v. Chicago Bridge & Iron Co.*, 514 F.2d 1082, 3 OSHC 1056 (7th Cir. 1975). See also *Stephenson Enters. v. Marshall*, 578 F.2d 1021, 6 OSHC 1860 (5th Cir. 1978); *Todd Shipyards Corp. v. Secretary of Labor*, 566 F.2d 1327, 6 OSHC 1227 (9th Cir. 1977).

[15]*Craig D. Lawrenz & Assocs.*, 4 OSHC 1464 (1976) (19-day delay). There were three opinions in this case. Chairman Barnako held that the citation should not have been vacated because there was no demonstration of prejudicial delay or "unreasonableness." Commission Cleary agreed with the result but disagreed that an unreasonable delay, absent employer prejudice, would warrant the vacation of a citation. Dissenting Commissioner Moran felt that the citation should be vacated because it had been issued more than 72 hours after the violation and because there had been no showing of exceptional circumstances by the Secretary of Labor. Commissioner Moran prevailed with this view of the Secretary's "unconscionable delay" in *Jack Conie & Sons*, 4 OSHC 1378 (Rev. Comm'n 1976) (125-day delay). Chairman Barnako concurred with this result, but not with its reasoning, holding that the delay was "patently unnecessary and unjustifiable." The views of Commissioner Moran on the Secretary's unconscionable delay were explicitly rejected by the Commission in *Stearns-Roger, Inc.*, 8 OSHC 2180 (Rev. Comm'n 1980), although employers have continued to urge their adoption. See *National Indus. Constructors*, 10 OSHC 1081 (Rev. Comm'n 1981).

The determination of what constitutes such prejudicial delay is made factually. Actual elapsed time therefore is only one factor to be considered, and delays of up to several months have been held nonprejudicial.[16] On the other hand, the unavailability of witnesses and equipment is a factor that affects the employer's ability to prepare its defense, and may occur during any delay related to the issuance of a citation. Such facts have constituted prejudice in several cases.[17]

Although Section 9(c) of the Act provides an absolute limitation of 6 months (180 days) within which a citation may be issued, the Commission has held that a citation issued 177 days after inspection was reasonably prompt under the circumstances in the *Stripe-A-Zone* case,[18] despite the fact that the employer had made a showing that two key witnesses were unavailable. The Commission reasoned:

> "The inspection in this case occurred in November 1978 and the citations were issued in April 1979. Yet Respondent has failed to establish any connection between the delay in the issuance of the citations and the asserted prejudice resulting from either the *** illness or the unavailability of [the witness] ***. Respondent represented to this Commission in January 1980—in its motion to reconsider a Commission order that proceedings be expedited—that 'the availability of the relevant witnesses and documents has not changed since the date the original citation was issued.' "[19]

Because the employer failed to demonstrate any prejudice causally related to the delay in issuing citations, the Commission refused to dismiss the citations.[20]

III. Service of Citations

The citation must be served in any manner reasonably calculated to provide the employer with adequate notice of the

---

[16]*Havens Steel Co. v. OSHRC*, 738 F.2d 397, 11 OSHC 2057 (10th Cir. 1984) (9-week delay); *Bethlehem Steel Corp. v. OSHRC*, 607 F.2d 871, 7 OSHC 1802 (3rd Cir. 1979) (7-week delay); *Stearns-Roger, Inc., supra* note 15 (4-week delay); and cases cited in note 14, *supra*.

[17]*E.C. Ernst, Inc.*, 2 OSHC 1468 (Rev. Comm'n 1975) (affected employee had left job and could not be located); *Southwire Co.*, 9 OSHC 2034 (Rev. Comm'n 1981) (1-month delay for forklift violation held prejudicial where forklifts had been returned to lessor and could not be examined for alleged defects).

[18]*Stripe-A-Zone, Inc.*, 10 OSHC 1694, 1695 (Rev. Comm'n 1982).

[19]*Id.* at 1695.

[20]*Id.* In a concurring opinion, Chairman Rowland noted his disagreement with the majority's holding that the Commission should not consider the "justifiability of the delay," noting that they "should also give relief to an employer in situations

violations and an opportunity to determine whether to abate
or contest. Proper service commences the 15-working-day period for filing a notice of contest.[21]

In *Buckley & Co. v. Secretary of Labor*, the Third Circuit
ruled that service of a citation by mail on the superintendent
at an employer's jobsite where the alleged violations occurred
was ineffective because it was not made upon a corporate
officer "who has the authority to disburse corporate funds to
abate the alleged violation, pay the penalty, or contest the
citation or proposed penalty."[22] Subsequently, the Commission refused to follow this rule, first distinguishing the *Buckley*
case in 1978 and then abandoning it in 1979.[23]

In *B.J. Hughes, Inc.*, the Commission held that service is
proper if it is "reasonably calculated to provide an employer
with knowledge of the citation and proposed penalty and an
opportunity to abate or contest."[24] It is now well established
that service at the employer's place of business on any employee who accepts delivery of certified mail constitutes proper
service as of the date of receipt. In this regard, the Commission
has held that the Secretary could presume that the mailed
citation would reach an appropriate official.[25]

IV. Amending Citations

Before a notice of contest is filed, a citation may be unilaterally amended by the Secretary.[26] After a notice of contest
is filed, a citation may still be amended before the hearing,
at the hearing, or after the hearing. In general, a lack of

---

where the delay is patently unreasonable, unnecessary and unjustified." *Id.* at 1696.
This language hearkens back to that used by former Chairman Barnako in *Jack Conie & Sons, supra* note 15.

[21]Section 10(a) of the Act, 29 U.S.C. §659(a).

[22]*Buckley & Co. v. Secretary of Labor*, 507 F.2d 78, 81 (3rd Cir. 1975). See also *Otis Elevator Co.*, 6 OSHC 1515 (Rev. Comm'n 1978) (Judge's Decision).

[23]*Joseph Weinstein Elec. Corp.*, 6 OSHC 1344 (Rev. Comm'n 1978); *B.J. Hughes, Inc.*, 7 OSHC 1471 (Rev. Comm'n 1979).

[24]*B.J. Hughes, Inc., supra* note 23. This decision was clarified in *Capital City Excavating Co.*, 8 OSHC 2008 (Rev. Comm'n 1980).

[25]*Henry C. Beck Co.*, 8 OSHC 1395 (Rev. Comm'n 1980). See also *Robert F. Wilson, Inc.*, 11 OSHC 1543 (Rev. Comm'n 1983). See Chapter 13 (Commencement of Pro-ceedings).

[26]Field Operations Manual, ch. V G (Oct. 21, 1985), OSHR [Reference File 77:2914].

prejudice to the employer has been the standard used by the Commission to determine whether to permit such amendments.[27]

The Commission approves the granting of prehearing amendments where respondents make an insufficient showing that they would be prejudiced in their case preparation or presentation. In one case, the Commission approved a prehearing amendment to a complaint alleging a general Section 5(a)(1) violation after the employer had moved to dismiss the complaint on grounds that a specific standard, of which no violation was alleged, governed the proceeding. The Commission held that no prejudice resulted to the employer from allowing an amendment substituting that specific standard.[28] In another case, the Commission upheld the granting of a prehearing amendment where the factual bases of the charge were the same, and where the employer declined an offered opportunity for a continuance at the hearing.[29] In *P.A.F. Equipment Co. v. OSHRC*, the Tenth Circuit rejected the employer's contention that the Secretary's postcontest amendment of a citation from serious to willful had a "chilling effect" on the employer's exercise of its due process right to contest the citation. The court found this case controlled by its prior holding that the Secretary, like the Commission, can increase proposed penalties after a notice of contest has been filed.[30]

Amendments at the hearing are also governed by this test of prejudice to the employer. In one case, the administrative law judge permitted an amendment at the hearing to plead alternatively a specific standard violation because the employer, in contesting the general duty clause violation, had introduced its applicability and was thereby not prejudiced.[31] In another case, the trial judge amended a citation on his own motion to include a general duty charge in addition to the

---

[27]*Cornell & Co. v. OSHRC*, 573 F.2d 820, 6 OSHC 1436 (3d Cir. 1978). See also Chapter 14 (Hearing Procedure) for a discussion of amended citations and pleadings.

[28]*Miller Brewing Co.*, 7 OSHC 2155 (Rev. Comm'n 1980).

[29]*Brown & Root, Inc.*, 8 OSHC 1055 (Rev. Comm'n 1980), citing with approval *Miller, supra* note 28.

[30]*P.A.F. Equip. Co. v. OSHRC*, 637 F.2d 741, 9 OSHC 1037 (10th Cir. 1980); *Clarkson Constr. Co. v. OSHRC*, 531 F.2d 451, 3 OSHC 1880 (10th Cir. 1976).

[31]*General Motors Corp.*, 8 OSHC 1412 (Rev. Comm'n 1980).

original specific standard violation.[32] Again, the judge found that no prejudice existed where both parties had impliedly tried the important issues of recognized hazard and employee exposure.

Such hearing amendments have been upheld on appeal to the Commission and the courts. The Commission has upheld an amendment at hearing to allege that an "investigation" as well as an inspection had been conducted the day after the accident where no prejudice was demonstrated by the employer.[33] The Commission reversed an administrative law judge's denial at hearing of an amendment that would have substituted a different respondent as the proper party. There, where the accident involved employees of a subcontractor of the cited employer, the judge had dismissed citations on the grounds that the Secretary had "inexcusably neglected" to ascertain the proper respondent and that the 6-month period for such an amendment under Section 9(c) of the Act had run. The Commission, however, found that the two owners of the cited employer also owned the subcontractor, and that they were both officers of the two corporations which worked closely together so that no prejudice would exist.[34] Inexcusable neglect was also not, the Commission held, an appropriate consideration under F. R. Civ. P. 15(c), governing amendments. Additionally, the Commission held that Section 9(c) is a statute of limitations relating only to the issuance, and not the amendment, of citations, citing *Duane Smelser Roofing Co.*[35] In the *Smelser* case, the Sixth Circuit upheld the Commission's reversal of a citation dismissal both because the proposed amendment (to change the violation date) related back to the original citation, and because the employer had consented to the amendment by submitting evidence of work conditions on the correct accident date and was thereby not prejudiced.[36]

Posthearing amendments are allowed only when the employer expressly or impliedly consents or is not prejudiced by

---

[32]*ASARCO, Inc.*, 8 OSHC 2076 (Rev. Comm'n Nos. 79-261, 79-2225, & 79-5269, 1980).

[33]*H.B. Zachry Co.*, 7 OSHC 2202 (Rev. Comm'n 1980).

[34]*CMH Co.*, 9 OSHC 1048 (Rev. Comm'n 1980).

[35]*Duane Smelser Roofing Co.*, 617 F.2d 448, 8 OSHC 1106 (6th Cir. 1980).

[36]*Id.*

such changes.[37] The Commission has itself struggled at length with this issue and remanded several cases for further determinations on possible prejudice to the employer.[38] In a recent case, the Commission split over the propriety of a *sua sponte* amendment on review. Because the employer had raised as a defense evidence which indicated the violation of another standard, the majority held that it was not prejudiced by an amendment on review encompassing such a violation.[39] On the other hand, an administrative law judge's *sua sponte* amendment after the close of the hearing which changed the citation to violation of another regulation was held to be prejudicial by the Tenth Circuit.[40]

V. Posting Citations

Employers are required to post prominently and promptly notices of citations, but not proposed penalties, at or near the place of the violations or where they will be readily observable by affected employees.[41] Violations of this posting requirement may subject employers to civil penalty and result in the dismissal of their notices of contest.[42]

The purpose of the posting requirement is to inform employees of the alleged hazard. The requirement has been up-

---

[37]Compare *C.E. Avery, Div. of Combustion Eng'g*, 8 OSHC 1417 (Rev. Comm'n 1980), where amendment was allowed by the judge, with *Vanco Constr. Co.*, OSAHRC No. 80/56/A2 (1980), where it was not.

[38]See *Foster & Kleiser*, 8 OSHC 1639 (Rev. Comm'n 1980); *Texaco, Inc.*, 8 OSHC 1677 (Rev. Comm'n 1980).

[39]*Turner Welding & Erection Co.*, 8 OSHC 1561 (Rev. Comm'n 1980). Commissioners Cleary and Cottine amended a citation on review which had been dismissed by the judge after the determination that the cause of the violation was unpreventable employee misconduct. In dissent, Commissioner Barnako argued that since such evidence was only raised as a defense, the employer should not be held to have expressly or impliedly consented to a new charge of violating a different specific standard. The majority did permit the employer 10 days to request reconsideration on the grounds of prejudice, but the employer did not do so and the decision became final.

[40]*R.A. Pohl Constr. Co. v. Marshall*, 640 F.2d 266, 9 OSHC 1224 (10th Cir. 1981).

[41]Section 9(b) of the Act, 29 U.S.C. §658(b); 29 C.F.R. §1903.16(a). Citations must remain posted until violations are abated or for 3 working days, whichever is longer. 29 C.F.R. §1903.16(b); see Chapter 5 (Employer Obligations to Obtain, Maintain, and Disseminate Information).

[42]Section 17(i) of the Act, 29 U.S.C. §666(i). See *C & H Erection Co.*, 3 OSHC 1293 (Rev. Comm'n 1975).

held when challenged by the employer on constitutional grounds.[43] Posting of slips of paper, at various locations, on which the employer had noted the cited violation did not comply with this requirement.[44] The requirement was also not met by posting a citation in a warehouse to which employees reported once a week or posting it in an unused latrine.[45] On the other hand, posting a citation so that only its first page was visible, with its other pages taped behind it, was held valid.[46]

[43]*Cullen Indus.*, 6 OSHC 2177 (Rev. Comm'n 1978).

[44]*K. Monkiewicz, Inc.*, 1 OSHC 3259 (Rev. Comm'n 1973).

[45]*Kesler & Sons Constr. Co.*, 3 OSHC 1589 (Rev. Comm'n 1975); *San Juan Constr. Co.*, 3 OSHC 1445 (Rev. Comm'n 1975).

[46]*Bay State Smelting Co.*, 11 OSHC 2254 (Rev. Comm'n 1984).

9

Types and Degrees of Violations

I. Introduction

The OSH Act provides for a scaled system by which an employer's violative conduct may be classified for penalty purposes. The types and degrees of violations within this scaled system are based on the gravity of the employer's culpability. A violation may be classified as "serious," "other than serious" (sometimes referred to as "nonserious"), or "de minimis." Additionally, serious and other than serious violations may be further classified as "willful" or "repeated" violations. The Act also contemplates that there may be a condition so immediately hazardous to employee safety or health as to constitute an "imminent danger."

II. Serious Violations

Section 17(k) of the Act provides that

"a serious violation shall be deemed to exist in a place of employment if there is a substantial probability that death or serious physical harm could result from a condition which exists, or from one or more practices, means, methods, operations, or processes which have been adopted or are in use, in such place of employment unless the employer did not, and could not with

257

the exercise of reasonable diligence, know of the presence of the violation."[1]

As the Act's definition indicates, a serious violation contains two basic elements, (a) that the employer has actual or constructive knowledge of the hazardous condition, and (b) that exposure to the condition could result in the substantial probability of death or serious physical harm.

A. Actual or Constructive Knowledge

Actual or constructive knowledge of the potentially harmful condition by the employer is, therefore, essential to the finding of a serious violation, and it is the Secretary's burden to establish satisfaction of the knowledge criterion as part of his *prima facie* case. Thus, the Ninth Circuit admonished the Secretary for failing to satisfy his knowledge burden of proof, stating:

> "There is no evidence in the record tending to show that the employer had any knowledge respecting these instances of employee disobedience of its established instructions. No effort was made to establish that the employer had any ongoing practice of permitting its instructions to be disregarded by its employees with impunity. The Secretary alleged employer knowledge and [the employer] denied it. As the case reaches us, the absence of employer knowledge must be considered one of the established facts."[2]

The court further rejected the Secretary's argument that, in the context of a serious violation, he need not prove the existence of employer knowledge but, rather, the employer must prove its absence.[3]

To satisfy the knowledge element of a serious violation, the Secretary need only show that the employer knew, or with reasonable diligence could have known, that a hazardous con-

---

[1] 29 U.S.C. §666(k) (1985).

[2] *Brennan v. OSHRC & Raymond Hendrix*, 511 F.2d 1139, 1140, 2 OSHC 1646, 1647 (9th Cir. 1975). See also *Puffer's Hardware v. Donovan*, 742 F.2d 12, 11 OSHC 2197 (1st Cir. 1984); *St. Joe Minerals Corp. v. OSHRC & Marshall*, 647 F.2d 840, 9 OSHC 1646 (8th Cir. 1981); *Bunge Corp. v. Secretary of Labor*, 638 F.2d 831, 9 OSHC 1312 (5th Cir. 1981); *Horne Plumbing & Heating Co. v. OSHRC*, 528 F.2d 564, 3 OSHC 2060 (5th Cir. 1976); *Prestressed Sys.*, 9 OSHC 1864 (Rev. Comm'n 1981); *MCC of Fla.*, 9 OSHC 1895 (Rev. Comm'n 1981); *D.R. Johnson Lumber Co.*, 3 OSHC 1124 (Rev. Comm'n 1975); *Cam Indus.*, 1 OSHC 1564 (Rev. Comm'n 1974).

[3] *Brennan v. OSHRC & Raymond Hendrix, supra* note 2.

dition existed. The Secretary need not establish that the employer knew that a particular OSHA standard prohibited a certain condition. Thus, in a 1977 Review Commission decision, an employer who allowed employees to perform ceiling and drywall work without the approved safety belts and lifelines argued that it did not know that such a practice was prohibited. The Commission rejected the argument, noting that the knowledge element for a serious violation refers to knowledge of the physical conditions and not to knowledge of the applicable law.[4] Similarly, the Secretary need not establish that the employer knew that a condition could result in harm to an employee.[5]

Actual knowledge of the cited hazard is not required for a serious violation. Rather, the Secretary will satisfy his burden of proving the existence of employer knowledge by showing that the employer could have known of the hazardous condition with the exercise of reasonable diligence.[6] Thus, in *North American Rockwell Corp.* the Commission expressly stated: "Knowledge of the existence of a violation, either actual or constructive, is an essential element of any violation of the Act. [Citation omitted.] Therefore, the evidence must establish that the respondent knew or, with reasonable diligence, should have known of the existence of the violation."[7] The Commission has also consistently held that a supervisor's knowledge of a hazardous condition will be imputed to the employer and will satisfy the Secretary's burden of proving knowledge.[8] The Third Circuit has held, however, that a supervisor's knowledge will not "end the supervisor's inquiry into foreseeability"; it is only evidence that an employer could

[4]See *Southwestern Acoustics & Specialty,* 5 OSHC 1091 (Rev. Comm'n 1977).

[5]See, e.g., *RSR Corp. v. Donovan & OSHRC,* 747 F.2d 294, 12 OSHC 1073 (5th Cir. 1984), *aff'g* 9 OSHC 2099 (Rev. Comm'n 1981); *Sunoutdoor Advertising,* 5 OSHC 1159 (Rev. Comm'n 1977).

[6]*Bunge Corp. v. Secretary of Labor, supra* note 2. See *St. Joe Minerals Corp. v. OSHRC & Marshall, supra* note 2; *Mahone Grain Corp.,* 10 OSHC 1275 (Rev. Comm'n 1981). But see *Texas Util. Generating Co.,* 7 OSHC 2112 (Rev. Comm'n J. 1979).

[7]*North Am. Rockwell Corp.,* 2 OSHC 1710, 1711 (Rev. Comm'n Nos. 2692 & 2875, 1975), *aff'd,* 540 F.2d 1283, 4 OSHC 1606 (6th Cir. 1976).

[8]See, e.g., *Wright & Lopez, Inc.,* 3 OSHC 1261 (Rev. Comm'n 1980); *Georgia Elec. Co.,* 5 OSHC 1112 (Rev. Comm'n 1977); *Structural Steel Erectors,* 2 OSHC 1506 (Rev. Comm'n 1975). But see *Ocean Elec. Corp. v. Secretary of Labor,* 594 F.2d 396, 7 OSHC 1149 (4th Cir. 1979); *Mountain States Tel. & Tel. Co. v. OSHRC,* 623 F.2d 155, 8 OSHC 1557 (10th Cir. 1980).

have foreseen and with reasonable diligence prevented the violation.[9]

B. Substantial Probability of Death or Serious Physical Harm

The second element of a serious violation is that exposure to the cited condition could result in the substantial probability of death or serious physical harm. As with the knowledge criterion, it is the Secretary's burden of proof to establish the probability of death or serious physical harm as part of his *prima facie* case.[10]

The substantial probability of death or serious physical harm required by the Act is *not* a substantial probability that an accident will, in fact, result, but only that, if the accident were to occur, there would be a substantial probability that death or serious physical harm would result.[11] Thus, for example, in *California Stevedore & Ballast Co. v. OSHRC*, although the court acknowledged that it was unlikely that an unsecured hatch beam would dislodge, the court nonetheless found that, if it were to dislodge, death or serious injury would result.[12]

Normally, the probability that the accident will occur will figure only into the calculation of the appropriate penalty.[13] Nevertheless, where the possibility of injury is *extremely* remote, the citation may be reduced to other than serious. For example, in *John W. McGowan*, where it was determined that there was one chance in 10 million that static electricity could set off an explosion of a storage tank, the Secretary's allegations of a serious violation failed.[14]

---

[9]*Pennsylvania Power & Light Co. v. OSHRC & Donovan*, 737 F.2d 350, 11 OSHC 1983 (3d Cir. 1984).

[10]See, e.g., *Crescent Wharf & Warehouse Co.*, 1 OSHC 1219 (Rev. Comm'n 1973).

[11]See, e.g., *Georgia-Pac. Corp.*, 9 OSHC 1460 (Rev. Comm'n 1981); *Usery v. Hermitage Concrete Pipe Co.*, 584 F.2d 127, 6 OSHC 1886 (6th Cir. 1979); *Dorey Elec. Co. v. OSHRC*, 553 F.2d 357, 5 OSHC 1285 (4th Cir. 1977); *Shaw Constr. v. OSHRC*, 534 F.2d 1183, 4 OSHC 1427 (5th Cir. 1976); *Brady-Hamilton Stevedore Co.*, 3 OSHC 1925 (Rev. Comm'n 1976).

[12]*California Stevedore & Ballast Co. v. OSHRC*, 517 F.2d 986, 3 OSHC 1174 (9th Cir. 1975).

[13]*Baltz Bros. Packing*, 1 OSHC 1118 (Rev. Comm'n 1973). See Chapter 11 (Assessment of Civil Penalties).

[14]*John W. McGowan*, 5 OSHC 2028 (Rev. Comm'n 1977). See also *Diamond Int'l Corp.*, 4 OSHC 1821 (Rev. Comm'n 1976).

OSHA's Field Operation Manual (FOM) contains instructions to its compliance officers with respect to the citation of serious violations.[15] The FOM sets forth a four-step process by which a compliance officer may determine if it is appropriate to cite a specific condition as serious.[16] The FOM also provides the compliance officer with a definition of the term "serious physical harm" which states that "serious physical harm" probably will be present where a condition could result in:

"*1* Impairment of the body in which part of the body is made *functionally useless* or is *substantially reduced in efficiency* on or off the job. Such impairment may be permanent or temporary, chronic or acute. Injuries involving such impairment would usually require treatment by a medical doctor. Examples of injuries which constitute such harm include:
"*a* Amputation (loss of all or part of a bodily appendage which includes the loss of bone).
"*b* Concussion.
"*c* Crushing (internal, even though skin surface may be intact).
"*d* Fracture, simple or compound.
"*e* Burn or scald, including electric and chemical burns.
"*f* Cut, laceration, or puncture involving significant bleeding and/or requiring suturing.
"*2* Illnesses that could shorten life or significantly reduce physical or mental efficiency by inhibiting the normal function of a part of the body. Some examples of such illnesses include cancer, silicosis, asbestosis, byssinosis, hearing impairment, central nervous system impairment and visual impairment."[17]

Although the Secretary has the burden of proof with respect to employer knowledge and the probability of serious injury, the Secretary need not establish that the employer had

---

[15]FOM ch. IV (violations), ch. V (citations), ch. VI (penalties) (Oct. 21, 1985), OSHR [2 Reference File 77:2701–3301].

[16]"The four elements the [compliance officer] shall consider are as follows:
"(1) *Step 1.* The type of accident or health hazard exposure which the violated standard or the general duty clause is designed to prevent.
***
"(2) *Step 2.* The type of injury or illness which could reasonably be expected to result from the type of accident or health hazard exposure identified in Step (1).
***
"(3) *Step 3.* Whether the types of injury or illness identified in Step (2) could include death or a form of serious physical harm.
***
"(4) *Step 4.* Whether the employer knew, or with the exercise of reasonable diligence, could have known of the presence of the hazardous condition."
FOM ch. IV B.1.b(1)–(4) (Oct. 21, 1985), OSHR [2 Reference File 77:2708–2709].
[17]FOM ch. IV B.1.b(3)(a)1–2 (Oct. 21, 1985), OSHR [2 Reference File 77:2709].

knowledge that a cited condition would have the potential for serious injury. In other words, the employer "knowledge necessary for a serious violation is knowledge only of the violative condition, and not of whether that condition presents a substantial probability of causing death or serious harm."[18]

With respect to serious violations of air contaminant standards, the Commission originally had maintained that the Secretary failed to meet his burden of showing that there was a substantial probability of death or serious physical harm in cases involving silica exposure[19] and lead exposure[20] because, the Commission concluded, the evidence did not establish that the nature, amount, and duration of exposure in the cases before it would lead to the diseases associated with such exposures. However, in *Anaconda Aluminum Co.*, the Commission, with prodding from the courts, reversed its prior holdings in the earlier silica and lead cases, reasoning that those decisions required the Secretary to prove that the degree of exposure to the contaminant *would* lead to a serious disease, whereas Section 17(k) of the Act only requires a showing that a substantial probability of death or serious harm *could* result from a violation.[21]

C. Mandatory Civil Penalties

Section 17(b) of the Act requires mandatory civil penalties for serious violations of up to $1000 for each such violation. Good faith efforts to abate a hazard may reduce the dollar amount of the penalty although not the degree of violation. For instance, in *Continental Steel Corp.* an employer failed to protect employees against excessive noise levels.[22] Although the citation was characterized as serious, the penalty was only $1.00 owing to the employer's good faith efforts to reduce noise exposure.[23]

---

[18]*Anaconda Aluminum Co.*, 9 OSHC 1460, 1474 n.37 (Rev. Comm'n 1981).

[19]*Hermitage Concrete Pipe Co.*, 3 OSHC 1920 (Rev. Comm'n 1976), *rev'd*, 584 F.2d 127, 6 OSHC 1886 (6th Cir. 1978).

[20]*Hydrate Battery Corp.*, 2 OSHC 1719 (Rev. Comm'n 1975).

[21]*Anaconda Aluminum Co., supra* note 18, at 1476.

[22]*Continental Steel Corp.*, 3 OSHC 1410 (Rev. Comm'n 1975), *rev'd sub nom. Penn-Dixie Steel v. OSHRC*, 553 F.2d 1078, 5 OSHC 1315 (7th Cir. 1977).

[23]For a more complete discussion of the penalties associated with serious violations, see Chapter 11 (Assessment of Civil Penalties).

III. Other Than Serious Violations

Unlike the serious violation, the "other than serious" or "nonserious" violation has no statutory definition.[24] Nevertheless, this void has been filled by case law. Thus in *Crescent Wharf & Warehouse Co.* the Commission stated that "a nonserious violation is one in which there is a direct and immediate relationship between the violative condition and occupational safety and health but not of such relationship that a resultant injury or illness is death or serious physical harm."[25] The FOM also offers a definition for what it refers to as "other than serious violations": "This type of violation shall be cited in situations where an accident or illness results from a hazardous condition that would probably not cause death or serious physical harm but would have a direct or immediate relationship to the safety or health of employees."[26]

Although there is no express statutory requirement of *scienter* with an other than serious violation, the Commission and most courts that have addressed the issue have held that the Secretary has the burden of showing that the employer had actual, constructive, or imputed knowledge of the condition underlying such a violation.[27] Thus, where a serious citation fails for lack of proof on the issue of probability of death or serious physical harm, a nonserious violation generally will be found.[28] However, where a serious citation fails for lack of proof of employer knowledge of the condition, an other than serious violation will not be appropriate, because the Commission as well as the courts requires *scienter* for nonserious as well as serious violations.[29]

---

[24]Section 17(c) of the Act merely refers to the penalty assessment for violations "specifically determined not to be of a serious nature." 29 U.S.C. §666(c) (1985).

[25]*Crescent Wharf & Warehouse Co.*, 1 OSHC 1219, 1222 (Rev. Comm'n 1973).

[26]FOM ch. IV B.2 (Oct. 21, 1985), OSHR [2 Reference File 77:2710].

[27]See, e.g., *Prestressed Sys.*, 9 OSHC 1864 (Rev. Comm'n 1981); *MCC of Fla.*, 9 OSHC 1895 (Rev. Comm'n 1981); *Green Constr. Co.*, 4 OSHC 1808 (Rev. Comm'n 1976); *Wally Taylor Constr. Co.*, 4 OSHC 1890 (Rev. Comm'n 1976); *Horne Plumbing & Heating Co. v. OSHRC*, 528 F.2d 564, 3 OSHC 2060 (5th Cir. 1976); *Dunlop v. Rockwell Int'l*, 540 F.2d 1283, 4 OSHC 1606 (6th Cir. 1976), *aff'g North Am. Rockwell Corp.*, 2 OSHC 1710 (Rev. Comm'n Nos. 2692 & 2875, 1975); *Brennan v. OSHRC & Raymond Hendrix*, 511 F.2d 1139, 2 OSHC 1646 (9th Cir. 1975).

[28]See, e.g., *Fulton Instrument Co.*, 2 OSHC 1366 (Rev. Comm'n 1974).

[29]See, e.g., *Wally Taylor Constr. Co.*, *supra* note 27.

OSHA has taken the position, with Commission approval, that a serious citation may issue for a group of individual related violations which, if cited individually, would be nonserious but when cited collectively are serious because in combination they present a substantial probability of injury resulting in death or serious physical harm.[30]

Section 17(c) makes penalties for nonserious violations discretionary, but sets an upward limit of $1000 for each such violation. However, as a matter of policy, penalties for other than serious violations will not normally exceed $300.[31]

IV. Repeated Violations

The statutory basis for the repeated violation is found in brief reference to "[a]ny employer who *** repeatedly violates the requirements" in Section 17(a) of the Act.

A. "Substantial Similarity"

The Commission has addressed itself in numerous decisions to the repeated violation concept, from which has evolved its current position on what constitutes a repeated violation, as articulated in *Potlatch Corp.*[32] The Secretary must show that, at the time of the alleged repeated violation, there was a Commission final order against the same employer for a substantially similar violation.[33] The Secretary may satisfy his burden of establishing "substantial similarity" in several ways. In cases arising under Section 5(a)(2) of the Act, the Secretary may show that the prior and present violations are for failure to comply with the same standard. If the same

---

[30]See *CTM, Inc.*, 4 OSHC 1468 (Rev. Comm'n 1976) (employer failed to both shore and slope trench to safe angle while soil contained large clods of earth which could roll creating trench collapse possibility, a single serious condition; although violation of one standard was nonserious and other was serious, together they constituted single serious violation).

[31]See Chapter 11 (Assessment of Civil Penalties).

[32]*Potlatch Corp.*, 7 OSHC 1061 (Rev. Comm'n 1979).

[33]See, e.g., *MICA Constr. Co.*, 10 OSHC 1381 (Rev. Comm'n 1982) (repeated citation vacated because Secretary failed to prove final order of Commission was entered against employer for prior substantially similar violation).

standard in both the prior and present citations is a specific standard (i.e., a standard applicable to particular equipment or operations), an employer will have an extremely difficult task in rebutting the Secretary's *prima facie* showing of similarity. However, if the same standard in both the prior and present citations is a general standard (i.e., a standard applicable to a variety of workplace settings), the Secretary's *prima facie* similarity showing may be challenged by the employer's evidentiary showing that disparate conditions and hazards were associated with the earlier and later violations. For example, a general construction standard such as 29 C.F.R. §1926.28(a) (1985), requiring the wearing of appropriate personal protective equipment where there is an exposure to hazardous conditions or where the wearing of such equipment will reduce hazards, may be used to cite an employer for failing to protect employees from fall hazards by requiring the use of safety belts. This same standard may also be used to cite an employer for failing to require employees to use seat belts in an earth-moving vehicle. An employer could rebut the Secretary's *prima facie* showing of similarity underlying a repeated violation of this standard by presenting evidence of the different conditions and hazards associated with these two violations of the same standard.[34]

If the prior and present citations do not involve the same standard, the Secretary's burden of establishing that the violations are substantially similar in nature is significantly increased. The Commission expressly held in *Potlatch* that the Act did not require that a repeated violation be predicated upon the same standard. Thus, "[a] section 5(a)(2) violation may *** be found to be repeated on the basis of either a prior section 5(a)(1) or 5(a)(2) violation and a section 5(a)(1) violation may similarly be found to be repeated on the basis of either a prior section 5(a)(1) or section 5(a)(2) violation. There must, of course, be evidence of substantial similarity between the prior and present violations."[35]

---

[34]See *Potlatch Corp., supra* note 32. See also *Kent Nowlin Constr. Co.,* 9 OSHC 1306 (Rev. Comm'n 1981).

[35]*Potlatch Corp., supra* note 32, at 1064. See *J.L. Foti Constr. Co. v. OSHRC & Donovan,* 687 F.2d 853, 10 OSHC 1937 (6th Cir. 1982). The Sixth Circuit affirmed the Commission's position in *Potlatch* that a repeated violation citation may be issued for a violation of one standard when it is preceded by a final order for violation of another standard. However, because the *Foti* proceeding commenced prior to *Potlatch*

The Commission further concluded in *Potlatch* "that one prior violation may support a finding of repeated."[36]

B. Qualifying Factors, OSHA "Uniform Policy," and Penalties

With respect to qualifying factors such as employer attitude, commonality of supervisory control over the violative condition, geographic distance of the violations, time lapse between violations, and number of prior violations, the Commission concluded that such factors would not bear on the classification of the violation as repeated but rather would be considered in assessing a penalty.[37]

Among the elements of a repeated violation is the requirement that the same employer be involved in the antecedent and present citation. However, the Commission has given an expansive interpretation to this requirement and has held in several cases that a change in corporate ownership may not preclude issuance of a repeated violation citation.[38]

Over the years, OSHA, the Commission, and the courts have faced recurring problems with respect to proper application of the repeated violation classification. Once such problem has involved the issue of multiple worksites. This issue can be a matter of different locations at the same large worksite,[39] different worksites of the same employer, or temporary

---

and because in *Potlatch* the Commission for the first time reached a consensus on this issue, the Sixth Circuit concluded it would not apply *Potlatch* retroactively as the employer would have had no reason to suspect it was vulnerable to citation for a repeated violation of one standard based on a previous final order being entered for violation of another standard.

[36] *Potlatch Corp., supra* note 32, at 1064. In so holding, the Commission rejected the Third Circuit's position in *Bethlehem Steel v. OSHRC*, 540 F.2d 157, 4 OSHC 1451 (3d Cir. 1976), and *Jones & Laughlin Steel Corp. v. Marshall*, 636 F.2d 32, 8 OSHC 2217 (3d Cir. 1980), that the term "repeatedly" means "more than twice" and involves a showing that there was a deliberate "flaunting" of the Act by the employer. *Bethlehem Steel*, 540 F.2d at 162, 4 OSHC at 1455. The Third Circuit's position has generally been rejected by other appellate courts. See, e.g., *Kent Nowlin Constr. Co. v. OSHRC, supra* note 34; *Todd Shipyards v. Secretary of Labor*, 566 F.2d 1327, 6 OSHC 1227 (9th Cir. 1977).

[37] *Potlatch Corp., supra* note 32, at 1064.

[38] See, e.g., *Turnbull Millwork Co.*, 6 OSHC 1148 (Rev. Comm'n 1976). But see *Universal Aluminal Extrusion Corp.*, 6 OSHC 1423 (Rev. Comm'n 1978) (bona fide change in corporate ownership and name between the prior and current citations precluded application of repeated classification).

[39] See, e.g., *J.M. Martinac Shipbuilding Corp.*, 6 OSHC 1645 (Rev. Comm'n 1978); *Bethlehem Steel Corp., supra* note 36; *Todd Shipyards Corp.*, 3 OSHC 1813 (Rev. Comm'n 1975).

versus permanent worksites.[40] Another problem has involved the time lapse between citations and its impact on the determination of whether a repeated violation citation should be issued.

In an attempt to clarify its position and to treat such issues in a consistent manner, on June 28, 1982, OSHA's "uniform policy" for citations for repeated violations became effective. This uniform policy is a guide for compliance officers in determining when to issue citations for repeated violations. Among its directives, the uniform policy, as set out in the FOM, provides:

1. That to be cited as a repeated violation, the cited condition must occur within 3 years of the date that the antecedent citation became a final order of the Commission or within 3 years of the final correction date, whichever is later.[41]

2. That a distinction be made between fixed and nonfixed worksites.[42]

3. That fixed worksites be defined as "a single physical location where business is conducted or where services or industrial operations are performed."[43]

4. That multifacility employers be cited for a repeated violation only if the same violative condition occurs at the same fixed worksite, i.e., the same facility, as the antecedent violation.[44]

5. That nonfixed worksites be defined as all geographical sites or locations within a single OSHA Area Office jurisdiction. Accordingly, where, for example, an em-

---

[40]For an employer with a fixed establishment, a repeated violation citation is issued if the subsequent violation occurred in the same establishment. For an employer with no fixed establishment, a repeated violation citation is issued if the subsequent violation occurred anywhere in the same Area Office jurisdiction. FOM ch. IV B.5.c(1–2) (Oct. 21, 1985), OSHR [2 Reference File 77:2711–2712]; *George Hyman Constr. Co. v. OSHRC*, 582 F.2d 834, 6 OSHC 1855 (4th Cir. 1978); *Desarrolos Metropolitanos, Inc. v. OSHRC*, 551 F.2d 874, 5 OSHC 1135 (1st Cir. 1977).

[41]See FOM ch. IV B.5.d (Oct. 21, 1985), OSHR [2 Reference File 77:2712]:
"Although there are no statutory limitations upon the length of time that a citation may serve as the basis for repeated violation, or [sic] order to ensure uniformity, 3 years from the date that the earlier citation became a final order or 3 years from the final statement date of that citation time period within which another violation of the same standard may be classified as repeated."

[42]See *supra* note 40.

[43]See FOM ch. IV B.5.c(1) (Oct. 21, 1985) (as defined in 29 C.F.R. §1903.2(b) (1985)), OSHR [2 Reference File 77:2711].

[44]See FOM ch. IV B.5.a, B.5.c(1), (Oct. 21, 1985), OSHR [2 Reference File 77:2711].

ployer has two construction sites, one in the jurisdiction of one Area Office and the other in the jurisdiction of another Area Office, a violation in one cannot be the basis for a repeated violation in the other.[45]
6. That with respect to longshoring operations, stevedoring activities within any single port area will be subject to repeated violations.[46]

Section 17(a) of the Act states that there shall not be a penalty of more than $10,000 for *each* repeated violation. In practice, the penalty for a repeated violation is derived by assigning a gravity-based penalty for each new violation. This gravity-based penalty is then doubled for the first repeat, quadrupled for the second repeat, and multiplied by 10 for the third repeat. In assigning the underlying gravity-based penalty, the factors discussed in *Potlatch Corp.* of employer attitude, commonality of supervisory control of the violative condition, geographical proximity, and time lapse may be considered.[47] However, with the uniform policy, there may be less discretion with respect to the latter two factors.[48]

V. Failure to Abate Violations

Under Section 17(b) of the Act, the Secretary is empowered to cite an employer for "failure to abate" or "failure to correct" a previously cited violation. Although a citation for failure to abate is conceptually similar to a citation for a repeated violation, the two are in fact vastly different. A failure to abate is established only when a cited condition has continued uncorrected; once corrective action is taken by an employer, there can be no citation for failure to abate.[49] A repeated citation, however, occurs when an employer commits an entirely new violation which is substantially similar to a previously cited condition in a final Commission order.[50]

---

[45]See FOM ch. IV B.5.c(2), (Oct. 21, 1985), OSHR [2 Reference File 77:2712].
[46]See FOM ch. IV B.5.c(3), (Oct. 21, 1985), OSHR [2 Reference File 77:2712].
[47]*Potlatch Corp.*, 7 OSHC 1061, 1064 (Rev. Comm'n 1979).
[48]See Chapter 11 (Assessment of Civil Penalties).
[49]*Braswell Motor Freight Lines*, 5 OSHC 1469 (Rev. Comm'n 1977). See also *Edward B. Fitzpatrick Jr. Assocs. & Schiavone Constr. Co., A Joint Venture*, 5 OSHC 2004 (Rev. Comm'n J. 1977).
[50]At least one Commission administrative law judge has ruled that owing to the fundamental difference between a repeated violation and a failure to abate an existing

The Secretary can establish his *prima facie* case of failure to abate where the original citation is not contested and has become a final Commission order, and upon reinspection the condition found is identical to the one for which the employer had been originally cited.[51] If the original citation requires immediate abatement, and the Secretary reinspects the employer prior to expiration of the 15-day contest period, the employer nevertheless may be cited for a failure to abate if it had not filed a notice of contest prior to the reinspection.[52]

An employer can rebut the Secretary's *prima facie* case by showing that the cited condition was corrected, or if not corrected, that the employer has prevented the exposure of its employees to the cited condition.[53] The Commission has also permitted an employer to rebut a *prima facie* case by showing that the condition for which the employer was cited was not violative of the Act either at the time of the first inspection or the reinspection,[54] that it is impossible to comply with the cited standard,[55] or that the employer has taken reasonable steps to abate the hazard, even though the employer is still in technical violation of the Act.[56]

Section 17(d) of the Act provides for penalties "of not more than *$1,000 for each day* during which such failure [to abate] or violation continues" (emphasis added). However, normally a single total proposed penalty for a failure to abate a particular violation is issued.[57]

VI. Willful Violations

The statutory predicate for a willful violation is Section 17(a) of the Act which refers to "[a]ny employer who willfully

---

violation, a particular condition may not be cited alternatively as a repeated violation and a failure to abate. See *Montgomery AMC Drilling*, 5 OSHC 1629 (Rev. Comm'n J. 1977).

[51]*Braswell Motor Freight Lines, supra* note 49.

[52]See, e.g., *Brennan v. OSHRC (Kesler & Sons Constr.)*, 513 F.2d 553, 2 OSHC 1668 (10th Cir. 1975); *Dunlop v. Haybuster Mfg. Co.*, 524 F.2d 222, 3 OSHC 1594 (8th Cir. 1975); *George T. Gerhardt Co.*, 4 OSHC 1351 (Rev. Comm'n 1976).

[53]*Braswell Motor Freight Lines, supra* note 49.

[54]*Id.* See also *York Metal Finishing Co.*, 1 OSHC 1655 (Rev. Comm'n 1974).

[55]*Bradley Plywood Corp.*, 9 OSHC 1103 (Rev. Comm'n J. 1980); *Robert's Sheet Metal Co.*, 5 OSHC 1659 (Rev. Comm'n Nos. 76-2806 & 76-3147, 1977).

[56]*Hudson Fabricating Co.*, 8 OSHC 1647 (Rev. Comm'n J. 1980).

[57]See Chapter 11 (Assessment of Civil Penalties).

*** violates the requirements" of the Act. The willful violation is subject to one of the most severe monetary sanctions prescribed by the Act.

An employer can be cited for a willful violation under the Act's general duty clause contained in Section 5(a)(1) of the Act[58] as well as for a violation of a standard promulgated under Section 5(a)(2) of the Act.[59] Further, a willful citation can be predicated upon an underlying violation which is serious or other than serious in gravity. However, absent an assertion in the citation that the alleged violation is also serious, the Commission will infer that the willful violation is other than serious.[60] Nevertheless, the Commission may find a serious violation where the Secretary has cited an employer only for a willful violation but fails to establish willfulness, if the serious violation was tried by express or implied consent.[61]

A. Willful Violation Standard

There is basic uniformity among the courts, the Commission, and OSHA with regard to the standard to be used in finding a willful violation.[62] The evidentiary burden to establish satisfaction of this standard rests with the Secretary.[63]

---

[58]See, e.g., *Hayes Albion Corp.*, 12 OSHC 1273 (Rev. Comm'n 1985).

[59]See, e.g., *Dic-Underhill, A Joint Venture*, 5 OSHC 1251 (Rev. Comm'n 1977).

[60]See, e.g., *Gates & Fox Co.*, 12 OSHC 1093 (Rev. Comm'n 1984); *Toler Excavating Co.*, 3 OSHC 1420 (Rev. Comm'n 1975); *Amulco Asphalt Co.*, 3 OSHC 1396 (Rev. Comm'n 1975).

[61]*National Realty & Constr. Co. v. OSHRC*, 489 F.2d 1257, 1264 n.41, 1 OSHC 1422, 1429 (D.C. Cir. 1973); *Emerson Elec. Co., Edwin L. Wiegand Div.*, 7 OSHC 1441 (Rev. Comm'n J. 1979); *Amulco Asphalt Co., supra* note 60; *Royster Co.*, 6 OSHC 1144 (Rev. Comm'n 1977); *Toler Excavating Co., supra* note 60.

[62]See, e.g., *F.X. Messina Constr. Corp. v. OSHRC*, 505 F.2d 701, 2 OSHC 1325 (1st Cir. 1974); *U.S. v. Dye Constr. Co.*, 510 F.2d 78, 2 OSHC 1510 (10th Cir. 1975); *Intercounty Constr. Co. v. OSHRC*, 522 F.2d 777, 3 OSHC 1337 (4th Cir. 1975), *cert. denied*, 423 U.S. 1072 (1976); *Detroit Steel Corp., Empire Detroit Steel Div. v. OSHRC*, 579 F.2d 378, 6 OSHC 1993 (6th Cir. 1978); *Western Waterproofing Co. v. Marshall*, 576 F.2d 139, 6 OSHC 1550 (8th Cir. 1978), *cert. denied*, 439 U.S. 965 (1978); *Cedar Constr. Co. v. OSHRC*, 587 F.2d 1303, 6 OSHC 2010 (D.C. Cir. 1978); *Georgia Elec. Co. v. Marshall*, 595 F.2d 309, 7 OSHC 1343 (5th Cir. 1979); *National Steel & Shipbuilding Co. v. OSHRC*, 607 F.2d 311, 7 OSHC 1837 (9th Cir. 1979); *Kent Nowlin Constr. Co. v. OSHRC*, 593 F.2d 368, 7 OSHC 1105 (10th Cir. 1979); *A. Schonbek & Co. v. Donovan*, 646 F.2d 799, 9 OSHC 1562 (2d Cir. 1980); *C.N. Flagg & Co.*, 2 OSHC 1195 (Rev. Comm'n 1974); *Amulco Apshalt Co., supra* note 60; *Kent Nowlin Constr.*, 5 OSHC 1051 (Rev. Comm'n 1977).

[63]*Stone & Webster Eng'g Corp.*, 8 OSHC 1753 (Rev. Comm'n 1980). See *Graven Bros. & Co.*, 7 OSHC 2228 (Rev. Comm'n 1976), *modified*, 5 OSHC 1074 (10th Cir. 1977). But see *Acme Fence & Iron Co.*, 4 OSHC 1045 (Rev. Comm'n 1980).

The Commission's basic articulation of the willful violation standard has been set forth in numerous cases. For example, in *C.N. Flagg & Co.* the Commission stated "that when used in the civil sense 'willful' means intentional, knowing, or voluntary as distinguished from accidental conduct and may be characterized as conduct marked by careless disregard."[64] This standard has been utilized by various courts of appeals as well, as is typified by the court's statement in *A. Schonbek & Co. v. Donovan*: "Both the Commission and the Occupational Safety and Health Administration define a 'willful' violation as one done either with an intentional disregard of, or plain indifference to, the statute. We join those other Courts of Appeals that have approved the administrative definition of willfulness."[65]

1. Employer Knowledge Criterion

As the definitions set out above indicate, employer knowledge of an OSHA standard or other requirement of the Act is a crucial criterion to the establishment of a willful violation. This requirement goes beyond mere knowledge of the existence of the violative condition; rather, the employer must have knowledge that the condition violated the Act.

The Secretary can satisfy his burden of proof with respect to this requirement by several means. First, he can impute a foreman's or supervisor's knowledge to the employer.[66] Second, the Secretary can show requisite knowledge by establishing that the employer had been previously cited for violations of the same standard.[67] Although mere prior citations for the same standard will not alone make out a willful violation, it can be used to prove employer knowledge.[68] Third, the Sec-

[64]*C.N. Flagg & Co., supra* note 62.

[65]*A. Schonbek & Co. v. Donovan, supra* note 62, at 800.

[66]See, e.g., *C.N. Flagg & Co., supra* note 62; *Fanning & Doorley Constr. Co.,* 2 OSHC 1195 (Rev. Comm'n 1977).

[67]See, e.g., *Branciforte Builders,* 6 OSHC 1251 (Rev. Comm'n J. 1980); *B&L Excavating Co.,* 8 OSHC 1739 (Rev. Comm'n J. Nos. 76-4419 & 76-4420, 1977); *F.X. Messina Constr. Corp., supra* note 62.

[68]See, e.g., *Cedar Constr. Co. v. OSHRC, supra* note 62; *Georgia Elec. Co. v. Marshall,* 595 F.2d 309 (5th Cir. 1979); *Dic-Underhill, supra* note 59. But see *Wright-Lopez, Inc.,* 10 OSHC 1108 (Rev. Comm'n 1981) (absence of prior citation no bar to finding of willful violation); *National Steel & Shipbuilding Co. v. OSHRC,* 607 F.2d 311 (9th Cir. 1979) (citation under different standard for similar condition will sometimes suffice).

retary can satisfy his knowledge evidentiary burden by showing that the employer's knowledge of a violative condition
constituted a knowing, voluntary disregard of employee safety.[69]
For example, in *Stabilized Pigments,* although eight employees had suffered from lead poisoning, the employer failed to
take action to reduce the lead oxide dust in its plant.[70] Similarly, in *National Steel & Shipbuilding Co. v. OSHRC,* despite
repeated warnings by the general contractor to the respondent
subcontractor at weekly meetings that unguarded scaffolds
were unsafe, no action was taken by the subcontractor to guard
the scaffolding.[71]

Just as a supervisor's knowledge will be imputed to the
employer to satisfy the knowledge requirement of a willful
violation, so will a supervisor's misconduct be attributed to
the employer to show that a violation has in fact occurred.[72]

2. State-of-Mind Criterion

Unlike any of the other violation classifications discussed
above, the willful violation involves the Secretary in establishing that the cited employer had a certain state of mind,
that is, the employer acted or failed to act "intentionally,"
"knowingly," "voluntarily," or "with plain indifference to the
Act."[73] This state-of-mind criterion has been interpreted by
the Third Circuit[74] to mean that the Secretary must show that
the employer engaged in "deliberate flaunting" of the Act. In
reaching this determination, the appellate court in *Frank Irey,
Jr. v. OSHRC* reasoned that "willfulness connotes such defi-
ance or such reckless disregard of the consequences as to be

---

[69]See, e.g., *Cedar Constr. Co. v. OSHRC, supra* note 62; *American Stair Corp.,*
6 OSHC 1899 (Rev. Comm'n J. 1978); *Amulco Asphalt Co., supra* note 60. See also
FOM ch. IV B.4.c (Oct. 21, 1985), OSHR [2 Reference File 77:2710–2711].

[70]*Stabilized Pigments,* 8 OSHC 1160 (Rev. Comm'n J. 1980).

[71]*National Steel & Shipbuilding Co. v. OSHRC,* 607 F.2d 311, 7 OSHC 1837 (9th
Cir. 1979). See also *Kus-Tum Builders, Inc.,* 10 OSHC 1128 (Rev. Comm'n J. 1981).

[72]See, e.g., *F.X. Messina Constr. Corp. v. OSHRC, supra* note 62; *Gates & Fox
Co.,* 8 OSHC 1548 (Rev. Comm'n J. 1980); *General Motors Corp., Central Foundry
Div.,* 8 OSHC 1352 (Rev. Comm'n J. 1980); *Georgia Elec. Co.,* 5 OSHC 1112 (Rev.
Comm'n 1977).

[73]See, e.g., *C. N. Flagg & Co., supra* note 62; *Kent Nowlin Constr.,* 5 OSHC 1051
(Rev. Comm'n 1977); *Georgia Elec. Co. v. Marshall, supra* note 62; *Intercounty Constr.
Co. v. OSHRC, supra* note 62. See also Chapter 11 (Assessment of Civil Penalties).

[74]*Universal Auto Radiator Mfg. Co. v. Marshall,* 631 F.2d 20 (3d Cir. 1980);
Frank Irey, Jr., Inc. v. OSHRC, 519 F.2d 1200, 2 OSHC 1283 (3d Cir. 1975).

equivalent to a knowing, conscious and deliberate flaunting of the Act. Willfulness means more than merely voluntary action or omission—it involves an element of obstinate refusal to comply."[75]

However, the Commission, as well as all the other courts of appeals that have addressed this issue, has rejected the Third Circuit's analysis. Thus, in *Intercounty Construction Co. v. OSHRC*, the Fourth Circuit expressly declined to follow *Frank Irey*, arguing that to require a bad purpose would be too serious a restriction on OSHA's authority to utilize strong sanctions to enforce the law.[76] The Sixth Circuit in *Empire Detroit Steel Division v. OSHRC* agreed, noting that it would be virtually impossible to establish willfulness were the Third Circuit's reasoning adopted.[77] The D.C. Circuit sought to reconcile the Third Circuit's approach to the majority view in *Cedar Construction Co. v. OSHRC* by asserting that there was not any real theoretical difference between an intentional disregard or plain indifference to the Act's requirements and an "obstinate refusal to comply" as defined by *Frank Irey*.[78] In *Kent Nowlin Construction Co.* and *Tri-City Construction Co.*, the Commission expressly declined to adopt the Third Circuit's *Frank Irey* test.[79]

B. Defenses

In defending against a willful violation, an employer has several options. First, the employer can attempt to rebut the Secretary's requisite showing of knowledge.[80]

Second, the employer can demonstrate that it tried to eliminate the hazard. The Commission has held that employer attempts to abate the cited hazard will defeat the willful classification.[81] Thus, in *Connecticut Natural Gas Corp.* no willful

---

[75]*Frank Irey, Jr., Inc. v. OSHRC, supra* note 74, at 1204.

[76]*Intercounty Constr. Co. v. OSHRC*, 522 F.2d 777, 3 OSHC 1377 (4th Cir. 1975), *cert. denied*, 423 U.S. 1072 (1976).

[77]*Detroit Steel Corp., Empire Detroit Steel Div. v. OSHRC*, 579 F.2d 378, 6 OSHC 1693 (6th Cir. 1978).

[78]*Cedar Constr. Co. v. OSHRC*, 587 F.2d 1303, 6 OSHC 2010 (D.C. Cir. 1978).

[79]*Kent Nowlin Constr. Co.*, 5 OSHC 1051 (Rev. Comm'n 1977); *Tri-City Constr. Co.*, 8 OSHC 1567 (Rev. Comm'n 1980).

[80]See *supra* note 63.

[81]See, e.g., *Randall Bearings, Inc.*, 9 OSHC 1100 (Rev. Comm'n J. 1980); *Quillian Pipe Co.*, 5 OSHC 1151 (Rev. Comm'n 1977); *Connecticut Natural Gas Corp.*, [1976–

violation resulted although a foreman intentionally violated trench shoring requirements because the employer had instructed its foreman to shore all trenches over five feet, had sent out instructions covering shoring, and had retained a soil expert to investigate potentially dangerous trenches.[82] Similarly, in *W.H. Allgood & Co.*, because the subcontractor was attempting to implement safety belt and lifeline procedures at the time of the OSHA inspections, the Commission reduced the citation from willful to serious.[83] In *Randall Bearings, Inc.*, because an employer, who had previously been cited for willful violations, had spent $125,000 trying to reduce lead concentration in its plant, the Commission reclassified the violation from willful to serious.[84] However, postcitation abatement efforts will not reduce a citation from willful to serious.[85]

A third possible defense to a willful citation can be a showing by the employer that it has a good faith dispute as to the existence of the violation.[86] For example, in *General Electric Co.* an employer's good faith challenge that standards for ladders were inapplicable to stator frames resulted in the Commission reducing a violation from willful to serious.[87]

Finally, the employer may defeat a willful classification by showing that compliance with the Act would create a greater hazard than noncompliance.[88]

C. Penalties

Section 17(a) of the Act provides penalties for willful or repeated violations of up to $10,000 per violation. As with other violations the penalties for willful violations are determined in accordance with the criteria in Section 17(j) of the

---

1977] OSHD ¶20,298 (Rev. Comm'n J. 1976), *reviewed on other points,* 6 OSHC 1796 (Rev. Comm'n 1978); *W.H. Allgood & Co.,* 5 OSHC 1640 (Rev. Comm'n J. 1977).

[82]*Connecticut Natural Gas Corp., supra* note 81.

[83]*W.H. Allgood & Co., supra* note 81.

[84]*Randall Bearings, Inc., supra* note 81.

[85]*Acme Fence & Iron Co.,* 7 OSHC 2228 (Rev. Comm'n 1980).

[86]See, e.g., *C.N. Flagg & Co.,* 2 OSHC 1539 (Rev. Comm'n 1975); *General Elec. Co.,* 3 OSHC 1031 (Rev. Comm'n 1975), *vacated in part,* 540 F.2d 67, 4 OSHC 1512 (2d Cir. 1976). But see *Western Waterproofing Co.,* 5 OSHC 1064 (Rev. Comm'n 1977).

[87]*General Elec. Co., supra* note 86.

[88]See *Acme Fence & Iron Co., supra* note 85 (employer's decision not to comply not in good faith.)

Act. In practice, a gravity-based penalty is determined and then that figure is multiplied by a number from 1 to 10, representing the "degree of willfulness." Generally, a high number, i.e., a high degree of willfulness, indicates intentional disregard of the Act; a middle number, i.e., a moderate degree of willfulness, indicates indifference to the Act; and a low number, i.e., a low degree of willfulness, indicates employer negligence.

The Act also provides for criminal penalties, including fines, imprisonment, or both, when a willful violation results in death.[89]

VII. *De Minimis* Violations

Section 9(a) of the Act defines *de minimis* violations as those that violate a standard not immediately or directly related to employee safety or health. Originally, notice of a *de minimis* violation had to be issued; this practice has now been abandoned.[90] No penalties are assessed and no abatement date is designated for *de minimis* violations.[91]

An example of a *de minimis* violation is found in *Hana Shoe Corp.* where an employer with an inadequately guarded sewing machine drive belt was cited for violation of a belt-guarding standard most often applied to large industrial machines.[92] Since nothing in the Act limited application of the standard to large machines, the employer was in technical violation of the standard. However, because it was concluded that the potential hazards were "trifling," only a *de minimis* citation resulted. Similarly, in *National Rolling Mills Co.* the Commission modified an other than serious violation to *de minimis* where only a negligible relationship to employee safety and health existed, and it was inappropriate to order abatement or assess a penalty.[93]

---

[89]See *infra* Chapter 11 (Assessment of Civil Penalties).

[90]FOM ch. IV B.6, (Oct. 21, 1985), OSHR [2 Reference File 77:2712]: "Whenever de minimis conditions are found during an inspection, they shall be documented in the same way as any other violation, but shall not be included on the citation."

[91]*Dravo Corp.*, 7 OSHC 2095 (Rev. Comm'n 1980).

[92]*Hana Shoe Corp.*, 4 OSHC 1635, 1637 (Rev. Comm'n 1976).

[93]*National Rolling Mills Co.*, 4 OSHC 1719 (Rev. Comm'n 1976). However, the Commission held in *Thunderbolt Drilling, Inc.*, 10 OSHC 1981 (Rev. Comm'n 1982), that an employer's failure to post a notice informing employees of their rights and

VIII. Imminent Danger Enforcement Proceedings

Section 13(a) of the Act provides that the Secretary may seek an injunction from the U.S. district courts "to restrain any conditions or practices *** which are such that a danger exists which could reasonably be expected to cause death or serious physical harm immediately or before the imminence of such danger can be eliminated through the enforcement procedures otherwise provided by [the] Act."[94] Not surprisingly, imminent danger enforcement proceedings are rarely used.[95]

As soon as a compliance officer discovers an imminent danger, he is required to inform the employer and employees of the condition and that he is recommending to the Secretary that relief be sought.[96]

---

employer duties under the Act was properly an other violation and had been mischaracterized by the administrative law judge as *de minimis*. For a more complete discussion of recordkeeping, see Chapter 5 (Employer Obligations to Obtain, Maintain, and Disseminate Information).

[94]29 U.S.C. §662(a) (1985).

[95]See generally *Usery v. Whirlpool Corp.*, 416 F. Supp. 30, 4 OSHC 1391 (N.D. Ohio 1976) (congressional intent underlying 29 U.S.C. §662 requires judicial due process prior to employer shutdown), *rev'd sub nom. Marshall v. Whirlpool Corp.*, 593 F.2d 715, 7 OSHC 1075 (6th Cir. 1979), *aff'd*, 445 U.S. 1, 8 OSHC 1001 (1980).

[96]See 29 U.S.C. §662(c).

10

Abatement

I. Determination of Abatement Dates

Section 9(a) of the OSH Act requires that each citation "fix a reasonable time for the abatement of the violation."[1] The OSHA Field Operations Manual directs the compliance officer to require abatement within "the shortest interval within which the employer can *reasonably* be expected to correct the violation."[2] During the closing conference the compliance officer considers abatement periods suggested by the employer.[3] Usually the citation fixes a specific date for abatement rather than a number of days within which abatement is to be accomplished.

In determining the appropriate abatement period, the Field Operations Manual directs the compliance officer to consider "[a]ll pertinent factors."[4] Among the factors considered are the gravity of the alleged violation; the availability of needed equipment, material, and/or personnel; the time necessary to complete delivery, installation, modification, or construction; and the training of personnel.[5] All of these factors are to be

---

[1]29 U.S.C. §658(a) (1980).
[2]Interim Field Operations Manual, ch. III E.1, OSHR [Reference File 77:2529] (emphasis in original).
[3]*Id.*, ch. III E.2.a, OSHR [Reference File 77:2529].
[4]*Id.*, ch. III E.2.b., OSHR [Reference File 77:2529].
[5]*Id.*, ch. III E.2.b(1)–(4), OSHR [Reference File 77:2529].

considered together, although one or more may be more significant or given more weight, depending on the circumstances.

The Field Operations Manual does not restrict the compliance officer's discretion for abatement periods of 30 days or less. If the abatement period exceeds 30 days, however, the compliance officer must include in the file an explanation for the length of time granted. Furthermore, the Field Operations Manual provides that abatement periods in excess of one year require prior approval from the Regional Administrator.[6]

Employers may seek and generally are granted an opportunity to submit an abatement plan with suggested dates, particularly for alleged violations that would require significant engineering or administrative controls. An employer may wish to discuss such a plan at the closing conference, since a citation issued "with reasonable promptness," as required by Section 9(a) of the OSH Act, would prescribe a particular abatement scheme which would preempt any plan the employer might have proposed.[7]

If significant changes in the employer's operations are required, abatement may be made or required in steps. Each step is assigned a specific date for completion. Interim measures are required while feasible engineering controls are developed or implemented. Short-range abatement dates may also be set to provide employee personal protections pending development or implementation of feasible long-range engineering or administrative controls. The types of steps required and the abatement dates are set forth in the citation or in the plan developed in consultation with the employer.[8]

II. Methods of Abatement

There are three general methods for accomplishing abatement: (1) engineering controls, (2) administrative controls, and (3) work practice controls. The compliance officer is required to

---

[6]*Id.*, ch. III E.3, OSHR [Reference File 77:2529].

[7] See *id.*, ch. III E.7, OSHR [Reference File 77:2531].

[8]*Id.*, ch. III E.8, OSHR [Reference File 77:2531–2532]. The citation must still be issued with "reasonable promptness" within the meaning of §9(a) of the Act.

discuss during the closing conference any control methodology he believes to be appropriate.[9]

A. Engineering Controls

Engineering controls, consisting of substitution, isolation, ventilation, or modification of equipment, are designed to remove the cause of the hazardous or potentially hazardous conditions. Implementation of engineering controls may range from placing a guard on a machine to reengineering a facility.[10]

B. Administrative Controls

Administrative controls may be required in addition to engineering controls or in place of engineering controls when implementation is not feasible.

Administrative controls are any procedure which significantly limits employee exposure by control or manipulation of the work schedule or manner in which the work is performed.[11] For example, if noise is a problem which cannot be corrected through engineering controls, the employer may rotate employees from one task to another to reduce exposure to high noise levels.

C. Work Practice Controls

Work practice controls are a type of administrative control whereby the employer modifies the manner in which work is performed, such as improving work habits or employee hygiene.[12]

D. Personal Protective Equipment

Personal protective equipment is not permitted as a substitute for feasible engineering, administrative, or work practice controls. Thus, personal protective equipment is not

---

[9] *Id.*, ch. III E.6, OSHR [Reference File 77:2529].

[10]*Id.*, ch. III E.6.a, OSHR [Reference File 77:2530].

[11]*Id.*, ch. III E.6.b, OSHR [Reference File 77:2530].

[12]*Id.*, ch. III E.6.c, OSHR [Reference File 77:2530].

considered a means of administrative control.[13] Personal protective equipment may be required when other controls are not technically or economically feasible or when other controls would not provide sufficient protection from exposure to potential hazards.[14] This control method consists of providing employees with devices which they wear, such as ear plugs or earmuffs, respirators, or protective clothing to protect them from potential hazards.

III. Methods of Challenging Abatement Dates

An employer may challenge the reasonableness of the abatement period set by the compliance officer in two different ways: (1) An employer may file a notice of contest asserting that the period for abatement set by the compliance officer is unreasonable under Section 9(a) of the Act or that compliance is impossible within the abatement period owing to factors beyond the employer's control, notwithstanding its good faith efforts.[15] (2) An employer may file a petition for modification of abatement (PMA) challenging the abatement date.[16] The two methods differ significantly, as is discussed below.[17]

A. Notice of Contest

In a contested citation proceeding, where the abatement date has been challenged the burden of proof rests with the Secretary to demonstrate the reasonableness of the abatement period.[18] In determining whether the Secretary meets the bur-

---

[13]*Id.*, ch. III E.6.b, OSHR [Reference File 77:2530].

[14]*Id.*, ch. III E.6.d(3)(c), OSHR [Reference File 77:2531].

[15]See 29 C.F.R. §2200.33.

[16]See 29 C.F.R. §2200.37; 29 C.F.R. §1903.14a. The Commission's jurisdiction is exclusive over PMAs. *H.K. Porter Co.*, 1 OSHC 1600, 1605 (Rev. Comm'n 1974). This includes jurisdiction to determine not only the reasonableness of the abatement period but also whether abatement has occurred. *American Cyanamid Co.*, 8 OSHC 1346, 1350 (Rev. Comm'n 1980).

[17]See *Gilbert Mfg. Co.*, 7 OSHC 1611, 1613–1615 (Rev. Comm'n 1979), for a discussion of the differences between the two methods. For a discussion of settlement of a notice of contest or a PMA, see Chapter 17 (Employee and Union Participation in Litigation Under the OSH Act).

[18]Cf. 29 C.F.R. §2000.36(b)(9) (1987) (Secretary to plead reasonableness of abatement date); but see §2200.30(f) (pleading rules not determinative on burden of persuasion). Details regarding the form of any pleading filed with the Commission are set forth in 29 C.F.R. Part 2200, Subpart C. See Chapter 13 (Commencement of Proceedings) for a discussion of notice of contest proceedings.

den of proving reasonableness, the Commission considers the same factors the compliance officer considers in setting the abatement date: gravity of the violation; number of exposed employees; extent of their exposure; availability of needed equipment, material, or personnel; necessary training of personnel; time required for delivery, installation, modification, or construction.[19]

Affected employees having a representative may also initiate a contest of the abatement period. They may file a notice of contest within the same 15 working-day period applicable to an employer's notice of contest.[20] Within 10 days after such notice is filed, the Secretary must file a clear and concise statement of the reasons the abatement period prescribed is not unreasonable. Within 10 days after receipt of the Secretary's statement, the affected employees or their representative may file a response.[21] The burden of proof is on the Secretary in the expedited proceeding that results.[22]

B. Petition for Modification of Abatement

A PMA may be filed with the Area Director who issued the citation anytime during the abatement period. The filing must occur, however, no later than the last business day following the expiration of the abatement period.[23] Unlike a notice of contest, a PMA does not generally affect the underlying citation and does not stay the abatement period. The purpose of a PMA is to obtain an extension of time within which to accomplish abatement.

In *Gilbert Manufacturing Co.*,[24] the Review Commission held that a notice of contest directed only at the abatement date, whether or not filed within 15 days after receipt of the citation, will be treated as a PMA rather than as a notice of

---

[19]*Supra* note 5.

[20]29 U.S.C §§10(a) and (c).

[21]29 C.F.R. §§2200.38(a) and (b).

[22]The contest is handled as an expedited proceeding. §2200.38(c). Cf. §2200.103 (expedited hearing procedure).

[23]29 C.F.R. §2200.37(c); 29 C.F.R. §1903.14a(c); *M.P. Kirk & Sons*, 1 OSHC 3267, 3268 (Rev. Comm'n J. 1974). If the petition is filed late, it must be accompanied by the employer's statement of exceptional circumstances. §1903.14a(c); §2200.37(c).

[24]7 OSHC 1611 (Rev. Comm'n 1979). Cf. *Philadelphia Coke Div.*, 2 OSHC 1171 (Rev. Comm'n 1974); *Bushwick Comm'n Co.*, 8 OSHC 1653, 1654 n.6 (Rev. Comm'n 1980).

contest. The Commission favored the use of the PMA proceeding over the notice of contest because a PMA does not
stay the employer's obligation to begin abatement. Accordingly, the abatement period is not automatically tolled when
an employer files for an extension of the abatement date.[25]

In a PMA proceeding the employer can satisfy the requirement that employees be notified of an application for
modification of abatement by posting a copy of the PMA at
the workplace. If no employee or authorized employee representative files any objections to the PMA with the Secretary
within 10 days after the posting, the Secretary may grant the
request without involving the Commission.[26] If the Secretary,
any of the employees, or the authorized employee representative objects to the PMA, the PMA and any objections are
forwarded to the Commission within 10 working days after
expiration of a 15-day period from the date of filing of the
PMA.[27] Although the Commission is directed to process such
a PMA as it would any other contested case, hearings are to
be handled in an expeditious fashion.[28]

The caption used in a PMA names the employer as the
"Petitioner" versus the Secretary of Labor as "Respondent."[29]
The PMA must be in writing and must contain the following
information: (1) all actions and the dates thereof taken to abate
within the abatement period; (2) the additional time necessary
to achieve abatement; (3) the reason such additional time is
necessary, e.g., unavailability of professional or technical personnel or impossibility of completing necessary construction
or alteration of facilities by the original abatement date; (4) all
available interim steps being taken to safeguard employees;
and (5) certification that the PMA has been posted for employees.[30] Once the PMA is filed, the burden is on the employer
to justify extension of the abatement period.[31]

---

[25]See, e.g., *Gilbert Mfg. Co., supra* note 24, overruling *Eastern Knitting Mills,* 1
OSHC 1677, 1679 n.4 (Rev. Comm'n 1974).

[26]See 29 C.F.R. §§2200.37(c)(2) and (3); *Bushwick Comm'n Co., supra* note 24.

[27]§2200.37(d)(1).

[28]§2200.37(d)(2). Cf. 29 C.F.R. §2200.103 (expedited hearing procedures).

[29]§2200.31(b).

[30]§2200.37(b) and (c); *Waugh & Co.,* 9 OSHC 1386 (Rev. Comm'n J. 1981).

[31]See §10(c) of the OSH Act, 29 U.S.C. §659(c); *Amoco Chem. Corp.,* 8 OSHC
1085, 1087 (Rev. Comm'n 1980); 29 C.F.R. §2200.37(d)(3).

C. Circumstances Supporting Change in Abatement Date

It is well established that a reasonable amount of time must be allowed to formulate and implement plans for abating cited violations.[32] Whether the time specified is reasonable depends on a myriad of factors, such as the present unavailability of feasible engineering or administrative controls specified by the compliance officer; good faith efforts made toward compliance; efforts made by the employer to reduce or eliminate the alleged violation in the absence of feasible controls; novelty, complexity, and magnitude of the modifications; weather conditions; financial hardship; construction delays; and strikes by employees.

1. Production or Design Problems

One of the most common reasons for extending an abatement period is unforeseen design or procurement problems.[33] Multiple extensions as a result of such problems are not uncommon.[34]

A closely related problem is the present unavailability of feasible engineering or administrative controls. The Commission or administrative law judge, in determining whether to extend the abatement period, considers what efforts have been made by the employer to attempt to reduce or eliminate the alleged violation in the absence of other controls.[35] For example, in *Mueller Brass Co.*, the administrative law judge, in granting a one-year extension, stressed that the employer had contacted 20 consultants, narrowed the field to six, and was still working on the problem.[36] In *Asbestos Textile Co.*, the

---

[32]See, e.g., *Matthews & Fritts, Inc.*, 2 OSHC 1149 (Rev. Comm'n J. 1974) (same day requirement did not allow sufficient time). But see *Murro Chem. Co.*, 2 OSHC 1268 (Rev. Comm'n J. 1973) (period of 5 days to provide accessible toilet facilities and 15 days to provide adequate eating facilities for employees reasonable).

[33]For example, in *American Metal Works*, 5 OSHC 1955 (Rev. Comm'n J. 1977), the administrative law judge found that a one-year abatement extension was justified where unforeseen design problems existed. See *Marshfield Steel, Inc.*, 7 OSHC 1741 (Rev. Comm'n J. Nos. 79-839-E & 79-1253-P, 1979) (six-month extension granted).

[34]See, e.g., *Columbus Marble Works*, [1974–1975] OSHD ¶19,152 (Rev. Comm'n J. 1974) (third extension); *Owens-Corning Fiberglas Co.*, 5 OSHC 1040 (Rev. Comm'n J. 1976) (third extension granted).

[35]See 29 C.F.R. §§2200.37(b)(1) and (4).

[36]8 OSHC 1776 (Rev. Comm'n J. 1980).

employer obtained a 9-month extension of the abatement period set for initiating engineering controls to limit employee exposure to excessive concentration of asbestos fibers. Although the employer did not necessarily prove that abatement was beyond its reasonable control, it was able to demonstrate that meeting the exposure limits was "not an easy task." The employer showed that it had visited a number of asbestos plants to see how they handled their dust collecting systems in order to determine what methods might be effective at the employer's facility. The company president also attended seminars on dust control and was able to show that a contractor had indicated it would be difficult to obtain effective fans and filters.[37] An employer's willingness to try different methods of abatement, however unsuccessful, has also helped to persuade an administrative law judge to extend an abatement date.[38]

Employers have also received extensions because of the novelty or complexity involved in installing or implementing the modifications.[39]

2. Financial Problems

Although more difficult to establish as a ground for extension of an abatement period, financial hardship has been found to justify an extension of time.[40] However, the impact on production caused by prescribed abatement measures has generally not been a valid ground for extension of an abatement period, notwithstanding the severe financial impact of

---

[37]1 OSHC 3302 (Rev. Comm'n J. 1974).

[38]*U.S. Ring Binder Corp.*, 2 OSHC 3312 (Rev. Comm'n J. 1975) (one-year extension granted where employer had shown good faith in attempting to comply, including moving facilities to larger building, but had encountered substantial engineering and technical difficulties).

[39]*Stearns & Foster Co.*, 6 OSHC 1881 (Rev. Comm'n J. 1978). See also *Florida Power & Light Co.*, 5 OSHC 1277 (Rev. Comm'n J. 1979) (six-month extension to test new devices); *Shenango Inc.*, 6 OSHC 1405 (Rev. Comm'n J. 1979) (three-year extension to reduce silica dust concentration); *Specialty Insulation Div. of Cantamount Mfg.*, 7 OSHC 1192 (Rev. Comm'n J. 1978) (two-year extension granted to test guards on hydraulic presses); *U.S. Ring Binder Corp., supra* note 38; *Prestolite Co., Div. of Eltra Co.*, 1 OSHC 3190 (Rev. Comm'n J. 1973); *Otis Elevator Co.*, 1 OSHC 3101 (Rev. Comm'n J. 1972). Cf. *Selma Timber Co.*, 1 OSHC 3083 (Rev. Comm'n J. 1973).

[40]*Serval Slide Fasteners*, 5 OSHC 1938 (Rev. Comm'n J. 1977) (two-year extension granted).

such measures.[41] Even the filing of a bankruptcy petition has been held insufficient in at least one case to stay an abatement period.[42]

In 1984, the Review Commission held an employer would not be required to implement administrative and engineering controls unless the Secretary could demonstrate that the benefits to be derived from implementation of the controls justifies the costs involved.[43] An employer may still, however, be required to supply personal protective equipment in order to achieve compliance.

3. Good Faith Efforts

Neither the Commission nor administrative law judges are inclined to grant extensions where the employer has not demonstrated a good faith abatement effort.[44] In *Mueller Brass Co.*, the administrative law judge granted a second extension of time to control noise based on the employer's good faith efforts to comply. The employer had surveyed other companies on noise control possibilities, inspected other plants, and hired a noise expert. The company had also experienced a three-month strike and the company's president had been incapacitated and replaced.[45] Similarly, in *Hudson Fabricating Co.*,[46]

---

[41]*Robertson Metal Fabricators*, [1980] OSHC ¶24,795 (Rev. Comm'n J. 1980) (guard on machine); *Markey Bronze Corp.*, 11 OSHC 301 (Rev. Comm'n J. 1981) (cost of protective clothing). But cf. *B.F. Goodrich Co.*, 2 OSHC 3035 (Rev. Comm'n J. 1975) (extension granted to allow employer opportunity to abate at lowest possible cost and with least disruptive impact on production activities).

[42]*Winfrey Structural Concrete Co.*, 10 OSHC 1270, 1271 (Rev. Comm'n J. 1981): "The employer's argument, that its filing of a petition of bankruptcy bars occupational safety and health administration action against it, is rejected. Although the petition might affect the secretary's ability to collect penalties from the employer, it is clear that its filing does not operate as a stay of OSHA proceedings."

[43]*Sherwin-Williams Co.*, 11 OSHC 2105, 2111 (Rev. Comm'n 1984), overruling *Sun Ship*, 11 OSHC 1028 (Rev. Comm'n 1982).

[44]See, e.g., *ITT Grinnell Corp.*, 11 OSHC 1464 (Rev. Comm'n 1983) (extension denied even though no feasible controls available); *Castings Div., Beloit Corp.*, 11 OSHC 1687 (Rev. Comm'n J. 1983) (employer failed to demonstrate good faith). Cf. *IBEW, Local 1031 (Stewart-Warner)*, 7 OSHC 1674 (Rev. Comm'n J. 1979), *vacated*, 8 OSHC 1316 (Rev. Comm'n 1980).

[45]8 OSHC 1776, 1776–1777 (Rev. Comm'n J. 1980).

[46]8 OSHC 1647 (Rev. Comm'n J. 1980). See also *Local 550, United Auto Workers (Chrysler Corp.)*, 11 OSHC 2238 (Rev. Comm'n J. 1984) (good faith effort where employer needed additional time to comply with Secretary's recommendations).

the administrative law judge dismissed a failure-to-abate charge because the employer had taken reasonable steps to correct the alleged violation by moving its operation and making other changes in the physical arrangement of the production process. The employer also installed other equipment to reduce the exposure to employees and thus, even though a technical violation of the standard remained, the administrative law judge held it was *de minimis*.

4. Miscellaneous Problems

Weather conditions have also served as a basis for an extension of an abatement period. In *Auto Sun Products Co.*, the administrative law judge granted the employer a second one-year extension to abate a noise violation. The employer showed that it had been unable to move to a larger, quieter building because of severe winter weather and power company use restrictions.[47]

Although legitimate construction delays may serve as a basis for extension of an abatement period,[48] delivery delays alone may not.[49] A strike by employees has, on the other hand, provided sufficient basis for an extension of time in at least one instance.[50]

IV. Stay of the Abatement Period

An employer may obtain judicial review of a Commission final order, by filing, within 60 days of the issuance of the order, an appropriate petition in the U.S. court of appeals for

---

[47]6 OSHC 1300 (Rev. Comm'n J. 1977). Cf. *Selma Timber Co., supra* note 39 (wet grounds).

[48]*Gates Rubber Co.*, 1 OSHC 3090 (Rev. Comm'n J. 1973). But see *Midwest Steel Fabricators*, 5 OSHC 2068 (Rev. Comm'n 1977) (third extension denied because not apparent new plant would be completed on time, appropriate protective equipment commercially available, and employer had not determined effectiveness of personal protective equipment).

[49]*Sigmund Sommer Constr. Co.*, 1 OSHC 3075, 3075–3076 (Rev. Comm'n J. 1973).

[50]*North Am. Royalties*, 1 OSHC 3088, 3089 (Rev. Comm'n J. 1973) (additional time granted to conduct engineering controls tests after employees' strike prevented tests). See also *Mueller Brass Co.*, 8 OSHC 1776 (Rev. Comm'n J. 1980) (3-month strike among other factors beyond employer's control causing delay).

the circuit in which the violation is alleged to have occurred or where the employer has its principal place of business, or in the Court of Appeals for the District of Columbia.[51] The filing of the petition, however, does not, in and of itself, operate as a stay of the Commission's order. Therefore, the employer is still bound by the abatement date.

Although the Act does not specifically provide a method for employers to obtain a stay of the abatement period pending judicial review, both the courts and the Commission have recognized this procedure. Commission Rule 94 provides that a party may seek a stay from the Commission as to any "final order" while a matter is within the jurisdiction of the Commission.[52] Therefore, an employer may file for a stay before commencing an appeal in the court of appeals.

Once an appeal is filed, however, the Commission loses its jurisdiction. The employer may then petition the court of appeals for a stay.[53] An employer seeking a stay must state the reasons for the stay and the length of time sought.[54] If a stay is not obtained, the abatement period commences the day after the Commission's order becomes "final," and failure to abate at the end of the period subjects the employer to a penalty for each day of nonabatement.[55]

An abatement period may also be stayed pending determination on an application for a variance.[56]

---

[51]29 U.S.C. §660(a); *Lance Roofing Co.*, 343 F. Supp. 685, 688 (N.D. Ga. 1972).

[52]29 C.F.R. §2200.94.

[53]29 U.S.C. §660(a).

[54]29 C.F.R. §2200.94. See *Lance Roofing Co., supra* note 51.

[55]29 U.S.C. §§659(b), 666(b).

[56]For a further discussion of variances, see Chapter 21 (Other Issues Related to Standards).

11

Assessment of Civil Penalties

In enacting the Occupational Safety and Health Act of 1970, Congress expressly authorized the Secretary of Labor to assess and collect civil penalties from employers who violate applicable OSHA standards.[1] These penalties, although initially proposed by the Secretary's Area Director, are subject to *de novo* review by the Review Commission[2] and are thereafter subject to traditional judicial review by a U.S. court of appeals.[3]

The statutory scheme of penalties contains several anomalies, but it has generally been applied so that intentional and truly dangerous violations are penalized more severely than minor matters. Peculiarly, the Act requires penalties for serious violations[4] and for violating posting requirements,[5] but only *permits* penalties for willful or repeated violations,[6] for nonserious violations,[7] including recordkeeping and reporting failures,[8] and for failures to abate within the periods prescribed by citation and applicable law.[9] On the other hand,

---

[1] 29 U.S.C. §659, §666(k) (1985).
[2] 29 U.S.C. §666(i).
[3] §660.
[4] §666(b).
[5] §666(i).
[6] §666(a).
[7] §666(c).
[8] See Section VII, *infra*.
[9] §666(d).

288

the amount of penalties escalates with increasing severity from "up to $1,000" for nonserious and serious violations to "up to $1,000 per day" for failure to abate and "up to $10,000" for willful and repeated violations.

I. Applicable Statutory Scheme for Assessment of Civil Penalties

A. Statutory Authority for Imposition of Civil Penalties

The assessment of penalties by the Secretary against an employer is expressly authorized by Section 10(a) of the Act, which provides:

> "If, after an inspection or investigation, the Secretary issues a citation under section 9(a), he shall, within a reasonable time after the termination of such inspection or investigation, notify the employer by certified mail of the penalty, if any, proposed to be assessed under section 17***."

1. Mandatory Penalties for Serious Violations

Section 17(b) provides that an employer who receives a citation for a serious violation of an OSHA standard *shall* be assessed a civil penalty of up to $1000 for each such violation.[10] The language of this section is mandatory and it has been held by the Commission that a penalty must be imposed for a serious violation.[11]

2. Discretionary or No Penalties for Nonserious Violations

In contrast to Section 17(b), Section 17(c) provides that any employer who has received a citation for a nonserious violation of an applicable OSHA standard *may* be assessed a

---

[10]§666(b).

[11]*Laster & Fingeret, Inc.*, 2 OSHC 1588 (Rev. Comm'n 1975). The Commission will, however, in an appropriate case impose only a nominal penalty to comply with the mandatory language of §17(b). *Hodgdon Shingle & Shake Co.*, 2 OSHC 1215 (Rev. Comm'n 1974) ($1 penalty substituted for administrative law judge's assessed penalty of $0 in view of mandatory nature of §17(b) of Act). See also *Varflex Corp.*, [1978] OSHD ¶22,784 (Rev. Comm'n J. 1978) (penalty for serious violation reduced from $700 to $1).

civil penalty of up to $1000 for each such violation.[12] Despite the fact that the range of penalties for serious and nonserious violations is the same, a larger penalty is likely to be imposed for a serious violation. Because the language of Section 17(c) is permissive, no penalty need be imposed for a nonserious violation. Indeed, in OSHA's appropriations bills Congress has repeatedly imposed limitations on the Secretary's ability to impose fines for nonserious violations. Most recently, the 1986 OSHA appropriations bill passed by Congress expressly provided that no funds may be spent to collect penalties imposed for fewer than 10 nonserious violations.[13]

3. *Discretionary but Significant Penalties for Violations Which Are Willful or Repeated or Not Abated in Time*

The amount of civil penalties that may be imposed by the Secretary increases dramatically where violations contain an element of intention, disregard, or repetition. Surprisingly, in these cases the language of Section 17(a) (willful and repeated violations) and Section 17(d) (failure to abate) is permissive; thus there is no statutory mandate that a penalty be imposed. However, it is far more likely that a substantial civil penalty will be imposed in these cases than in the case of an initial citation for a nonserious violation.

The clearest example of such a violation is a failure to abate a cited violation. The employer, having been cited, is on notice that there is a violation and has been given a time to correct, but fails to heed the warning. Section 17(d) of the Act provides that an employer who fails to abate a cited violation within the abatement period may be assessed a civil penalty of not more than $1000 per day until the violation is abated.[14]

Of even greater monetary concern are the provisions of Section 17(a), which provide that any employer who willfully or repeatedly violates its statutory duty under Section 5 of the Act to "furnish to each of his employees employment and a place of employment which are free from recognized hazards

---

[12]29 U.S.C. §666(c).

[13]Pub. L. No. 97-377, 96 Stat. 1830 (1986). *Texaco, Inc.*, 8 OSHC 1758 (Rev. Comm'n 1980); *GAF Corp.*, 5 OSHC 1583 (Rev. Comm'n J. 1977) (Commission noted appropriations limit).

[14]29 U.S.C. §666(d).

that are causing or likely to cause death or serious physical harm to his employees"[15] or who violates any applicable OSHA standard may be assessed a civil penalty of not more than $10,000 for each violation.[16]

4. *Mandatory Penalty for Posting Violations*

The final civil penalty provided by the Act is contained in Section 17(i), which provides that any employer who violates any of the OSHA posting requirements shall be assessed a civil penalty of up to $1000 for each violation.[17] As in the case of serious violations, the language of this subsection is mandatory and a civil penalty must be imposed, but it is not likely that a large penalty will be proposed.

5. *Notification of Proposed Civil Penalties*

The Area Director, acting in the Secretary's name, proposes penalties for violations and includes them on the citation form, together with the citation itself and the proposed abatement date, which is required to be issued to the employer by Section 9 of the Act.[18] Proposed penalties must be assessed against the employer within 6 months following the latest occurrence of any violation, since under Section 9(c) of the Act no citation may be issued after the expiration of 6 months following the occurrence of any violation.[19]

B. Contesting the Proposed Civil Penalties

Upon receipt of a citation with proposed penalties, an employer has 15 working days within which to notify the Secretary that it desires to contest the proposed penalties (as well as the citation).[20] The form can be as simple as "the employer contests each and every penalty" and the penalty would be contested if the employer "contests each and every

---

[15] §654.
[16] §666(a).
[17] §666(h).
[18] 29 U.S.C. §§658(a) and 659(a).
[19] §658(c).
[20] §659(a).

provision of the citation." Failure to contest within 15 working days results in the proposed penalty being deemed a final order of the Commission and not subject to judicial review.[21] When the employer contests timely, the Solicitor of Labor issues a complaint bringing the matter before the Commission via a hearing before an administrative law judge.[22] In such a hearing, the burden of proof is on the Secretary to establish the appropriateness of the proposed penalties.[23] The Commission's powers and duties regarding the assessment of penalties is derived from Section 17(j) of the Act, which provides that

> "[t]he Commission shall have authority to assess all civil penalties provided in this section, giving due consideration to the appropriateness of the penalty with respect to the size of the business of the employer being charged, the gravity of the violation, the good faith of the employer, and the history of previous violations."[24]

Based on the language of Section 17(j), the Commission, as affirmed by several U.S. courts of appeals, has consistently held that it is for the Commission to determine, *de novo*, the appropriateness of the penalty to be imposed for violation of the Act or an OSHA standard.[25]

The Act requires OSHA and the Commission to give "due consideration to the appropriateness of the penalty" after considering and evaluating (1) "the gravity of the violation," (2) "the good faith of the employer," (3) "the size of the business," and (4) "the history of previous violations."[26]

OSHA'S method of calculating civil penalties can best be characterized as an inflexible and mechanical mathematical formula. OSHA employs numerous arbitrary numerical values and maximum percentages which are brought into operation, by and large, by the rote application of easily identifiable objective facts, such as the number of employees and the frequency of exposure to a hazard.

---

[21]*Id.*

[22]§659(c).

[23]*Colorado Fuel & Iron Steel Corp.*, 2 OSHC 1295 (Rev. Comm'n 1974).

[24]§666(i).

[25]*Long Mfg. Co. v. OSHRC*, 554 F.2d 903 (8th Cir. 1977); *Clarkson Constr. Co. v. OSHRC*, 531 F.2d 451 (10th Cir. 1976); *Dan J. Sheehan Co. v. OSHRC*, 520 F.2d 1036 (5th Cir. 1975), *cert. denied*, 424 U.S. 965 (1976); *California Stevedore & Ballast Co. v. OSHRC*, 517 F.2d 986 (9th Cir. 1975); *Bomac Drilling*, 9 OSHC 1681 (Rev. Comm'n 1981); *Delaware & H. Ry.*, 8 OSHC 1252 (Rev. Comm'n 1980); *P.A.F. Equip. Co.*, 7 OSHC 1209 (Rev. Comm'n 1979), *aff'd*, 637 F.2d 741 (10th Cir. 1980).

[26]§666(i).

The Commission, showing greater concern for the statutory requirement, has criticized and refused to follow OSHA's mechanical formula for calculating penalties as not being permitted by the requirements of Section 17(j). The Commission has repeatedly declined to affirm penalties based on a routine application of the instructions in the OSHA Field Operations Manual (FOM) to Area Directors for calculating penalties.[27] As a result, although the Commission has generally applied the same criteria as the Area Directors in calculating and assessing penalties, it evaluates these criteria less rigidly and more on a case-by-case basis.

II. How OSHA Calculates and Proposes Penalties for Serious and Nonserious Violations

A. Range of Penalties for Serious and Nonserious Violations

In calculating proposed penalties, the first determination made by OSHA is whether the violation is serious or nonserious.[28] Once this determination has been made, procedures promulgated by the Secretary in the FOM are utilized to calculate the proposed penalty. According to the FOM, penalties for serious violations *shall* range from $300 to $1000 for each violation.[29] Penalties for nonserious violations normally range from no penalty to $300 for each violation.[30] As a matter of OSHA policy, penalties for nonserious violations will not normally exceed $300, except in exceptional cases.[31]

---

[27]*Nacirema Operating Co.*, 1 OSHC 1001 (Rev. Comm'n 1972); *Rose Acre Farms*, [1981] OSHD ¶25,611 (Rev. Comm'n J. 1981); *American Airlines*, 1 OSHC 3249 (Rev. Comm'n J. 1973); *Rogge Lumber Sales*, [1971–1973] OSHD ¶15,267 (Rev. Comm'n J. 1972); *aff'd*, 1 OSHC 1408 (Rev. Comm'n 1973).

[28]Field Operations Manual (FOM) ch. VI A.1, OSHR [Reference File 77:3101]. A serious violation is defined in §17(k):
"[A] serious violation shall be deemed to exist in a place of employment if there is a substantial probability that death or serious physical harm could result from a condition which exists, or from one or more practices, means, methods, operations, or processes which have been adopted or are in use, in such place of employment unless the employer did not, and could not with the exercise of reasonable diligence, know of the presence of the violation."

[29]Ch. VI A.2, OSHR [Reference File 77:3101].

[30]*Id.*

[31]*Id.*

B. Determining the Proposed Penalty

1. Gravity-Based Penalty

In both serious and nonserious cases, OSHA initially chooses an unadjusted penalty based on the "gravity" of the violation.[32]

OSHA considers two factors in determining gravity: (1) the severity of the injury or illness which could result from the alleged violation (the "severity quotient"), and (2) the degree of probability that an injury or illness could occur as a result of the alleged violation (the "probability quotient").[33]

Following the calculation of both the severity quotient and the probability quotient, the Area Director averages the severity quotient and the probability quotient, resulting in a "probability/severity quotient," which OSHA then uses to determine the unadjusted gravity-based penalty by consulting either Penalty Table A (Serious Violations) or Penalty Table B (Other-Than-Serious Violations).[34]

2. Reduction of Gravity-Based Penalty by Application of Adjustment Factors

The unadjusted gravity-based penalty may be adjusted downward by OSHA by as much as 80 percent, depending on the employer's good faith (up to 30%), size of business (up to 40%), and history of previous violations (up to 10%).[35] But in cases of serious violations, the adjustments for good faith and history are limited to a combined total of 30 percent if the probability quotient is 6, 20 percent if the probability quotient is 7, and no reduction if the probability quotient is 8.[36]

Moreover, a small business which is cited for one or more serious violations of high probability or a number of moderate probability serious violations—either alternative indicating

---

[32]Ch. VI A.2.d OSHR [Reference File 77:3101].
[33]Ch. VI A.2.e–.f, OSHR [Reference File 77:3101–3102].
[34]Ch. VI A.2.h(4), OSHR [Reference File 77:3103]. See tables *infra.*
[35]Ch. VI A.2.j, OSHR [Reference File 77:3104].
[36]*Id.*

a serious disregard for worker safety and health—may be denied any reduction in penalty based on size of business.[37]

(a) Size of the business. In calculating the percentage of reduction for the size of the business, OSHA measures size solely on the basis of the number of employees in all of the employer's workplaces.[38] Utilizing this information, the rates of reduction applied by OSHA are as follows:[39]

Employees	Reduction
10 or fewer	40%
11 to 25	30%
26 to 60	20%
61 to 100	10%
More than 100	None

OSHA generally does not consider an employer's ability to pay a penalty in determining the penalty reduction for size of business. However, where an employer presents convincing evidence of an inability to pay the penalty, OSHA may determine that a further penalty reduction or an elimination of the penalty is appropriate.[40]

(b) Good faith of the employer. With regard to good faith, a maximum penalty reduction of 30 percent is permitted by OSHA in recognition of an employer's good faith.[41] OSHA seeks to quantify an employer's good faith by utilizing the following criteria:

(1) Evidence of genuine and effective safety and health efforts prior to the inspection; and
(2) Evidence of a desire to comply with the Act during and after an inspection.

The extent to which an employer can satisfy these criteria determines the actual percentage reduction that will be given by OSHA. A 30 percent reduction will be given for prompt and aggressive initiation of abatement of cited violations during the inspection or (by commitment) as soon as is practical,

---

[37]Ch. VI A.2.j(5)(a)3, OSHR [Reference File 77:3105].
[38]Ch. VI A.2.j(5)(a), OSHR [Reference File 77:3104].
[39]*Id.*
[40]Ch. VI A.2.j(5)(a)2, OSHR [Reference File 77:3105].
[41]Ch. VI A.2.j(5)(b), OSHR [Reference File 77:3105].

or for generally thorough and effective safety and health efforts. A 20 percent reduction will be given for reasonably prompt but slightly reluctant initiation of abatement, or for moderately thorough and effective safety and health efforts. A 10 percent reduction will be given for initiation of abatement with barely acceptable promptness or with some reluctance, or for fairly thorough and effective safety and health measures with a number of deficiencies. No reduction will be given for obvious reluctance to initiate correction of violations with indications that complete correction may not be made, or for little safety and health effort, with little effectiveness.[42]

OSHA recognizes, however, that many employers will not fit exactly these general criteria for penalty reduction based on good faith. Accordingly, OSHA has directed its Area Directors and Field Inspectors to exercise professional judgment in balancing the important factors in determining an appropriate rate of reduction for a particular employer.[43]

(c) History of prior violations. OSHA permits a maximum 10 percent reduction in the proposed penalty for an employer who lacks a significant history of previous violations.[44] Normally, OSHA only takes into account previous violations resulting from earlier OSHA or OSHA-funded inspections. OSHA permits the 10 percent reduction in the proposed penalty only where no previous inspection was made under any state or federal safety and health statute and there is no history of serious violations, no finding of a new serious violation or repeated violation of a serious nature, and no finding of a number of repeated violations of a nonserious nature.[45]

3. Calculating the Adjusted Proposed Civil Penalty

After the amount of percentage reduction for each violation is determined, the final adjusted penalty is determined by reference to the appropriate penalty table:

---

[42]*Id.*
[43]Ch. VI A.2.j(5)(b)4, OSHR [Reference File 77:3105].
[44]Ch. VI A.2.j(5)(c), OSHR [Reference File 77:3105].
[45]Ch. VI A.2.j(5)(c)3, OSHR [Reference File 77:3106].

Serious Violations

P/S Quo-tient	GBP	Percent Reduction							
		10	20	30	40	50	60	70	80
1	100	90	80	70	60	50	40	30	20
2	200	180	160	140	120	100	80	60	40
3	300	270	240	210	180	150	120	90	60
4	400	360	320	280	240	200	160	120	80
5	500	450	400	350	300	250	200	150	100
6	600	540	480	420	360	300	240	180	120
7	700	630	560	490	420	350	280	210	140
8	800	720	640						
9	900	810							
10	1000								

Other-Than-Serious Violations

P/S Quo-tient	GBP	Percent Reduction							
		10	20	30	40	50	60	70	80
1									
2									
3									
4									
5	300	270	240	210	180	150	120	90	60

The previously determined gravity-based penalty is now reduced by the percentage reduction allowed for each violation. The resulting penalty is found in the column under the amount of the percentage reduction on the chart. If this adjusted penalty amounts to less than $60, no penalty will be proposed by OSHA for that violation.[46]

---

[46]Ch. VI A.2.c. OSHR [Reference File 77:3101]. OSHA employs the same criteria for the calculation and assessment of civil penalties for violations of the general duty clause contained in §5(a)(1) of the Act and in serious and nonserious situations of imminent danger. FOM Ch. VII. If the violations involved are repeated or willful, OSHA calculates civil penalties in the same manner as in the case of a repeated or willful violation of a specific standard. Ch. VI A.6, 7. OSHR [Reference File 77:3107].

III. How the Commission Sets Penalties

In assessing penalties, the Commission, like the Secretary, determines the gravity of the violation by taking into account the probability of an accident occurring and the seriousness of the injury if an accident occurs. The Commission does not, however, first calculate an unadjusted gravity-based penalty and then apply the so-called penalty adjustment factors, as does OSHA, to determine whether that penalty should be reduced. Rather, the Commission considers all of these factors together in assessing penalties. In fact, the Commission has held that the criteria to be considered cannot always be given equal weight and that no single factor is controlling in assessing penalties.[47]

A. Gravity Factor

In examining the gravity of a violation, the Commission relies almost exclusively on the probability of an accident occurring. But, like the Area Director, the Commission also looks to the number of employees exposed and the duration of the exposure in determining the probability of an occurrence.

1. Duration of Employee Exposure

While the Commissioner has rejected the Secretary's attempt to quantify the frequency of employee exposure to the cited danger, as a practical matter the Commission reduces penalties where the employee exposure is only occasional or infrequent.[48]

Thus, an administrative law judge reduced a $600 penalty to $75 where a safety belt violation existed for only 15 minutes and a guardrail violation lasted for less than one hour.[49] Similarly, where an employee working in a trench was exposed to a backhoe for only 15 minutes an administrative law judge

---

[47]*Tech-Steel, Inc.*, 2 OSHC 1266 (Rev. Comm'n 1974); *Nacirema Operating Co.*, 1 OSHC 1001 (Rev. Comm'n 1972).

[48]See cases cited in note 27, *supra*.

[49]*Abbott Bldg. Co.*, [1977–1978] OSHD ¶21,955 (Rev. Comm'n J. 1977).

reduced the proposed $550 penalty to $75.[50] In another case, the Commission, affirmed by the Fourth Circuit, reduced a $10,000 penalty to $5000 where only two employees were exposed intermittently to the risk of a cave-in during a total of not more than 4 to 5 hours in a 2-day period.[51]

In contrast, where employee exposure to the danger cited is continuous or frequent, the Secretary's proposed penalty will be sustained or increased by the Commission. Thus, where from 4 to 24 employees were exposed to the hazard of uncovered and unelevated electric cables for a period of from 2 to 6 weeks, the Commission upheld penalties totaling $6000.[52] Similarly, where an employer failed to bring an elevator into compliance with the state elevator code and continuously used the elevator during the existence of the hazardous condition, the Commission imposed the maximum penalty of $10,000.[53]

2. *Number of Employees Exposed*

Although the Commission has likewise not adopted the Secretary's fixed method of determining and evaluating the number of employees exposed,[54] the fact is that the number of employees exposed to a violation figures into the Commission's consideration of gravity.

Thus where only one, two, or even three employees are exposed to a hazard, the Commission's, like the Secretary's, ultimate assessment of gravity is likely to be moderate or low. For example, a $500 proposed penalty was reduced to a $200 penalty for violation of an OSHA standard caused by suspending a 3000-pound wrecking ball from a crane load line by a used rubber truck tire, because the gravity of the in-

---

[50]*Max J. Kuney, Inc.*, [1976–1977] OSHD ¶21,003 (Rev. Comm'n J. 1976), *aff'd*, 5 OSHC 1569 (Rev. Comm'n 1977).

[51]*Intercounty Constr. Co.*, 1 OSHC 1437 (Rev. Comm'n 1973), *aff'd*, 522 F.2d 777 (4th Cir. 1975), *cert. denied*, 423 U.S. 1072 (1976). See also *Williams Enters.*, 10 OSHC 2224 (Rev. Comm'n 1982); *Niagara Mohawk Power Corp.*, 7 OSHC 1447 (Rev. Comm'n 1979); *Hullenkremer Constr. Co.*, 6 OSHC 1469 (Rev Comm'n 1978); *Thunderbird Coos Bay*, 3 OSHC 1904 (Rev. Comm'n 1976); *Deering Milliken Inc.*, 3 OSHC 1317 (Rev. Comm'n 1975); *National Precast Concrete*, 10 OSHC 1720 (Rev. Comm'n J. 1982); *Murphey Well Serv.*, 10 OSHC 1455 (Rev. Comm'n J. 1982); *Cleveland Metal Abrasive*, [1981] OSHD ¶25,108 (Rev. Comm'n J. 1981).

[52]*Morrison-Knudsen & Assocs.*, 8 OSHC 2231 (Rev. Comm'n 1980).

[53]*Kus-Tum Builders*, 10 OSHC 1128 (Rev. Comm'n 1981); *Ford Motor Co.*, 5 OSHC 1765 (Rev. Comm'n 1977). See also *St. Joe Minerals Corp.*, 10 OSHC 1023 (Rev. Comm'n 1981).

[54]See cases cited in note 27, *supra*.

fraction was determined to be low to moderate where only two employees were exposed to the hazard.[55] Similarly, a penalty imposed for the serious violation of sending an employee into a ventilation duct with a bucket of kerosene to clean an air filter where the employee could have been overcome by kerosene vapors was reduced where only the single employee was exposed for a limited time.[56]

In contrast, where numerous employees are exposed to the hazard in question, the Commission is more likely to sustain or increase the proposed OSHA penalty. Thus, where from 4 to 24 employees were exposed to the hazard of uncovered and unelevated electric cables, proposed penalties of $6000 were affirmed by the Commission.[57]

3. Partial Precautions and Other Factors

Beyond these two factors, a number of other factors, also considered by the Secretary, are looked to by the Commission in its determination of probability.

The Commission will normally find a lower probability of accident where an employer has taken partial precautions against the potential accident or injury.[58] The inherent physical qualities of machinery used by employees is also a factor relied upon by the Commission in determining the probability of accidents. Thus where a bacon press at a food processing plant could be operated by one lever of a defective two-lever control device, which was a serious violation because of the

---

[55]*Ed Miller & Sons*, 2 OSHC 1132 (Rev. Comm'n 1974). See also *Vaughn Roofing Co.*, 2 OSHC 1683 (Rev. Comm'n 1975) (gravity low to moderate where only one employee exposed to hazard); *Metal Recycling Co.*, 10 OSHC 1730 (Rev. Comm'n J. 1982) (proposed $1600 penalty reduced to $1 where only 3 or 4 employees exposed to lead); *Murphey Well Serv.*, *supra* note 51 (violation low in gravity where only four employees exposed to hazard); *Messick Constr. Co.*, [1976–1977] OSHD ¶20,988 (Rev. Comm'n J. 1976) (proposed $550 penalty reduced to $200 where only one employee exposed to fall from I-beam); *Maskel Constr. Co.*, 2 OSHC 3150 (Rev. Comm'n J. 1974) (proposed $1900 penalty reduced to $500 where only one employee working in improperly shored trench).

[56]*ALMCO Steel Prods. Corp.*, 6 OSHC 1532 (Rev. Comm'n J. 1978). See also *Psaty & Fuhrman, Inc.*, 2 OSHC 1006 (Rev. Comm'n 1974).

[57]*Morrison-Knudsen & Assocs.*, *supra* note 52. See also *Jerry's Restaurant*, 2 OSHC 3227 (Rev. Comm'n J. 1974) (where 15 employees exposed to hazards of slipping and falling, proposed penalties warranted).

[58]*Pratt & Whitney Aircraft, Div. of United Aircraft Corp.*, 2 OSHC 1713 (Rev. Comm'n 1975); *Cadillac Malleable Iron Co.*, 2 OSHC 1039 (Rev. Comm'n 1974); *St. Louis County Water Co.*, 1 OSHC 1295 (Rev. Comm'n 1973); *W.J. Lazynski, Inc.*, 1 OSHC 1203 (Rev. Comm'n 1973). But see *Frank Irey, Jr., Inc.*, 5 OSHC 2030 (Rev. Comm'n 1977); *Brand Insulation Inc.*, 4 OSHC 1700 (Rev. Comm'n 1976).

possible resulting injury, the probability of an accident occurring was found to be low since the press took 12 seconds to complete a full cycle.[59] In contrast, where these mitigating factors are not present, such as where the employer has failed to take any precautions to prevent injury or where the nature of the machinery involved is such that defective control or maintenance can result in immediate injury to the employee, the probability of an accident will be found by the Commission to be high.[60]

Although the seriousness of an injury which can occur if an accident takes place is one of the factors the Area Director deems relevant to a determination of the gravity of the violation, the Commission has rarely used this factor in determining gravity. Rather, the Commission appears to allow for this factor in its determination of whether a violation is serious or nonserious and, as a result, has usually ignored the Secretary's calculation of a severity quotient.[61]

B. Size of the Business

The Commission has declined to follow the Secretary's mathematical formula.[62] Rather than permitting a fixed reduction in the amount of the penalty of between 0 to 40 percent depending upon the number of employees employed by the employer at all of the employer's worksites, the Commission's adjustments for size are determined by considering both the total number of persons employed at all the employer's worksites and the employer's gross annual dollar volume.[63]

C. Employer Good Faith

The Commission, as mandated by the Act, likewise considers the good faith of the employer in determining the ap-

---

[59]*Henry's Hickory House*, [1979] OSHD ¶23,290 (Rev. Comm'n 1978). See also *Specialty Insulation Div. of Cantamount Mfg.*, 7 OSHC 1192 (Rev. Comm'n J. 1978); *Kennedy Houseboats, Inc.*, [1975–1976] OSHD ¶20,611 (Rev. Comm'n J. 1976), *aff'd*, 4 OSHC 2050 (Rev. Comm'n 1977).

[60]*St. Joe Minerals Corp.*, *supra* note 53; *Emerick Constr.*, 5 OSHC 2048 (Rev. Comm'n 1977).

[61]But see *Astra Pharmaceutical Prods.*, 10 OSHC 2070 (Rev. Comm'n 1982) (severity of injury relied upon as factor in assessing penalty).

[62]See cases cited in note 27, *supra*.

[63]*Jasper Constr.*, 1 OSHC 1269 (Rev. Comm'n 1973).

propriateness of the penalty. It has declined, however, to apply the Secretary's mathematical formula in determining the effect of the employer's good faith on the ultimate penalty.[64] Instead, the Commission, as in the case of business size, has treated each case on its own merit and has arrived at various results based on the employer's good faith, including in some cases vacating the penalty entirely.

The Commission has determined that the showing of any one of a number of factors will be sufficient to demonstrate the employer's good faith and justify a lesser penalty. Foremost among these criteria for a demonstration of good faith are early or immediate abatement of the cited violation,[65] an active and ongoing safety program,[66] cooperation with OSHA and its inspectors,[67] compliance with an industry practice or consensus,[68] requests for consultation or inspection by OSHA or state health and safety agencies,[69] the employment of industrial engineers or safety experts to ensure safety for employees,[70] and an honest belief that a particular standard does not apply to the employer.[71] Other factors include the employer's safety record (as distinguished from its history of prior violations),[72] the fact that the employer was unaware of the

---

[64]See cases cited in note 27, *supra.*

[65]*Spaghetti Importing & Warehouse*, [1977–1978] OSHD ¶21,655 (Rev. Comm'n 1977); *Oklahoma Beverage Co.*, 11 OSHC 1191 (Rev. Comm'n J. 1983); *Holman Mfg. Co.*, [1981] OSHD ¶25,180 (Rev. Comm'n J. 1981); *Bethlehem Steel Corp.*, 7 OSHC 1575 (Rev. Comm'n J. 1979); *Hoover Aircraft Prods.*, [1979] OSHD ¶23,362 (Rev. Comm'n J.1979); *Earl Durden Signs*, [1979] OSHD ¶23,314 (Rev. Comm'n J. 1978); *Visbeen Constr. Co.*, [1978] OSHD ¶22,831 (Rev. Comm'n J. 1978).

[66]*Williams Enters.*, *supra* note 51; *Lundell Mfg. Co.*, 1 OSHC 1530 (Rev. Comm'n 1974); *Oklahoma Beverage Co.*, *supra* note 65; *Horvitz Co.*, [1982] OSHD ¶25,966 (Rev. Comm'n J. 1982); *Hydraform Prods. Corp.*, 7 OSHC 1995 (Rev. Comm'n J. 1979); *Koller Craft Plastic Prods.*, 7 OSHC 1409 (Rev. Comm'n J. 1979); *55th Street Taxi Garage*, 6 OSHC 1673 (Rev. Comm'n J. 1978); *Truesdale Constr. Co.*, [1976–1977] OSHD ¶20,766 (Rev. Comm'n J. 1976); *Harry Rich Affiliates*, [1975–1976] OSHD ¶20,616 (Rev. Comm'n J. 1976); *Marketing Specialties*, [1975–1976] OSHD ¶20,420 (Rev. Comm'n J. 1976); *Standard Elec. Co.*, [1974–1975] OSHD ¶19,706 (Rev. Comm'n J. 1975).

[67]*Marino Dev. Corp.*, 2 OSHC 1260 (Rev. Comm'n 1974); *Apex Granite Co.*, [1977–1978] OSHD ¶22,182 (Rev. Comm'n J. 1977); *Hensel Optical Co.*, [1977–1978] OSHD ¶21,728 (Rev. Comm'n J.), *aff'd*, 5 OSHC 1772 (Rev. Comm'n 1977).

[68]*Hammond Masonry, Inc.*, [1981] OSHD ¶25,793 (Rev. Comm'n 1981).

[69]*Omark-CCI, Inc.*, [1979] OSHD ¶23,761 (Rev. Comm'n J. 1979); *Riteway Mfg. Co.*, [1979] OSHD ¶23,253 (Rev. Comm'n J. 1978); *Nip-Co. Mfg.*, [1976–1977] OSHD ¶21,264 (Rev. Comm'n J. 1976), *aff'd*, 5 OSHC 1632 (Rev. Comm'n 1977); *John Knoell & Son*, [1973–1974] OSHD ¶17,654 (Rev. Comm'n J. 1974).

[70]*Connecticut Natural Gas Corp.*, 6 OSHC 1796 (Rev. Comm'n 1978).

[71]*Thorlief Larsen & Son*, 1 OSHC 3274 (Rev. Comm'n 1973).

[72]*Marshall v. M.W. Watson, Inc.*, 652 F.2d 977 (10th Cir. 1981); *Boonville Div. of Ethan Allan*, 6 OSHC 2169 (Rev. Comm'n 1978); *Westinghouse Elec. Corp.*, 6 OSHC

Act or its requirements at the time of the inspection,[73] the fact that the employer had relied on prior inspection by OSHA during which a violation was not cited,[74] and the fact that the employer has ceased its business operations.[75]

The absence of these factors may cast an employer's good faith into doubt. Where affirmative evidence is present that an employer failed to abate or begin abatement of a violation, had no active safety program, or had an extensive previous accident history, the Commission may in some cases find bad faith and assess a higher penalty.[76]

The Commission has held that the following conduct by the employer, although not constituting affirmative evidence of good faith, does not demonstrate bad faith and may not be relied upon as a justification for denial of credit for good faith: an employer's demand for a search warrant prior to permitting an OSHA inspection,[77] an employer's criticism or disagreement with the OSHA inspector's conclusions,[78] an employer's attempt to abate the violation where correction proved to be impossible,[79] and an employer's failure to comply if the nec-

---

1095 (Rev. Comm'n J. 1977); *Pease Co., Builder Div.*, 5 OSHC 1901 (Rev. Comm'n J. 1977); *Charles Beck Mach. Corp.*, [1976–1977] OSHD ¶21,480 (Rev. Comm'n J. 1976); *Consolidated Metal Prods.*, 2 OSHC 3152 (Rev. Comm'n J. 1974); *Flagler Tribune, Inc.*, 1 OSHC 3001 (Rev. Comm'n J. 1973).

[73]*Reliable Drywall Co.*, [1975–1976] OSHD ¶20,635 (Rev. Comm'n J. 1976); *Keystone Constr. Co.*, 3 OSHC 1904 (Rev. Comm'n J. 1975); *C&L Trucking*, 1 OSHC 3057 (Rev. Comm'n J. 1974); *Sunpruf Aluminum Prods.*, [1971–1973] OSHD ¶15,213 (Rev. Comm'n J. 1972), *aff'd*, 1 OSHC 1568 (Rev. Comm'n 1974).

[74]*Columbian Art Works*, 10 OSHC 1132 (Rev. Comm'n 1981); *Anchor Roofing*, 6 OSHC 1238 (Rev. Comm'n J. 1977); *Wall Trends, Inc.*, [1975–1976] OSHD ¶20,581 (Rev. Comm'n J. 1976); *Herbert Seidler Agencies*, 2 OSHC 3247 (Rev. Comm'n J. 1974); *Ideal of Idaho*, 2 OSHC 3171 (Rev. Comm'n J. 1974).

[75]*Custom Painting Co.*, 1 OSHC 1233 (Rev. Comm'n 1973); *Tom Bean Grain & Fertilizer Corp.*, [1979] OSHD ¶23,722 (Rev. Comm'n J. 1979); *Karner, Inc.*, 6 OSHC 1615 (Rev. Comm'n J. 1978); *L.R. Ward Steel Prods.*, 5 OSHC 1931 (Rev. Comm'n J. 1977); *Allied Equip. Co.*, [1976–1977] OSHD ¶20,747 (Rev. Comm'n J. 1976), *aff'd*, 5 OSHC 1401 (Rev. Comm'n 1977); *Recreational Components*, 2 OSHC 3095 (Rev. Comm'n J. 1974).

[76]*St. Joe Mineral Corp.*, 10 OSHC 1023 (Rev. Comm'n 1981) (failure to abate); *Scullin Steel Corp.*, 6 OSHC 1764 (Rev. Comm'n 1978) (failure to abate); *Anchorage Plastering Corp.*, [1979] OSHD ¶23,783 (Rev. Comm'n J. 1979) (failure to abate); *Anderson-Tully Co.*, [1979] OSHD ¶23,418 (Rev. Comm'n J. 1979) (weakly enforced safety program); *Ellanef Mach. Tool Co.*, 6 OSHC 1853 (Rev. Comm'n 1978) (proposed $320 penalty increased to $1000, in part because employer had 14 accidents in prior year); *Associated Chemists*, 2 OSHC 3258 (Rev. Comm'n J. 1975) (failure to abate); *Stack Constr. Co.*, [1971–1973] OSHD ¶15,323 (Rev. Comm'n J. 1972), *aff'd*, 1 OSHC 1786 (Rev. Comm'n 1974) (failure to abate).

[77]*Walter H. Kessler Co.*, 7 OSHC 1401 (Rev. Comm'n J. 1979).

[78]*State Painting Co.*, [1978] OSHD ¶22,678 (Rev. Comm'n J. 1978).

[79]*Hecfer Constr. Corp.*, 2 OSHC 1217 (Rev. Comm'n 1974).

essary corrective equipment or materials have been ordered but did not arrive in time.[80]

Similarly, the Commission has held that the mere fact that an employer is cited for a large number of violations does not automatically disqualify that employer from availing itself of the good faith credit,[81] although the nature of the violations and the employer's conduct regarding those violations may in itself negate or reduce the existence of good faith.[82] So too, the good faith credit is available, according to the Commission, where the employer has only partially complied with the applicable OSHA standard.[83] The rationale for this appears to be that the employer at least made some effort to comply with the standard.

D. History of Prior Violations

The Commission has declined to follow the Secretary's maximum 10 percent reduction where the employer does not have a substantial history of previous violations.[84] The Commission's view, consistent with that taken with regard to the other statutory factors, is that an adjustment of more or less than 10 percent may be granted depending upon the facts of the particular case.

Thus, an administrative law judge reduced proposed penalties of $1800 to $35 based, in part, on his finding that the employer did not have a history of prior violations.[85] In contrast, the Commission has sustained proposed penalties where the employer has had a significant history of prior violations. These prior violations may be of the same standard for which the employer is currently cited or for violations of other standards.[86]

---

[80]*Kay-Townes, Inc.*, [1971–1973] OSHD ¶15,596 (Rev. Comm'n J. 1973), *aff'd*, 2 OSHC 1088 (Rev. Comm'n 1974).

[81]*Omaha Body & Equip. Co.*, 1 OSHC 3066 (Rev. Comm'n J. 1973).

[82]*St. Louis County Water Co.*, 1 OSHC 1295 (Rev. Comm'n 1973) (full credit for good faith disallowed owing in part to large number of violations).

[83]*Williams Constr. Co.*, 6 OSHC 1093 (Rev. Comm'n 1977).

[84]See cases cited in note 27, *supra*.

[85]*Harry Rich Affiliates*, [1975–1976] OSHD ¶20,616 (Rev. Comm'n J. 1976). See also *55th Street Taxi Garage*, 6 OSHC 1673 (Rev. Comm'n J. 1978); *Koller Craft Plastic Prods.*, 7 OSHC 1409 (Rev. Comm'n J. 1979).

[86]*Alpha Masonry, Inc.*, [1978] OSHD ¶22,846 (Rev. Comm'n J. 1978); *Continental Heller Corp.*, 1 OSHC 3067 (Rev. Comm'n J. 1973).

The Commission, like the Secretary, also considers it appropriate to examine an employer's record under other federal or state safety and health statutes, where there is evidence of continuing poor performance under those statutes. In cases where such poor performance has been shown, the Commission will not order a reduced penalty simply because there is an absence of prior violations of the Act itself.[87]

E. Inability to Pay

The Commission has adopted the Secretary's practice of adjusting penalties downward where the employer has presented convincing evidence of its inability to pay owing to its financial circumstances. Thus, in one case an administrative law judge reduced proposed penalties totaling $1920 to $7 on the ground of inability to pay where the employer had been out of operation for 4 months before the hearing and the imposition of the penalty raised the possibility that the employer would be forced to go into receivership.[88] Similarly, in another case a $500 proposed penalty was reduced to $1 by an administrative law judge where the employer had lost $10,000 in the 9 months prior to the inspection.[89] In fact, severe operating loss is the objective fact most commonly cited by the Commission as justifying a reduction in penalties.

IV. Penalties for Failure to Abate Violations

As previously noted, Section 17(d) of the Act provides that a civil penalty of not more than $1000 may be assessed for failure to abate a violation for each day that violation continues.[90] If upon reinspection OSHA determines that a previously cited violation has not been corrected by the abatement date

---

[87]*Buck Kreihs Co.*, [1971–1973] OSHD ¶15,418 (Rev. Comm'n J.), *aff'd*, 2 OSHC 1119 (Rev. Comm'n 1974).

[88]*Gulf Marine Servs.*, [1981] OSHD ¶25,571 (Rev. Comm'n J. 1981).

[89]*J.F. Higgins & Co.*, [1977–1978] OSHD ¶21,736 (Rev. Comm'n J. 1977). See also *Pittsburgh Brewing Co.*, 7 OSHC 1099 (Rev. Comm'n J. 1978); *Perry Foundry, Inc.*, 6 OSHC 1690 (Rev. Comm'n J. 1978); *Willse Indus.*, 6 OSHC 1424 (Rev. Comm'n J. 1978).

[90]29 U.S.C. §666(d).

set forth in the citation (as extended by contest or petition for modification of abatement period) or specified in a final order of the Commission, a new gravity-based penalty will be calculated based upon the facts determined during the reinspection.[91] OSHA will then multiply the newly computed penalty by the number of calendar days that the previously cited violation continued uncorrected.[92]

In cases where no penalty was initially proposed, the minimum gravity-based penalty for failure to correct upon reinspection will be, as a matter of OSHA policy, $100 per day.[93] The total proposed penalty for failure to correct a particular violation will not normally exceed 10 times the amount of the daily additional penalty proposed.[94] Higher penalties will be proposed by OSHA in unusual circumstances, where the gravity of the violation is especially high or the employer has exhibited a high degree of negligence in failing to correct the violation.[95]

Adjustment factors for size, good faith, and history are applied on the basis of circumstances noted during the reinspection.[96] For example, OSHA will give credit where an employer promptly ordered parts or materials necessary to correct the alleged violation by the abatement date but has not yet received them, even though the employer failed to inform OSHA of its correction difficulties or failed to petition for a modification of the abatement period.[97]

After the proposed penalty for failure to correct a violation has been determined, a separate notice of the proposed failure-to-abate penalty will be sent to the employer[98] and may be contested in the same manner as the original citation.[99]

In determining the amount of the penalty for failure to abate a previously cited violation, the Commission, like the Secretary, considers the four statutory factors of gravity, good faith, size, and prior violations. In addition, the Commission

---

[91]FOM ch. VI A.5.d, OSHR [Reference File 77:3106].
[92]Ch. VI A.5.d(3), OSHR [Reference File 77:3107].
[93]Ch. VI A.5.d(1), OSHR [Reference File 77:3106].
[94]Ch. VI A.5.d(3), OSHR [Reference File 77:3107].
[95]*Id.*
[96]Ch. VI A.5.d(2), OSHR [Reference File 77:3106].
[97]*Id.*
[98]Ch. VI A.5.a, OSHR [Reference File 77:3106].
[99]29 U.S.C. §659(b).

has adopted the requirement that there be a reasonable relationship between the nonabatement penalties and the penalties assessed for the original citation. Thus, in a case where only nominal penalties had initially been assessed, the Commission reduced the nonabatement penalties to $375 from the proposed $3250.[100]

An employer is most successful in obtaining a reduction in failure-to-abate penalties before the Commission by establishing its good faith as evidenced by efforts to abate during the abatement period. Thus, where an employer had promptly installed temporary safeguards and erroneously believed that it had been granted an informal extension of time, the Commission reduced a $5000 proposed penalty to $1000.[101] Similarly, in cases where materials have been ordered but not yet delivered or installed, or where specialized workmen needed to correct violations are unavailable during the abatement period, the Commission has generally found the employer to have demonstrated good faith and has reduced the nonabatement penalties.[102] Reductions in proposed nonabatement penalties may also be obtained where the employer successfully demonstrates that most but not all of the violations have been abated within the prescribed basic period.[103]

Where the employer failed, however, to take any affirmative action during the abatement period to correct the violation, the Commission and the appeals court have affirmed or increased the penalties proposed by the Secretary. Thus, the Eighth Circuit affirmed a $5000 penalty assessed by the Commission against an employer who failed to abate violations regarding two press brakes during the abatement period, even though the employer claimed that the penalty was excessive because it had acted in good faith and with reasonable

---

[100]*Empire Art Prods. Co.*, 2 OSHC 1230 (Rev. Comm'n 1974). See also *Serval Slide Fasteners*, 2 OSHC 3287 (Rev. Comm'n J. 1975); *AAA Mattress Co.*, 2 OSHC 3256 (Rev. Comm'n J. 1975).

[101]*Cross & Brown Co.*, 9 OSHC 1412 (Rev. Comm'n 1981).

[102]*Everett Lemke*, [1977–1978] OSHD ¶21,909 (Rev. Comm'n J. 1977); *Concrete Technology Corp.*, [1976–1977] OSHD ¶21,193 (Rev. Comm'n J. 1976), *aff'd*, 5 OSHC 1751 (Rev. Comm'n 1977); *C&D Lumber Co.*, [1971–1973] OSHD ¶15,162 (Rev. Comm'n J. 1972).

[103]*Atlas Tag Co.*, 3 OSHC 1264 (Rev. Comm'n J. 1975) (reduction in $1000 proposed penalty to $650 where 36 of 80 violations abated); *Congress Inc.*, 2 OSHC 3268 (Rev. Comm'n J. 1975) (where 5 of 7 cited violations abated during abatement period, proposed $210 penalties for failure to abate remaining two violations reduced to $70).

diligence and because most of the employer's difficulties resulted from lack of guidance from the Department of Labor.[104] Similarly, an administrative law judge doubled a proposed penalty of $750 to $1500 where an employer cited for violations of the noise standard took no noise control steps during the abatement period.[105]

Another example of employer bad faith was found where the employer falsely certified that it had abated the cited violation, when in fact it had merely ordered but not yet received the radial saw guard that it was required to install.[106] As a result of this false certification by the employer, an administrative law judge affirmed a $500 proposed penalty for the employer's failure to abate.

The remaining statutory factors of gravity, size, and prior history of violations are also considered by the Commission in the failure-to-abate situations as they are in the initial citation situation, except that the employer that has failed to abate a citation has at least that violation as a prior violation.[107] But the prior uncorrected violation may not be determinative on the issue of history of previous violations, since only a single nonserious violation may be involved.

In addition to the statutory considerations, the Commission has reduced the proposed additional penalties in two other situations: (1) where abatement is impossible, and (2) where the employer honestly misunderstood the abatement requirements.

Thus, where the employer's employees went on a general strike during the abatement period, which made abatement difficult if not impossible, a proposed $63,970 penalty for failure to abate 37 violations was reduced to $1 for each violation.[108] Similarly, a proposed $150 penalty for violation of an exit sign requirement was vacated where it was established that the sign specified to be installed in the original citation was unavailable by the end of the abatement period.[109]

[104]*Long Mfg. Co. v. OSHRC*, 554 F.2d 903 (8th Cir. 1977).

[105]*Allen Clark Inc.*, 1 OSHC 3047 (Rev. Comm'n J. 1973).

[106]*Schuessler Lumber Co.*, [1979] OSHD ¶23,930 (Rev. Comm'n J. 1979).

[107]*Allway Tools*, 5 OSHC 1094 (Rev. Comm'n 1977) (size of business and low gravity); *Perry Foundry, Inc.*, 6 OSHC 1690 (Rev. Comm'n J. 1978) (financial condition and size of business); *Abaddon Prods. Co.*, [1975–1976] OSHD ¶20,423 (Rev. Comm'n J. 1976) (low gravity and prior history).

[108]*San Juan Shipyard*, [1980] OSHD ¶24,944 (Rev. Comm'n J. 1980).

[109]*S&W Framing Supplies*, 3 OSHC 1514 (Rev. Comm'n J. 1975).

In a related context, where the employer has genuinely misunderstood the abatement requirements, the Commission has reduced the proposed penalty. Thus, the Commission reduced a $4230 penalty to $1000 where the employer's failure to abate the violation of floor hole guarding requirements occurred because the employer had misunderstood the standard and the Secretary's personnel could have realized this, but failed to, from correspondence with the employer.[110]

Finally, an employer may not avoid responsibility for failure to abate because management personnel failed to carry out instructions necessary to correct the violation. Thus, the Commission held the employer responsible for the failure of its plant manager to act, where the plant manager had been instructed to abate the violation but had failed to do so, even though he had been discharged for failing to perform these duties.[111]

V. Penalties for Repeated Violations

Section 17(a) of the Act provides that for a repeated violation of the Act, an employer may be assessed a civil penalty of not more than $10,000 for each violation.[112] For every repeated violation, both serious and nonserious, the gravity-based penalty is recalculated and a new proposed penalty determined by OSHA.[113] Even though an alleged violation is a repetition of a previous violation, it may differ in one or more factors relating to gravity, requiring that gravity be reassessed for each repetition.[114]

Once the gravity-based penalty is arrived at, OSHA will normally double that amount for the first repeated violation

---

[110]*Caldwell Lace Leather Co.*, [1971–1973] OSHD ¶15,221 (Rev. Comm'n J. 1972), *aff'd*, 1 OSHC 1302 (Rev. Comm'n 1973). See also *Command Trucking & Warehouse Corp.*, [1976–1977] OSHD ¶21,006 (Rev. Comm'n J. 1976), *aff'd*, 5 OSHC 1517 (Rev. Comm'n 1977); *Siggins Co.*, 3 OSHC 1562 (Rev. Comm'n J. 1975); *David Weber Co.*, 1 OSHC 3081 (Rev. Comm'n J. 1972); *Utah-Idaho Sugar Co.*, 1 OSHC 3227 (Rev. Comm'n J. 1973).

[111]*DeSotacho, Inc.*, 3 OSHC 1460 (Rev. Comm'n J. 1975). But see *General Drywall Contractor*, 10 OSHC 2068 (Rev. Comm'n J. 1982) (violation of superintendent's instructions by foremen given consideration in determining employer's good faith).

[112]29 U.S.C. §666(a).

[113]FOM ch. VI A.6.a, OSHR [Reference File 77:3107].

[114]*Id.*

and quadruple that amount if the violation has been cited and repeated twice. If a third repetition of a previous violation occurs, OSHA will multiply the unadjusted gravity-based penalty by 10. In the event of four or more repetitions of a previously cited violation, the Area Director will consult with the Regional Administrator who may in turn consult with the Regional Solicitor to determine the amount of the proposed penalty.[115] If there is no initial proposed penalty, the minimum unadjusted gravity-based penalty OSHA will impose will be $100 per repeated violation.[116]

Despite the fact that the employer has been cited for a repeated violation, it may still obtain reduction on the basis of the statutorily mandated adjustment factors.[117] In the case of a repeated violation, the size of the business factor remains unchanged.[118] However, the maximum reduction for the good faith and prior history factor will be less, because good faith is frequently negated by repetition and a prior history of the same violation can eliminate any historical credit.[119] Thus, OSHA invariably gives an employer a significantly lower percentage reduction for a repeated violation.

The Commission requires that the employer's attitude, the commonality of supervisory control of the violative condition, geographic proximity or time lapse between violations, and the number of prior violations be considered in determining penalties for repeated violations.[120] The Commission also considers the good faith of the employer, the size of the business, and the employer's safety record as mitigating factors in reviewing proposed penalties.[121]

VI. Penalties for Willful Violations

Section 17(a) of the Act provides that for a willful violation an employer may be assessed a civil penalty of not more

---

[115]*Id.*

[116]Ch. VI A.6.c, OSHR [Reference File 77:3107].

[117]Ch. VI A.6.a, OSHR [Reference File 77:3107].

[118]*Id.*

[119]*Id.*

[120]*Potlatch Corp.*, 7 OSHC 1061 (Rev. Comm'n 1979); *Becker Elec. Co.*, 7 OSHC 1519 (Rev. Comm'n J. 1979).

[121]*T. Smith & Son*, 5 OSHC 2032 (Rev. Comm'n 1977); *Jonas Haies & Co.*, [1980] OSHD ¶24,897 (Rev. Comm'n J. 1980); *Ceco Corp.*, [1979] OSHD ¶23,812 (Rev. Comm'n J. 1979).

than $10,000 for each violation.[122] But before a violation may be classified as willful, OSHA attempts to ascertain if the employer had "specific knowledge" of the standard or its requirements or of the particularly hazardous working conditions to which the employees were exposed.[123] Each violation is first evaluated by OSHA as serious or nonserious and, following that determination, an unadjusted gravity-based penalty corresponding to the probability/severity quotient is determined.[124] In cases of willful violations, OSHA always proposes a minimum unadjusted penalty of at least $100.[125]

In willful, repeated, high gravity serious, and failure-to-abate violations, an additional factor of up to the number of violation instances may be applied to the gravity-based penalty.[126] Penalties calculated with this additional factor cannot be proposed without the permission of the Assistant Secretary of Labor for Occupational Safety and Health.[127] Usually, no credit for good faith will be given by OSHA in situations involving willful violations.[128]

The Commission applies the same statutory criteria as does the Secretary to determine penalties to be imposed for a willful violation of the Act. Although the evaluation and weight to be ascribed to each of the statutory criteria in a particular case has varied between the Secretary and the Commission, the analysis applied by both is essentially the same. One exception has been the Commission's holding that the $10,000 maximum penalty for a willful violation under Section 17(a) of the Act permits a civil penalty of not more than $10,000 for each employee death or injury covered by that section. Accordingly, a Commission judge reversed a proposed penalty by the Secretary of $20,000 for three willful violations of trenching standards which resulted in the death of a single employee.[129] The judge held that Congress did not intend to multiply penalties for hazards leading to the death of a single employee to an amount above the $10,000 limit.

---

[122]29 U.S.C. §666(a).
[123]FOM ch. IV B.3, OSHR [Reference File 77:2710].
[124]Ch. VI A.7.a, OSHR [Reference File 77:3107].
[125]Ch. VI A.7.b, OSHR [Reference File 77:3107].
[126]Ch. VI A.2.i(4).
[127]*Id.*
[128]Ch. VI A.2.j(4), OSHR [Reference File 77:3104].
[129]*James Tull*, 6 OSHC 1426 (Rev. Comm'n J. 1978).

VII. Civil Penalties for Violations of Posting, Recordkeeping, and Reporting Requirements

Penalties will also be imposed by OSHA for failure to comply with the posting, recordkeeping, and reporting requirements contained in 29 C.F.R. §§1903 and 1904. The Act requires that penalties be imposed only for violation of posting requirements.[130] However, the Secretary has determined that, despite the permissive language of Section 17(c) of the Act, violations of the recordkeeping and reporting requirements also carry mandatory, fixed unadjusted penalties. But OSHA imposes penalties only if the employer has received a copy of the "Recordkeeping Requirements" booklet or had knowledge of the requirements. In cases where the employer has not received the booklet and does not have knowledge of it, citations without proposed penalties will be issued to the employer.[131]

The adjustment factors of good faith, size of business, and prior history of violations are applicable to these violations and may result in a reduced proposed penalty.[132] The only exception is the $200 penalty for violation of the requirement that the employee representative be notified of an inspection, which will not be adjusted by OSHA.[133]

To guard against the pyramiding of penalties in cases where violations of posting and recordkeeping requirements involve the same document, such as Form 102, the higher unadjusted penalty will be proposed and the lower unadjusted penalty will be disregarded.[134]

A. Failure to Comply With OSHA Posting Requirements

Pursuant to Section 17(i) of the Act, an employer who violates any of the posting requirements shall be assessed a civil penalty of up to $1000 for each violation.[135] But in prac-

---

[130]29 U.S.C. §666(i).
[131]FOM ch. VI A.8.a(1), OSHR [Reference File 77:3107].
[132]Ch. VI A.8.a(2), OSHR [Reference File 77:3107].
[133]Ch. VI A.8.e, OSHR [Reference File 77:3108].
[134]Ch. VI A.8.d, OSHR [Reference File 77:3108].
[135]29 U.S.C. §666(h).

tice a nonwillful violation of Section 17(i) will not lead to such a large penalty.

1. Failure to Post the Official OSHA Notice

If the notice furnished by OSHA is not posted as required by 29 C.F.R. §1903.2(a), the unadjusted penalty for this violation will be $100.[136]

2. Failure to Post Citation and Notification of Penalty

The failure to post a citation as prescribed in 29 C.F.R. §1903.16 will result in the imposition of a mandatory unadjusted penalty of $500 for each citation not posted.[137]

3. Failure to Post Annual Summary of Occupational Injuries and Illnesses

A mandatory unadjusted penalty of $200 will be proposed if an employer fails to post the portion of the OSHA-200 form in accordance with 29 C.F.R. §1904.5.[138]

B. Failure to Comply With OSHA Recordkeeping, Record Maintenance, and Reporting Requirements

Pursuant to the provisions of Section 17(c) of the Act, violations of the recordkeeping and reporting requirements may result in civil penalties of up to $1000 for each violation.[139] In practice, however, the Secretary ignores the permissive language of Section 17(c) and proposes mandatory unadjusted penalties for violation of these requirements.

If an employer fails to maintain the "Log of Occupational Injuries and Illnesses," Form 100, or the "Supplementary Record," Form 101, as required by 29 C.F.R. §1904, an unadjusted penalty of $100 will be proposed for each OSHA form not maintained.[140] An employer lucky enough to have no re-

---

[136]Ch. VI A.8.b(1), OSHR [Reference File 77:3108].
[137]Ch. VI A.8.b(3), OSHR [Reference File 77:3108].
[138]Ch. VI A.8.b(2), OSHR [Reference File 77:3108].
[139]Ch. VI A.8.c, OSHR [Reference File 77:3108].
[140]Ch. VI A.8.c(1), OSHR [Reference File 77:3108].

cordable injuries or illnesses will escape imposition of this
penalty; OSHA will consider these forms as being maintained,
because no entries would appear on them.[141]

C. Failure to Report Reportable Accidents to OSHA

Employers are required to report to the nearest Area Office within 48 hours any occurrence of an employment accident
that is fatal to one or more employees or that results in hospitalization of five or more employees. Failure to comply with
this regulation will result in a proposed unadjusted penalty
of $400.[142]

D. Failure to Notify Employee Representative of Inspection

An employer who receives advance notice of an inspection
but fails to notify the authorized employee representative,
most commonly the union or unions which represent the employees, as required by 29 C.F.R. §1903.6 will be assessed a
proposed penalty of $200.[143] Moreover, this penalty will not
be subject to reduction because of good faith, size of business,
or prior record.

E. The Commission's Views

In considering the appropriateness of penalties for an employer's failure to post OSHA notices and citations, to maintain records of illnesses and injuries, and to submit reports of
the fatality of one or more employees and hospitalizations
involving five or more employees, the Commission has generally imposed penalties of lesser magnitude than those provided for in the FOM. Indeed, only where an employer fails
to post a notice of citation has the Commission held that a
penalty is mandatory under the Act.[144] And even in this situation, the Commission has imposed mere nominal penalties

---

[141]*Id.*
[142]Ch. VI A.8.c(2), OSHR [Reference File 77:3108].
[143]Ch. VI A.8.f, OSHR [Reference File 77:3108].
[144]*C & R Cabinet Co.*, 3 OSHC 1206 (Rev. Comm'n 1975).

upon the employer,[145] particularly where the employer has in good faith believed that it was not required to post the OSHA citation.[146]

The Commission has, in fact, consistently taken the position that the failure to maintain an illness and injury log is generally a *de minimis* violation requiring no penalty, and it has vacated penalties imposed by the Secretary for violation of this recordkeeping requirement.[147] So too, the Commission has assessed only nominal, if any, penalties for failure by an employer to post the OSHA informational poster.[148] Similar results have been directed by the Commission where the employer has failed to post the annual summary.[149]

VIII. *De Minimis* Violations

Section 9(a) of the Act provides, in relevant part, that "[t]he Secretary may prescribe procedures for the issuance of a notice in lieu of a citation with respect to *de minimis* violations which have no direct or immediate relationship to safety or health."[150] In practice, OSHA does not issue citations for *de minimis* violations and will take no enforcement action even though an employer fails to correct the *de minimis* violation.[151] Consequently, no penalties are proposed by OSHA for *de minimis* violations and no abatement date is established.[152] Likewise, the Commission will not impose any penalties for a *de minimis* violation.[153]

---

[145]*Intracoastal Shipyard*, 1 OSHC 3276 (Rev. Comm'n J. 1974) (where all employees knew of cited conditions and were aware of this correction by employer, $500 penalty reduced to $10).

[146]*Atcheson, T. & S.F. Ry.*, 18 OSAHRC 276 (Rev. Comm'n J. 1975); *Southern Stevedoring Co.*, 2 OSHC 3246 (Rev. Comm'n J. 1974).

[147]*Jenny Indus.*, 2 OSHC 3308 (Rev. Comm'n J. 1975); *Mantua Mfg. Co.*, 1 OSHC 3070 (Rev. Comm'n J. 1973).

[148]*Ecklar-Moore Express*, 2 OSHC 3209 (Rev. Comm'n J. 1974); *Trojan Builders, Inc.*, 1 OSHC 3263 (Rev. Comm'n J. 1973); *Holland Constr. Co.*, 1 OSHC 3199 (Rev. Comm'n J. 1973).

[149]*Stowe Canoe Co.*, 4 OSHC 1012 (Rev. Comm'n 1976).

[150]29 U.S.C. §658(a).

[151]FOM ch. IV B.6, OSHR [Reference File 77:2712].

[152]*Id.*

[153]E.g., *Jenny Indus.*, *supra* note 147; *Mantua Mfg. Co.*, *supra* note 147.

IX. Settlement of Penalties Assessed by the Secretary of Labor

The Secretary has indirect authority under the Act to settle any penalty proposed under the Act, provided that the Secretary states the reason for such settlement in the *Federal Register*.[154] The Secretary has direct authority to settle the proposed penalties assessed pursuant to the provisions of the Compromise and Collection of Federal Claims Act.[155]

In practice, it is the general experience that an employer may obtain a reduction in the Secretary's proposed penalty of 50 percent in settling an OSHA citation. OSHA's willingness to almost uniformly grant a 50 percent reduction in the proposed penalty as part of an overall settlement is no doubt motivated, in significant part, by (1) the employers' success rate (around 80–90%) in obtaining a substantially reduced penalty before the Commission in a contest proceeding; (2) the high expense of litigation that OSHA must incur in a contest proceeding; (3) the amount of the proposed penalty, which is often small and rarely high enough to justify expending those funds; and (4) OSHA's greater interest in obtaining abatement and improving safety than in punishing employers. Accordingly, at least with regard to the proposed penalty portion of an OSHA citation, the Secretary has been extremely willing to negotiate significant reductions in the amount of the proposed penalties provided there is assurance of abatement.

As a result, the employer has three significant opportunities to obtain a reduction in penalties: (1) from the Area Director in settlement of the OSHA citation; if such negotiations are unsuccessful, (2) by filing a notice of contest and seeking a settlement through the Solicitor of Labor, and (3) by actually trying the case before an administrative law judge and ultimately appealing to the Commission and the courts.

X. Collection of OSHA Penalties

After the final penalty imposed on an employer has been assessed after a contest proceeding, in a settlement agree-

---

[154]29 U.S.C. §655(e). *Dale M. Madden Constr. Co. v. Hodgson*, 502 F.2d 278 (9th Cir. 1974).
[155]31 U.S.C. §§951–953 (1983).

ment, or through acceptance by the employer of the original proposed penalty, those penalties become due and owing to the United States and accrue to the United States. The Act specifically provides in Section 17(l) that civil penalties owed under the Act may be recovered in a civil action in the name of the United States brought in the U.S. district court where either the violation is alleged to have occurred or the employer has its principal office.[156]

An employer is required to pay the proposed penalty immediately after the expiration of the 15-working-day period for filing a notice of contest.[157] If the employer files a notice of contest, payment is due after a final order of the Commission has been entered and the 60-day period for taking an appeal to the U.S. court of appeals has expired.[158] If an appeal is taken to the courts, the penalties are payable upon decision by the court.[159]

An employer who contests some but not all of the penalties proposed in a citation obtains the extended periods of time only with regard to those penalties that are contested and is required to pay the uncontested penalties immediately upon the expiration of the contest filing period.[160] Failure to pay can result in late-payment penalties and interest charges.[161]

OSHA will commence a civil action to recover unpaid penalties where the employer has failed to pay them within the applicable time period and has ignored the subsequent notification which OSHA sends by certified mail informing the employer of the overdue payment.[162]

In the collection proceeding, the court will generally look only to whether or not there is an appropriate final order directing the employer to pay a penalty and will not permit an employer to contest the merits of either the citation itself or the penalty assessed.[163]

---

[156]29 U.S.C. §666(k).

[157]FOM ch. VI C.2(1), OSHR [Reference File 77:3108–3109].

[158]Ch. VI C.2(2), OSHR [Reference File 77:3109].

[159]*Id.*

[160]Ch. VI C.2(3), OSHR [Reference File 77:3109].

[161]Ch. VI D.1, OSHR [Reference File 77:3110–3112].

[162]Ch. VI D.3, OSHR [Reference File 77:3112].

[163]*United States v. Fornea Road Boring Co.,* 565 F.2d 1314 (5th Cir. 1978); *United States v. Jan Hardware Mfg. Co.,* 463 F. Supp. 732 (E.D.N.Y. 1979); *United States v. Tomco Stud Co.,* 420 F. Supp. 470 (E.D. Wis. 1976); *United States v. E. Fultz & M. Fultz,* [1975–1976] OSHD ¶19,919 (N.D. Ind. 1975); *United States v. F.A. Weber*

XI. Criminal Penalties

Criminal penalties are provided in Sections 17(e)–(h) of the OSH Act.[164] These penalties may be imposed only by the U.S. district courts, after trial, and may not be imposed by either the Commission or OSHA.[165] In situations where there is reason to believe that a violation may involve a crime, the Regional Solicitor of Labor, after consultation with the appropriate OSHA officials, refers the matter to the appropriate federal agencies, such as the Department of Justice or the U.S. Attorney's Office, for further action.[166]

& *Sons*, [1974–1975] OSHD ¶19,777 (W.D. Wis. 1975); *United States v. J.M. Rosa Constr. Co.*, [1971–1973] OSHD ¶15,643 (D. Conn. 1973). But see *United States v. Crown Kitchen Cabinets Corp.*, [1975–1976] OSHD ¶20,546 (E.D.N.Y. 1976).

[164]29 U.S.C. §666(e)–(g); 18 U.S.C. §1114 (1984). Under §17(e) of the Act, 29 U.S.C. §666(e), an employer is subject to a fine of not more than $10,000 and imprisonment for not more than 6 months where it willfully violates any OSHA standard and that violation causes the death of an employee. In the event of a repeat willful violation which causes the death of an employee, the punishment is raised to a fine of not more than $20,000 or imprisonment for not more than one year or both. Other criminal penalties are provided in §17(f) and (g) of the Act. Section 17(f), 29 U.S.C. §666(f), provides that any person who gives advance notice of any inspection without authority from the Secretary of Labor shall, upon conviction, be punished by a fine of not more than $1000 or by imprisonment for not more than 6 months, or by both. Section 17(g), 29 U.S.C. §666(g), provides that whoever knowingly makes any false statement, representation, or certification in any application, record, report, plan, or other document filed or required to be maintained pursuant to the Act shall, upon conviction, be punished by a fine of not more than $10,000, or by imprisonment for not more than 6 months, or by both. Section 17(h), 18 U.S.C. §1114, provides that any person who kills, assaults, or resists OSHA personnel shall, upon conviction, be subject to a fine of $5,000, or imprisonment for not more than 3 years, or both, for assaulting or resisting such personnel; a fine of $10,000 or imprisonment of not more than 10 years, or both, for assaulting such personnel with a deadly weapon; and a fine of $10,000 and life imprisonment for the murder of such personnel.

[165]FOM ch. VI B.2, OSHR [Reference File 77:3108].

[166]Ch. IV B.2.c, OSHR [Reference File 77:2711].

12

Criminal Enforcement of Occupational Safety and Health Laws

I. Introduction

The enforcement scheme of the Act is essentially civil in nature. Sections 8, 9, 10, and 11[1] set forth detailed procedures for investigations and inspections, the issuance of citations, the contesting of citations, and Commission and court review of contested citations. However, Section 17(e) of the Act gives rise to possible criminal liability in the event of a willful violation of a standard, rule, order, or regulation which causes the death of an employee.[2]

In addition, certain state legislatures and prosecutors have chosen to increase the extent to which criminal sanctions are looked to for enforcement of occupational health and safety laws. In such jurisdictions, state law tends to expand the scope and severity of the criminal sanctions beyond that available under the Act.

---

[1]29 U.S.C. §§657, 658, 659, and 660.

[2]§666(e). The Act also includes other criminal provisions which, however, are not as central to the enforcement scheme of the Act as is §17(e): §666(f) (giving advance notice of an inspection); §666(g) (making a false statement, representation, or certification in a document filed or maintained under the Act); §666(h)(2) (killing an OSHA investigator).

II. Elements of a Criminal Willful Violation

Section 17(e) provides as follows:

"[A]ny employer who willfully violates any standard, rule, or order promulgated pursuant to section 6 of this Act, or of any regulations prescribed pursuant to this Act, and that violation caused death to any employee, shall, upon conviction, be punished by a fine of not more than $10,000 or by imprisonment for not more than six months, or both; except that if the conviction is for a violation committed after a first conviction of such person, punishment shall be by a fine of not more than $20,000 or by imprisonment for not more than one year, or by both."

Thus, criminal sanctions can be imposed for violations of any Section 6(a) or 6(b)[3] standard, any variance rule or order issued pursuant to Section 6(b)(6)(A)[4] or Section 6(d),[5] and any rule or regulation prescribed pursuant to Section 8(c)(1) or 8(g)(2).[6] Since the language of Section 17(e) clearly limits its applicability to violations of "any standard, rule, or order promulgated pursuant to Section 6 of this Act or of any regulations prescribed pursuant to this Act," criminal liability cannot be founded on the basis of a violation of Section 5(a)(1) of the Act, the general duty clause of the statute.[7]

A. Willfulness

In order to establish a violation of Section 17(e), the violation of the standard, rule, order, or regulation must be willful. While the Act itself does not define the term "willful," there has been considerable litigation, both at the Review Commission and in the courts, concerning the definition of the term. Currently, the Commission and the courts are in substantial agreement that "willful" refers to an action taken deliberately and with knowledge of the standard or with plain indifference to its requirement.[8]

---

[3]§655(a), (b).

[4]§655(b)(6)(A).

[5]§655(d).

[6]§657(c)(1), (g)(2).

[7]§654(a)(1).

[8]*National Steel & Shipbuilding Co.*, 6 OSHC 1680 (Rev. Comm'n 1978), *aff'd*, 607 F.2d 311, 7 OSHC 1837 (9th Cir. 1979); *Cedar Constr. Co. v. OSHRC*, 587 F.2d

Willfulness therefore may be established by showing that an employer knew of the requirements of a standard and, as evidenced by its action or inaction, chose not to comply with it.[9] It is not necessary to show malicious intent on the part of the employer.[10]

The employer's knowledge of a standard's requirements may be shown in a number of ways. Knowledge can be demonstrated by proving that the employer had previously received a citation for violating the identical standard.[11] Knowledge also may be established by prior warnings from employees or others who inform the employer of the requirements of the standard[12] or by the presence of a supervisor or corporate official at the location of the violation.[13]

B. Death of an Employee

While civil willful violations may be established regardless of whether an employee's death occurred, in order to es-

---

1303, 6 OSHC 2010 (D.C. Cir. 1978); *Kent Nowlin Constr. Co. v. OSHRC*, 648 F.2d 1278, 1281, 9 OSHC 1709, 1711 (10th Cir. 1981); *Intercounty Constr. Co. v. OSHRC*, 522 F.2d 777, 3 OSHC 1337 (4th Cir. 1975), *cert. denied*, 423 U.S. 1072 (1976); *A. Schonbek & Co. v. Donovan*, 646 F.2d 799, 9 OSHC 1562 (2d Cir. 1981); *F.X. Messina Constr. Co. v. OSHRC*, 505 F.2d 701, 2 OSHC 1325 (1st Cir. 1974). For a more detailed discussion of willfulness for civil violations of the Act, see Chapter 9 (Types and Degrees of Violations). The only court that arguably deviated from this definition was the Third Circuit in *Frank Irey, Jr., Inc. v. OSHRC*, 519 F.2d 1200, 2 OSHC 1283 (3rd Cir. 1974), *aff'd en banc*, 519 F.2d 1215, 3 OSHC 1329. Since *Irey*, however, the Third Circuit reevaluated its position in *Universal Auto Radiator Mfg. Co. v. Marshall*, 631 F.2d 20, 8 OSHC 2026 (3rd Cir. 1980), and seemingly adopted the Tenth Circuit standard of willfulness as set forth in *United States v. Dye Constr. Co.*, 510 F.2d 78, 2 OSHC 1510 (10th Cir. 1975). In *Dye Constr. Co.*, the court adopted the following definition of "willful":

> "The failure to comply with a safety standard under the Occupational Safety and Health Act is willful if done knowingly and purposely by an employer who, having a free will or choice, either intentionally disregards the standard or is plainly indifferent to this requirement. An omission or failure to act is willfully done if done voluntarily and intentionally."

510 F.2d at 81, 2 OSHC at 1512–1513.

[9]For cases in which an employer's failure to act contributed to a finding of a willful violation, see, e.g., *Williams Enters.*, 11 OSHC 1410, 1420 (Rev. Comm'n 1983) (supervisors repeatedly told of need to have employees wear safety belts yet employer took no action to require their use); *Daniel Int'l Corp. v. Donovan*, 705 F.2d 382, 387, 11 OSHC 1305, 1308–1309 (10th Cir. 1983) (employer had reason to know of Act's requirements but made no effort to comply with them).

[10]*United States v. Dye Constr. Co.*, 510 F.2d 78, 2 OSHC 1510 (10th Cir. 1975). See *Speis v. United States*, 317 U.S. 492 (1943).

[11]See generally cases cited *supra* note 8.

[12]*Intercounty Constr. Co. v. OSHRC*, *supra* note 8 (warning given by OSHA personnel); *Ford Motor Co.*, 5 OSHC 1765 (Rev. Comm'n 1977) (warning by state agency); *Tri-City Constr. Co.*, 8 OSHC 1567 (Rev. Comm'n 1980) (warning by power company to construction company of hazard caused by overhead lines.

[13]See generally cases cited *supra* note 8.

tablish a violation of Section 17(e) it is necessary to prove that the violation "caused death to any employee."[14] Nearly all of the Section 17(e) cases that have been prosecuted involved a direct, causal relationship between a violation of a standard, rule, order, or regulation, and the employee's death. For example, in *United States v. Dye Construction Co.*,[15] the employer's failure to shore or slope a trench caused the trench to collapse and killed the employee working there. However, a criminal violation may also be found if the violative condition was only a contributing cause of the employee's death. In *United States v. Youngstown Sheet & Tube*,[16] the defendant was found guilty where its failure to maintain the regular brakes on a tractor, in violation of a safety standard, was only a contributing factor to the death since the tractor operator was unaware of the location of the emergency brake and therefore failed to use it.

III. Sanctions for Criminal Violations

Section 17(e) provides for 6 months' imprisonment or a $10,000 fine, or both, for the first conviction and one year's imprisonment or a $20,000 fine, or both, for any second or subsequent convictions.[17] Unlike civil willful citations, criminal willful violations do not impose abatement requirements.[18]

Although Section 17(e) initially uses the term "employer" to identify entities subject to conviction for violations of its provisions, it subsequently indicates that "conviction of such

---

[14]As noted previously, civil willful violations can be based on §5(a)(1), whereas criminal willful violations cannot arise under the general duty clause.

[15]*Supra* note 10. See also *United States v. Nichols Contracting Co.*, Cr. No. 80–0169A (N.D. Ga. 1980) (failure to shore or slope trench caused collapse killing employee); *United States v. Brown Steel Contractors*, Cr. No. 80–G–0254–S (N.D. Ala. 1980) (allowing employee to ride crane load killed employee when load fell); *United States v. Port Allen Marine Serv.*, Magis. No. 81–77A U.S. (M.D. La. 1981) (failure to protect employee working on dry dock with guardrails on dock or, in alternative, flotation device, caused employee's drowning death); *United States v. Newton Roofing & Sheet Metal*, Cr. No. 4–82–028 (N.D. Tex. 1982) (failure to provide guardrail around open skylight killed employee when he fell through opening).

[16]Cr. No. HCR–82–14 (N.D. Ind. 1982).

[17]Only $10,000 in penalties may be imposed for civil willful violations under §17(a), 29 U.S.C. §666(a).

[18]Cf. §17(d), 29 U.S.C. §666(d) ("civil penalty of not more than $1,000 for each day during which *** violation continues").

person" for a second violation shall bring a fine of not more than $20,000 or imprisonment for one year. Section 3(4) of the Act defines a person as "one or more individuals, partnerships, associations, corporations, businesses, trusts, legal representatives, or any organized group of persons."[19] This gives rise to the question of whether Section 17(e) may apply not only to corporate and other business entities for the actions of their authorized agents, but also to individuals who may be employed by such entities.

The issue of individual as opposed to corporate liability was addressed in *United States v. Pinkston-Hollar, Inc.*[20] There the court held that "responsible corporate officials who have the power to prevent or correct violations *** may be prosecuted"[21] under Section 17(e). However, in *United States v. Dye Construction Co.*,[22] the district court dismissed the charges against the corporate president on the grounds that only the corporation may be charged as an "employer" under Section 17(e).

The determination of whether an indictment will issue under Section 17(e) is a matter for a grand jury, with the standards for indictment varying widely depending on the applicable facts and circumstances. In one instance, for example, individuals were indicted based on the number of standards violated multiplied by the number of employees killed as a result of such violations. Thus, in *United States v. Jones & Caulder*,[23] violations of two standards allegedly contributed to the deaths of 18 employees. Each defendant was therefore charged with 36 separate violations, which could have resulted in a maximum of 36 6-month prison terms, i.e., 18 years, and $360,000 in penalties. The case went to trial in 1981 and a hung jury resulted. The case was thereafter dismissed.

Individuals involved in possible violations of Section 17(e) also may be indicted pursuant to 18 U.S.C. §2(a) providing that "[w]hoever commits an offense against the United States

---

[19]29 U.S.C. §652(4).

[20]4 OSHC 1697, Cr. No. 76–33–CR6 (D. Kan. 1976).

[21]*Id.* at 1699. See *United States v. Turcon Co.*, Cr. No. 72–0–239 (D. Neb. 1972) (court orally denied motion to dismiss charges against corporate president); *United States v. Jones & Caulder, infra* note 23 (only individual supervisors indicted; corporation not charged); *United States v. S.O. Jennings Constr. Corp.*, Cr. No. 80–200 (E.D. Va. 1980) (both corporation and corporate officers indicted).

[22]510 F.2d 78, 2 OSHC 1510 (10th Cir. 1975).

[23]Cr. No. G–80–11 (S.D. Tex. 1980).

or aids, abets, counsels, commands, induces or procures its commission is punishable as a principal."[24]

IV. The *Miranda* Rule in OSHA Cases

Since OSHA's compliance officers do not have the authority to arrest or otherwise restrict the freedom of an individual, the *Miranda* rules do not apply to any Section 17(e) investigation conducted by OSHA personnel.[25]

The *Miranda* rule also does not apply to Section 17(e) cases where only a corporation is charged: "It is settled that a corporation is not protected by the constitutional privilege against self-incrimination. A corporate officer may not withhold testimony or documents on the ground that his corporation would incriminated."[26]

V. Search Warrant Requirements[27]

Only civil probable cause need be shown to obtain a warrant unless "the investigation changes from purely administrative to criminal in nature."[28] However, "if the investigating officials find probable cause to believe that *** [a criminal violation] has occurred and require further access to gather evidence for a possible prosecution, they may obtain a warrant only upon a traditional showing of probable cause applicable

---

[24]*Id.*

[25]In *Miranda v. Arizona*, 384 U.S. 436 (1966), the Supreme Court held that "the prosecution may not use statements, whether exculpatory or inculpatory, stemming from custodial interrogation of the defendant unless it demonstrates the use of procedural safeguards effective to secure the privilege against self-incrimination. By custodial interrogation, we mean questioning initiated by law enforcement officers after a person has been taken into custody or otherwise deprived of his freedom of action in any significant way."
384 U.S. at 444. In determining whether an interrogation has occurred in a custodial setting the courts will examine whether (a) probable cause to arrest has arisen; (b) the subjective intent of the interrogating officer was to hold the defendant in custody; (c) the defendant subjectively believed that his freedom was significantly restricted; and (d) the investigation was focused on the defendant. See, e.g., *United States v. Henry*, 604 F.2d 908 (5th Cir. 1979); *United States v. Bridwell*, 583 F.2d 1135 (10th Cir. 1978); *United States v. Miller*, 643 F.2d 713 (10th Cir. 1981).

[26]*Curcio v. United States*, 354 U.S. 118, 122 (1957); *Bellis v. United States*, 417 U.S. 85 (1974).

[27]Chapter 7 (Inspections) discusses the general search warrant requirements for OSHA investigations.

[28]*Burkart Randall Div. of Textron v. Marshall*, 625 F.2d 1313, 1318, 8 OSHC 1467, 1469–1470 (7th Cir. 1980).

to searches for evidence of crime."[29] As yet, there has been no case where OSHA has had to obtain a search warrant based upon a criminal probable cause showing.

VI. Concurrent Investigation and Litigation of Related Civil and Criminal Cases

Although Section 9(c)[30] requires that civil citations must be issued within 6 months following the occurrence of any violation, criminal charges under Section 17(e) may be brought at any time up to 5 years following the violation.[31] Generally, OSHA will not issue the civil citations until "(1) the criminal case has been referred to the Department of Justice, (2) a decision has been made not to refer the criminal case, or (3) the 6-month statute of limitations for the issuance of the civil citations has nearly expired, whichever occurs first."[32] Under these circumstances, OSHA will have the maximum time available to consider a possible criminal referral to the Department of Justice prior to the issuance of civil citations.[33] Once the civil citation has been issued and contested, OSHA may seek a stay of the civil litigation until the criminal proceeding is completed,[34] e.g., no indictment is obtained or a final verdict is rendered. Such a procedure limits any possible adverse effect that conducting simultaneous civil and criminal litigation might have on the related cases.

VII. State Criminal Sanctions

State criminal penalties associated with unsafe working conditions exist both within and outside of the context of the Act.

---

[29]*Michigan v. Tyler*, 436 U.S. 499, 511–512 (1978).

[30]29 U.S.C. §658(c).

[31]18 U.S.C. §3282.

[32]OSHA Instruction CPL 2.39 (Oct. 9, 1979), 1 Employment Safety & Health Guide ¶4012.

[33]However, if employees continue to be exposed to the hazard because the condition has not been abated, OSHA will issue the citation as soon as possible. *Id.*

[34]See *Research Cottrell, Inc.*, OSHRC No. 78–2927 (April 5, 1979) (unpublished order staying civil proceedings).

Under Sections 18 and 23 of the Act,[35] a state may assume
responsibility for implementing occupational health and safety
laws, provided that prior approval by OSHA of the state's
occupational health and safety plan has been obtained. State
plans must include standards "at least as effective" as comparable OSHA standards.[36] Twenty-three states currently have
in place approved plans which provide for criminal enforcement.[37] Several of the OSHA-approved state plans contain
criminal provisions significantly broader than Section 17(e)
of the Act. Perhaps the broadest such statute is Minnesota's,
which permits assessment of fines of up to $10,000, and/or
imprisonment of up to 6 months, for any willful or repeated
violation of the state plan.[38] California and Washington, in
addition to enacting statutes roughly parallel to Section 17(e),[39]
also provide for lesser criminal penalties against employers
who have committed certain serious or dangerous violations.[40]
Indiana supplements the traditional OSHA criminal sanctions
with a catchall misdemeanor statute.[41] Alaska and Hawaii
apparently have dispensed entirely with Section 17(e)'s *mens*

---

[35]29 U.S.C. §§667, 672.

[36]§667(c)(2).

[37]See ALASKA STAT. §§18.60.10–18.60.105 (1986); ARIZ. REV. STAT. ANN. §§23–
401–23–433 (1983 & Supp. 1986); CAL. LAB. CODE §§140–149, 6300–6308 (1976 &
Supp. 1987); HAWAII REV. STAT. ch. 396 (1976 & Supp. 1982); IND. CODE §§22–8–1.1
et seq. (1986); IOWA CODE §§88.1–88.21 (West 1984); KY. REV. STAT. §§338.011–
338.911 (1983 & Supp. 1986); MD. ANN. CODE art. 89, §§28–49A (1985 & Supp. 1986);
MICH. COMP. LAWS §§408.1001–408.1094 (1985 & Supp. 1986); MINN. STAT. ANN. ch.
182 (Supp. 1987); N.M. STAT. ANN. §§50–9–1–50–9–25 (1978 & Supp. 1985); NEV.
REV. STAT. ch. 618 (1986); N.C. GEN. STAT. §§95–126–95–155 (1985); ORE. REV. STAT.
§§654.001–654.991 (1985); P.R. LAWS ANN. tit. 29, §§361–362 (1985); S.C. CODE ANN.
§§41–15–10–41–15–640 (Law Co-op. 1986 & Supp. 1986); TENN. CODE ANN. §§50–
3–101–50–3–918 (1983 & Supp. 1986); UTAH CODE ANN. tit. 35, ch. 9 (1974 & Supp.
1986); VT. STAT. ANN. tit. 21, §§201–231 (1978 & Supp. 1986); scattered sections of
VA. CODE tit. 40.1 (1986); V.I. CODE ANN. tit. 24, ch. 2 (Supp. 1985); WASH. REV. CODE
ANN. §§49.17.010–49.17.910 (Supp. 1987); WYO. STAT. §§27–11–101–27–11–114 (1977).

[38]MINN. STAT. §182.667. See also VA. CODE §40.1–49.2, which contains language
parallel to that of §17(e) of the OSH Act, but which also provides for a fine of up to
$10,000 for any repeat or willful violation.

[39]CAL. LAB. CODE §6425; WASH. REV. CODE §49.17.190. Note that the California
statute addresses willful violations that cause "permanent or prolonged impairment
of the body of any employee," as well as those that cause death.

[40]See CAL. LAB. CODE §6423, providing that any employer or other person having
the direction of a place of employment or other employee who knowingly or negligently commits a serious violation, who repeatedly violates any standard, order or
special order, or who fails to abate a hazard which creates a real and apparent hazard
to employees is guilty of a misdemeanor; WASH. REV. CODE §49.17.190, providing for
a $10,000 fine and/or 6 months of imprisonment for violation of any order issued by
the superior court or the State Director of the Department of Labor and Industries,
which restrains a serious violation or a situation of imminent danger.

[41]IND. CODE §22–8–1.1–49.

rea requirement, since they provide for $10,000 fines and/or 6-month jail sentences for employers whose willful *or repeated* violations cause employee deaths.[42] Several state laws implicitly or explicitly permit criminal proceedings for violations of statutory general duties, as well as for violations of specific orders, standards, or regulations.[43]

In addition to these state legislative initiatives, state and local prosecuting attorneys in at least seven states have shown a willingness to file charges under more general state criminal statutes against employers they feel are responsible for worksite fatalities.[44] Interest has been shown by prosecutors in states having OSHA-approved state plans as well as by those in states that do not. In Illinois, a nonplan state, three executives of Film Recovery Systems Inc., a silver reclamation facility, received 25-year prison terms for murder when an employee died from exposure to cyanide gas.[45] The case is presently on appeal. However, the outcome on appeal is questionable since the Appellate Court of Illinois affirmed the dismissal of indictments brought against the Chicago Magnet Wire Corp. for allegedly exposing workers to toxic gases.[46] In Texas, another nonplan state, prosecutors obtained indictments against supervisory employees for criminally negligent homicide in connection with two construction trench cave-ins.[47] California, a state with a plan that emphasizes criminal enforcement of worker safety laws,[48] prosecuted a corporation for manslaughter following an industrial accident,[49] and also

---

[42]ALASKA STAT. §18.60.095; HAWAII REV. STAT. §396–10.

[43]E.g., WASH. REV. CODE §49.17.190.

[44]See Pond, *Local Officials Are More Inclined to Seek Prosecution for Workplace Hazards*, 32 CONSTR. LAB. REP. 196 (April 16, 1986).

[45]*Illinois v. Film Recovery Sys.*, [1984–1985] OSHD ¶27,356 (Cook County Cir. Ct. 1985).

[46]*Illinois v. Chicago Magnet Wire Corp.*, 13 OSHC 1337 (Ill. App. Ct. 1987).

[47]*Id.* at 196, 197–198.

[48]See CAL. LAB. CODE §§6311, 6314, 6313.5, 6314, 6321, 6322, 6323, 6325, and 6326. The multiplicity of California's criminal penalties has raised double jeopardy issues. See *People v. Lockheed Shipbuilding & Constr. Co.*, 69 Cal. App.3d Supp. 1, 138 Cal. Rptr. 445 (Cal. App. Dep't Super. Ct. 1977). It has also prompted a California court to hold that in California an inspection warrant based on probable cause must be obtained once an employer refuses to consent to a safety inspection. *Salwasser Mfg. Co. v. Municipal Court*, 94 Cal. App.3d 223, 156 Cal. Rptr. 292 (1979).

[49]*Granite Constr. Co. v. Superior Court*, 149 Cal. App.3d 465, 197 Cal. Rptr. 3 (1984).

convicted the superintendent of a water reclamation plant for both state plan violations and manslaughter following asphyxiation of two employees.[50]

Similar criminal prosecutions have been brought or contemplated in Michigan, Connecticut, Texas, and Washington.[51] Courts disagree as to whether murder and manslaughter charges may be lodged against a corporation;[52] the issue may well turn on the wording of the particular criminal statute.

---

[50]*People v. Gaglione*, 138 Cal. App.3d 52, 187 Cal. Rptr. 603 (1982).

[51]Pond, *supra* note 44.

[52]Compare *Granite Constr. Corp. v. Superior Court*, *supra* note 49, with *Vaughan & Sons v. State*, 649 S.W.2d 677 (Tex. App. 1983), *review granted*, June 15, 1983.

Part IV

Enforcement Proceedings Under the Occupational Safety and Health Act

13

Commencement of Proceedings

I. Statute and Regulations Governing the Notice of Contest

A contest of a citation, penalty, abatement period, or notification of failure to correct a prior violation is initiated by the filing of a notice of contest within 15 working days from receipt of the notice in question. Notices of contest are provided for in Section 10(c) of the OSH Act, which states:

> "If an employer notifies the Secretary that he intends to contest a citation issued under section 9(a) or notification issued under subsections (a) or (b) of this section, or if, within fifteen working days of the issuance of a citation under section 9(a), any employee or representative of employees files a notice with the Secretary alleging that the period of time fixed in the citation for the abatement of the violation is unreasonable, the Secretary shall immediately advise the Commission of such notification and the Commission shall afford an opportunity for a hearing *** ."[1]

Failure to file a timely notice of contest will convert the Secretary's citation into a final order, which is not subject to review. Thus, the second sentence of Section 10(a) provides:

> "If, within fifteen working days from the receipt of the notice [of proposed penalty] issued by the Secretary the employer fails to notify the Secretary that he intends to contest the citation

---

[1] 29 U.S.C. §659(c) (1970).

331

or proposed assessment of penalty, and no notice is filed by any employee or representative of employees under subsection (c) within such time, the citation and the assessment, as proposed, shall be deemed a final order of the Commission and not subject to review by any court or agency."

The second sentence of section 10(b) contains a similar provision making final a notification of failure to correct a prior violation if not contested within 15 working days.[2]

Regulations promulgated by the Secretary of Labor at 29 C.F.R. §§1903.17 and 1903.18(b) provide that a notice of contest is to be in writing and filed with the OSHA Area Director. The regulations also state that a notice of contest will be postmarked within 15 working days of receipt of the notice of proposed penalty, and a notice of a contest of a citation will specify whether it is directed to the citation, the proposed penalty, or both.

The Commission's rule at 29 C.F.R. §2200.32 provides that, within 7 days of receipt of a notice of contest, the Secretary will transmit it and copies of all relevant documents to the Commission. The Commission's rule at 29 C.F.R. §2200.7 requires that a notice of contest be posted at the worksite if there are any affected employees who are not represented by an "authorized employee representative" (i.e., union). The rule also requires that the notice of contest be served on the union if the affected employees are represented by a union.

II. Formal Requirements of the Notice of Contest

A. Form: Oral or Written

The act does not explicitly state whether a notice of contest must be in writing. This point, however, has been the subject of a number of decisions by the Commission.

Several rulings have upheld oral notices of contest. In *Wood Products Co.*,[3] the employer timely informed OSHA's Area Director of its intent to contest and then filed an untimely written notice of contest. The employer filed the notice

---

[2]§659(b).
[3]4 OSHC 1688 (Rev. Comm'n 1976).

of contest shortly after being told by the Area Director that notices of contest must be in writing. Because the employer had been confused by its dealings with the Area Director, which had included a number of meetings and telephone conversations, a divided Commission held the notice of contest valid.

In *H-E Loudermilk Co.*,[4] at a hearing on one item of a citation, the employer orally contested another item after being informed that a $65 penalty was proposed for the latter item. The notice of proposed penalty that the employer had received was imprecise and appeared to impose a penalty only for the item originally contested. The oral notice of contest given at the hearing was held valid because it was made promptly upon clear notice of the proposed penalty.

In *P & M Sales*,[5] the employer filed a written notice of contest addressed to the proposed penalties and contemporaneously notified OSHA orally that it wished to contest the merits of the citation. The oral notice, together with the written one, was held to be a valid notice contesting both the merits and the penalties of the citation.

The line of cases holding oral notices of contest to be valid was interrupted by *Merritt-Meridian Corp.*[6] There the employer timely informed OSHA of its intent to contest, but, because of a clerical error, did not file a written notice of contest until 17 days after expiration of the contest period. The late written notice of contest was not saved by the earlier oral notice. The Commission held that clerical error was insufficient to justify variation from the 15-working-day time limit.

After *Merritt-Meridian*, the Commission subsequently upheld an oral notice of contest in *Florida Power & Light Co.*[7] The employer's safety director in that case prepared a written notice of contest but forgot to bring it to a meeting with OSHA's Area Director. That meeting took place on a Friday, three days before expiration of the contest period. At the meeting, the safety director informed the Area Director that the employer was going to contest the citation. The Area Director

---

[4]1 OSHC 1663 (Rev. Comm'n 1974).
[5]4 OSHC 1158 (Rev. Comm'n 1976).
[6]4 OSHC 2025 (Rev. Comm'n 1977).
[7]5 OSHC 1277 (Rev. Comm'n 1977).

told the safety director that he could mail the notice of contest on Monday, which was the final day of the contest period. A safety emergency arose on Monday, however, and prevented the safety director from mailing the notice. He tried to telephone the Area Director after 5:00 p.m. but could not reach him. When he contacted the Area Director on Tuesday, he was told that he could bring the notice of contest to the Area Office in the next day or two. The notice of contest was hand delivered on Thursday. A divided Commission held the notice of contest valid because the employer had acted in a reasonable, nondilatory manner without bad faith.

In 1979, the Commission changed course on the oral notice of contest issue in *Keppel's, Inc.*[8] Although finding that no oral notice of contest had been given in that case, the Commission went on to hold that a notice of contest must be in writing and must be filed within 15 working days from receipt of the citation. *Keppel's* expressly overruled *Wood Products Co.* and *Florida Power & Light Co.*, in which timely oral notices of contest followed by untimely written notices of contest had been held valid. The only stated exception to the rule in *Keppel's* was a showing by the employer that a delay in filing was caused by the Secretary's deception or failure to follow proper procedures.[9] The decision of the Commission in *Keppel's* was followed in *B.J. Hughes, Inc.*,[10] *Henry C. Beck Co.*,[11] *National Roofing Co.*,[12] and *Harris Brothers Roofing Co.*[13]

It should be noted, however, that many of the cases after *Keppel's* involved application of the *Atlantic Marine* exception.[14] The exception has been applied broadly in conjunction with a long-stated Commission policy in favor of allowing employers an opportunity for a full hearing on the merits.[15]

---

[8]7 OSHC 1442 (Rev. Comm'n 1979).

[9]See *Atlantic Marine, Inc. v. OSHRC*, 524 F.2d 476, 3 OSHC 1755 (5th Cir. 1975).

[10]7 OSHC 1471 (Rev. Comm'n 1979).

[11]8 OSHC 1395 (Rev. Comm'n 1980).

[12]8 OSHC 1916 (Rev. Comm'n 1980).

[13]9 OSHC 1074 (Rev. Comm'n 1980).

[14]See Section II.D.4, *infra*.

[15]*Elmer Constr. Co.*, 12 OSHC 1002 (Rev. Comm'n 1984); *reaffirmed after remand*, 12 OSHC 2051 (Rev. Comm'n 1986); *Barretto Granite Corp.*, 12 OSHC 1088 (Rev. Comm'n 1984), *reaffirmed after remand*, 12 OSHC 2051 (Rev. Comm'n 1986), *rev'd*, 830 F.2d 396 (1st Cir. 1987). The Commission reaffirmed the results after the remands in both *Elmer* and *Barretto*, but relied on its holding in *Pav-Saver Mfg. Co.*, 12 OSHC 2001 (Rev. Comm'n 1986), that oral notices of contest were valid. The First Circuit disagreed with *Pav-Saver* in *Barretto*. It relied on the OSHA regulation at 29 C.F.R. §1903.17(a) and held that notices of contest must be written.

These cases could be construed as a relaxation of the general rule in *Keppel's*.[16]

B. Liberal Construction of Possible Notices of Contest

Case law consistently has favored a liberal interpretation of documents that might be deemed notices of contest. In *Brennan v. OSHRC (Bill Echols Trucking Co.)*,[17] within 15 working days of receiving a citation, the employer sent the Secretary a letter stating that "the signaling device has been installed" and requesting that the penalty be "abated." The Secretary delayed forwarding the letter to the Commission because he did not believe it to be a notice of contest. The Commission, however, held that the letter constituted a notice of contest.[18] On appeal, the Fifth Circuit also found that the letter contested the penalty, stating that "the Commission properly gave a liberal interpretation of the letter."[19]

In *Eastern Knitting Mills*,[20] the Commission held that a letter from an employer stating that it had been unable to devise a way to abate one item of a citation and requesting advice on how to do so constituted a notice of contest as to that item. The lead opinion stated that writings filed with the Secretary during the 15-working-day period after receipt of a citation are to be given a liberal interpretation.

In *Tice Industries*,[21] the Commission held that the statement "I protest any penalty" written by the employer across the notice of proposed penalty constituted a notice of contest as to the penalty. Further, the employer's makeshift answer, filed within the 15-day contest period and comprised of a note written across item 1 of the complaint asserting that it was "no hazard," was deemed effective to contest item 1 of the citation.

In *Harris Brothers Roofing Co.*,[22] the employer sent a letter 35 days after receipt of a citation requesting that its

---

[16]See also *Con-Lin Constr. Co.*, 11 OSHC 1757 (Rev. Comm'n 1983).
[17]487 F.2d 230, 1 OSHC 1398 (5th Cir. 1973).
[18]1 OSHC 1107 (Rev. Comm'n 1973).
[19]*Supra* note 17, 487 F.2d at 234, 1 OSHC at 1401.
[20]1 OSHC 1677 (Rev. Comm'n 1974).
[21]2 OSHC 1489 (Rev. Comm'n 1975).
[22]*Supra* note 13.

case be "reopened" and admitting that the letter was not filed within the 15-working-day contest period. Although untimely, the letter was construed by the Commission to be a notice of contest because it showed the employer's intent to contest the citation even though it spoke in terms of reopening the case. The Commission remanded the case to the administrative law judge to determine whether the employer is able to establish that the untimely notice of contest was the result of deception or a failure to follow proper procedures on the part of the Secretary.

Despite the "liberal construction" doctrine, a letter from an employer was held not to constitute a notice of contest in *Branciforte Builders*.[23] The letter, sent 34 days after receipt of the citation and notice of proposed penalties, stated that "there has been little activity" at the inspected job site and that construction would resume "within several weeks." Because the letter was not sent during the contest period and the employer never asserted that the letter was intended to be a notice of contest, the Commission ruled that it could find no basis for holding the letter to be a notice of contest.

C. Multiple Documents as a Notice of Contest

A few cases have held that more than one document may serve as a notice of contest. As noted above, in *Tice Industries*, the Commission found one document effective to contest the penalties and another document, the answer to the complaint, to contest one item of the citation where both documents were filed during the contest period. *Atlantic & Gulf Stevedores*[24] similarly held the employer's answer to constitute an amendment to its notice of contest when both were filed within the 15-working-day contest period. In *Chesapeake & Ohio Railway*,[25] the Commission held that both letters filed by the employer during the contest period had to be taken into account in determining what issues the employer included in its contest.

---

[23] 9 OSHC 2113 (Rev. Comm'n 1981).
[24] 3 OSHC 1003 (Rev. Comm'n 1975), *aff'd*, 534 F.2d 541, 4 OSHC 1061 (3d Cir. 1976).
[25] 3 OSHC 1762 (Rev. Comm'n 1975).

D. Timely Filing

1. *Start of the Contest Period*

To be timely, a notice of contest must be filed within 15 working days of the employer's receipt of the notification of proposed penalties.[26] When the notice of proposed penalties is received by a low-level employee, sent to a worksite other than the employer's main office, or not served by certified mail as required by Section 10(a) of the Act, it has been questioned whether such service is sufficient to trigger the start of the contest period. Courts of appeals, as well as the Commission, have ruled on these issues, with varying results.[27]

An early ruling in this area came in *Buckley & Co. v. Secretary of Labor*,[28] where the citation and notice of proposed penalties were sent to the superintendent of the employer's maintenance shop which was at a different location from that of the corporate headquarters. The superintendent apparently did not inform corporate headquarters of the citation or pro-

---

[26]The Act's language concerning the time period for filing a notice of contest is not uniform for employers, employees, and employee representatives. Section 10(a) provides that an employer must "notify" the Secretary that it intends to contest a citation or proposed penalty within 15 working days of *receipt* of the Secretary's notice of proposed penalty. Section 10(b) provides that, in the case of a notification of failure to correct a prior violation, the employer must "notify" the Secretary that it intends to contest the notification of "failure to correct" or the proposed penalty within 15 working days of receipt of the notification of "failure to correct" or the proposed penalty. Section 10(c) permits an employee or employee representative to contest a citation's abatement period by "fil[ing] a notice" with the Secretary within 15 working days of *issuance* of the citation. Section 10(a) refers to §10(c)'s provision for an employee or employee representative to contest an abatement period, but §10(b), which provides for notifications for failure to correct a prior violation, makes no provision for contests by employees or employee representatives.

Thus, §10 requires an employer to "notify" the Secretary that it intends to contest within 15 working days of *receipt* of the penalty notice or "failure to correct" notice, while an employee or employee representative who wishes to contest the period for abatement must file a notice within 15 working days of issuance of the citation. Since the contest period for employees and employee representatives runs from issuance of the citation, rather than receipt of the penalty notice, it is effectively shorter than that provided for employers. However, the Secretary's regulation at 29 C.F.R. §1903.7(b) purports to allow an employee or employee representative 15 working days from the employer's receipt of the penalty notice to file a notice of contest. Contests by employees or employee representatives are infrequent, and there are no Commission or court cases addressing the apparent discrepancy in contest periods.

Although the contest period for employers runs from receipt of the notification of proposed penalty, this notification normally accompanies the citation, and cases discussing the timeliness of an employer's notice of contest often speak in terms of the number of days elapsing between receipt of the citation and filing of the notice of contest. This usage also is employed at times in this chapter.

[27]See also Chapter 8 (Citations), Section III.

[28]507 F.2d 78, 2 OSHC 1432 (3d Cir. 1975).

posed penalties. The court of appeals set aside the Commission's dismissal of the employer's late notice of contest, stating:

> "[N]otification must be given to one who has the authority to disburse corporate funds to abate the alleged violation, pay the penalty, or contest the citation or proposed penalty. As to this corporate employer, this means, at the very least, a notice to the officials at the corporate headquarters, not the employee in charge at the particular worksite."[29]

Thus, mailing of the notice to the maintenance shop superintendent was held insufficient to trigger running of the contest period.

The Commission adopted a variety of positions with respect to the court's holding in *Buckley & Co.* In *Womack Construction Co.*,[30] the Commission held that a citation and notice of proposed penalties sent to an employer's corporate headquarters, where they were received by a bookkeeper who later lost them, was sufficient to start the contest period. The employer's notice of contest, filed four months later, was dismissed as untimely. The Commission distinguished *Buckley & Co.* on the basis that there the notice had not been sent to the corporate headquarters. The Commission's decision was affirmed without opinion.[31]

Without expressly saying so, the Commission followed the holding of *Buckley & Co.* in *Joseph Weinstein Electric Corp.*[32] The citation and notice of proposed penalties in that case were mailed to a construction worksite. The Commission observed that the court in *Buckley & Co.* had not held the citation invalid because the citation and penalty notice had been served incorrectly; rather, the *Buckley & Co.* decision suggested merely that the contest period begins to run only when an appropriate corporate official has received notice of the proposed penalty. Thus, although the date the citation was received by an employee of the cited employer at the job site was mentioned in the decision, the Commission remanded the case for determination of the date the employer's president received actual notice of the notification of proposed penalties, stating that it

---

[29]*Id.* at 81, 2 OSHC at 1435.

[30]6 OSHC 1125 (Rev. Comm'n 1977).

[31]No. 78–1178, 7 OSHC 1312 (9th Cir. Mar. 2, 1979).

[32]6 OSHC 1344 (Rev. Comm'n 1978). The Commission also tacitly followed *Buckley & Co.* in *Imperial Lumber Co.*, 4 OSHC 1908 (Rev. Comm'n 1976), and *Norkin Plumbing Co.*, 5 OSHC 2062 (Rev. Comm'n 1977).

was the latter date that would determine whether the notice of contest was timely.

Less than two years later, however, the Commission explicitly rejected *Buckley & Co.* In *B.J. Hughes, Inc.*,[33] the citation and penalty notice were sent to the employer's Louisiana district office and a copy was sent to the employer's California corporate headquarters and received there later. The Commission held that the contest period began to run when the penalty notice was received in the Louisiana district office. The Commission stated that the test for whether service of a penalty notice is proper is whether the service is reasonably calculated to provide an employer with knowledge of the citation and notification of proposed penalty and an opportunity to determine whether to abate or contest. The Commission noted that service on an employee who will know to whom in the corporate hierarchy to forward the documents will satisfy this test. Service on an employee in charge of a local worksite usually will be appropriate. Rather than explaining its change of position concerning *Buckley & Co.* and explicitly overruling *Joseph Weinstein Electric Corp.*, the Commission simply observed that it had never before expressly adopted or rejected the court decision in *Buckley & Co.*

The Commission has continued to follow *B.J. Hughes.* In *Henry C. Beck Co.*,[34] the Commission found the employer's receipt of a citation and notice of proposed penalties at its Atlanta worksite was sufficient to trigger the running of the contest period, even though the employer's main office was in Dallas. The Commission reiterated its holding in *B.J. Hughes* that mailing a citation to a company employee in charge of a local worksite where the alleged violation occurred generally is valid service.

On a related issue, in *Capital City Excavating Co.*,[35] the Commission relied on *B.J. Hughes* and held that the contest period ran from the date the penalty notice and citation were received by a clerical employee at the employer's main office, rather than two days later when the notice was forwarded to the employer's corporate secretary. As a result, the employer's notice of contest was dismissed as untimely. Although not

---

[33]7 OSHC 1471 (Rev. Comm'n 1979).

[34]8 OSHC 1395 (Rev. Comm'n 1980).

[35]8 OSHC 2008 (Rev. Comm'n 1980), *aff'd*, 679 F.2d 105, 10 OSHC 1625 (6th Cir. 1982).

mentioned by the Commission, this decision directly followed its earlier *Womack Construction Co.* precedent. The Commission's decision in *Capital City Excavating Co.*, like *Womack*, was upheld on court review. Although affirming the decision, the Ninth Circuit noted its disagreement with the Commission's holding in *B.J. Hughes*, stating rather that it agreed with the Third Circuit's ruling in *Buckley & Co.* that delivery of a citation and penalty notice to a worksite other than the employer's main office is not adequate service.[36]

Between its rulings in *B.J. Hughes* and *Henry C. Beck*, the Commission had occasion to rule on the sufficiency of service of a citation and penalty notice in a somewhat different context. In *P & Z Co.*,[37] a citation and a penalty notice were addressed to two corporations that were participants in a joint venture. The joint venture itself filed a notice of contest, and the Secretary moved to amend the caption of the citation to name the joint venture and the two participants individually. The three entities argued that none of them received adequate service of the citations and penalty notice. Rejecting this argument, the Commission analogized a citation and penalty notice to an Internal Revenue Service deficiency notice and relied on a Third Circuit ruling that mailing of a technically perfect deficiency notice is not a prerequisite to the jurisdiction of the Tax Court. Finding that delivery of a citation and penalty notice does not constitute service of process and need not meet the requirements for such service, the Commission concluded that if an employer receives actual notice of a citation, it is immaterial to the exercise of the Commission's jurisdiction that the manner in which the citation was sent was not technically perfect.

The Commission's decision in *P & Z Co.* seemed to erode an earlier Commission decision which had held that when a citation and notice of proposed penalty were delivered by means other than certified mail, the delivery must comply with the service of process requirements of Fed. R. Civ. P. 4(d). In *Donald K. Nelson Construction*,[38] the citation and notice of proposed penalty were personally delivered to the home of the employer's president, where they were received by his wife.

---

[36]*Id.*, 679 F.2d at 110 n.4, 10 OSHC at 1628.
[37]7 OSHC 1589 (Rev. Comm'n 1979).
[38]3 OSHC 1914 (Rev. Comm'n 1976).

Because this form of delivery did not comply with Fed. R. Civ. P. 4(d)(3), it was held to be invalid service. On this basis, the Commission vacated the citation, rather than merely holding the notice of contest to be timely. In a subsequent case, the Commission held that a citation and penalty notice sent by certified mail, as authorized by Section 10(a) of the Act, need not be addressed to a person authorized to receive service of process under Fed. R. Civ. P. 4(d).[39] Although the Commission then held in *P & Z Co.* that an employer's actual notice of a citation and notification of proposed penalty is sufficient to cure any technical defect in service, it later reiterated in *Capital City Excavating Co.*[40] that the federal rule on service of process is not applicable when the citation and penalty notice are sent by certified mail, thus implying that the service rule still is applicable to delivery by other means. In reviewing *Capital City Excavating Co.*, the Ninth Circuit broadened the strict service requirement, stating that the citation and penalty notice need not be served in accordance with the federal rule only when they are sent to the employer's corporate headquarters by certified mail. Thus, the Commission's holding in *P & Z Co.* that actual notice cures defective service has been largely ignored.

Finally, in establishing the date when the citation and penalty notice sent by certified mail were received by an employer, the Commission had held that the dated return receipt is *prima facie* evidence of receipt by the employer. Thus, in *Kerr-McGee Chemical Corp.*,[41] the return receipt date was deemed dispositive in the absence of evidence to support the employer's claim that the person who signed the receipt was not its employee.

2. Definition of "Working Days"

Although Section 10 of the Act requires a notice of contest to be filed within 15 working days of receipt of the penalty notice, the Act does not define "working days." The Secretary's regulation at 29 C.F.R. §1903.21(c) fills this void, defining working days as "Mondays through Fridays but *** not ***

---

[39]*Joseph Weinstein Elec. Corp.*, 6 OSHC 1344 (Rev. Comm'n 1978).
[40]*Supra* note 35.
[41]4 OSHC 1739 (Rev. Comm'n 1976).

Saturdays, Sundays, or Federal holidays."[42] This regulation was followed by the Commission in *B.J. Hughes, supra*, where it found a notice of contest to be one day late, rejecting the employer's argument that Good Friday, a company holiday but not a federal holiday, should not be counted as a working day.

3. What Constitutes Filing

For a notice of contest to be timely, an employer must "notify" the Secretary that it intends to contest a citation within 15 working days of the employer's receipt of the notice of proposed penalty, or an employee or employee representative must "file a notice" with the Secretary within 15 working days of issuance of the citation.[43] Questions have arisen concerning whether such a notification is effective at the time it is made or the time it is received. As discussed above, a notice of contest generally must be in writing. If the notice of contest is transmitted to the Secretary by mail, the Commission has held that the notice of contest need not be received by the Secretary within the 15-working-day contest period, but need only be mailed within that period.[44] Although the Secretary's regulations at 29 C.F.R. §§1903.17 and 1903.18(b) require that the notice of contest be postmarked within 15 working days of receipt of the notice of proposed penalty, the Commission had held that the postmark is only *prima facie* evidence of the mailing date and evidence may be introduced to show that the notice was mailed on a date different from that shown on the postmark.[45] Thus, in *Newport News Shipbuilding & Drydock Co.*,[46] an affidavit of a secretary of the employer that she mailed the notice of contest on a date that was the last day of the contest period was held to overcome both the postmark date and the proffered testimony of the postmaster that any

---

[42]The definition of "working day" in the Commission's rules of procedure is essentially the same: " 'Working day' means all days except Saturdays, Sundays, and Federal Holidays." 29 C.F.R. §2200.1(1).

[43]29 U.S.C. §§659(a) & (c). See note 26, *supra*.

[44]*Transamerica Delavel, Inc.*, 9 OSHC 1938 (Rev. Comm'n 1981); *Newport News Shipbuilding & Drydock Co.*, 9 OSHC 1085 (Rev. Comm'n 1980); *J.D. Blum Constr. Co.*, 4 OSHC 1255 (Rev. Comm'n 1976); *Electrical Contractors Assocs.*, 2 OSHC 1627 (Rev. Comm'n 1975).

[45]*Stone Container Corp.*, 9 OSHC 1832 (Rev. Comm'n 1981); *Transamerica Delavel, Inc.*, *supra* note 44; *J.D. Blum Constr. Co.*, *supra* note 44. But see *Kerr-McGee Chem. Corp.*, 4 OSHC 1739 (Rev. Comm'n 1976).

[46]*Supra* note 44.

letter postmarked on the date in question must have been mailed on that date.

4. Exceptions to the Timely Filing Requirement

The filing of a timely notice of contest is the action that invokes the jurisdiction of the Commission.[47] Because it is a jurisdictional issue, the timeliness of a notice of contest may be challenged at any time.[48] Even though a timely notice of contest is a prerequisite to Commission jurisdiction, in some situations the running of the contest period has been deemed tolled or a facially untimely notice of contest nevertheless has been held valid.

One circumstance in which an untimely notice of contest does not preclude Commission review is when the Secretary's deception or failure to follow proper procedure is responsible for the late filing. This exception to the timely filing requirement was first suggested by the Fifth Circuit in *Atlantic Marine, Inc. v. OSHRC*,[49] alluded to by the Commission in *Florida Power & Light Co.*,[50] and fully embraced by the Commission in *Keppel's Inc.*[51] and *B.J. Hughes, Inc.*[52] The Sixth Circuit cited *Keppel's* approvingly in holding that the notice of contest period may be extended under proper circumstances.[53]

An employer's late filing of a notice of contest has been excused because of the Secretary's deception or failure to follow proper procedures in numerous cases. In *Seminole Distributors*,[54] a notice of contest filed over two months after receipt of a notification of failure to correct violations was held effective because, through a printing error, language was omitted from the notification that would have informed the cited employer of the 15-working-day time limit for filing a

---

[47]See *Capital City Excavating Co. v. Donovan*, 8 OSHC 2008 (Rev. Comm'n 1980), *aff'd*, 679 F.2d 105, 10 OSHC 1625 (6th Cir. 1982); *Dan J. Sheehan Co. v. OSHRC*, 520 F.2d 1036, 3 OSHC 1573 (7th Cir. 1975); *Brennan v. OSHRC (Bill Echols Trucking Co.)*, 487 F.2d 230, 1 OSHC 1398 (5th Cir. 1973); *Willamette Iron & Steel Co.*, 9 OSHC 1900 (Rev. Comm'n 1981); *Harris Bros. Roofing Co.*, 9 OSHC 1074 (Rev. Comm'n 1980).

[48]*Chanal Plastics Corp.*, 9 OSHC 1844 (Rev. Comm'n 1981); *Stone Container Corp., supra* note 45.

[49]524 F.2d 476, 3 OSHC 1755 (5th Cir. 1975).

[50]5 OSHC 1277 (Rev. Comm'n 1977).

[51]7 OSHC 1442 (Rev. Comm'n 1979).

[52]7 OSHC 1471 (Rev. Comm'n 1979).

[53]*Capital City Excavating Co. v. Donovan, supra* note 47.

[54]6 OSHC 1194 (Rev. Comm'n 1977).

notice of contest. In *B.J. Hughes, Inc.*,[55] a notice of contest filed one day late was held effective because the cited employer relied on the OSHA Area Director's erroneous statements about the filing deadline. In *Henry C. Beck Co.*,[56] the OSHA Area Director's failure to correct the cited employer when it stated an incorrect deadline for filing of a notice of contest was held a sufficient basis for deeming the late notice of contest to be effective. In *Merritt Electric Co.*,[57] a late notice of contest was held effective because the Area Director did not dispel the cited employer's impression that an informal conference would serve to contest the citation. The Area Director failed to emphasize to a small contractor with no prior exposure to OSHA that oral disagreement at the informal conference was insufficient and that a written contest within the 15-working-day time period was required.

The most recent cases at the Commission review level present facts and results similar to those in *Merritt Electric*. These cases are *Con-Lin Construction Co.*,[58] *Elmer Construction Corp.*,[59] and *Barretto Granite Corp.*[60] Common features include oral disagreements at one or more informal conferences between the employer and OSHA, lack of legal representation for the employer, and weak evidence of clear notice to the employer at the meetings of what is required and when it is due. In *Elmer* and *Barretto*, the Commission found the result to be consistent with a long-standing policy to favor an opportunity to allow employers a full hearing on the merits.

The cases in which a late filing has not been excused are fewer and older. In *Keppel's, Inc.*,[61] the employer's late filing of its notice of contest was held not to have been caused by deception or failure to follow proper procedures on the Secretary's part. There, an employer brought a written notice of contest to an informal conference with the OSHA Area Director, but failed to present the notice because it felt intimi-

---

[55]*Supra* note 52.

[56]8 OSHC 1395 (Rev. Comm'n 1980).

[57]9 OSHC 2088 (Rev. Comm'n 1981).

[58]11 OSHC 1757 (Rev. Comm'n 1983).

[59]12 OSHC 1002 (Rev. Comm'n 1984), *reaffirmed after remand*, 12 OSHC 2051 (Rev. Comm'n 1986).

[60]12 OSHC 1088 (Rev. Comm'n 1984), *reaffirmed after remand*, 12 OSHC 2051 (Rev. Comm'n 1986) (upholding oral notice contest), *rev'd*, 830 F.2d 396 (1st Cir. 1987) (notices of contest must be written).

[61]*Supra* note 51.

dated by the sheer mass of government resources arrayed against it. The Commission stated that the determination of whether the Secretary's conduct was deceptive or improper is based on objective criteria, and it found no such deception or improper conduct to have occurred despite the employer's subjective feeling of intimidation. Similarly in *Capital City Excavating Co. v. Donovan*,[62] the Sixth Circuit recognized that running of the contest period could be tolled in some circumstances but noted that the employer in that case had not pointed to any exceptional circumstances to justify tolling of that period.

Clearly, the exception that permits a late-filed notice of contest where the late filing has resulted from a deception or a failure to follow proper procedures by the Secretary has been applied generously to favor the equities. It can be questioned whether the facts in *Keppel's* and *Capital City Excavating* would yield the same result today. The cases also suggest an obligation by the Secretary at informal conferences with employers without legal counsel to make clear the filing requirements for a notice of contest. It cannot be said, however, that the cases herald a return to a pre-*Keppel's* approach.

Finally, when a party files a late notice of contest and asserts that its lateness was caused by the Secretary's deception or violation of proper procedures, the party is entitled to a hearing on those allegations. A judge's dismissal of the notice of contest without a factual inquiry is error.[63]

A party who files a late notice of contest also may obtain Commission review by establishing grounds for relief from judgment or order under Fed. R. Civ. P. 60(b). That rule authorizes a motion for relief from a judgment or order to be granted on a number of grounds, including mistake, inadvertence, surprise, excusable neglect, newly discovered evidence, fraud, misrepresentation, misconduct of an adverse party, and "any other reason justifying relief from the operation of the judgment." The Commission initially held in *Plessy, Inc.*[64] that it lacked authority to consider a Rule 60(b) motion filed by an employer whose notice of contest was untimely. The

---

[62]8 OSHC 2008 (Rev. Comm'n 1980), *aff'd*, 679 F.2d 105, 10 OSHC 1625 (6th Cir. 1982).

[63]*National Roofing Co.*, 8 OSHC 1916 (Rev. Comm'n 1980); *Harris Bros. Roofing Co.*, 9 OSHC 1074 (Rev. Comm'n 1980).

[64]2 OSHC 1302 (Rev. Comm'n 1974).

Third Circuit, however, found the Commission to have authority to consider a Rule 60(b) motion in *J.I. Hass Co. v. OSHRC*.[65] The Commission subsequently agreed with the Third Circuit in *Branciforte Builders*,[66] noting that the grounds for relief under Rule 60(b) would encompass the Secretary's deception or failure to follow proper procedures, which the Commission previously had held to excuse a tardy notice of contest.

The Commission ruled on a Rule 60(b) motion concerning an untimely notice of contest in *P & A Construction Co.*[67] There an employer's attorney dictated a notice of contest letter, and his secretary told him that it had been mailed. When the employer received a request from OSHA for payment of the penalty, the attorney discovered that the notice of contest had not been mailed. He promptly mailed it and sought relief under Rule 60(b) from the citation and penalty which had become final orders under Section 10(a) of the Act. The Commission held that the untimely filing of the notice of contest resulted from "mistake, inadvertence, or excusable neglect" and found that the employer alleged that a defense to the citation could be shown. Accordingly, the Commission granted the Rule 60(b) motion and remanded the case for a hearing.

The Commission also has held in two cases that the absence of company officials from the United States might excuse untimely filing of a notice of contest.[68] These decisions, however, are grounded on the Third Circuit's holding in *Buckley & Co.* that the citation and penalty notice must be delivered to someone who has authority to disburse corporate funds to abate the alleged violation, to pay the penalty, or to contest the citation or proposed penalty. As discussed in Section II.B.1 above, the Commission later rejected *Buckley & Co.* in *B.J. Hughes, Inc.*

In a case that is *sui generis*, the Commission vacated a judge's dismissal of an untimely notice of contest and remanded for a hearing as to the employer's assertion that its notice of contest was late because it never received one of two citations issued to it.[69]

[65]648 F.2d 190, 9 OSHC 1712 (3d Cir. 1981).
[66]9 OSHC 2113 (Rev. Comm'n 1981).
[67]10 OSHC 1185 (Rev. Comm'n 1981).
[68]*Norkin Plumbing Co.*, 5 OSHC 2062 (Rev. Comm'n 1977); *Imperial Lumber Co.*, 4 OSHC 1908 (Rev. Comm'n 1976).
[69]*Safeway Stores*, 6 OSHC 2055 (Rev. Comm'n 1978).

When two citations resulting from a single inspection are issued at different times, it is sometimes unclear if a subsequent notice of contest is timely as to both citations. In *Owens-Illinois, Inc.*,[70] an employer received two citations three or four days apart. It contested only the earlier citation, but the notice of contest was timely only as to the later citation. The Commission held that the later citation supplemented the earlier one, and therefore the contest period for both citations ran from receipt by the employer of the later one. Accordingly, the notice of contest as to the first citation was held to be timely. In *Dic-Underhill*,[71] however, the Commission held that, even construed most favorably to the employer, a notice of contest expressly contesting a particular six-item citation could not be read also to contest a separate citation issued 10 days earlier.

Several other arguments made to excuse late notices of contest have been uniformly unsuccessful. In *American Airlines*,[72] the Commission held that the waiver provision of its rules of procedure[73] could not be applied to waive the requirement that a notice of contest be filed within 15 working days of receipt of the citation, because the notice of contest requirement is statutory. In *Womack Construction Co.*,[74] the Ninth Circuit affirmed without opinion the Commission's holding that the loss of the citation by the employer's bookkeeper did not excuse a notice of contest filed several months late. In that case, however, the employer did not move to set aside the final order because of mistake, inadvertence, or excusable neglect under Fed. R. Civ. P. 60(b). Arguments that the Secretary was not prejudiced by the late filing of a notice of contest and that the employer proceeded in good faith in filing the notice have met with no success.[75] Finally, the Commission's docketing of a notice of contest and the assignment of it to an administrative law judge does not constitute waiver of the timely filing requirement.[76]

---

[70]4 OSHC 1250 (Rev. Comm'n 1976).

[71]4 OSHC 1766 (Rev. Comm'n 1976).

[72]2 OSHC 1326 (Rev. Comm'n 1974).

[73]29 C.F.R. §2200.108.

[74]6 OSHC 1125 (Rev. Comm'n 1977).

[75]*Branciforte Builders, supra* note 66; *Fitchburg Foundry,* 7 OSHC 1516 (Rev. Comm'n 1979); *City Mills Co.,* 5 OSHC 1129 (Rev. Comm'n 1977).

[76]*Branciforte Builders, supra* note 66; *Norkin Plumbing Co., supra* note 68.

E. Posting

As noted in Section I above, the Commission's rules of procedure at 29 C.F.R. §2200.7(g) require an employer who files a notice of contest to post a copy of the notice at the location where the citation is required to be posted if any affected employees are not represented by an authorized employee representative. The practical effect of this provision is to require posting of virtually all notices of contest, because, even if a plant is unionized, supervisory personnel who are not represented by a union typically are exposed to alleged hazardous conditions located at the worksite. Several cases have addressed the consequences of an employer's failure to post a notice of contest. In *Caribtow Corp.*,[77] an employer in Puerto Rico that believed that it was not covered by the Act expressly refused to post its notice of contest. The Commission administrative law judge dismissed the notice of contest and affirmed the citation because, by refusing to post the notice of contest, the employer was depriving the affected employees of their statutory right of an opportunity to participate in the proceedings. The Commission summarily affirmed, and its decision was upheld by the court of appeals, which found that the employer was subject to the Act's requirements.

In subsequent cases, the employer's failure to post its notice of contest or to certify that it posted it usually has been by mistake and, once rectified, has not precluded the employer from proceeding with its contest. Dismissals of notices of contest by administrative law judges were reversed on this basis by the Commission in *Colonnade Cafeteria*[78] and *Browar Wood Products Co.*[79] In *Daron R. Dickherber*,[80] the Commission similarly reversed a judge's dismissal of a notice of contest for failure to provide proof of posting of the notice of contest. An employer's letter, stating that his business at the cited location had closed, was construed by the Commission to mean that the employer was unable to post the notice of contest because of the closing of his place of business. The Commission stated that, under these circumstances, another method of notifica-

---

[77]1 OSHC 1503 (Rev. Comm'n 1973), *aff'd*, 493 F.2d 1064, 1 OSHC 1592 (1st Cir. 1974).

[78]7 OSHC 2234 (Rev. Comm'n 1980).

[79]7 OSHC 1165 (Rev. Comm'n 1979).

[80]6 OSHC 1972 (Rev. Comm'n 1978).

tion, such as mailing copies of the notice of contest to the affected employees, could be used. Finally, in *Car & Truck Doctor*,[81] an administrative law judge dismissed an employer's notice of contest for failure to provide proof of posting of the notice even though the Secretary had moved to withdraw the citations. The Commission reversed the dismissal on the grounds that the employer may have thought that no action was necessary once the Secretary filed his withdrawal motion. On remand, the employer was to submit proof of posting of the notice of contest before the Secretary's withdrawal motion could be granted.

III. Transmittal of Notice of Contest

Section 10(c) of the Act[82] provides that if an employer notifies the Secretary that it intends to contest a citation or penalty or if an employee or representative of employees files a notice with the Secretary challenging the reasonableness of the time of abatement, "the Secretary shall immediately advise the Commission of such notification." This requirement is implemented by Commission Rule 32.[83] Rule 32 provides: "The Secretary shall, within 7 days of receipt of a notice of contest, transmit the original to the Commission, together with copies of all relevant documents."

It is established Commission precedent that a delay by the Secretary in forwarding a notice of contest to the Commission does not warrant dismissal of the citation in the absence of a showing of prejudice to the employer.[84] Furthermore, the Fifth Circuit held in *Bill Echols Trucking*[85] that imposition of the sanction of dismissal is not authorized without a hearing first being provided for. It is the burden of the employer to claim and establish prejudice.[86] To show prejudice,

---

[81]8 OSHC 1767 (Rev. Comm'n 1980).

[82]29 U.S.C. §659(c).

[83]29 C.F.R. §2200.32.

[84]*J. Dale Wilson, Builder*, 1 OSHC 1146 (Rev. Comm'n 1973); *ADM Grain Co.*, 1 OSHC 1148 (Rev. Comm'n 1973).

[85]487 F.2d 230, 235, 1 OSHC 1398, 1401 (5th Cir. 1973).

[86]*Rollins Outdoor Advertising Corp.*, 4 OSHC 1861 (Rev. Comm'n 1976); *Southern Ry.*, 3 OSHC 1657 (Rev. Comm'n 1975), *withdrawn*, No. 75–2493 (6th Cir. June 4, 1976).

it must be shown that the procedural irregularity or delay denied the employer an opportunity to prepare or present its case.[87] Thus, a delay of 79 days in *Rollins* because the notice of contest was "misplaced" resulted in no sanction because no prejudice was shown.

Similarly, in *Jensen Construction Co. v. OSHRC*,[88] the Tenth Circuit affirmed a judge's finding that the company did not demonstrate actual prejudice in a case involving a delay of 28 days. The company claimed that it was unable to secure necessary witnesses, but it had made no effort to subpoena witnesses known to be in the area.

More recently, in *Pennsylvania Electric Co.*, the Commission has held that a delay in transmittal of the notice of contest can be based on a finding of contumacious conduct on the part of the Secretary.[89] Contumacious conduct, however, was found not to be present. Although the delay amounted to nearly seven months, the case involved an amended citation that the Secretary was attempting to include in the contest of the original citation. In *Texas Masonry*,[90] the Commission found that the Secretary's conduct was not contumacious but remanded to allow the employer to show prejudice. A delay in transmittal of about 45 days resulted from efforts in the Area Office to settle the case. The Commission found that the lack of diligence in pursuing the settlement did not rise to the level of contumacious conduct.

IV. Issues Contested

A. Effect of Contest of the Penalty on the Citation

Notices of contest, as provided for in Section 10(c) of the Act, may place in issue the citation or notification of proposed penalty or both. The regulation issued by the Secretary of Labor at 29 C.F.R. §1903.17 also provides that the notice of contest "shall specify whether it is directed to the citation or to the proposed penalty, or both."

[87]*Pennsylvania Elec. Co.*, 11 OSHC 1235 (Rev. Comm'n 1983).
[88]597 F.2d 246, 7 OSHC 1283 (10th Cir. 1979).
[89]*Supra* note 87.
[90]11 OSHC 1835 (Rev. Comm'n 1984).

The unequivocal contest of the proposed penalty places only the amount of penalty in issue before the Review Commission. The Commission does not acquire jurisdiction to review the underlying citation. The abatement period set out in the citation is not tolled. The Commission so held in its decision in *Florida East Coast Properties*.[91]

The *Florida East Coast Properties* decision relied on the earlier decision of the Fifth Circuit in *Brennan v. OSHRC (Bill Echols Trucking Co.)*.[92] In *Bill Echols*, the notice of contest had represented that the cited conditions had been remedied and requested that "the penalty be abated" because of the timely corrective action. The Fifth Circuit held that the particular notice of contest was limited to a contest of the proposed penalty.[93]

B. Effect of Contest of the Citation on the Penalty

The Commission consistently has held that once a notice of contest has been filed as to a citation, the Commission has jurisdiction to assess a civil penalty.[94] The Commission reaffirmed this view in *Danco Construction Co.*[95]

C. *Turnbull Millwork* Rule

In 1975, the Commission retreated from its ruling in *Florida East Coast Properties*. In *Turnbull Millwork Co.*,[96] the Commission held that a notice of contest by a *pro se* employer limited solely to penalty will be construed to include a contest of the citation if subsequent communications indicate that the actual intent was to contest the citation as well.[97]

---

[91] 1 OSHC 1532 (Rev. Comm'n 1974).

[92] *Supra* note 85.

[93] The Fifth Circuit reaffirmed this view in *Dan J. Sheehan Co. v. OSHRC*, 520 F.2d 1036, 3 OSHC 1573 (5th Cir. 1975), *cert. denied*, 424 U.S. 965 (1976), but suggested that it might agree with the Commission if the Commission adopted another position in a consistent course of administrative interpretations.

[94] *Thorleif Larsen & Son*, 1 OSHC 1095 (Rev. Comm'n 1973), *appeal dismissed*, No. 73–1232 (7th Cir. Jan. 22, 1974).

[95] 3 OSHC 1114 (Rev. Comm'n 1975).

[96] 3 OSHC 1781 (Rev. Comm'n 1975).

[97] The holding of the Commission in *Turnbull* was conditioned on the failure of the Secretary to implement a suggestion of the Fifth Circuit in its decision in *Bill Echols Trucking*:

The notice of contest in *Turnbull* unequivocally challenged only the proposed penalty. Three months later, the employer filed a second letter that stated in part:

> "We contend that the manner in which we use these saws makes the use of a blade guard very inconvenient and in many ways impossible and unsafe. I now realize that my first letter of April 2, 1974, in which I notified you that I desired to contest the proposed penalty did not meet all the requirements of the Commission as I did not state therein my reasons for contesting the penalty. I am sorry that I was negligent in this matter, and trust that this will serve to meet all requirements."[98]

Shortly after the filing of the second letter, the Secretary submitted a request for admissions in which the employer was asked to admit that only penalties were in issue. In response, the employer stated that the notice of contest placed in issue the penalty and the citation.

On these facts, the Commission found that its *Florida East Coast Properties* rule operated harshly on employers acting under pressures of time and often without the benefit of counsel. Thus, the Commission stated it would construe notices of contest that are limited to the penalty to include the contest of the citation, if the employer "indicates at a later time that it was its intent to also contest the citation."[99]

D. *Turnbull Millwork* Rule Endorsed

The rule announced by the Commission in *Turnbull Millwork* has been upheld by the courts of appeals in two circuits.

> "This Court claims no bureaucratic rules for the Secretary or the Commission. We would be remiss, however, if, in the interest of minimizing the need of these parties for the future assistance of this Court, we did not suggest a simple means of eliminating ambiguity in notice of contest. If each citation or notification of proposed penalty sent to an employer were accompanied by a reply form on which the employer could check boxes indicating intent to contest the citation or proposed penalty, or neither or both, with space for listing reasons or making comments, no confusion need ever again arise on the part of either the Secretary or the Commission."

487 F.2d 230, 234 n.7, 1 OSHC 1398, 1401 (5th Cir. 1973). The Commission noted in *Turnbull* that the Secretary had not implemented the suggestion in *Bill Echols* during the two years that intervened between the two decisions. The Commission adopted its position in *Turnbull* "in the absence of the Secretary's implementation" of the *Bill Echols* suggestion. Even today, the suggestion remains unadopted by the Secretary. It is not clear what effect on the *Turnbull* rule would be had by the implementation by the Secretary of the *Bill Echols* suggestion.

[98]*Supra* note 96, at 1781.

[99]*Id.* at 1782.

In *Penn-Dixie Steel Corp. v. OSHRC*,[100] the Seventh Circuit
applied the *Turnbull Millwork* rule to a case that the Commission had reviewed under the *Florida East Coast Properties*
rule prior to the *Turnbull* decision. About one and one-half
years later, the Eighth Circuit denied a petition for review by
the Secretary in the case of *Marshall v. Gil Haugen Construction Co.*[101] The court agreed with the Seventh Circuit in *Penn-
Dixie Steel* in upholding the *Turnbull Millwork* rule, while
noting that "the Secretary may easily remedy its alleged dilemma if it so chooses."[102]

E. Application of *Turnbull Millwork* Rule

1. *Rule Applied Even If Respondent Is Not* Pro Se

The employer in the *Turnbull Millwork* case was not represented by counsel. The Commission placed some emphasis
on this fact in its decision in *Turnbull Millwork*. Nevertheless,
the Commission did not explicitly limit the rule to *pro se*
employers.

About eight months later, the rule was applied in *Nilson
Smith Roofing & Sheet Metal Co.*,[103] a case in which the employer was represented by counsel in the filing of the notice
of contest and all subsequent proceedings. The Commission
decision in *Nilson Smith*, however, made no mention of the
fact that the employer was represented by counsel, but adopted
the decision by the judge who had rejected an argument by
the Secretary urging that the *Turnbull Millwork* rule was
applicable only to *pro se* employers.

The application of the *Turnbull Millwork* rule to an employer who has been represented by counsel from the notice
of contest stage of the proceeding has not been upheld in the
courts of appeals. In *Gil Haugen Construction Co.*, the employer was not represented by counsel before the Commission.
The Eighth Circuit did not limit explicitly its affirmance to
pro se employers but noted that the unfairness that the rule

---

[100]553 F.2d 1078, 5 OSHC 1315 (7th Cir. 1977).
[101]586 F.2d 1263, 6 OSHC 2067 (8th Cir. 1978).
[102]*Id.* at 1266, 6 OSHC at 2069. See note 97, *supra.*
[103]4 OSHC 1765 (Rev. Comm'n 1976).

was implemented to alleviate was more pronounced with a *pro se* employer.

Penn-Dixie Steel Corp.[104] also involved an employer who was *pro se* at the earlier stages of the proceeding. The Seventh Circuit did not limit its holding to *pro se* employers; indeed, the court noted that the rule was not limited to *pro se* employers, citing *Nilson Smith*; on the other hand the court placed emphasis on the layman status of Penn-Dixie's representative in determining Penn-Dixie's intent at the time of the notice of contest. It appears to be a fair conclusion that representation by counsel at the notice of contest stage will increase the difficulty of showing that an employer's actual intent differed from the express language in the notice of contest.

2. *Other Applications*

The principal feature of the cases in which the *Turnbull Millwork* rule has been applied and the underlying citation has been found to be in issue is the existence of ambiguity in the notice of contest. In those cases in which the notice of contest is clear and specific in challenging only the penalties, the citation has been found not to be in issue. The notice of contest is read in the light most favorable to the employer. Most of the cases involve *pro se* employers, but the Commission has not emphasized this fact in deciding the applicability of the rule. It is clear that communications from the employer after the notice of contest must support a finding that the employer intended to challenge the citation at the time of the notice.[105]

In *Data Electronic Co.*,[106] the employer received a document entitled "Amended Notification of Proposed Penalty" on which all alleged violations were specified. In response, it filed *pro se* a notice of contest that stated, "We, herewith, contest your amended notification of proposed penalty ***." Subsequently, the employer answered with a general denial the

---

[104]*Supra* note 100.

[105]For other cases holding the citation to be in issue, see *Hougan Constr. Co.*, 5 OSHC 1956 (Rev. Comm'n 1977); *State Home Improvement Co.*, 6 OSHC 1249 (Rev. Comm'n 1977). Other cases holding that the citation is not in issue are *Acme Metal*, 3 OSHC 1932 (Rev. Comm'n 1976); *F.H. Sparks of Md.*, 6 OSHC 1356 (Rev. Comm'n 1978).

[106]5 OSHC 1077 (Rev. Comm'n 1977).

Secretary's complaint, in which all violations were realleged. Finally, when an admission at the hearing was sought by the Secretary regarding facts underlying the disputed citation, the employer specifically denied the admission. At the time of the hearing, an attorney represented the employer. A unanimous Commission found the citation in issue based on these facts.

Similarly, in *Big 6 Drilling Co.*,[107] a unanimous Commission found the underlying citation in issue. The *pro se* employer filed a notice of contest that stated in pertinent part:

> "We have taken immediate steps to have all of these alleged violations corrected to the standards requested by your investigator. We believe we have everything in order.
> "We do, however, feel that your proposed penalties of $805 are excessive, and we would like to present our case at a hearing in an attempt to have this violation penalty waived. Our company is taking great pride in providing safe working conditions, and each month we spend several hundred dollars for safety awards to our employees ***."[108]

Thereafter, the Secretary filed the complaint that stated that the violations were not in issue and had become a final order pursuant to Section 10(a) of the Act. The employer, still representing itself, responded with a letter stating in part: "This letter is to ask for a hearing so that we may express our objections in connection with the inspection conducted June 16, 1975 and the resulting penalties. We provide a safe place for our employees to work and feel the penalty is not appropriate."[109] Later, the employer was represented by counsel, who disputed the finality of the citations. The citations were found to be in issue.

The notice of contest was construed to place in issue only the penalties in *J & H Livestock Co. & Karler Meat Packing.*[110] The essentially identical notices of contest stated:

> "This is to notify you *** that we wish to contest the penalty portion of that citation. We do not contest the safety hazards listed, the time for abatement allocated, nor any action by your office or staff. Our contest is limited strictly to the penalty assessments, and we specifically waive any and all time extensions normally attributable to a contest."[111]

---

[107]5 OSHC 1683 (Rev. Comm'n 1977).

[108]*Id.* at 1684.

[109]*Id.*

[110]5 OSHC 1742 (Rev. Comm'n Nos. 76–324 & 76–326, 1977).

[111]*Id.* at 1743.

Thereafter, the complaints filed by the Secretary alleged specifically that only the penalties were in issue. The employers did not deny the allegations, nor did they respond to the judge's prehearing order that indicated that only penalties were in issue. At the hearing, the employers stated that they wished to contest the citations also. The Commission concluded that the only reasonable conclusion to be drawn from the facts is that the employers did not intend to contest the citations when the notices of contest were filed.

In *Frank C. Gibson*,[112] the issue was whether item 2 of a three-item citation was contested. The *pro se* notice of contest objected with specificity to item 3. It did not mention item 2. In his complaint, the Secretary alleged that item 2 was not contested and became a final order. Gibson was represented by counsel when its answer was filed. In the answer, Gibson denied that item 2 was not contested. The Commission agreed with the judge, who had found that the notice of contest was unambiguous and did not include item 2.

F. Contest of the Abatement Period

1. *Effect on the Citation*

In an early case, the filing of a notice of contest challenging the reasonableness of the abatement date or the method of abatement was held to place the citation in issue.[113] An employer filed a letter within the contest period that stated an inability to abate an item and requested advice. The Commission held that the letter constituted a notice of contest as to the citation and the abatement period. The holding was reaffirmed in *Philadelphia Coke Division, Eastern Ass'n Coal Corp.*[114]

2. *Effect of Abatement Extension Request on the Citation*

About five years later, the Commission overruled, to some extent, *Philadelphia Coke Division* and *Eastern Knitting Mills*

---

[112]6 OSHC 1557 (Rev. Comm'n 1978).
[113]*Eastern Knitting Mills*, 1 OSHC 1677 (Rev. Comm'n 1974).
[114]2 OSHC 1171 (Rev. Comm'n 1974).

in *Gilbert Manufacturing Company*.[115] In *Gilbert*, the employer requested additional time to abate six violations in a letter sent during the 15-day contest period. The Secretary was agreeable to the requested extensions. The Commission held that procedurally a simple request for an extension of the abatement period filed during the contest period should be treated as a petition for modification of abatement. Where there is no objection to the extension, as in *Gilbert*, the Secretary may grant the extension pursuant to Commission Rule 34.[116] If a dispute over the extension occurs, the Commission stated that the case would be docketed and a hearing afforded. The employer bears the burden of justifying the extension. The Commission also noted that if the request for relief by the employer is ambiguous regarding whether the citation also is contested, or if the employer disputes the Secretary's interpretation of the request for relief, the matter is to be referred to the Commission "for appropriate action."

The Commission applied *Gilbert Manufacturing* in *Bushwick Commission Co.*[117] It should be noted, however, that *Bushwick* involved an unambiguous request for an extension of the abatement date and the Secretary granted the extension.

3. Turnbull Millwork Rule Applied

The Commission addressed the issue of an ambiguity in a letter that is filed within the 15-day contest period and that appeared to contest only the abatement dates in *Maxwell Wirebound Box Co.*[118] The Commission applied the *Turnbull Millwork* test to determine the intent of the employer at the time of filing the request by examining later "pleadings." If the employer's intent, when the ambiguous request for extension was filed, was to contest the citation, then the merits of the citation are in issue. In *Maxwell*, the employer's vigorous contest of the citation in its answer was the basis for the Commission concluding that the merits were in issue.

---

[115] 7 OSHC 1611 (Rev. Comm'n 1979).
[116] 29 C.F.R. §2200.34.
[117] 8 OSHC 1653 (Rev. Comm'n 1980).
[118] 8 OSHC 1995 (Rev. Comm'n 1980).

14

Hearing Procedure

As discussed in Chapter 13, the employer's notice of contest is sent to the OSHA Area Director, and it triggers the adjudicatory procedure of the Review Commission under Section 10 of the OSH Act.[1] The Area Director, acting for the Secretary of Labor, must transmit the notice of contest to the Commission within 7 days of receipt.[2] The Review Commission's Executive Secretary, who acts much as the clerk of a court, then assigns a docket number to the case, and on behalf of the Commission sends to the employer a notice requiring it to inform employees or their authorized representative that a notice of contest has been filed.[3] A form certifying that such notification has occurred must be completed by the employer and returned to the Review Commission. The case's adjudicatory posture then essentially moves into two general phases: the prehearing stage followed by the hearing itself.

---

[1] 29 U.S.C. §659 (1982); 29 C.F.R. §1903.17 (1985).

[2] 29 C.F.R. §2200.32 (1985).

[3] The form of this notification is supplied by the Commission and states:
"Your employer has been cited by the Secretary of Labor for violation of the Occupational Safety and Health Act of 1970. The citation has been contested and will be the subject of a hearing before the Occupational Safety and Health Review Commission. Affected employees are entitled to participate in this hearing as parties under terms and conditions established by the Occupational Safety and Health Review Commission in its Rules of Procedure. Notice of intent to participate should be sent to the Occupational Safety and Health Review Commission, 1825 K Street, N.W., Washington, D.C. 20006. All papers relevant to this matter may be inspected at [a place reasonably convenient to employees, preferably at or near the workplace.]"

I. Prehearing Proceedings

A. Rules Governing Procedure—An Overview*

Proceedings before the Review Commission are governed by the Rules of Procedure promulgated at 29 C.F.R. §§2200.1–.211 (Rules 1–211).[4] The Commission's rules cover the preparation and trial of OSHA cases, including the form, timing, and service of various notices to parties; pleadings; discovery; scheduling and conduct of hearings; Commission review of judges' decisions; and settlement. In addition, Commission Rules 200–211 contain a special set of provisions providing for the expeditious and less expensive handling of certain categories of cases, and these special provisions are called "simplified proceedings."

The rules also incorporate the Federal Rules of Civil Procedure and the Federal Rules of Evidence. Commission Rule 2(b) provides that in the absence of a specific provision in the Commission's rules, procedure shall be governed by the Federal Rules of Civil Procedure. Commission Rule 72 requires "insofar as practicable" that hearings be governed by the rules of evidence applicable in the U.S. district courts.[5]

B. Initial Pleadings

1. Notice of Contest

Although technically it is not a pleading, the notice of contest triggers the initial pleadings. Therefore, the two types of notices of contest must be distinguished. First, where affected employees or their authorized representative contests the proposed abatement period for correcting the alleged violation, the Secretary has 10 days from receipt of this notice

---

*Editors Note: As this book entered the final production process, the Occupational Safety and Health Review Commission totally revised their rules of procedure. Rather than delay the entire book while this chapter was revised, it was decided to include the new procedural rules as an appendix (see Appendix B) and to include appropriate revisions in the first supplement. As a result, many of the references to Commission Rules in this chapter are not up to date. Accordingly, the reader should take care to cross-reference any reference to the Commission's Rules to Appendix B.

[4]The Commission's rules are identified by numbers and letters corresponding to the location of each rule within 29 C.F.R. §§2200.1–211 (1985). Thus Commission Rule 35(a) is found at 29 C.F.R. §2200.35(a).

[5]The rules of evidence in the federal district courts are codified as the Federal Rules of Evidence. 28 U.S.C. (1975).

to file a statement with the Review Commission supporting the reasonableness of the proposed abatement period.[6] A response by the contesting party then must be filed 10 days after receipt of the Secretary's statement.[7]

Second, where an employer files a notice of contest, the Secretary has 20 days from his receipt of the notice of contest to file a complaint with the Review Commission and serve a copy of the complaint on the employer.[8] The Secretary's complaint is in a form similar to complaints in any court of law.[9] It is important to note that the Secretary cannot rely on the citation as the complaint because, as held by the Review Commission, the complaint serves important purposes independent of the citation by defining and limiting the issues in the case.[10]

With respect to petitions to modify the proposed abatement period under Rule 34 and the difference in the form of pleadings with respect to the same under Rules 31 and 34, see Chapter 10 (Abatement), Section III.B.

2. *The Complaint*

The typical OSHA complaint usually (a) contains the requisite jurisdictional allegations, namely that the Commission has jurisdiction over the case, that the employer is an "employer" within the meaning of the Act, and that the employer's business affects interstate commerce;[11] (b) repeats the substance of the charges contained in the citation, including any requisite allegation of knowledge by the employer about the existence of the hazard; and (c) alleges that employees were exposed to the alleged hazardous conditions. However, sometimes the Secretary seeks in the complaint to alter the cita-

---

[6]Commission Rule 35(a).

[7]Commission Rule 35(b).

[8]Commission Rule 33(a)(1).

[9]In cases before the Review Commission resulting from an employer's notice of contest, the Secretary of Labor is referred to as the Complainant and the employer is referred to as the Respondent. Commission Rule 31(a). The heading of the case as specified in Commission Rule 31(a) must appear on the initial page of every document filed with the Commission. Rule 31(c). As to other requirements concerning form, see Rule 31.

[10]See, e.g., *ASARCO, El Paso Div.*, 8 OSHC 2156, 2162 (Rev. Comm'n 1980). It should be noted, however, that a case is unlikely to be dismissed if the Secretary fails to file a complaint in a timely fashion. See *Boring & Tunneling Co. of Am. v. OSHRC*, 670 F.2d 13, 10 OSHC 1409 (5th Cir. 1982).

[11]29 U.S.C. §652(5).

tion's allegations in order to make the citation's charges complete or accurate. Rule 33(a)(3) governs this procedure, requiring the Secretary to state the reasons for any change. Alterations to pleadings at this early stage in the proceeding are routinely allowed as encompassed under Rule 15(a) of the Federal Rules of Civil Procedure unless they cause prejudice to the defense of the case.[12]

3. The Answer

The employer must file its answer within 15 days after service of the Secretary's complaint.[13] Although Commission Rule 33(b)(2) merely requires the employer to file "a short and plain statement denying those allegations in the complaint which the party intends to contest," the employer also should use the answer to raise each affirmative defense it may seek to pursue.[14] In pleading any affirmative defense, however, the employer should draft a short statement of the essential elements of the defense in order to put the Secretary on notice that the defense is being raised.

4. Amendments

Amendments to either the complaint or the answer prior to the hearing also are normally permitted unless the proposed amendment causes prejudice to the opposing party in presenting its case.[15] Obviously, as the hearing date approaches, the possibility that an amendment may prejudice a party in the presentation of its case increases; and the likelihood of amendment approval correspondingly decreases, because the party against whom the amendment is offered may not have the opportunity to fully prepare a defense to the new issue.[16]

---

[12]*George Barry*, 9 OSHC 1264 (Rev. Comm'n 1981); *Western Mass. Elec. Co.*, 9 OSHC 1940 (Rev. Comm'n 1981); *Brown & Root, Inc.*, 8 OSHC 1055 (Rev. Comm'n 1980); *Leisure Resources Corp.*, 7 OSHC 1485 (Rev. Comm'n 1979); *Henkels & McCoy, Inc.*, 4 OSHC 1502 (Rev. Comm'n 1976).

[13]By operation of Commission Rule 4(b), when the complaint is served by mail the employer must mail its answer to the Commission's Executive Secretary no later than 18 days after the date of mailing of the complaint.

[14]See FED. RS. CIV. P. 8(c) and 12. For commonly raised defenses in OSHA cases, see Chapter 15 (Defenses).

[15]*Cornell & Co. v. OSHRC*, 573 F.2d 820, 6 OSHC 1436 (3d Cir. 1978); *Brown & Root, Inc.*, *supra* note 12; *General Motors Corp.*, 10 OSHC 1293 (Rev. Comm'n 1982).

[16]See *Cornell & Co. v. OSHRC*, *supra* note 15 (Secretary's request to amend complaint 9 days before hearing with new theory of liability prejudiced employer because of lack of opportunity to secure testimony to rebut new issue).

C. Computation of Time, Filing Papers, and Service

The mechanics of practice before the Review Commission, such as time computations and the filing and serving of papers, are principally governed by Commission Rules 4, 7, and 8.

Commission Rule 4 concerns computation of time, providing that in computing any period of time, the day the period begins to run is excluded, with all days counted consecutively until and including the last day of the time period, unless it falls on a Saturday, Sunday, or federal holiday. If that is the case, the last day is the next business day. However, if the time period for response is less than 7 days, intermediate Saturdays, Sundays, and federal holidays are not counted. Commission Rule 4(b) provides that when a document is served by mail, 3 days are added to the time for filing a response.

With respect to the actual filing of papers, Commission Rule 8 governs. For example, prior to the assignment of a case to an administrative law judge (A.L.J.), papers are filed with the Commission's Executive Secretary. After the A.L.J. is assigned and until he issues a decision, all papers are filed with that A.L.J. Once the A.L.J.'s decision is filed (docketed) with the Review Commission, all filing is again with the Executive Secretary. Filing is accomplished by first class mail and is deemed effective at time of mailing.

Commission Rule 7 addresses service of all papers. The rule requires that all pleadings and other papers required to be filed with the Review Commission are to be served on every other party or intervenor in the case. If a party appears through a representative, service shall be made only on that representative. Service of papers on a party can be effected either by first class mail, postage prepaid, or by personal delivery. Each document filed with the A.L.J. or the Commission also must be accompanied by proof of service, which is deemed made on the date of mailing (if by mail) or on the date of delivery (if by delivery). However, as noted earlier, 3 days are added to the response time for any document served by mail.

D. Administrative Law Judge

The Review Commission assigns an A.L.J. to each case very early in the proceeding, usually shortly after the complaint and answer are filed. Once an A.L.J. is assigned to the case, he controls the progress of the proceedings.

The first communication the parties usually receive from the A.L.J. is either a prehearing order setting the ground rules for hearing preparation or an order setting a hearing date. Typical examples of matters covered in prehearing orders include requiring that the parties meet to explore settlement at an early date;[17] granting leave to conduct discovery without further specific requests to the judge or setting forth the ground rules under which discovery will be permitted; setting forth the procedures for submitting discovery requests or any motions regarding the same; establishing ground rules on hearing procedures, including such matters as marking and exchanging exhibits and witness lists; and, perhaps, asking the parties to admit routine facts such as the correct legal name of the employer, uncontested jurisdictional facts, size of the employer's business, and history of the employer's compliance with the Act.

With respect to the notice of the hearing, Commission Rule 60 requires that it be given at least 10 days in advance of the hearing date. Normal practice is for the A.L.J. to set an initial hearing shortly after being assigned the case. That date is usually a few months in the future. Commission Rule 61 prescribes circumstances for postponement of the hearing date, which will not normally be allowed, absent good cause, because it is usually set so far in advance. The normal considerations of good cause that govern extensions of time in any court case govern extensions of an OSHA hearing date. Typically, the A.L.J. wants to know whether the parties are working diligently to prepare for trial, whether additional time is reasonably necessary, and what work the parties intend to do in the extended time period. In any case, extensions exceeding 30 days must receive full Review Commission approval.

The Commission Rules also provide for prehearing conferences. Specifically, Rule 51(a) states that at any time before a hearing the Review Commission or the A.L.J., *sua sponte*

---

[17]Settlement is governed by Commission Rule 100. The necessary elements of any settlement are specified in part (b) of the rule. Rule 100(c) provides that proposed settlements must be served on affected employees or their representatives in the manner in which notices of contest are served under Commission Rule 7. However, settlements need only be served on employees or authorized representatives who have elected party status. *American Bakeries Co.*, 11 OSHC 2024 (Rev. Comm'n 1984).

or on the motion of a party, may direct the parties to exchange information or to participate in a prehearing conference to consider settlement or matters which will tend to simplify issues or expedite the hearing. In complicated cases a prehearing conference before the A.L.J. may prove quite useful, particularly in limiting, narrowing, or defining the issues that will be tried. This is especially important in many OSHA cases which often involve contests of many citations, and where each item of the citation concerns different factual and legal questions. A.L.J.s also use prehearing conferences to ferret out items really at issue, to work out a general outline of the proof and the approach each party intends to follow with respect to a particular issue, and sometimes to prod the parties toward settlement.

Caution must be noted, however, that because the Review Commission is an adjudicatory body, *ex parte* communications are not permitted with respect to the merits of a pending case between an officer or employee of the Review Commission involved in the decisional process and any of the parties.[18] This rule is directed at communications with the Commissioners and their staffs as well as the A.L.J.s.

E. Motions

The Review Commission has no express provisions on motions, and thus resort must be made to the Federal Rules of Civil Procedure. Under Fed. R. Civ. P. 7 an application for an order is by motion. For prehearing matters the motion must be in writing and state "with particularity" the relief sought and the grounds thereof. Responses to motions, however, are governed as to time by Commission Rule 37, which allows the filing of responses 10 days from service.

To expedite matters, a motion may be decided without oral argument and upon brief written statements in support of or in opposition to the motion.[19] Most motions are decided without hearing. When a motion is based on facts not appearing in the record, it may be ordered that the matter be

---

[18]Commission Rule 103.
[19]FED. R. CIV. P. 78.

heard on affidavits presented by the parties or in whole or part upon oral testimony or deposition.[20]

Turning to specific examples of motions filed in OSHA cases, the following appear most frequently:

1. Motions to Dismiss

Fed. R. Civ. P. 12 permits a party to raise certain defenses to an action by motion before any responsive pleading is filed, including (1) lack of jurisdiction over the subject matter, (2) lack of jurisdiction over the person, (3) improper venue, (4) insufficiency of process, (5) insufficiency of service of process, (6) failure to state a claim upon which relief can be granted, or (7) failure to join a party. The Commission's standards for deciding whether to grant a motion to dismiss are set forth in *Spector Freight System*.[21] There, dismissal was requested for lack of jurisdiction over the subject matter, i.e., the condition cited as an OSHA violation was allegedly exempt from OSHA jurisdiction under Section 4(b)(1) of the Act.[22] In determining whether to grant the motion to dismiss, the Commission stated that it must take the well-pleaded allegations of the complaint as true and that it is improper to dismiss a complaint unless it appears beyond a reasonable doubt that the Secretary can prove no set of facts in support of a claim entitling relief. Here, the citation related to the operation of a truck, which the employer claimed was subject to certain Department of Transportation (DOT) motor vehicle safety regulations and, therefore, exempt from OSHA regulations. However, the Commission refused to grant dismissal because it perceived that the Secretary might be able to prove that DOT regulations were inapplicable to the condition involved.

2. Motions for More Definite Statement

Fed. R. Civ. P. 12(e) provides for motions for more definite statement when "a pleading to which a responsive pleading is permitted is so vague or ambiguous that a party cannot

---

[20]FED. R. CIV. P. 43(e).

[21]3 OSHC 1233 (Rev. Comm'n 1974).

[22]29 U.S.C. §653(b)(1). This section exempts working conditions over which other federal agencies exercise authority to prescribe or enforce standards affecting occupational safety or health.

reasonably be required to frame a responsive pleading." Although a motion for a more definite statement is appropriate when a complaint lacks sufficient particulars, the thrust of the motion is to strike at unintelligibility rather than want of detail.[23] A motion for more definite statement is not a substitute for discovery.[24] The Commission will not grant a motion for more definite statement if the complaint states the jurisdictional grounds of the claim; identifies the sections of the law and standards allegedly violated; and sets forth the time, location, place, and circumstances of each such alleged violation.[25]

3. Motions for Summary Judgment

Under Fed. R. Civ. P. 56, summary judgment is appropriate where the pleadings, discovery, and any affidavits presented by the parties demonstrate that there is no genuine issue of material fact and that the moving party is entitled to a judgment as a matter of law.[26] As an example of the application of the summary judgment test in OSHA cases, the Commission has held that summary judgment was appropriate in a case where prior precedent established that a standard regulating cotton dust exposure did not apply to the company's process.[27] In that case, affidavits accompanying the summary judgment motion uncontestably established that the employer was engaged solely in the processing of cotton seed, a matter not covered by the standard.

4. Motions for Consolidation

Commission Rule 9 permits cases to be consolidated "where there exist common parties, common questions of law or fact, or both, or in such other circumstances as justice and the administration of the Act require." Consolidation can be ordered on the motion of any party or on motion of the A.L.J. or the Commission. This is a beneficial procedure in handling

---

[23]*Cement Asbestos Prods. Co.*, 8 OSHC 1151 (Rev. Comm'n 1980).

[24]*American Can Co.*, 10 OSHC 1305, 1313–1314 (Rev. Comm'n 1982).

[25]*Cement Asbestos Prods. Co.*, *supra* note 23.

[26]See *Poller v. Columbia Broadcasting Sys.*, 368 U.S. 464 (1962); *United States Steel Corp.*, 9 OSHC 1527 (Rev. Comm'n 1981).

[27]*Traders Oil Mill Co.*, 2 OSHC 1508 (Rev. Comm'n 1975).

cases efficiently and in the least costly manner when they involve the same party or issue. The consolidation may be either for prehearing purposes or for purposes of all proceedings, including the hearing. The Commission and its judges have readily issued consolidation orders pursuant to motions of one or both of the parties, but consolidation typically will not occur unless one or more of the parties seek it. The primary purpose of consolidation is to refrain from duplicating efforts on common issues and to avoid inconsistent results. Probably the most comprehensive guidebook for handling consolidated matters is the Manual for Complex Litigation used by the federal district courts in cases consolidated under 28 U.S.C. §1407.

An example of how OSHA cases can be consolidated to avoid both duplication of efforts and inconsistent results is found in *Continental Can Co.*[28] There, numerous citations were issued against different manufacturing plants of the employer, many of which used the same equipment to make the same product. Three citations were consolidated for hearing. After the hearing but before the decision issued, the parties agreed that the evidence would govern five additional citations, which in effect consolidated eight cases for decision. However, while the employer prevailed on the merits in the consolidated cases, OSHA continued to issue citations against other plants of the company. A federal court issued an injunction against OSHA from prosecuting these subsequent citations.[29]

F. Discovery

In any court-litigated proceeding, the four principal methods of discovery are (1) requests for admissions, (2) requests for production of documents or entry onto land, (3) interrogatories, and (4) depositions. Under the Federal Rules of Civil Procedure those discovery methods are freely available without prior leave of court.

On their face the Commission Rules on discovery are much more limited than the Federal Rules. The only discovery method that is specifically available without first obtaining leave from

---

[28]4 OSHC 1541, 1541 n.2 (Rev. Comm'n 1976).

[29]*Continental Can Co., U.S.A. v. Marshall*, 603 F.2d 590, 7 OSHC 1521 (7th Cir. 1979).

the A.L.J. is requests for admissions.[30] Depositions and interrogatories are only allowed upon a special order of the judge.[31] In deciding whether to permit such discovery, the A.L.J. must balance the need of the party seeking discovery for the information sought against the burden the discovery would impose on the opposing party.[32] The A.L.J. should "look with disfavor on broad, open-ended discovery requests."[33] The Commission's rules do not mention requests for production of documents or discovery by entry on land. These forms of discovery are therefore governed by Fed. R. Civ. P. 34.[34] (For explanation of the role of the authorized employee representative in the discovery process, see Chapter 17 (Employee and Union Participation in Litigation Under the OSH Act)).

1. Scope of Discovery

When discovery is allowed, the Review Commission has held that the scope of discovery is governed by the Federal Rules of Civil Procedure because the Commission has no rule delineating the permissible scope.[35] The appropriate Federal Rule governing the scope of discovery is Fed. R. Civ. P. 26(b)(1). Under that rule, discovery is permitted "if the information sought appears reasonably calculated to lead to the discovery of admissible evidence." A compliance officer's file, including notes and memoranda, has been ruled to be within the scope of allowable discovery.[36] It should be noted, however, that objections to discovery under the Federal Rules of Civil Procedure may be invoked in OSHA proceedings. Such objections include, for example, attorney-client privilege, work product, relevancy, undue burden, trade secrets, confidential information, or informer's privilege.[37]

---

[30]Commission Rule 52.

[31]Commission Rule 53.

[32]*N.L. Indus.*, 11 OSHC 2156, 2159 (Rev. Comm'n 1984).

[33]*Id.*

[34]*West Point-Pepperell, Inc.*, 9 OSHC 1784, 1789 (Rev. Comm'n 1981); *Reynolds Metals Co.*, 3 OSHC 1749, 1750 (Rev. Comm'n 1975).

[35]*Quality Stamping Prods. Co.*, 7 OSHC 1285 (Rev. Comm'n 1979); *Newport News Shipbuilding & Drydock Co.*, 9 OSHC 1120 (Rev. Comm'n 1980); *West Point-Pepperell, Inc.*, *supra* note 34.

[36]*Frazee Constr. Co.*, 1 OSHC 1270 (Rev. Comm'n 1973); *Gulf & W. Food Prods. Co.*, 4 OSHC 1436 (Rev. Comm'n 1976).

[37]See FED. R. CIV. P. 26. See also *Continental Oil Co.*, 9 OSHC 1737 (Rev. Comm'n 1981); *Massman-Johnson (Luling)*, 8 OSHC 1369 (Rev. Comm'n 1980); *Quality Stamping*

Discovery of facts and opinions known by experts retained by the opposing party is governed by Fed. R. Civ. P. 26(b)(4). This rule differentiates between an expert the opponent expects to call as a witness and an expert not expected to be called as a witness. Discovery is permitted from the former, but is only permitted from the latter upon the showing of exceptional circumstances under which the party seeking discovery cannot obtain facts or opinions on the same subject by any other means.

2. Requests for Admission

Commission Rule 52 allows any party to request any other party to admit facts under oath. The procedure established in the rule requires each admission to be set forth separately. The party served with the request must respond within 15 days or such longer or shorter time as the A.L.J. may order. If no response is filed within the allotted time, the matter is deemed admitted under the rule.

Fed. R. Civ. P. 36 sets forth more detailed requirements as to what the content of the response to a request for admission must contain. Specifically, (1) the answer shall deny the matter or set forth in detail the reasons why the answering party cannot truthfully admit or deny the matter; (2) a denial shall fairly meet the substance of the requested admission; and (3) when good faith requires that a party qualify its answer or deny only a part of the matter of which an admission is requested, so much of it as is true must be specified and the remainder must either be qualified or denied.

3. Requests for Production of Documents or for Entry on Land

OSHA cases are similar to many other kinds of litigation in that the opponent's files may contain relevant documents. In such instances access to the documents is very helpful in preparing for the hearing because the documents can give insight into the details or inconsistencies of the opponent's case.

Prods. Co., supra note 35; *Stephenson Enters.*, 2 OSHC 1080 (Rev. Comm'n 1974), *aff'd*, 578 F.2d 1021, 6 OSHC 1860 (5th Cir. 1978). And see Section II.C.3, *infra*.

An example of the type of case where one or both of the parties may wish to obtain documents or OSHA may wish to have its expert enter the plant is a case involving a noise citation under 29 C.F.R. §1910.95. In noise cases, the charge typically is that the employer failed to implement feasible engineering or administrative controls to reduce sound levels. The employer may have records in its files that detail work it performed or studies that its own engineers or outside consultants may have prepared. OSHA may have analyses of possible engineering controls in its compliance officer's file or reports relating to other plants that use similar equipment to manufacture the same product. Further, OSHA's compliance officer may have issued the citation solely on the basis of sound level measurements without any analysis of what engineering controls might be available to reduce those sound levels. In this type of case, the Commission has held that discovery is necessary and critical, particularly to prepare expert testimony.[38]

Fed. R. Civ. P. 34 provides the vehicle in OSHA cases for seeking the production of documents or entry into the workplace in question.[39] Under Rule 34, any party may without leave of court

> "serve on any other party a request (1) to produce and permit the party making the request *** to inspect and copy any designated documents *** or to inspect and copy, test, or sample any tangible things which constitute or contain matters *** which are in the possession, custody or control of the party upon whom the request is served; or (2) to permit entry upon designated land or other property in the possession or control of the party upon whom the request is served for the purpose of inspection and measuring, surveying, photographing, testing, or sampling ***."

The request must describe each item and category sought with "reasonable particularity." The request also must specify a reasonable time, place, and manner for making the production. The person served with the request must respond within 30 days and must state with respect to each request whether production or objection will be made.

---

[38]*Federated Metals*, 9 OSHC 1906, 1911 (Rev. Comm'n 1981).

[39]*West Point-Pepperell, Inc.*, *supra* note 34; *Forte Bros.*, 9 OSHC 1065 (Rev. Comm'n 1980); *Wheeling-Pittsburgh Steel Corp.*, 4 OSHC 1578 (Rev. Comm'n 1976); *Reynolds Metals Co.*, *supra* note 34.

4. Depositions and Interrogatories

Commission Rule 53 does not permit, as referenced earlier, the use of depositions or interrogatories unless the party seeking to utilize those discovery tools obtains an order from the A.L.J. The Commission has stated that the judge, in deciding whether to permit depositions or interrogatories, "should consider the need of the moving party for the information sought, any undue burden on the party from whom the discovery is sought and in balance, any undue delay in the proceeding which may occur."[40]

In cases involving complex factual issues, the Commission typically permits depositions and interrogatories. Indeed, some judges even issue a blanket order permitting depositions and interrogatories before any party seeks such authority. If the judge allows a party to take depositions, the procedures for requiring the attendance of a witness are set out in Commission Rules 55 and 63 and Fed. Rs. Civ. P. 30 and 45. These rules are particularly important in the case of a nonparty witness whose attendance at a deposition can be compelled only by subpoena.[41]

In that regard, Fed. R. Civ. P. 45(c) provides the mechanics for service of a subpoena since the Commission's rules are silent on the method of service. Service may be made by any person over 18 years of age who is not a party by delivering a copy to the prospective deponent. Commission Rule 63, similar to Fed. R. Civ. P. 45(c), also requires that the subpoena be accompanied with the same witness fees and mileage paid to witnesses in the U.S. courts.[42]

Finally, a party served with a subpoena has 5 days under Commission Rule 55(b) to move for revocation or modification of the subpoena.[43] Proceedings seeking relief for failure to comply with a subpoena may be brought in the "appropriate" district court if, in the Commission's judgment, enforcement

---

[40]*KLI, Inc.*, 6 OSHC 1097, 1098 (Rev. Comm'n 1977). See also *N.L. Indus.*, *supra* note 32; *Del Monte Corp.*, 9 OSHC 2136 (Rev. Comm'n 1981).

[41]Commission Rule 55(a) provides that once a case is assigned to a judge, the judge shall grant an application for subpoena which can be made *ex parte*. The Commission has subpoena forms which can be used for this purpose.

[42]See 28 U.S.C. §1821 (1982).

[43]See *Lee Way Motor Freight*, 3 OSHC 1843 (Rev. Comm'n 1975).

of the subpoena would be consistent with law and the policies of the Act.[44]

5. Sanctions for Failure to Make Discovery

Commission Rule 54 merely provides that if a party fails to comply with an order permitting discovery, the A.L.J. may issue an appropriate order. The Commission therefore looks to Fed. R. Civ. P. 37 for dealing with failures to make discovery.[45]

Under Fed. R. Civ. P. 37, if a party fails to respond to discovery or answers incompletely or evasively, the court may order the party to respond properly. If the party fails to comply with the court's order, the court "may make such orders in regard to the failure as are just," including

"(A) An order that the matters regarding which the order was made or any other designated facts shall be taken to be established for the purposes of the action in accordance with the claim of the party obtaining the order.

"(B) An order refusing to allow the disobedient party to support or oppose designated claims or defenses, or prohibiting him from introducing designated matters in evidence.

"(C) An order striking out pleadings or parts thereof, or staying further proceedings, until the order is obeyed, or dismissing the action or proceeding or any part thereof, or rendering a judgment by default against the disobedient party."[46]

6. Trade Secrets and Discovery

In complicated cases, particularly those involving hygiene and general duty issues, the Secretary will retain outside experts to assist in the preparation of a case and to testify at the hearing.[47] As part of the Secretary's hearing preparation, the expert will inspect the worksite pursuant to Fed. R. Civ. P. 34 while Commission Rule 11 permits the A.L.J. upon application to issue appropriate orders to protect confidentiality;

---

[44]Commission Rule 55(d). See *Equitable Shipyards*, 12 OSHC 1288 (Rev. Comm'n 1985) (judge erred in refusing to enforce subpoena against witness whose testimony would be relevant to case).

[45]*Federated Metals, supra* note 38.

[46]FED. R. CIV. P. 37(b)(2).

[47]See, e.g., *Newport News Shipbuilding & Drydock Co.*, 9 OSHC 1085 (Rev. Comm'n 1980); *Noranda Aluminum*, 9 OSHC 1187 (Rev. Comm'n 1980); *Circle T. Drilling Co.*, 8 OSHC 1681 (Rev. Comm'n 1980).

employers have nevertheless objected to these inspections on the grounds that the expert will have access to valuable trade secrets which the expert could appropriate for his benefit to the employer's detriment. The Commission's decision in *Owens-Illinois, Inc.*,[48] permits an expert to inspect a facility containing trade secrets, but only on condition that a protective order is entered containing the following provisions:

(1) the Secretary must submit the resume of the expert to the employer and the employer may challenge the use of that particular expert on the grounds that he or she is closely aligned with a competitor;

(2) the expert must sign a confidentiality oath not to disclose information established by the employer to be a trade secret;

(3) the Secretary must include a confidentiality oath in his contract with the expert and make the employer the third-party beneficiary of the provision.

G. Simplified Proceedings

As briefly mentioned in Section I.A above, in an effort to make it easier for parties to have their day in court, the Commission has adopted rules for simplified proceedings in certain categories of cases.[49] Cases involving the general duty clause and certain health regulations are ineligible for simplified proceedings,[50] and other cases may also be too complicated for the invocation of these procedures. The purpose of simplified proceedings is to make the resolution of cases faster, to make it easier for those appearing before the Commission to proceed without an attorney, to reduce paperwork, and to lessen the expense of litigation.

The principal differences between simplified proceedings and the procedures described earlier in this chapter are that there are no pleadings (i.e., complaint and answer), discovery, or interlocutory appeals.[51] Furthermore, if the case goes to a

---

[48]6 OSHC 2162, 2167 (Rev. Comm'n 1978).

[49]See Commission Rules 200–211. Employee or employee representative notices of contests to the proposed abatement period are governed under the expedited procedures as set forth in Rule 101.

[50]Commission Rule 202.

[51]Rules 205, 210, and 211.

hearing, the formal rules of evidence are not followed. Simplified proceedings also encourage the A.L.J. to take an active role in the case in such matters as trying to narrow the issues and to assist in settling the case.[52]

A party seeking to obtain simplified proceedings must request them in writing, with service on other parties, within 10 days after the notice of docketing is received from the Commission.[53] Any party may object to a request for simplified proceedings within 15 days after service. If an objection is filed, the institution of simplified proceedings is precluded.[54] If no objection is filed and the case is eligible, simplified proceedings are instituted. However, even where instituted, a party may move to discontinue such proceedings any time prior to the commencement of the hearing.[55] The motion shall be granted if all parties consent or if sufficient reason is shown for application of the regular rules of procedure.[56]

H. Interlocutory Appeals

A party may appeal an A.L.J.'s order that is not a final order to the Commission if the appeal involves questions of discovery or decisions on other pretrial motions. This is called an interlocutory appeal. Interlocutory appeals are not allowed as a matter of right, but are governed by Commission Rule 75.

The procedure set forth in Rule 75 requires the party seeking the interlocutory appeal to seek a certification of the appeal from the A.L.J. within 5 days after receipt of the A.L.J.'s ruling. The rule provides that "the judge shall certify an interlocutory appeal when the ruling involves an important question of law or policy about which there is substantial ground for difference of opinion and an immediate appeal of the ruling may materially expedite the proceedings."

Even if the A.L.J. certifies the interlocutory appeal, however, the Commission may decline to accept it.[57] If the A.L.J.

---

[52]Rules 206 and 207.
[53]Rule 203(a).
[54]Rule 203(b).
[55]Rule 204.
[56]*Id.*
[57]Rule 75(b)(2).

denies certification, then the party may file a petition for interlocutory appeal directly with the Commission within 5 days after the order denying certification. The Commission will not grant a direct petition for interlocutory review unless the Commission believes that the petition satisfies the required criteria and that there is a substantial probability that the A.L.J.'s order will be reversed. A party who has not been granted the right of interlocutory appeal may later seek to have the same question reviewed after the A.L.J.'s final decision on the merits of the case.[58]

When a party files a request for interlocutory appeal or even if an interlocutory appeal is granted, the prehearing proceedings are not stayed unless otherwise ordered by the A.L.J. or Commission.[59] There is one exception to that rule. When a trade secret is involved, an interlocutory appeal automatically stays a proceeding until the A.L.J. denies the request or, if the request is granted, until the Commission rules on the appeal or declines to accept the certification.[60] Furthermore, when trade secrets are involved, if the A.L.J. denies certification, the A.L.J. upon motion by the requesting party must also stay the effect of his ruling for 5 days to enable the party to petition the Commission. If such a petition is filed, the A.L.J.'s ruling concerning the trade secret is stayed until the Commission denies the petition or rules on the appeal.[61]

II. The Hearing

The requirement for a formal hearing stems from Section 10(c) of the Act,[62] which expressly provides that where there is a contest before the Commission an opportunity for a hearing shall be provided in accordance with the Administrative Procedure Act (APA).[63] A formal administrative hearing under the APA is generally the equivalent of a civil trial without a jury.[64] As to the degree of formality contemplated by the

---

[58]Rule 75(c).
[59]Rule 75(e)(2).
[60]Rule 75(e)(1).
[61]*Id.*
[62]29 U.S.C. §659(c).
[63]5 U.S.C. §554 (1982).
[64]See *Opp Cotton Mills v. Wage & Hour Div.*, 312 U.S. 126 (1941).

Act, Section 12(f)[65] prescribes, after vesting the Commission
with conventional rulemaking authority, that Commission
proceedings shall be in accordance with the Federal Rules of
Civil Procedure unless the Commission adopts a different rule.

A. Elements of a Hearing

1. Burden of Proof

Commission Rule 73 assigns to the Secretary of Labor the
burden of proof in any case begun by a notice of contest. When
a case is started by a "petition for modification of the abatement period" by an employer under Rule 34, the burden of
establishing the necessity for the change lies with the petitioner. The burden of proving an issue must be met by a preponderance of the evidence.

The pleadings required by the Commission (complaint
and answer in the case of employer contests) normally establish the issues to which the burden applies. However, the
pleadings alone may not necessarily control the proof. Also
considered will be any express or implied amendments of the
pleadings, including amendments to conform to the evidence
and the content of prehearing conferences. A matter of official
notice, which is discussed in Section II.C below, may also influence the quantum of proof.

Normally, in cases involving an alleged violation of a
standard, the Secretary must establish the following elements:
(1) there was a violation of the standard's requirement; (2)
one or more employees were actually exposed or had access
to the hazard; and (3) the employer had actual or constructive
knowledge of the violation.[66] In addition, under the test of
some standards, e.g., 29 C.F.R. §1910.95 (noise standard), the
Secretary also has the burden of proving the feasibility of
abatement practices. The Secretary has the same burden in
cases arising under the general duty clause. Finally, with
respect to affirmative defenses of an employer (e.g., impossibility of compliance, isolated employee misconduct, compliance would create a greater hazard), the employer has the
burden of going forward.

---

[65]29 U.S.C. §661(f).
[66]See, e.g., *Anning-Johnson Co.*, 4 OSHC 1193 (Rev. Comm'n 1976).

2. *Oral and Documentary Evidence*

While a more extensive analysis of evidentiary matters is contained in Section II.C below, it should be noted that submittal of evidence is, of course, an element of Review Commission hearings. The APA[67] provides that any oral or documentary evidence may be received, but the agency (i.e., the Commission) as a matter of policy shall provide for the exclusion of irrelevant, immaterial, or unduly repetitious evidence. Affidavits may be used as documentary evidence under Commission Rule 69 if matters therein are otherwise admissible and the parties agree to admission.

Moreover, a sanction may not be imposed or rule or order issued except on consideration of the whole record or those parts of the record cited by a party and supported by and in accordance with the reliable, probative, and substantial evidence. Thus, the APA's text permits evidence to be admitted without regard to technical rules of evidence except for the exclusion of "irrelevant, immaterial, or unduly repetitious" evidence.

In addition, Commission Rule 72 provides, as mentioned in Section I.A above, that "insofar as practicable" hearings shall be governed by the Federal Rules of Evidence. While certain circumstances under which the Commission has found it not practicable to apply the Federal Rules of Evidence are discussed in Section II.C below, Commission Rule 72 parallels closely the last sentence in Section 10(b) of the Labor Management Relations Act of 1947.[68] Under that statute similar wording has been construed to require the NLRB to rest its rulings upon facts rather than conjecture and presumed expertness, thereby foreclosing expert inferences not based on evidence as well as the "wholesale use" of hearsay evidence.[69] Courts have also noted that the phrase "so far as practicable" was intended to authorize departure from the rules of evidence when necessary because of the peculiar characteristics of administrative hearings.[70]

---

[67]5 U.S.C. §556.
[68]29 U.S.C. §160(b).
[69]*Pittsburgh S.S. Co. v. NLRB*, 180 F.2d 731 (6th Cir. 1950).
[70]*General Eng'g v. NLRB*, 341 F.2d 367 (9th Cir. 1965).

3. Right to Counsel

Under the APA, a party is entitled to appear in person, or to be represented by counsel or "other duly qualified representative."[71] The term "other qualified representative" refers to someone with expertise who is not an attorney, such as an accountant appearing in matters in this field.[72] Commission Rule 22 provides only that any party may appear in person or through a representative and states expressly that nothing in the rule is to be construed to require that any representative be an attorney. However, the Commission in *Yaffe Iron & Metal Co.*[73] implied that a person may be unqualified by reason of ability or experience to represent a party (but noted that there was nothing in the *Yaffe Iron* record indicating that the representative in that case, an engineer, was unqualified).

4. Speedy Proceedings

Speedy hearing proceedings are contemplated in adjudications under the OSH Act.[74] The Commission's procedural rules are laced with devices encouraging relatively quick proceedings, and in fact admonish presiding judges to avoid delay in adjudication.[75] For example, postponement of scheduled hearings is not ordinarily allowed. In addition, while exceptions are made for "extreme emergency" or "unusual circumstances," postponement for more than 30 days is not allowed without Commission approval.[76] The Commission has a consonant duty under the APA to proceed to conclude any matter presented to it within a "reasonable time" and with due regard for the convenience of the parties.[77]

---

[71] 5 U.S.C. §555(b).

[72] ATTORNEY GENERAL'S MANUAL ON THE APA 62 (1947).

[73] 5 OSHC 1057, 1058 (Rev. Comm'n 1977).

[74] *Atlas Roofing Co. v. OSHRC*, 430 U.S. 442, 461, 5 OSHC 1105, 1111–1112 (1977).

[75] Commission Rules 66, 61, and 75.

[76] Commission Rule 61. The Commission's power under the rule for postponements exceeding 30 days has been delegated to the Chief Administrative Law Judge.

[77] 5 U.S.C. §555(b).

B. Administrative Law Judge's Hearing Functions and Duties

While an exhaustive listing of all functions and duties the A.L.J. possesses is beyond the scope of this section, certain matters are noteworthy. First, the A.L.J. has a duty to conduct a "fair" and "impartial" hearing to assure that the facts are fully elicited, to adjudicate all issues, and to avoid delay.[78] Cases contested before the Commission are assigned in rotation "so far as practicable" to its judges.[79] Some practicalities that may result in a variation in assignments include referring cases to an A.L.J. for resolution of issues of consolidation, travel costs, calendar arrangements, and a reasonable distribution of caseload among the Commission's regional offices and its individual judges.

The A.L.J. who presides at the reception of evidence in a formal Commission hearing must make an initial decision unless that A.L.J. becomes "unavailable."[80] This requirement is met if the original judge presides at a prehearing conference and issues a prehearing order even though a substituted judge presides at the formal evidentiary hearings.[81] "Unavailability" has been found where the A.L.J. has died[82] or retired,[83] or where bias or prejudice exists.[84] While there must be an opportunity for taking additional evidence when there is a substitution of the A.L.J., in general this is done only when the demeanor of witnesses is critical to a decision.[85] When demeanor is not important, no unfairness results from a decision made without augmenting the record.[86]

The A.L.J. may also have to consider the issue of disqualification. Commission Rule 67 provides that the A.L.J. may disqualify himself on the grounds of bias or prejudice on the filing of an affidavit setting forth facts sufficient for disqualification. Hearings, however, are not required on every

---

[78]Commission Rule 66.

[79]5 U.S.C. §3015.

[80]5 U.S.C. §554(d).

[81]*Ace Books, Inc.*, 17 Pike & Fischer, Ad. L (2d) 555 (FTC 1965).

[82]*Pigrenet v. Boland Marine & Mfg. Co.*, 656 F.2d 1091 (5th Cir. 1981).

[83]*Gamble-Skogmo v. FTC*, 211 F.2d 106 (8th Cir. 1954).

[84]*NLRB v. Dixie Shirt Co.*, 176 F.2d 969 (4th Cir. 1949).

[85]2 K. Davis, ADMINISTRATIVE LAW TREATISE §11.18 at 113 (1st ed. 1955).

[86]*New England Coalition on Nuclear Pollution v. NRC*, 582 F.2d 87 (1st Cir. 1978).

such charge. If the affidavit is not sufficient on its face, it may be dismissed summarily,[87] but the record shall reflect the grounds of the judge's ruling.

There have been no full Commission decisions on the personal bias or disqualification of an A.L.J., although the issue has arisen with respect to a Commission member.[88] The standards considered in that case in rejecting the claim were the ABA Code of Judicial Conduct, 28 U.S.C. §§144 and 455, along with controlling case law on the subject.[89]

Another crucial function of the A.L.J. concerns the scheduling of hearings. At least 10 days' notice must be given of the time, place, and nature of any hearing, unless a member of the Review Commission has ordered that "expedited proceedings" are to be used.[90] If that occurs, a shorter period of notice is permissible, so long as it is "timely" under the APA.[91]

Timeliness may conceivably also require that there be no unreasonable delay in giving notice of hearing; i.e., there is a right to a hearing within a reasonable time.[92] The convenience and necessity of the parties or their representatives is to be given "due regard" in setting the time and place of hearings.[93] In *Bethlehem Steel Corp.*,[94] the presiding A.L.J. scheduled the hearing in Philadelphia although the case arose out of inspections at a Bethlehem, Pennsylvania, plant. The employer, the Secretary of Labor, and the Steelworkers Union expected to call more than 20 witnesses, all of whom were employed or resided in the Bethlehem-Allentown area. In addition, federal and county court facilities were available in the Allentown area, some 50 miles from Philadelphia. Upon interlocutory appeal, the Commission held that the A.L.J. had abused his discretion in locating the hearing in Philadelphia.

---

[87]ATTORNEY GENERAL'S MANUAL ON THE APA 73 (1947).

[88]*National Mfg. Co.*, 8 OSHC 1435 (Rev. Comm'n 1980).

[89]Disqualification is warranted where there is prejudgment of the merits of a particular case or application of the law to the operative facts. An expression of views on general legal issues is not enough. *Cinderella Career & Finishing Schools v. FTC*, 425 F.2d 583 (D.C. Cir. 1970); *Texaco, Inc. v. FTC*, 336 F.2d 754 (D.C. Cir. 1964). See also *Faultless Div., Bliss & Laughlin Indus. v. Secretary of Labor*, 674 F.2d 1177, 10 OSHC 1481 (7th Cir. 1982) (evidence did not raise any substantial doubt about A.L.J.'s impartiality).

[90]Commission Rule 60.

[91]5 U.S.C. §554(b).

[92]Cf. *White v. Mathews*, 559 F.2d 852, 858 (2d Cir. 1977).

[93]5 U.S.C. §554(b).

[94]6 OSHC 1912 (Rev. Comm'n 1978).

Another important function of the A.L.J. concerns postponement or continuance of hearings. As briefly mentioned earlier, Commission Rule 61 expressly provides that a postponement of hearing will not ordinarily be allowed. "Extreme emergency" or "unusual circumstances" are required. No postponement by the presiding officer may exceed 30 days without Commission approval. The Commission has, however, condoned postponements by an A.L.J. for longer periods when the A.L.J.'s action was "harmless error."[95] Matters of continuance or postponement are within the sound discretion of the presiding officer or the agency.[96]

Sometimes postponement or continuance arises from the failure of a party or representative to appear at the hearing. While a failure to appear will normally act as a waiver of a party's rights except as to receiving the A.L.J.'s decision or seeking review,[97] the Commission or the A.L.J. may excuse such failure upon a showing of "good cause." Examples of "good cause" have included the fact that the employer had been injured in an accident the previous evening, and his attorney had tried unsuccessfully five times to contact the A.L.J.;[98] the employer had met with a Department of Labor official several days before the hearing to attempt to settle the case, and assumed that it would be settled outside of the adjudicative process;[99] counsel failed to appear on behalf of the employer as a result of a misunderstanding of instructions.[100]

With respect to specific duties A.L.J.s are empowered to take, Commission Rule 66 follows closely the enumerated powers of the presiding officers under the APA. Specifically, A.L.J.s are granted the following powers with respect to hearings:

- To administer oaths and affirmations
- To issue subpoenas authorized by law
- To rule upon offers of proof and receive relevant evidence
- To regulate the course of the hearing
- To dispose of procedural requests or similar matters

---

[95]*Maxwell Wirebound Box Co.*, 8 OSHC 1995 (Rev. Comm'n 1980).
[96]*Ralston Purina Co.*, 7 OSHC 1730 (Rev. Comm'n 1979).
[97]Commission Rule 62. See also *Bob Bolles*, 7 OSHC 1580 (Rev. Comm'n J. 1979).
[98]*Simpson Roofing Co.*, 5 OSHC 1836 (Rev. Comm'n 1977).
[99]*Duquesne Electric & Mfg. Co.*, 5 OSHC 1843 (Rev. Comm'n 1977).
[100]*Ribblesdale, Inc.*, 5 OSHC 1179 (Rev. Comm'n 1977).

- To make initial decisions or recommend decisions
- To take other consistent action authorized by the Commission's rules

1. Oaths and Affirmations

"Oath" or "affirmation" is normally taken before any testimony is given.[101] These terms signify a knowledge of the solemnity of testimony and the consequences of perjury to the extent they may be dealt with under administrative law.

2. Subpoenas

Commission Rule 55 empowers any Commission member on an application of any party to issue a subpoena requiring the attendance of a witness or the production of any evidence. Subsequent to the assignment of a case to an A.L.J., an application for subpoena is to be filed with the A.L.J. The A.L.J. shall grant the application on behalf of any member of the Commission. An application for subpoena may be made *ex parte*. In practice, subpoenas are issued routinely upon application, and commonly bear the name of the Commission's chairperson. They are issued in blank. A party fills in the blanks before service for both subpoenas *ad testificandum* and *duces tecum*.

The APA[102] provides that when an agency possesses statutory subpoena power, it shall issue a subpoena at a party's request "and when required by rules of procedure, on a statement or showing of general relevance and reasonable scope of the evidence sought." A person served with a subpoena may move to quash or modify it.[103] This is similar to the practice in federal district courts.

---

[101]FED. R. EVID. 603 reads: "Before testifying, every witness shall be required to declare that he will testify truthfully, by oath or affirmation administered in a form calculated to awaken his conscience and impress his mind with his duty to do so." See also *M.K. Binkley Constr. Co.*, 5 OSHC 1411 (Rev. Comm'n 1977) (A.L.J. did not err in permitting employer's representative to take "retroactive" oath when witness repeated previous unsworn statements while under oath).

[102]5 U.S.C. §555(d). See also W. Gellhorn, C. Byse, & P. Strauss, ADMINISTRATIVE LAW 684 (7th ed. 1979). For an example where a subpoena on discovery matters was modified, see *Bethlehem Steel Corp.*, 9 OSHC 1321, 1326 (Rev. Comm'n 1981).

[103]Commission Rule 66(c).

3. Offers of Proof and Receipt of Relevant Evidence

A party has the right to make an offer of proof when evidence is excluded.[104] In addition, Fed. R. Evid. 103, pertaining to offers of proof, applies to Commission proceedings. That provision requires that where evidence is excluded either the substance of the evidence must be made known to the court by offer or the substance must be apparent from the context within which questions were asked. Fed. R. Evid. 103(b) then permits the court to add any further statement to show the character of the evidence, the form offered, objection made, and the ruling.

It should be noted that the Federal Rules of Evidence fuse the concept of relevancy with that of materiality. Also, exclusion of repetitious evidence is permitted under Fed. R. Evid. 403.

4. Disposition of Procedural Requests or Similar Matters

This power has been broadly construed by the Commission to include a power to rule on matters upon which judges have "traditionally" ruled.[105] The policy announced by the Commission is that on routine procedural matters not expressly covered by the Commission's rules of procedure, there should be recourse to a ruling by the presiding A.L.J. rather than the extraordinary measure of seeking a waiver in the specific case from the full Commission. It should also be noted, as mentioned in Section I.D above, that the A.L.J.'s power to hold conferences for the settlement or simplification of the issues is set forth in Commission Rule 51. The Commission has equated the purposes of Rule 51 with Fed. R. Civ. P. 16.[106] However, in *Usery v. Marquette Cement Manufacturing Co.*,[107] the Second Circuit observed that the Commission Rule, unlike Fed. R. Civ. P. 16, does not provide that the prehearing order will control the subsequent course of the action.

The Commission itself considers the A.L.J. to have much discretion. For example, it has sustained discipline employed

---

[104]Commission Rule 74(b).

[105]*Carhar Contracting Co.*, 9 OSHC 1237 (Rev. Comm'n 1981). See also *ASARCO, El Paso Div.*, 8 OSHC 2156 (Rev. Comm'n 1980).

[106]*Duquesne Light Co.*, 8 OSHC 1218, 1221 (Rev. Comm'n 1980).

[107]568 F.2d 902, 5 OSHC 1793 (2d Cir. 1977).

by an A.L.J. for breach of a prehearing order to provide a witness list;[108] held that a modification of a witness list determined at prehearing need not be accepted by an A.L.J.;[109] and ruled that an A.L.J. is empowered to refuse a party's request to allow an unscheduled witness to testify.[110]

5. *Power to Make Initial Decisions*

The A.L.J.'s power under the APA to render a decision must be read with the OSH Act itself, which provides that the A.L.J. hear and render a "determination," and make a "report" of the "determination" to the full Commission. The "report" of the presiding A.L.J. will become a final order of the Commission, unless it is ordered by a Commission member for review.[111]

C. Evidence

1. *Commission Rules*

Commission Rule 72 provides, as mentioned previously, that Commission hearings are to be in accord with the APA's evidentiary requirements and "insofar as practicable" shall be governed by the Federal Rules of Evidence. The Commission, however, has some specific rules dealing with evidentiary matters that warrant the practitioner's attention.

(a) Objections. Rule 74(a)[112] deals expressly with objections to the introduction of evidence. The rule, however, appears to fall short of covering the scope of matters covered

---

[108]*Hoerner Waldorf Corp.*, 4 OSHC 1836 (Rev. Comm'n 1976).

[109]*Williams Enters.*, 4 OSHC 1663 (Rev. Comm'n 1976).

[110]*Fleming Foods of Neb.*, 6 OSHC 1233 (Rev. Comm'n 1977).

[111]29 U.S.C. §661(j). See Chapter 16 (The Occupational Safety and Health Commission), Section II.A.

[112]Rule 74(a) reads:

"(a) Any objection with respect to the conduct of the hearing, including any objection to the introduction of evidence or a ruling by the judge, may be stated orally or in writing, accompanied by a short statement of the grounds for the objection, and shall be included in the record. No such objection shall be deemed waived by further participation in the hearing.

"(b) Whenever evidence is excluded from the record, the party offering such evidence may make an offer of proof, which shall be included in the record of the proceeding."

under Fed. R. Evid. 103(a)(1) and therefore the two should be considered together. For example, although Rule 74(a) provides that when an oral objection is made it is to be accompanied by a short statement of the grounds for objection, the actual practice is to follow the approach of Fed. R. Evid. 103 that no specific ground for objection need be assigned when the ground is apparent from the context. This saves the time of the counsel and the A.L.J. An exception cannot be taken to an A.L.J. ruling if there has been no objection before the A.L.J.[113]

(b) Offers of proof. Paragraph (b) of Commission Rule 74 provides that a party may make an offer of proof of excluded evidence. Offers of proof, as discussed earlier, are also expressly contemplated by 5 U.S.C. §556(c), and provided for under Fed. R. Evid. 103. Since Rule 74 is silent as to the content of offers of proof, Fed. R. Evid. 103 should also be considered, i.e., that the substance of the offer be made known to the court unless it is apparent from the context of the questions asked.

(c) Examination of witnesses. Commission Rule 68 provides for examination of witnesses and the opposing party's right to cross-examine, but contains no other directives. Since the Secretary has the burden of persuasion, he normally presents the government's case first, followed by the union if it is a party. However, the A.L.J. is empowered to manage the course of the hearing to serve the objectives of ascertaining the truth of the allegations, avoiding needless consumption of time, and protecting witnesses from harassment or undue embarrassment;[114] to these ends, the A.L.J. can control the mode and order of interrogating witnesses and presenting evidence.

As to cross-examination, the APA entitles a party to conduct such cross-examination as is required for a full and true disclosure of the facts.[115] Fed. Rs. Evid. 611(b) and (c) permit leading questions on cross-examination and, at the A.L.J.'s discretion, cross-examination on matters not brought up on direct examination.

Leading questions are not normally permitted on direct examination, except where the A.L.J. deems them necessary

---

[113]See *Williams Enters., supra* note 109, at 1665–1666.

[114]FED. R. EVID. 611(a).

[115]5 U.S.C. §556(d).

to develop the witness's testimony.[116] However, when a party calls a hostile witness or a witness identified with an adverse party, the rule allows interrogation by leading questions.

(d) Exhibits. Commission Rule 71 requires exhibits offered into evidence to be marked in a manner similar to that of any court proceeding. Unless an objection is raised, exhibits are normally admitted into evidence, except that the A.L.J. may exclude an exhibit that is irrelevant, immaterial, or unduly repetitious.[117] Copies of admitted exhibits are usually given to opposing parties, and all rejected exhibits are placed in a separate file labeled as such.

(e) Calling and interrogating witnesses. Commission Rule 66(j) permits the A.L.J. to call and examine witnesses and to introduce evidence into the record.[118] While judicial activism on a jurisdictional issue has been approved,[119] there is nevertheless a need for caution in order to preserve the appearance of impartiality.

2. Federal Rules of Evidence

Certain evidentiary matters are not provided for under the Commission's rules. Consequently, direction on such matters is obtained by reference to the Federal Rules of Evidence.

(a) Official notice (Rule 201). Official notice is the administrative counterpart of judicial notice of adjudicative facts. The last sentence of 5 U.S.C. §556(e) provides that "[w]hen an agency decision rests on official notice of a material fact not appearing in the evidence on the record, a party is entitled, on timely request, to an opportunity to show the contrary." This is the only statement in the APA regarding "official notice." It does not deal with admissibility. It concerns reliance by an A.L.J. or the Commission upon information not supported by the evidence "on the record" of a "material fact," and provides that the parties in the adjudication must be given

---

[116]FED. R. EVID. 611(c).

[117]*Id.* See also 5 U.S.C. §554.

[118]See also FED. R. EVID. 614; *Noblecraft Indus.*, 3 OSHC 1727, 1728 n.3 (Rev. Comm'n 1975).

[119]*Brennan v. OSHRC (John J. Gordon Co.)*, 492 F.2d 1027, 1 OSHC 1580 (2d Cir. 1974).

an opportunity to present contrary information. Agency recognition of facts outside the record is not limited to indisputable facts. Official notice extends to a broader range of information that may be useful in deciding the adjudication, so long as elementary fairness (notice and opportunity to be heard) is provided. Generally, however, Fed. R. Evid. 201 is looked to for guidance. That provision allows official notice of facts not subject to reasonable dispute that are generally known within the territorial jurisdiction of the court or are capable of accurate and ready determination by reasonably unquestionable sources.

It is cautioned, however, that the concept of official notice does not allow a major circumvention on matters of proof. For example, official notice may not serve as a substitute for expert evidence by one of the parties. In *National Realty & Construction Co. v. OSHRC*,[120] the Commission had sought to cure deficiencies in the record on methods of abatement of the hazard involved by itself suggesting several methods. The court noted that this was a matter for expert evidence, and that the Secretary should have called his own expert. Significantly, the court did not consider the use of official notice as a possibility.

Examples of facts given official notice include that the San Francisco Bay is a navigable waterway of the United States,[121] and that the eye is a delicate organ and any foreign material in the eye can be potentially injurious.[122] The Commission has also taken official notice of the scientific data contained in *Dangerous Properties of Industrial Materials* concerning the properties of hydrogen cyanide.[123] However, in one decision[124] a divided Commission declined to take official notice of scientific studies on exposure levels of airborne lead reported in a NIOSH document. In the same case, however, the majority without reference to official notice considered as a fact that excessive exposure to dangerous levels of airborne lead can cause serious physical harm.

---

[120]489 F.2d 1257, 1267, 1 OSHC 1422, 1428 (D.C. Cir. 1973).

[121]*Cable Car Advertisers*, 1 OSHC 1446 (Rev. Comm'n 1973).

[122]*Stearns-Rogers, Inc.*, 7 OSHC 1919 (Rev. Comm'n 1979).

[123]*Pratt & Whitney Aircraft*, 8 OSHC 1329, 1331 n.4 (Rev. Comm'n 1980) (citing N. Sax, DANGEROUS PROPERTIES OF INDUSTRIAL MATERIALS 822 (5th ed. 1979)).

[124]*Hydrate Battery Corp.*, 2 OSHC 1719 (Rev. Comm'n 1975).

(b) Presumptions (Rule 301). Under Fed. R. Evid. 301 a presumption imposes on a party against whom it is directed the burden of going forward with evidence to rebut it, but the ultimate burden of proof remains with the party benefiting from the presumption. In OSHA cases, the existence of a hazard is presumed when the requirements of a specification standard are not met.[125] For example, the standard in question in this case provided that oxygen cylinders in storage be separated from fuel-gas cylinders or combustible materials by at least 20 feet or by a noncombustible barrier of specified dimensions. Proof that such materials were not separated by the required distance resulted in a presumption of the existence of a hazard. In contrast, when a standard is general and by its terms requires protection from a hazard, the Secretary of Labor must prove the existence of hazard.[126]

(c) Relevancy (Rules 401–411). The rules on relevancy are practical rules designed to restrict evidence to reasonable bounds. The APA and Fed. Rs. Evid. 401–411 govern relevancy. Several cases show typical rulings. For example, an evidentiary offer of an industry study designed to determine noise controls for typical machines in an industry was properly refused because the study did not include the employer's own machinery.[127] On the other hand, advisory National Fire Protection Association (NFPA) standards were determined relevant in construing industry recognition of an electrical hazard under the OSH Act's general duty clause.[128] In *Williams Enterprises,*[129] it was held that the A.L.J. erred in not admitting into evidence testimony that following a fatal accident a large counterweight was secured to a crane because social legislation rather than negligence at common law was involved. This evidence was deemed relevant, primarily because it demonstrated the feasibility of preventive or corrective measures.

---

[125]*Ormet Corp.,* 9 OSHC 1055, 1060 (Rev. Comm'n 1980).

[126]*General Motors Corp., GM Parts Div.,* 11 OSHC 2062 (Rev. Comm'n 1984), aff'd, 764 F.2d 32, 12 OSHC 1377 (1st Cir. 1985). See also *Hermitage Concrete Pipe Co.,* 10 OSHC 1517, 1519–1520 (Rev. Comm'n 1982) (exposures to silica dust measured during OSHA inspection presumed to be typical of normal concentrations in workplace).

[127]*Bethlehem Steel Corp.,* 9 OSHC 1321 (Rev. Comm'n 1981).

[128]*Kansas City Power & Light Co.,* 10 OSHC 1417 (Rev. Comm'n 1982).

[129]4 OSHC 1663 (Rev. Comm'n 1976).

(d) Writing used to refresh memory (Rule 612). Fed. R. Evid. 612 provides that if a witness uses a writing to refresh his memory for the purpose of testifying, and if the court determines such use is necessary in the interests of justice, the adverse party is entitled to examine the writing at the hearing, use it for cross-examination, or introduce into evidence those portions that are related to the testimony. In *Massman-Johnson (Luling)*,[130] the Commission required previous statements of witnesses to be made available for cross-examination. While *Massman-Johnson (Luling)* makes no reference to Rule 612, the rule is not limited to statements of witnesses, but to any writing used to refresh the memory of any witness. For example, Rule 612 has been used for the production of a penalty worksheet used by a compliance officer witness.[131]

(e) Prior statements (Rule 613). Fed. R. Evid. 613 provides that a witness being examined concerning his prior statement need not be shown the statement or its content, although opposing counsel has a right to see it upon request. However, the admissibility of the prior statement should not be confused with the weight of the evidence. In *Ellanef Machine Tool Co.*,[132] the A.L.J. found that a prior statement to the police within two hours after a fatal accident was more reliable than a subsequent contradictory statement.

(f) Opinion testimony by lay witnesses (Rule 701). Fed. R. Evid. 701 permits opinion testimony of nonexpert witnesses which is rationally based on the perception of the witness and helpful to the clear understanding of the testimony of fact in issue. For example, in *Harrington Construction Corp.*[133] the testimony of a compliance officer, who was the Secretary of Labor's only witness, was not admitted because he had not been qualified as an expert on the subject on which he was asked to testify. According to an offer of proof, he would have testified as to the texture of the soil in a trenching case. With-

---

[130]8 OSHC 1369 (Rev. Comm'n 1980). See also *Pratt & Whitney Aircraft*, 9 OSHC 1653, 1657–1658 (Rev. Comm'n 1981) (A.L.J. erred in not requiring compliance officer's statement to be made available during cross-examination, but error not prejudicial).

[131]*Blakeslee-Midwest Prestressed Concrete Co.*, 5 OSHC 2036 (Rev. Comm'n 1977).

[132]6 OSHC 1853 (Rev. Comm'n 1978).

[133]4 OSHC 1471 (Rev. Comm'n 1976).

out reaching the question of whether the compliance officer was qualified as an expert, the Commission held that the excluded testimony was admissible under Rule 701 because it was helpful in the resolution of a material issue and was based upon the witness's personal knowledge. The Commission held that the compliance officer could render an opinion on the texture of the soil and could draw inferences as to its stability.

In another case, *Ray Evers Welding Co. v. OSHRC*,[134] a court of appeals was persuaded by the expert testimony of the employer's witnesses: an engineer with many years of experience in steel erection work; the company president, a man with much demonstrated experience in steel erection; and a former union officer. Weighed against this was the testimony of a compliance officer, not qualified as an expert, and who had no special familiarity with steel erection and whose education was in history.

(g) Testimony by experts (Rule 702). Under Fed. R. Evid. 702 opinion testimony of experts is allowed as to their scientific, technical, or otherwise special knowledge to understand the evidence. With respect to qualifying a witness as an expert, in *York Heel of Maine*,[135] the Commission reversed an A.L.J. ruling that a compliance officer was not an expert on machine guarding (a mechanical press) because he had neither an engineering degree nor mechanical engineering experience. The Commission noted that Rule 702 allows great flexibility in qualifying an expert. The witness had extensive experience in design drafting; was a member of several associations of safety professionals; was a graduate of a technical institute and had taken postgraduate courses in his field; and had much experience in inspecting a variety of mechanical presses.

Expert testimony, even if uncontradicted, may lack persuasiveness.[136] Without expert testimony on the elements of a case to assist the A.L.J., the Secretary may be unable to sustain his burden of proof.[137]

---

[134]625 F.2d 726, 8 OSHC 1271 (6th Cir. 1980).

[135]9 OSHC 1803 (Rev. Comm'n 1981).

[136]*Connecticut Natural Gas Corp.*, 6 OSHC 1796 (Rev. Comm'n 1978).

[137]See *Cape & Vineyard Div. v. OSHRC*, 512 F.2d 1148, 2 OSHC 1628 (1st Cir. 1975).

(h) Bases of opinion testimony by experts (Rule 703). Reflecting Fed. R. Evid. 703, the Commission indicated in *York Heel of Maine*[138] that as an expert, the witness need not base his testimony on personal knowledge, but may base it on facts "perceived or made known to him at or before the hearing." Concerning hypothetical questions, the Commission ruled in *Pipe-Rite Utilities* that if the facts that are hypothesized are not proved, the expert opinion is reduced to speculation.[139] In that case, which involved an alleged excavation violation, the expert's opinion on moving ground was based upon the presence of moist excavation walls, but other independent persuasive evidence indicated that the walls were dry.

(i) Opinion on ultimate issue (Rule 704). Fed. R. Evid. 704 allows otherwise admissible testimony even though it embraces an ultimate issue of fact. *York Heel of Maine*[140] illustrates an application of Rule 704. The Secretary's expert was permitted to render an opinion that the machine involved a "point-of-operation hazard," one of the ultimate issues of the case. Testimony in the form of opinions on ultimate issues is now common.[141] Obviously, ultimate issues in any case will depend on the charges, but normally such issues concern "recognized hazards," "serious physical harm," "substantial probability," or "reasonable diligence."

(j) Admissions (Rule 801(d)). It is common in OSHA cases to have testimony by a compliance officer concerning what employees told him about working conditions. Statements made by an employee to a compliance officer during an inspection are not hearsay under Fed. R. Evid. 801(d)(2) when the statements are about a matter within the scope of the employee's employment and are made during the existence of the employment relationship.[142] Also, an admission by an employer's

---

[138]*Supra* note 135.

[139]10 OSHC 1289 (Rev. Comm'n 1982).

[140]*Supra* note 135.

[141]See *Pipe-Rite Util., supra* note 139.

[142]*Astra Pharmaceutical Prods.*, 9 OSHC 2126, 2131 n.19 (Rev. Comm'n 1981), *aff'd*, 681 F.2d 69, 10 OSHC 1697 (1st Cir. 1982); *Power Sys. Div., United Technologies Corp.*, 9 OSHC 1813, 1817 (Rev. Comm'n 1981); *H-30, Inc.*, 5 OSHC 1715 (Rev. Comm'n 1977), *rev'd on other grounds*, 597 F.2d 234, 7 OSHC 1253 (10th Cir. 1979); *A.J. McNulty & Co.*, 4 OSHC 1097, 1099 (Rev. Comm'n 1976).

vice president to a compliance officer was ruled an admission by a party opponent under Rule 801(d)(2).[143]

(k) Hearsay rule (Rule 802). The test of Fed. R. Evid. 802 provides, among other things, that hearsay is not admissible except pursuant to an Act of Congress. However, the APA[144] does not preclude the use of hearsay evidence in formal administrative adjudications. Consequently, hearsay evidence is admissible in Commission proceedings and may be used as probative evidence, although the weight to be assigned to the evidence depends upon the degree of reliability.[145] The traditional hearsay exceptions may nevertheless be useful in assigning probative force to evidence.[146] For example, in *Power Systems Div., United Technologies Corp.,*[147] at the outset of the hearing the A.L.J. expressed doubt about the validity of the standard involved (provisions of the National Electrical Code) because of remarks of the Secretary of Labor reported in a labor relations publication that "many of the provisions of the National Electrical Code do not apply to worker's safety and health, because it is written for architects and engineers rather than employers and employees." The Commission ruled that the A.L.J. erred in assigning probative value to the reported remarks and that the reported remarks were properly characterized as hearsay. A more common situation concerns the weight to be assigned to a manufacturer's instructions or warnings that accompany a product. A manufacturer's warning has been ruled to be probative evidence as to the existence of a hazard.[148]

The Commission has also followed the "residuum" rule in regard to admittance of hearsay evidence in support of a factual finding.[149] However, the Commission has broadly con-

---

[143]*Prestressed Sys.*, 9 OSHC 1864 (Rev. Comm'n 1981). See also *Stephenson Enters.*, 4 OSHC 1702 (Rev. Comm'n 1976), *aff'd*, 578 F.2d 1021, 6 OSHC 1860 (5th Cir. 1978) (plant manager's admission during walkaround probative evidence admissible under Rule 801(d)(2)).

[144]5 U.S.C. §556.

[145]*Power Sys. Div., United Technologies Corp., supra* note 142; *Tri-City Constr. Co.*, 8 OSHC 1567, 1569 (Rev. Comm'n 1980); *Hurlock Roofing Co.*, 7 OSHC 1867, 1872–1873 (Rev. Comm'n 1979).

[146]Weinstein, *The Probative Force of Hearsay*, 146 IOWA L. REV. 331 (1961).

[147]*Supra* note 142.

[148]*Young Sales Corp.*, 7 OSHC 1297 (Rev. Comm'n 1979).

[149]*Paramount Plumbing & Heating Co.*, 5 OSHC 1459, 1461 (Rev. Comm'n 1977); *Metro-Mechanical*, 3 OSHC 1350 (Rev. Comm'n 1975); *B & K Paving Co.*, 2 OSHC 1173 (Rev. Comm'n 1974).

strued the exceptions to the hearsay rule, thereby narrowing greatly the residuum rule's use.[150]

(l) Exceptions to the hearsay rule. There are 24 exceptions under Fed. R. Evid. 803 to the general rule that hearsay is inadmissible. These exceptions do not depend on the availability of the declarant. Among them are *present sense impression and excited utterance* (Rules 803(1) and (2)), referring to statements of a witness describing an event or condition made while either perceiving the event or condition or shortly thereafter, as well as statements made while under stress or excitement to a startling event or condition. For example, a statement by an employer's foreman when questioned by the compliance officer that he had not worn a body belt in an aerial lift was considered admissible by the A.L.J. under Rules 803(1) and (2).[151]

Another kind of admissible hearsay is *records of regularly conducted activity* (Rule 803(6)), i.e., in general, regularly kept records of events, conditions, etc., made at or near the time of the event and kept in the normal practice of the business. One application of the rule in OSHA proceedings has been the admissibility of accident reports.[152]

There is also a "catchall" exception, *other exceptions* (Rule 803(24)), under which specific OSHA cases are lacking. However, in measuring the trustworthiness of statements it should be considered whether (1) the trier's evaluation of the statement would likely be different if cross-examination were available; (2) whether there is any testimonial inferiority in the statement with respect to sincerity, perception, and memory; and (3) whether the evidence is inferior because of a lack of cross-examination and a lack of opportunity to observe the demeanor of the statement giver.[153]

3. Privileges

As mentioned in Section I.F.1 above, privileges may be invoked to prevent discovery in prehearing proceedings. They

---

[150]*B & K Paving Co., supra* note 149.
[151]*Asplundh Tree Expert Co.*, 6 OSHC 1951 (Rev. Comm'n 1978).
[152]See *American Airlines*, 6 OSHC 1252 (Rev. Comm'n 1977).
[153]See Tribe, *Triangulating Hearsay*, 87 HARV. L. REV. 97 (1974).

may also be raised to reject evidence at trial. Examples include the following:

(a) Doctor-patient. An employer's argument that the names and social security numbers on employee health records were protected from disclosure by a doctor-patient privilege under state law was rejected.[154] The Commission noted that federal law did not recognize a physician-patient privilege and held that the federal law of privilege applied because federal substantive law supplied the rules of decision under the Act. However, as custodian of the medical records, the employer had standing in an enforcement action to assert the constitutional right of patient-employees to privacy in the disclosure of their medical records. But the right to privacy must be balanced against the competing interests of the enforcing agency under the federal statute involved.[155]

(b) Informer's privilege. The Commission indicated generally in *Pratt & Whitney Aircraft*[156] that the Secretary was not obligated to reveal in any material or testimony of his witnesses the identity of confidential informants.

(c) Financial information. In *West Point-Pepperell, Inc.,*[157] profit-and-loss statement for a textile mill asserted to be a trade secret was protected from disclosure. The Commission reasoned that in considering the economic feasibility of measures to control cotton dust, the disclosure as to the particular mill was not necessary, because a determination of feasibility would be based on the financial viability of the entire company rather than the mill.

(d) Attorney-client and work product. In *Wheeling-Pittsburgh Steel Corp.,*[158] photographs taken by an employer's agents at a workplace during an OSHA inspection were held to be not within the attorney-client privilege, since there was no intention that the photographs be considered confidential. The case also involved a assertion of the qualified privilege of

[154]*West Point-Pepperell, Inc.,* 9 OSHC 1784 (Rev. Comm'n 1981).

[155]See, e.g., *Whalen v. Roe,* 429 U.S. 589 (1977); *E.I. du Pont de Nemours & Co. v. Finklea,* 442 F. Supp. 821, 6 OSHC 1167 (S.D. W. Va. 1977).

[156]9 OSHC 1653, 1657 (Rev. Comm'n 1981). See also *Massman-Johnson (Luling),* 8 OSHC 1369 (Rev. Comm'n 1980).

[157]*West Point-Pepperell, Inc., supra* note 154.

[158]*Wheeling-Pittsburgh Steel Corp.,* 4 OSHC 1578 (Rev. Comm'n 1976).

"work product." Privilege was refused because the Secretary's photographs had not developed as a result of a malfunctioning camera and the photographs could not be retaken. Hence, there was a showing of "substantial need" for production.

D. Miscellaneous Issues

1. Oral Argument

As a matter of right, any party may request oral argument before the A.L.J. following the taking of evidence.[159] In practice, some parties waive the right to oral argument and rely exclusively upon briefs. Others, particularly small employers and local unions, often make oral argument and waive the right to file a brief.

2. Transcripts

All hearings are transcribed verbatim.[160] Copies of the transcript are available to the parties from the reporter at rates established by the agency. Daily copy of the transcript is usually available but at a higher cost. Under the Federal Advisory Committee Act,[161] the Commission and other agencies are required to make copies of the transcript available to any person at the actual cost of reproduction. The time for filing proposed findings of fact and conclusions of law runs from the availability of the transcript.

3. Proposed Findings of Fact and Conclusions of Law

Proposed findings of fact and conclusions of law may be filed within a reasonable period fixed by the A.L.J., but not exceeding 20 days from the availability of the transcript. There are no cases on whether the Commission's liberal interpretation of the broad power of the A.L.J. under Commission Rule 66 permits the A.L.J. to allow a greater period when reasonable.[162]

---

[159]Commission Rule 76.

[160]Commission Rule 65.

[161]5 U.S.C. app. 1, ¶11 (1982).

[162]Cf. *Carhar Contracting Co.*, 9 OSHC 1237 (Rev. Comm'n 1981) (A.L.J.s generally have authority to rule on procedural requests).

The A.L.J. may require, however, that proposed findings and conclusions be supported by precise citations to the record or legal authorities, as the case may be.[163] In proposing findings of fact on matters requiring credibility evaluation, the parties should consider the Commission's decisions requiring the A.L.J. to support findings with some specificity.[164]

Under *Universal Camera Corp. v. NLRB*,[165] an agency's findings must also be justified by "a fair estimate of the worth of the testimony of witnesses or its informed judgment on matters within its special competence, or both, after a review of the record as a whole." In short, there must be substantial evidence to support the conclusions reached. Under *Securities & Exchange Commission v. Chenery Corp.*,[166] the reasons for agency action must also be stated in order to permit judicial review.[167]

4. Form of Decision

Fed. R. Civ. P. 52 controls the form of the A.L.J.'s decisions since no specific Commission rule exists. The A.L.J. is to find the facts and state separately his conclusions of law. If an opinion or memorandum decision is filed, it is sufficient if the findings of fact and conclusions of law appear in the opinion or memorandum. In actual practice most A.L.J.'s do make separate and specific findings of fact and state separately their conclusions of law.

---

[163]ATTORNEY GENERAL'S MANUAL ON THE APA 85 (1947).

[164]*Asplundh Tree Expert Co.*, 7 OSHC 2074 (Rev. Comm'n 1979); *P & Z Co.*, 6 OSHC 1189 (Rev. Comm'n 1977); *Butler Lime & Cement Co.*, 5 OSHC 1370 (Rev. Comm'n 1977).

[165]340 U.S. 474 (1951).

[166]318 U.S. 80 (1943).

[167]See also *Brennan v. Gilles & Cotting, Inc.*, 504 F.2d 1255, 2 OSHC 1243 (4th Cir. 1974); *General Elec. Co. v. OSHRC*, 540 F.2d 67, 70 n.3, 4 OSHC 1512, 1515 (2d Cir. 1976).

15

Defenses

I. Overview

A. Nature of Defenses

The burden of proving that an employer violated the OSH Act rests with the Secretary.[1] To carry his burden, the Secretary must prove all essential elements of a violation.[2]

An employer's substantive duties are found in Section 5(a) of the Act. Section 5(a)(1), the general duty clause, requires each employer to furnish to its employees a place of employment free from recognized hazards that are causing or are likely to cause death or serious physical harm. Section 5(a)(2) requires that employers comply with the occupational safety and health standards promulgated under the Act.

To establish a violation of Section 5(a)(1) of the Act, the Secretary must prove

"(1) the employer failed to render its workplace free of a hazard, (2) the hazard was recognized either by the cited employer or generally within the employer's industry, (3) the hazard was

---

[1]*Mountain States Tel. & Tel. Co. v. OSHRC*, 623 F.2d 155, 8 OSHC 1577 (10th Cir. 1980); *Ocean Elec. Corp. v. Secretary of Labor*, 594 F.2d 396, 7 OSHC 1149 (4th Cir. 1979); *Brennan v. OSHRC (Alsea Lumber Co.)*, 511 F.2d 1139, 2 OSHC 1646 (9th Cir. 1975).

[2]*Mountain States Tel. & Tel. Co. v. OSHRC, supra* note 1.

causing or was likely to cause death or serious physical harm, and (4) there was a feasible means by which the employer could have eliminated or materially reduced the hazard."[3]

In order to established that an employer violated Section 5(a)(2) of the Act, the Secretary must show by a preponderance of the evidence that

> "(1) the cited standard applies, (2) there was a failure to comply with the cited standard, (3) employees had access to the violative condition, and (4) the cited employer either knew or could have known of the condition with the exercise of reasonable diligence."[4]

If the Secretary, during his case-in-chief, establishes a *prima facie* case of a violation, the employer can avoid liability either by rebutting an element of the Secretary's case[5] or by establishing a reason why the employer should be entitled to prevail despite the proven instance of noncompliance. It is convenient to distinguish between these two ways of avoiding liability by referring to the former as "rebuttal" and to the latter as "affirmative defenses." This chapter deals with both means of avoiding liability under the general category of defenses.[6]

Defenses generally fall into three categories: coverage, substantive, and procedural. Coverage issues involve whether the Act applies to the cited employer or to the working conditions that are the subject of the citation. An example is an exemption from the Act that exists because the cited working

---

[3]*Jones & Laughlin Steel Corp.*, 10 OSHC 1778, 1781 (Rev. Comm'n 1982). For further discussion, see Chapter 4 (The General Duty Clause), Section III.

[4]*Astra Pharmaceutical Prods.*, 9 OSHC 2126, 2129 (Rev. Comm'n 1981), *aff'd*, 681 F.2d 69, 10 OSHC 1697 (1st Cir. 1982).

[5]For example, in *Cargill, Inc., Nutrena Feed Div.*, 10 OSHC 1398 (Rev. Comm'n 1982), the Commission concluded that the employer had successfully rebutted the Secretary's showing of a feasible means of abatement in a §5(a)(1) case. See also *Royal Logging Co.*, 7 OSHC 1744, 1751 (Rev. Comm'n 1979), *aff'd*, 645 F.2d 822, 9 OSHC 1755 (9th Cir. 1981) (employer can rebut feasibility showing under §5(a)(1) by proving proposed means of abatement would produce greater hazards).

[6]There is not always a sharp distinction between "rebuttal" and "affirmative defenses." As will be discussed later in this chapter, the defense of unpreventable employee misconduct recognized by the Review Commission is closely tied to the question of an employer's actual or constructive knowledge, which is one of the elements the Secretary must prove to establish a violation of a standard. Similarly, although proof of the applicability of a standard is listed as an element of the Secretary's proof, the Commission may not consider such an issue if the employer does not raise it. See *Bechtel Power Corp.*, 4 OSHC 1005, 1010 n.13a (Rev. Comm'n 1976).

condition is regulated by another federal agency.[7] Substantive defenses involve matters directly related to the merits of the violations, e.g., whether noncompliance with a standard resulted from unpreventable employee misconduct.[8] Procedural defenses permit an employer to avoid liability because of the Secretary's failure to comply with procedural requirements in enforcing the Act, such as his failure to follow constitutional, statutory, and regulatory requirements in conducting inspections.[9] This chapter also discusses certain defenses that do not fit comfortably into these three categories.

B. When and How Defenses Must Be Raised

The Review Commission has designated certain defenses as affirmative defenses and has required that such defenses be raised in the answer or by way of amendment to the answer.[10] However, because pleadings are liberally construed and easily amended (see Chapter 14, Hearing Procedure), the Commission will generally consider an affirmative defense fairly presented during the trial of the case, regardless of whether the defense has been pleaded.[11] In extraordinary circumstances, the Commission will permit defenses to be raised for the first time on review.[12]

---

[7]See, e.g., *Northwest Airlines,* 8 OSHC 1982 (Rev. Comm'n 1980), *appeals dismissed,* No. 80-4218 (2d Cir. Feb. 18, 1981), No. 80-4222 (2d Cir. Mar. 31, 1981).

[8]See, e.g., *Texland Drilling Corp.,* 9 OSHC 1023 (Rev. Comm'n 1980).

[9]See, e.g., *Sarasota Concrete Co.,* 9 OSHC 1608 (Rev. Comm'n 1981), *aff'd,* 693 F.2d 1061, 11 OSHC 1001 (11th Cir. 1982).

[10]Commission Rule 36(b)(1), 29 C.F.R. §2200.36(b)(1) (1987); *General Motors Corp., Chevrolet Motor Div.,* 10 OSHC 1293 (Rev. Comm'n 1982).

[11]*Id.* See *Daniel Int'l Corp.,* 9 OSHC 1980 (Rev. Comm'n 1981), *rev'd on other grounds,* 683 F.2d 361, 10 OSHC 1890 (11th Cir. 1982) (affirmative defense of unpreventable employee misconduct tried by consent).

[12]See Commission Rule 92(d), 29 C.F.R. §2200.92(d). In *A. Prokosch & Sons Sheet Metal,* 8 OSHC 2077 (Rev. Comm'n 1980), the Commission considered the argument, raised for the first time on review, that a §5(a)(1) citation was inappropriate because of the applicability of a standard to the cited condition. The Commission noted that cases involving the same issue with respect to other employers on the same worksite had been disposed of on that basis and reasoned that the uniform application of the Act required it to consider the defense. In *B.J. Hughes, Inc.,* 10 OSHC 1545 (Rev. Comm'n 1982), the employer argued, for the first time on review, that the cited working condition was exempt from the Act pursuant to §4(b)(1). Although the Commission had previously labeled such a claim an affirmative defense, it remanded to permit the employer the opportunity to pursue the issue. However, in another case in which a §4(b)(1) argument was raised for the first time on review, the Commission refused to consider it, holding that the employer had waived the defense before the administrative law judge. *Allegheny Airlines,* 9 OSHC 1623 (Rev. Comm'n 1981), *rev'd sub nom. U.S. Air v. OSHRC,* 689 F.2d 1191, 10 OSHC 1721 (4th Cir. 1982). In reversing the Commission's decision, the Fourth Circuit held that preemption

In some cases, the Commission has held that it would not consider various defenses because they were not raised in a timely manner. In *Willamette Iron & Steel Co.*,[13] the employer argued that a citation was issued in violation of an operational agreement between the Secretary and the state of California. The Commission held that this issue did not relate to its subject matter jurisdiction and would therefore not be considered when raised for the first time on review. Other cases in which the Commission refused to consider issues first raised on review include *Gulf Stevedore Corp.*,[14] *Huber, Hunt & Nichols, Inc.*,[15] and *John F. Beasley Construction Co.*[16] In *River Terminal Railway*,[17] the Commission held that the argument that a standard was unenforceably vague was untimely when raised for the first time in the employer's posthearing submissions to the administrative law judge.

Thus, the failure to raise a defense in a timely manner may result in the Commission's refusal to consider the defense. However, in special circumstances, the Commission will consider the defense despite the lack of timeliness.

II. Coverage Defenses

A. "Affecting Commerce"

The Act imposes duties on "employers," and Section 3(5) defines "employer" as "a person engaged in a business affecting commerce who has employees, but does not include the United States or any State or political subdivision of a State."[18]

---

under §4(b)(1) is not an affirmative defense but is a jurisdictional limitation on OSHA's authority to issue a citation. Hence, the court concluded that the issue could be raised at any time. Accord *Columbia Gas v. Marshall*, 636 F.2d 913, 9 OSHC 1135 (3d Cir. 1980).

[13]9 OSHC 1900 (Rev. Comm'n 1981).

[14]5 OSHC 1625 (Rev. Comm'n 1977) (claim that inspection invalid because employer's senior representative at site did not accompany compliance officer).

[15]4 OSHC 1406 (Rev. Comm'n 1976) (validity of standards).

[16]2 OSHC 1086 (Rev. Comm'n 1974) (citation not issued with reasonable promptness).

[17]3 OSHC 1808 (Rev. Comm'n 1975).

[18]Although §3(2) refers to "employees," the Commission has held that an organization with one employee falls within the Act's coverage. *Poughkeepsie Yacht Club*, 7 OSHC 1725 (Rev. Comm'n 1979).

In a number of cases, organizations that received citations contended that they were not "employers" within the meaning of this definition. These contentions generally have proven unsuccessful.

In using the phrase "affecting commerce," Congress intended to exercise its full powers under the Commerce Clause of the Constitution.[19] Thus, virtually any nexus with interstate commerce has been sufficient for the courts and the Commission to find that an enterprise was engaged in a business affecting commerce. In *Avalotis Painting Co.*,[20] the Commission concluded that an employer's use of goods produced out of state established that the business affected commerce. A private yacht club was found to affect commerce because the club dispensed fuel for boats, members' boats were manufactured out of state, and the club was situated on a navigable waterway used by members to travel between two states.[21] The Ninth Circuit held that the clearing of land for the growing of grapes to be used in wine making affected commerce because the wine business was interstate in nature.[22] The same court subsequently held that an individual who hired approximately 40 workers for the construction of a 15-unit apartment building was engaged in a business affecting commerce as a matter of law because such an activity was within the class of activities that Congress sought to regulate.[23]

B. Political Subdivisions of States

Several cited organizations have contended that they were not employers within the meaning of Section 3(2) because they were political subdivisions of states. The University of Pittsburgh argued that it fell within this category because of its relationship to the state of Pennsylvania. Prior to 1966, the

---

[19]*Usery v. Franklin R. Lacy*, 628 F.2d 1226, 8 OSHC 2060 (9th Cir. 1980).

[20]9 OSHC 1226 (Rev. Comm'n 1981). See also *Atlanta Forming Co.*, 11 OSHC 1667 (1983). But see *Vak-Pak, Inc.*, 11 OSHC 2094 (Rev. Comm'n 1984) (Secretary failed to prove employer used goods purchased or manufactured out of state).

[21]*Poughkeepsie Yacht Club, supra* note 18.

[22]*Godwin v. OSHRC*, 540 F.2d 1013, 4 OSHC 1603 (9th Cir. 1976).

[23]*Usery v. Franklin R. Lacy, supra* note 19. Accord *Clarence M. Jones*, 11 OSHC 1529 (Rev. Comm'n 1983). But see *Austin Road Co. v. OSHRC*, 683 F.2d 905, 10 OSHC 1943 (5th Cir. 1982) (Secretary failed to prove contractor engaged in building residential streets, storm drains, sanitary sewers, and water transmission lines engaged in business affecting commerce).

Occupational Safety and Health Law

University of Pittsburgh was unquestionably a private institution. In that year, because the university had encountered financial problems, the state enacted a law declaring the university to be "state-related," granting the university substantial financial assistance, and providing for 12 of the university's 36 trustees to be appointed by state officials. Despite these actions, the Commission concluded that the university retained its "fundamental characteristics as a private institution of higher learning" and had not become "an instrumentality of the state."[24]

The Commission also rejected the argument that a firm which contracted with the state of New York to provide architectural and construction manager services for the construction of a state building was not an employer within the meaning of Section 3(2).[25] The Commission noted that such an employer was not an instrumentality of the state but was a private employer that had entered into an arm's length transaction with New York.[26]

C. Exemption Under Section 4(b)(1)[27]

Section 4(b)(1) of the Act provides that the Act does not apply to "working conditions" over which other federal agencies "exercise statutory authority to prescribe or enforce stan-

---

[24]*University of Pittsburgh*, 7 OSHC 2211, 2217–2219 (Rev. Comm'n 1980).

[25]*Bertrand Goldberg Assocs.*, 4 OSHC 1587 (Rev. Comm'n 1976).

[26]The applicability of the Act to an Indian tribal enterprise was at issue in *Navajo Forest Prods. Indus.*, 8 OSHC 2094 (Rev. Comm'n 1980), *aff'd*, 692 F.2d 709, 10 OSHC 2159 (10th Cir. 1982). Although Navajo Forest Products Industries was unquestionably engaged in a business affecting commerce, the Commission held that the Act did not apply to it because application of the Act would be inconsistent with an 1868 treaty between the Navajo Tribe and the United States. The treaty reserved the general right of sovereignty, or self-government, to the tribe, and Navajo Forest Products Industries had been established by the tribe for governmental purposes: to promote the tribe's general welfare and economic well-being. Therefore, the Commission concluded, applying the Act to the enterprise would be inconsistent with the treaty and there was no indication that Congress had intended for the Act to override existing Indian treaties. In affirming the Commission's decision, the Tenth Circuit stated that Indian tribes whose reservations were created by executive order as well as by treaty could exercise the inherent attributes of tribal sovereignty, including the right to exclude non-Indians from the reservation. The Commission subsequently relied on this reasoning in concluding that an Indian tribal enterprise operating on land established as reservation land by executive order was not subject to the Act. *Coeur d'Alene Tribal Farm*, 11 OSHC 1705 (Rev. Comm'n 1983), *rev'd*, 751 F.2d 1113, 12 OSHC 1169 (9th Cir. 1985).

[27]See also discussion in Chapter 27 (Other Federal Legislation and Regulation Affecting Workplace Safety and Health), Section I.

dards or regulations affecting occupational safety or health."
This section has been the subject of considerable controversy
and litigation.

1. *Statutory Purpose Underlying Other Agency's Rules*

The defense requires initially that an agency other than
the Secretary of Labor issue standards or regulations pursuant
to a statute that has as a policy or purpose the protection of
occupational safety or health. Thus, regulations issued under
the Wholesome Meat Act,[28] which is intended to protect consumers against tainted meat, do not preempt OSHA standards
even though they incidentally affect the safety and health of
workers in a meat packing plant.[29] However, it is not necessary that the other agency's enabling legislation or regulations have the exclusive purpose of protecting employees. It
is sufficient if the safety and health protection they seek to
provide includes the protection of employees while on the job.[30]
When the purpose of another agency's enabling legislation is
at issue, the Commission will give substantial deference to
the other agency's interpretation.[31]

2. *Extent of Exemption—"Working Conditions"*

The Section 4(b)(1) exemption extends to those "working
conditions" over which the other agency has actually exercised
its authority. In a series of early cases, employers in the railroad industry contended that their industry was entirely exempt from the Act because the Federal Railroad Administration
(FRA) possessed authority under the Federal Railway Safety

---

[28]21 U.S.C. §§601–695 (1982).

[29]*Fineberg Packing Co.*, 1 OSHC 1598 (Rev. Comm'n 1974). See also *Haas &
Haynie Corp.*, 4 OSHC 1911 (Rev. Comm'n 1976) (GSA regulations issued under
procurement statutes do not preempt OSHA standards); *Gearhart-Owen Indus.*, 2
OSHC 1568 (Rev. Comm'n 1975) (Department of Defense procurement regulations
do not preempt OSHA standards). Accord *Ensign-Bickford Co. v. OSHRC*, 717 F.2d
1419, 11 OSHC 1657 (D.C. Cir. 1983), *cert. denied*, 466 U.S. 937 (1984).

[30]*Northwest Airlines*, 8 OSHC 1982 (Rev. Comm'n 1980) (airline maintenance
workers within class of persons intended to be protected under Federal Aviation Act
of 1958); *Texas E. Transmission Corp.*, 3 OSHC 1601 (Rev. Comm'n 1975) (workers
in liquid natural gas storage facility within class protected by Natural Gas Pipeline
Safety Act of 1968); *Organized Migrants in Community Action v. Brennan*, 520 F.2d
1161, 3 OSHC 1566 (D.C. Cir. 1975) (Federal Environmental Pesticide Control Act
of 1972 encompasses farm worker exposure to pesticides).

[31]*Northwest Airlines, supra* note 30, at 1988–1989.

Act of 1970[32] to regulate safety and health throughout the industry and had issued some regulations pursuant to that authority. However, the citations issued by the Secretary concerned working conditions in railroad maintenance shops, and the FRA had not issued any regulations governing conditions in such shops. The Commission and the reviewing courts uniformly rejected the "industrywide exemption" argument, holding that an exemption only existed for those working conditions over which the other agency had actually exercised its authority.[33]

The cases, however, have not evidenced such universal agreement over what Section 4(b)(1) means by "working conditions." Although the courts rejected the railroad industry's contention that the term encompasses the entire industry, they also rejected the Secretary's argument that only those specific hazards regulated by the other agency were exempted, an argument sometimes referred to as the nook-and-cranny theory of regulation. The Fourth Circuit defined "working conditions" as the "environmental area in which an employee customarily goes about his daily tasks."[34] The Third Circuit has expressed agreement with this test.[35] The Fifth Circuit said that "working conditions" embraces both "surroundings" and "hazards."[36] The First Circuit has also adopted the Fifth Circuit's interpretation.[37]

Inasmuch as the FRA had not promulgated *any* regulations applicable to the maintenance shop areas that were the subject of many OSHA citations, both the environmental area test of the Fourth Circuit and the surroundings and hazards test of the Fifth Circuit led to the same result. Subsequently, however, the Commission was faced with cases in which the

---

[32]45 U.S.C. §§421–442 (1982).

[33]*Southern Pac. Transp. Co.*, 2 OSHC 1313 (Rev. Comm'n 1974), *aff'd*, 539 F.2d 386, 4 OSHC 1693 (5th Cir. 1976), *cert. denied*, 434 U.S. 874 (1977); *Southern Ry.*, 2 OSHC 1396 (Rev. Comm'n 1974), *aff'd*, 539 F.2d 335, 3 OSHC 1940 (4th Cir.), *cert. denied*, 429 U.S. 999 (1976); *Baltimore & O.R.R. v. OSHRC*, 548 F.2d 1052, 4 OSHC 1917 (D.C. Cir. 1976).

[34]*Southern Ry. v. OSHRC*, 539 F.2d 335, 339, 3 OSHC 1940 (4th Cir.), *cert. denied*, 429 U.S. 999 (1976). The court subsequently reaffirmed this holding in *U.S. Air v. OSHRC*, 689 F.2d 1191, 10 OSHC 1721 (4th Cir. 1982).

[35]*Columbia Gas v. Marshall*, 636 F.2d 913, 9 OSHC 1135 (3d Cir. 1980).

[36]*Southern Pac. Transp. Co. v. Usery*, 539 F.2d 386, 4 OSHC 1693 (5th Cir. 1976), *cert. denied*, 434 U.S. 874 (1977).

[37]*PBR, Inc. v. Secretary of Labor*, 643 F.2d 890, 9 OSHC 1721 (1st Cir. 1981).

choice of test did make a difference. In *Allegheny Airlines*,[38] the Commission expressed a preference for the surroundings and hazards interpretation. In that case, the airline was cited for maintaining locked exit doors in an airport terminal departure lounge. The employer contended that the OSHA citation was preempted by Federal Aviation Administration (FAA) regulations issued under the FAA Act of 1958[39] to prevent skyjacking and which required that access to the departure area from the lounge be limited. The Commission held that there was no preemption because the FAA regulations were directed at an entirely different hazard than was the OSHA citation. The Commission explicitly rejected the Fourth Circuit's environmental area test; but it noted that it did not disagree entirely with the Fourth Circuit's analysis, because the court had apparently contemplated a situation in which the other agency had adopted *comprehensive* regulations governing a certain environmental area.[40] The Commission hinted that it might find a particular environmental area entirely exempt in the face of another agency's comprehensive scheme of occupational safety and health regulation applicable to that area.[41]

A series of cases in which employers relied on Coast Guard regulations to support exemptions from the Act add another factor to be considered in the interpretation of Section 4(b)(1): whether the exemption depends on the job classification of the affected employees. The Coast Guard has general statutory authority to promote and regulate safety on vessels operating on navigable waters, except to the extent such authority is delegated by law to other agencies.[42] The Longshoremen's and Harbor Workers' Compensation Act[43] removed authority over longshoring activities from the Coast Guard and placed it with the Secretary of Labor. Thus, early Commission decisions held

---

[38]9 OSHC 1623 (Rev. Comm'n 1981), *rev'd sub nom. U.S. Air v. OSHRC*, 689 F.2d 1191, 10 OSHC 1721 (4th Cir. 1982).

[39]49 U.S.C. §§1301–1542 (1982).

[40]*Allegheny Airlines, supra* note 38, at 1629.

[41]The alleged violation in *Allegheny* occurred at JFK International Airport in New York, which is within the geographical confines of the Second Circuit. However, the employer was able to appeal the Commission's decision to the Fourth Circuit, because the employer's principal office is located there. Thus, the employer was ultimately able to have the citation vacated by the application of the Fourth Circuit's environmental area test. *U.S. Air v. OSHRC, supra* note 34.

[42]14 U.S.C. §2 (1982).

[43]33 U.S.C. §§901–950 (1976 & Supp V 1981).

that the OSH Act applied to a hazard affecting longshoremen despite a Coast Guard standard similar in scope and directed at the same hazard as the cited OSHA standard.[44] In a similar vein, where employees brought negligence suits to recover for injuries suffered while working on vessels on navigable waters, courts held that OSHA standards were directed exclusively at the protection of longshoremen and Coast Guard standards were directed exclusively at the protection of seamen.[45]

In *Puget Sound Tug & Barge*,[46] several employers claimed Section 4(b)(1) exemptions from citations alleging OSHA violations aboard vessels inspected and certificated by the Coast Guard. Certain of the employees affected by the alleged violations were seamen, while others were longshoremen and construction workers. The employers made two basic arguments: Relying on the cases holding that the Coast Guard exercises exclusive jurisdiction over seamen, they contended that the OSHA citations were preempted to the extent that they were directed at hazards affecting seamen. They also made the environmental area argument that the working conditions of all employees aboard vessels inspected and certificated by the Coast Guard were exempt from the Act because the Coast Guard was enforcing a reasonable and comprehensive program to promote health and safety on such vessels. The Commission viewed the "seamen" argument as seeking an industrywide exemption from the Act and rejected it on the basis of the decisions involving the railroad industry, holding that Section 4(b)(1) does not create industrywide exemptions. The Commission did not discuss whether preemption should be found because the Coast Guard comprehensively regulated safety and health in the "environmental area" of inspected vessels. The lack of discussion of this issue was particularly noteworthy in light of the Commission's suggestion in *Allegheny Airlines*, a case decided the day before *Puget Sound*, that an environmental area exemption might properly

[44]*California Stevedore & Ballast Co.*, 1 OSHC 1757 (Rev. Comm'n 1974). See also *T. Smith & Son*, 2 OSHC 1177 (Rev. Comm'n 1974).

[45]*Taylor v. Moore-McCormick Lines*, 621 F.2d 88, 8 OSHC 1277 (4th Cir. 1980) (longshoremen not within class of persons intended to be protected by Coast Guard regulations); *Clary v. Ocean Drilling & Exploration Co.*, 609 F.2d 1120, 7 OSHC 2209 (5th Cir. 1980) (seamen not within class of persons intended to be protected by OSHA standards).

[46]9 OSHC 1764 (Rev. Comm'n 1981), *appeals dismissed*, Nos. 81-7405 & 81-7406 (9th Cir. Mar. 21, 1983), No. 81-4243 (5th Cir. July 5, 1983).

be found if another agency's regulation of such an area was sufficiently comprehensive.[47]

In a dissenting opinion, Commissioner Cleary criticized what he termed the majority's adoption of the "nook-and-cranny" approach. He would have found that the citations directed at the working conditions of seamen were preempted because the Coast Guard comprehensively regulates their working conditions on vessels on navigable waters. He would have found no exemption, however, for the citations involving longshoremen and construction workers because those classes of workers do not fall within the Coast Guard's jurisdiction.[48]

Shortly after *Puget Sound* was decided, another case involving an OSHA citation directed at a hazard to seamen came before the Commission. In the meantime, there had been a membership change on the Commission, and the result in this new case was different. The newly appointed Chairman, Robert A. Rowland, joined Commissioner Cleary in holding that Coast Guard regulation of the working conditions of seamen aboard vessels on navigable waters created, in effect, an industrywide exemption from the Act.[49] To the extent it held otherwise, *Puget Sound* was overruled. However, the part of *Puget Sound* finding no exemption for the working conditions of longshoremen and construction workers was not at issue and was not overruled.[50]

---

[47]*Supra* note 38, 9 OSHC at 1629.

[48]*Puget Sound Tug & Barge, supra* note 46, at 9 OSHC 1782–1783 (Cleary, Comm'r, dissenting).

[49]*Dillingham Tug & Barge Corp.*, 10 OSHC 1859 (Rev. Comm'n 1982), *appeal dismissed*, No. 82-7552 (9th Cir. Nov. 4, 1982). The vessel in *Dillingham* was a towboat that was not subject to Coast Guard inspection and certification. The Commission did not consider the absence of Coast Guard inspection and certification significant, because the Coast Guard's statutory authority and certain of its regulations extended to uninspected vessels as well as inspected and certificated vessels. Subsequent to the decision in *Dillingham*, OSHA and the Coast Guard entered into a Memorandum of Understanding in which OSHA "concluded that it may not enforce the OSH Act with respect to the working conditions of seamen aboard inspected vessels." 48 FED. REG. 11,366 (1983). The Memorandum stated that nothing in it pertains to uninspected vessels. Subsequently, the Second Circuit disagreed with *Dillingham* and held that §4(b)(1) did not create an industrywide exemption for seamen aboard uninspected vessels. *Donovan v. Red Star Marine Servs.*, 739 F.2d 774, 11 OSHC 2049 (2d Cir. 1984), *cert. denied*, 470 U.S. 1003 (1985). The court held that the Secretary could enforce his noise standard aboard such vessels because the Coast Guard's statutory authority over uninspected vessels did not extend to noise hazards.

[50]The Memorandum of Understanding between OSHA and the Coast Guard, *supra* note 49, provides that OSHA has the authority to enforce the Act's antidiscrimination provision, 29 U.S.C. §660(c) (1982), with respect to seamen. The Fifth Circuit has rejected OSHA's attempt to exercise such authority, endorsing the Com-

3. Exercise of Authority by Other Agency

Another issue that has created difficulty is whether the other agency has "exercised" its authority in a manner that creates a Section 4(b)(1) exemption. In *Pennsuco Cement & Aggregates*,[51] the Secretary issued a citation following an investigation of an explosion in a cement kiln. It was undisputed that the Interior Department's Mining Enforcement and Safety Administration (MESA) had the statutory authority at that time to regulate kilns and had in fact adopted regulations applicable to the cited working conditions. However, MESA had temporarily suspended enforcement of those regulations pending the resolution of a dispute with OSHA over the extent of each agency's authority in the cement making industry. The Secretary contended that a Section 4(b)(1) exemption could not exist when the other agency was not actively enforcing its regulations. The Commission rejected the argument, holding that MESA had exercised its statutory authority through the promulgation of regulations and through enforcing those regulations both before and after the incident that led to Pennsuco being cited by OSHA. The Commission viewed MESA's temporary suspension of enforcement as an act of agency discretion whose wisdom was outside the scope of a permissible inquiry under Section 4(b)(1). The Commission further noted that the public had been given notice of the MESA regulations and of the applicability of those regulations to cement kilns but had not been given notice of the temporary suspension of MESA enforcement. Therefore, during the period when MESA enforcement was suspended, employers still had to comply with MESA requirements and had no reason to believe they also had to comply with OSHA requirements.[52]

In *Northwest Airlines*,[53] the Commission was faced with the question of whether the Federal Aviation Administration

---

mission's holding in *Dillingham Tug & Barge, supra* note 49, that the working conditions of seamen are entirely exempt from the Act. *Donovan v. Texaco, Inc.*, 720 F.2d 825, 11 OSHC 1721 (5th Cir. 1983). Because it was holding part of the Memorandum of Understanding invalid, the court said that no part of the Memorandum would be recognized as authoritative in the Fifth Circuit. 720 F.2d at 827–828 n.3, 11 OSHC at 1722–1723.

[51] 8 OSHC 1378 (Rev. Comm'n 1980).

[52] Authority over mine safety and health has been transferred from MESA to the Mine Safety and Health Administration of the Department of Labor. See Federal Mine Safety & Health Act of 1977, 30 U.S.C. §§801–960 (1982).

[53] 8 OSHC 1982 (Rev. Comm'n 1980).

had exercised authority to regulate the safety and health of ground personnel in the airline industry. The FAA had promulgated regulations requiring airlines to submit and comply with maintenance manuals for each type of aircraft they flew, and had established a procedure whereby an airline could change a provision in a manual by submitting the change to the FAA and having it become effective automatically if the FAA did not disapprove. Generally, the airlines would adopt as their manuals the maintenance manuals developed by the manufacturer of the aircraft, with whatever changes an airline believed were necessary to adapt the manual to its specific operations. In effect, then, the FAA regulations required the airlines to comply with the manufacturers' maintenance manuals but provided a procedure by which the airlines could make changes to the manuals. Moreover, although the manuals were primarily intended to assure that the airplanes were properly maintained so that they would be safe in flight, the manuals also contained numerous provisions intended to protect the safety of personnel performing maintenance on the planes. The Commission held that this scheme fulfilled the requirement of Section 4(b)(1) that the other agency exercise its statutory authority to promulgate "standards or regulations" affecting working conditions.[54] The Commission interpreted "standards or regulations" to mean rules having the force and effect of law, and concluded that the FAA regulatory scheme met this requirement because airlines could be penalized for failing to comply with provisions in the maintenance manuals.[55] The Commission further noted that the FAA's scheme was similar to OSHA regulations that required employers to comply with manufacturers' specifications or recommendations.[56]

The meaning of the phrase "standards or regulations" was also at issue in a series of cases involving the Consolidated

---

[54]*Id.* at 1989–1993.

[55]*Id.* at 1990 n.22.

[56]*Id.* at 1992 & nn.27 & 28. The California Supreme Court has concluded that the FAA's regulatory scheme does not preempt the California Division of Occupational Safety and Health from enforcing the state's occupational safety and health plan established under §18 of the Act, 29 U.S.C. §667, with respect to the working conditions of airline ground maintenance personnel. *United Air Lines v. Occupational Safety & Health Review Appeals Bd.*, 654 P.2d 157 (1982). The court noted that the preemption language in the California statute substantially differed from that in §4(b)(1) of the Act and relied on that difference in reaching its decision.

Rail Corp. (Conrail). After the decisions holding that the railroad industry was not totally exempt from the Act, the Federal Railroad Administration initiated a series of rulemaking actions ultimately intended to bring the industry entirely under FRA regulation. However, after it had issued notices of proposed rulemaking, the FRA determined that its resources did not permit such an endeavor, and it terminated the rulemaking proceedings. In doing so, it published a policy statement[57] which listed the OSHA standards that the FRA believed should and should not apply in the railroad industry. Among other things, the FRA said that OSHA regulations requiring guarding of open pits and ditches should not apply to inspection pits in locomotive or car repair facilities.

The Secretary subsequently cited Conrail for failing to guard inspection pits in a locomotive maintenance shop, and Conrail argued that the policy statement preempted the citation. The Commission rejected the argument, holding that the policy statement did not have the force and effect of law because it was not promulgated pursuant to the rulemaking procedures of the Administrative Procedure Act.[58] The Commission relied on its holding in *Northwest Airlines*[59] that "standards or regulations" under Section 4(b)(1) means rules having the force and effect of law. In a dissent, Commissioner Cleary argued that the formality required for rules having the force and effect of law should not be required when another agency seeks to displace an OSHA rule by the type of "negative exercise" represented by the FRA's policy statement.

Once again, the Commission's change in membership resulted in a dissent by Commissioner Cleary becoming the majority rule. In a subsequent Conrail case, the Commission reversed its earlier decision and held that OSHA standards are preempted to the extent that the FRA policy statement said that OSHA standards should not apply in the railroad industry.[60] Thus, under the second Conrail case, another agency can preempt the application of the OSH Act without the necessity of engaging in statutory rulemaking proceedings if it

[57]43 FED. REG. 10,583–590 (1978).

[58]*Consolidated Rail Corp.*, 9 OSHC 1258 (Rev. Comm'n 1981), *appeal dismissed*, No. 81-4192 (2d Cir. Sept. 17, 1982).

[59]*Supra* note 53.

[60]*Consolidated Rail Corp.*, 10 OSHC 1577 (Rev. Comm'n 1982), *appeal dismissed*, No. 82-3302 (3d Cir. Nov. 16, 1982).

has the statutory authority to regulate a particular subject and issues a formal statement that OSHA standards should not apply to that subject.

III. Substantive Defenses

A. Multiemployer Worksite Defense

Section 5(a)(1) of the Act places a duty on employers to provide their own employees with workplaces free from serious recognized hazards. The question arises under Section 5(a)(2) of the Act whether an employer's duty to "comply" with standards similarly runs only to the employer's own employees. This question becomes important when it is necessary to apportion liability for violations of standards when employees of more than one employer occupy the same worksite.

Multiemployer worksites are most common in the construction industry, and most, but not all, of the cases involving the liability of employers on multiemployer worksites involve construction work. In early cases, the Commission held that liability under Section 5(a)(2) paralleled that under Section 5(a)(1): each employer was responsible for the safety of its own employees and no others. Thus, an employer who created a hazardous condition was not responsible for the exposure of employees of another employer to the hazard.[61] Employers were, however, liable for the exposure of their own employees to hazardous conditions even if they did not create or control those conditions.[62]

Reviewing courts rejected the idea that liability under Section 5(a)(2) should be based solely on the employment relationship. In *Brennan v. OSHRC (Underhill Construction Corp.)*,[63] the employer stored material closer to the edge of a floor than permitted by a standard. This exposed workers on lower floors to the danger that the material would fall on them. However, the workers so endangered were employees of other employers. The Commission found that the employer was not

---

[61]*Martin Iron Works*, 2 OSHC 1063 (Rev. Comm'n 1974).
[62]*Robert E. Lee Plumbers*, 3 OSHC 1150 (Rev. Comm'n 1975).
[63]513 F.2d 1032 (2d Cir. 1975).

liable because its own employees were not shown to be exposed to the hazard. The Second Circuit reversed, holding that an employer who controlled an area and created a hazard within that area in contravention of a standard was responsible for the exposure to the hazard of any employees on the worksite regardless of whose employees they were.

In *Anning-Johnson Co. v. OSHRC*,[64] the Seventh Circuit was faced with a situation in which a subcontractor on a multiemployer construction site had been found liable by the Commission for certain violations it neither created nor controlled but to which its employees were exposed. The court rejected this basis for liability insofar as other than serious violations were involved. Comparing the language of Section 5(a)(2) with Section 5(a)(1), the court concluded that Congress intended to make employers responsible for the exposure of their employees to serious hazards but that the purposes of the Act would not be served by having employers held liable for the correction of nonserious conditions that lay within the control of other employers.

1. Elements of the Defense

The Commission reconsidered its early decisions in light of these court cases and, in two cases decided on the same day, adopted rules for apportioning liability on multiemployer construction sites that conformed, in most respects, to the decisions in *OSHRC (Underhill Construction Corp.)* and *Anning-Johnson Co.* The Commission held that each employer had primary responsibility for the safety of its own employees and would generally be held responsible for violations to which its employees were exposed, even if another employer was contractually responsible for providing the necessary protection.[65] However, the Commission created a two-pronged affirmative defense by which such an employer could avoid liability. An employer that did not create or control a violation could avoid liability by proving either that it took whatever

---

[64]516 F.2d 1081 (7th Cir. 1975).

[65]*Anning-Johnson Co.*, 4 OSHC 1193 (Rev. Comm'n 1976); *Grossman Steel & Aluminum Corp.*, 4 OSHC 1185 (Rev. Comm'n 1976). See *Central of Ga. R.R. v. OSHRC*, 576 F.2d 620, 6 OSHC 1784 (5th Cir. 1978); *Dun-Par Engineered Form Co.*, 8 OSHC 1044 (Rev. Comm'n 1980), *aff'd*, 676 F.2d 1333, 10 OSHC 1561 (10th Cir. 1982).

steps were reasonable under the circumstances to protect its employees against the hazard or that it lacked the expertise to recognize the condition as hazardous.[66] The Commission stated that these rules applied equally to serious and other than serious violations, rejecting the distinction between these classes of violation drawn by the Seventh Circuit in *Anning-Johnson Co. v. OSHRC.*[67]

Reviewing courts have approved the Commission's approach of holding noncreating and noncontrolling employers responsible for the safety of their own employees but permitting such employers to defend on the basis that they acted reasonably under the circumstances.[68]

2. Application of the Defense

One means of which noncreating and noncontrolling employers can protect their employees is to complain to the responsible contractor in an attempt to have the violative conditions corrected. They can also instruct their employees to avoid the areas where the hazards exist insofar as the employees' work assignments do not require them to be in such areas. A combination of such instructions coupled with reasonably zealous complaints to the responsible employer will generally establish the defense in the case of minor violations.[69] Where, however, employees are unavoidably exposed to a serious hazard over a long period of time and alternative physical means of protection are possible, even sincere and repeated requests to the responsible contractor will not be sufficient.[70]

Lewis & Lambert Metal Contractors[71] illustrates a situation in which complaints to the responsible contractor were

---

[66]*Anning-Johnson Co., supra* note 65; *Grossman Steel & Aluminum Corp., supra* note 65.

[67]See text accompanying note 64, *supra.*

[68]*Electric Smith, Inc. v. Secretary of Labor*, 666 F.2d 1267, 10 OSHC 1329 (9th Cir. 1982); *Bratton Corp. v. OSHRC*, 590 F.2d 273, 7 OSHC 1004 (8th Cir. 1979); *De Trae Enters. v. Secretary of Labor*, 645 F.2d 103, 9 OSHC 1425 (2d Cir. 1980).

[69]*Electric Smith, Inc. v. Secretary of Labor, supra* note 68; *Dutchess Mechanical Corp.*, 6 OSHC 1795 (Rev. Comm'n 1978).

[70]*Data Elec Co.*, 5 OSHC 1077 (Rev. Comm'n 1977). Cf. *Novak & Co.*, 11 OSHC 1783 (Rev. Comm'n 1984) (noncreating subcontractor not in violation when violative condition not present for enough time to permit subcontractor to complain to responsible contractor).

[71]12 OSHC 1026 (Rev. Comm'n 1984).

sufficient to establish the defense. The employer, a sheet metal subcontractor, was issued citations because of inadequate guardrail protection at various locations on the worksite. Due to union jurisdictional rules, the employer could not abate the hazard itself. The employer's foreman had, however, complained about inadequate guardrail protection on several occasions to the general contractor. The Commission concluded that, under the circumstances, complaining to the general contractor was the only realistic alternative available to the employer and that the complaints made were sufficiently persistent to establish the defense. The Commission also, however, found the employer in violation for failure to use ground-fault circuit interrupters. The Commission noted that the ground-fault devices were commercially available and easily installed, and that union jurisdictional rules would not preclude a sheet metal subcontractor from obtaining and installing the equipment. Thus, the employer had the ability to comply with the ground-fault standard.

In *Weisblatt Electric Co.*,[72] the employer was cited for exposure of its employees to certain violations it did not create or control. The employer proved that its employees had arrived at the worksite shortly before the OSHA compliance officer, and that their reason for being there was to check on conditions at the site. Also, the employer had previously taken its employees off the worksite on six or seven previous occasions owing in part to unsafe conditions, and had complained to the general contractor about those conditions. The Commission held that the employer's total course of conduct was reasonable to protect its employees and vacated the citations.

A noncreating, noncontrolling employer also may defend on the basis that it neither knew nor could have known, with the exercise of reasonable diligence, that a violative condition was hazardous. Thus, a subcontractor on a construction site may avoid liability for violations that lie outside its expertise and which it therefore cannot recognize as hazardous. Applying this defense, the Commission held that a plumbing contractor was not liable for nonobvious electrical hazards.[73]

---

[72]10 OSHC 1667 (Rev. Comm'n 1982).

[73]*4 G Plumbing & Heating*, 6 OSHC 1528 (Rev. Comm'n 1978). See also *A.A. Will Sand & Gravel Corp.*, 4 OSHC 1442 (Rev. Comm'n 1976).

3. Liability of the Controlling Employer

At the same time it established the multiemployer worksite defense, which is available to an employer whose employees are exposed to a hazard it does not create or control, the Commission adopted the holding of *Brennan v. OSHRC (Underhill Construction Corp.)* that an employer who creates or controls a hazard is liable even if its own employees are not exposed.[74] The Commission also held that a general contractor on a construction site is responsible for violations it controls by virtue of its supervisory authority over the site.[75] Thus, an employer that erected defective scaffolding on a construction site was found liable even though none of its own employees worked on the scaffold.[76] A general contractor was held liable for the lack of guardrails on a scaffold that was erected and used by a subcontractor because the general contractor could have had the subcontractor correct the condition.[77] In the same case, however, the general contractor was held not liable for a structural defect in the scaffold which it could not reasonably have detected. An employer who does not have the title "general contractor" but who nevertheless exercises supervisory control over a worksite will be held responsible for violations it could reasonably be expected to detect and prevent.[78] It is the functions an employer performs, and not its title, that determine liability under the Act.[79]

The early cases apportioning liability on multiemployer worksites involved construction sites. The Commission subsequently concluded that there was no valid distinction between construction sites and nonconstruction sites and held that all employers were responsible for hazards they created or controlled even if their own employees were not exposed.[80]

---

[74]*Grossman Steel & Aluminum Corp., supra* note 65, at 1188; *Anning-Johnson Co., supra* note 65, at 1199.

[75]*Id.*

[76]*Beatty Equip. Leasing,* 4 OSHC 1211 (Rev. Comm'n 1976), *aff'd,* 577 F.2d 534, 6 OSHC 1699 (9th Cir. 1978). See also *F.L. Heughes & Co.,* 11 OSHC 1391 (Rev. Comm'n 1983).

[77]*Knutson Constr. Co.,* 4 OSHC 1759 (Rev. Comm'n 1976), *aff'd,* 566 F.2d 596, 6 OSHC 1077 (8th Cir. 1977).

[78]*Red Lobster Inns of Am.,* 8 OSHC 1762, 1763 & n.3 (Rev. Comm'n 1980).

[79]See *Cauldwell-Wingate Corp.,* 6 OSHC 1619, 1621 (Rev. Comm'n 1978).

[80]*Harvey Workover, Inc.,* 7 OSHC 1687 (Rev. Comm'n 1979). See also *Plains Coop. Oil Mill,* 11 OSHC 1370 (Rev. Comm'n 1983).

The Commission has not explicitly extended the multiemployer worksite defenses to nonconstruction sites, but the rationale underlying the defense and the rejection of any distinction between construction and nonconstruction sites suggests that the defense would apply to nonconstruction sites.

4. *The Loaned Worker Situation*

Cases have arisen in which workers employed by one employer are assigned to work at a site controlled by another employer. For example, when a crane is leased, the company supplying the crane will sometimes furnish an operator as well.[81] The question arises whether the company who thus "loans" an employee to another employer retains responsibility for violations to which the loaned worker is exposed.

In general, if an employer who loans an employee retains control over the activities of the employee or if the violation is of the type that could be prevented by adequate training and supervision of the employee, the employer will be responsible for the employee's exposure to hazards. Thus, where the lessor of a crane also supplies an operator, that lessor company is responsible for deficiencies in the operation of the crane resulting from inadequate training of the operator.[82] Where the borrowing employer's purpose is to obtain the expertise of the employees, the loaning employer retains responsibility for the actions of the employees relevant to their area of expertise.[83] If the employees have no particular expertise and the borrowing employer controls their activities and is in the better position to provide for their safety, the loaning employer is not liable.[84] Moreover, as discussed later in this chapter, an employer is not responsible for violations of which it lacks actual or constructive knowledge, and the

---

[81]See *Frohlick Crane Serv.*, 2 OSHC 1011 (Rev. Comm'n 1974), *aff'd*, 521 F.2d 628, 3 OSHC 1431 (10th Cir. 1975).

[82]*Lidstrom, Inc.*, 4 OSHC 1041 (Rev. Comm'n 1976); *Frohlick Crane Serv., supra* note 81. Cf. *Sasser Elec. & Mfg. Co.*, 11 OSHC 2133 (Rev. Comm'n 1984), *aff'd*, 12 OSHC 1445 (4th Cir. 1985) (borrowing employer not responsible for crane operator's action in contacting power line).

[83]*Del-Mont Constr. Co.*, 9 OSHC 1703 (Rev. Comm'n 1981); *Sam Hall & Sons*, 8 OSHC 2176 (Rev. Comm'n 1980); *Gordon Constr. Co.*, 4 OSHC 1581 (Rev. Comm'n 1976). The loaning employer may, however, raise a multiemployer worksite defense with respect to hazards it did not create or control. *Gordon Construction Co.*, 4 OSHC at 1583.

[84]*MLB Indus.*, 12 OSHC 1525 (Rev. Comm'n 1985).

loaning employer may well lack such knowledge as to violations outside its area of expertise.[85]

B. Impossibility of Compliance

It is often stated that the OSH Act imposes on employers only duties that are achievable.[86] This principle suggests that an employer is not required to comply with the Act when to do so would prevent the employer from performing a task required by its business. Thus, an employer may defend on the basis that it is impossible to comply with a standard while performing required work. An early application of this principle occurred when the Commission vacated a citation alleging the absence of guardrails at the perimeter of an elevator shaft when the guardrails had been removed in order to erect the wall around the shaft.[87]

To establish the impossibility defense,[88] the employer must show not only that compliance was impossible, but that alternative means of protection either were being used or were unavailable.[89] Thus, an employer faced with a situation in which it is unable to comply with a standard cannot ignore

---

[85]In some cases, the Commission has analyzed "loaned employee" cases by determining whether an employer-employee relationship exists between the cited employer and the employee(s) exposed to the hazard. In making this determination, the Commission has looked to factors such as (1) whom the employee considers to be his or her employer; (2) who pays the employee's wages; (3) who is responsible for controlling the employee's activities; (4) who has the power, as opposed to the responsibility, to control the employee; and (5) who has the power to fire the employee or to modify the employee's employment conditions. *MLB Indus., supra* note 84, and cases cited therein. However, in view of the Commission's decisions holding employers that create or control hazards responsible for the exposure of any workers, and providing a defense for employers that do not create or control hazards, the importance of whether an employer-employee relationship exists is considerably reduced.

[86]See, e.g., *National Realty & Constr. Co. v. OSHRC*, 489 F.2d 1257, 1 OSHC 1422 (D.C. Cir. 1973).

[87]*W.B. Meredith, II, Inc.*, 1 OSHC 1782 (Rev. Comm'n 1974).

[88]The Commission labels impossibility of compliance an affirmative defense and places the burden on the employer to prove the defense. *General Motors Corp., Chevrolet Motor Div.*, 10 OSHC 1293 (Rev. Comm'n 1982). Insofar as specification standards are concerned, the courts are in agreement with this allocation of the burden of proof. See, e.g., *Quality Stamping Prods. v. OSHRC*, 709 F.2d 1093, 11 OSHC 1550 (6th Cir. 1983); *Ace Sheeting & Repair Co. v. OSHRC*, 555 F.2d 439, 5 OSHC 1589 (5th Cir. 1977). However, some courts have placed on the Secretary the burden of identifying and proving the feasibility of a means of compliance with a performance standard. *Modern Drop Forge Co. v. Secretary of Labor*, 683 F.2d 1105, 10 OSHC 1825 (7th Cir. 1982); *Ray Evers Welding Co. v. OSHRC*, 625 F.2d 726, 733, 8 OSHC 1271, 1275 (6th Cir. 1980); *Diebold, Inc. v. Marshall*, 585 F.2d 1327, 1333, 6 OSHC 2002, 2005–2006 (6th Cir. 1978); *Bristol Steel & Iron Works v. OSHRC*, 601 F.2d 717, 723–724, 7 OSHC 1462, 1466 (4th Cir. 1979).

[89]*M.J. Lee Constr. Co.*, 7 OSHC 1140 (Rev. Comm'n 1979).

the safety or health hazard to which noncompliance will expose its employees but must provide whatever protection it reasonably can. Such alternative protection might require the employer to perform its work in a manner different from that which renders compliance with the standard impossible.[90]

The Commission has stressed that the defense requires the employer to prove that compliance is functionally impossible, and that proof of impracticality or infeasibility is not sufficient.[91] The courts of appeals, however, have generally stated that an employer may defend on the basis that compliance is "infeasible."[92] However, in none of these court decisions was it found that the employer actually established that compliance was infeasible, so whether there is any practical difference between the defense of infeasibility discussed by the courts and the defense of impossibility recognized by the Commission remains uncertain. If compliance with a standard would be so expensive as to render compliance economically infeasible, the Commission might conclude that the employer's duty is not an achievable one and that the defense of impossibility has been established.[93]

In most cases in which the defense is raised, the employer's problem is more one of difficulty of compliance rather than impossibility. Thus, the Commission has generally not found the defense to be proven.[94]

In *M.J. Lee Construction Co.*,[95] the employer was cited for failing to shore or slope the walls of an excavation in which its employees were working. The employer proved that it could not slope the excavation because of the presence of a building

---

[90]*Cleveland Consol. v. OSHRC*, 649 F.2d 1160, 9 OSHC 2043 (5th Cir. 1981); *F.H. Lawson Co.*, 8 OSHC 1063 (Rev. Comm'n 1980).

[91]*Duane Smelser Roofing Co.*, 9 OSHC 1530 (Rev. Comm'n 1981); *Research-Cottrell, Inc.*, 9 OSHC 1489 (Rev. Comm'n 1981); *Dun- Par Engineered Form Co.*, 8 OSHC 1044 (Rev. Comm'n 1980), *aff'd*, 676 F.2d 1333, 10 OSHC 1561 (10th Cir. 1982).

[92]*Faultless Div., Bliss & Laughlin Indus. v. Secretary of Labor*, 674 F.2d 1177, 10 OSHC 1487 (7th Cir. 1982); *A.E. Burgess Leather Co. v. OSHRC*, 576 F.2d 948, 6 OSHC 1661 (1st Cir. 1978); *Ace Sheeting & Repair Co. v. OSHRC, supra* note 88; *Atlantic & Gulf Stevedores v. OSHRC*, 534 F.2d 541, 4 OSHC 1061 (3d Cir. 1976). But see *Cleveland Consol. v. OSHRC, supra* note 90; *PBR, Inc. v. Secretary of Labor*, 643 F.2d 890, 9 OSHC 1721 (1st Cir. 1981).

[93]See *Consolidated Aluminum Corp.*, 9 OSHC 1144, 1158 (Rev. Comm'n 1980).

[94]See, e.g., *Tube-Lok Prods.*, 9 OSHC 1369 (Rev. Comm'n 1981); *American Luggage Works*, 10 OSHC 1678 (Rev. Comm'n 1982); *Duane Smelser Roofing Co., supra* note 91.

[95]*Supra* note 89.

near one of the walls. Moreover, it could not shore the walls in strict compliance with the standard because the process of installing shoring would have destabilized the wall. The Commission nevertheless found that the impossibility defense was not established because an alternative means of protection, a system of piles, could have been used to partially support the excavation walls. The Commission stressed that an employer faced with a compliance problem cannot ignore the hazard but must provide whatever protection it can.[96]

The employer prevailed on the impossibility defense in *Alberici-Koch-Laumand*.[97] The citation involved employees working on steel beams while connecting structural steel. The employees wore safety belts, but they did not tie off the lanyards in circumstances where they needed to retain mobility in order to avoid sudden unpredictable movements of the beams. The employer used safety nets on the job where conditions permitted, but at the time of the alleged violation, nets could not have been used because a crane was lifting girders through the area the nets would occupy, and the crane could not be repositioned. The Commission found that the use of nets was impossible, and that employees would have been subjected to a greater hazard from moving beams if they had tied off their safety belts. The Commission further found that other means of protection could not have been used. Thus, the Commission vacated the citation on the basis that the employer had established a combined defense of impossibility and greater hazard.

A similar combined impossibility and greater hazard defense was raised in *H.S. Holtze Construction Co.*[98] The employer was erecting a wooden frame apartment building, and its employees were working on a floor 19½ feet above ground level that was not protected by a guardrail. The employees, however, were working at the interior of the structure, fabricating sections of the exterior walls. When a section was completed, it would be moved to the edge of the floor, raised into place, nailed down, and braced. Thus, under the construction method being used, the first work that was performed on each floor was the erection of the framework for the exterior

---

[96]See also *Cleveland Consol. v. OSHRC, supra* note 90.
[97]5 OSHC 1895 (Rev. Comm'n 1977).
[98]7 OSHC 1753 (Rev. Comm'n 1979).

walls. The employer contended that it was in the process of complying with the guardrail standard because the framework for the exterior walls was the equivalent of a guardrail and was erected before any other work was done on the floor. The employer also raised a greater hazard defense, arguing that employees would have had to spend more time near the perimeter of the floor to erect and dismantle guardrails than to raise the walls in the manner they were doing. The Commission rejected the employer's arguments, essentially finding that the employees could have been afforded fall protection of some type during the wall-raising operation.

The Commission's decision was reversed on appeal by the Eighth Circuit. The court essentially held that the combination of problems raised by the employer rendered further remedial measures unreasonable:

> "The Commission would require petitioner to build a guardrail along the 171-foot edge of a building under construction although the reason for the employees working on this third level is to erect an exterior wall which will serve as the functional equivalent of the standard guardrail. *** While we are mindful of the broad scope and remedial purpose of the Occupational Safety and Health Review [sic] Act, we are of the opinion that some modicum of reasonableness and common sense is implied. *** [S]ome demarcation line must be drawn between that which is genuinely aimed at the promotion of safety and health and that which, while directed at such aims, is so imprudent as to be unreasonable."[99]

In sum, an employer must prove considerably more than mere difficulty of compliance in order to establish the impossibility defense. The employer must demonstrate that it has not ignored the hazard to its employees that the standard is intended to protect against, but has provided whatever protection against the hazard is reasonable under the circumstances.[100] The defense does, however, provide a manner in which an employer who genuinely cannot literally comply with a standard can avoid liability.

---

[99]*H.S. Holtze Constr. Co. v. Marshall*, 627 F.2d 149, 151–152, 8 OSHC 1785, 1787 (8th Cir. 1980).

[100]See *PBR, Inc. v. Secretary of Labor, supra* note 92.

C. Greater Hazard

An employer is excused from strict compliance with a standard if compliance would result in a greater hazard to its employees than noncompliance. As with the impossibility defense, however, the defense of greater hazard requires that the employer show that alternative means of protection either were in use or were unavailable.[101] The employer also must show that a variance application under Section 6(d) of the Act is inappropriate.[102] This last requirement forces the employer to raise its greater hazard claim in a variance proceeding before exposing its employees to the hazard the standard is intended to address, rather than bypassing the variance procedure and waiting until an enforcement proceeding occurs to raise the claim.[103] The requirement that a variance application be inappropriate limits the defense to transitory situations, since a variance application will generally be appropriate if the violative condition is of a continuing or recurring nature.[104]

The defense often involves a contention that installing the means of protection required by a standard would expose employees to a hazard for a longer period of time than employees would otherwise be exposed to the hazard. This argument has been made when employees are exposed to fall hazards on a construction site for a relatively short period of time and when the act of installing guardrails, safety nets, or lifelines would expose employees to a fall hazard. As discussed in the previous section, the Eighth Circuit's decision in *H.S. Holtze Construction Co. v. Marshall*[105] was partly based on its acceptance of such an argument. In other cases, however, the contention has been rejected because the Commission or court

[101]*M.J. Lee Constr. Co.*, 7 OSHC 1140 (Rev. Comm'n 1979).

[102]*Id.; General Elec. Co. v. Secretary of Labor*, 576 F.2d 558, 6 OSHC 1541 (3d Cir. 1978).

[103]*General Elec. Co. v. Secretary of Labor, supra* note 102, at 561, 6 OSHC at 1542. See also *Modern Drop Forge Co. v. Secretary of Labor*, 683 F.2d 1105, 10 OSHC 1825 (7th Cir. 1982).

[104]See *Morton Bldgs.*, 7 OSHC 1792 (Rev. Comm'n 1979) (variance application would have been appropriate when hazard arose during standard construction technique). The rationale underlying the requirement that a variance application be inappropriate would appear to apply equally to the defense of impossibility of compliance. However, to date this requirement has not been imposed on the impossibility defense.

[105]*Supra* note 99.

was unpersuaded that the protection could not be installed safely or that it would indeed take longer to install a protective device than the length of time employees would be exposed to falling without the device.[106]

As with the impossibility defense, an employer may be required to change its method of operation in order to avoid a hazard that compliance would cause. Thus, where the greater hazard envisioned by the employer can be avoided by changes in the employer's work practices, the defense will fail.[107]

As discussed in the previous section on impossibility, the Commission upheld a combined defense of impossibility and greater hazard in *Alberici-Koch-Laumand*. In an earlier case, the Commission had also vacated a citation involving fall hazards to steel connectors because of the greater hazard associated with being unable to escape moving steel beams.[108] However, these decisions only provide a complete defense where no means of fall protection can be used without creating a greater hazard.[109]

D. Unpreventable Employee Misconduct

In numerous cases, employers have contended that violations of standards resulted from the failure of employees to obey instructions. Such a claim is often made in situations where compliance with a standard requires positive employee action, such as tying off a safety belt. Although Section 5(b) of the Act provides that employees shall comply with all of the Act's requirements applicable to their own actions and conduct, the Act contains no provision for enforcement against employees.[110] Thus, if the Act is to be enforced at all when violations result from derelictions of employees, it must be through citations issued to their employers.

---

[106]*United States Steel Corp. v. OSHRC*, 537 F.2d 780, 4 OSHC 1424 (3d Cir. 1976); *Martin-Tomlinson Roofing Co.*, 7 OSHC 2125 (Rev. Comm'n 1980); *Hurlock Roofing Co.*, 7 OSHC 1108 (Rev. Comm'n 1979).

[107]*Voegele Co. v. OSHRC*, 625 F.2d 1075, 8 OSHC 1631 (3d Cir. 1980); *Martin-Tomlinson Roofing Co., supra* note 106.

[108]*Industrial Steel Erectors*, 1 OSHC 1497 (Rev. Comm'n 1974).

[109]See *Holman Erection Co.*, 5 OSHC 2078 (Rev. Comm'n 1977).

[110]*Atlantic & Gulf Stevedores v. OSHRC*, 534 F.2d 541, 553, 4 OSHC 1061, 1069–1071 (3d Cir. 1976); *Nacirema Operating Co.*, 4 OSHC 1393, 1397–1398 (Rev. Comm'n 1976).

The Commission and courts have uniformly held that employers are not strictly liable for their employees' conduct, but that an employer has some duty to promote compliance by its employees. The Commission takes this problem into account by recognizing the affirmative defense of unpreventable employee misconduct, or, as it is sometimes called, the "isolated incident" defense. In order to establish this defense, the employer must show that the violation resulted from employee misconduct that violated a company work rule that was effectively communicated and uniformly enforced.[111]

Reviewing courts have generally agreed that an employer has the duty to establish, communicate, and enforce work rules in order to avoid liability for its employees' actions that violate standards. The Eighth Circuit has stated that an employer "cannot fail to properly train and supervise its employees and then hide behind its lack of knowledge concerning their dangerous working practices."[112] Several courts, however, have rejected the Commission's characterization of the issue as an affirmative defense, and have viewed the issue of safety training and supervision as bearing on an employer's knowledge of a violation, with the burden of proof on the Secretary to prove that the employer either knew or could have, with reasonable diligence, known of the violation.[113]

Regardless of where the burden of proof lies, the employer can avoid liability for the misconduct of its employees if it maintains an effective, ongoing safety program designed to achieve compliance with OSHA standards. In finding the employer was not liable in *Capital Electric Line Builders v. Marshall*, the court noted:

"It is undisputed that Capital Electric provided its employees with all the safety equipment required by the standard. The company regularly disseminated safety information to its employees, conducted weekly tailgate safety meetings at which the subject of 'covering up' was discussed, and held a tailgate meet-

[111]*Texland Drilling Corp.*, 9 OSHC 1023 (Rev. Comm'n 1980); *H.B. Zachry Co.*, 7 OSHC 2202 (Rev. Comm'n 1980), *aff'd*, 638 F.2d 812, 9 OSHC 1417 (5th Cir. 1981).

[112]*Danco Constr. Co. v. OSHRC*, 586 F.2d 1243, 1247, 6 OSHC 2039, 2041 (8th Cir 1978).

[113]*Pennsylvania Power & Light Co. v. OSHRC*, 737 F.2d 350, 11 OSHC 1985 (3d Cir. 1984); *Capital Elec. Line Builders v. Marshall*, 678 F.2d 128, 10 OSHC 1593 (10th Cir 1982); *Mountain States Tel. & Tel. Co. v. OSHRC*, 623 F.2d 155, 8 OSHC 1577 (10th Cir. 1980); *Ocean Elec. Corp. v. Secretary of Labor*, 594 F.2d 396, 7 OSHC 1149 (4th Cir. 1979); *Danco Constr. Co. v. OSHRC, supra* note 112; *Brennan v. Butler Lime & Cement Co.*, 520 F.2d 1011, 3 OSHC 1461 (7th Cir. 1975).

ing immediately preceding every job. Capital Electric enforces its safety program by reprimands, suspensions, and, if necessary, firings. *** Thus, this is not a case in which the employer took inadequate safety precautions by not supplying the required equipment or by not strictly requiring its use."[114]

Similarly, in a case where an employee working on an electric utility line failed to wear rubber insulating gloves, the Commission found the employer was not liable because

"The record reveals that Respondent's rubber glove rule was repeatedly and effectively communicated to employees, that violations of the rule were rare, that adequate means to discover such violations were taken, and that when violations were discovered, the rule was enforced through disciplinary measures. Under these circumstances, Respondent could not have prevented the violation and can therefore not be held responsible."[115]

Where the record reveals a pattern of violations over a period of time, the employer's safety program generally will be found inadequate even though work rules were established and employees knew of the rules.[116] On the other hand, infrequency of violations is strong evidence of an effective safety program.[117] Similarly, the stringency of the steps an employer must take to enforce its rules depends on the frequency with which violations are detected. Verbal reprimands will be sufficient if violations are isolated and infrequent.[118] Further disciplinary steps are required, however, if an employer knows that an employee continues to violate a work rule despite oral reprimands.[119]

Lack of knowledge of a work rule on the part of employees is strong evidence that the rule is not effectively communicated.[120] Moreover, a work rule must be at least as stringent as the requirements of the standard the employer is charged with violating for the defense to be established.[121] An em-

---

[114]*Supra* note 113, at 130, 10 OSHC at 1594.

[115]*Utilities Lines Constr. Co.*, 4 OSHC 1681, 1685 (Rev. Comm'n 1976).

[116]*Jensen Constr. Co.*, 7 OSHC 1477 (Rev. Comm'n 1979). See also *Daniel Int'l Corp.*, 9 OSHC 1980 (Rev. Comm'n 1981), *rev'd*, 683 F.2d 361, 10 OSHC 1890 (11th Cir. 1982).

[117]*Texland Drilling Corp., supra* note 111.

[118]*Asplundh Tree Expert Co.*, 7 OSHC 2074 (Rev. Comm'n 1979).

[119]*Wallace Roofing Co.*, 8 OSHC 1492 (Rev. Comm'n 1980).

[120]*New England Tel. & Tel. Co.*, 8 OSHC 1478, 1490 (Rev. Comm'n 1980).

[121]*Iowa Southern Util. Co.*, 5 OSHC 1138 (Rev. Comm'n 1977).

ployer cannot generally avoid liability on the basis that it did not know its employees would be exposed to certain hazards in the course of their work: "An employer must make a reasonable effort to anticipate hazards to which its employees may be exposed in the course of their scheduled work and must be given specific, appropriate instructions to prevent exposure to these hazards."[122]

E. Special Problems Under Section 5(a)(1)

In general, an employer's substantive duties are the same under Section 5(a)(1) (the general duty clause) as under Section 5(a)(2).[123] Differences exist, however, in the allocation of the burden of proof for certain issues. Under Section 5(a)(1), the Secretary must show a feasible means by which the employer could eliminate or materially reduce the hazard.[124] Factors such as the impossibility of using a particular means of protection or greater hazards that protective devices might create, which are regarded as defenses to Section 5(a)(2) violations, are relevant to the feasibility of the means of abatement under Section 5(a)(1).[125] This does not mean, however, that the Secretary must anticipate all problems his proposed means of abatement may cause. Once the Secretary proves the existence of an abatement method that would provide protection against the hazard, the burden "shifts to the employer to produce evidence showing or tending to show that use of the method or methods *** will cause consequences so adverse as to render their use infeasible."[126] The employer need not, however, meet the additional elements of its burden of proof under the impossibility and greater hazard defenses, such as the unavailability of other means of protection or the inappropriateness of a variance application.[127]

---

[122]*Merritt Elec. Co.*, 9 OSHC 2088, 2092 (Rev. Comm'n 1981). Cf. *Hogan Mechanical*, 6 OSHC 1221 (Rev. Comm'n 1979) (citation vacated because employer could not have anticipated its employee would be present at worksite where violations occurred).

[123]See *Western Mass. Elec. Co.*, 9 OSHC 1940, 1944–1945 (Rev. Comm'n 1981).

[124]*Jones & Laughlin Steel Corp.*, 10 OSHC 1778 (Rev. Comm'n 1982). And see discussion in Chapter 4 (The General Duty Clause), Section III.D.

[125]*Royal Logging Co.*, 7 OSHC 1744, 1751 (Rev. Comm'n 1979), *aff'd*, 645 F.2d 822, 9 OSHC 1755 (9th Cir 1981).

[126]*Id.; Cargill, Inc., Nutrena Feed Div.*, 10 OSHC 1398, 1401 (Rev. Comm'n 1982).

[127]*Royal Logging Co., supra* note 125, at 1751.

For some Section 5(a)(1) violations, the Secretary alleges that the feasible means to abate the hazard lies in improved training and supervision of employees. In such a case, the Secretary bears the burden of proving that improved training and supervision are feasible and useful methods of reducing the hazard.[128] Thus, where the Secretary alleges that the employer's safety program is inadequate to eliminate a recognized hazard, the Secretary must demonstrate the inadequacies in the program.[129] As discussed in the previous section on unpreventable employee misconduct, in cases arising under Section 5(a)(2) the employer has the burden of proving the adequacy of its safety program to establish the defense of unpreventable employee misconduct. Thus, the burden of proof with respect to the adequacy of an employer's safety program differs between Section 5(a)(1) and Section 5(a)(2) cases.

F. Invalidity of a Standard

OSHA standards have the force and effect of law and, like all laws, their validity may be challenged on the basis either that they were established according to improper procedures or that the requirements they seek to impose are unenforceable. The former challenges are referred to as procedural and the latter as substantive.

1. Procedural Invalidity

A standard does not have the force and effect of law unless the Secretary followed the procedural requirements of the OSH Act in promulgating the standard.[130] Thus, in an enforcement proceeding, the employer may contend that the standard under which it has been cited is unenforceable because of procedural defects in its promulgation. The Commission and the courts generally will entertain such challenges.[131]

---

[128]See *Champlin Petroleum Co. v. OSHRC*, 593 F.2d 647, 7 OSHC 1241 (5th Cir. 1979).

[129]See *General Dynamics Corp. v. OSHRC*, 599 F.2d 453, 458, 7 OSHC 1373 (1st Cir. 1979); *National Realty & Constr. Co. v. OSHRC*, 489 F.2d 1257, 1265–1266, 1 OSHC 1422, 1426–1427 (D.C. Cir. 1973); *Western Mass. Elec. Co., supra* note 123, at 1944–1945.

[130]See *Northwest Airlines*, 8 OSHC 1982, 1989 (Rev. Comm'n 1980).

[131]*Daniel Int'l Corp. v. OSHRC*, 656 F.2d 925, 9 OSHC 2102 (4th Cir. 1981); *Deering Milliken, Inc. v. OSHRC*, 630 F.2d 1094, 9 OSHC 1001 (5th Cir. 1980);

During the first two years of the Act's existence, Section 6(a) authorized the Secretary to issue national consensus standards and established federal standards as OSHA standards without conducting notice-and-comment rulemaking proceedings. The Secretary was not authorized to make any substantive changes in such standards when adopting them under this summary procedure. Therefore, the validity of Section 6(a) standards may be challenged on the basis that the Secretary made a substantive change to the source standard in promulgating it under the Act, such as by converting advisory language in the source standard into a mandatory requirement.[132] A Section 6(a) standard is not, however, invalid simply because its language differs from that of its source standard.[133] Changes in language that do not affect an employer's substantive duties are permissible.[134]

The Commission will not entertain a challenge to the validity of a standard based on an alleged procedural defect in the promulgation of the established federal standard that is the source of the OSHA standard.[135] In refusing to permit challenges to the procedural validity of such ancestor standards, the Commission noted "the public interests in finality and in avoiding the burden that continuous challenges would impose upon the Secretary's enforcement program and the Commission's adjudicating process."[136]

2. Substantive Invalidity

Challenges to the substantive validity of standards include contentions that standards are unenforceable either because they are overly vague or because the requirements they impose are arbitrary and capricious. In *Santa Fe Trail Transportation Co.*,[137] the Commission found unenforceably vague

---

Marshall v. Union Oil Co., 616 F.2d 1113, 8 OSHC 1169 (9th Cir. 1980); *Rockwell Int'l Corp.*, 9 OSHC 1092 (Rev. Comm'n 1980). But see *National Indus. Constructors v. OSHRC*, 583 F.2d 1048, 6 OSHC 1914 (8th Cir. 1978) (preenforcement proceeding under §6(f) is exclusive means for challenging procedural validity of standards).

[132]*Senco Prods.*, 10 OSHC 2091 (Rev. Comm'n 1982); *Kennecott Copper Corp.*, 4 OSHC 1400 (Rev. Comm'n 1976), *aff'd*, 577 F.2d 1113, 6 OSHC 1197 (10th Cir. 1977).

[133]*American Can Co.*, 10 OSHC 1305, 1310–1311 (Rev. Comm'n 1982).

[134]*Id.; Senco Prods., supra* note 132, at 2095.

[135]*American Can Co., supra* note 133, at 1309–1310; *General Motors Corp.*, 9 OSHC 1331 (Rev. Comm'n 1981).

[136]*General Motors Corp., supra* note 135, at 1337.

[137]1 OSHC 1457 (Rev. Comm'n 1974).

a standard[138] requiring that each worksite have a person or persons trained in first aid unless a hospital, infirmary, or clinic was in "near proximity" to the worksite. The Commission concluded that in the absence of evidence regarding customary practice or any other guidance to employers, the standard did not convey a "sufficiently definite warning" of the meaning of "near proximity." This decision was reversed by the Tenth Circuit.[139] The court held that regulations promulgated pursuant to remedial legislation must be construed in light of the conduct to which they are applied and, as so construed, the standard in question was not overly vague. The court stated that the words "near proximity" were given meaning because they were used in a context that emphasized the desirability of prompt assistance when an injury occurs. The court further noted that, in cases of certain severe injuries, first aid, to be effective, would have to be rendered within three minutes. Thus, the standard's purpose gave sufficient meaning to the term "near proximity" so that it was not unenforceably vague.

In *Ryder Truck Lines v. Brennan*,[140] the court rejected a challenge to the vagueness of 29 C.F.R. §1910.132(a), the general industry personal protective equipment standard.[141] The court stated:

> "The regulation appears to have been drafted with as much exactitude as possible in light of the myriad conceivable situations which could arise and which would be capable of causing injury. Moreover, we think inherent in that standard is an external and objective test, namely, whether or not a reasonable person would recognize a hazard of foot injuries *** which would warrant protective footwear. So long as the mandate affords a

---

[138]29 C.F.R. §1910.151(b).

[139]*Brennan v. OSHRC (Santa Fe Trail Transp. Co.)*, 505 F.2d 869, 2 OSHC 1274 (10th Cir. 1974).

[140]497 F.2d 230, 2 OSHC 1075 (5th Cir. 1974).

[141]29 C.F.R. §1910.132(a) provides:
"Protective equipment, including personal protective equipment for eyes, face, head, and extremities, protective clothing, respiratory devices, and protective shields and barriers, shall be provided, used and maintained in a sanitary and reliable condition wherever it is necessary by reason of hazards of processes or environment, chemical hazards, radiological hazards, or mechanical irritants encountered in a manner capable of causing injury or impairment in the function of any part of the body through absorption, inhalation or physical contact."
Similar claims of vagueness have been raised with respect to 29 C.F.R. §1926.28(a), the construction industry counterpart of §1910.132(a). See *S & H Riggers & Erectors v. OSHRC*, 659 F.2d 1273, 10 OSHC 1057 (5th Cir. 1981) and cases cited therein at n. 12.

reasonable warning of the proscribed conduct in light of common understanding and practices, it will pass constitutional muster."[142]

In *Kropp Forge Co. v. Secretary of Labor*,[143] the court declared unenforceably vague a standard requiring that "a continuing, effective hearing conservation program shall be administered."[144] The Secretary contended that the employer violated the standard by not administering a program containing annual audiometric tests, referral of employees to a physician, retests of employees with significant threshold shifts, selection and use of hearing protection, training of employees in use of hearing protectors, and enforcement of proper wearing of hearing protectors. The court stated that the standard afforded no warning to employers that hearing conservation programs had to include all of these elements, and further noted that certain OSHA publications, as well as the opinion of the compliance officer who had inspected the employer's plant, indicated that the standard did not require all of these elements. The court further noted that it was not the custom and practice in the forging industry to include all of these elements in a hearing conservation program.

In an enforcement proceeding an employer may challenge whether a standard is arbitrary and capricious, but beyond that cannot challenge the wisdom of a standard.[145] The burden is on the employer to show why the standard, as applied to it, is arbitrary, capricious, unreasonable, or contrary to law.[146] No standard has yet been declared invalid on this basis. The Commission has rejected the argument that the noise standard[147] is arbitrary and capricious in preferring administrative or engineering controls to personal protective equipment, stating that the employer "has not shouldered its heavy bur-

---

[142]497 F.2d at 233, 2 OSHC at 1077. The Fifth Circuit subsequently applied the same analysis in rejecting a vagueness challenge to the eye and face protection standard at 29 C.F.R. §1926.102(a)(1). *Vanco Constr. v. Donovan*, 723 F.2d 410, 11 OSHC 1772 (5th Cir. 1984). Accord *Cape & Vineyard Div. v. OSHRC*, 512 F.2d 1148, 2 OSHC 1628 (1st Cir. 1975); *Gold Kist, Inc.*, 7 OSHC 1855 (Rev. Comm'n 1979).

[143]657 F.2d 119, 9 OSHC 2133 (7th Cir. 1981).

[144]The standard was published at 29 C.F.R. §1910.95(b)(3). It has since been superseded by a new standard specifying in detail the requirements a hearing conservation program must have. See 29 C.F.R. §§1910.95(c)–(p) (1985).

[145]*Borg-Warner Corp.*, 6 OSHC 1939 (Rev. Comm'n 1978).

[146]*Atlantic & Gulf Stevedores v. OSHRC*, 534 F.2d 541, 551–552, 4 OSHC 1061, 1068 (3d Cir. 1976).

[147]29 C.F.R. §1910.95(b)(1) (1985).

den of proving this standard to be manifestly lacking in any rational justification."[148]

3. *Validity Challenges to Section 6(b) Standards*

Many standards promulgated under the notice-and-comment rulemaking procedures of Section 6(b) have been the subject of preenforcement validity challenges under Section 6(f) of the Act. Chapter 20 (Judicial Review of Standards) discusses many of these challenges. In general, it would appear that validity issues resolved in a Section 6(f) challenge could not later be raised in an enforcement proceeding.[149] The Commission has stated that it has the discretion to consider a validity challenge in an enforcement proceeding while a parallel challenge under Section 6(f) is pending in a court of appeals.[150] However, an employer may be barred from raising in an enforcement proceeding issues that could have been but were not raised in a Section 6(f) challenge.[151]

One of the most common validity challenges to health standards raised in Section 6(f) proceedings is to the Secretary's finding that the exposure limit established in a standard can feasibly be met by the affected industries.[152] In upholding the validity of the lead standard, the District of Columbia Circuit indicated that an employer could later challenge the feasibility of the standard in an enforcement proceeding even if the court upheld the Secretary's feasibility finding on preenforcement review.[153] The court viewed the Secretary's finding of feasibility during the standard setting process as establish-

---

[148]*Turner Co.,* 4 OSHC 1554, 1563 (Rev. Comm'n 1976), *rev'd on other grounds,* 561 F.2d 82, 5 OSHC 1790 (7th Cir. 1977) (footnote omitted).

[149]See *RSR Corp.,* 11 OSHC 1163, 1166 (Rev. Comm'n 1983), *aff'd,* 764 F.2d 355, 12 OSHC 1413 (5th Cir. 1985) (parties to enforcement proceeding agreed that resolution by courts of pending §6(f) challenge to validity of medical removal protection provision of lead standard was binding).

[150]*Phelps Dodge Corp.,* 11 OSHC 1441, 1445 n.7 (Rev. Comm'n 1983), *aff'd,* 725 F.2d 1237, 1769 (9th Cir. 1984), 11 OSHC 1769 (9th Cir. 1984).

[151]*RSR Corp. v. Donovan,* 747 F.2d 294, 302, 12 OSHC 1073, 1079 (5th Cir. 1984) (where employer participated in rulemaking and raised certain validity arguments in §6(f) proceeding, employer cannot later raise other validity challenges in enforcement proceeding where such challenges could have been raised in §6(f) proceeding).

[152]See *American Textile Mfrs. Inst. v. Donovan,* 452 U.S. 490, 9 OSHC 1913 (1981) (interpreting meaning of "feasible" in §6(b)(5)).

[153]*United Steelworkers v. Marshall,* 647 F.2d 1189, 1270, 8 OSHC 1810, 1868 (D.C. Cir. 1980), *cert. denied,* 453 U.S. 913 (1981).

ing a presumption of feasibility that an employer could later rebut in an enforcement proceeding if the compliance experience in the employer's industry demonstrated the standard's infeasibility. The court stressed, however, that any showing of economic infeasibility in an enforcement proceeding would have to encompass the experience of an entire industry, not only of the particular employer bringing the challenge.[154]

G. Exception to a Standard's Requirements

An employer may defend against an alleged violation of a standard on the basis that the employer's conduct falls within an exception to the standard's requirements. The burden lies with the employer to prove that it is entitled to the benefit of the exception.[155] Thus, where a standard generally required certain electrical tools to be grounded, but excepted such tools from the requirement if they were double insulated,[156] the Commission affirmed a citation where the evidence showed a lack of grounding and the employer failed to prove the tools were double insulated.[157]

Certain standards applicable to vehicles used for the transportation of ammonia contain an exception saying that the standards do not apply to "farm vehicles."[158] In *Durant Elevator*,[159] the Commission found that the employer proved that trailers it used to transport ammonia from a storage facility to farms fell within the exception, and vacated a citation alleging that the trailers did not have adequate brakes.

IV. Procedural Defenses

A. In General

Several of the defenses that allege that the Secretary failed to follow proper procedures are discussed in other chapters.

---

[154]*Id.*, 647 F.2d at 1270 & n.119, 8 OSHC at 1868.
[155]*Finnegan Constr. Co.*, 6 OSHC 1496 (Rev. Comm'n 1978).
[156]29 C.F.R. §1926.401(a) (1985).
[157]*Finnegan Constr. Co., supra* note 155.
[158]29 C.F.R. § 1910.111(f) (1986).
[159]8 OSHC 2187 (Rev. Comm'n 1980).

The most important such defense involves the failure of the Secretary to conform to constitutional and statutory requirements in conducting inspections of worksites. This is fully discussed in Chapter 7 (Inspections). Other procedural defenses—improper service of a citation, failure to issue a citation with reasonable promptness, and failure of a citation to state the alleged violation with particularity—are discussed in Chapter 8 (Citations).

B. Statute of Limitations

Section 9(c) of the Act provides that a citation may not be issued more than 6 months after the occurrence of a violation. In several cases, employers have argued that the Secretary violated this requirement. Most of these cases involve citations issued within the 6-month period, but where the Secretary seeks amendments outside that time. In *Bloomfield Mechanical Contracting v. OSHRC*,[160] the Secretary sought to amend a citation to name a joint venture as the cited employer in place of one of the companies that was a part of the joint venture. The court held that the amendment was improper because the statute of limitations had run, and there had therefore been no service of process upon the joint venture within the 6-month period. The court noted, however, that Rule 15(c) of the Federal Rules of Civil Procedure provides that amendments changing the name of the party would relate back to the date of the original pleading if certain conditions were satisfied, and the court remanded to the Commission to determine whether that rule applied in Commission proceedings and, if so, whether it should be applied to allow the amendment in this case.

Rule 15(c) also provides for the relation back of amendments when "the claim or defense asserted in the amended pleading arose out of the conduct, transaction, or occurrence set forth or attempted to be set forth in the original pleading." The Commission has applied this rule to permit amendments outside the 6-month period that change the standard an employer is charged with violating.[161]

---

[160]519 F.2d 1257, 3 OSHC 1403 (3d Cir. 1975).

[161]*Structural Painting Corp.*, 7 OSHC 1682 (Rev. Comm'n 1979). Accord *Southern Colo. Prestress Co. v. OSHRC*, 586 F.2d 1342, 6 OSHC 2032 (10th Cir. 1978).

In *Yelvington Welding Service*,[162] the employer was issued a citation more than 6 months after an employee suffered a fatal injury. The citation alleged two violations: failure to report the fatality to OSHA within 48 hours, and a violation related to the conditions that led to the fatality. The Commission held that the first violation was not barred by the statute of limitations because the failure to report the fatality did not end with the 48-hour period in which reporting was required, but was a continuing violation that existed until the Secretary learned of the fatality. The Commission also held that the second violation was not barred. Although acknowledging that the violation occurred outside the 6-month period, the Commission concluded that Congress did not intend the limit to apply when the Secretary's failure to discover a violation resulted from an employer's failure to report a fatality. The Commission stated: "To hold otherwise would reward an employer for violation of the Act in that by failing to report he would escape the attention of the agency designated to enforce the statute."[163] Where, however, the Secretary failed to issue a citation within 6 months of the time he learned all the facts surrounding the alleged violation, the Commission held that the citation was barred by the statute of limitations.[164]

V. Attacks on the Applicability of the Provision Allegedly Violated

A. Inapplicability of the Cited Standard

When the Secretary alleges that an employer violated a standard, the burden lies with the Secretary to prove that the standard applies to the facts of the case.[165] Employers have often been successful in challenging this aspect of the Secretary's case.

---

[162]6 OSHC 2013 (Rev. Comm'n 1978).

[163]*Id.* at 2016.

[164]*Sun Ship, Inc.*, 12 OSHC 1185 (Rev. Comm'n 1985).

[165]*United Geophysical Corp.*, 9 OSHC 2117 (Rev. Comm'n 1981), *aff'd mem.*, 683 F.2d 415, 10 OSHC 1936 (5th Cir. 1982).

1. Vertical and Horizontal Standards

OSHA standards generally can be categorized as either vertical standards, which apply to a particular industry or activity, or horizontal standards, which impose requirements that apply without regard to the industry or activity in which an employer is engaged. The nature of vertical standards limits their applicability to a particular industry or activity. An employer cited under such a standard may defend on the basis that it is not engaged in that industry or activity. In *B. J. Hughes, Inc.*,[166] an employer engaged in oil drilling operations was cited under a standard applicable to the construction industry. The Commission held that oil drilling was not construction work and the standard therefore did not apply. Similarly, an architectural and engineering firm that designed a building and inspected its construction to assure that the design specifications were being met was held to not be engaged in construction work.[167] However, a construction manager who performs no construction work itself but supervises the work of other contractors is subject to the construction standards.[168]

An employer cited under a horizontal standard may contend that the standard's applicability is preempted by a more specific vertical standard. In *New England Telephone & Telegraph Co.*,[169] the employer was cited under a general industry standard requiring that cranes maintain a distance of 10 feet from energized power lines. The employer contended that the general industry standard did not apply because a standard specifically applicable to telecommunications work required the maintenance of only a 3-foot clearance between cranes and power lines. The Commission agreed with the employer that the work in which it was engaged at the time of the alleged violation was subject to the telecommunications standard and concluded that the general industry standard therefore did not apply. Similarly, the Commission has held that

---

[166]10 OSHC 1545 (Rev. Comm'n 1982).

[167]*Skidmore, Owings & Merrill*, 5 OSHC 1762 (Rev. Comm'n 1977).

[168]*Bechtel Power Corp.*, 4 OSHC 1005 (Rev. Comm'n 1976), *aff'd*, 548 F.2d 248, 4 OSHC 1963 (8th Cir. 1977). See also *Dravo Corp.*, 10 OSHC 1651 (Rev. Comm'n 1982) (standards applicable to "shipbuilding" do not apply to land-based shop in shipyard).

[169]8 OSHC 1478 (Rev. Comm'n 1980).

an employer engaged in construction work cannot be cited under general industry standards where the hazard is addressed by construction standards.[170] However, a construction employer may be cited under a general industry standard where the specific hazard is not addressed by a construction standard.[171]

A controversy involving the requirements for fall protection during steel erection illustrates that it is not always certain whether a vertical standard addresses a specific hazard to an extent sufficient to preempt a horizontal standard. The standards in Subpart R of the construction standards, which apply to steel erection, contain certain provisions addressed to fall protection, but generally do not require protection for fall distances of less than 30 feet. Upon being cited under general construction standards when their employees were exposed to falls, employers contended that the Subpart R standards preempted the application of the more general standards. The Commission has accepted the argument.[172] The courts, however, have held that the more general standards are not preempted, stressing the inadequacy of the fall protection that would be provided to employees by the steel erection standards alone.[173]

2. Scope of Cited Standard

A standard that applies to an employer's industry may nonetheless not apply to the specific circumstances alleged to be in violation. Considerable litigation surrounded OSHA's attempt to require guardrails on roofs under a standard requiring guardrails around open-sided "floors." The Commission originally held that the standard applied, but its decision was reversed by the Fifth Circuit.[174] The court held that a

---

[170]*Daniel Constr. Co.*, 10 OSHC 1549 (Rev. Comm'n 1982).

[171]*Western Waterproofing Co.*, 7 OSHC 1499 (Rev. Comm'n 1979).

[172]*Adams Steel Erection*, 11 OSHC 2073 (Rev. Comm'n 1984), *rev'd*, 766 F.2d 804, 12 OSHC 1393 (3d Cir. 1985). In *Adams Steel*, the Commission overruled *Williams Enters.*, 11 OSHC 1410 (Rev. Comm'n 1983), *aff'd on this point*, 744 F.2d 170, 11 OSHC 2241 (D.C. Cir. 1984).

[173]*Brock v. L.R. Willson & Sons*, 773 F.2d 1377, 12 OSHC 1499 (D.C. Cir. 1985); *Donovan v. Adams Steel Erection*, 766 F.2d 804, 12 OSHC 1393 (3d Cir. 1985); *Donovan v. Daniel Marr & Son*, 763 F.2d 477, 12 OSHC 1361 (1st Cir. 1985); *Donovan v. Williams Enters.*, 744 F.2d 170, 11 OSHC 2241 (D.C. Cir. 1984); *Bristol Steel & Iron Works v. OSHRC*, 601 F.2d 717, 7 OSHC 1462 (4th Cir. 1979).

[174]*Diamond Roofing Co. v. OSHRC*, 528 F.2d 645, 4 OSHC 1001 (5th Cir. 1976), *rev'g S.D. Mullins Co.*, 1 OSHC 1364 (Rev. Comm'n 1973).

standard must afford fair warning to employers of the conduct the standard prohibits or requires, and should therefore "be construed to give effect to the natural and plain meaning of its words."[175] The court concluded that the natural and plain meaning of "floor" did not include "roof" and that the standard therefore did not apply to roofs. Ultimately, the Commission acquiesced in the court's ruling.[176]

The roofing cases illustrate that there is a certain tension between applying standards broadly so as to promote the remedial purpose of the Act and giving standards an interpretation that accords with the plain meaning of their words. Other cases illustrate the same dichotomy. In some cases, the remedial purpose of the Act prevails.[177] In other cases, the Commission places greater stress on the plain meaning of words or on giving the standard a reasonable interpretation.[178]

The question has arisen whether the applicability of a standard is limited by headings to the standard. In *Wray Electric Contracting*,[179] the cited standard[180] required employees working from an aerial lift to wear a body belt and lanyard. The standard was found in a section headed "Extensible and articulating boom platforms." The citation involved an employee in an aerial lift that was not an extensible and articulating boom platform. The Commission held that titles and headings can be used to resolve an ambiguity but cannot limit the plain meaning of a standard. It therefore held that the standard applied. In a later case, the Commission reaffirmed this holding but noted that titles and headings can be used to indicate or characterize the subject matter of a standard.[181]

---

[175]*Id.*, 528 F.2d at 649, 4 OSHC at 1004.

[176]See, e.g., *Dun-Par Engineered Form Co.*, 8 OSHC 1044 (Rev. Comm'n 1980), *aff'd*, 676 F.2d 1333, 10 OSHC 1561 (10th Cir. 1982).

[177]See, e.g., *Borton, Inc.*, 10 OSHC 1462 (Rev. Comm'n 1982), *rev'd*, 734 F.2d 508, 11 OSHC 1921 (10th Cir. 1984); *Seattle Crescent Container Serv.*, 7 OSHC 1895 (Rev. Comm'n 1979).

[178]See, e.g., *Bechtel Power Corp.*, 12 OSHC 1509 (Rev. Comm'n 1985), *aff'd*, 803 F.2d 999, 12 OSHC 2169 (9th Cir. 1986); *Globe Indus.*, 10 OSHC 1596 (Rev. Comm'n 1982); *Berglund-Cherne Gen. Contractors*, 10 OSHC 1644 (Rev. Comm'n 1982), *aff'd*, No. 82-1768 (10th Cir., Oct. 24, 1983).

[179]6 OSHC 1981 (Rev. Comm'n 1978), *aff'd*, 633 F.2d 220, 9 OSHC 1077 (6th Cir. 1980).

[180]29 C.F.R. §1926.556(b)(2)(v).

[181]*Everglades Sugar Ref.*, 7 OSHC 1410 (Rev. Comm'n 1979), *aff'd*, 658 F.2d 1076, 10 OSHC 1035 (5th Cir. 1981).

B. Preemption of Section 5(a)(1)

From the earliest days of the OSH Act, the Commission has held that Section 5(a)(1), the general duty clause, cannot properly be cited when a standard applies to the cited condition.[182] The rule, however, is more easily stated than applied. In some cases, the Commission has held that citation under Section 5(a)(1) is proper when a standard provides insufficient protection against a hazard. For example, the Commission affirmed a Section 5(a)(1) citation in which the Secretary sought to have employees vaccinated against anthrax despite also affirming a citation under 29 C.F.R. §1910.132(a), the general personal protective equipment standard, requiring employees to be protected against anthrax by means of respirators and protective clothing. The Commission stated: "[W]hen no specific standard entirely protects against the hazard alleged, citation under Section 5(a)(1) is proper."[183]

In other cases, the Commission's approach has been different. Where the Secretary sought to have an employer erect canopies at the entrances to a building under construction to protect employees against falling objects, the Commission held that the citation was preempted by a combination of standards designed to protect employees against falling objects even though the standards would not have provided complete protection to employees.[184] Similarly, the Commission vacated a Section 5(a)(1) citation in which the Secretary sought to have an area of a construction site at ground level barricaded against entry by employees because various tools and materials associated with overhead steel erection work could have fallen.[185] The Commission relied on the existence of specific standards designed to prevent certain objects, particularly the girders being erected, from falling and concluded that requiring further precautions under Section 5(a)(1) would circumvent the decisions the Secretary had made in rulemaking concerning the proper means of protection against the hazard.

---

[182]*Brisk Waterproofing Co.*, 1 OSHC 1263 (Rev. Comm'n 1973).

[183]*Peter Cooper Corp.*, 10 OSHC 1203, 1211 (Rev. Comm'n 1981). See also *Con Agra, Inc.*, 11 OSHC 1141 (Rev. Comm'n 1983); *Ted Wilkerson, Inc.*, 9 OSHC 2012 (Rev. Comm'n 1981).

[184]*John T. Brady & Co.*, 10 OSHC 1385 (Rev. Comm'n 1982), *rev'd*, No. 82-4082 (2d Cir., Oct. 14, 1982).

[185]*Daniel Int'l*, 10 OSHC 1556 (Rev. Comm'n 1982).

The rule that Section 5(a)(1) is preempted by a standard applies even when the standard is advisory and cannot serve as the basis for a citation.[186] The Commission noted that Congress intended that the objective of safe and healthful workplaces could best be achieved through the promulgation of specific standards and concluded that advisory standards serve the congressional purpose because they

> "represent the considered judgment of the Secretary, after receiving input from safety experts and persons who will be affected by the standards, of the proper means to guard against particular hazards. They promote the Act's objective by placing employers on notice of the steps they must take to provide safe and healthful workplaces to their employees."[187]

VI. Miscellaneous Defenses

A. *Res Judicata* and Collateral Estoppel

In general, the principles of *res judicata* and collateral estoppel apply to cases litigated under the Act.[188] These doctrines generally prevent a party from relitigating issues on which it lost in an earlier litigation. *Continental Can Co.*[189] illustrates a situation in which an employer was able to use *res judicata* and collateral estoppel as a defense to OSHA citations. The Secretary had issued a number of noise citations to Continental Can, alleging failure to install feasible engineering and administrative controls at a number of the company's can manufacturing facilities. The cases were consolidated, and the Commission ultimately held that the Secretary had failed to meet the burden of proving the feasibility of controls.[190] The Secretary continued to issue noise citations to Continental Can concerning other plants with similar noise problems and the company sought an injunction against the Secretary pursuing the enforcement actions with respect to the other plants. The Seventh Circuit held that the

---

[186]*A. Prokosch & Sons Sheet Metal*, 8 OSHC 2077 (Rev. Comm'n 1980).

[187]*Id.* at 2082.

[188]See *International Harvester Co. v. OSHRC*, 628 F.2d 982, 8 OSHC 1780 (7th Cir. 1980).

[189]*Continental Can Co., USA v. Marshall*, 603 F.2d 590, 7 OSHC 1521 (7th Cir. 1979).

[190]*Continental Can Co.*, 4 OSHC 1541 (Rev. Comm'n 1976).

subsequent enforcement actions were barred by *res judicata* and collateral estoppel and affirmed the district court's granting of the injunction.[191] The court concluded that the issues presented by the Secretary's continuing enforcement actions were precisely the same as had been presented in the earlier case and that the relevant issues had been fully litigated and resolved in the earlier case. The court also indicated that, if the Secretary could show that a particular citation involved a significantly different noise problem than had been litigated in the earlier case, the Secretary could pursue such a citation.

International Harvester Co. v. OSHRC[192] also involved citations alleging violations of the noise standard. In defending against a citation alleging failure to implement feasible engineering and administrative controls, International Harvester contended that an earlier citation, to which it had withdrawn its notice of contest, was *res judicata*. The court noted that no issues involving the earlier citation had been litigated on the merits and there was therefore "no *res judicata* effect in this case where the feasibility issue has been contested by Harvester and decided by the OSHRC on its merits after compilation of an extensive record."[193]

In *Consolidated Aluminum Corp.*,[194] the Secretary cited the employer for failing to guard certain machinery. The employer contended that the citation was barred by an earlier settlement agreement disposing of a similar citation. In the earlier agreement, the Secretary had withdrawn the citation in return for the company's promise to take certain abatement measures. In the subsequent citation, the Secretary contended that the precautions the company had agreed to take were inadequate and that the standard required further protective measures. The Commission held that the earlier agreement did not establish a binding abatement program because the citation had been withdrawn, thereby negating any abatement requirement. Accordingly the subsequent action was not barred. Although the issue was not presented in *Consolidated Aluminum*, the decision implies that a settlement agreement

---

[191]*Continental Can Co., USA v. Marshall*, 603 F.2d 590, 7 OSHC 1521 (7th Cir. 1979).

[192]*Supra* note 188.

[193]628 F.2d at 984, 8 OSHC at 1781.

[194]9 OSHC 1144 (Rev. Comm'n 1980).

in which a citation was affirmed would be binding on the Secretary in a later action.

B. Classification of Violation as *de Minimis*

When an employer violates an OSHA standard but the hazard resulting from the violation is minimal or nonexistent, the violation may be classified as *de minimis*. Such a classification is tantamount to the vacation of the citation, for a *de minimis* violation carries no penalty, need not be abated, and cannot be used as evidence of a history of violations for purposes of penalty assessments for future violations.[195] Because a *de minimis* finding carries no adverse consequence for the cited employer, the Commission will not normally permit an employer to seek review of an administrative law judge's finding of a *de minimis* violation.[196]

In *Alton Box Board Co.*,[197] the Commission found that an unenclosed chain and sprocket drive was a *de minimis* violation of a standard requiring all such mechanisms not more than 7 feet above the floor to be enclosed.[198] The particular drive in that case was underneath a machine and was recessed in such a manner that the danger of accidental contact was negligible. The absence of a railing on a stairway of five risers was found to be *de minimis* where the ground adjacent to the stairway was sloped alongside the stairway, thereby eliminating the hazard of a vertical drop from the open sides of the risers.[199]

The presence of unapproved electrical equipment near a spray booth was found to be *de minimis* in *Clifford B. Hannay & Son.*[200] The OSHA standard that was violated[201] had been drawn from the 1971 edition of the National Electrical Code, which OSHA had adopted as a national consensus standard. However, another version of the National Electrical Code, issued in 1975 but not adopted by OSHA, permitted the em-

---

[195]*Westburne Drilling, Inc.*, 5 OSHC 1457 (Rev. Comm'n 1977).
[196]*Id.*
[197]9 OSHC 1846 (Rev. Comm'n 1981).
[198]29 C.F.R. §1910.219(f)(3) (1985).
[199]*Continental Oil Co.*, 7 OSHC 1432 (Rev. Comm'n 1979).
[200]6 OSHC 1335 (Rev. Comm'n 1978).
[201]29 C.F.R. §1910.107(c)(6) (1985).

ployer's electrical equipment to be near a spray booth if certain other precautions, which the employer had taken, were maintained. The Commission noted that it was required to enforce the OSHA standard, not the 1975 National Electrical Code; but it concluded that the fact that the employer was in compliance with the 1975 code, coupled with expert testimony that the employer's installation presented no hazard, established the *de minimis* nature of the violation.

A *de minimis* classification is inappropriate where the likelihood of an accident is small but the consequences could be severe. In *Bethlehem Steel Corp.*, [202] the employer argued that its failure to have a 36-inch-high guardrail around a 15-foot-deep pit was *de minimis*. The pit was surrounded by a wall 23 to 36 inches high and 25 inches deep, and the employer contended that the wall would prevent a fall into the pit. The Commission concluded that the wall would not be as effective in preventing a fall as a 36-inch guardrail. The Commission noted that, at the time of the violation, there was snow on the ground and that this increased the likelihood that an employee could slip and fall over the wall. Similarly, the Commission rejected a *de minimis* classification for a violation involving the storage together of oxygen and acetylene cylinders. The hazard involved was the intensification of an existing fire if the two cylinders should simultaneously be exposed to a fire. Although such a possibility was slight, the Commission concluded that the hazard bore more than a negligible relationship to employee safety.[203]

---

[202]9 OSHC 2177 (Rev. Comm'n 1981), *aff'd mem.*, 688 F.2d 818, 10 OSHC 1824 (3d Cir. 1982).

[203]*Belger Cartage Serv.*, 7 OSHC 1233 (Rev. Comm'n 1979). A further discussion on the *de minimis* classification is found in Chapter 9 (Types and Degrees of Violations).

16

The Occupational Safety and Health Review Commission

I. Legislative History[1]

One of the principal issues that concerned Congress when the OSH Act was under consideration was whether the Act should follow the usual administrative model of vesting authority for rulemaking, enforcement, and adjudication[2] in a single agency, or whether these powers should be divided between three separate agencies. As often happens during the legislative process, Congress ultimately resolved the conflict by a series of compromises. The primary result was the creation of the Occupational Safety and Health Review Commission. The struggle to reach this compromise, however, provides essential insight into the role envisioned by Congress for the Review Commission.

A. In the Senate

The original bill introduced in the Senate by Senator Williams, S. 2193, gave authority for all three functions—rule-

---

[1] For the legislative history of the OSH Act as a whole, see Chapter 2.

[2] As used in this chapter, rulemaking means the establishment of substantive occupational safety and health standards; enforcement refers to the investigation and prosecution of alleged violations of the Act; and adjudication means the resolution of disputes arising out of such enforcement actions.

making, enforcement, and adjudication—to the Secretary of Labor.[3] This bill was considered by the Senate Committee on Labor and Public Welfare together with two other bills, S. 2788 and S. 4404, both of which contained provisions dividing the three functions between separate agencies. S. 2788, introduced by Senator Javits, would have given the Secretary of Labor enforcement authority only, with rulemaking and adjudicatory responsibilities lodged in a five-member National Occupational Safety and Health Board.[4] The other bill, S. 4404, introduced by Senator Dominick, contained a more far-reaching separation-of-powers proposal: enforcement powers would reside in the Secretary, standards-setting authority would be given to a five-member National Occupational Safety and Health Board, and the adjudicatory function would be assigned to a three-member Occupational Safety and Health Appeals Commission.[5]

The Committee on Labor and Public Welfare favorably reported S. 2193, the Williams bill, to the Senate floor. The Committee Report explained that a separate board for setting standards had been rejected because

> "the committee believes that a sounder program will result if responsibility for the formulation of rules is assigned to the same administrator who is also responsible for their enforcement and for seeing that they are workable and effective in their day-to-day application, thus permitting cohesive administration of a total program."[6]

The Committee gave the following reasons for rejecting the proposal for establishing a separate adjudicatory agency:

> "[S]ounder policy would be to place the responsibility and accountability for administration of the total program in the Secretary of Labor, rather than to establish a new agency and create an unnecessary division of responsibility. While the argument has been made that due process considerations would be better served if the investigative and adjudicative functions were separated between two different agencies, the fact is that the pro-

---

[3]S. 2193, 91st Cong., 2d Sess. §§3(b)–(f), 5, and 6(a)(1) (1970), *reprinted in* Subcommittee on Labor of the Senate Comm. on Labor and Public Welfare, 92d Cong., 1st Sess., LEGISLATIVE HISTORY OF THE OCCUPATIONAL SAFETY AND HEALTH ACT OF 1970, at 4–7, 10–12 (Comm. Print 1971) (hereafter cited as LEGIS. HIST.).

[4]S. 2788, 91st Cong., 2d Sess. §§4–7 (1970), *reprinted in* LEGIS. HIST. at 36–49.

[5]S. 4404, 91st Cong., 2d Sess. §§6 and 9–11 (1970), *reprinted in* LEGIS. HIST. at 80–86, 92–106.

[6]S. REP. NO. 1282, 91st Cong., 2d Sess. 8 (1970), *reprinted in* LEGIS. HIST. at 148.

visions of the Administrative Procedure Act insure that under the bill as reported by the committee there will be a separation of functions within the Department of Labor between those subordinates of the Secretary who are engaged in investigation and prosecution, and those who are engaged in adjudication. The overwhelming majority of other regulatory programs are administered in just this fashion, and the requirements of due process are fully observed."[7]

The proponents of independent standards-setting and adjudicatory panels carried their fight to the floor of the Senate. In attempting to substitute his own bill, S. 4404, for the Williams bill on the Senate floor, Senator Dominick argued in favor of the complete separation of the three powers of rulemaking, enforcement, and adjudication.[8] Senator Williams countered the Dominick argument by stressing the "time-honored" structure (all three powers in one agency) employed in his bill, S. 2193.[9] Senator Dominick's effort was defeated in a 41 to 39 vote.[10]

Senator Javits, however, continued to press for an independent adjudicatory body and also offered an amendment to that effect. Senator Javits stressed the increased confidence that the business community would have in the Act if adjudication were placed in a body independent of the Secretary of Labor:

> "The important thing is to inspire confidence in the community that we expect to obey this law ***. [T]he community will be considerably reassured in the difficult, and one might say dangerous situation, by the adoption of this amendment.
> ***
> "This is a situation which can disturb very seriously and be very costly to the business community. I feel very strongly that a great element of confidence will be restored in how this very new and very wide-reaching piece of legislation will be administered if the power to adjudicate violations is in the hands of an autonomous body, more than one man, and more than in the Department of Labor itself. *** We have a difficult piece of legislation reaching the whole of American business, involving millions of employees and tens of thousands of employers. This will give them a greater measure of confidence."[11]

---

[7]*Id.* at 15, LEGIS. HIST. at 155.
[8]Senate debate on OSH Act of 1970 (Nov. 16, 1970), LEGIS. HIST. 420.
[9]*Id.* at 435.
[10]*Id.* at 449.
[11]Senate debate on OSH Act of 1970 (Nov. 17, 1970), LEGIS. HIST. 469–470.

Speaking in support of Senator Javits's amendment, Senator Holland stressed the concern among employers that decisions by the Department of Labor would tend to favor organized labor in disputes between labor and management, adding: "[W]hen we are setting up a body to judge the controversies between the employers and the labor groups, we certainly should require the setting up of an agency that will be respected and is capable, impartial, and objective in its approach."[12]

The Senate adopted the Javits amendment by a vote of 43 to 38.[13] Thus, the bill ultimately adopted by the Senate placed authority for rulemaking and enforcement in the Secretary of Labor, but established an independent three-member panel, called the Occupational Safety and Health Review Commission, to adjudicate disputes arising out of the Secretary's enforcement actions.

B. In the House of Representatives

The proceedings in the House of Representatives largely paralleled those in the Senate with respect to the separation-of-powers issue. The House Committee on Education and Labor favorably reported a bill introduced by Representative Daniels, H.R. 16785, that vested rulemaking, enforcement, and adjudication authority in the Secretary of Labor.[14] The Committee considered and rejected two other bills: H.R. 13373, which was introduced by Representative Ayers on behalf of the administration, and H.R. 19200, introduced by Representative Steiger. H.R. 13373 proposed to establish an independent board that would have authority for both setting standards and adjudication,[15] while the Steiger bill would have established separate and independent agencies for these two purposes.[16]

---

[12]*Id.* at 476.

[13]*Id.* at 478–479.

[14]H.R. 16785, 91st Cong. 2d Sess. §§6, 7, and 11 (1970), *reprinted in* LEGIS. HIST. at 727–732, 739–742.

[15]H.R. 13373, 91st Cong., 2d Sess §§4, 5, and 7 (1970), *reprinted in* LEGIS. HIST. at 684–693, 695–696.

[16]H.R. 19200, 91st Cong., 2d Sess., §§6, 10, and 11 (1970), *reprinted in* LEGIS. HIST. at 770–776, 785–796.

On the floor of the House, Representative Steiger offered H.R. 19200 as a substitute (referred to as the Steiger-Sikes substitute) for the Committee bill. Although there were other differences between the two bills, Representative Steiger referred to the separation-of-powers provisions as the "most basic and most important difference."[17] As in the Senate, the relative virtues of vesting all authority in the Secretary of Labor as opposed to dividing the functions between different agencies were extensively debated. The proponents of the Steiger-Sikes substitute stressed the danger of too much concentration of power in the Secretary of Labor, particularly in view of the perceived pro-labor bias on the part of the Secretary.[18] The representatives who spoke in favor of H.R. 16785, the Committee bill, were primarily concerned that fragmenting the powers would unnecessarily complicate enforcement of the Act and lead to a lack of accountability.[19] When the question came to a vote, the Steiger-Sikes substitute prevailed by a vote of 220 to 172.[20]

Thus, the bills passed by both houses of Congress provided for a three-member independent agency to adjudicate disputes arising out of enforcement actions brought by the Secretary of Labor. The House bill also established an independent board to set standards, while the Senate version gave the power to establish standards to the Secretary.

C. Conference Committee

A conference committee was convened to resolve the differences between the two bills. The committee adopted the provision in the Senate bill that vested authority for setting standards in the Secretary of Labor rather than an independent board.[21] The committee retained, however, the provision contained in both bills for an independent adjudicatory agency

---

[17]House of Representatives debate on OSH Act of 1970 (Nov. 23, 1970), LEGIS. HIST. 989.

[18]*Id.* at 981 (Rep. Anderson); 991 (Rep. Steiger); 1014 (Rep. Scherle); 1050 (Rep. Michel).

[19]House of Representatives debate on OSH Act of 1970 (Nov. 24, 1970), LEGIS. HIST. 1074 (Rep. Perkins); 1079 (Rep. Pucinski); 1090–1091 (Rep. Randall).

[20]*Id.* at 1112–1113.

[21]H.R. REP. No. 1765 (Conference Report), 91st Cong., 2d Sess. 33 (1970), *reprinted in* LEGIS. HIST. at 1186.

and gave it the name that appeared in the Senate bill—the Occupational Safety and Health Review Commission.[22] Thus, the proponents of separation of powers achieved a partial victory, and a new independent agency—whose only function was to adjudicate contested enforcement actions under the Occupational Safety and Health Act—was born.

II. Review Commission's Structure and Function

A. Overview

The Commission's authority derives entirely from Sections 10(c) and 12 of the Act.[23] If an employer contests a citation, a notification of failure to abate, or a proposed penalty, or if an employee or a representative of employees contests the reasonableness of an abatement date contained in a citation, the Commission must afford the contesting party an opportunity for a hearing. According to Section 10(c), "The Commission shall [after providing the opportunity for a hearing] issue an order, based on findings of fact, affirming, modifying, or vacating the Secretary's citation or proposed penalty, or directing other appropriate relief ***."

Section 10(c) goes on to provide that, if an employer has made a good faith effort to comply with the abatement requirements of a citation but abatement has not been completed owing to factors beyond the employer's reasonable control, "the *Secretary*, after an opportunity for a hearing *** shall issue an order affirming or modifying abatement requirements in such citation." (Emphasis added.) The Review Commission, relying on the legislative history showing that Congress created the Commission for the purpose of resolving disputed issues arising out of the enforcement actions brought by the Secretary, held that this use of the word "Secretary" in Section 10(c) was an inadvertent mistake and that the Commission has jurisdiction over petitions by employers for modification of the abatement period.[24] The Secretary apparently agrees,

---

[22]*Id.* at 39, Legis. Hist. at 1192.

[23]29 U.S.C. §§659(c) and 661 (1982).

[24]*H.K. Porter Co.*, 1 OSHC 1600 (Rev. Comm'n 1974).

for he has never challenged the Commission's jurisdiction over such petitions.[25]

The Commission is composed of three members who are appointed by the President with the advice and consent of the Senate, with one member designated by the President to be chairman.[26] The members serve six-year staggered terms, so that a vacancy occurs every two years.[27] Two members of the Commission constitute a quorum, and the Commission can take official action only on the affirmative vote of at least two members.[28] The Review Commission is specifically authorized to adopt rules of procedure.[29] Unless it has adopted a different rule, the Federal Rules of Civil Procedure apply to its proceedings.[30]

Any proceeding before the Commission must initially be heard and decided by an administrative law judge (A.L.J.).[31] The A.L.J.'s decision becomes the final order of the Review Commission unless, within 30 days of the docketing of the decision, a Commission member directs that the decision be reviewed.[32] Accordingly, a party aggrieved by an A.L.J.'s decision cannot obtain Commission review as a matter of right. A final order of the Commission, however, whether resulting from a decision by the Review Commission or from an A.L.J.'s decision that becomes final with Commission review, is appealable to an appropriate U.S. court of appeals.[33]

---

[25]See *ITT Grinnell Corp. v. Donovan*, 744 F.2d 344, 345–346, 11 OSHC 2257, 2258 (3d Cir. 1984) (Commission's authority to resolve petition for modification of abatement not challenged and presumably Commission has such authority). The Commission's current Rule 37, 29 C.F.R. §2200.37 (1987), which governs petitions for modification of the abatement period, provides that the Secretary can grant such a petition if neither he nor the affected employees object to the requested extension. Any dispute arising out of such a petition, however, must be transmitted to the Commission for resolution. See *Gilbert Mfg. Co.*, 7 OSHC 1611 (Rev. Comm'n 1979).

[26]OSH Act §12(a), 29 U.S.C. §661(a).

[27]OSH Act §12(b), 29 U.S.C. §661(b).

[28]OSH Act §12(f), 29 U.S.C. §661(f).

[29]OSH Act §12(g), 29 U.S.C. §661(g). See also discussion in Chapter 5 (Employer Obligations to Obtain, Maintain, and Disseminate Information), Section VII.

[30]*Id.*

[31]OSH Act §12(j), 29 U.S.C. §661(j).

[32]*Id.*

[33]OSH Act §11, 29 U.S.C. §660. A party other than the Secretary may appeal either in the circuit where the violation allegedly occurred or where the employer has its principal place of business, or in the District of Columbia Circuit. OSH Act §11(a), 29 U.S.C. §660(a). The Secretary may seek review or enforcement of a Commission order only in the circuit where the alleged violation occurred or where the employer has its principal place of business. OSH Act §11(b), 29 U.S.C. §660(b). Parties should note that a failure to seek Commission review may foreclose the

B. Decision of an Administrative Law Judge

1. Procedure for Issuance

The procedure by which an A.L.J.'s decision is issued has changed over the years. The current procedure under Commission Rule 90[34] is as follows: The A.L.J. mails copies of the decision to the parties but does not mail it to the Review Commission for docketing. The parties are informed that the decision will be filed with the Commission within 20 days unless the A.L.J. changes the decision. The A.L.J. then submits his final decision to the Commission's Executive Secretary, who dockets the decision, mails a notice of docketing to the parties, and circulates copies to the Commissioners. The docketing date is the date that begins the 30-day discretionary review period. The 20-day delay between the time the A.L.J. first mails his decision to the parties and the actual docketing of the decision with the Commission does not unlawfully extend the statutory 30-day review period.[35] After docketing and until the A.L.J.'s decision is directed for review or becomes a final order, the A.L.J. has authority to correct clerical errors and errors arising through oversight or inadvertence.[36]

2. Adequacy of the A.L.J.'s Decision

The A.L.J.'s decision must conform to Section 8 of the Administrative Procedure Act[37] (APA) even in simplified proceedings cases.[38] Section 8(b) of the APA states that judges' decisions are part of the record, and requires that they include "a statement of *** findings and conclusions, and the reasons

---

possibility of seeking judicial review unless there are extraordinary circumstances excusing the failure. *Keystone Roofing Co. v. OSHRC*, 539 F.2d 960, 4 OSHC 1481 (3d Cir. 1976), *cited in* Commission Rule 91(f), 29 C.F.R. §2200.91(f) (1987).

[34]29 C.F.R. §§2200.90 and 2200.91. The Commission's Rules of Procedure are identified by numbers and letters corresponding to the location of each rule within 29 C.F.R. §2200.1–211 (1987). Hereafter the Commission rules will be cited without reference to 29 C.F.R. pt. 2200.

[35]*Gulf & W. Food Prods. Co.*, 4 OSHC 1436 (Rev. Comm'n 1976); *Robert W. Setterlin & Sons*, 4 OSHC 1214 (Rev. Comm'n 1976); *Northwest Airlines*, 8 OSHC 1982, 1983 n.5 (Rev. Comm'n 1980); *H.S. Holtze Constr. Co. v. Marshall*, 627 F.2d 149, 8 OSHC 1785 (9th Cir. 1980), *approving in pertinent part* 7 OSHC 1753, 1755 (Rev. Comm'n 1979). But see *Herriott Printing Co.*, 2 OSHC 1702 (Rev. Comm'n 1975) (Cleary, Comm'r, concurring) (criticizing rules).

[36]Commission Rule 90(b)(3).

[37]5 U.S.C. §557 (1982).

[38]Commission Rules 67(i), 90(a), and 209(a).

or basis therefor, on all the material issues of fact, law or discretion represented on the record."[39]

The Review Commission has set aside decisions that do not comply with these requirements.[40] It has required that the A.L.J. make findings on subsidiary factual issues as well as "ultimate" factual issues,[41] that the A.L.J. consider the entire record in making a decision,[42] and that the A.L.J.'s decision "show on its face what evidence has been considered in reaching his findings and conclusions."[43] Furthermore, credibility evaluations must be adequately explained; the A.L.J. should identify conflicting testimony and give reasons for crediting one witness over another, or for rejecting the unimpeached, uncontradicted testimony of a witness.[44]

If the A.L.J.'s decision is inadequate, the Review Commission may either remand it to the judge for the making of a new decision or decide the case itself and enter the necessary findings, conclusions, and reasons.[45] The Commission has stated that it is a sounder practice to let the A.L.J. make any credibility evaluation in the first instance.[46] Therefore, the Review Commission generally remands if a credibility evaluation is necessary, although it may decide a credibility question if it considers the answer to be clear.[47]

3. Unreviewed A.L.J.'s Decision

An unreviewed decision of an A.L.J. is binding on the parties as a final order of the Review Commission, but it is not a precedent that must be followed by the Commission or its other judges.[48] Unreviewed decisions, however, may be

---

[39]5 U.S.C. §557(c).

[40]E.g., *Syntron, Inc.*, 10 OSHC 1848 (Rev. Comm'n 1982) (lack of findings and reasons); *Thunderbolt Drilling*, 10 OSHC 1981 (Rev. Comm'n 1982) (recitation of conflicting evidence; unexplained resolution).

[41]*P & Z Co.*, 6 OSHC 1189 (Rev. Comm'n 1977); *Evansville Materials*, 3 OSHC 1741 (Rev. Comm'n 1975).

[42]*P & Z Co., supra* note 41; *C. Kaufman, Inc.*, 6 OSHC 1295 (Rev. Comm'n 1978); *Asplundh Tree Expert Co.*, 6 OSHC 1951 (Rev. Comm'n 1978).

[43]*P & Z Co., supra* note 41, 1191–1192.

[44]*Id.* at 1192. See also *C. Kaufman, Inc., supra* note 42.

[45]See *Vampco Metal Prods.*, 8 OSHC 2178 (Rev. Comm'n 1980).

[46]*Evansville Materials, supra* note 41.

[47]Compare *C. Kaufman, Inc., supra* note 42 (no remand, credibility question open to only one answer), with *Asplundh Tree Expert Co., supra* note 42 (remand; credibility question unclear).

[48]*Leone Constr. Co.*, 3 OSHC 1979, 1981 (Rev. Comm'n 1976); *RMI Co. v. Secretary of Labor*, 594 F.2d 566, 571 n.13, 7 OSHC 1119, 1122 (6th Cir. 1979); *Willamette Iron*

cited by the Review Commission and be accorded the weight that the soundness of the reasoning warrants.[49] In addition, it should be noted that some courts of appeals have ascribed significance to the Review Commission's failure either to direct review of an A.L.J.'s decision or to harmonize apparent inconsistencies between Review Commission and unreviewed A.L.J. decisions.[50]

Decisions of A.L.J.s, both reviewed and unreviewed, may be important in a different respect. An employer who argues that a standard is too vague to afford fair notice of its requirements may be able to point to varying interpretations of the standard by different judges as evidence of the standard's lack of clarity.[51]

C. Review of a Judge's Decision by the Review Commission

1. Petition for Discretionary Review

Review must be directed, if at all, within 30 days of the date the A.L.J.'s decision is docketed by the Review Commission's Executive Secretary.[52] The review period begins on the day after the docketing date and ends on the thirtieth day thereafter, unless the thirtieth day falls on a Saturday, Sun-

---

& *Steel Co. v. Secretary of Labor*, 604 F.2d 1177, 1180, 7 OSHC 1641, 1642 (9th Cir. 1979), *cert. denied*, 445 U.S. 942 (1980); *Donovan v. Anheuser-Busch, Inc.*, 666 F.2d 315, 326, 10 OSHC 1193, 1200 (8th Cir. 1981); *Faultless Div. v. Secretary of Labor*, 674 F.2d 1177, 10 OSHC 1481 (7th Cir. 1982); *Fred Wilson Drilling Co. v. Marshall*, 624 F.2d 38, 40, 8 OSHC 1921, 1922 (5th Cir. 1980).

[49]*Havens Steel Co.*, 6 OSHC 1740, 1742 & n.7 (Rev. Comm'n 1978) (relying on unreviewed A.L.J.'s decision; collecting cases). See *Fred Wilson Drilling Co. v. Marshall, supra* note 48.

[50]See *L.R. Willson & Sons v. OSHRC*, 698 F.2d 507, 515, 11 OSHC 1097, 1103 (D.C. Cir. 1983); *Kropp Forge Co. v. Secretary of Labor*, 657 F.2d 119, 9 OSHC 2133 (7th Cir. 1981); *Ocean Elec. Corp. v. Secretary of Labor*, 594 F.2d 396, 399–400, 7 OSHC 1149, 1151 (4th Cir. 1979) (on rehearing); *United Parcel Serv. v. OSHRC*, 570 F.2d 806, 809, 6 OSHC 1347, 1348–1349 (8th Cir. 1978) (denial of discretionary review constituted adoption of A.L.J.'s legal conclusions); *Brennan v. Gilles & Cotting, Inc.*, 504 F.2d 1255, 1264–1266, 2 OSHC 1243, 1250–1252 (4th Cir. 1974).

[51]See *Diebold, Inc. v. Marshall*, 585 F.2d 1327, 1335, 6 OSHC 2002, 2008 (6th Cir. 1978). But cf. *Faultless Div. v. Secretary of Labor, supra* note 48, at 1188 n.20, 10 OSHC at 1489 (single inconsistent A.L.J.'s decision; not binding on Commission).

[52]Commission Rules 90(b)(2) and (d). See also *Angel Constr. Co.*, 1 OSHC 1749 (Rev. Comm'n 1974) (A.L.J.'s decision made when received by Commission, not when signed by A.L.J.); *Gurney Indus.*, 1 OSHC 1376 (Rev. Comm'n 1973) (same).

day, or federal holiday, in which case it is extended to the next working day.[53] If, however, the A.L.J.'s decision has not been circulated to the Commission members because of a clerical error, the review period does not begin until the decision is circulated.[54]

Although the Commission members have the authority to direct review on their own motion (*sua sponte* review), review is normally directed only when a party seeks review by filing a petition for discretionary review. Commission Rule 91 prescribes the procedure for filing such a petition. A petition may only be filed by a party "adversely affected or aggrieved" by the A.L.J.'s decision.[55] A party who merely disagrees with a statement in the decision but not the disposition is not necessarily aggrieved.[56]

Petitions for discretionary review may be filed with the A.L.J. during the 20-day reconsideration period provided by Commission Rule 90(b)(2). The effect of filing a petition with the A.L.J. is to ask the A.L.J. to reconsider his decision during the 20-day period between the mailing of the decision to the parties and its docketing by the Commission's Executive Secretary. After the Executive Secretary dockets the decision, the parties have 20 days to file petitions, and 27 days to file cross-petitions, directly with the Commission.[57]

Commission Rule 91(d) states that a petition "should concisely state the portions of the decision for which review is sought." The rule also prohibits the incorporation by reference of a brief or legal memorandum. The petition should also identify the legal issues because, if a petition is granted, review may be limited to the issues in the petition.[58] To assist parties appearing without legal counsel, the Commission no longer requires that the petition be supported by citations to the record and by authorities.[59] For example, in *Hullenkremer*

---

[53]See *Consolidated Freightways*, 9 OSHC 1822, 1825 (Rev. Comm'n 1981).

[54]*Allway Tools, Inc.*, 5 OSHC 1094 (Rev. Comm'n 1977).

[55]Commission Rule 91(b).

[56]Cf. *Cagle Constr Co.*, 6 OSHC 1330 (Rev. Comm'n 1978) (A.L.J.'s decision affirmed without review of issues raised in petition); *PPG Indus.*, 8 OSHC 2003 (Rev. Comm'n 1980) (employer not aggrieved by finding of *de minimis* violation; A.L.J.'s decision affirmed without review).

[57]Commission Rules 91(b) and (c).

[58]Commission Rule 92(a).

[59]Rule 91(d) (last sentence); 43 FED. REG. 4604 (1978) (amending prior rule).

Construction Co.,[60] the Commission accepted a petition from a *pro se* employer that stated only: "We request a consideration for a review of the above case."

Three copies and the original of the petition must be filed.[61] Statements in opposition to the petition may be filed in the same manner as a petition.[62] Commission Rule 92(d) lists factors that the Commissioners will consider in deciding whether to grant a petition: a finding of material fact is not supported by a preponderance of the evidence; legal error was committed; the A.L.J.'s decision raises an important question of law, policy, or discretion; Commission review will resolve a conflict in A.L.J. decisions; and a prejudicial error of procedure was committed. A party whose petition has been granted may move that the petition be withdrawn and that the direction for review be vacated.[63]

Even if a direction for review does not mention all of the issues raised in a petition, the Commission may still decide the unspecified issues.[64] Although the Commission now notifies parties that a petition for discretionary review has not been granted and that the A.L.J.'s decision is final, the failure to receive such a notice does not necessarily extend the time for seeking judicial review.[65]

2. *Direction for Review*

A Commissioner who directs that an A.L.J.'s decisions be reviewed issues a "direction for review." The direction may be issued in response to a petition for discretionary review or on the Commissioner's own motion. The direction for review is effective when it is received by the Commission's Executive Secretary, not when it is signed.[66] A Commissioner may not

---

[60]6 OSHC 1469 n.3 (Rev. Comm'n 1978).

[61]Commission Rule 91(h).

[62]Commission Rule 91(g).

[63]Commission Rule 92(b), codifying *Hamilton Die Cast, Inc.,* 12 OSHC 1797 (Rev. Comm'n 1986). This changed prior practice. See *Emery Smiser Constr. Co.,* 3 OSHC 1794 (Rev. Comm'n 1975).

[64]*Austin Bldg. Co.,* 8 OSHC 2150 (Rev. Comm'n 1980); *S & S Diving Co.,* 8 OSHC 2041 n.1 (Rev. Comm'n 1980).

[65]See *United States v. Fornea Road Boring Co.,* 565 F.2d 1314, 1316, 6 OSHC 1232, 1233 (5th Cir. 1978). See also *Consolidated Andy, Inc. v. Donovan,* 642 F.2d 778, 9 OSHC 1525 (5th Cir. 1981) (notice dated after final order date; judicial review period not extended).

[66]*Gurney Indus.,* 1 OSHC 1376 (Rev. Comm'n 1973).

unilaterally withdraw a direction for review once the review period has expired.[67] When review is directed, the Commission notifies the parties. This notice need not be sent during the 30-day review period.[68]

If a direction for review has been issued, the Commission will ordinarily review the A.L.J.'s decision and affirm, modify, or reverse it. In unusual circumstances, however, the Commission may either vacate the direction for review or affirm the judge's decision without review. In *A.C. & S., Inc.*,[69] the Commission vacated the directions for review of a large number of consolidated cases arising out of a construction project. The Review Commission held that full-scale review was "improvident" because it had already issued opinions in other cases on most of the legal questions involved, the violations were not serious, the proposed penalties were low or zero, abatement was not at issue because the building at issue had been completed, and new briefs would be necessary. The Commission did, however, afford the parties an opportunity to have their cases reinstated if they did, in fact, desire further review.[70]

The Review Commission may affirm a judge's decision without review if there is a lack of party interest in review and the case presents no issue of compelling public interest.[71] The Commission usually finds a lack of party interest in review if no party has petitioned for review (review of the case was directed by a Commissioner *sua sponte*) and the party aggrieved by the A.L.J.'s decision does not file a brief or otherwise urge reversal.[72]

Whether the Commission vacates the direction for review or affirms the A.L.J.'s decision without review, the effect is

[67]*Thorleif Larsen & Son*, 2 OSHC 1256 (Rev. Comm'n 1974).

[68]*H.S. Holtze Constr. Co.*, 7 OSHC 1753, 1755–1756 n.5 (Rev. Comm'n 1979), *aff'd in part and rev'd in part on other grounds*, 627 F.2d 149, 8 OSHC 1785 (8th Cir. 1980).

[69]4 OSHC 1529 (Rev. Comm'n 1976).

[70]See also *P & Z, Inc.*, 10 OSHC 1427 (Rev. Comm'n 1982). The Commission has also vacated directions for review that were issued solely to assure publication of the A.L.J.'s decision, holding that the directions did not present an issue capable of adjudication. *Francisco Tower Serv.*, 3 OSHC 1952 (Rev. Comm'n 1976).

[71]*Abbott-Sommer, Inc.*, 3 OSHC 2032 (Rev. Comm'n 1976). See also *Lone Star Steel Co.*, 10 OSHC 1228 (Rev. Comm'n 1981) (lead and dissenting opinions).

[72]Compare *Tunnel Elec. Constr. Co.*, 8 OSHC 1961 (Rev. Comm'n 1980) (*sua sponte* direction; brief filed), with *Cargill, Inc.*, 7 OSHC 2045 (Rev. Comm'n 1979) (no brief filed).

the same. The judge's decision stands as the final order of the Review Commission but is not precedent binding on the Commission.[73]

As noted earlier, a Commissioner may direct that an A.L.J.'s decision be reviewed even if no party asks for review.[74] Such *sua sponte* review has become rare, however. Commission Rule 92(b) states that a Commissioner will normally not direct *sua sponte* review unless the case raises novel questions of law or policy or questions involving conflict in Administrative Law Judge's decisions." Directions for review may raise only issues that could have been raised by a party.[75] The Commission will ordinarily review only issues raised before the A.L.J.[76] As mentioned above, however, even if *sua sponte* review is ordered, the Commission may nevertheless decline to review an A.L.J.'s decision if there is neither party interest in review nor issues of compelling public interest to be addressed.[77]

3. *Issues on Review, Briefs, Oral Argument, and Amendments*

The Review Commission ordinarily will decline to pass upon issues that are not properly presented for review. To be properly presented, an issue must have been both timely raised before the A.L.J.[78] and raised in either the granted portions of a petition for discretionary review or in the direction for review itself.[79] Jurisdictional issues, however, may be raised

[73]See *Water Works Installation Corp.*, 4 OSHC 1339 (Rev. Comm'n 1976); *Cargill, Inc., supra* note 72; *Texaco, Inc.*, 8 OSHC 1758 (Rev. Comm'n 1980).

[74]Commission Rule 92(a); see *GAF Corp.*, 8 OSHC 2006 (Rev. Comm'n 1980).

[75]See *Francisco Tower Serv., supra* note 70.

[76]Commission Rule 92(c); see *Puterbaugh Enters.*, 2 OSHC 1030, 1031–1032 (Rev. Comm'n 1974) (reasonable promptness and vagueness first raised in direction; no review); *Scientific Coating Co.*, 2 OSHC 1339, 1340 n.2 (Rev. Comm'n 1974) (compliance with §8(a) of the Act).

[77]See *supra* notes 71 and 72.

[78]Commission Rule 92(c). See *J.L. Manta Plant Servs. Co.*, 10 OSHC 2162 (Rev. Comm'n 1982) (Rule 15(a) motion to amend not made before A.L.J.; no extraordinary circumstances); *A. Prokosch & Sons Sheet Metal*, 8 OSHC 2077, 2079 (Rev. Comm'n 1980) (need for uniformity of decisions on same worksite; extraordinary circumstance); *Consolidated Freightways*, 9 OSHC 1822, 1826 (Rev. Comm'n 1981) (Secretary abandoned prosecution before A.L.J.; no revival on review); *River Terminal Ry.*, 3 OSHC 1808 (Rev. Comm'n 1975) (vagueness raised in posthearing brief to A.L.J. too late). See also *John T. Brady & Co.*, 10 OSHC 1385, 1386 (Rev. Comm'n 1982) (extraordinary circumstances because of new case law), *rev'd on other grounds*, No. 82-4082 (2d Cir. Oct. 14, 1982).

[79]Commission Rule 92(a) (review limited to issues in granted petition unless direction expressly provides differently) and Rule 92(b) (issues raised in *sua sponte*

at any time.[80] The Commission may find that an issue raised in a petition and on which review has been directed has been abandoned by a failure to brief the issue or otherwise indicate interest in review.[81]

After the A.L.J.'s decision has been directed for review, the Commission usually will invite briefs from the parties prior to considering the case. Commission Rule 93(a) states that the Commission will "ordinarily" request briefs. When it first adopted the rule, the Commission stated that "[w]hile the Commission expects that it will request briefs in almost all cases, there are cases in which briefs are not necessary."[82] Commission Rule 93 sets out the requirements for briefs and the usual briefing schedules.

As noted, the petitioning party's failure to brief an issue raised in its petition for review may be considered to be an abandonment of the issue.[83] If review has been directed on a Commissioner's own motion, the failure of the party aggrieved by the A.L.J.'s decision to file a brief may cause the Commission and any reviewing court to decline further review.[84]

The Review Commission rarely hears oral argument. Commission Rule 95(a) states: "Oral argument before the Commission ordinarily will not be allowed." Rule 95 goes on to provide that, if oral argument is ordered, the Review Commission will give the parties at least 10 days' notice of the argument, and the notice of argument will advise the parties not only of the date but also of the "hour, place, time allotted, and scope of such argument."

The Review Commission will deny a motion under Rule 15(a) of the Federal Rules of Civil Procedure to amend the pleadings while the case is on review, unless the request was first made to the A.L.J. during the hearing stage or there are extraordinary circumstances excusing a party's failure to do

---

direction). See *Austin Bldg. Co.*, 8 OSHC 2150 (Rev. Comm'n 1980); *S & S Diving Co.*, 8 OSHC 2041 (Rev. Comm'n 1980).

[80]See *U.S. Air v. OSHRC*, 689 F.2d 1191, 10 OSHC 1721 (4th Cir. 1982); *Willamette Iron & Steel Co.*, 9 OSHC 1900, 1904 (Rev. Comm'n 1981); *Stone Container Corp.*, 9 OSHC 1832 (Rev. Comm'n 1981).

[81]See *StanBest, Inc.*, 11 OSHC 1222, 1224–1225 n.4 (Rev. Comm'n 1983).

[82]44 FED. REG. at 70,107–108 (1979) (ancestor of current rule).

[83]See *StanBest, Inc., supra* note 81.

[84]See *John W. McGowan*, 5 OSHC 2028 (Rev. Comm'n 1977), *aff'd*, 604 F.2d 885, 7 OSHC 1842 (5th Cir. 1979).

so.[85] A motion under Fed. R. Civ. P. 15(b) to amend the pleadings to conform to the issues actually tried by the parties during the hearing may be made at any time, even on review, and the Commission may so amend on its own motion.[86]

4. Standard of Review

The OSH Act prescribes no standard of review of A.L.J. decisions. The Commission is, however, subject to Section 8(a) of the APA,[87] which provides that unless the agency provides otherwise by rule or notice, it has all the powers it would have in making the initial decision in the case. The Review Commission, therefore, has held, and the courts have agreed, that it may freely disagree with the A.L.J.'s decision on questions of law and, giving due regard to the A.L.J.'s credibility findings, it may disagree on questions of fact.[88] Thus, the Commission is not required to leave an A.L.J.'s finding undisturbed if it is supported by substantial evidence.[89] The Fifth Circuit held that "the statutory scheme contemplates that the *Commission* is the fact-finder, and that the judge is an arm of the Commission for that purpose."[90] Thus, while the A.L.J.'s findings must be considered by the Commission because they are part of the record, the Commission may disagree with the findings and substitute its own.[91]

---

[85]*J.L. Manta Plant Servs. Co., supra* note 78.

[86]*Farmers Coop. Grain & Supply Co.*, 10 OSHC 2086 (Rev. Comm'n 1982); *Rogers Mfg. Co.*, 7 OSHC 1617 (Rev. Comm'n 1979). Note that a Rule 15(b) amendment sought on review normally will be denied if reopening of the record is necessary to avoid prejudice. *Texaco, Inc.*, 8 OSHC 1677 (Rev. Comm'n 1980). Cf. *Copperweld Steel Co.*, 11 OSHC 2235, 2237 (Rev. Comm'n 1984) (remand to A.L.J. to consider Rule 15(b) motion made on review).

[87]5 U.S.C. §557(b).

[88]The lack of a special statutory standard for review, and hence the reliance on the Administrative Procedure Act, distinguishes the Occupational Safety and Health Review Commission from the Federal Mine Safety and Health Review Commission (FMSHRC) and the Labor Department's Benefits Review Board. These tribunals are required by their enabling statutes to apply a substantial evidence test. See *Donovan v. Phelps Dodge Corp.*, 709 F.2d 86 (D.C. Cir. 1983) (FMSHRC); *Atlantic & Gulf Stevedores v. Director, OWCP*, 542 F.2d 602, 608 (3d Cir. 1976), *cert. denied*, 439 U.S. 818 (1978) (Benefits Review Board).

[89]*Little Beaver Creek Ranches*, 10 OSHC 1806, 1810 (Rev. Comm'n 1982).

[90]*Accu-Namics, Inc. v. OSHRC*, 515 F.2d 828, 834, 3 OSHC 1299, 1302 (5th Cir. 1975), *cert. denied*, 425 U.S. 903 (1976) (emphasis in original).

[91]*Astra Pharmaceutical Prods.*, 9 OSHC 2126, 2131 n.18 (Rev. Comm'n 1981), *aff'd in pertinent part*, 681 F.2d 69, 10 OSHC 1697 (1st Cir. 1982) (no credibility question; Commission may draw different factual inferences from A.L.J.'s); (Pratt & Whitney Aircraft v. Secretary of Labor, 649 F.2d 96, 105, 9 OSHC 1554, 1561 (2d Cir. 1981) (A.L.J.'s findings entitled to some weight; should not be disturbed without explanation).

The Review Commission ordinarily will accept the A.L.J.'s evaluation of the credibility of witnesses, "for it is the judge who has lived with the case, heard the witnesses, and observed their demeanor."[92] If the A.L.J. does not adequately explain his credibility findings, however, the Commission may decline to accept them.[93] When the Commission overturns an A.L.J.'s credibility finding, it must adequately explain why it disagrees with the A.L.J.[94]

Factual findings by the Commission and its A.L.J.s must be supported by a preponderance of the evidence.[95] The "substantial evidence" test used by the courts of appeals must not be applied by the Commission or its A.L.J.s as a substitute for the "preponderance" test.[96] A preponderance of the evidence is that quantum of evidence sufficient to convince the trier of fact that the facts asserted are more probably true than false.[97]

D. Disqualification of a Commission Member

Commission Rule 68 governs the disqualification of an A.L.J., but there is no Commission rule governing the disqualification of a Commissioner. Section 7(a) of the APA[98] requires that "[t]he functions of presiding employees and of employees participating in decisions in accordance with [5 U.S.C. §557] shall be conducted in an impartial manner." The statute also provides that "[a] presiding or participating employee may at any time disqualify himself" and that "[o]n the filing in good faith of a timely and sufficient affidavit of personal bias or other disqualification of a presiding or participating employee the agency shall determine the matter as a part of the record and decision in the case."

In *National Manufacturing Co.*,[99] the Commission held that it would leave the resolution of a disqualification motion

---

[92]*C. Kaufman, Inc.*, 6 OSHC 1925, 1297 (Rev. Comm'n 1978), and cases cited.

[93]*P & Z, Inc.*, 6 OSHC 1189, 1192 (Rev. Comm'n 1977).

[94]*General Dynamics Corp. v. OSHRC*, 599 F.2d 453, 463, 7 OSHC 1373, 1379 (1st Cir. 1979).

[95]*Astra Pharmaceutical Prods., supra* note 91; *Armor Elevator Co.*, 1 OSHC 1409 (Rev. Comm'n 1973).

[96]*Olin Constr. Co. v. OSHRC*, 525 F.2d 464, 3 OSHC 1526 (2d Cir. 1975).

[97]*Astra Pharmaceutical Prods., supra* note 91, 2131 & n.17.

[98]5 U.S.C. §556(b).

[99]8 OSHC 1435 (Rev. Comm'n 1980).

to a member sought to be disqualified. It stated that the APA does not require it to rule on the disqualification of presidentially appointed Commissioners in the same way that it passes on the disqualification of A.L.J.s. The Commission also stated that even if it had the power to disqualify one of its members, it declined to exercise that power under the circumstances. It observed that the disqualification motion before it required a subjective inquiry that was best left to the person who best knew the facts underlying the motion.

Moreover, it is not clear what standards should be used to decide such disqualification questions. The judicial code provisions at 28 U.S.C. §§144 and 455 (1982) apply by their terms only to justices, judges, magistrates, and bankruptcy referees. They may, however, be applied by analogy.[100] Two former members of the Commission who, before being appointed to the Commission, had formed views on certain legal issues, declined to disqualify themselves from considering cases involving those issues.[101]

E. Deference to the Courts of Appeals[102]

The Review Commission generally has stated that, while it gives due deference to the decisions of the courts of appeals, it is not bound to acquiesce in them.[103] The Commission has

---

[100]See *National Mfg. Co.*, 8 OSHC 1435, 1439–1440 (Rev. Comm'n 1980) (separate opinion of Comm'r Cottine).

[101]*Id.* at 1441–1444; *United States Steel Corp.*, 5 OSHC 1289, 1297 (Rev. Comm'n 1977) (memorandum of former Comm'r Barnako). The memorandum opinions of former Commissioner Cottine in *National Mfg. Co.* and *American Cyanamid Co.*, 9 OSHC 1596, 1605–1608 (Rev. Comm'n 1981), hold that various personal and professional associations with a party and with employees of a party are insufficient to warrant disqualification. In *Sun Petroleum Prods. Co.*, 7 OSHC 1306 (Rev. Comm'n 1979), *remanded*, 622 F.2d 1176, 8 OSHC 1422 (3d Cir.), *cert. denied*, 449 U.S. 1061 (1980), Mr. Cottine disqualified himself because he had appeared as counsel at the hearing. Mr. Barnako disqualified himself from cases involving Bethlehem Steel Corp. because of his prior employment with and financial interest in the company. *United States Steel Corp.*, 5 OSHC at 1297.

[102]See also discussion in Chapter 18 (Judicial Review of Enforcement Proceedings), Section III.D.

[103]See *Mobil Oil Corp.*, 10 OSHC 1905, 1907–1908 n.4 (Rev. Comm'n 1982), *rev'd without consideration of point*, 713 F.2d 918, 11 OSHC 1609 (2d Cir. 1983); *Farmers Export Co.*, 8 OSHC 1655 (Rev. Comm'n 1980); *S & H Riggers & Erectors*, 7 OSHC 1260 (Rev. Comm'n 1979), *rev'd on other grounds*, 659 F.2d 1273, 10 OSHC 1057 (5th Cir. 1981). But see *Davis Metal Stamping*, 12 OSHC 1259, 1261 (Rev. Comm'n 1985) (warrant invalid because Fifth Circuit would inevitably find warrant invalid on appeal), *aff'd*, 800 F.2d 1351, 12 OSHC 2129 (5th Cir. 1986).

also stated that the national scope of the OSH Act and its uniform and orderly administration require the Commission, rather than its A.L.J.s, to determine whether to follow an appellate court precedent.[104] The Commission has, on occasion, reconsidered and revamped its precedent in light of court of appeals decisions.[105]

Two courts of appeals have unequivocally stated that the Commission is bound to follow their holdings in cases arising in their circuits.[106] However, because Commission decisions are appealable to more than one circuit, a potential problem would exist if two different circuits to which a particular case could be appealed had conflicting precedents.[107]

F. Effect of Split Decisions

Split Review Commission decisions may occur when three Commissioners cast three differing votes or, as has occurred more frequently, when a member's term has expired or a member has been disqualified, leaving only two members to vote on a case. The result announced by the Commission has been consistent, though the rationale has changed: the A.L.J.'s decision becomes the final order of the Review Commission but is accorded the precedential value of an unreviewed A.L.J. decision.

The problem first arose after the expiration of former Commissioner Van Namee's term in April 1975. The two remaining members agreed to dispose of several cases in which they were divided by announcing that "[t]he judge's decision is affirmed by an equally divided Commission."[108] Two courts of appeals held that the Commission could not dispose of a

---

[104]*Grossman Steel & Aluminum Corp.*, 4 OSHC 1185, 1188 (Rev. Comm'n 1975).

[105]*E.g., Potlatch Corp.*, 7 OSHC 1061 (Rev. Comm'n 1979); *Grossman Steel & Aluminum Corp., supra* note 104; *Anning-Johnson Co.*, 4 OSHC 1193 (Rev. Comm'n 1976).

[106]See *Smith Steel Casting Co. v. Donovan*, 725 F.2d 1032, 1035, 11 OSHC 1785, 1786–1787 (5th Cir. 1984); *Jones & Laughlin Steel Corp. v. Marshall*, 636 F.2d 32, 8 OSHC 2217 (3d Cir. 1980); *Babcock & Wilcox Co. v. OSHRC*, 622 F.2d 1160, 1166, 8 OSHC 1317, 1321 (3d Cir. 1980).

[107]*Raybestos Friction Materials Co.*, 9 OSHC 1141, 1143 (Rev. Comm'n 1980) (case distinguishable; appealable to three circuits); *Bethlehem Steel Corp.*, 9 OSHC 1346, 1349 n.12 (Rev. Comm'n 1981) (conflicting court precedent; Commission noted "dilemma," followed own precedent).

[108]E.g., *Garcia Concrete, Inc.*, 3 OSHC 1211 (Rev. Comm'n 1975).

case in this manner on the ground that the affirmances of the A.L.J.s' decisions were supported by only one vote and thus did not comply with the requirement in Section 12(f) of the OSH Act that "official action can be taken only on the affirmative vote of at least two members."[109]

After these court decisions, the Review Commission issued several divided decisions in which it announced that, because it could take no official action, the A.L.J.'s decision "becomes the final order of the Commission."[110] Two courts held that such orders were final and therefore judicially reviewable.[111]

The Commission adopted a different approach shortly thereafter. In *Life Science Products Co.,*[112] the two members sitting on the case disagreed on the correctness of the A.L.J.'s decision. To resolve the impasse, they *both* voted to affirm the A.L.J.'s decision but to accord it no binding precedential value. The Commission stated that this approach was different from an affirmance by an equally divided Commission because both members were explicitly voting to affirm the decision.

The courts are divided on the *Life Science* approach. In *Willamette Iron & Steel Co. v. Secretary of Labor,*[113] the Ninth Circuit rejected the approach on the basis that it had the same effect as the affirmance by an equally divided Commission, which the court had previously rejected. However, in *Marshall v. Sun Petroleum Products Co.,*[114] the Third Circuit disagreed with *Willamette* and held that the Commission's order was judicially reviewable.

In response to the continuing disagreement between the circuits, the Commission has announced a new method of disposing of split decisions: vacating the direction for review. The Commission explained that vacating a direction for review

[109]*Shaw Constr. v. OSHRC,* 534 F.2d 1183, 1185–1186, 4 OSHC 1427, 1428 (5th Cir. 1976); *Cox Brothers v. Secretary of Labor,* 574 F.2d 465, 6 OSHC 1484 (9th Cir. 1978).

[110]E.g., *Bethlehem Steel Corp.,* 5 OSHC 1025 (Rev. Comm'n 1976), *rev'd on another ground without consideration of point,* 573 F.2d 157, 6 OSHC 1440 (3d Cir. 1978).

[111]*George Hyman Constr. Co. v. OSHRC,* 582 F.2d 834, 6 OSHC 1855 (4th Cir. 1978); *Marshall v. L.E. Myers Co.,* 589 F.2d 270, 6 OSHC 2159 (7th Cir. 1978).

[112]6 OSHC 1053 (Rev. Comm'n 1977), *aff'd without discussion of point sub nom. Moore v. OSHRC,* 591 F.2d 991, 7 OSHC 1031 (4th Cir. 1979).

[113]604 F.2d 1177, 7 OSHC 1641 (9th Cir. 1979), *cert. denied,* 445 U.S. 942 (1980).

[114]622 F.2d 1176, 8 OSHC 1422 (3d Cir. 1980), *cert. denied,* 449 U.S. 1061 (1982).

has the same effect as affirming the A.L.J.'s decision and according it no binding precedential value.[115]

G. Effective Date of Final Commission Order

The date a Commission order becomes final is important because any abatement date established by the order does not begin to run until the order is final, as long as the employer filed the contest in good faith.[116] Section 10(c) of the Act states that a Commission order "shall become final thirty days after its issuance." If an A.L.J.'s decision is not directed for review, it becomes a final order at the expiration of the 30-day review period.[117]

The Commission has issued many decisions in which the Commission's order will take effect only if a stated contingency does not occur.[118] In such cases, the date of the Commission's final order remains uncertain until the time limit placed on the contingency has passed.

H. Stay of Commission Orders

Section 11(a) of the OSH Act[119] states that the filing of a petition for judicial review does not, unless ordered by the court, stay the Commission's order. Section 10(d) of the APA[120] permits the Commission to postpone the effective date of its order pending judicial review when "justice so requires." Rule 18 of the Federal Rules of Appellate Procedure, entitled "Stay Pending Review," requires that stay motions "shall ordinarily be made in the first instance to the agency." Com-

---

[115]*Texaco, Inc.*, 8 OSHC 1758 (Rev. Comm'n 1980). See also *Baldwin Indus.*, 10 OSHC 1572 (Rev. Comm'n 1982).

[116]OSH Act §10(b), 29 U.S.C. §659(b). Additionally, a finding that a prior Commission order has become final is a necessary prerequisite to the findings of a repeated violation. *Otis Elevator Co.*, 8 OSHC 1019, 1026 (Rev. Comm'n 1980).

[117]OSH Act §12(j), 29 U.S.C. §661(j). See *Monroe & Sons*, 4 OSHC 2016, 2018 & n.11 (Rev. Comm'n No. 6031, 1977), *aff'd*, 615 F.2d 1156, 8 OSHC 1034 (6th Cir. 1980).

[118]E.g., *Pima Constr. Co.*, 4 OSHC 1620, 1625 (Rev. Comm'n 1976) (citation amended unless good cause to contrary shown within 20 days); *Anning-Johnson Co.*, 4 OSHC 1193, 1200 (Rev. Comm'n 1976) (citation items affirmed unless remand motion filed within 10 days).

[119]29 U.S.C. §660(a) (1982).

[120]5 U.S.C. §705.

mission Rule 94 permits aggrieved parties to file a motion for a stay. The Review Commission has generally declined to stay the abatement requirement of an affirmed citation but will usually stay the penalty assessment.[121] Once the record has been filed in a court of appeals, the court has exclusive jurisdiction to grant a stay.[122]

I. Reconsideration of Final Orders

There are three ways in which Commission orders become final: by operation of law if a citation is not contested; by an A.L.J.'s decision not being directed for review within the 30-day review period; and upon issuance of a Commission decision following a direction for review. The authority of the Commission to reconsider an order depends on how it became final.

Sections 10(a) and (b) of the Act[123] state that a citation and notification of proposed penalty that are not contested within 15 working days "shall be deemed a final order of the Commission and not subject to review by any court or agency." At first the Commission held, with some equivocation, that such final orders could not be reexamined at all.[124] The Commission has, however, now followed the Third Circuit and held that reconsideration is permissible if Fed. R. Civ. P. 60(b), Relief from Judgment or Order, is satisfied.[125]

The Commission has held that it may reconsider an A.L.J.'s decision not directed for review within the 30-day review period if Fed. R. Civ. P. 60(b) is satisfied and the record has not

---

[121]E.g., *Ryder Truck Lines*, 1 OSHC 1326 (Rev. Comm'n 1973), *aff'd*, 497 F.2d 230, 2 OSHC 1075 (5th Cir. 1974).

[122]OSH Act §11(a), 29 U.S.C. §660(a).

[123]29 U.S.C. §§659(a) and (b).

[124]See *American Airlines*, 2 OSHC 1326 (Rev. Comm'n 1974); *Plessey, Inc.*, 2 OSHC 1302, 1306 (Rev. Comm'n 1974) (Secretary moved to vacate uncontested citation; motion denied); *Penn Cent. Transp. Co.*, 2 OSHC 1379 (Rev. Comm'n 1974) (citation final despite possibly meritorious exemption argument), *aff'd*, 535 F.2d 1249, 3 OSHC 2059 (4th Cir. 1976); *Phoenix, Inc.*, 1 OSHC 1011 (Rev. Comm'n 1972) (Secretary moved to vacate uncontested citation because of meritorious exemption argument; motion granted).

[125]*Branciforte Builders, Inc.*, 9 OSHC 2113 (Rev. Comm'n 1981), following *J.I. Hass Co. v. OSHRC*, 648 F.2d 190, 9 OSHC 1712 (3d Cir. 1981). See also *Atlantic Marine, Inc. v. OSHRC*, 524 F.2d 476, 478, 3 OSHC 1755, 1756 (5th Cir. 1975) (suggesting statute does not establish "impenetrable barrier" in face of "deceptive practices" by OSHA), followed in *Henry C. Beck Co.*, 8 OSHC 1395 (Rev. Comm'n 1980).

been filed in a court of appeals.[126] The Commission also has reopened an A.L.J.'s decision under Fed. R. Civ. P. 60(a) when the decision became final as a result of clerical error by the Commission.[127]

The Review Commission can reconsider its own decisions within 30 days of their issuance under Section 10(c) of the OSH Act, although the Commission has no specific rule on reconsideration. It has reconsidered its decisions on motion of a party.[128] The Commission's power to reconsider its decisions expires once the record in a case is filed in a court of appeals.[129]

J. Attorney's Fees and Costs

In early cases, the Commission held that it lacked authority to assess costs, including attorney's fees, against a party.[130] These decisions reflected the so-called American Rule, under which each party bears it own costs of litigation, including attorney's fees, in the absence of statutory authorization to shift those costs to another party.[131]

In 1980, Congress enacted the Equal Access to Justice Act (EAJ Act),[132] which authorizes agencies that conduct admin-

---

[126]*Monroe & Sons, supra* note 117. Contra *Brennan v. OSHRC (S.J. Otinger, Jr., Constr. Co.)*, 502 F.2d 30, 2 OSHC 1218 (5th Cir. 1974).

[127]*Voegele Co.*, 7 OSHC 1713, 1714 n.2 (Rev. Comm'n 1979), *aff'd without consideration of point*, 625 F.2d 1075, 8 OSHC 1631 (3d Cir. 1980).

[128]E.g., *George A. Hormel & Co.*, 2 OSHC 1281 (Rev. Comm'n 1974), *reconsidering* 2 OSHC 1190 (1974).

[129]See OSH Act §11(a), 29 U.S.C. §660(a) (court of appeals has jurisdiction once record is filed with court).

[130]*John W. McGowan*, 5 OSHC 2028 (Rev. Comm'n), *aff'd without consideration of point*, 604 F.2d 885, 7 OSHC 1842 (5th Cir. 1979); *General Elec. Co.*, 3 OSHC 1031, 1049 (Rev. Comm'n 1975), *rev'd in part on other grounds*, 540 F.2d 67, 4 OSHC 1512 (2d Cir. 1976).

[131]See *Donovan v. Nichols*, 646 F.2d 190, 9 OSHC 1818 (5th Cir. 1981) (district court erred in awarding attorney's fees against Secretary of Labor; statutory authority lacking for such an award even if prosecution of case in bad faith).

[132]Pub. L. No. 96-481, 94 Stat. 2325 (1980). As originally enacted, the EAJ Act applied to adjudications pending on Oct. 1, 1981. See *S & H Riggers & Erectors v. OSHRC*, 672 F.2d 426, 10 OSHC 1495 (5th Cir. 1982) (appeal of Commission decision pending on Oct. 1, 1981; Commission proceeding, however, not pending on that date, so no award for fees and expenses relating to Commission proceedings could be made). The original version of the EAJ Act was effective for the period from Oct. 1, 1981, to Oct 1, 1984. It expired on the latter date pursuant to a repealer included in the statute. An amended version of the EAJ Act was enacted and made effective on Aug. 5, 1985. Pub. L. 99-80 (1985). The provisions of the EAJ Act that apply to fees in agency adjudications are codified at 5 U.S.C. §504 (1982), and those that apply to court proceedings are found at 28 U.S.C. §2412 (1982). The Commission's rules implementing the EAJ Act are found at 29 C.F.R. §§2204.101-.311.

istrative adjudications and courts to assess attorney's fees and other litigation expenses against the government in limited circumstances. Awards can only be made to parties who prevail in the litigation and who meet certain size restrictions.[133] No award can be made, however, if the government's position in the litigation was substantially justified or if special circumstances make an award unjust.[134]

An application for an EAJ Act award based on proceedings before the Commission must be filed with the Commission no later than 30 days after the Commission's final disposition of the proceeding.[135] If the Commission decision on which the application is based is subsequently appealed, the Commission must dismiss the application.[136] Following the resolution of the appeal, an eligible party may apply for an EAJ Act award to the court that decided the appeal, which can award fees and expenses incurred in connection with the Commission proceedings as well as those incurred in the appeal.[137] If the Commission dismisses an application owing to the filing of an appeal and the appeal is subsequently withdrawn, the applicant may reinstate its application before the Commission within 30 days of the withdrawal.[138]

In order for a party to be regarded as having prevailed in the litigation, it is not necessary that the party have won an unqualified victory. It is sufficient if the party prevailed in a discrete, substantive portion of the case.[139] In a case resolved by a settlement agreement that deleted one willful item of a citation, downgraded two others from willful to serious, and reduced the penalties from $11,200 to $2,080, the Commission concluded that the employer qualified as a prevailing party under the EAJ Act.[140]

---

[133]In order to be eligible for an award, a business must have a net worth of not more than $7 million, and not more than 500 employees. An individual must have a net worth of no more than $2 million to be eligible. 5 U.S.C. §504(b)(1)(B).

[134]5 U.S.C. §504(a)(1).

[135]29 C.F.R. §2204.302(a).

[136]§2204.302(c); *Federal Clearing Die Casting Co.*, 11 OSHC 1157 (Rev. Comm'n 1983).

[137]§2412(d)(1) and 2412(d)(3).

[138]§2204.302(c).

[139]§2204.106(a); *H.P. Fowler Contracting Corp.*, 11 OSHC 1841, 1845 (Rev. Comm'n 1984).

[140]*H.P. Fowler Contracting Corp.*, *supra* note 139, at 1846.

A prevailing party may be denied an award if the Secretary's position in the litigation was substantially justified. In *S & H Riggers & Erectors v. OSHRC*,[141] the court reversed the Commission's affirmance of citations alleging noncompliance with 29 C.F.R. §1926.28(a), the general personal protective equipment standard for construction work. The court held that the language of the standard was so general that due process required a showing either that the employer's conduct failed to conform to the custom and practice of its industry or that the employer had actual knowledge that personal protective equipment was necessary to protect its employees. The court rejected the Commission's application of a "reasonable person" test under the standard. Having disagreed with the Commission's interpretation of the standard, however, the Fifth Circuit assigned substantial significance to that interpretation in denying a subsequent EAJ Act application.[142] The court noted that its interpretation of Section 1926.28(a) in cases preceding *S & H Riggers*[143] was premised in part on the lack of an authoritative interpretation of the standard sufficient to give employers notice of what conduct the standard required. As the Commission's decision in *S & H Riggers*[144] represented the first definitive administrative interpretation of the standard, the court concluded that the Secretary was justified in attempting to persuade the court to approve the Commission's interpretation.

A citation under 29 C.F.R. §1926.28(a) was also involved in an application before the Commission in *Hocking Valley Steel Erectors*.[145] The A.L.J. had affirmed a citation alleging that employees were not using safety belts to protect against falls. The employer's petition for review was granted by the

---

[141]659 F.2d 1273, 10 OSHC 1057 (5th Cir. 1981), *rev'g S & H Riggers & Erectors*, 7 OSHC 1260 (Rev. Comm'n 1979).

[142]*S & H Riggers & Erectors v. OSHRC*, 672 F.2d 426, 10 OSHC 1495 (5th Cir. 1982). In addition to applying for an EAJ Act award against the Secretary, the employer also sought an award against the Commission. The court noted that the Commission had played no role in prosecuting the appeal before it and found this to be a special circumstance sufficient to render an award unjust. *Id.* at 429, 10 OSHC at 1497. The court declined to rule on the Commission's argument that, as an independent adjudicatory agency, it could never suffer an award under the EAJ Act.

[143]E.g., *B & B Insulation v. OSHRC*, 583 F.2d 1364, 6 OSHC 2062 (5th Cir. 1978).

[144]7 OSHC 1260 (Rev. Comm'n 1979), *supra* note 141.

[145]11 OSHC 1492 (Rev. Comm'n 1983).

Commission, but the Secretary moved to withdraw the citation while review was pending. Upon examining the facts of the case, the Commission denied the fee application, concluding that the Secretary's case was reasonable in fact and law under Commission precedent and that his position was therefore substantially justified. The Commission criticized the Secretary for not explaining why he chose to withdraw the citation, but concluded that the unexplained withdrawal did not detract from the justification for his earlier position.

17

Employee and Union Participation in Litigation Under the OSH Act

I. Overview

The OSH Act confers upon employees and their representatives (i.e., their unions[1]) two express rights to participate in proceedings before the Review Commission. The first sentence of Section 10(c) provides, *inter alia*, that after a citation

---

[1]The Act does not define the term "representative of employees." However, Commission Rule 20, 29 C.F.R. §2200.20(b) (1987), refers to the filing of a notice of contest by an "*authorized* employee representative" (emphasis added), and Rule 1(g), 29 C.F.R. §2200.1(g), defines "authorized employee representative" as "a labor organization which has a collective bargaining relationship with the cited employer and which represents affected employees." Thus, the Commission's rules appear to contemplate that only affected employees or their union may file a notice of contest.

Commission Rule 22(c), 29 C.F.R. §2200.22(b), provides that where affected employees are "represented by an authorized employee representative" (i.e., a union), they may not appear in Commission proceedings through any other representative. However, where affected employees are not represented by a union, the Commission's rules apparently allow them to appear in Commission proceedings through any representative they may select. See Rule 22(a). (Rule 22(a), like all the Commission's Rules of Procedure, is found in the corresponding section of 29 C.F.R. pt. 2200, i.e., §2200.22(a). Hereafter Commission rules will be cited without reference to 29 C.F.R. pt. 2200.) In addition, Rule 21, governing intervention in Commission proceedings, is not limited to employees and unions.

Accordingly, there are instances in which an affected employee may appear in Commission proceedings through some representative other than a union. But, as a practical matter, unions are the only entities that have participated significantly in Commission proceedings as employee representatives. Consequently, in this chapter the terms "employee representatives" and "unions" are used interchangeably.

has been issued, any employee or representative of employees may file a notice of contest "alleging that the period of time fixed in the citation for the abatement of the violation is unreasonable," and the Commission shall then "afford an opportunity for a hearing."[2] The final sentence of Section 10(c) provides an additional right to employees and their representatives, by stating that "[t]he rules of procedure prescribed by the Commission shall provide affected employees or representatives of affected employees an opportunity to participate as parties to hearings under this subsection."[3]

The legislative history of the OSH Act provides little guidance concerning the relationship between these two provisions, except to indicate that they create separate and distinct rights. Thus, the Senate report states:

> "If the employer decides to contest a citation or notification, or proposed assessment of penalty, the [Commission] must afford an opportunity for a formal hearing * * *. * * * The procedural rules prescribed by the [Commission] for the conduct of such hearings must make provision for affected employees or their representatives to participate as parties.
>
> "Section 10(c) *also* gives an employee or representative of employees a right, whenever he believes that the period of time provided in a citation for abatement of a violation is unreasonably long, to challenge the citation on that ground."[4]

The fact that the first sentence of Section 10(c) refers only to employee or union complaints concerning the *abatement period* specified in a citation, while the final sentence refers more generally to "an opportunity to participate as parties," has given rise to a host of controversies concerning the scope of employee and union rights to participate in Commission proceedings.

II. Issues That May Be Litigated Where an Employee or Union Files a Notice of Contest

One of the first issues to arise concerning employee and union rights in Commission proceedings was whether, in a

---

[2]29 U.S.C. §659(c) (1976).

[3]*Id.*

[4]S. REP. NO. 1282, 91st Cong., 2d Sess. 14–15 (1970), *reprinted in* Subcommittee on Labor of the Senate Comm. on Labor and Public Welfare, 92nd Cong., 1st Sess., LEGISLATIVE HISTORY OF THE OCCUPATIONAL SAFETY AND HEALTH ACT OF 1970, at 154–155 (Comm. Print 1971) (emphasis added).

proceeding initiated by a union's notice of contest, the union
is entitled to challenge not only the length of time provided
for abatement (the issue specifically mentioned in the first
sentence of Section 10(c)), but the substance of the required
abatement as well. In *Local 588, UAW*, the union objected to
the substantive provisions of an "abatement plan" that had
been submitted by the employer after the issuance of a citation. The Commission held that "the union is not entitled to
seek, and the Commission is not empowered to grant, a modification in the abatement plan."[5] The Commission stated that
the union was entitled to show that abatement could be accomplished in less time than was provided in the plan, and
that as part of such a showing "the union may properly adduce
evidence to show that the employer's abatement plan does not
include all feasible abatement methods or controls that are
presently available," because "[s]uch specific methods or controls and the time required for their implementation are determinative of the length of time needed for abatement."[6] But
the Commission held that the language of the first sentence
of Section 10(c) precludes a union from *directly* challenging
any aspect of an abatement plan other than the length of time
allowed for abatement: "Because the clear language of the
statute limits employee contests to the reasonableness of the
period of time which is fixed in the citation for abatement, we
hold that the sufficiency of the so-called abatement plan cannot be *directly* contested by the union."[7] On appeal, the Seventh Circuit adopted the Commission's decision on this point.[8]

In its 1982 decision in *Mobil Oil Corp.*[9] the Commission
overruled *Local 588, UAW*. Commissioner Cottine, writing the
lead opinion, explained that the reason why the first sentence
of Section 10(c) refers only to the time for abatement rather
than the method of abatement is simply that under Section 9
of the Act the citations issued by the Secretary need only
specify the time, and not the method, of abatement. From this
Commissioner Cottine reasoned that "inasmuch as the manner for achieving abatement is not ordinarily set forth in the

[5]*Local 588, UAW*, 4 OSHC 1243, 1244 (Rev. Comm'n 1976), *aff'd*, 557 F.2d 607,
5 OSHC 1525 (7th Cir. 1977).
[6]*Supra* note 5, 4 OSHC at 1244.
[7]*Id.* (emphasis in original).
[8]*Supra* note 5, 557 F.2d at 610, 5 OSHC at 1527.
[9]*Mobil Oil Corp.*, 10 OSHC 1905 (Rev. Comm'n 1982), *rev'd*, 713 F.2d 918, 11
OSHC 1609 (2d Cir. 1983).

citation, it would be impossible for employees to file a notice of contest regarding the abatement method," and consequently, the fact that Section 10(c) does not refer to the filing of a notice of contest with respect to abatement methods does not necessarily mean that Congress wished to foreclose unions and employees from being heard with respect to that subject in cases where the Secretary has specified the proposed abatement method.[10]

On appeal, the Second Circuit reversed the Commission's order in *Mobil Oil*, expressly disagreeing with the reasoning of Commissioner Cottine and concluding that "the limitation on the scope of employee contests in §10(c) * * * demonstrates a congressional intent to limit the rights of employees."[11] Subsequently, in a 1984 decision, *Pan American World Airways*, the Commission overruled *Mobil Oil*.[12] In *Pan American* the Commission did not expressly address the portion of *Mobil Oil* that had overruled *Local 588, UAW*, but it seems clear that as a result of *Pan American* the Commission's current view has returned to what was stated in *Local 588, UAW*. Furthermore, several courts of appeals have approved the reasoning and result of *Local 588, UAW*.[13] At this writing, then, it appears to be established that when a Commission proceeding is initiated by a notice of contest filed by an employee or a union, the first sentence of Section 10(c) limits the issues the Commission may consider to the reasonableness of the time allowed for abatement, and does not permit the Commission to consider any other abatement issues.

III. Employee and Union Rights in Contested Cases Initiated by an Employer

By its terms, the language in the first sentence of Section 10(c) referring to the period of time for abatement applies only

---

[10]*Supra* note 9, 10 OSHC at 1919. The same view was adopted by Judge Pollak in dissent in *Marshall v. Sun Petroleum Prods., Co.*, 622 F.2d 1176, 1189 n.2, 8 OSHC 1422, 1432 (3d Cir.), *cert. denied*, 449 U.S. 1061 (1980).

[11]*Donovan v. OSHRC (Mobil Oil Corp.)*, 713 F.2d 918, 928, 11 OSHC 1609, 1617 (2d Cir. 1983).

[12]*Pan Am. World Airways*, 11 OSHC 2003 (Rev. Comm'n 1984).

[13]See *Donovan v. Allied Indus. Workers*, 722 F.2d 1415, 1419, 11 OSHC 1737, 1739–1740 (8th Cir. 1983); *Donovan v. OSHRC (Mobil Oil Corp.)*, *supra* note 11, at 929, 930, 11 OSHC at 1617, 1618; *Oil, Chem. & Atomic Workers v. OSHRC (American Cyanamid)*, 671 F.2d 643, 650, 10 OSHC 1345, 1349–1350 (D.C. Cir. 1982), *cert. denied*, 459 U.S. 905 (1982).

to cases initiated by an employee or union notice of contest. Where an *employer* files the notice of contest, the operative provision is the final sentence of Section 10(c), which states that affected employees or their representatives must be given "an opportunity to participate as parties to hearings under this subsection."

Commission Rule 20(a), adopted to implement that provision, states that "[a]ffected employees . . . may elect party status" at least 10 days before hearing unless good cause is shown for a later request for party status. Rule 20 authorizes either affected employees or their union to elect automatic party status; but the Commission has held that where an affected employee is a member of a union, the employee may elect party status only if the union has not done so. Once the union elects party status, its members are precluded from participating independently as parties.[14] In addition, the Commission's regulations define "affected employee" as "an employee of a cited employer who is exposed to or has access to the hazard arising out of the allegedly violative circumstances, conditions, practices or operations."[15] Thus, for example, employees on a construction site who are exposed to the hazard described in a citation are not "affected employees" if they are employed by a subcontractor and the Labor Department cites only the general contractor, because they are not "employee[s] of [the] cited employer." Such employees are not entitled to elect party status under Commission Rule 20, nor is their union,[16] although interventions will generally be allowed in such instances under Commission Rule 21.[17]

---

[14]*United States Steel Corp.*, 11 OSHC 1361 (Rev. Comm'n 1983) (overruling prior cases).

[15]Commission Rule 1(e).

[16]*Brown & Root, Inc.*, 7 OSHC 1526 (Rev. Comm'n 1979) (decided under old rules).

[17]In *Brown & Root, supra* note 16, the union representing the subcontractor's employees was granted "unconditional" leave to intervene, and was accorded the same right it would have had if it had been granted party status. The three Commissioners who decided *Brown & Root* took three separate approaches to the issues presented. Chairman Cleary concluded that Commission Rule 21 was applicable, and that where an applicant for intervention has more than a minimal interest in the proceeding, leave to intervene should be freely granted unless the intervenor's participation would unduly hinder the efficient resolution of the case. Commissioner Cottine concluded that the applicable rule was not Commission Rule 21, which he characterized as dealing only with permissive intervention, but Rule 24(a) of the Federal Rules of Civil Procedure, under which, in his view, the union representing the exposed employees could intervene as of right. Commissioner Barnako agreed with Chairman Cleary that Commission Rule 21 was controlling, but he concluded

Employees or unions electing party status may participate in prehearing discovery, present their own witnesses, and exercise the other procedural rights of a party.[18] What is less certain, however, is whether employees and unions have a right to be heard with respect to all issues that arise in an employer-initiated case, or only with respect to the length of the abatement period.

The Commission has stated that "[an] employee representative electing party status has the right to litigate all the issues raised by the citation and complaint."[19] Where an employer is continuing to contest a citation (i.e., where the employer has not reached a settlement with the Labor Department), the courts have generally concluded that "Congress did not intend to limit the interest assertable by the union in an employer-initiated proceeding to the length of the abatement period * * *. The employees' request for party status confers jurisdiction on the Commission to entertain the employees' objections on all matters relating to the citation in question."[20] The Labor Department, which had previously taken the position that employees and unions have a right to be heard on matters other than the abatement date in cases where an employer is contesting a citation, reversed its position on that question in 1983.[21] Furthermore, *dicta* in some court of appeals decisions may suggest that those courts regard employee and union participation rights in all Commission proceedings to be limited to the question of the reasonableness of the abatement period.[22] However, in cases where an employer is con-

---

that intervention was not warranted because the exposed employees had no legally protected interest in the case, as the cited employer owed them no duty under the Act. Cf. *Pennsylvania Truck Lines*, 7 OSHC 1722 (Rev. Comm'n 1979) (employer not cited entitled to intervene where violation occurred on its property and involved activities it controlled and violation might affect safety of its own employees).

[18] See *Donovan v. OSHRC (Mobil Oil Corp.)*, *supra* note 11, at 927 n.13, 11 OSHC at 1616; *Marshall v. OSHRC (IMC Chem. Group)*, 635 F.2d 544, 552, 9 OSHC 1031, 1036 (6th Cir. 1980).

[19] *Southwestern Bell Tel. Co.*, 5 OSHC 1851, 1852 (Rev. Comm'n 1977).

[20] *Oil, Chem. & Atomic Workers v. OSHRC (American Cyanamid)*, *supra* note 13, at 648, 10 OSHC at 1348. Accord *Donovan v. Oil, Chem. & Atomic Workers (American Petrofina)*, 718 F.2d 1341, 1349, 11 OSHC 1689, 1694–1695 (5th Cir. 1983), *cert. denied*, 466 U.S. 971 (1984); *Donovan v. Allied Indus. Workers*, *supra* note 13, at 1419, 11 OSHC at 1739–1740. But see note 22, *infra*.

[21] See *Donovan v. Oil, Chem. & Atomic Workers (American Petrofina)*, *supra* note 20, at 1347–1348 n.25, 11 OSHC at 1693.

[22] In *Donovan v. Oil, Chem. & Atomic Workers (American Petrofina)*, *supra* note 20, the Fifth Circuit stated that three circuits—the Second, Third, and Sixth—have taken the position that even in employer-initiated cases, employees and unions may

tinuing to contest the Secretary's citation, as a practical matter employees and unions are generally heard with respect to all matters in dispute, and the theoretical question whether they have a right to be heard has generated little discussion other than in *dicta*. Much more important and controversial has been the question of employee and union rights in cases where, subsequent to the filing of a notice of contest, a dispute ceases to exist between the employer and the Secretary, and the only party opposing a proposed resolution of the case is an affected employee or union. To that question we now turn.

IV. Employee and Union Rights When the Secretary and the Employer Wish to Settle or Withdraw

More than 90 percent of all cases disposed of by Commission administrative law judges are terminated prior to a hearing by way of settlement, withdrawal by the Secretary of the citation, or withdrawal by the employer of the notice of contest.[23] These include numerous cases where the Secretary and the employer negotiate terms on which they propose to resolve a citation and, as well, cases where the Secretary simply decides that he no longer wishes to prosecute a citation. In either situation, it is often the case that the affected employees or unions are the only parties opposed to the proposed termination of the proceedings. Questions as to the rights of employees and unions in such circumstances have been the subject of much litigation.

---

be heard only on the issue of the length of time allowed for abatement. 718 F.2d at 1348, 1352, 11 OSHC at 1693–1694, 1697 citing *Marshall v. Sun Petroleum Prods.*, 622 F.2d 1176, 8 OSHC 1422 (3rd Cir.), *cert. denied*, 449 U.S. 1176 (1980); *Marshall v. Oil, Chem. & Atomic Workers (American Cyanamid)*, 647 F.2d 383, 9 OSHC 1584 (3d Cir. 1981); *Donovan v. OSHRC (Mobil Oil Co.)*, 713 F.2d 918, 11 OSHC 1609 (2d Cir. 1983); *Marshall v. OSHRC (IMC Chem. Group)*, *supra* note 18. However, in each of those cases the question presented was whether an employee or union could challenge matters other than the abatement date *after the employer had ceased to contest the citation* as a result of a settlement between the employer and the Secretary of Labor. The cited decisions of the Second, Third, and Sixth Circuits do not directly address the scope of employee and union participation rights in cases where the employer continues to contest a citation. The courts that have addressed that question have concluded that in such cases employees and unions may be heard on all issues. See cases cited in note 20, *supra*.

[23]See *Mobil Oil Corp.*, 10 OSHC 1905, 1932 n.18 (Rev. Comm'n 1982) (Rowland, Chmn., dissenting).

From its earliest days the Commission's rules have provided that once the Commission asserts jurisdiction over a case, any proposed settlement or withdrawal of notice of contest must be presented to the Commission by motion.[24] In a 1972 decision, *Dawson Brothers-Mechanical Contractors*, the Commission announced its general approach with respect to such motions:

> "The Commission will give hospitable consideration to stipulated withdrawals of notice of contest where the record reflects (1) the date on which abatement has been or will be accomplished; (2) assurance by the respondent of continuing compliance; (3) tender of payment of the penalty proposed by the Secretary of Labor; and (4) evidence that the affected employees or their authorized representatives have been afforded an opportunity to participate in the proceedings."[25]

For more than a decade thereafter, the Commission revisited various aspects of this problem, and many courts of appeals had occasion to review these questions.

---

[24]Both the interim rules issued by the Commission on Aug. 31, 1971, and the permanent regulations first promulgated on Sept. 28, 1972, provided for the Commission to review proposed settlements and motions to withdraw notices of contest. See 36 FED. REG. 169 (1971) (Interim Rules 23 and 11); 37 FED. REG. 20,239 (1972) (Permanent Rules 50 and 100). As amended, those regulations are now codified as Commission Rules 100 (Settlement) and 102 (Withdrawal of Notice of Contest).

[25]*Dawson Bros.-Mechanical Contractors*, 1 OSHC 1024, 1025 (Rev. Comm'n 1972). In subsequent cases the Commission consistently held that it has jurisdiction to review at least some aspects of a proposed settlement or withdrawal. See, *e.g., Raybestos Friction Materials Co.*, 9 OSHC 1141 (Rev. Comm'n 1980); *Nashua Corp.*, 9 OSHC 1113 (Rev. Comm'n 1980); *Weldship Corp.*, 8 OSHC 205 (Rev. Comm'n 1980); *Farmers Export Co.*, 8 OSHC 1655 (Rev. Comm'n 1980); *American Airlines*, 2 OSHC 1391 (Rev. Comm'n 1974); *Blaisdell Mfg.*, 1 OSHC 1406 (Rev. Comm'n 1973). But the Commission changed its position several times on the question of whether it may properly entertain employee or union objections that deal with matters other than the length of time allowed for abatement of a hazard.

Prior to 1977 the Commission assumed it had the authority to entertain union and employee objections to proposed settlements that involved matters other than the period of time allowed for abatement, and in fact the Commission regularly directed administrative law judges to consider such objections. See, e.g., *American Airlines, supra; Gurney Indus.*, 1 OSHC 1218 (Rev. Comm'n 1973). Then, in a 1977 decision the Commission held that unions and employees do not have a right to object to a proposed settlement, except insofar as they wish to challenge the length of the abatement period. *United States Steel Corp.*, 4 OSHC 2001 (Rev. Comm'n 1977). The Commission asserted that this result was compelled by its previous decision in *Local 588, UAW*, 4 OSHC 1243 (Rev. Comm'n 1976). In that same year (1977), the Commission held in *Southern Bell Tel. & Tel. Co.*, 5 OSHC 1405 (1977), that employees and unions have no right to object to the Secretary's motion to withdraw a citation.

Thereafter, in *IMC Chem. Group*, 6 OSHC 2075 (Rev. Comm'n 1980), the Commission overruled *Southern Bell*, and in *Mobil Oil Corp.*, 10 OSHC 1905 (Rev. Comm'n 1982), the Commission overruled *United States Steel*.

After a change in its composition, the Commission decided a series of cases in 1984 in which it turned away completely from the views it had expressed in *IMC*

The Supreme Court effectively put these matters to rest in *Cuyahoga Valley Railway v. United Transportation Union.*[26] There, the Secretary had cited the company for a violation of the OSH Act. The company contested the citation, whereupon the Secretary filed a complaint with the Commission and the company then filed its answer. The union moved for and was granted intervention by the Commission. At the hearing before the administrative law judge, the Secretary *sua sponte* moved to vacate the citation on legal grounds—namely, that the Federal Railway Administration had exclusive jurisdiction over the relevant safety conditions at issue. The union objected but the administrative law judge granted the Secretary's motion and vacated the citation. The Commission directed review of that order and then remanded the case to the administrative law judge for consideration of the union's objections. The Secretary noticed an appeal from the Commission's order. The Sixth Circuit affirmed the Commission's holding that it had authority to review the Secretary's decision to withdraw a citation.[27] The court recognized that under the scheme of the OSH Act the Secretary "has the sole authority to determine whether to prosecute" a violation of the Act.[28] However, the court found that the Secretary "had already made the decision to prosecute by filing a complaint and that [the] complaint already had been answered by the time the Secretary attempted to withdraw the citation."[29] Because the adversarial process was well advanced at the time the Secretary attempted to withdraw the citation, the court reasoned that the Commission, "as the adjudicative body, had control

---

and *Mobil Oil.* Thus, in *American Bakeries Co.*, 11 OSHC 2024 (Rev. Comm'n 1984), the Commission held, contrary to *IMC*, that "a union-party may not object to a motion by the Secretary to withdraw a citation." Subsequently, in *Copperweld Steel Co.*, 11 OSHC 2235 (Rev. Comm'n 1984), the Commission made clear that it will not entertain objections to the withdrawal of a citation even in a case that has been fully litigated and has been decided on the merits by an administrative law judge. In *Pan Am. World Airways*, 11 OSHC 2003 (Rev. Comm'n 1984), the Commission overruled *Mobil Oil*, holding "that a union lacks the right to object to the adequacy of the abatement methods specified in a settlement agreement between the Secretary and an employer, and that a union may object only to the reasonableness of the abatement period specified by the agreement." 11 OSHC at 2004. See also *Willamette Iron & Steel Co.*, 11 OSHC 1955 (Rev. Comm'n 1984).

[26]474 U.S. 3, 12 OSHC 1521 (1985).

[27]*Donovan v. United Transp. Union*, 748 F.2d 340, 12 OSHC 1057 (6th Cir. 1984).

[28]*Id.* at 343, 12 OSHC at 1059.

[29]*Id.*

over the case and the authority to review the Secretary's withdrawal of the citation."[30]

The Supreme Court granted the petitions for certiorari filed by the Secretary and the company, and simultaneously issued a *per curiam* opinion summarily reversing the Sixth Circuit. The Court initially observed that the Sixth Circuit's position previously had been rejected by eight other courts of appeals which held that the Secretary has unreviewable discretion to withdraw a citation against an employer.[31] Reasoning from the premise that the Secretary alone is charged with the responsibility of protecting employee rights and enforcing the Act and that the Secretary alone has the authority to determine if a citation should be issued to an employer for unsafe working conditions, the Court's conclusionary analysis reduced to one point: "A necessary adjunct of that power is the authority to withdraw a citation and enter into settlement discussions with the employer."[32] What this means, the Court held, is that the Commission has no authority to entertain an employee or union request to review the Secretary's decision not to issue or to withdraw a citation. The Court observed that its statutory analysis effectuated an important policy consideration as well: allowing the Commission to review the Secretary's decision to withdraw a citation would discourage the Secretary from seeking voluntary settlements with employers, thus unduly hampering enforcement of the Act.[33]

---

[30]*Id.*

[31]The Court stated:

"Vacating the citation thus did not rest solely on jurisdictional grounds. Nor did the Court of Appeals' decision sustaining the Commission's order focus on jurisdiction. Its holding would permit review by the Commission of the Secretary's withdrawal of any citation, whatever the reason, provided the adversarial process was sufficiently advanced to vest control in the Commission. For these reasons and because the issue relates to the statutory division of authority between the Secretary and the Commission, rather than the question of judicial review of administrative action, the case does not pose the question whether an agency's decision, resting on jurisdictional concerns, not to take enforcement action is presumptively immune from judicial review under §701(a)(2) of the Administrative Procedure Act. See *Heckler v. Chaney*, 470 U.S. 821, 833, n.4 (1985)."
Cuyahoga Valley Ry. v. United Transp. Union, supra note 26 at n.1.

[32]*Id.* at 7.

[33]*Id.*

V. Employee and Union Rights to Obtain Judicial Review of Commission Decisions

Section 11(a) of the Act provides that "[a]ny person adversely affected or aggrieved by an order of the Commission * * * may obtain a review of such order in * * * the court of appeals."[34] In *Oil, Chemical & Atomic Workers v. OSHRC (American Cyanamid)*,[35] the District of Columbia Circuit addressed the question whether a union that has elected party status in a Commission case initiated by an employer's notice of contest has a right to judicial review of a Commission order dismissing the citation, where the Secretary does not seek review.

The employer argued in *American Cyanamid* that the union could not appeal, because "the statute precludes the union from being heard on matters other than the reasonableness of the abatement period."[36] The District of Columbia Circuit rejected that argument because the court disagreed with its premise. The court concluded that in an employer-initiated case that is resolved by a Commission decision on the merits rather than by a settlement, employees and unions who elect party status have a right to be heard on "all matters relating to the citation in question."[37] The court therefore held that the union was "aggrieved" by the Commission's order dismissing the citation, and had a right to judicial review. The court added:

> "The union's right to appeal OSHRC decisions where it has participated as a party in the commission proceedings is, however, subject to two conditions, derived from the general statutory scheme and purpose of the Act. First, the union must give the Secretary notice of its intention to appeal and must serve him with copies of all of the pleadings. * * * * * Second, the case may become moot in those instances when the Secretary, participating in the appeal as an amicus curiae or as an intervenor, provides th[e] court with a clear and unconditional statement that he will not prosecute the claim regardless of the disposition of the appeal by th[e] court. The prosecutorial dis-

---

[34]29 U.S.C. §660(a).

[35]671 F.2d 643, 10 OSHC 1345 (D.C. Cir. 1982), *cert. denied*, 459 U.S. 905 (1982).

[36]*Id.* at 646, 10 OSHC at 1346.

[37]*Id.* at 648, 10 OSHC at 1348. As previously noted, it is not clear whether all courts would agree with the District of Columbia Circuit's position on this issue. See note 22 and accompanying text, *supra*.

cretion with which the Secretary is vested empowers him not to renew his prosecution effort even if this court were to find that the citation dismissed by the OSHRC asserted a violation under the Act."[38]

VI. Other Litigation Rights of Employees and Unions

The foregoing discussion has focused on employee and union rights in Commission proceedings initiated by a notice of contest with respect to a citation, and the right to judicial review of orders issued in such proceedings. There are certain other proceedings under the Act in which employees and unions may become parties; these warrant brief mention, although they have not engendered any disputes as to the scope of employee and union litigation rights.

A. Petitions for Modification of Abatement Dates

Where an employer petitions for modification of an abatement date,[39] affected employees and their union have a right to object and to have their objections heard by the Commission.[40]

B. Mandamus Actions in Situations of Imminent Danger

Section 13 of the Act authorizes the Secretary to obtain an injunction in federal court "to restrain any conditions or practices in any place of employment which are such that a danger exists which could reasonably be expected to cause death or serious physical harm immediately or before the imminence of such danger can be eliminated through the enforcement procedures otherwise provided by th[e] Act."[41] Sec-

---

[38]*American Cyanamid, supra* note 35, at 650–651, 10 OSHC at 1350.

[39]See Chapter 10 (Abatement), Section III.

[40]See previous Commission Rules 34(c)(2) and 34(d). See also *Auto Bolt & Nut Co.*, 7 OSHC 1203 (Rev. Comm'n 1979); *Aspro, Inc.*, 6 OSHC 1980 (Rev. Comm'n 1978) (decided under the old rules).

[41]29 U.S.C. §612.

tion 13(d) provides that an affected employee or his union may bring a mandamus action to compel the Secretary to seek such an injunction "if the Secretary arbitrarily or capriciously fails [to do so]." There are no reported cases under this provision.

C. Standards and Variances

The rights of employees and unions with respect to the promulgation and review of OSHA standards and variances are discussed in Chapter 19 (The Development of Occupational Safety and Health Standards) and Chapter 21 (Other Issues Related to Standards).

18

Judicial Review of Enforcement Proceedings

I. Conditions Precedent to Review

A. Statutory Provisions

Sections 11(a) and (b) of the OSH Act[1] set forth the basic requirements for judicial review of Review Commission orders in enforcement proceedings. In general, Section 11(a) of the Act contains provisions covering who may file for an appeal of a Commission order, in what court(s) the appeal may be filed, and the time limits for the filing of an appeal. It also establishes the relief courts may grant, the weight to be given findings of fact by the Commission, and the procedure for raising additional evidence not considered by the Commission during the enforcement proceedings. Section 11(b) deals with appeals by the Secretary of Labor from a Commission order, as well as with the ramifications of not filing an appeal within the specified time limit.

B. Who May Appeal (Parties)

Section 11(a) of the Act provides:

Any person adversely affected or aggrieved by an order of the Commission issued under subsection (c) of section 10 may obtain

---

[1]29 U.S.C. §§660(a) and (b) (1976).

a review of such order in any United States court of appeals for the circuit in which the violation is alleged to have occurred or where the employer has its principal office, or in the Court of Appeals for the District of Columbia Circuit ***."

The Secretary under Section 11(b) may also appeal a ruling of the Commission, but is limited to the court of appeals for the circuit in which the violation is alleged to have occurred or where the employer has its principal office.[2]

The primary participants to enforcement proceedings, the employer and the Secretary of Labor, are clearly parties "adversely affected or aggrieved" who may initiate proceedings for the review of a Commission final order in the courts.[3]

With respect to the Commission, two courts of appeals have recognized its right to appear as a party in the courts in order to defend its policies and decisions.[4] However, the majority of circuit courts have held that the Commission does not have the authority to participate as a party in judicial proceedings reviewing the Commission's own decisions.[5] The rationale for this viewpoint is clearly expressed in *Marshall v. Sun Petroleum Products*:[6]

---

[2]The timely filing of a notice of contest, the conduct of a hearing, and the issuance of a Commission order are prerequisites to the seeking of judicial review. See 29 U.S.C. §§659 (b) and (c) and Chapters 14 (Hearing Procedure) and 16 (The Occupational Safety and Health Review Commission).

[3]The rights of employees and unions to participate in judicial proceedings under the Act are discussed in Chapter 17 (Employee and Union Participation in Litigation Under the OSH Act).

[4]In *Brennan v. Gilles & Cotting, Inc.*, 504 F.2d 1255, 1267, 2 OSHC 1243, 1252 (4th Cir. 1974), the Fourth Circuit held that " it is appropriate for the Commission to appear in the courts of appeal to defend the policies Congress empowered it to adopt in adjudication." In *Diamond Roofing Co. v. OSHRC*, 528 F.2d 645, 648, n.8, 4 OSHC 1001, 1003 (5th Cir. 1976), the court noted that the Commission is "properly a party similar to other administrative agencies in suits by the Secretary or private parties to review its order."

[5]*Oil, Chem. & Atomic Workers v. OSHRC (American Cyanamid)*, 671 F.2d 643, 652, 10 OSHC 1345, 1351 (D.C. Cir. 1982), *cert. denied*, 459 U.S. 905 (1982); *Marshall v. OSHRC*, 635 F.2d 544 (6th Cir. 1980); *Dale Madden Constr. v. Hodgson*, 502 F.2d 278, 2 OSHC 1101 (9th Cir. 1974); *Marshall v. Sun Petroleum Prods.*, 622 F.2d 1176, 8 OSHC 1422 (3rd Cir.), *cert. denied*, 449 U.S. 1061 (1980); *General Elec. Co. v. OSHRC*, 583 F.2d 61, 6 OSHC 1868 (2nd Cir. 1978); *Brennan v. OSHRC* (Santa Fe Trail Transportation Co.), 505 F.2d 869, 2 OSHC 1274 (10th Cir. 1974). See also *Cuyahoga Valley Ry. v. United Transp. Union*, 474 U.S. 3, 12 OSHC 1521 (1985).

[6]622 F.2d 1176, 1184, 8 OSHC 1422, 1428 (3rd Cir. 1980). The court in *Oil, Chem. & Atomic Workers v. OSHRC (American Cyanamid)*, *supra* note 5, at 652, 10 OSHC at 1351, also stated:

"The OSHRC has no enforcement power comparable to the FCC, FERC or the NLRB. We believe the OSHRC should never be considered a proper statutory respondent under 19 U.S.C. §660(a). The commission was envisioned by its creators to be similar to a district court. It was established to settle disputes between employers and the Secretary of Labor over citations issued by the Secretary's

"[T]he Act grant the Secretary exclusive authority to enforce Commission decisions in the courts of appeal, 29 U.S.C. §600(b). Section 14 of the Act authorizes only the Secretary to conduct OSHA civil litigation, subject to the direction and control of the Attorney General. 29 U.S.C. §663. The Commission is not so entitled. Nor can it claim to participate as a party in appellate proceedings by virtue of being 'adversely affected or aggrieved' by its own decision. See 29 U.S.C. §600(a). At every turn, then, the statute denies the Commission the authority to appear in the court of appeals. We therefore conclude that the Review Commission was designed strictly as an independent adjudicator with no rule-making authority other than procedural rules for hearings, no direct policy role in administering the Act, and, accordingly, no right to independent presentation in judicial review procedures before this court."

Once a party has shown itself to be entitled to party status in judicial review proceedings, the party must then prove itself to be "adversely affected or aggrieved" by the Commission's final order before review will be initiated. Interpreting this phrase, the court in *Savina Home Industries v. Secretary of Labor*[7] stated that, to be adversely affected or aggrieved, a challenger must show that he has been injured by a Commission order, or suffered a chilling effect.

C. When an Appeal May Be Taken

Parties with standing to appeal a final order of the Review Commission must, within 60 days of the filing of the final order, file a petition with the appropriate court of appeals requesting that the order be modified or set aside.

---

inspectors. The commission, like a district court, has no duty or interest in defending its decision on appeal. As a purely adjudicative entity, it has no stake in the outcome of the litigation. Accordingly, the OSHRC may not participate in this court to review the commission decision as a party respondent in proceedings initiated by any party to the commission hearing. The role of the commission closely parallels the functions of the Benefits Review Board. This court has treated the Board as an independent adjudicator, similar to a district court and has dismissed it as a respondent. McCord v. Benefits Review Board, 514 F.2d 198 (D.C. Cir. 1975)."

[7]594 F.2d 1358, 1366, 7 OSHC 1154, 1160 (10th Cir. 1979). Cf. *Dan J. Sheehan Co. v. OSHRC*, 520 F.2d 1036, 1040, 3 OSHC 1573, 1575 (5th Cir. 1975). In *Savina* the employer lacked standing to raise objections to the Commission's authority to increase penalties because it had not shown it was aggrieved by a Commission decision, which had reduced the penalty. In *RSR Corp. v. Donovan*, 733 F.2d 1142, 11 OSHC 1953 (5th Cir. 1984), even though the case was remanded by the Commission to the A.L.J. for further findings, the fact that a penalty was imposed on the employer demonstrated it was affected and aggrieved.

Timely filing of a petition for review is a necessary prerequisite for obtaining appellate jurisdiction. In *Midway Industrial Contractors v. OSHRC*,[8] an employer's failure to file a written petition for review within the required 60 days from the date of filing the final order required the court to dismiss the petition for lack of jurisdiction. The time limit is strictly adhered to and may not be extended by pleading neglect; unlike the application of Fed. R. Civ. P. 4(a) in administrative proceedings, Fed. R. App. P. 15 does not permit late filing under such circumstances.[9] The date of filing with the court of appeals is the date the petition for judicial review is received at the court clerk's office, not the date on which it was mailed.[10]

If no timely petition for review is filed within 60 days after the service of the Commission's final order, the Commission's findings of fact, conclusions of law, and resultant orders are deemed conclusive in connection with any enforcement petition filed by the Secretary of Labor after the expiration of the 60-day period. Furthermore, the final orders of the Commission are considered uncontested, and the clerk of the court will, unless otherwise directed by the court, enter a decree enforcing the final order (since Commission orders are not self-enforcing), and send copies of the decree to the Secretary and the employer named in the petition.

D. How an Appeal Is Taken

The Federal Rules of Appellate Procedure, specifically Rules 15–20, set forth the applicable rules for the review and enforcement of the decisions of administrative agencies, boards, and commissions. The general rules for the maintenance of such actions are contained in Rules 25–48, and include, *inter alia*, page limitations, filing and service, motions, briefs, oral arguments, and costs. Additionally, each of the federal appellate courts issues its own local rules which supplement the federal rules.

---

[8]616 F.2d 346, 8 OSHC 1076 (7th Cir. 1980). See also *Hoerner Waldorf Pan Am. Bag Co. v. OSHRC*, 614 F.2d 795, 7 OSHC 2210 (1st Cir. 1980) (time period for review of unreviewed A.L.J. decision starts from date it becomes final order); *Consolidated Andy, Inc. v. Donovan*, 642 F.2d 778, 9 OSHC 1525 (5th Cir. 1981).

[9]*Consolidated Andy, Inc. v. Donovan, supra* note 8.

[10]*Heath & Stitch Inc. v. Donovan*, 641 F.2d 338, 9 OSHC 1448 (5th Cir. 1981).

Within 40 days after the filing of the petition for review with the court, the Commission must file with the court the record of the administrative proceedings. The record includes (unless the parties stipulate otherwise) the order sought to be reviewed or enforced; the findings or report on which it is based; and the pleadings, testimony, and proceedings before the Commission.[11] Upon filing the record, the Court's jurisdiction over the proceeding is asserted, giving it the power to grant temporary relief or a restraining order, and to make and enter upon the record a decree affirming, modifying, or setting aside, in whole or in part, the Commission's order, as well as enforcing the order to the extent such an order is affirmed or modified.[12]

The filing of a petition for review and the commencement of proceedings in the court of appeals does not operate as a stay of the Commission's final order, unless specifically ordered by the court. The aggrieved party should ordinarily seek an application for a stay of the final order from the Commission pursuant to Commission Rule 94[13]; however, if such a request is not practical or has been denied, an application for a stay of the final order can be requested from the court under Fed. R. App. P. 18. If the court grants a stay, it may condition the grant of such relief upon the filing of a bond or other appropriate security. If a stay is not obtained from either the Commission or the court, the Secretary of Labor may legitimately collect penalties or cite the employer for failure to abate.

II. Exhaustion of Administrative Remedies

A. Compliance With Statutory Requirements and Commission Procedure

Before the court will consider the questions, issues, and objections raised by a petitioning party on review, the party must have first exhausted the available administrative remedies and obtained a final order from the Review Commis-

---

[11]FED. RS. APP. P. 16 & 17.

[12]*Atlantic Drydock Corp. v. OSHRC*, 524 F.2d 467, 3 OSHC 1755 (5th Cir. 1975).

[13]29 C.F.R. §2200.94 (1985).

sion.[14] In *Continental Can Co. U.S.A. v. Marshall*,[15] the court explained that the basic purpose of the exhaustion of the administrative remedies is "to allow the administrative agency to perform functions within its special competence—to make a factual record, to apply its expertise and to correct its own errors so as to moot judicial controversies."

Under the Act, the petitioning party is required to avail himself of the established administrative procedures before obtaining judicial review. For instance, any employer who wishes to file a petition for court review must initially have filed a notice of contest of the citations and/or penalty with the Secretary within 15 working days after receiving notification of the citation.[16] If an employer does not contest the Secretary's citation within 15 working days, the citation will become a final order of the Commission and will not be subject to court review.[17] The notice of contest should indicate what the employer intends to contest—the citation (all or part), the penalty, or the abatement; for a notice of contest to one portion of the citation does not relieve the employer of its obligations to comply with the uncontested portion of the citation (which become binding upon the employer after the contest period expires).[18]

Once a citation has been timely contested, the petitioning party must obtain a hearing with an administrative law judge (A.L.J.). Once the ALJ issues a decision, a party must then seek administrative review. To obtain administrative review, the petitioning party must file a petition for discretionary review (PDR) with the Commission pursuant to Commission Rule 91.[19]

---

[14]This limitation appears not to be applicable to contentions raised by the responding parties. *RMI Co. v. Secretary of Labor*, 594 F.2d 566, 7 OSHC 1119 (6th Cir. 1979).

[15]603 F.2d 590, 597, 7 OSHC 1521, 1525 (7th Cir. 1979).

[16]However, some courts have held that such a requirement does not apply to a union contesting the reasonableness of the Secretary's order. *Marshall v. Oil, Chem. & Atomic Workers (American Cyanamid)*, 647 F.2d 383, 9 OSHC 1584 (3rd Cir. 1981).

[17]Section 10(a) of the Act; *Penn Cent. Transp. Co. v. OSHRC*, 535 F.2d 1249, 3 OSHC 2059 (4th Cir. 1976); *Brennan v. Winters Battery Mfg. Co.*, 531 F.2d 317, 3 OSHC 1775 (6th Cir. 1975).

[18]*Dan J. Sheehan Co. v. OSHRC, supra* note 7; *Penn Dixie Steel Corp. v. OSHRC*, 533 F.2d 1078, 5 OSHC 1315 (7th Cir. 1977).

[19]29 C.F.R. §2200.91. For greater discussion of the adminstrative procedures involved, see Chapter 16 (The Occupational Safety and Health Review Commission), Section II.C.1.

If the entire administrative procedure is not followed, the court will not grant judicial review. In *Keystone Roofing Co. v. OSHRC*,[20] the employer bypassed the Commission and petitioned for review of an A.L.J. decision directly with the court of appeals. The court held the employer had failed to exhaust its administrative remedies as it had not filed a PDR, and therefore the unreviewed A.L.J.'s decision became a final unreviewable order of the Commission. Even if an aggrieved party does not file a PDR, but a Commission member directs review *sua sponte*, the aggrieved party is required to respond and follow the Commission's procedural rules, including its invitation to file briefs,[21] or it will be precluded from judicial review of the Commission's final order.

B. Existence of Final Order

Even if a decision has been issued by the Commission following its review of the A.L.J. findings, the case may not yet be ready for judicial review. The threshold issue is whether the Commission's action constitutes an appealable agency ruling. Under the terms of the Act, for the court to have jurisdiction to review, the Commission is required to hold a hearing and thereafter issue an order which must affirm, modify, or vacate the Secretary of Labor's citation of proposed penalty, or direct other appropriate relief. Otherwise, it is not a final order subject to judicial review. In *Stripe-A-Zone Inc. v. OSHRC*,[22] it was stated:

> "We agree with the Fourth Circuit's analysis of Sections 10(c) and 11(a) of the Act. The Commission's order vacating and remanding [the A.L.J.'s] decision certainly did not affirm, modify or vacate the Secretary's proposed penalty, and we believe the phrase 'directing other appropriate relief' can refer only to the OSHRC decisions which order remedial measures after a de-

---

[20]539 F.2d 960, 4 OSHC 1481 (3rd Cir. 1976). See also *McGowan v. Marshall*, 604 F.2d 885, 7 OSHC 1842 (5th Cir. 1979).

[21]*McGowan v. Marshall, supra* note 20. If review is not directed, the report of the A.L.J. becomes the final order of the Commission, i.e., the necessary predicate for judicial review. *Modern Drop Forge Co. v. Secretary of Labor*, 683 F.2d 1105, 10 OSHC 1825 (7th Cir. 1982).

[22]643 F.2d 230, 233, 9 OSHC 1587, 1589 (5th Cir. 1981). See also *Noranda Aluminum v. OSHRC*, 650 F.2d 934, 9 OSHC 1894 (8th Cir. 1981); *Fieldcrest Mills v. OSHRC*, 545 F.2d 1384, 4 OSHC 1845 (4th Cir. 1976). But see *RSR Corp. v. Donovan*, 733 F.2d 1142, 11 OSHC 1953 (5th Cir. 1984).

termination on the merits of the allegations that the Act has been violated. Accordingly, the Commission's order remanding the case was not an order issued under Section 10(c) of the Act, and therefore it is not reviewable under Section 11(a) of the Act."

Thus, for example, a Commission interlocutory order which upholds the validity of an inspection or resolves a discovery dispute would not be a final order.[23] Nor would a Commission decision remanding a case to an A.L.J. for disposition normally be a final order.[24]

However, there are exceptions to the "final order" rule. The Supreme Court has held that if an order is collateral to the proceedings, a final order need not be issued for the court to grant judicial review:

"To come within the 'small class' of decisions exempted from the final judgment [rule articulated in *Cohen v. Beneficial Finance*, 337 U.S. 541 (1949)], the order must conclusively determine the disputed question, resolve an important issue completely separate from the merits of the action, and be effectively unreviewable on appeal from the final judgment."[25]

All three criteria identified in *Cohen* must be satisfied for the collateral order doctrine to apply. An appealable collateral order may not be incomplete, informal, or tentative, for *Cohen* does not apply to decisions that may be reconsidered or revised.

Several of the circuit courts have applied the *Cohen* tests and found a Commission order from which appeal was taken to be collateral since it disposed of the matter. Such appealable orders have included, *inter alia*, a Commission order which remanded a proposed settlement agreement to an A.L.J. for consideration of the union's objections to the method of abatement;[26] and a Commission order denying the Secretary's mo-

---

[23]*Robberson Steel Co. v. OSHRC*, 645 F.2d 22, 9 OSHC 1165 (10th Cir. 1980).

[24]*Fieldcrest Mills v. OSHRC, supra* note 22; *Chicago Bridge & Iron Co. v. OSHRC*, 514 F.2d 1082, 3 OSHC 1056 (7th Cir. 1975). But see *RSR Corp. v. Donovan*, 733 F.2d 1142, 11 OSHC 1953 (5th Cir. 1984) (Review Commission order remanding cause for further proceedings to determine amounts owed employees under medical removal protection is final order for purposes of judicial review).

[25]*Coopers & Lybrand v. Livesay*, 437 U.S. 463, 468 (1978). See also *Englehard Indus. v. OSHRC*, 713 F.2d 45, 11 OSHC 1630 (3d Cir. 1983).

[26]*Donovan v. OSHRC (Mobil Oil Corp.)*, 713 F.2d 918, 11 OSHC 1609 (2d Cir. 1983). See also *Marshall v. Oil, Chem. & Atomic Workers (American Cyanamid)*, *supra* note 16; *Donovan v. Allied Indus. Workers*, 722 F.2d 1415, 11 OSHC 1737 (8th Cir. 1983).

tion to vacate an order of review.[27] Other courts in applying *Cohen* have not found the Commission's action to fully determine a collateral claim of right. In one case, for example, where the Commission merely reserved and deferred ruling on a motion to withdraw a citation and petition for review, the action did not fully dispose of a claimed right.[28]

C. Raising Issues and Objections

Section 11(a) of the Act provides that "no objection that has not been argued before the Commission shall be considered by the court, unless the failure or neglect to urge such objection shall be excused because of extraordinary circumstances." This statutory requirement is similar to the procedural requirement of filing a PDR in that, unless raised during the exhaustion of administrative remedies, issues will be considered uncontested and hence unreviewable. In *Felton Construction Co. v. OSHRC*,[29] the A.L.J. rejected material documents proffered during the hearing. The employer did not raise an objection to the rejection of this evidence before the Commission, nor did it file a PDR. The A.L.J.'s decision became a final order. Thereafter, the employer filed a petition for review with the court of appeals and asked that the hearing be reopened so that additional evidence could be presented. The court rejected the petition, stating that the employer could have raised its objections at any time prior to the A.L.J.'s rendering of the decision, even in a brief to the A.L.J. Thus, Section 11(a) precluded the court's consideration of the employer's petition. A similar result was reached in *GAF Corp. v. OSHRC*,[30] where an employer's objection was barred from judicial review because it had not been raised before the Commission.

---

[27]*Donovan v. Oil, Chem. & Atomic Workers (American Petrofina)*, 718 F.2d 1341, 11 OSHC 1689 (5th Cir. 1983).

[28]*Donovan v. United Steelworkers, Local 2243*, 731 F.2d 345, 11 OSHC 1895 (6th Cir. 1984). See also *Englehard Indus. v. OSHRC*, 713 F.2d 45, 11 OSHC 1630 (3d Cir. 1983).

[29]518 F.2d 49, 3 OSHC 1269 (9th Cir. 1985). Although, as a general rule, the Review Commission has no jurisdiction over its final orders, it does have the authority to reconsider a final order pursuant to FED. R. CIV. P. 60(b), as long as the case has not come within the jurisdiction of the reviewing court. *J.I. Hass Co. v. OSHRC*, 648 F.2d 190, 9 OSHC 1712 (3d Cir. 1981).

[30]561 F.2d 913, 5 OSHC 1555 (D.C. Cir. 1977). See also *Stockwell Mfg. Co. v. Usery*, 536 F.2d 1306, 4 OSHC 1332 (10th Cir. 1976).

The term "objection" includes issues that would arise at administrative enforcement hearings, such as constitutional challenges, jurisdictional challenges, and affirmative defenses. The Commission has authority to rule on such objections and is, according to the majority of the courts,[31] the proper forum to address such subjects as:

- Fourth Amendment challenges and questions relating to the exclusion of evidence gathered in an inspection[32]
- the validity of OSHA's rulemaking procedure in the promulgation of standards[33]
- the particularity of citations and the availability of affirmative defenses[34]
- jurisdictional issues, both constitutional and statutory (such as §4(b)(1).[35]

Judicial intervention prior to an agency's determination of an issue is appropriate only where there is clear evidence that the exhaustion of administrative remedies will result in irreparable injury; where the agency's jurisdiction or competence is plainly lacking; and/or where the agency's special expertise would be of no help, would prove unavailing, or would be futile.[36]

---

[31]*Marshall v. Union Oil Co.*, 616 F.2d 1113, 8 OSHC 1169 (9th Cir. 1980); *Cape & Vineyard Div., New Bedford Gas & Edison Light Co. v. OSHRC*, 512 F.2d 1148, 2 OSHC 1628 (1st Cir. 1975); *Marshall v. Whittaker Corp., Berwick Forge & Fabricating Div.*, 610 F.2d 1141, 7 OSHC 1888 (3d Cir. 1979); *American Airlines v. Secretary of Labor*, 578 F.2d 38, 6 OSHC 1691 (2d Cir. 1978); *Marshall v. Central Mine Equip. Co.*, 608 F.2d 719, 7 OSHC 1907 (8th Cir. 1979).

[32]*Todd Shipyards Corp. v. Secretary of Labor.* 586 F.2d 683, 6 OSHC 2122 (9th Cir. 1978); *In re Worksite Inspection of Quality Prods*, 592 F.2d 611, 7 OSHC 1093 (1st Cir. 1979); *Babcock & Wilcox Co. v. Marshall*, 610 F.2d 1128, 7 OSHC 1880 (3d Cir. 1979).

[33]*Deering Milliken, Inc. v. OSHRC*, 630 F.2d 1094, 9 OSHC 1001 (5th Cir. 1980); *Marshall v. Union Oil Co.,supra* note 31; *American Airlines v. Secretary of Labor, supra* note 31. But see *National Indus. Constructors v. OSHRC*, 583 F.2d 1048, 6 OSHC 1914 (8th Cir. 1978).

[34]The failure to contest the merits of a citation or penalty will also bar the right to challenge a later civil collection action brought by the Secretary of Labor. *Marshall v. Painting by CDC*, 497 F. Supp. 653, 8 OSHC 2085 (E.D.N.Y. 1980); *Marshall v. Church Drilling Inc.*, 9 OSHC 1391 (W.D. Mo. 1981).

[35]*Marshall v. Able Contractors*, 573 F.2d 1055, 6 OSHC 1317 (9th Cir. 1978); *Marshall v. Burlington N.*, 595 F.2d 511, 7 OSHC 1314 (9th Cir. 1979); *Marshall v. Northwest Orient Airlines*, 574 F.2d 119, 6 OSHC 1481 (2d Cir. 1978).

[36]*Continental Can Co. U.S.A. v. Marshall*, 603 F.2d 590, 7 OSHC 1521 (7th Cir. 1979).

D. Extraordinary Circumstances

Even where the Commission is without authority to rule on an issue or objection such as fundamental challenge to the Act's constitutionality, the issue should nevertheless be raised at the administrative level so as to properly preserve it for judicial review.[37] Some courts have taken the contrary position that where the Commission has no authority to determine an issue, it need not be raised at the administrative level.[38] However, the majority of courts have found that "extraordinary circumstances" may excuse the failure to raise the issue—a viewpoint taken by the court in *McGowan v. Marshall*[39] (relying on its previous decision in *Buckeye Industries v. Secretary of Labor*[40]):

> "In *Buckeye* the court was faced with the question of whether it could consider the Secretary's collateral estoppel defense to the appellant's constitutional attack on Section 8(a) of the Act, although the defense had not been raised in proceedings before the Commission. ***
> "The court thus held the essence of Commission authority to rule on the constitutional question was, within the meaning of Section 660(a), an 'extraordinary circumstance' that excused the failure to urge the objection before the Commission.
> "This holding is especially applicable here. It would have been utterly fruitless for the appellant to have petitioned for the Commission review of his constitutional attacks on the Act since the Commission could not have ruled on these claims. Accordingly, his failure to seek discretionary review of these circumstances is excused by extraordinary circumstances within the meaning of section 660(a)."

---

[37]*Buckeye Indus. v. Secretary of Labor*, 587 F.2d 231, 6 OSHC 2181 (5th Cir. 1979).

[38]*Weyerhaeuser Co. v. Marshall*, 592 F.2d 373, 7 OSHC 1090 (7th Cir. 1979).

[39]604 F.2d 885, 892, 7 OSHC 1842, 1846 (5th Cir. 1979).

[40]587 F.2d 231, 235, 6 OSHC 2181 (5th Cir. 1979). In this case the court held that the objections limitation applied only to the adversely affected or aggrieved party who was seeking review of the Commission's order. It does not speak to contentions raised by the secondary party. In *RMI Co. v. Secretary of Labor*, 594 F.2d 566, 7 OSHC 1119 (6th Cir. 1979), because the Secretary had the burden of proving economic feasibility, the employer was not barred from raising the issue before the court, even though it was not raised before the A.L.J. or the Commission. See also, *Todd Shipyards Corp. v. Secretary of Labor*, 586 F.2d 683, 6 OSHC 2122 (9th Cir. 1978).

The court in *Lloyd C. Lockrem Co. v. United States*[41] also found "extraordinary conditions" where there had been two prior reviews of the same case by the Commission. Such a finding justified the employee's failure to file a second PDR. However, in *Continental Can Co. U.S.A. v. Marshall*,[42] the court stated that the fact that the Review Commission has previously rejected the argument which is being raised does not in itself constitute an extraordinary circumstance.[43] Unless it can be shown that extraordinary circumstances exist, the failure to raise issues and objections during the administrative hearing may preclude later review of those issues.

III. Standards of Judicial Review

A. General Overview

Judicial review of enforcement proceedings ensures that the administrative review and adjudication process is consistent with the mandate of the Act as well as with procedural and substantive due process. As the court in *Frank Irey Jr., Inc. v. OSHRC*[44] stated,

> "When the delegation of legislative authority to an administrative agency is broad in scope, the courts have a greater role to play to prevent or correct disparate treatment of those subjected to legislation. Furthermore, it is the duty of the courts to interpret the statute under which the agency functions and to determine whether the agency is acting within the Congressional purpose."

Although the Act itself does not contain standards for judicial review as to issues of law, Section 10(e) of the Administrative Procedure Act[45] provides that a reviewing court shall hold unlawful and set aside an agency decision if it is

---

[41]609 F.2d 940, 7 OSHC 1999 (9th Cir. 1979).

[42]603 F.2d 590, 7 OSHC 1521 (7th Cir. 1979).

[43]*Power Plant Div., Brown & Root, Inc. v. Donovan*, 659 F.2d 1291, 10 OSHC 1066 (1981), *rehearing*, 672 F.2d 111, 10 OSHC 1529 (5th Cir. 1982).

[44]519 F.2d 1200, 1206, 2 OSHC 1283, 1288 (3d Cir. 1975).

[45]5 U.S.C. §706(2) (1966).

(a) arbitrary, capricious, an abuse of discretion, or otherwise not in accordance with law;
(b) contrary to constitutional right, power, privilege, or immunity;
(c) in excess of statutory jurisdiction, authority, or limitations; or short of statutory right;
(d) without observance of procedure required by law;
(e) unsupported by substantial evidence in a case subject to Sections 556 and 557 of Title 5 or otherwise reviewed on the record of an agency hearing provided by statute; or
(f) unwarranted by the facts to the extent that the facts are subject to trial by *de novo* by the reviewing court.

In addition, courts have general supervisory authority to review administrative agency decisions.[46]

B. Evidentiary and Factual Findings

Section 11(a) of the Act provides that the findings of the Review Commission with respect to factual questions (and the inferences derived from them) shall be conclusive, provided that such findings are supported by *substantial evidence* on the record considered as a whole. "Substantial evidence" has been defined by the courts as such relevant evidence as a reasonable mind might accept as adequate to support a conclusion.[47] The rationale for applying the substantial evidence test to evidentiary and factual findings on judicial review is that the Commission has technical expertise and experience in the area of occupational safety and health. Furthermore, the Commission through the A.L.J. has lived with the case, conducted the hearing, heard the witnesses, and observed their

---

[46]*Greater Boston Television Corp. v. FCC*, 463 F.2d 268 (D.C. Cir. 1971); *General Dynamics Corp. v. OSHRC*, 599 F.2d 453, 7 OSHC 1373 (1st Cir. 1979).

[47]*Marshall v. Knutson Constr. Co.*, 566 F.2d 596, 6 OSHC 1077 (8th Cir. 1977); *Martin Painting & Coating Co. v. Marshall*, 629 F.2d 437, 8 OSHC 2173 (6th Cir. 1980). Substantial evidence must be enough to warrant denial of a motion for directed verdict in a civil case tried by a jury. *Dunlop v. Rockwell Int'l*, 540 F.2d 1283, 4 OSHC 1606 (6th Cir. 1976). The evidence a reasonable mind might accept as adequate to support a conclusion is surely less in a case where it stands entirely unrebutted in the record by a party having full possession of all the facts than in a case where there is contrary evidence to detract from its weight. *Astra-Pharmaceutical Prods. v. Donovan*, 681 F.2d 69, 10 OSHC 1697 (1st Cir. 1982).

demeanor; it has firsthand knowledge of the case and veracity and credibility of the witnesses. Thus, the Commission's determination of factual matters is accorded substantial deference.[48]

Although all 12 of the circuit courts have applied the substantial evidence test in reviewing the Commission's findings on evidentiary and factual matters, they have varied in their interpretation of the test. For example, in *Super Excavators v. Secretary of Labor*,[49] the court affirmed the finding of a violation of the trenching standard, 29 C.F.R. §1926.652, even though it doubted that if it were the trier of fact it would have believed the Secretary's opinion over that of the employer. However, in *Austin Road Co. v. OSHRC*,[50] the court found the Commission's finding regarding employer coverage and impact on interstate commerce to be so speculative and conclusive that it overturned the decision as not supported by substantial evidence.

Use of the substantial evidence test permeates the entire judicial review process. For example, an important initial determination during review of Commission decisions is whether the issue or finding in question should be characterized as legal or factual.[51] The courts have held that if the Secretary of Labor fails to meet his burden of providing all elements of a violation, the Commission cannot establish the missing elements by opinion and conjecture and cannot characterize issues as findings of fact when they are actually conclusions of law.[52]

With respect to alleged violations of the general duty clause, Section 5(a)(1) of the Act, where evidence of the employee's actual knowledge is relied upon as proof that a hazard is "recognized," the Secretary of Labor has the burden of demonstrating by substantial evidence that the employer's safety precautions were unacceptable in its industry or a relevant

[48]*Accu-Namics, Inc. v. OSHRC*, 515 F.2d 838, 3 OSHC 1299 (5th Cir. 1975); *General Dynamics Corp. v. OSHRC*, 599 F.2d 453, 7 OSHC 1373 (1st Cir. 1979); *National Steel & Shipbuilding Co. v. OSHRC*, 607 F.2d 311, 7 OSHC 1837 (9th Cir. 1979).

[49]674 F.2d 592, 10 OSHC 1369 (7th Cir. 1981).

[50]683 F.2d 905, 10 OSHC 1943 (5th Cir. 1982).

[51]*Electric Smith, Inc. v. Secretary of Labor*, 666 F.2d 1267, 10 OSHC 1329 (9th Cir. 1982).

[52]*Marshall v. CFI Steel Corp.*, 576 F.2d 809, 6 OSHC 1543 (10th Cir. 1978).

industry as measured by the standard of a reasonably conscientious safety expert familiar with the pertinent industry.[53] Where constructive knowledge is the basis for showing the existence of a recognized hazard, substantial evidence must include evidence of the custom and practice in the employer's industry relative to the hazard involved.[54] For instance, in *General Electric Co. v. OSHRC*,[55] the Second Circuit found that unsupported generalizations on the part of the Commission could not support a finding of employer violation of the general duty clause. In *Southern Ohio Building Systems v. Marshall*,[56] the Sixth Circuit reversed a Commission decision, stating that an obvious danger is not enough to establish that a hazard is recognized unless there is substantial evidence to demonstrate that either the employer or the industry in general knew of the particular hazard.

Construction and application of the substantial evidence requirements also vary when dealing with violations of specific hazards. In *Bethlehem Steel Corp. v. OSHRC*,[57] the employer attacked the accuracy of an A.L.J.'s findings regarding the existence of a serious violation on the grounds that the A.L.J. failed to specify the evidentiary basis upon which a violation may be found. The court agreed:

> "The Administrative Procedure Act provides in relevant part: all decisions, including initial *** decisions are part of the record and shall include a statement of (A) findings and conclusions, and the reasons or basis therefor, on all material issues of fact, law, or discretion prosecuted on the record.*** 5 U.S.C. Section 557(e).
>
> "At a minimum, the ALJ's findings in this case should have indicated the evidentiary basis for his conclusion.*** Yet examination of the ALJ's decision reveals that absolutely no findings were made with respect to the alleged seriousness of the violation.
>
> "We believe the ALJ's findings do not comport with the minimum requirements of the Administrative Procedure Act,

---

[53]*Magma Copper Co. v. Marshall*, 608 F.2d 373, 7 OSHC 1893 (9th Cir. 1979); *General Dynamics Corp. v. OSHRC*,599 F.2d 453, 7 OSHC 1373 (1st Cir. 1979) (application of harmless error rule).

[54]*Usery v. Marquette Cement Mfg. Co.*, 568 F.2d 902, 5 OSHC 1793 (2d Cir. 1977); *National Realty & Constr. Co. v. OSHRC*, 489 F.2d 1257, 1 OSHC 1422 (D.C. Cir. 1973).

[55]540 F.2d 67, 4 OSHC 1512 (2d Cir. 1976).

[56]649 F.2d 456, 9 OSHC 1848 (6th Cir. 1981). But see *Donovan v. Missouri Farmers Ass'n*, 674 F.2d 690, 10 OSHC 1460 (8th Cir. 1982).

[57]607 F.2d 1069, 7 OSHC 1833 (3d Cir. 1979).

[and] we hold that the ALJ's findings of fact are inadequate to sustain the government charges."[58]

In most cases, however, the substantial evidence rule is not an issue. If the Commission sets forth the basis for its factual findings and such findings are supportable by the record, the reviewing court allows the Commission considerable latitude and discretion in arriving at those findings and drawing inferences from them,[59] even if the reviewing court would have reached a different result *de novo*.[60] The courts generally will not substitute their judgment for that of the Commission.

C. Questions of Law and Policy

Generally, the courts show great deference to the Commission's interpretation of the Act, and a lesser, but still significant, deference to its interpretation of OSHA standards. In reviewing questions of law or legislative policy, the courts apply the "arbitrary and capricious" or "abuse of discretion" test, which is highly deferential to agency actions by presuming such actions to be valid. In its application, courts will normally not inquire into the wisdom of an agency's exercise of discretion[61] and will uphold the agency's decision unless it finds it to be arbitrary and capricious, not in accordance with law, or in excess of the authority granted to the Commission under the Act.

Under the "arbitrary and capricious" standard, the scope of review is narrower than that employed under the substantial evidence test, since a reviewing court may only consider whether the Commission's decision was based upon a consideration of the relevant factors and whether there has been a clear error of judgment. Although this inquiry into the facts is to be searching and careful, the ultimate standard of review

---

[58]*Id.* at 1073, 7 OSHC at 1836.

[59]Factual determinations based on the credibility of witnesses are basically considered nonreviewable, unless contradicted by incontrovertible documentary evidence or physical facts. *International Harvester Co. v. OSHRC*, 628 F.2d 982, 8 OSHC 1780 (7th Cir. 1980); *Olin Constr. Co. v. OSHRC*, 525 F.2d 464, 3 OSHC 1526 (2d Cir. 1975).

[60]*A. Schonbek & Co. v. Donovan*, 646 F.2d 799, 9 OSHC 1562 (2d Cir. 1981); *Donovan v. Daniel Constr. Co.*, 692 F.2d 818, 10 OSHC 2188 (1st Cir. 1982).

[61]*Everglades Sugar Ref. v. Donovan*, 658 F.2d 1076, 10 OSHC 1035 (5th Cir. 1981).

is a narrow one. Minor procedural violations are commonly overlooked in order to ensure that cases are decided on their merits. However, the Commission must articulate a rational connection between the facts found and the choice made.[62]

Although in theory greater deference is accorded to the Commission's interpretation of the Act than to its interpretation of standards and regulation,[63] it appears that the difference in judicial decisions on this question depends not so much on whether an interpretation of the Act or a standard is in question, but rather whether the Commission's interpretation is reasonable, provides fair notice, and is consistent with the Act's purpose.[64] Thus, there should be no greater weight accorded, for instance, to the Commission's interpretation of the general duty clause than to its interpretation of 29 C.F.R. §1926.28(a) or 29 C.F.R. §1910.105(a) (the personal protective equipment standards), as both involve questions of the recognition of a hazard, custom and practice in the industry, and feasibility. Furthermore, actual court decisions suggest support of such a view.[65]

It also appears that deference to the Commission and application of the "arbitrary and capricious" standard depends in large part on whether the Commission's interpretation of the Act or a standard is in agreement with that of the Secretary of Labor. In such a situation the court would be reluctant to impose its own interpretation.[66] The court in *California*

---

[62]*Marshall v. Knutson Constr. Co.*, 566 F.2d 596, 6 OSHC 1077 (8th Cir. 1977), quoting from *Bowman Transp. v. Arkansas Best Freight Sys.*, 419 U.S. 281 (1974). Factual findings include such matters as the existence of a violation and the appropriateness of a penalty.

[63]Although in many cases the courts have deferred to the Commission's statutory interpretations, e.g., *Intercounty Constr. Co. v. OSHRC*, 522 F.2d 777, 3 OSHC 1337 (4th Cir. 1975) (definition of a willful violation), in other cases there has been no such deferring, e.g., *Bethlehem Steel Corp. v. OSHRC*, 540 F.2d 157, 4 OSHC 1451 (3d Cir. 1976)(definition of repeated violation).

[64]*General Elec. Co. v. OSHRC*, 583 F.2d 61, 6 OSHC 1868 (2d Cir. 1978); *Wray Elec. Constr. v. Secretary of Labor*, 638 F.2d 914, 9 OSHC 1077 (6th Cir. 1980); *L.R. Willson & Sons v. Donovan*, 685 F.2d 664, 10 OSHC 1881 (D.C. Cir. 1982); *J.L. Foti Constr. Co. v. OSHRC*, 687 F.2d 853, 10 OSHC 1937 (6th Cir. 1982).

[65]See *B & B Insulation v. OSHRC*, 583 F.2d 1364, 6 OSHC 2062 (5th Cir. 1978); *Cape & Vineyard Div., v. OSHRC*, 512 F.2d 1148, 2 OSHC 1628 (1st Cir. 1975); *Voegele Co. v. OSHRC*, 625 F.2d 1075, 8 OSHC 1631 (3d Cir. 1980); *United Technologies Corp., Pratt & Whitney Aircraft Div. v. Secretary of Labor*, 649 F.2d 96, 9 OSHC 1554 (2nd Cir. 1981); *Fry's Tank Serv. & Cities Serv. Oil Co. v. Marshall*, 577 F.2d 126, 6 OSHC 1631 (10th Cir. 1978).

[66]*Clarkson Constr. Co. v. OSHRC*, 531 F.2d 451, 3 OSHC 1880 (10th Cir. 1976); *Wisconsin Elec. Power Co. v. OSHRC*, 567 F.2d 735, 6 OSHC 1137 (7th Cir. 1977); *Lloyd C. Lockrem Co. v. United States*, 609 F.2d 940, 7 OSHC 1999 (9th Cir. 1979).

Stevedore & Ballast Co. v. OSHRC[67] deferred to the Secretary of Labor's definition of a serious violation primarily because the Commission also agreed with that interpretation. However, if those interpretations are not reasonable and consistent with the purpose of the Act, then the fact that the Commission and the Secretary are in agreement will have no bearing on whether the court would defer to their interpretation.[68]

If the Secretary and the Commission disagree on the interpretation of a statutory provision or standard, the courts have differing approaches to the question of deference.[69] For example, in *Usery v. Hermitage Concrete Pipe Co.*,[70] the Sixth Circuit stated: "Where, as here, the Secretary of Labor and the Commission differ over the construction of the Act, we have indicated that the Commission's ruling is entitled to great deference." Most courts are in accord with *Hermitage Concrete Pipe* and give a more significant degree of deference to the Commission's interpretation (rather than the Secretary's) in those cases where its position is different from that of the Secretary's.[71] However, in *Brennan v. OSHRC (Kessler & Sons Construction)*,[72] the Tenth Circuit accepted the Secretary's interpretation of Section 10 of the Act rather than the Commission's because a question of enforcement of the provision was involved.

Intertwined with the application of the "arbitrary and capricious" and "abuse of discretion" standards of review is

---

[67]517 F.2d 986, 3 OSHC 1174 (9th Cir. 1975). See also *Western Waterproofing v. Marshall*, 576 F.2d 139, 6 OSHC 1550 (8th Cir. 1978).

[68]*Wisconsin Elec. Power Co. v. OSHRC, supra* note 66; *General Elec. Co. v. OSHRC, supra* note 64.

[69]The "rules" of deference do not apply when a decision of an A.L.J. is involved. In *Donovan v. Anheuser-Busch, Inc.*, 666 F.2d 315, 10 OSHC 1193 (8th Cir. 1981), the court refused to defer to the A.L.J.'s findings since (1) the report was only a recommended decision until adopted by the Commission; (2) the findings had no precedential value since they were not reviewed by the Commission; and (3) there was no authoritative interpretation of the standard involved available. See also *CCI, Inc. v. OSHRC*, 688 F.2d 88, 10 OSHC 1718 (10th Cir. 1982); *Accu-Namics, Inc. v. OSHRC*, 515 F.2d 828, 3 OSHC 1299 (5th Cir. 1975), *cert. denied*, 425 U.S. 903 (1976).

[70]584 F.2d 127, 132, 6 OSHC 1886, 1889 (6th Cir. 1978).

[71]*Budd Co. v. OSHRC*, 513 F.2d 201, 2 OSHC 1698 (3d Cir. 1975); *Diebold Inc. v. Marshall*, 585 F.2d 1327, 6 OSHC 2002 (6th Cir. 1978); *Marshall v. Anaconda Co.*, 596 F.2d 370, 7 OSHC 1382 (9th Cir. 1979); *Donovan v. Castle & Cooke Foods*, 692 F.2d 641, 10 OSHC 2169 (9th Cir. 1982).

[72]513 F.2d 553, 2 OSHC 1668 (10th Cir. 1975). See also *Marshall v. Western Elec.*, 565 F.2d 240, 5 OSHC 2054 (2d Cir. 1977); *GAF Corp. v. OSHRC*, 561 F.2d 913, 5 OSHC 1555 (D.C. Cir. 1977).

the additional basis for review under the Administrative Procedure Act of "otherwise not in accordance with law." This standard is a catchall developed to ensure that a correct legal standard was used, that correct legal conclusions were reached, that overall legal consistency was followed, and a correct legal formulation was applied.[73]

D. Effect of Judicial Decision Making[74]

In reaching its decisions the Commission has taken the position that it will consider the views of the courts but, unless reversed by the Supreme Court, feels obligated to establish its own precedent and is not bound to acquiesce to the views of courts whose decisions are in conflict with the Commission's.[75] Although the Commission may generally decline to follow the holdings of courts with which it disagrees, it appears that it is required to adhere to the holdings established by a court if the case to be decided is within the jurisdiction of the court. In *Jones & Laughlin Steel Corp. v. Marshall*,[76] the court held that in dealing with matters within the court's jurisdiction, an agency is not free to apply its own views in contravention of the court's mandated precedent. Moreover, since the principle that an administrative agency is required to follow the court's instructions on remand is well established, the Commission's failure to follow a court's instructions in that circuit would be reversible error.[77]

---

[73]5 K. Davis, ADMINISTRATIVE LAW 332–344 (1984).

[74]See also discussion in Chapter 16 (The Occupational Safety and Health Review Commission), Section II.E.

[75]*Western Mass. Elec. Co.*, 9 OSHC 1940 (Rev. Comm'n 1981); *Farmers Export Co.*, 8 OSHC 1665 (Rev. Comm'n 1980); *B.J. Hughes, Inc.*, 7 OSHC 1471 (Rev. Comm'n 1979).

[76]636 F.2d 32 (3d Cir. 1980). But see *S & H Riggers & Erectors v. Donovan*, 659 F.2d 1273, 10 OSHC 1057 (5th Cir. 1981), in which it was stated that "we assume without deciding that the Commission is free to decide not to follow decisions of the courts of appeal with which it disagrees, even in cases arising in those circuits."

[77]*Butler Lime & Cement Co. v. OSHRC*, 658 F.2d 544, 9 OSHC 2169 (7th Cir. 1981).

Part V

Development and
Promulgation of Standards

19

The Development of Occupational Safety and Health Standards

Ethylene oxide (EtO), a colorless gas with a characteristic ether-like odor, has been manufactured in the United States since 1925.[1] OSHA's regulation of EtO began in 1971 with the agency's adoption of the established federal standard for EtO, which had originally been issued under the Walsh-Healey Public Contracts Act.[2] In 1981, the Public Citizen Health Research Group petitioned OSHA for an emergency temporary standard for EtO on the ground that the 1971 standard did not protect employees adequately against cancer and other serious illness. OSHA denied the petition and was sued in the district court; in 1983, a court of appeals ordered OSHA to commence a rulemaking proceeding within 30 days to issue a standard lowering the permissible exposure level and affording employees fuller protection from EtO-induced cancer and other chronic health risks. OSHA conducted the rulemaking and a final standard was issued in June 1984, 14 months after the proposal was published.

The publication of the final standard did not end the controversy, however. The standard reduced the 8-hour permis-

---

[1] OSHA's final standard on EtO, published on June 22, 1984, contains a background section entitled "Physical Properties, Manufacture and Uses of Ethylene Oxide," 49 FED. REG. 25,734 (1984).

[2] 36 FED. REG. 10,466, 10,504, Table G-1 (1971). OSHA's authority to adopt established federal standards without rulemaking is discussed *infra* Section I.

sible exposure limit (PEL)[3] for the chemical from 50 parts per million (ppm) to 1 ppm. A short-term exposure limit, that is, the average EtO level permitted in any 15-minute period, was deleted shortly before publication of the standard, in response to "reservations" expressed by the Office of Management and Budget (OMB) to OSHA's "final draft standard."[4] OSHA then undertook further rulemaking on the issue of the short-term exposure limit;[5] and on January 2, 1985, it issued a Supplemental Statement of Reasons for the Final Rule, in which it determined that a short-term limit for EtO was not warranted.[6] OSHA's decision not to include a short-term exposure limit was challenged in the District of Columbia Circuit and the court remanded the proceeding to OSHA, concluding that substantial evidence did not support the agency's determination.[7] On July 21, 1985, just short of one year after the remand, the court of appeals issued a decision asserting that OSHA would be found in contempt if it did not make a final decision on the short-term exposure level issue by March 1988.[8]

This case study on OSHA's regulation of EtO will seek to illuminate the often labyrinthine corridors of OSHA's rulemaking activity.[9] This is not to suggest that this proceeding is typical of OSHA rulemaking; on the contrary, it is far more complex than most proceedings. The regulation of EtO was selected as a case study because it encompasses such an unusual variety of important regulatory issues. Thus, there was both a court suit *to compel OSHA to initiate rulemaking*, and,

---

[3]The PEL is normally expressed as a time-weighted average (TWA) and represents the amount of exposure to a toxic substance, averaged over an 8-hour workday, that is permitted by the standard. This and related concepts are discussed in B. Mintz, OSHA: HISTORY, LAW, AND POLICY 90–91 (1984).

[4]See 49 FED. REG. at 25,775 (1984). OMB authority to review OSHA regulatory actions derives from Executive Order 12,291 (1981). See discussion *infra* Section IV.B.3.a.

[5]49 FED. REG. 36,659 (1984).

[6]50 FED. REG. 64 (1985).

[7]*Public Citizen Health Research Group v. Tyson*, 796 F.2d 1479, 12 OSHC 1905 (D.C. Cir. 1986).

[8]*Public Citizen Health Research Group v. Brock*, ___ F.2d ___, 13 OSHC 1362 (D.C. Cir. 1987).

[9]Prof. Kenneth Culp Davis, a widely respected expert in the field of administrative law, has referred to rulemaking under the Administrative Procedure Act as "one of the greatest inventions of modern Government." Davis, ADMINISTRATIVE LAW TEXT §6.03 (3d ed. 1972). Prof. Steven Kelman, who has written extensively on the activities of OSHA, referred to its standards process as "Byzantine in its complexity." Kelman, *Occupational Safety and Health Administration*, in THE POLITICS OF REGULATION at 236 (J. Wilson ed. 1980).

later, another court proceeding to determine *whether the standard issued by OSHA was valid.* During the rulemaking proceeding, there was active involvement by the judicial, legislative, and executive branches of the federal government, with OMB playing a major role in the late stages of the administrative decision-making process. As in most health standards rulemaking, this proceeding raised critical questions regarding the scope and limits of agency discretion in providing protection against cancer-causing substances; the role of various branches of the federal government in constraining the broad reach of administrative discretion; and the legal, policy, and evidentiary bases for agency decision-making. In addition to these broad issues of regulatory and administrative policy, the EtO rulemaking raised several specific "science-policy" issues:[10] whether the standard should impose a short-term exposure limit in addition to the 8-hour PEL; whether the elements of the medical surveillance program should be mandatory; and whether OSHA itself should identify operations where it is likely that engineering controls will be infeasible.

In sum, this proceeding, dramatically and in vivid detail, portrays the intricacies of the relationship among administrative agencies, Congress, the courts, and the President in policy rulemaking in the field of public health. In the following pages, these issues, the roles of the major actors, and the history of other OSHA proceedings are interwoven to provide a background necessary to an understanding of both the EtO case history in particular, and OSHA rulemaking in general.

I. 1971: Initial Promulgation

Section 5(a)(2) of the OSH Act requires employers to comply with standards issued by the Secretary of Labor.[11] By defining employer compliance obligations primarily by means of substantive general rules with prospective effect (occupa-

---

[10]The term has been used by Prof. Thomas McGarity to describe the major issues in OSHA and Environmental Protection Agency rulemaking. McGarity, *Substantive and Procedural Discretion in Administrative Resolution of Science Policy Questions: Regulating Carcinogens in EPA and OSHA*, 67 GEO. L.J. 729, 731–747 (1979).

[11]29 U.S.C. §654(a)(2) (1982). Employers are also required to comply with the general duty clause (§5(a)(1)). The Secretary of Labor has delegated his responsibilities under the Act to the Assistant Secretary for OSHA.

tional safety and health standards), Congress achieved two salutary results: (1) it provided a mechanism for defining the obligations of covered employers in advance, and thereby avoided the imposition of new duties after the fact on a case-by-case basis; and (2) it permitted the agency to benefit from public views obtained in rulemaking proceedings before prescribing the obligations of employers.[12] The Act expressly conferred on OSHA authority to issue three different types of standards: Section 6(b) standards, promulgated after rulemaking;[13] emergency temporary standards (Section 6(c) standards), promulgated without rulemaking after OSHA finds that there is "grave danger" to employees and that the standard is needed to protect employees from the danger;[14] and Section 6(a), or start-up, standards, issued without rulemaking during the first two years after the Act's effective date. The most important reasons for authorizing the promulgation of start-up standards was the need to provide OSHA with a comprehensive group of standards for a speedy start for its enforcement program.[15] Start-up standards consisted of those established federal standards that had previously been adopted by the Department of Labor under pre-OSHA statutes such as the Walsh-Healey Public Contracts Act and the Longshoremen's and Harbor Workers' Compensation Act, and standards that had previously been adopted by certain private organizations in accordance with procedures allowing at least some measure of public comment before they were issued (national consensus standards).[16]

Although the Act gave OSHA a period of two years to issue start-up standards, the agency promulgated those standards quickly, motivated at least in part by the desire to satisfy substantial public and congressional pressure for the agency to impose immediate and specific compliance requirements on

---

[12]See *National Petroleum Refiners Ass'n v. FTC*, 482 F.2d 672 (D.C. Cir. 1973), *cert. denied*, 415 U.S. 951 (1974).

[13]Rulemaking procedures are set forth in detail in §6(b)(1) through (5).

[14]Emergency temporary standards are discussed *infra* Section II.

[15]The delays in OSHA's promulgation of standards after rulemaking are discussed generally in REPORT FROM THE OFFICE OF THE CHAIRMAN, ADMINISTRATIVE CONFERENCE OF THE UNITED STATES, TO THE ASSISTANT SECRETARY FOR OCCUPATIONAL SAFETY AND HEALTH ON OSHA RULEMAKING PROCEDURES (1987).

[16]"National consensus standard" is defined in §3(9) of the Act. The legislative history expressly refers to the American National Standards Institute and the National Fire Protection Association as the major organizational sources of national consensus standards. S. REP. NO. 1282, 91st Cong., 2d Sess. at 6 (1970).

employers for the protection of employees. A substantial corpus of these standards, both national consensus standards and established federal standards, was issued by OSHA on May 28, 1971, only a month after the Act became effective.[17] OSHA's enforcement of the national consensus safety standards, most of which were not intended by the voluntary standards groups to be legally enforceable and which were issued by OSHA after only minimal review, has been the subject of continuing criticism. Indeed, Lane Kirkland, President of the AFL-CIO and a strong exponent of OSHA enforcement, told an oversight committee of Congress in 1980 that "[i]n retrospect [OSHA] made a major mistake in 1971 when it hastily issued, en masse, a ramshackle collection of so-called 'national consensus' standards" and that "this hodge-podge collection of standards and OSHA's early efforts to enforce them, probably did more to damage the initial acceptance of the entire program than any other single action."[18] OSHA eventually deleted approximately 600 of these national consensus standards, almost all in the safety area, after it had determined that these standards were unenforceable or that their revocation would not jeopardize employee safety in the workplace.[19] In addition, OSHA has undertaken to review all of its 1971 national consensus safety standards and to revise them to the extent necessary, including updating and simplifying them. Two comprehensive revisions have already been completed: those pertaining to electrical hazards and to fire protection requirements.[20]

OSHA's major difficulty with start-up safety standards was that in some cases they were unnecessarily stringent and detailed and unrelated to employee safety. The health standards adopted by OSHA in 1971, on the other hand, were often inadequate to provide necessary protection from serious illnesses that were job-related. The Senate Labor and Public

---

[17]36 FED. REG. 10,466 (1971). Most of the standards adopted were safety standards, reflecting the fact that more was known at the time about safety hazards and precautions than about health hazards and controls. However, as discussed below, some health standards were also adopted in 1971. In general, OSHA established a 90-day delayed effective date for the standards issued.

[18]*Oversight on the Administration of the Occupational Safety and Health Act, 1980: Hearings Before the Senate Comm. on Labor and Human Resources*, 96th Cong., 2d Sess. 730 (1980).

[19]43 FED. REG. 49,726 (1978).

[20]45 FED. REG. 60,656 (1980) (codified at 29 C.F.R. §§1910.155–.165) (1983) (fire protection); 46 FED. REG. 4034 (1981) (codified at 29 C.F.R. §§1910.301–.308) (1983) (electrical hazards).

Welfare Committee acknowledged this when it reported the bill that later became the Act; the Committee said that the Section 6(a) standards "may not be as effective or as up-to-date as is desirable" and that they would provide only a "minimum level of health and safety."[21] The health standards issued in 1971 were mostly established federal standards previously adopted under the Walsh-Healey Public Contracts Act; the ultimate source of these standards was the so-called Threshold Limit Values (TLVs) contained in the health standards adopted in 1968 by the American Conference of Governmental Industrial Hygienists (ACGIH). In addition, some health standards were national consensus standards that had been issued by the American National Standards Institute (ANSI). Many of the permissible exposure limits in these 1971 standards were only guidelines. Even in cases where the permissible exposure limit was believed to be sufficiently protective when issued in 1971, it often became clear on the basis of newly discovered evidence, such as new studies establishing the carcinogenicity of the substances, that the standard was far from adequate a decade later.

The 50-ppm PEL in the standard for EtO issued in 1971 was derived from the ACGIH 1968 TLV, which had been adopted in 1969 by the Department of Labor under the Walsh-Healey Act. This TLV was based on the then available evidence on the toxic effects of EtO caused by cutaneous contact with solutions of the compound. The effects were primarily irritation and burns of the skin. Chronic health effects caused by EtO had not been reported up to that time.[22] However, major new animal and human studies conducted during the 1970s and published beginning in 1979 established the carcinogenicity, mutagenicity, and possible reproductive effects of EtO.[23] These findings led directly to the filing of the petition for an emergency temporary standard in 1981.

---

[21]S. REP. No. 1282, 91st Cong., 2d Sess. at 6 (1970). The committee statement was also true to a more limited extent about the §6(a) safety standards.

[22]See 48 FED. REG. at 17,286–287 (1983) (preamble to proposed EtO standard).

[23]The studies are summarized in the proposed EtO standard, 48 FED. REG. at 17,287–292 (1983), and in the final standard, 49 FED. REG. at 25,734, 25,738–755 (1984). ACGIH itself lowered its TLV to 1 ppm, effective in 1984, and NIOSH recommended that EtO be treated as a human carcinogen. 49 FED. REG. at 25,736. The Environmental Protection Agency, acting under the Federal Insecticide, Fungicide and Rodenticide Act (FIFRA), 7 U.S.C. §§135 *et seq.* (1947), has published notices providing guidance on methods of reducing EtO exposures to manufacturers of certain EtO sterilants used in hospitals and health care facilities, see 49 FED. REG. at 25,737.

II. Petition for Emergency Standard

In August 1981, the Public Citizen Health Research Group (HRG), a public interest organization, petitioned OSHA to issue an emergency temporary standard (ETS) for EtO under Section 6(c). According to HRG, new studies demonstrated that the existing PEL was inadequate to protect employees. In September 1981, OSHA denied the petition, stating that "the available evidence did not indicate that an emergency situation existed."[24] OSHA recognized, however, the need to reevaluate the existing EtO standard by means of rulemaking, and four months later it published a notice requesting public data and views on the issue.[25]

OSHA's decision to reject the petition for an ETS should not be viewed in isolation. As of August 1981, when this petition was received, OSHA had issued eight ETSs, four of which were either stayed, vacated, or both, by courts of appeals.[26] Three of the others were not challenged, and were ultimately replaced by permanent standards.[27] One emergency standard, for acrylonitrile, was challenged, but the court of appeals denied a stay and the challenge was withdrawn.[28] Subsequently, in November 1983, OSHA issued a second ETS for asbestos, which was later indefinitely stayed by the Fifth Circuit.[29]

---

[24]See 48 FED. REG. at 17,287 (1983). Section 6(c) of the Act requires OSHA to issue an "emergency temporary standard" if it determines that "employees are exposed to grave danger from exposure to substances or agents determined to be toxic or physically harmful or from new hazards" and that the emergency standard is "necessary" to protect employees from the danger. While standards issued under §6(c) are characterized as "emergency" standards, the provision does not state explicitly that a finding of an "emergency situation" is necessary.

[25]See discussion *infra* Section II.B on advance notices of proposed rulemaking.

[26]ETS for organophosphorous pesticides, 38 FED. REG. 10,715 (1973), vacated in *Florida Peach Growers Ass'n v. Department of Labor*, 489 F.2d 120, 1 OSHC 1472 (5th Cir. 1974); ETS for ethyleneimine (EI) and 3,3' dichlorobenzidine (DCB) 38 FED. REG. 10,929 (1973), vacated in *Dry Color Mfrs. Ass'n v. Department of Labor*, 486 F.2d 98, 1 OSHC 1331 (3d Cir. 1973); ETS for diving operations, 41 FED. REG. 24,272 (1976), stayed in *Taylor Diving & Salvage Co. v. Department of Labor*, 537 F.2d 819, 4 OSHC 1511 (5th Cir. 1976); ETS for benzene, 42 FED. REG. 22,516 (1977), stayed in *Industrial Union Dep't v. Bingham*, 570 F.2d 965, 6 OSHC 1107 (D.C. Cir. 1977). A second ETS on asbestos was issued, and vacated, subsequent to the EtO petition. See *infra* note 29.

[27]Asbestos, 37 FED. REG. 11,318 (1972); vinyl chloride, 39 FED. REG. 35,890 (1974); DBCP, 42 FED. REG. 45,536 (1977).

[28]*Vistron Corp. v. OSHA*, 6 OSHC 1483 (6th Cir. 1978). The ETS appears at 43 FED. REG. 2586 (1978).

[29]48 FED. REG. 51,086 (1983), stayed in *Asbestos Information Ass'n v. OSHA*, 727 F.2d 415, 11 OSHC 1817 (5th Cir. 1984).

The bases for the refusals by the courts of appeals to permit challenged ETSs to become effective have varied. Only in the case of the pesticide standard and the second asbestos ETS did the Fifth Circuit expressly hold that no grave danger existed that warranted the issuing of an ETS.[30] The other ETSs were vacated and stayed essentially for "procedural" reasons.[31] But, apart from the specific rationale for the decisions, the courts of appeals have repeatedly admonished OSHA to utilize its emergency authority cautiously because no public procedures precede the imposition of obligations on the employer. For example, in the *Dry Color* case,[32] the Third Circuit warned that "emergency temporary standards should be considered an unusual response to exceptional circumstances. The courts should not permit temporary emergency standards to be used as a technique for avoiding the procedural safeguards of public comment and hearing ***."[33] Significantly, then Assistant Secretary Eula Bingham, who had earlier spoken strongly in favor of OSHA's use of the emergency authority,[34] ultimately endorsed a more constrained view. In the Carcinogens Policy, issued toward the close of her administration, she noted that even in dealing with carcinogenic substances, "the level of the Agency's resources including compliance, legal and technical personnel, at the given time, may suggest that employee health may be more effectively protected by concentrating those resources in work on permanent standards."[35] However, as OSHA's experience with vinyl chloride and DBCP[36] clearly demonstrated, where there is a substantial consensus among the affected groups on the need for an emergency standard, the swift issuance of an ETS can perform an important function in publicizing the seriousness of the

---

[30]*Florida Peach Growers Ass'n, supra* note 26, at 120, 1 OSHC at 1479–1482; *Asbestos Information Ass'n, supra* note 29, at 415, 11 OSHC at 1825.

[31]E.g., the EI and DCB standards were vacated because of the inadequacy of OSHA's statement of reasons. *Dry Color Manufacturers' Ass'n, supra* note 26, at 98, 1 OSHC at 1336–1337. However, the Third Circuit in that case also expressed reservations as to whether a grave danger existed.

[32]*Supra* note 26.

[33]*Supra* note 26, at 104–106 n.9a.

[34]E.g., *Performance of the Occupational Safety and Health Administration: Hearing before the Subcomm. on Manpower and Housing of the House Comm. on Government Operations*, 95th Cong., 1st Sess. 77–78 (1977) (testimony of Dr. Bingham).

[35]45 FED. REG. 5002, 5215–5216 (1980).

[36]1,2-Dibromo-3-chloropropane.

hazard and in defining the manner in which employees can be protected from grave danger.

III. 1981–1983: Litigation

A. District Court Litigation

Even while its request for an ETS was being reviewed by OSHA, HRG brought an action in the U.S. District Court for the District of Columbia, seeking as a remedy that the court order OSHA to issue an EtO standard. The EtO suit was not brought under Section 6(f), which expressly provides for court of appeals review of standards that have already been issued. Rather, the suit was brought under the Administrative Procedure Act,[37] with HRG seeking affirmative relief from OSHA and arguing that OSHA unreasonably delayed action in regulating EtO.[38] While the District of Columbia Circuit has held that agency denials of requests for rulemaking are not exempt from judicial review, it has also made clear that "judicial intrusion into an agency's exercise of discretion in the discharge of its essentially legislative rulemaking functions should be severely circumscribed."[39] This very limited review is in contrast to court of appeals review of standards already issued, which, under Section 6(f) of the Act, is conducted on the basis of the "substantial evidence" test.[40]

During the past decade, there have been a number of suits, mostly brought by unions and public interest groups, to compel OSHA to issue standards. These suits often reflected public

---

[37]5 U.S.C. §706(1) (1982).

[38]These delays have been the subject of extensive comment. See, e.g., Comptroller General of the United States, Report to Congress, DELAYS IN SETTING WORKPLACE STANDARDS FOR CANCER-CAUSING AND OTHER DANGEROUS SUBSTANCES, HRD-77-71 (May 10, 1977).

[39]*WWHT, Inc. v. FCC*, 656 F.2d 807 (D.C. Cir. 1981). The Supreme Court recently held that a decision by an agency not to enforce a statute is "presumptively unreviewable." *Heckler v. Chaney*, 470 U.S. 821, 832 (1985). The Court majority left open the issue of judicial review of an agency refusal to initiate rulemaking. 470 U.S. at 825 n.2. See also, Note, *Judicial Review of Administrative Inaction*, 83 COLUM. L. REV. 627, 680–689 (1983) (discussing particularly court review of OSHA's decision not to issue a pesticide standard). Under a recent court of appeals decision, *Telecommunications Research & Action Center v. FCC*, 750 F.2d 70 (D.C. Cir. 1984), the suit to compel OSHA to issue a standard would now have to be initiated in the court of appeals and not in the district court.

[40]The scope of court review under the substantial evidence test was first adumbrated by the District of Columbia Circuit in reviewing OSHA's asbestos standard. *Industrial Union Dep't v. Hodgson*, 499 F.2d 467 (D.C. Cir. 1974). See discussion in

impatience with the slow pace of standards development at OSHA, particularly in protecting employees against exposure to carcinogenic substances. None of these suits, at least not until the EtO case, resulted in a court order to issue a standard; however, in several cases—e.g., the ETS standard for pesticides—the pendency of the suit may have been a contributing factor to OSHA's decision to promulgate a standard.[41]

The lengthiest of these lawsuits was *National Congress of Hispanic American Citizens v. Marshall,*[42] which was commenced in 1973. The action sought to compel OSHA to issue a standard that would require sanitation facilities for field workers. OSHA resisted issuing the standard,[43] partly on the basis of its claim that it had more pressing priorities.[44] The district court ordered OSHA several times to proceed promptly to issue the standard.[45] OSHA appealed in each instance and the court of appeals twice reversed the district court, holding that more deference should be paid to the Secretary's discretion in establishing priorities among standards.[46] However, the court of appeals insisted that OSHA exhibit good faith in establishing these priorities and required that it submit to the district court a timetable that would indicate when the field sanitation standard would in fact be issued. The district

---

OSHA and the Courts: An Uneasy Partnership, in Mintz, *supra* note 3, at 165–206. For an article arguing for less court deference to administrative agencies regarding their scientific and technical rulemaking determinations, see Stever, *Deference to Administrative Agencies in Federal Environmental, Health and Safety Litigation—Thoughts on Varying Judicial Application of the Rule,* 6 W. NEW ENG. L. REV. 35 (1983).

[41]*See* discussion in Mintz, *supra* note 3, at 97–107.

[42]626 F.2d 882, 7 OSHC 2029 (D.C. Cir. 1979).

[43]In 1976, OSHA issued a proposed field sanitation standard, 41 FED. REG. 17,576 (1976), but was subjected to strong criticism from Congress in large part because of this proposal. See, e.g., 122 CONG. REC. 20,367 (1976) (statement of Rep. Skubitz). OSHA never acted on the proposal. See discussion in Mintz, *supra* note 3, at 19.

[44]Section 6(g) states that, in determining standards priorities, OSHA should give "due regard" to the "urgency of the need" for mandatory safety and health standards for particular industries, trades, crafts, occupations, businesses, workplaces, or work environments, and to the recommendations of NIOSH. OSHA's priorities are discussed in *National Congress of Hispanic Am. Citizens v. Marshall, supra* note 42, at 886, 7 OSHC at 2034 (D.C. Cir. 1979).

[45]E.g., *National Congress of Hispanic Am. Citizens v. Dunlop,* 425 F. Supp. 900 (1975).

[46]*National Congress of Hispanic Am. Citizens v. Usery,* 554 F.2d 1196, 5 OSHC 1255 (D.C. Cir. 1977); *National Congress of Hispanic Am. Citizens v. Marshall, supra* note 42, at 882, 7 OSHC at 2033.

court rejected the timetable,[47] but the lengthy litigation was settled in 1982. However, after completing further rulemaking, OSHA, on April 16, 1985, made a "Final Determination" that it would not issue a field sanitation standard. This determination was challenged and the Court of Appeals for the District of Columbia Circuit directed OSHA to issue a final standard within 30 days, describing the proceeding as a "disgraceful chapter of legal neglect."[48] The final standard was issued on April 28, 1987.[49]

In the EtO case, the district court concluded that the record before OSHA "presented a solid and certain foundation showing that workers are subjected to grave health dangers from exposure to ethylene oxide at levels within the currently permissible range" and therefore that OSHA's decision not to issue an ETS for ethylene oxide was an "abuse of discretion."[50] The court accordingly ordered OSHA to issue an ETS within 20 days. OSHA quickly appealed the decision, obtaining a stay from the court of appeals of the requirement that the ETS be issued.

B. Court of Appeals Litigation

The court of appeals considered the case on an expedited schedule and in February 1983 unanimously decided the case in a *per curiam* opinion. While reversing the decision of the district court on the issuance of the ETS, the court nonetheless ordered OSHA to issue an EtO proposal within 30 days, and directed OSHA to proceed on a "priority, expedited basis, and to issue a permanent standard as promptly as possible, well in advance of the current, latter part of 1984 estimate."[51]

OSHA conceded before the court that the standard for EtO then in effect was inadequate to protect against carcinogenic risk. It argued, however, that the "average" exposure

---

[47]*National Congress of Hispanic Am. Citizens v. Donovan*, No. 2142–73 (D.D.C. Oct. 30, 1981).

[48]*Farmworker Justice Fund v. Brock*, 811 F.2d 613, 13 OSHC 1049 (D.C. Cir. 1987).

[49]52 FED. REG. 16,053 (1987).

[50]*Public Citizen Health Research Group v. Auchter*, 554 F. Supp. 242, 251, 11 OSHC 1049, 1055 (D.D.C. 1983), *rev'd in part*, 702 F.2d 1150, 11 OSHC 1209 (D.C. Cir. 1983).

[51]*Id.*, 702 F.2d at 1159, 11 OSHC at 1215. The court rejected OSHA's argument that OSHA's authority to regulate EtO had been preempted under §4(b)(1) of the Act by EPA's regulation of EtO under FIFRA. 702 F.2d at 1156 n.23, 11 OSHC at 1213.

of most employees to EtO was at the 10 ppm level and at that level the danger was not sufficiently grave to warrant the issuance of an ETS. In deciding that OSHA's conclusion on this issue was "rational," the court of appeals emphasized the absence of evidence in the record contradicting OSHA's estimated "average" level and the "absence of clear scientific evidence" on the degree of harm from exposure at 10 ppm and below. Although noting that OSHA's evidence was "impressionistic," the court stated that it was "hesitant to *compel* the Assistant Secretary to grant extraordinary relief."[52]

At the same time, the court of appeals concluded—particularly in light of the fact that "some workers" are exposed to levels above the 10 ppm average and therefore "encounter a potentially grave danger to both their health and the health of their progeny"—that OSHA's unaccounted for delay in initiating a rulemaking on EtO was action "unreasonably delayed," within the meaning of the Administrative Procedure Act.[53] The court of appeals noted that in denying the petition for an ETS in September 1981 OSHA emphasized its intention to commence rulemaking proceedings under Section 6(b), but that *"no notice or proposed rulemaking has yet been issued."*[54] Instead, the court said, OSHA in January 1982 published an Advance Notice of Proposed Rulemaking (ANPR) announcing its intention to conduct the reevaluation of the EtO standard and asking interested parties to submit data, views, and comments to aid in the process.[55] In the ANPR, OSHA posed a number of specific questions connected with the regulation of EtO on such issues as the appropriate PEL, the feasibility of available control methods, the numbers of workers exposed, and methods of monitoring the substance. Although the comment period on the notice ended in March 1982, with OSHA receiving about 50 written responses to the ANPR, OSHA had not taken any further steps toward issuing a standard.

ANPRs are not required or even mentioned in the Administrative Procedure Act or the OSH Act, which contemplate that rulemaking will typically be initiated by the publication of a proposed rule. Federal administrative agen-

---

[52]*Id.* at 1156–1157, 11 OSHC at 1214 (emphasis by court).
[53]*Id.* at 1158, 11 OSHC at 1215.
[54]*Id.* at 1157, 11 OSHC at 1214 (emphasis by court).
[55]47 FED. REG. 3566 (1982).

cies have increasingly utilized ANPRs to facilitate agency collection of data to be used in formulating a proposed rule. OSHA's use of ANPRs became more extensive when, beginning in 1977, the agency discontinued its reliance on standards advisory committees as a source of recommendations for proposed standards.[56] OSHA has also solicited the comments of interested persons—employers, unions, experts—on drafts of proposed standards.[57] Indeed, the EtO proceeding docket contains "informal comments received from interested parties who reviewed the March 30, 1983 draft ethylene oxide proposal."[58]

IV. 1983: Proposed Standard

A. Hybrid Rulemaking

The EtO proposal, published by OSHA on April 21, 1983, invited all interested parties to submit written data, views, and comments during a 60-day period ending June 17, 1983.[59] In the notice, OSHA also scheduled an "informal public hearing" to be held on July 18, 1983, during which members of the public would have an opportunity to present oral testimony.[60]

---

[56]Section 6(b) allows but does not require OSHA to convene an advisory committee to make recommendations on proposed standards. If an advisory committee is established, it must act in accordance with statutory procedures and time frames. See *Synthetic Organic Chem. Mfrs. Ass'n v. Brennan*, 506 F.2d 385 (3d Cir. 1974), *cert. denied*, 423 U.S. 830 (1975) (vacating OSHA's MOCA (4,4' methylene bis (2-chloroaniline)) standard). The composition of the advisory committee is governed by §7(b). Early OSHA health standards, such as those for coke oven emissions and vinyl chloride, were preceded by comprehensive recommendations of advisory committees.

[57]To achieve the same purpose, OSHA has also scheduled preproposal public information-gathering meetings. See, e.g., *Procedure for Revision of Safety Standards*, 41 Fed. Reg. 17,100 (1976).

[58]Exhibit 7, Attachment VI, in the EtO docket. The proposal was issued April 21, 1983. At the earliest stage of a rulemaking proceeding, OSHA establishes at its Technical Data Center in the Department of Labor in Washington, D.C., a docket which constitutes the record for the proceeding. The EtO docket number is H-200. All documents related to the proceeding are placed in the record and are given an exhibit number. This record is the basis for OSHA's decision in issuing a final standard and for any resulting court review. The record in the proceeding is available for public inspection and copying. On the issue of records in informal rulemaking, see Pedersen, *Formal Records and Informal Rulemaking*, 85 Yale L.J. 38 (1975).

[59]48 Fed. Reg. 17,284, 17,309 (1983).

[60]*Id.* The EtO hearing is discussed *infra* Section V.B.

Section 6(b) of the OSH Act sets forth the procedures that OSHA must follow in adopting, revising, and modifying standards. These basic steps are the publication of a proposed standard; an opportunity for public comment; a public hearing, if requested; and a final standard, with a statement of reasons explaining the action,[61] or a determination, accompanied by a statement of reasons, that a standard should not be issued.

In its preamble to the EtO proposal, OSHA, as it always does, characterized its rulemaking proceedings as "informal." Under traditional principles of administrative law, rulemaking is either formal or informal. Formal rulemaking is similar to adjudication and generally involves more rigid, time consuming, and trial-like, adversarial procedures. Informal rulemaking, on the other hand, is merely so-called "notice and comment" rulemaking, with the minimal procedural requirements given in the previous paragraph: publication of a proposal, opportunity for written comment, and publication of a final rule, with a brief statement of reasons.[62] As government rules became more controversial and burdensome, affecting broad sectors of the economy with increased impact, the minimum informal procedures under the Administrative Procedure Act were not adequate to encompass the complexities of the rulemaking proceedings that were being conducted. In some cases, as in the OSH Act, an increased formalization of informal rulemaking took place as a result of action by Congress; in other cases, it resulted from judicial decisions.[63] This more formalized informal rulemaking has often been referred to as "hybrid rulemaking."[64] The OSH Act added several procedural requirements to the minimum informal rulemaking steps required under the APA. In particular, the Act required that a public hearing be held, if requested.[65] In addition, the

---

[61]The requirement for a statement of reasons is set forth in §6(e) and is discussed in *Dry Color Mfrs. Ass'n v. Department of Labor*, 486 F.2d 98, 1 OSHC 1331 (3d Cir. 1973).

[62]The "classic" model for informal rulemaking is prescribed in the Administrative Procedure Act, 5 U.S.C. §553 (1982). See DeLong, *Informal Rulemaking and the Integration of Law and Policy*, 65 VA. L. REV. 257 (1979).

[63]See, e.g., *Ethyl Corp. v. EPA*, 541 F.2d 1 (D.C. Cir.), *cert. denied*, 426 U.S. 941 (1976). For a discussion of OSHA rulemaking, see Kestenbaum, *Rulemaking Beyond APA: Criteria for Trial-Type Procedures and the FTC Improvement Act*, 44 GEO. WASH. L. REV. 679, 682–685 (1976).

[64]A number of articles dealing with the evolution of informal rulemaking are cited in DeLong, *supra* note 62, at 260 n.22.

[65]§6(c)(3), 29 U.S.C. §655(c)(3).

Act provided that the "substantial evidence" rule, normally applied in court review of formal rulemaking and adjudication, would be the basis of court review of OSHA standards.[66] On the basis of the statutory language and legislative history of the Act, OSHA promulgated regulations in 1971[67] on standards rulemaking that confirmed that OSHA's standards rulemaking process would be informal but that more than the bare procedural essentials of informal rulemaking would be used. Specifically, the regulations required that a hearing examiner (now an administrative law judge) preside at the hearing; that there be an opportunity for cross-examination on crucial issues; and that a verbatim transcript of the hearing be kept. In the asbestos case,[68] the District of Columbia Circuit commended OSHA for adopting rulemaking regulations that went beyond the minimum procedural requirements of the statute by providing for evidentiary hearings that would furnish a record for court of appeals review under Section 6(f). The basic procedures for OSHA standards rulemaking established in 1971 have remained in effect to the present; as will be discussed,[69] however, a number of changes have taken place in the mechanics of the hearings.

B. EtO Proposal Preamble

1. General Features of the Preamble

The preamble to the EtO proposal filled 25 printed pages in the *Federal Register*.[70] Preambles to OSHA standards are currently far more extensive than those published in OSHA's early history. For example, the preamble to the final asbestos standard issued in 1972 was two pages long and dealt only cursorily with the issues in the proceeding.[71]

---

[66]§6(f), 29 U.S.C. §655(f). The "substantial evidence" test for court review is discussed in *Universal Camera Corp. v. National Labor Relations Bd.*, 340 U.S. 474 (1951).

[67]29 C.F.R. §1911.15 (1984).

[68]*Industrial Union Dep't v. Hodgson*, 499 F.2d 467, 1 OSHC 1631 (D.C. Cir. 1974) (asbestos standard upheld for most part).

[69]See discussion *infra* Section V.B.

[70]49 FED. REG. 25,734 (1984). The preamble to the final EtO standard was considerably longer. See *infra* Section V. The preamble to the final lead standard was more than 200 pages. 43 FED. REG. 52,952 (1978); 43 FED. REG. 54,354 (1978).

[71]37 FED. REG. 11,318 (1972).

There are several related reasons for the increasing length of preambles. OSHA's early attempts to limit the length of its statement of reasons may be attributed in part to its attempt to issue standards as quickly as possible. However, several courts, relying particularly on Section 6(e)[72] of the OSH Act, vacated standards, or portions of standards, because of the inadequacy of the statement of reasons.[73] The court of appeals in the asbestos case made it clear that the need for a full statement of reasons was no less pressing where, as in the case of OSHA, the agency is obliged to make policy judgments and no factual certainties exist.[74] In another OSHA case,[75] the District of Columbia Circuit emphasized that a full statement of reasons for a regulatory action enables the courts to perform their review function more effectively; helps the public to understand and react to the action of the agency; and helps the agency to develop an orderly process of decision making.

To satisfy the requirements of Section 6(e) as construed by the courts, OSHA now routinely explains in its preambles the reasons for its numerous "decisions"—i.e., resolution of issues—in the standard, as well as the reasons for its rejection of contrary views. These lengthy discussions of OSHA determinations reflect the increased scope, complexity, and controversial nature of standards proceedings, particularly in regard to the health effects of the regulated substance and the feasibility of the standard.

The major components of the preamble to all occupational health standards proposals, including the EtO proposal, are the following: a section on the "health effects" associated with the substance being regulated; a summary of the regulatory impact of the standard; and a summary and explanation of

---

[72]Section 6(e) provides in relevant part: "Whenever the Secretary promulgates any standard, *** he shall include a statement of the reasons for such action, which shall be published in the Federal Register."

[73]*Dry Color Mfrs. Ass'n v. Department of Labor, supra* note 61; *Associated Indus. v. Department of Labor,* 487 F.2d 342 (2d Cir. 1973) (vacating in part OSHA's lavatory standard). Many other court decisions underscore the importance of agency articulation of the reasons for its action. See, e.g., *Amoco Oil Co. v. EPA,* 501 F.2d 722 (D.C. Cir. 1974).

[74]*Industrial Union Dep't v. Hodgson, supra* note 68, at 476, 1 OSHC at 1637–1638.

[75]*AFL-CIO v. Marshall,* 617 F.2d 636, 651–652, 7 OSHC 1775, 1783 (D.C. Cir. 1979) (affirming for the most part OSHA's cotton dust standard), *aff'd sub nom. American Textile Mfrs. Inst. v. Donovan,* 452 U.S. 490, 9 OSHC 1913 (1981).

the proposed standard.[76] Customarily, the text of the proposed standard and various appendices follow the preamble.

2. Health Effects Section

The health effects section of the EtO proposal preamble discusses the bases for OSHA's determination that the substance must be regulated and for the specific PEL selected. Following the Supreme Court decision in *Industrial Union Department v. American Petroleum Institute*,[77] in which the Court decided that OSHA must show by "substantial evidence" that a "significant risk" is being addressed by a proposed standard and that the standard would eliminate or substantially reduce the risk, OSHA began its present practice of routinely performing risk assessments in developing health standards. Risk assessments are a statistical method used to extrapolate from risk at relatively high exposure levels to risk at lower exposure levels; the latter are typically the levels at which employees are likely to be exposed.[78] Prior to the benzene decision,[79] OSHA, as a matter of policy, refused to conduct quantitative risk assessments in regulating carcinogenic substances. Instead, the agency had adopted a "lowest feasible level" policy, taking the position that, given present scientific knowledge, *no* level of exposure to a carcinogen could be considered safe for a given population; therefore, prudent health policy to assure that "no employee" will suffer material impairment of health[80] requires that the permissible level set

---

[76]Preambles to final standards have essentially the same basic format. Thus, the preamble to the final EtO standard discussed the health effects of EtO, 49 FED. REG. at 25,737–755 (1984); summarized the final Regulatory Impact and Regulatory Flexibility Analysis that was placed in the record, *id.* at 25,766–773; and explained the reasons for its adoption of the individual requirements of the EtO standard, *id.* at 25,773–796. See discussion of final standard *infra* Section V.

[77]448 U.S. 607, 8 OSHC 1586 (1980). The plurality consisted of Chief Justice Burger and Justices Stevens, Stewart, and Powell. There were several concurring opinions and a dissent by four Justices.

[78]See Congress of the United States, Office of Technology Assessment, ASSESS-MENT OF TECHNOLOGIES FOR DETERMINING CANCER RISKS FROM THE ENVIRONMENT 157–172 (1981). Data in both animal and human studies typically relate to exposures at levels which are much higher than the level at which regulation is contemplated. Where animal data are used, OSHA must also make the determination that a carcinogenic risk to humans is present. See, e.g., *Synthetic Organic Chem. Mfrs. Ass'n v. Brennan*, 506 F.2d 385, 2 OSHC 1402 (3d Cir. 1974) (standard for 13 carcinogens affirmed), *cert. denied*, 423 U.S. 830 (1975).

[79]*Supra* note 77.

[80]See §6(b)(5) of the Act.

by OSHA be the lowest feasible.[81] The Supreme Court, however, rejected OSHA's policy.[82]

In the EtO proposal, OSHA described the two different types of studies which, in its view, established the carcinogenicity of EtO. These were epidemiological studies, comparing the cancer incidence among groups of individuals, some exposed and some not exposed to the substance; and animal bioassays, i.e., controlled experiments on laboratory animals.[83] OSHA concluded that EtO was a "potential occupational carcinogen," mainly on the basis of positive findings of cancer in a two-year chronic inhalation study performed on three groups of rats at the Bushy Run Research Center. These findings were supported, OSHA noted, by the "strongly suggestive" findings of two epidemiological studies. OSHA also determined that the human and animal studies indicated that EtO causes genetic damage and spontaneous abortions.[84]

OSHA selected the inhalation study performed on rats at Bushy Run as a basis for its risk assessment, stating that this study provided "by far the best quantitative information available" on EtO. In that study, the rats had been exposed to EtO concentrations of 100, 33, or 10 ppm. Since OSHA was considering regulation to levels as low as 0.5 ppm, the purpose of the risk assessment was to determine on the basis of the animal data what the cancer risk was to exposed employees at levels of 50, 10, 5, 1, and 0.5 ppm. After discussing a number of the difficult issues that arise when performing risk analysis,[85] OSHA indicated that it would use a "conservative" estimate "allowing error on the side of maximum worker protection";[86] and it concluded that EtO presents an excess risk of 100–152 cases of cancer per 1000 workers with regular exposure for a working lifetime to a 50-ppm EtO level, and an excess risk of two to three cases of cancer per 1000 workers at a 1-ppm level.

---

[81]See, e.g., the preamble to the benzene standard. 43 FED. REG. 5918, 5946–5947 (1978).

[82]See note 77, *supra*, and accompanying text.

[83]See Merrill, "Federal Regulation of Cancer-Causing Chemicals," Report to the Administrative Conference of the United States, 47–67 (April 1, 1982) (draft), for a discussion of various types of studies establishing carcinogenicity.

[84]48 FED. REG. at 17,292 (1983).

[85]*Id.* at 17,292–296.

[86]*Id.* at 17,295.

The crucial questions that remained were whether the 50-ppm level presented a "significant risk" and whether the new PEL would reduce that risk. The Supreme Court in the benzene case had said that OSHA would be accorded substantial deference in its determination as to what level or degree of risk should be deemed "significant." As a guide, the plurality opinion suggested that one chance in a thousand of an individual contracting cancer could be deemed a significant risk.[87] On the basis of its risk assessment, OSHA concluded preliminarily that the 50-ppm level presented a significant risk that would be substantially reduced by the new level.[88]

3. Summary of Regulatory Impact

The preamble next discusses the regulatory impact of the proposed standard. Section 6(b)(5) requires that OSHA, in promulgating health standards, set a standard that would provide maximum protection to employees "to the extent feasible." In the 1974 asbestos case, the court of appeals stated: "Congress does not appear to have intended to protect employees by putting their employers out of business—either by requiring protective devices unavailable under existing technology or by making financial viability generally impossible."[89] The Supreme Court embraced that view in 1981 in its cotton dust decision.[90] Thus, the Court, in an opinion by Justice Brennan for the majority, accepted OSHA's view in that case that a standard was economically feasible if it would allow the industry as a whole to maintain "long-term profitability and competitiveness."[91]

In order to obtain data with which to make feasibility determinations, OSHA has for many years contracted for studies on the impact of a proposed standard.[92] As feasibility be-

---

[87]*Industrial Union Dep't v. American Petroleum Inst.*, 448 U.S. 607, 655, 8 OSHC 1586, 1603 (1980).

[88]48 FED. REG. at 17,296 (1983).

[89]*Industrial Union Dep't v. Hodgson*, 499 F.2d 467, 478, 1 OSHC 1631, 1639 (D.C. Cir. 1974).

[90]*American Textile Mfrs. Inst. v. Donovan*, 452 U.S. 490, 9 OSHC 1913 (1981).

[91]*Id.* at 530–531 n. 55, 9 OSHC at 1928–1929.

[92]This practice began with the vinyl chloride standard, which was issued in 1974. 39 FED. REG. 35,890 (1974). That standard was affirmed by the Second Circuit, which announced the "technology-forcing" doctrine. *Society of Plastics Indus. v. OSHA*, 509 F.2d 1301, 2 OSHC 1496 (2d Cir. 1975), *cert. denied*, 421 U.S. 992 (1975). The court there emphasized that OSHA may lawfully issue standards that "require improvements in existing technologies or which require the development of new technology."

came a more controversial issue in OSHA standards proceedings, the feasibility studies introduced into the rule-making record became more extensive; and, as in the coke oven emissions and cotton dust standards proceedings,[93] parties often offered their own feasibility studies. In addition, in a series of executive orders, beginning with the order issued by President Gerald Ford requiring "inflation impact" statements, federal regulatory agencies have been required to develop economic (or, as they were sometimes called, regulatory) impact studies before taking actions that could have a major impact on the economy. The stated purpose of these studies was to improve the quality of regulatory decision making and to assure that the agencies would consider costs and other impacts on the economy in implementing their statutory goals.[94]

(a) Executive Order 12,291. The most recent executive order, No. 12,291,[95] was issued by President Ronald Reagan shortly after he took office in 1981. Executive Order 12,291 requires agencies to prepare preliminary and final regulatory impact analyses which contain, among other things, a description of the potential benefits of a rule, the potential costs of the rule, and a determination of the potential net benefits. The Executive Order also requires a discussion of alternative approaches that could achieve the same regulatory goals and an explanation as to why the agency did not adopt these alternatives. Executive Order 12,291 also requires cost-benefit analysis in the development of agency regulations, to the extent permitted by law.[96] In *American Textile Manufacturers Institute v. Donovan*,[97] the major case involving review of OSHA's cotton dust standard, the Supreme Court held that cost-benefit analysis in OSHA health standards proceedings was inconsistent with the requirements of Section 6(b)(5).[98]

---

[93]41 FED. REG. 46,742 (coke oven emissions); 43 FED. REG. 47,350 (cotton dust).

[94]These executive orders are discussed in Baram, *Cost-Benefit Analysis: An Inadequate Basis for Health, Safety, and Environmental Regulatory Decisionmaking*, 8 ECOLOGY L.Q. 473, 502–515 (1980). Some critics have suggested that a significant purpose of these orders is to slow down the pace of government regulation.

[95]3 C.F.R. §127 (1982).

[96]On Jan. 4, 1985, President Reagan issued Executive Order 12,498, 50 FED. REG. 1036 (1985), providing for oversight by the Office of Management and Budget of the regulatory planning process of Executive Department agencies.

[97]*Supra* note 90.

[98]See Mintz, *supra* note 3, at 322 (1984).

Since then, OSHA has not conducted formal cost-benefit analyses in its standards proceedings.

(b) Assessment of risks and benefits. The regulatory impact section of the preamble to the EtO proposal is a summary of the Preliminary Regulatory Impact Analysis (RIA) that OSHA submitted on April 15, 1983, as required by Executive Order 12,291.[99] The major portion of the EtO Preliminary RIA dealt, separately, with the costs and benefits of the proposed standard. After noting that any attempt to "quantify" the benefits would "seriously underestimate" the impact of the illness and death associated with EtO exposure, the Preliminary RIA utilized the OSHA risk assessment, which was based on the inhalation study performed on rats, and estimated that reducing the PEL to 1 ppm would reduce the number of excess EtO-induced cancers from 958 to 95 among EtO-exposed workers in five industry sectors, a reduction of 90 percent.[100] The Preliminary RIA went on to discuss the possibility of "monetizing" the benefits, i.e., placing a dollar value on the potential reduction in injuries, illnesses, and deaths attributable to the new standard. OSHA initially took the position that it could not reliably assign monetary values to benefits derived from saving lives and avoiding injury and illness; this position was first articulated by OSHA in the preamble to the final coke oven standards, which stated that any such attempt would involve "insuperable obstacles."[101] However, in light of Executive Order 12,291 and, more particularly, the interim regulatory analyses guidance provided by the Office of Management and Budget in June 1981,[102] OSHA has sought in some stan-

---

[99]The full title of the document is PRELIMINARY REGULATORY IMPACT AND REGULATORY FLEXIBILITY ASSESSMENT OF THE PROPOSED STANDARD FOR ETHYLENE OXIDE, U.S. Dep't of Labor, OSHA (April 15, 1983) (hereinafter cited as PRELIMINARY RIA). This impact analysis itself is based on a longer document prepared by an OSHA contractor, JRB Associates, entitled "Economic and Environmental Impact Study of Ethylene Oxide." A comprehensive study of the use of regulatory analysis was recently published by the Administrative Conference of the United States. McGarity, THE ROLE OF REGULATORY ANALYSIS IN REGULATORY DECISIONMAKING, FINAL REPORT TO THE ADMINISTRATIVE CONFERENCE OF THE UNITED STATES (May 1985). The report discusses in detail the role of regulatory analysis at OSHA. McGarity, FINAL REPORT at 190–247.

[100]PRELIMINARY RIA, at V-14. See also the parallel discussion in the preamble to the proposal, 48 FED. REG. at 17,293–296 (1983).

[101]Preamble to coke oven standard, 41 FED. REG. 46,742, 46,750–751 (1976).

[102]The OMB "Guidance" is reprinted in *Hearings on the Role of OMB in Regulation Before the Subcomm. on Oversight and Investigation of the House Comm. on Energy and Commerce*, 97th Cong., 1st Sess. 360 (1981).

dards proceedings to quantify the benefits of the standard in monetary terms.[103]

(c) Methods of "monetization." Two methods of estimating the dollar value of the benefits were discussed in the Preliminary RIA. The first is the "human capital" approach, which estimates the "direct and indirect" costs of cancer. Using this method, one estimate of the cost of treating a case of cancer is $16,473, and the annual loss to the Gross National Product (GNP) of an employee leaving the work force because of impaired health ranges from $13,936 in the spice industry to $23,400 in the EtO-producer sector.[104] These amounts, the Preliminary RIA said, represent the "minimum estimate" of monetizable benefits because they ignore such factors as the value of non-market-related production, the productivity loss of other potentially affected persons, such as relatives or co-workers, and the associated pain and suffering.

The Preliminary RIA then described the second method of monetization as the

"willingness to pay method. This method estimates the theoretical amount that the beneficiaries of a regulatory action would be willing to pay in order to obtain the benefits of the regulatory action. In an occupational safety and health setting, this does not represent what an individual worker would pay to avoid certain health impairment, but rather what society or a group of workers would pay to reduce the probability of this impairment. Willingness to pay estimates reported in the literature range from $500,000 to $7 million per worker."[105]

(d) Feasibility of the standard. Without reaching any conclusions on the dollar value of benefits or on the proper technique for making the calculation,[106] the Preliminary RIA then

---

[103]See, e.g., hazard communications standard, 48 FED. REG. 53,280, 53,327–330 (1983).

[104]The variation in amounts is due to differences in average earnings in these sectors.

[105]PRELIMINARY RIA at V-16–17.

[106]In OSHA's final regulatory impact analysis, OSHA calculated benefits based on an updated version of Prof. W.K. Viscusi's willingness-to-pay studies. See W. Viscusi, EMPLOYMENT HAZARDS: AN INVESTIGATION OF MARKET PLACE PERFORMANCE (1979), and Viscusi, *Labor Market Valuation of Life and Limb: Empirical Estimates and Policy Implications*, 26 PUB. POL'Y 359 (1978). Viscusi estimated the value of a life as the "marginal increase in annual income per worker associated with an increase in a 'death risk' variable, extrapolated to the labor force as a whole." OSHA used a midpoint, $3.5 million, of Viscusi's range as the implied value of a life. REGULATORY IMPACT AND REGULATORY FLEXIBILITY ANALYSIS OF THE FINAL STANDARD FOR ETHYLENE OXIDE V-14–17, U.S. Dep't of Labor, OSHA (June 18, 1984) (hereinafter

considered the feasibility of the standard. The estimate of total annualized costs, which included an allowance for capital costs as well as for annual operating costs, would be $72.4 million for the proposed 1 ppm standard for the five major industry sectors studied. This cost, the Preliminary RIA concluded, would not have a "negative impact on the viability of firms in these five industries."[107] The Preliminary RIA also concluded that the proposed standard was technologically feasible, utilizing "conventional technology" that is "commonly known and presently used by the affected industries."[108]

4. Summary and Explanation of the Standard

As indicated earlier, the preambles to all proposed and final OSHA standards in both the safety and health areas typically also contain a "Summary and Explanation of the [Proposed] Standard." This serves as a full explanation of OSHA's reasons for having selected certain provisions rather than others, and includes a discussion of the significant arguments presented by participants to the proceeding.[109] This section also constitutes OSHA's first interpretation of the standard, which is often relied upon by OSHA, the Review Commission, and the courts in proceedings for enforcement of the final standard.[110] The EtO Summary and Explanation section[111] deals with the following major topics in the proposed

---

cited as FINAL RIA). OMB, commenting on OSHA's final standard, noted that, using OSHA's highest benefit estimate, the benefits of the final EtO standard "could approach" OSHA's cost estimate if "some" methods of discounting future costs and benefits were applied and "high willingness to pay values [were used]." However, OMB sharply disagreed with OSHA's estimate of the numbers of cancer cases avoided by the standard. OMB's views are discussed *infra* Section V.B.

[107]PRELIMINARY RIA, at I-4–5. This refers, of course, to the test for economic feasibility set forth in the asbestos case and confirmed by the Supreme Court, *supra* notes 89–91 and accompanying text.

[108]PRELIMINARY RIA at VI-1–5. See also preamble to proposal, 48 FED. REG. at 17,298 (1983). Thus, OSHA did not rely on "technology-forcing" principles.

[109]OSHA's lavatory standard was set aside by the Second Circuit in part because OSHA had failed to present "*some* reasoned explanation" for rejecting certain arguments against the proposal. *Associated Indus. v. Department of Labor*, 487 F.2d 342, 1 OSHC 1340 (2d Cir. 1973).

[110]See, e.g., *Phelps Dodge Corp. v. OSHRC*, 725 F.2d 1237, 1238, 11 OSHC 1769, 1770 (9th Cir. 1984), in which the court of appeals relied on OSHA's statement in the Summary and Explanation section of the arsenic standard in deciding that an employer is required to compensate employees for their time and travel in taking medical examinations after regular working hours.

[111]48 FED. REG. 17,301–308 (1983).

standard: scope and application; definitions; permissible exposure limit; exposure monitoring; regulated areas; methods of compliance; respiratory protection; medical surveillance; employee information and training; precautionary labels and signs; recordkeeping; and effective dates.

5. *Other Portions of the Preamble*

In accordance with the requirements of the Regulatory Flexibility Act,[112] OSHA certified in the preamble that the proposed standard "would not have a significant adverse economic impact on a substantial number of small entities."[113] Further, as required by the National Environmental Policy Act,[114] OSHA concluded that as a result of the proposal there would be "no significant impact on the general quality of the human environment external to the workplace."[115] The preamble also contained a summary of the comments OSHA had received on the ANPR published in January 1982.[116]

V. Public Comment and Hearings

A. Public Comment

The major purpose of publishing a proposed standard is to initiate the process of public participation. The EtO proposal invited interested persons to submit "written data, views, and arguments with respect to this proposed standard,"[117] and afforded 60 days for the filing of comments. The OSH Act provides that not less than 30 days shall be given for comment on proposals; however, OSHA almost always gives 60 days or longer for comment. The EtO proposal also scheduled an "in-

---

[112]5 U.S.C. §§601–612 (1982).

[113]48 FED. REG. at 17,298.

[114]42 U.S.C. §4321 (1982), at implemented by the Council on Environmental Quality, 40 C.F.R. §1500 (1984), and the Department of Labor Procedures, 29 C.F.R. §11 (1984).

[115]48 FED. REG. at 17,298.

[116]47 FED. REG. at 17,298–301 (1982). Forty-six comments were received and placed in the record (Attachment III to Exhibit IV in EtO docket, H-200). See *supra* note 58 for an explanation of the docket.

[117]48 FED. REG. at 17,309.

formal public hearing" in Washington, D.C., for July 18, 1983, one month after the end of the comment period, during which an opportunity to present oral testimony would be provided. Under Section 6(b)(3), a hearing must be scheduled if requested and if objections are raised to the proposal. However, to save time when requests for a hearing are expected, OSHA often schedules the hearing at the same time that it publishes the proposal. In the case of EtO, because of the court deadline, shortening of the time frame was particularly important, and there was little question that a hearing would be requested.

In the early history of OSHA, there were few procedural steps required of an individual seeking to testify at the informal hearing. However, as OSHA standards rulemaking became more complex, it became necessary for OSHA to formalize the process in certain respects. The EtO notice reflected these changes. It required persons desiring to participate in the hearing to file, no later than 30 days before the start of the hearing, a notice of intention to appear that includes, among other things, the amount of time requested, the specific issues that would be addressed, and a "detailed statement" of the position to be taken with respect to each issue. In addition, OSHA required persons requesting more than 10 minutes for presentation or who intended to submit documentary evidence to provide the complete text of the testimony and the documentary evidence by July 1, 1983.[118] These procedures were designed to assure that all hearing time would be used productively and to provide all participants an opportunity to prepare adequately to cross-examine witnesses at the hearing.

The major focus of such hearings as well as of the written comments is, of course, the proposed standard and OSHA's explanation in the preamble. In the EtO proposal, as in some other proposals, OSHA broadened the scope of public participation by posing specific issues for public comment; these issues related to EtO's health effects, the technological and economic feasibility of the proposed PEL, and additional provisions to be considered for inclusion in the final standard.[119] Thus, for example, OSHA asked: "Is a short-term or ceiling exposure limit for EtO exposure necessary for the PEL or action level in view of recent information regarding increased

---

[118]*Id.*
[119]*Id.* at 17,284.

spontaneous abortions and chromosome changes in workers exposed to EtO?"[120] Although it could be argued that the proposed standard, which contained an 8-hour time-weighted average PEL, implicitly raised the question of a short-term exposure limit, OSHA hoped, by raising the issue explicitly, to receive a broader range of public comment on the issue and, more importantly, to lessen the likelihood of a successful challenge on "lack of notice" grounds in the event that such a short-term limit were eventually adopted.[121]

B. The EtO Hearing and OSHA Hearing Procedure

OSHA received 158 written comments on its proposed EtO standard.[122] The hearing began on July 19, 1983, and lasted through July 28. OSHA received notices of intention to appear at the hearing from representatives of public interest groups, unions, individual employers, employer and employee associations, and university faculties. Major participants included the Public Citizen Health Research Group (HRG), which had petitioned for the EtO emergency standard in 1981; the American Federation of State, County and Municipal Employees (AFSCME) and the Service Employees International Union, which represents large groups of exposed employees; the Department of Occupational Safety and Health of the AFL-CIO; the Ethylene Oxide Industry Council, an association of companies that produces and uses EtO; and the American Hospital Association.[123] In addition, as is usually the situation, OSHA itself presented several expert witnesses who testified concerning the proposal.

---

[120]*Id.* This issue became critical as the rulemaking progressed. See discussion *infra* Section VIII.

[121]The challenge of the Lead Industries Association to OSHA's lead standard was based in part on the argument that OSHA failed to give adequate notice in proposing a PEL of 100 micrograms and eventually adopting a 50-microgram PEL. The District of Columbia Circuit, with one judge dissenting, rejected the argument, holding that the final PEL was a "logical outgrowth" of the proposal. *United Steelworkers v. Marshall,* 647 F.2d 1189, 1221, 8 OSHC 1810, 1829 (D.C. Cir. 1980), *cert. denied,* 453 U.S. 913 (1981). The notice issue in the lead litigation is discussed in Rochvarg, *Adequacy of Notice of Rulemaking Under the Federal Administrative Procedure Act—When Should a Second Round of Notice and Comment Be Provided?* 31 Am. U.L. Rev. 1 (1981).

[122]Attachment VII to Exhibit 11 in the EtO docket. See *supra* note 58 for an explanation of the docket.

[123]A significant portion of employee exposure to EtO takes place in health care facilities. Final RIA, *supra* note 106, at ch. IV.

In its early history, OSHA provided the public with an opportunity to present testimony but neither presented testimony itself nor asked questions of witnesses. In time, however, it became obvious to OSHA that, to better explain the basis for its proposal and to assist in the development of a full record for the purposes of court review, it had to arrange for its own expert witnesses who would present evidence relevant to the proposal. OSHA now routinely does so. Another important development in hearing procedure relates to the examination of witnesses. OSHA's original regulations on rulemaking procedures provided for an "opportunity for cross-examination on crucial issues."[124] Recently, the cross-examination has become far more extensive and more searching to allow participants more effectively to develop a record and to test the reliability and accuracy of the testimony of witnesses, particularly experts.[125]

In the EtO hearing, OSHA offered Dr. Leonard Vance, OSHA's Director of Health Standards Programs; a panel of experts from JRB Associates, OSHA's contractor, who testified on the regulatory impact analysis; and several other experts, including Dr. Marvin S. Legator, Professor and Director of the Department of Preventive Medicine and Community Health at the University of Texas at Galveston.[126]

Under OSHA's regulations,[127] an administrative law judge presides over the informal hearing. At the opening of the EtO hearing, the administrative law judge made a statement outlining the manner in which the hearing would be conducted:

> "This informal Public Hearing is being held by the Occupational Safety and Health Administration often referred to as OSHA, for the purpose of receiving data and testimony on the issues in order to make a complete record in this proceeding. I will not make or recommend a decision in the matter. My function is to conduct a thorough and orderly hearing and to expedite the proceedings.

---

[124]29 C.F.R. §1911.15(b)(2) (1984).

[125]For an example of cross-examination at an OSHA rulemaking hearing, see text accompanying note 130, *infra*.

[126]Altogether OSHA presented seven witnesses. This is average for a health standards proceeding. In the lengthy Carcinogens Policy proceeding, OSHA presented 46 of its own witnesses. 45 FED. REG. 5002, 5008 (1980).

[127]29 C.F.R. §1911.15 (1984).

* * *

"There is a list of the order of appearances of the witnesses with the time of their presentations available on the table at the back.

"The first witness will be a representative from the Occupational Safety and Health Administration, and he will be followed by other representatives and/or experts invited to testify by OSHA. Then, other governmental and public witnesses will appear as scheduled.

"I would like to emphasize that unduly repetitious testimony will not be permitted and to expand further on that, I would like to say that all written submissions of all participants, including statements, exhibits, scientific data and studies are made a part of the record and should have been submitted for docketing by July 1, 1983.

"Therefore, the presentation of a participant should be as brief as possible. Each participant should focus primarily on presenting the highlights or of summarizing their submissions and if necessary, briefly responding to other submissions.

"The written submissions are, of course, entitled to equal weight with the oral testimony. Therefore, participants may wish to identify and respond to their written submissions and then make themselves available for questions from other participants.

"The time period for witnesses will be limited generally to fifteen minutes. Questioning of witnesses will be permitted after a witness has completed his or her statements, and only parties who filed appearances are entitled to ask questions.

"After a witness has testified, I will ask for the names of those participants who wish to ask questions of the witnesses. When it is your turn to ask questions, please come forward to the podium which is on my right, identify yourself and your organization, if any, and state the amount of time that you would like for your questions.

"Normally, no more than fifteen minutes will be permitted for the questioning of each participant. Therfore, be sure to ask your most crucial questions first.

"The Occupational Safety and Health Administration as the initiating party will have the first opportunity to question witnesses of other parties unless there is reason to proceed differently. The order of questioning by others will vary. It will depend on how many there are primarily.

"In view of the number of witnesses, questions should be kept as brief as possible. The questions should be limited to the issues relating to this hearing, and they should not be repeated by other parties if at all possible.

"Questions should be designed to clarify a presentation and elicit information that is within the witness' area of competence or expertise. Counsel should try to avoid routine questioning of their own witnesses.

"The record of this proceeding along with written comments and studies will be available for inspection and copying at the OSHA Docket Office ***."[128]

The major purposes of informal rulemaking were to give the public an opportunity to participate in the decisional process and provide full information to the head of the agency, which would improve the quality of the final decision within the framework of a flexible and efficient process. As government rulemaking became more controversial and complex, and with the development of hybrid rulemaking,[129] public hearings were increasingly held in informal rulemaking, and at the hearings the questioning of witnesses in a more adversarial atmosphere became more prevalent. For example, the questioning of Dr. Vance by Margaret Seminario, representing the Safety and Health Department of the AFL-CIO, at the EtO hearing is typical of the tenor of the questioning that often takes place at OSHA health standards proceedings:

"Ms. SEMINARIO: If the Agency determines from information gathered in this rulemaking that it is feasible to reduce exposures below the proposed one part per million level, will the Agency in its final determination lower the proposed exposure limit?

"DR. VANCE: It will give it every consideration of doing that. It's impossible to say prior to the time that the evidence is received what in fact the Agency will do, but the Agency's mind is certainly not closed to that possibility.

"Ms. SEMINARIO: ***[H]ow does OSHA define feasibility ***? Are you looking at feasibility determination based across industries in all operations all the time? Is it a determination based upon feasibility *** in most of the industries, in most operations most of the time? What definition of feasibility is OSHA using ***?"

"DR. VANCE: *** There are about three [or] four different courts which have defined feasibility. OSHA itself doesn't have any kind of a generic rule defining feasibility in a systematic way for the Agency, so it's necessarily done on a case-by-case basis.

---

[128]OCCUPATIONAL EXPOSURE TO ETHYLENE OXIDE, U.S. Dep't of Labor, OSHA (July 19, 1983) (statement of Judge Rhea Burrow; transcript, 4–8). Mr. George Henschel of the Office of the Solicitor of Labor was project attorney for the EtO rulemaking. He explained the responsibility of the OSHA attorney at the hearing as follows: "Our role in this proceeding will be to help to develop a clear, complete and accurate record upon which the Assistant Secretary can make a well-informed decision in issuing the final standard." *Id.* at 9–10. Mr. Henschel was of great assistance in the preparation of this chapter.

[129]See discussion in text accompanying notes 62–66, *supra*.

"If you will note, there are two different aspects to feasibility, economic and technological. The possibility of doing feasibility determinations on an industry-by-industry basis is one that can never be foreclosed. OSHA very rarely does it. It's given some consideration [to] doing that in at least one case that I'm aware of, but whether or not OSHA would do that in this case is in open question.

"It's not highly likely, given our past track records that we would do that, but it's certainly not beyond all possibility.

"Ms. SEMINARIO: But the Agency has said in the Federal Register, 17296, that it believes that compliance of one ppm TWA is technologically and economically feasible based upon data. It says, 'Regarding feasibility of the compliance with the PEL of 0.5 ppm, however, OSHA does not have sufficient data to demonstrate that a substantial portion of the EtO industry could control exposure to this level.'

"So clearly, not just on a generic basis, but in this instance, the Agency has already made some preliminary determination *** using some kind of a definition of feasibility ***.

"DR. VANCE: Most of the time, OSHA uses a definition, technological feasibility—that *** which can be done in most of the places most of the time. This is some language that comes out of some litigation across the street in the District of Columbia Court of Appeals and is standard. And that is a definition which OSHA has applied, more often than not. As I say, each individual rulemaking proceeding that OSHA engages in, advances OSHA's ability to do rational feasibility analysis. *** [E]ach and every rulemaking probably changes *** as we get more and more information. [The] question [of how we will do feasibility analyses] arises in every rulemaking that we do."[130]

In closing the hearings on July 28, the administrative law judge gave 30 days for the submission of further information for the record, including both information that the participants had been requested to submit and other information. The administrative law judge gave the parties an additional 20 days for the submission of posthearing comments and briefs. The briefs typically stated the position of the various participants on the issues in the proceeding, with appropriate citations to the record.[131] The EtO record, consisting of about

---

[130]OCCUPATIONAL EXPOSURE TO ETHYLENE OXIDE, *supra* note 128 (statement of Dr. Vance at 31–40). For similar persistence in examination of a witness, see the questioning of Dr. Legator by Robert Barnard, attorney for the Ethylene Oxide Industry Council. *Id.* at 86–107. The fact that attorneys participate frequently and actively in OSHA rulemaking hearings undoubtedly contributes to the tone of OSHA hearings.

[131]E.g., the brief of the Ethylene Oxide Industry Council, Exhibit 53, comprised 113 pages and addressed itself to issues relating to exposure levels, medical sur-

1600 transcript pages and over 300 exhibits, was certified by the administrative law judge on November 7, 1983, for decision by the Assistant Secretary.[132]

C. Congressional Hearing

On November 1, 1983, after the close of the OSHA hearing and shortly before the certification of the record, a congressional subcommittee held a hearing on OSHA's EtO rulemaking proceeding.[133] According to the chairperson of the subcommittee, the hearing was precipitated by the fact that the subcommittee had obtained a series of internal OSHA memoranda[134] which suggested that Dr. Vance had "a private *ex parte* meeting with a Union Carbide representative in mid-June 1983," while the EtO rulemaking was in progress; and that as a result of that meeting, Dr. Vance "actively intervened" in the EtO proceeding to prevent OSHA staff from developing information which could lead to a short-term limit for EtO, which Union Carbide opposed.[135] Testifying at the hearing were Dr. Vance, several OSHA staff members,[136] and

---

veillance, industrial hygiene/compliance issues, and phase-in times. One of the major arguments made in the brief was that a short-term exposure limit was not necessary, pp. 45–61, and that the study by Hemminki on spontaneous abortions had methodological shortcomings and should not be relied on for regulatory purposes, pp. 10–22. In its final standard, OSHA discussed the study conducted by Hemminki and coworkers which examined the rate of spontaneous abortions among women exposed to EtO in sterilizing operations in Finnish hospitals as opposed to women not exposed. OSHA concluded that the study provided "additional support" for the regulation of EtO. 49 FED. REG. at 25,752 (1984).

[132]See 49 FED. REG. at 25,737 (1984). The decisional official in the Department of Labor on OSHA standards is the Assistant Secretary for Occupational Safety and Health. In reaching his decision, the Assistant Secretary is assisted by OSHA staff, staff in the Office of the Solicitor, and consultants. The relationship between the Assistant Secretary and staff in the decisional process is discussed in *United Steelworkers v. Marshall, supra* note 121, at 1213, 8 OSHC at 1825. See M. Asimow, *When the Curtain Falls: Separation of Functions in the Federal Administrative Agencies*, 81 COLUM. L. REV. 759 (1981).

[133]*Use and Control of Ethylene Oxide (EtO): Hearing before the Subcomm. on Labor Standards of the House Comm. on Education and Labor*, 98th Cong., 2d Sess. (1983) (hereinafter cited as *EtO Hearing*).

[134]These memoranda would not normally have been placed in the rulemaking docket. However, Dr. Vance testified that he placed the memoranda in the docket because he "felt confident that this [the EtO] standard, when it is promulgated, would be litigated."*EtO Hearing* at 246.

[135]*EtO Hearing* at 2.

[136]According to the chairperson at the subcommittee, Assistant Secretary of OSHA Thorne Auchter had originally taken the position that Dr. Peter Infante, then Director of OSHA Carcinogen Identification and Classification, and Dr. H. Robert

representatives of the Amalgamated Clothing and Textile Workers Union, the AFL-CIO, AFSCME, the National Union of Hospital and Health Care Employees, and the American Hospital Association.

Dr. Vance testified that he met with Mr. Arlin Voress of the Ethylene Oxide Industry Council, representing affected industries, in May or June 1983, but that Mr. Voress "didn't spend any protracted period of time" with him.[137] Dr. Vance further testified that there was "nothing unusual about a person who has an interest in an ongoing standard development appearing at OSHA, talking with the folks who are working on that standard."[138] Dr. Vance said that there is a "fair amount of conversation during the hearings" and that "it pretty much phases down right after the hearings."[139] Dr. Vance promised to show the committee his appointment logs, but they were never produced.[140] Dr. Beliles[141] testified that Mr. Voress spent at least an hour with him, and that Mr. Voress "persisted" in discussing several points regarding the EtO standard, the short-term limit, the PEL, and the medical surveillance provisions, despite Dr. Beliles's attempt to avoid such substantive discussion.[142] Mr. Voress later wrote to the subcommittee, stating his "recollection" of his meetings with Dr. Vance and Dr. Beliles.[143] He said in his letter that he had given Dr. Beliles a "very brief overview of the subject which would be included in EOIC's [Ethylene Oxide Industry Council] official filing" and that in order to "assist OSHA in planning the hearing," he "described" the testimony that EOIC would present.[144]

Dr. Vance was also questioned by the chairperson of the subcommittee on the agency view of *ex parte* meetings with interested parties during a rulemaking proceeding.[145] Accord-

---

Beliles, Director of Carcinogens Standards, would not be permitted to testify before the subcommittee. However, Mr. Auchter "reluctantly" authorized the appearance when advised that the subcommittee was prepared to compel the testimony through legal measures. *EtO Hearing* at 2.

[137]*EtO Hearing* at 243.

[138]*Id.*

[139]*Id.*

[140]It was reported tht Dr. Vance advised the committee that he threw the logs away because his dog had vomited on them in his car. N.Y. Times, April 3, 1984, at 30, col. 2.

[141]See note 136, *supra.*

[142]*EtO Hearing* at 244–245.

[143]*Id.* at 265–266.

[144]*Id.*

[145]*Id.* at 236.

ing to Dr. Vance, OSHA's view was that "any time the information is provided after the record closes, it is very important that we record the fact of the meeting, the nature of the information transmitted, and submit it for the record."[146] Dr. Vance's view was based, he said, on a memorandum from the Solicitor of Labor to the Departmental Executive staff dealing with *ex parte* communications.[147] The Solicitor's memorandum reached the following conclusion:

> "In light of the above discussion, we should probably recommend at the very least that in informal and hybrid rulemakings ex parte written comments conveying factual information centrally relevant to the rulemaking should be placed in the rulemaking file. Additionally, it would appear that ex parte oral comments conveying centrally relevant factual information should also be reduced to writing and be entered in the rulemaking file. Where the factual information in a post-comment submission or in a submission late in the comment period is of central importance and vial [sic] to the agency's support for its rule, it may be appropriate for the agency to reopen its comment period. Finally, to the extent that the Department may have employed formal rulemaking, the guidelines pertinent to the formal rulemakings should be followed by agencies, as addressed in this memorandum."[148]

The Solicitor's views were based in large measure on the court of appeals decision in *Sierra Club v. Costle.*[149]

During the course of the hearing, a concern was raised by Congressman Packard, an *ex officio* member of the subcommittee, that since OSHA staff was currently deliberating on the EtO standard, the hearing, which also dealt with the standard, would be "prejudicial" to those deliberations.[150] The chairperson of the subcommittee stated that it was "within the jurisdiction of this committee to determine the procedures by which those discussions [among OSHA staff] are taking place," and that witnesses could always inform the committee at any point "when they think a question is improper."[151]

---

[146]*EtO Hearing* at 242.

[147]The memorandum is reprinted in *EtO Hearing* at 237–242.

[148]*EtO Hearing* at 242.

[149]657 F.2d 298 (D.C. Cir. 1981), discussed *infra* in text at notes 153 & 154. The relevant portions of the decision appear at 657 F.2d at 408–409. The subcommittee hearing transcript also included a memorandum on *ex parte* communications prepared by the Congressional Research Service of the Library of Congress. *EtO Hearing* at 248–252. The issue of *ex parte* communications during rulemaking proceedings is also discussed *infra* Section VI.B and VIII.B.

[150]*EtO Hearing* at 227.

[151]*Id.*

While the subcommittee questioning related in large measure to the deliberative procedures that OSHA was following, there was also considerable testimony, and a number of submissions, on the merits of the EtO standard. Thus, for example, AFSCME submitted for the subcommittee record a copy of the lengthy testimony which it had previously offered at an OSHA EtO hearing.[152]

The subject of congressional involvement in administrative agency rulemaking has been the subject of court rulings and much scholarly comment. In *Sierra v. Costle*, the District of Columbia Circuit said:

> "We believe it entirely proper for Congressional representatives vigorously to represent the interests of their constituents before administrative agencies engaged in informal, general policy rulemaking, so long as individual Congressmen do not frustrate the intent of Congress as a whole as expressed in statute, nor undermine applicable rules of procedure. Where Congressmen keep their comments focused on the substance of the proposed rule—and we have no substantial evidence to cause us to believe Senator Byrd did not do so here—administrative agencies are expected to balance Congressional pressure with the pressures emanating from all other sources. To hold otherwise would deprive the agencies of legitimate sources of information and call into question the validity of nearly every controversial rulemaking."[153]

The court refused to invalidate the Environmental Protection Agency rule in *Sierra v. Costle* because of senatorial attempts to influence the contents of the rule.[154]

---

[152]*EtO Hearing* at 28–188. The exchange of memoranda between Dr. Vance and Dr. Infante also raised the question whether Dr. Vance improperly refused to allow OSHA staff to perform a quantitative risk assessment on reproductive hazards resulting from EtO exposure. In urging that a risk assessment be performed, Dr. Infante stated the view of OSHA "scientific staff" that a short-term exposure limit should be included in the standard. *EtO Hearing* at 229. Dr. Vance responded by refusing to permit the risk assessment to be done and stating that the fact that OSHA's scientific staff had reached conclusions on a major issue before the record was complete constituted prejudgment and denied interested parties a fair hearing. *EtO Hearing* at 230. Dr. Infante responded by saying that the staff has an "open mind not a blank mind" and that the staff scientists were not required by law to have an absence of scientific opinion concerning potential toxic agents while the rulemaking was in progress. *EtO Hearing* at 233. The issue of prejudgment by agency staff in rulemaking proceedings was discussed in *United Steelworkers v. Marshall*, 647 F.2d 1189, 8 OSHC 1810 (D.C. Cir. 1980), *cert. denied*, 453 U.S. 913 (1981) (refusing to disqualify Dr. Bingham from deciding the OSHA lead standard because of her views on medical removal protection and other issues).

[153]*Sierra Club v. Costle*, *supra* note 149, at 409–410.

[154]*Id.*

The continuing refusal of Congress to set specific limits to rulemaking authority in enabling statutes[155] and, of course, the fact that the Supreme Court has held legislative veto provisions to be unconstitutional[156] have played major roles in bringing about the continuing involvement by Congress in administrative rulemaking activity. As one commentator has noted, "Policy making is continuous and incremental and legislators pursuing their own mix of goals will often find it worth their while to intervene in administrative rulemaking."[157] Thus, the EtO subcommittee hearing may be understood as a function of Congress's continuing informal oversight role in policy rulemaking under the OSHA statute.[158]

VI. 1984: Final Standard

A. Introduction

The final standard on EtO was published in the *Federal Register* on June 22, 1984.[159] OSHA reaffirmed its preliminary determinations that EtO was carcinogenic and mutagenic and had cytogenic effects and that exposure to the chemical caused reproductive hazards and other serious health effects.[160] OSHA established a PEL of 1 ppm, determined as an 8-hour TWA, and an "action level" of 0.5 ppm (also as an 8-hour TWA) as the level above which employers must initiate periodic employee exposure monitoring, medical surveillance, and certain other compliance requirements.[161] OSHA "reserved decision"

---

[155]The courts have generally refused to find delegations of authority to administrative agencies unconstitutional on grounds of vagueness. Compare dissenting opinion of Rehnquist, J., in *American Textile Mfrs. Inst. v. Donovan*, 452 U.S. 490, 9 OSHC 1913 (1981).

[156]*Immigration & Naturalization Serv. v. Chadha*, 462 U.S. 919 (1983).

[157]J.M. Berry, Feeding Hungry People—Rulemaking in the Food Stamp Program 125 (1984).

[158]The issue of congressional involvement in administrative rulemaking is discussed *id.* at 105–126. See also R.B. Ripley, Congress—Process and Policy 362–382 (1983).

[159]49 Fed. Reg. 25,734 (1984).

[160]The health effects portion of the preamble appears in 49 Fed. Reg. at 25,737–755. There, OSHA discussed the available epidemiological and animal experimental data establishing the carcinogenic, mutagenic, cytogenetic, reproductive, and other health effects of EtO.

[161]The PEL and action level adopted in the final standard were the same as those originally proposed, and are discussed in the preamble to the final standard at 49

on the issue of the "short-term exposure limit" (STEL),[162] largely, as OSHA explained in the preamble, "in response to reservations expressed by the Office of Management and Budget (OMB) to STEL provisions in the draft final standard."[163] The final standard also modified the language of some provisions in the proposed standard in light of public comment; changes included such critical issues as engineering controls and medical surveillance,[164] and many provisions were rewritten for purposes of clarity. Finally, the entire preamble was greatly expanded, with a much fuller statement of reasons for OSHA's conclusions. In particular, as has been the case with all recent preambles to final standards, the preamble contained extensive analysis of the public comments and testimony, as well as explanations of why OSHA accepted or rejected particular views.[165]

B. Short-Term Exposure Limit; OMB Review of Regulatory Actions

While OSHA's EtO proposal contained an 8-hour TWA PEL and an "action level," it did not propose a short-term exposure limit.[166] The preamble to the proposal, however, raised

---

FED. REG. 25,775 and extensively in the portions of the preamble dealing with health effects and feasibility.

[162]See Section VI.B, *infra.*

[163]*Id.* The changes made by OSHA in response to OMB's comments are discussed *infra* Section VI.B.

[164]49 FED. REG. at 25,788–789 (engineering controls) and 49 FED. REG. at 25,784–789 (medical surveillance). See discussion *infra* Section VI.C.

[165]See, e.g., OSHA's preamble discussion of the union representatives' objections to OSHA's use of the data from the animal inhalation study as a basis for its quantitative risk assessment. See 49 FED. REG. at 25,761. The American Federation of State, County and Municipal Employees objected to OSHA's failure to rely as well on epidemiological data in determining the level of risk from EtO. In the preamble to the final standard, OSHA analyzed the relative merits of making predictions of risk from epidemiologic data and animal studies, reaffirming the conclusion in its preamble to the proposal that, based on animal experimental evidence, the "best estimate of risk" over a working lifetime is 634–1,093 excess deaths per 10,000 workers exposed to EtO at 50 ppm and approximately 12–23 excess deaths per 10,000 workers exposed for the same period at 1 ppm. OSHA added that in light of the predictions obtained on the basis of epidemiological data, these estimates do not overstate the risk and "may understate the risk." 49 FED. REG. at 25,762–764.

[166]The TWA exposure level is typically calculated for an 8-hour workday and a 40-hour workweek. Ceiling limits are usually defined as 15-minute TWA exposures designed to protect against high but brief excursions in the atmospheric concentration of the toxic substance. The term "short-term exposure limit" is sometimes used instead of "ceiling limit." See Congress of the United States, Office of Technology Assessment, PREVENTING ILLNESS AND INJURY IN THE WORKPLACE 259, Table 13-1 (1985).

the question: "Is a short-term or ceiling exposure limit for EtO exposure necessary for the PEL or action level in view of recent information regarding increased spontaneous abortions and chromosome changes in workers exposed to EtO?"[167] That issue was debated both in the public comments and during the hearing, and was discussed in OSHA's draft final standard,[168] submitted to the Office of Management and Budget on June 14, 1984, in accordance with Executive Order 12,291.[169] OMB responded on the same day,[170] stating that the draft standard was inconsistent with the Executive Order and, in particular, that the STEL was "unsupported by any reasonable risk assessment or inference from the available scientific evidence."[171] Based on OMB's objections, OSHA reserved decision on the STEL issue and submitted OMB's comments to a "num-

---

[167]48 FED. REG. at 17, 284 (1983). The studies on EtO-induced spontaneous abortions are discussed in the preamble to the final standard, 49 FED. REG. at 25,749–752 (1984), and chromosome damage is discussed at 49 FED. REG. 25,744–747.

[168]OSHA's June 14 draft of the EtO final standard contained an extensive discussion of the record evidence on the STEL (Draft, pp. 221–231). On the June 14 draft, see *infra* notes 169 and 224.

[169]Section 3, 3 C.F.R. §127 (1982), discussed *supra* Section IV.B.3.a. As noted, the EtO standard as published did not contain a STEL or a discussion of the record in relation to that issue. However, in the motion submitted on July 18, 1984, to the U.S. District Court for the District of Columbia by HRG and others for an order compelling OSHA to issue a "completed final standard," the plaintiffs attached portions of OSHA's June 14 draft which had been submitted to OMB (Exhibit B). (OSHA did not claim that the excerpts from the draft were inaccurate.) The draft would have established a STEL of 10 ppm of EtO averaged over any 15-minute period during the workday. The draft (p. 231) stated that the STEL was based on a "confluence of circumstances, including the nature of the health effects, the significance of the cancer risk, the nature of intermittent exposures in the affected industries, and the practical utility of the STEL as a control strategy as demonstrated by current industrial hygiene practice." The renewed district court litigation is discussed *infra* Section VII.B.

[170]The response was placed on the EtO record as Exhibit 162. Letter from Christopher DeMuth, Administrator for Information and Regulatory Affairs, Office of Management and Budget, to Francis X. Lilly, Solicitor, Dep't of Labor (June 14, 1984). There has been an ongoing debate on whether oral and written communications between OMB and the agencies during a rulemaking proceeding should be recorded and placed in the rulemaking file. See, e.g., S. REP. NO. 305, 97th Cong., 1st Sess. 67–72 (1981), and discussion *infra* note 183 and text accompanying notes 183 & 184.

[171]Letter, *supra* note 170, at 2. OMB also stated that the Final RIA "overstates the likely health benefits of the rule by a large margin"; that the rule, by requiring engineering controls when "feasible" and regardless of cost, "would needlessly prohibit more cost-effective means of achieving the rule's health objectives"; and that the standard's paperwork requirements had not been submitted to OMB for approval under the Paperwork Reduction Act of 1980, 44 U.S.C. §§3501–3520 (1982). In the final standard, OSHA postponed the effective date of those provisions requiring employers to record or report information because OMB had not approved these requirements. 49 FED. REG. at 25,734 (1984). On March 12, 1985, OSHA announced that OMB clearance had been obtained and that the information collection requirements were effective, with a variety of start-up dates. 50 FED. REG. 9800 (1985). The issue of engineering controls is discussed *infra* Section VI.C.2.

ber of scientifically qualified peer reviewers" for comment, analysis, and criticism and for public comment.[172]

Executive Order 12,291 established a system of regulatory review whose purpose was to "reduce the burdens of existing and future regulations, increase agency accountability for regulatory actions, provide for presidential oversight of the regulatory process, minimize duplication and conflict of regulations, and insure well-reasoned regulations."[173] Similar executive orders, intended to provide a procedure for assessment of the economic effects of federal regulatory actions, had been issued by Presidents Ford and Carter.[174] Christopher C. DeMuth, OMB's then Administrator for Information and Regulatory Affairs, testifying in 1982, emphasized the "substantial differences" between President Reagan's order and those of his predecessors. He said that the policies and procedures of Executive Order 12,291 were "much more ambitious and forthright" than those of Presidents Ford and Carter in that President Reagan's order sets forth specific policies that the agencies must follow to the extent permitted by law, requires that all regulations be reviewed "for consistency with those policies," and charges the Vice-President and a Cabinet-level task force with "general oversight of the Administration's regulatory reform efforts."[175] The principal procedures under the Executive Order were described by Mr. DeMuth at the same hearing:

> "Under the terms of the Executive Order, executive branch agencies submit all proposed and final regulations to OMB prior to publication in the *Federal Register*. To aid in the review and consultation process, and to ensure that agencies have a proper basis on which to make their most important regulatory decisions, agencies are required to prepare Regulatory Impact Analyses for rules that are defined as 'major' according to criteria established in the Order. The most important criterion for de-

---

[172]The notice requesting comment on OMB's letter was published in the *Federal Register* on Sept. 19, 1984, 49 FED. REG. 36,659.

[173]3 C.F.R. §127 (1982).

[174]EXEC. ORDER 11,821, 3 C.F.R. §926 (1971–1975 Comp.); EXEC. ORDER 12,044, 3 C.F.R. §152 (1979). OMB and the Council on Wage and Price Stability, as representatives of the President, had important roles in reviewing agency regulations under these orders. See COUNCIL ON WAGE AND PRICE STABILITY, BENEFIT COST ANALYSES OF SOCIAL REGULATION (J.C. Miller III & B. Yandle eds. 1979).

[175]*Office of Management and Budget Control of OSHA Rulemaking: Hearings Before a Subcommittee of the House Comm. on Government Operations*, 97th Cong., 2d Sess. 307–309 (1982) (hereinafter cited as *OMB-OSHA Hearings*).

termining a major rule is that it is expected to have an economic impact of $100 million or more. Any disagreement with OMB's views about the consistency of a proposed or final rule with the President's regulatory principles is taken up by the Task Force [on Regulatory Relief] or, if necessary, the President. The agencies, however, retain authority over the final decision pursuant to their governing statutes."[176]

This procedure has caused considerable controversy. The Department of Justice expressed the view in 1981 that the President would not exceed his constitutional or statutory authority by authorizing the task force and OMB "to supervise agency rulemaking" in the manner provided by the Executive Order.[177] The contrary view is that the Order, taken as a whole or separated into its procedural and substantive components, violates the constitutional separation of powers.[178] In particular, the legality and propriety of OMB's review of OSHA regulations has been questioned before Congress. In 1982, a subcommittee of the House Committee on Government Operations held a hearing on the issue of OMB involvement in OSHA standards rulemaking.[179] George H.R. Taylor, Director of the Department of Occupational Safety and Health, AFL-CIO, complained about OMB's "meddling" in OSHA rulemaking on the issue of hearing conservation and in other issues. Thorne Auchter, Assistant Secretary for OSHA, on the other hand, defended the procedures under the Executive Order, saying that "it is both necessary and appropriate that the President delegated day to day oversight of the regulatory process to OMB. The function OMB performs is useful to the agency, and essential to the President's economic recovery program." In December 1983, the full Government Operations

---

[176]*Id.* at 306–307. In practice, agencies have generally also prepared RIAs to determine if a regulation is likely to be "major" under the Executive Order. Thus, OSHA prepared a Preliminary and Final RIA for EtO and concluded in the Final RIA that the EtO standard was not a "major" action. Under the terms of the Order, agencies must also submit for review by OMB every notice of proposed rulemaking and final rule for other than major rules. EXEC. ORDER 12,291 §3(c)(3).

[177]Memorandum from Larry L. Simms, Acting Assistant Attorney General, Dep't of Justice (Feb. 13, 1981), *reprinted in OMB-OSHA Hearings, supra* note 175, at 323–335.

[178]Rosenberg, *Beyond the Limits of Executive Power: Presidential Control of Agency Rulemaking Under Executive Order 12,291*, 80 MICH. L. REV. 193 (1981). See also Horton, "Executive Order 12,291 and the Conflict Between the Legislative and Executive Branches of Government," in *OMB-OSHA Hearings, supra* note 175, at 103–121; Comments, *Capitalizing on a Congressional Void: Executive Order No. 12,291*, 31 AM. U.L. REV. 613 (1982).

[179]Note 175, *supra.*

Committee issued a report based on these hearings entitled "OMB Interference with OSHA Rulemaking."[180] Focusing particularly on the OSHA diving standard and the hazard communication standard, a majority of the committee concluded that the Executive Order had been "used as a back door, unpublicized, channel of access to the highest levels of political authority in the Administration for industry alone."[181] Twelve members of the committee vigorously dissented.[182]

A related controversial issue that has also caused disagreement is whether the communications between OMB and the various federal agencies during OMB review of regulatory actions must be placed in the rulemaking record, so that they would be available for public examination and rebuttal. The Administrative Conference of the United States recommended that there be no requirement that the contents of communications between presidential advisors and agency officials be placed in the record except where the communications contain "material factual information" relating to the rule or where the communications originated from persons outside the government.[183] On the other hand, the American Bar Association Commission on Law and the Economy recommended that memoranda exchanged between an agency and the President and his staff be made part of the public record; and that if there was an oral discussion, the fact that a discussion had taken place be made part of the public record but that the substance of the discussion remain private.[184] In all events, OSHA's practice, at least recently, as evidenced by the cotton dust and EtO rulemakings, has been to place copies of at least

---

[180]H.R. REP. NO. 583, 98th Cong., 1st Sess. (1983). Mr. Taylor's statement appears in *OMB-OSHA Hearings, supra* note 175, at 15, Mr. Auchter's statement in *OMB-OSHA Hearings* at 66, 82–83.

[181]H.R. REP. NO. 583, *supra* note 180, at 11.

[182]*Id.* at 15 and 24. A widely publicized dispute took place between OSHA and OMB over the scope of OSHA's reconsideration of its cotton dust standard, particularly respecting the use of engineering controls. The text of the OSHA-OBM correspondence is included in the record, Docket H-025, available in the OSHA Docket Office. See Wall St. J., May 20, 1984, at 4. OSHA ultimately decided to adhere to the priority given to engineering controls in the proposed revisions to the cotton dust standard. 48 FED. REG. 26,962, 26,964 (1983).

[183]Administrative Conference Recommendation No. 80–6, 1 C.F.R. §§305.80–.86 (1984). If the agency relies on "material factual information" not in the record in promulgating the rule, there could also be an issue whether the rule was supported by substantial evidence. See also Verkuil, *Jawboning Administrative Agencies: Ex Parte Contacts by the White House*, 80 COLUM. L. REV. 943 (1980).

[184]ABA Comm. on Law and the Economy, FEDERAL REGULATION: ROADS TO REFORM 81 (1979). See also *Sierra Club v. Costle*, 657 F.2d 298, 408 (D.C. Cir. 1981).

some of the communications exchanged with OMB in the rule-making record.

C. Other Changes in Final Standard

1. Medical Surveillance

As noted, in its final EtO standard OSHA made several changes to the proposal on the basis of public comment. There were two changes related to the medical surveillance program. Under Section 6(b)(7) of the Act, OSHA is required to prescribe in health standards "the type and frequency of medical examinations or other tests which shall be made available, by the employer or at his cost, to employees exposed to such hazards in order to most effectively determine whenever the health of employees is adversely affected by such exposure."[185] OSHA medical surveillance provisions typically state the frequency of the examinations, specify the employees to whom they are to be administered, and give other details of the program. In addition, the minimum protocol for the medical examination is usually included.[186] In the EtO proposal, however, OSHA did not specify the type of medical tests that must be made available to employees, stating that it had insufficient information upon which to base mandatory requirements, and that the examining physician was best qualified to make this judgment.[187] In the final standard, OSHA modified the provision, mandating certain elements for all medical examinations "to ensure uniformity of medical surveillance for all EtO workers."[188]

OSHA also deleted the nonmandatory proposed medical test for chromosome damage on the ground that the results

---

[185]On the evolution of OSHA's medical surveillance provisions, see Mintz, *Medical Surveillance of Employees Under the Occupational Safety and Health Administration*, 28 J. OCCUPATIONAL MED. 913 (1986). See also M.A. Rothstein, MEDICAL SCREENING OF WORKERS ch. 8 (1984).

[186]In its standard for 14 carcinogens, issued in 1973, OSHA did not include detailed criteria for medical examinations. On appeal, the Third Circuit remanded the case to OSHA for amplification of the standard on the issue so that the "worker [would] be assured of the benefits of medical procedures presently available and that the industry be advised of what is expected of it." *Synthetic Organic Chem. Mfrs. Ass'n v. Brennan*, 506 F.2d 385, 391, 2 OSHC 1402, 1407 (3d Cir. 1974), *cert. denied*, 423 U.S. 830 (1975).

[187]48 FED. REG. at 17,305 (1983).

[188]49 FED. REG. at 25,784 (1984).

of these tests, "as applied to an individual rather than a group," cannot be related to individual exposures of EtO.[189] The EtO proposal also asked for comments on whether a provision for medical removal protection (MRP) should be added to the standard.[190] The clause would provide for maintenance of earnings, seniority, and other benefits by employees removed from jobs involving EtO exposure because they wish to procreate.[191] The final standard discussed the comments received on this issue but did not include a MRP provision primarily, the preamble said, because the effects of EtO exposure were not highly reversible and thus the record contained insufficient evidence that temporary removal would provide long-term employee health benefits.[192]

2. Compliance Methods

OSHA's traditional policy on the methods of compliance with PELs in health standards is to require engineering controls in the first instance and only in limited situations to allow the use of other methods of compliance, such as personal protective equipment. Personal protective equipment may be used, according to OSHA, only in emergencies, or where other methods of control are not feasible, are not adequate to achieve the PEL, or have not yet been installed. The rationale for preferring engineering controls was stated in the preamble to the final EtO standard, as follows:

> "Respirators have traditionally been accorded the least preferred position in the hierarchy of controls because of the many problems associated with their use. *** The effective use of respirators requires that they be individually selected and fitted, conscientiously worn, carefully maintained, and replaced when necessary; these conditions may be difficult to achieve

---

[189]*Id.*

[190]48 FED. REG. at 17,285. OSHA's authority to provide MRP in the context of the lead standard was upheld by the District of Columbia Circuit in *United Steelworkers v. Marshall*, 647 F.2d 1189, 1228–1238, 8 OSHC 1810, 1834–1842 (D.C. Cir. 1980), *cert. denied*, 453 U.S. 913 (1981).

[191]It has been asserted that EtO causes reproductive damage, i.e., spontaneous abortions. See Note, *The Validity of Medical Removal Protection in OSHA's Lead Standard*, 59 TEX. L. REV. 1461 (1981).

[192]49 FED. REG. at 25,788. OSHA's decision to include MRP in the lead standard was based on the fact that employee blood lead levels could be reduced by removal from exposure to lead. 43 FED. REG. at 52,972 (1978). OSHA's recently issued benzene standard contained provisions on medical removal protection. 52 FED. REG. 34,569–572.

and to maintain consistently in many workplace environments."[193]

Since 1981, OSHA has been reviewing the health benefits and cost effectiveness of the traditional hierarchy of controls and the effectiveness of respirators in protecting employees.[194] In the proposed EtO standard, OSHA retained its preference for engineering controls; thus, the proposal required the employer to meet the PEL by engineering controls and work practices, "except to the extent that the employer can establish that such controls are not feasible."[195] In the preamble to the proposal, however, OSHA asked whether there were "conditions under which respirator use should be permitted in addition to those proposed."[196] In the final EtO standard, OSHA maintained its preference for engineering controls, but with several changes in language. The final standard states that the employer is required to institute engineering controls "except to the extent that such controls are not feasible."[197] This somewhat different language arguably would reduce to some extent the employer's burden of proof in establishing infeasibility in an enforcement proceeding.[198] In addition, for the first time OSHA added a provision to the standard specifically finding that engineering controls were "generally infeasible" in certain specific operations, such as removal of biological indicators and maintenance operations; for these operations, the standard requires engineering controls only where OSHA demonstrates that such controls are feasible.[199]

In its June 14 letter reviewing OSHA's draft final standard, OMB had sharply disagreed with the standard's reliance

---

[193]49 FED. REG. at 25,788. See also OSHA's Carcinogens Policy, 45 FED. REG. 5002, 5223 (1980).

[194]OSHA has issued two advance notices on the issue. 47 FED. REG. 20,803 (1982), and 48 FED. REG. 7473 (1983).

[195]48 FED. REG. at 17,310.

[196]*Id.* at 17,285.

[197]49 FED. REG. at 25,797.

[198]On the issue of the employer's burden in establishing infeasibility in an enforcement proceeding, see *United Steelworkers v. Marshall*, 647 F.2d 1189, 1269, 8 OSHC 1810, 1867–1868 (D.C. Cir. 1980), *cert. denied*, 453 U.S. 913 (1981).

[199]49 FED. REG. at 25,797. This provision is discussed in the preamble at 49 FED. REG. 25,778–779. As a general matter in health standards, OSHA does not require *specific* engineering controls or state specifically in which operations engineering controls are feasible or infeasible. Compare, however, OSHA's coke oven emission standard, 29 C.F.R. §1910.1029(f) (1984), where engineering controls are specified. These provisions were upheld in *American Iron & Steel Inst. v. OSHA*, 577 F.2d 825, 837–838, 6 OSHC 1451, 1460–1461 (3d Cir. 1978), *cert. denied*, 448 U.S. 917 (1980).

on engineering controls. OMB stated that in light of the variety of workplace conditions in which employees are exposed to EtO and the types of respirators available, respirators would be at least as effective as, and in some cases more effective than, engineering controls "in reducing in-lung exposure." And, since respirators are clearly less costly—involving a cost saving with a present value of $150 million—a performance standard allowing employers the choice of using respirators or engineering controls "results in an equally or more cost effective rule as employers seek the least cost approach appropriate to their own circumstances."[200] OSHA later asserted that it made no changes in the final standard on the engineering control-respirator issue on the basis of OMB's letter.[201]

3. Effective Dates

The effective date of various provisions of health standards, particularly the requirements relating to engineering controls, has always been a critical issue in rulemaking proceedings. In the proposed EtO standard, OSHA provided for a 30-day delayed effective date for the standard as a whole[202] and posed the issue of the "length of time *** needed for affected employers to reduce employee exposure to the proposed PEL through engineering and work practice controls."[203]

The final standard set a 60-day delayed effective date to allow affected parties "to familiarize themselves with this rather comprehensive document."[204] On the issue of the "start-up" dates[205] for engineering and work practice controls, OSHA

---

[200]Letter, *supra* note 170, at 3.

[201]OSHA's memorandum to the district court in the 1984 litigation, *infra* Section VII.B, stated that OSHA "considered, but ultimately did not agree with OMB reservations *** relating to *** the compliance strategy adopted in the standard." Government's Memorandum of Points and Authorities in Opposition to the Plaintiffs' Motion for an Order Directing Defendants to Issue Immediately a Complete Standard Governing Worker Exposure to Ethylene Oxide at 7 n. 4, *Public Citizen Health Research Group v. Rowland*, No. 81–2343 (D.D.C. 1984). This is a continuation of the action in the district court suit in which HRG had originally asked for an order requiring OSHA to initiate rulemaking (see *supra* Section III.A).

[202]48 FED. REG. at 17,313.

[203]*Id.* at 17,285.

[204]49 FED. REG. at 25,793.

[205]OSHA distinguishes between the "effective date" of the standard, applicable to the entire standard, and "start-up" dates for specific provisions. Section 6(b)(4) of the Act provides that the effective date of a standard may not be delayed longer than

summarized the extensive record evidence showing a broad
range of estimates on the length of time needed by various
covered industries to implement these requirements.[206] It ultimately concluded that a one-year start-up date[207] for engineering and work practice controls would provide adequate
time for affected employers, "with few exceptions," to meet
the PEL.[208] OSHA also provided an 80-day start-up date from
the effective date for all other requirements of the standard.[209]

VII. 1984: Renewed Litigation

A. Court of Appeals Litigation

Section 6(f) of the OSH Act permits any person "adversely
affected" by a standard issued by OSHA, within 60 days of
the issuance, to challenge the validity of the standard in a
court of appeals for the circuit in which the person resides or
has his place of business. The issue of the appropriate venue
for court review of OSHA standards has been litigated several
times.[210] But it did not arise in the review of the EtO standard,
since the earliest petition for review of the standard was filed
in the District of Columbia Circuit immediately after the standard was issued.[211] This petition for review was filed by the
Public Citizen Health Research Group (HRG), which had originally brought the suit to compel the issuance of the standard,[212] and by several labor organizations. The Association
of Ethylene Oxide Users, an employer association, sought to
intervene in the proceeding brought by HRG, and later filed

---

90 days; however, the legislative history indicates that this provision was not intended
to bar "graduated requirements to take effect progressively on specific dates exceeding
90 days." 116 CONG. REC. 42,206 (1970) (statement of Rep. Steiger).

[206]49 FED. REG. at 25,793–795.

[207]The start-up period begins to run from the effective date of the standard.

[208]49 FED. REG. at 25,795.

[209]*Id.* Effective dates and start-up dates for OSHA health standards are discussed
in Mintz, *supra* note 3, at 94–95 and n.96.

[210]See, e.g., *Industrial Union Dep't v. Bingham*, 570 F.2d 965, 6 OSHC 1107 (D.C.
Cir. 1977) (benzene emergency standard); *United Steelworkers v. Marshall*, 592 F.2d
693, 7 OSHC 1001 (3d Cir. 1979) (lead standard).

[211]*Public Citizen Health Research Group v. Rowland*, No. 84–1252 (D.C. Cir.
1984).

[212]See Section III.A, *supra*.

its own petition for review of the standard.[213] The court of appeals later consolidated the two proceedings.[214]

On July 31, 1984, the employer association asked the court of appeals to stay the effective date of the EtO standard, more particularly, the one-year start-up date for engineering controls and the 180-day start-up date for instituting the warning signs and labels requirement.[215] The Act provides that the filing of a petition for review in the court of appeals does not operate as a stay of the standard, unless otherwise ordered by the court.[216] Courts of appeals for some circuits, notably the Fifth Circuit, have often granted requests for stays of OSHA health standards.[217] Since litigation on the merits of OSHA health standards is usually lengthy, the granting of a stay significantly delays the implementation of the standard. With respect to emergency temporary standards, which under the statute remain in effect no longer than six months, the stay as a practical matter may prevent the ETS from ever becoming effective.[218] Unlike the petitioners in most health standard challenges, however, the employer association in the EtO case did not press the court of appeals to decide the stay issue immediately; on August 17, 1984, it filed a joint motion

---

[213]*Ethylene Oxide Users v. Rowland*, No. 84–1392 (D.C. Cir. 1984).

[214]*Public Citizen Health Research Group v. Tyson*, 796 F.2d 1479, 12 OSHC 1905 (D.C. Cir. 1986).

[215]In the final standard, OSHA, among other things, required that EtO regulated areas be demarcated through signs referring to EtO as a "Cancer Hazard and Reproductive Hazard" and that EtO containers be labeled with similar warning signs. 49 FED. REG. at 25,789–790 (1984). OSHA has separately promulgated a final generic rule on hazard communication. 29 C.F.R. §1910.1200 (1984). The standard was remanded to OSHA by the Third Circuit for further consideration on several issues. *United Steelworkers v. Auchter*, 763 F.2d 728, 12 OSHC 1337 (3d Cir. 1985). The court of appeals later directed OSHA to take action on its remand on the issue of expanding the scope of the standard, *United Steelworkers v. Pendergrass*, 819 F.2d 1263, 13 OSHC 1305 (3d Cir. 1987), and an expanded standard issued on August 24, 1987, 52 FED. REG. 31,852 (1987).

[216]Section 6(f).

[217]E.g., OSHA's permanent benzene standard was stayed by the Fifth Circuit. See *American Petroleum Inst. v. OSHA*, 581 F.2d 493, 6 OSHC 1959 (5th Cir. 1978), *aff'd sub nom. Industrial Union Dep't v. American Petroleum Inst.*, 448 U.S. 607, 8 OSHC 1586 (1980), and the standard never went into effect. OSHA's cotton dust standard was stayed by the District of Columbia Circuit, but the stay was later lifted by the court of appeals which upheld the standard in most respects. *AFL-CIO v. Marshall*, 617 F.2d 636, 7 OSHC 1775 (D.C. Cir. 1979). On occasion, the court of appeals will grant a partial stay of the more costly provisions of the standard. *United Steelworkers v. Marshall*, 647 F.2d 1189, 8 OSHC 1810 (D.C. Cir. 1980), *cert. denied*, 453 U.S. 913 (1981) (lead standard).

[218]E.g., in the case of the benzene emergency standard, the stay granted by the Fifth Circuit remained in effect for the entire six-month period because of the litigation over venue. See *Industrial Union Dep't v. Bingham*, *supra* note 210.

with OSHA requesting that the court hold the motion for stay in abeyance until after OSHA decided whether to grant the association's motion for an *administrative* stay of the standard.

Substantially the same grounds were advanced by the association for both the court and the administrative stays.[219] In substance, the association argued first that its members would suffer irreparable harm if they began to install engineering controls and these requirements were later modified or vacated by the court. Further, the association argued that if OSHA ultimately decided to include a STEL in the EtO standard, the STEL-oriented engineering controls in many instances would require modification or removal of the controls already begun or installed to comply with the time-weighted average requirement.[220] Finally, the association argued that the substantial evidence in the record did not support the requirement in the standard for signs and labels stating that EtO is a "Cancer Hazard and Reproductive Hazard" and that its members could suffer irreparable harm if the labeling requirements were implemented.[221]

OSHA did not answer the association's request for a stay on the merits but agreed that it be held in abeyance; OSHA also asked to be given 10 days after its decision on the motion for an administrative stay to file a response. HRG filed a brief strongly opposing the stay, arguing that the association's arguments against the standard were "unfounded," that the challenged provisions were "virtually certain" to be upheld, and that the association's nonparticipation in the rulemaking also weighed against granting the stay.[222] On November 5, 1984, the court of appeals ordered OSHA to respond to the association's request for a stay. After notifying the association

---

[219]The courts of appeals have generally applied the criteria for preliminary injunctions in determining whether to grant stays of OSHA standards. Thus, the courts have said that to obtain a preliminary injunction a party must show a substantial likelihood of prevailing on the merits; irreparable harm if the stay is denied; that the stay will not substantially harm other parties; and that the stay will not interfere with the public interest. See *Taylor Diving & Salvage Co. v. Department of Labor*, 537 F.2d 819, 821 n.8, 4 OSHC 1511, 1512 (5th Cir. 1976) (staying OSHA's ETS on diving).

[220]Memorandum of Petitioner Association of Ethylene Oxide Users in Support of Motion for Stay Pending Review, in *Ethylene Oxide Users v. Rowland, supra* note 213, at 28–31.

[221]Memorandum, *supra* note 220, at 6, 25–29. The association argued also that the 1-ppm PEL was not supported by substantial evidence and therefore there was a substantial likelihood that the association would prevail on the merits.

[222]Memorandum of Health Research Group, in *Ethylene Oxide Users v. Rowland, supra* note 213.

that OSHA was denying its request for a stay, OSHA opposed the stay in the court of appeals; and on January 23, 1985, the court of appeals denied the request on the grounds that it "would not be in the public interest." The court noted that the association had conceded that most employers had already taken steps to lower EtO exposure to the new permissible level, and ruled that "irreparable injury resulting from the denial of a stay had not been established."

B. District Court Litigation

At the same time that the court of appeals litigation was taking place, HRG renewed its suit in the U.S. District Court for the District of Columbia, asking the court for "an order directing defendants to issue immediately a completed final standard governing worker exposure to ethylene oxide."[223] The thrust of HRG's argument was that OSHA had violated its stipulation that the EtO rulemaking proceedings would be "completed" by June 15, 1984, by issuing an EtO standard that lacked a provision on the short-term exposure limit that OSHA "conceded to be a vital component of any regulatory action relating to ethylene oxide."[224] In response, OSHA argued that the Act provided OSHA with a "great deal of flexibility in exploring and exercising regulatory options" and the fact that it had retained the STEL issue on its regulatory calendar "for a brief period" did not negate the fact that the rulemaking had been "completed" and that a "comprehensive" EtO standard had been issued.[225] OSHA also argued in its memorandum that it had not acted in "bad faith" in concluding "that OMB had raised serious questions concerning the adequacy of the rulemaking record to support a STEL."[226] HRG's reply memorandum presented the additional argument that

---

[223]*Public Citizen Health Research Group v. Tyson*, No. 81–2343 (D.D.C. 1984). This is the same proceeding referred to in note 201, *supra*, except that Patrick J. Tyson, the new Deputy Assistant Secretary, was substituted as principal defendant.

[224]*Id.* at 2. The stipulation was entered into by OSHA when the agency had failed to issue a final standard one year after the original court of appeals decision. HRG attachments to the motion papers were excerpts from OSHA's draft final standard submitted to OMB (the June 14 draft) and excerpts from the text of the standard submitted by OSHA to the *Federal Register* with those portions dealing with the STEL blacked out.

[225]OSHA Memorandum, *supra* note 201, at 4–5.

[226]*Id.* at 7. The Association of Ethylene Oxide Users filed a brief *amicus curiae* in the district court in support of OSHA's position.

OMB's "involvement at this juncture raises serious questions about the integrity of OSHA's rulemaking process." According to OSHA's regulations on rulemaking at 29 C.F.R. §1911.15, the rulemaking must be "on the record," and, HRG argued, since OMB's "last minute, off the record interventions" were not part of the public record, "they cannot form the basis of OSHA's failure to include a short term exposure limit in its final rule."[227]

On August 30, 1984, the district court entered an order requiring OSHA to reach a final decision on the STEL issue by December 17, 1984, and to periodically report to the court on its progress in the rulemaking.[228] Having obtained a short extension of the December 17 deadline, OSHA notified the district court on December 21 that the adoption of a STEL was not warranted by the available health evidence, and was thus "not reasonable, necessary or appropriate for inclusion in the final EtO standard."[229] On January 2, 1985, OSHA published in the *Federal Register* a Supplemental Statement of Reasons to the Final Rule, containing its determination that no STEL would be established for EtO.[230]

VIII. 1985: Short-Term Exposure Limit

A. Supplemental Statement of Reasons

As noted, in its final EtO standard, OSHA reserved judgment on whether the standard should provide a STEL, largely in response to reservations expressed by the Office of Management and Budget.[231] OSHA indicated that its draft final

---

[227]Plaintiff's Memorandum in reply to Defendant's Opposition to Plaintiff's Motion, see *supra* note 201, at 3–4. The issue of the legality of presidential involvement in agency rulemaking is discussed *infra* Section VIII.B. The broad issue of *ex parte* communications in informal rulemaking has been the subject of several major court cases. *Home Box Office v. FCC*, 567 F.2d 9 (D.C. Cir. 1977), *cert. denied*, 434 U.S. 829 (1977); *Action for Children's Television v. FCC*, 564 F.2d 458 (D.C. Cir. 1977). It has also been the subject of much legal comment, e.g., Nathanson, *Report to the Select Committee on Ex Parte Communications in Informal Rulemaking Proceedings*, 30 AD. L. REV. 377 (1978).

[228]14 OSHR 315 (1984). OSHA outlined the steps it would have to take in issuing a STEL in its affidavit to the court.

[229]14 OSHR 563 (1985).

[230]50 FED. REG. 64 (1985).

[231]49 FED. REG. at 25,775 (1984). See discussion *supra* Section VI.B.

standard and OMB's comments would be submitted to a number of scientifically qualified "peer reviewers" for comment, analysis, and criticism.[232] These comments would be placed in the record, the preamble stated, together with any public comments on the peer review statements.[233]

OSHA thereafter requested peer review of the STEL data from 23 individuals and organizations, who were asked "to provide comment on those aspects of the STEL issues that were within their areas of expertise." These specialists included those in the fields of toxicology, epidemiology, and industrial hygiene.[234] Among the 23 peer reviewers were the 12 members of OSHA's National Advisory Committee on Occupational Safety and Health.[235] OSHA received comments from the peer reviewers, and in addition received 41 comments from the general public on the STEL issue.[236]

On January 2, 1985, OSHA published a Final Rule: Supplemental Statement of Reasons[237] determining, on the basis of the entire record—including the peer review and the comments received after the final standard was published—that "a STEL for EtO is not warranted by the available health evidence, and *** a STEL is not reasonably necessary or appropriate for inclusion in the final EtO standard."[238] In ex-

---

[232]The role of scientists in regulatory decision-making has been the subject of considerable debate. On the one hand, it has been suggested that "scientifically sophisticated outsiders" are in the best position to describe the current state of technical knowledge and to provide statements, founded on that knowledge, that will provide defensible, credible, technical bases for urgent policy decisions. *The Science Court Experiment: An Interim Report*, 193 SCIENCE 653 (1976). On the other hand, others have expressed the view that "the very concept of objectivity embodied in the word disinterested is now discredited." Bazelon, *Risk and Responsibility*, 205 SCIENCE 277 (1979). The desirability of giving important decision-making power to scientists has been seriously questioned. Talbott, *"Science Court": A Possible Way to Obtain Scientific Certainty for Decisions Based on Scientific "Fact"?* 8 ENVTL. L. 827 (1978). The District of Columbia Circuit spoke of the usefulness of peer review in its decision vacating OSHA's second asbestos emergency standard. *Asbestos Information Ass'n v. OSHA*, 727 F.2d 415, 420 n.12, 11 OSHC 1817, 1820 (5th Cir. 1984).

[233]49 FED. REG. at 25,775.

[234]See 50 FED. REG. 65 (1985).

[235]The composition and responsibilities of the committee, called NACOSH, are described in §7(a). See also OSHA's regulations on NACOSH. 29 C.F.R. §1912.5(a) (1984). Not all NACOSH members are expected to be experts in occupational health. See 29 C.F.R. §1912.9(b). The names and organizational affiliations of the NACOSH members at the time appear at 50 FED. REG. 66 (1985).

[236]The note asking for comment on the comments of the peer reviewers and the issues raised by the draft standard was published on Sept. 19, 1984. 49 FED. REG. 36,659.

[237]50 FED. REG. 64 (1985).

[238]*Id.*

plaining its conclusion, OSHA noted that "no new studies on the health effects of EtO have been submitted to the record" during the renewed comment period.[239] However, OSHA asserted that "there has been considerable new documentation of reasons why the existing health data do not necessitate that a STEL be established." This information, according to OSHA, "has provided the agency with clearer and more definitive findings than were available at the time of promulgation of the June 22 final rule."[240]

OSHA based its decision on three findings:

"First, the available health data do not demonstrate the risks from EtO exposure to be dose-rate dependent. In other words, the studies do not indicate that the risks from exposure to a given dose of EtO are greater when that dose is distributed at higher concentrations over a short period of exposure during a workday rather than at a lower concentration during a longer period of time. Second, since the effects of EtO are assumed to be dose dependent rather than dose-rate dependent, reduction of the total dose is the critical factor in dealing with the significant risks of EtO exposure. Therefore, the 1 ppm TWA is sufficient to minimize significant risk within the bounds of feasibility. Third, in terms of industrial hygiene and methods of controlling EtO, compliance with the TWA will necessitate the control of short-term exposures, particularly for employees whose exposure consists primarily of short-term bursts. Therefore, to the extent that good industrial hygiene practice calls for the reduction of short-term peak exposures, the low TWA of 1 ppm will result in the minimization of short-term exposures within the workday. Further, where burst-type exposures occur more than once per day, and where there are background levels of EtO between bursts, the TWA will place internal limitations on the levels and durations of such bursts during the workday to assure compliance with the TWA ***."[241]

OSHA further stated that it "believes that by reducing the total EtO dose through establishment of a 1 ppm TWA, and by including other ameliorative provisions in the standard [such as exposure monitoring and employee training], a substantial reduction of the adverse health effects of EtO will be achieved."[242]

---

[239]*Id.* at 73.

[240]*Id.*

[241]*Id.*

[242]*Id.* at 77. OSHA also indicated that it had asked NIOSH to fund additional studies on whether there is a dose-rate relationship for EtO and that OSHA would review those studies. *Id.* at 64.

B. Challenges in the Court of Appeals

At this point, the main focus of the EtO proceeding shifted to the Court of Appeals for the District of Columbia, where petitions for review of the OSHA EtO final standard were already pending.[243] HRG, one of the petitioners in the court proceedings, comprised a public interest group and four unions which represented a "major segment" of the more than 100,000 hospital and health care workers exposed to EtO. HRG's request for review was limited to the legality of OSHA's last-minute decision to "remove" the STEL from the final standard. The other petitioner was the Association of Ethylene Oxide Users, which supported the deletion of the STEL and also urged reversal of the 1 ppm PEL and the requirement that employees be warned that EtO is a cancer and reproductive hazard.

1. Health Research Group Brief

The brief of the HRG recognized that under applicable precedent, including OSHA cases,[244] OSHA's "expert" determinations in standards proceedings are entitled to some measure of deference. In this proceeding, however, HRG contended that the court should engage in a "heightened level of scrutiny for at least two reasons." First, HRG argued, "this proceeding has been marked by precisely the sort of abrupt shift in agency policy that is a 'danger signal' that the agency may be acting in an arbitrary fashion."[245] Second, HRG argued, the "determinative" decision on the STEL was made by OMB, "which has no technical expertise whatsoever," and not by OSHA, the expert agency. Therefore, "the normal reasons for according deference to agency expertise are entirely absent here."[246]

---

[243]*Public Citizen Health Research Group v. Rowland, supra* note 211; *Ethylene Oxide Users v. Rowland, supra* note 213. The court of appeals had previously denied the request for a stay pending review. See discussion *supra* Section VII.A.

[244]E.g., *Industrial Union Dep't v. Hodgson*, 499 F.2d 467, 1 OSHC 1631 (D.C. Cir. 1974).

[245]Brief of the Public Citizen Health Research Group at 33, *Public Citizen Health Research Group v. Rowland, supra* note 243. The brief cites, among several other cases on this point, *Public Citizen v. Steed*, 733 F.2d 93, 99 (D.C. Cir. 1984).

[246]HRG Brief at 33–34.

HRG referred to the central three findings upon which OSHA based its decision to eliminate the STEL[247] and argued that none of these findings "can withstand scrutiny." HRG also asserted that apart from the specific flaws in the findings, OSHA's explanation suffers from an "overriding error," namely, it "leaves totally unanswered OSHA's own justifications for including the STEL in the first place: that a 1 ppm TWA standard, by itself, does not reduce the significant risk of cancer to an acceptable level, and it does not reduce at all the risks of chromosome damage and spontaneous abortions, which OSHA concedes are caused by short-term exposures at TWA levels at or below 1 ppm."[248]

After seeking to refute each of OSHA's three basic findings in light of the record,[249] HRG turned to the role of OMB in the proceeding, and made the following major points:

(1) Since Congress directed that the Secretary of Labor—not the Director of OMB—determine the content of OSHA standards, OSHA's "abdication of [its] rulemaking responsibilities to OMB violated the language and the spirit of the OSH Act." Further, "because OMB simply parroted industries' concerns," which had already been submitted to the record, it is clear that it was not the "substance of the comments that influenced OSHA but rather it was their source—OMB—that was the determinative factor."[250]

(2) Whatever the scope of OMB's proper role under Executive Order No. 12,291 in the formulation of agency regulatory policy "as a general matter," that role, according to HRG, "surely does not extend to the formulation of the kind of complex scientific and technical judgments OSHA was called on to make regarding the need for a STEL."[251]

(3) The thrust of OMB's concern "relates almost exclusively to cost-benefit concerns," which are not a proper

---

[247]See text accompanying note 241, *supra.*

[248]HRG Brief, *supra* note 245, at 41.

[249]*Id.* at 42–51.

[250]*Id.* at 53–55. In dealing with the issue of intergovernmental communications in informal rulemaking, the Administrative Conference has distinguished between communications between presidential advisors and agency officials that originated from "persons outside the government" and other communications. See 1 C.F.R. §305.80–6 (1983).

[251]HRG Brief, *supra* note 245, at 56.

basis for OSHA health standards determinations under the cotton dust case.[252]

To remedy OSHA's unlawful actions, HRG asked the court to direct OSHA to issue the EtO standard containing a STEL for EtO as set forth in the June 14 draft.[253]

2. Employer Association Brief

In its brief to the court of appeals, the Association of Ethylene Oxide Users asserted that OSHA's finding that EtO was carcinogenic was based primarily on animal studies; that there are "significant uncertainties involved in attempts to translate animal health effect to possible risk in humans," particularly in estimating risks to humans exposed to low doses based upon data concerning animals exposed to high doses; that the human experimental data here do not "establish" cancer risk in humans; and that OSHA's reliance on evidence of chromosomal aberrations and sister chromatid exchanges "is contrary to the weight of reputable scientific thought" and its conclusions on EtO's reproductive hazards are based on unreliable and insubstantial evidence.[254] In sum, the association argued that OSHA's "prejudgment of crucial factual issues and the lack of substantial evidence to support the Secretary's findings require a determination that the Secretary has not established [that] the 1 ppm permissible exposure [limit] and the signs and labels requirement are necessary to eliminate a significant risk of harm."[255]

---

[252]*American Textile Mfrs. Inst. v. Donovan*, 452 U.S. 490, 9 OSHC 1913 (1981).

[253]HRG Brief, *supra* note 245, at 60. HRG refers to OSHA's action here as "essentially" an "unjustified or illegal rescission of rules by administrative agencies," citing, among other cases, *Public Citizen v. Steed*, *supra* note 245, at 93. HRG is presumably anticipating the argument that the deletion of the STEL should be judged by the court of appeals under the more deferential standard of review applied in cases of agency refusal *to initiate* rulemaking. See *WWHT Inc. v. FCC*, 656 F.2d 807 (D.C. Cir. 1981). The Supreme Court has held, however, that rescissions of rules are judged by the same review standard as decisions to adopt rules in the first place. *Motor Vehicle Mfrs. Ass'n v. State Farm Mut.*, 463 U.S. 29 (1983). But the *State Farm* case is not factually the same, since here the STEL was never promulgated; therefore there was no "rescission" and the question as to the standard of review to be applied may remain.

[254]Brief of the Association of Ethylene Oxide Users at 9–33, *Public Citizens Health Research Group v. Rowland*, No. 84–1252 (D.C. Cir. 1984), & No. 85–1014 (D.C. Cir. 1985).

[255]*Id.* at 60.

3. OSHA Brief

The OSHA brief was filed on May 29, 1985.[256] Addressing itself first to the issue raised by the Association of Ethylene Oxide Users relating to the legality of the 1 ppm PEL, the brief cited extensive record evidence supporting OSHA's conclusion that EtO poses a cancer threat to exposed workers; that it also poses a danger of mutagenic and cytogenetic effects and a hazard to the reproductive system of employees; that this risk is "significant" and that the new PEL will "substantially" reduce the level of risk faced by workers.[257] In answer to arguments of HRG, the brief contended that OSHA properly concluded that issuance of a STEL was not warranted on the record before OSHA. Although acknowledging the mandate of Section 6(b)(5) of the OSH Act, OSHA argued that the section "has never been interpreted, and cannot reasonably be interpreted, to require OSHA to adopt *every conceivable* feasible measure to reduce worker exposure to a regulated substance, no matter how limited the measure's effectiveness nor how attenuated the evidence suggesting its necessity."[258] Here, the brief argued, the evidence no more than "suggest[s], in a very tentative and hypothetical way," that there is a health need for a STEL. This is "simply not good enough to serve as the basis for regulation," particularly since, for a number of reasons, "benefits that would be achieved by a short-term limit will be almost entirely achieved by a PEL."[259]

---

[256]Brief for the Secretary of Labor, Nos. 84–1252, 84–1392, & 85–1014 (D.C. Cir. 1984 & 1985). The brief was signed both by attorneys from the Office of the Solicitor in the Department of Labor and by attorneys in the Civil Division of the Department of Justice. Court of appeals litigation involving the review of OSHA standards has normally been handled exclusively by Department of Labor attorneys since 1975, when the Solicitor of Labor and the Deputy Attorney General entered into an agreement governing the handling of OSHA court litigation. The background of this agreement is discussed in D.L. Horowitz, THE JUROCRACY: GOVERNMENT LAW-YERS, AGENCY PROGRAMS, AND JUDICIAL DECISIONS 109–114 (1977). The special role of the Department of Justice in the EtO litigation presumably related to the issue of the legality of OMB involvement in the rulemaking.

[257]OSHA Brief, *supra* note 256, at 17–45.

[258]*Id.* at 54. The brief cited *Vermont Yankee Nuclear Power Corp. v. Natural Resources Defense Council*, 435 U.S. 519, 551 (1978), for the principle that an environmental impact statement "cannot be found wanting simply because the agency failed to include every alternative device and thought conceivable by the mind of man."

[259]OSHA Brief, *supra* note 256, at 54–55.

OSHA then addressed HRG's arguments based on OMB's involvement in the proceeding.[260] The brief made the following major points on that issue:

(1) Executive Order 12,291 as applied to rulemaking under the OSH Act is valid. In the OSH Act, Congress vested policy-making responsibilities in the Secretary of Labor, who serves at the President's pleasure. Accordingly, it must be assumed that Congress intended the exercise by OSHA of responsibilities under the OSH Act "to be discharged pursuant to the President's supervisory authority."[261]

(2) The District of Columbia Circuit in *Sierra Club v. Costle*[262] "squarely rejected" the argument made by HRG that when the subject matter of a proposed agency regulation involves technical and "scientific" judgments regarding which agency expertise is particularly important, "Presidential guidance and influence is so inappropriate that Congress must be presumed to have forbidden it."[263] In that case, involving a complex engineering issue of pollution controls, the court recognized the "basic need" for the President to "monitor the consistency of executive agency regulations with Administration policy."[264]

(3) OMB did not "dictate" OSHA's decision deleting the STEL from the EtO standard. First, when OSHA sent its draft final EtO standard, which included a STEL, to OMB, the conclusion on the STEL was "at best provisionally final," unsigned by the Secretary and subject to OMB comments.[265] Secondly, OSHA did not precipitously reverse its decision on the basis of the OMB comments; rather, "in order to make a reasoned and deliberate response," OSHA sought additional comments and peer review, and only "after this extended consideration," made a "deliberate and mea-

[260]*Id.* at 70–80.
[261]*Id.* at 80.
[262]657 F.2d 298 (D.C. Cir. 1981).
[263]OSHA Brief, *supra* note 256, at 85.
[264]*Id.* at 86, quoting from *Sierra Club v. Costle, supra* note 262, at 405.
[265]OSHA Brief, *supra* note 256, at 89.

sured" decision that the evidence did not support a STEL.[266]

(4) Contrary to HRG's arguments, there was no "reversal" by OSHA of its policy and factual findings which would constitute "danger signals" warranting more stringent court review.[267] In the first place, in its supplemental decision, OSHA made a "new" finding on an issue previously unaddressed, namely, there was no basis for linking the intensity of the exposure with the health effects observed.[268] Nor was the absence of a STEL a departure from a "well-established pattern" of including STELs in health standards, since "in recent years" OSHA has regulated toxic substances, such as benzene, arsenic, and lead, without including a STEL.[269]

4. *Brief of* Amici Curiae

The position of HRG was supported in a brief filed on June 28 by five members of Congress who are chairpersons of congressional committees which, the representatives said, have "generated much of the substantive legislation pursuant to which Executive Branch agencies have promulgated rules."[270] In their statement of interest in their brief, the congresspersons "object[ed] to the systematic usurpation of legislative power by OMB" under Executive Order 12,291. OMB has used Executive Order 12,291, they argued, in a way that "severely

---

[266]*Id.* at 89–90.

[267]*Id.* at 91.

[268]*Id.* at 92.

[269]*Id.* at 92–93.

[270]Brief of Representatives John D. Dingell, Peter W. Rodino, Jr., Jack Brooks, Augustus F. Hawkins, and William D. Ford as *Amici Curiae* in Support of Petitioners at 2, *Public Citizen Health Research Group v. Rowland, supra* note 243. Rep. Dingell is chairperson of the House Committee on Energy and Commerce and its Subcommittee on Oversights and Investigations; Rep. Rodino is chairperson of the House Committee on the Judiciary; Rep. Brooks is chairperson of the House Government Operations Committee; Rep. Hawkins is chairperson of the House Education and Labor Committee; and Rep. Ford is chairperson of the House Committee on Post Office and Civil Service. The question whether the courts will entertain suits by members of Congress which seek to determine "political" questions, such as the "pocket veto" issue, has been the subject of sharp differences of opinion. See, e.g., *Barnes v. Kline*, 759 F.2d 21 (D.C. Cir. 1985), *dismissed as moot*, 107 S. Ct. 734 (1986), and the dissenting opinion of Judge Bork, 759 F.2d at 41. In this case, of course, the members of Congress filed a brief as *amici* and therefore no standing issue would arise.

dislocates the allocations of powers in our government." The Executive Order is the "cornerstone of a steadily growing Presidential apparatus, the effect of which is to contravene explicit Congressional delegations of authority, to subvert meaningful public participation in and judicial review of federal regulations, and to impose substantive standards on decisionmakers foreign to the statutes they administer." This activity, the congresspersons asserted, unless checked, will "fundamentally damage" the administrative process, the legislative system, and the separation of powers.[271]

This case, the congresspersons argued, illustrates their concerns about the role of OMB in the governmental process, since OMB, purportedly acting under Executive Order 12,291, "force[d] OSHA to drop a vital occupational health standard that was fully supported by the agency record."[272] This interference, they contended, is "far from an isolated example," since OMB has pervasively acted in contravention of the will of Congress. Several other examples were cited in the brief, involving, in addition to OSHA, the Environmental Protection Agency and the Food and Drug Administration.[273] The thrust of the congresspersons' legal argument was that once Congress has exercised its constitutional authority to delegate rulemaking authority to the agency in accordance with substantive and procedural criteria provided by Congress, "the responsibility falls to the Judiciary to ensure that agency rulemaking conforms to the Congressional mandate."[274] In the *Chadha* case,[275] the Supreme Court held that Congress cannot overrule the action of an agency in promulgating a regulation pursuant to its delegated power, except by passing new legislation, consistent with the requirements of bicameralism and the presentment clause. Similarly, OMB cannot "usurp" the legislative power delegated by the Congress to a specific Executive Department official.[276] Arguing that the OMB ac-

---

[271]Brief of *Amici Curiae, supra* note 270, at 7–8.

[272]*Id.* at 13–14.

[273]*Id.* at 13–16.

[274]*Id.* at 17.

[275]*Immigration & Naturalization Serv. v. Chadha*, 462 U.S. 919 (1983) (legislative veto case). See particularly the discussion regarding the relationship of Congress and Executive Department agencies, *id.* at 953 n. 16. The *Chadha* case is discussed critically in E.D. Elliott, *INS v. Chadha: Administrative Constitution, the Constitution, and the Legislative Veto,* in 1983 THE SUPREME COURT REVIEW at 125 (Kurland, Casper, & Hutchinson eds. 1984).

[276]Brief of *Amici Curiae, supra* note 270, at 16–18.

tions concerning EtO contravened Congress's stated intent in the OSH Act that standards be set by the Secretary of Labor, that standards be set following "meaningful" public participation and be subject to judicial review, and that they be based on the "substantive health and safety criteria" contained in the OSH Act, the congresspersons asked the court to "reinstate the EtO standard incorporating a 10 ppm 15-minute STEL."[277]

C. Court of Appeals Decision

On July 25, 1986, the District of Columbia Circuit decided the EtO case for the second time.[278] The court affirmed OSHA's 8-hour exposure limit, but concluded that the agency's STEL was not supported by the record and remanded for further consideration on the issue. The court did not reach the constitutional issues raised by HRG's challenge based on OMB interference with the rulemaking proceeding.[279]

The court first addressed the scope of court review. It restated the interpretation of the "substantial evidence" test previously set forth in the asbestos[280] and lead[281] cases, emphasizing that this "deferential" standard of review "in no way conflicts with the substantive legal requirements" established in the benzene[282] case. Thus, the court of appeals said, the benzene decision reversed OSHA on its legal interpretation of the Act, and imposed on OSHA the burden of establishing the existence of a "significant risk." At the same time, the Supreme Court asserted that this requirement was not a "mathematical straitjacket," and that the court in reviewing OSHA standards must give the agency "some leeway where its findings must be made on the frontiers of scientific knowl-

---

[277]*Id.* at 57.

[278]*Public Citizen Health Research Group v. Tyson*, 796 F.2d 1479, 12 OSHC 1905 (D.C. Cir. 1986).

[279]The panel consisted of Chief Judge Robinson, Senior Circuit Judge McGowan, and Circuit Judge (now Senior Circuit Judge) Wright. The opinion was written by Judge McGowan, who wrote the opinion for the same court in the original asbestos decision, *Industrial Union Dep't v. Hodgson*, 449 F.2d 467, 1 OSHC 1631 (D.C. Cir. 1974).

[280]*Industrial Union Dep't v. Hodgson*, *supra* note 279, at 472–474, 1 OSHC at 1635–1638.

[281]*United Steelworkers v. Marshall*, 647 F.2d 1189, 1206–1207, 8 OSHC 1810, 1816–1817 (D.C. Cir. 1980), *cert. denied*, 453 U.S. 913 (1981).

[282]*Industrial Union Dep't v. American Petroleum Inst.*, 448 U.S. 607, 8 OSHC 1586 (1980).

edge."[283] Thus, according to the court of appeals, the Supreme Court preserved "traditional deference to agency decision making."[284]

The court of appeals next dealt with the challenge of the Association of Ethylene Oxide Users (AEOU) to the 8-hour exposure limit. The court first upheld OSHA's finding that EtO has carcinogenic, cytogenetic, and mutagenic effects and poses reproductive hazards. While recognizing that some of the studies relied on by OSHA were not "models of scientific investigation" and that the court "might be hesitant to accept a single study" as substantial evidence, the "cumulative impact" of the studies is "compelling," said the court. In reaching this conclusion, the court stated that AEOU "fundamentally misconstrues the roles that OSHA and reviewing courts must play" in regulating toxic substances. If the evidence before the court "can be reasonably interpreted" as supporting the agency, the court must affirm the agency's conclusion "despite the fact that the same evidence is susceptible of another interpretation."[285]

AEOU next challenged OSHA's quantitative risk assessment, which was based on the animal data in the Bushy Run study. The major contention of AEOU was that the assumption in OSHA's risk assessment that there is no safe threshold for EtO[286] was inconsistent with the Supreme Court's rejection in the benzene case of OSHA's "lowest feasible level policy." The court found no merit in the argument, saying that AEOU "simply misconstrues" the benzene opinion. The Supreme Court vacated the benzene standard because the agency did not "even" attempt to construct a dose-response curve and its "bald" assertion that no level of benzene was safe was "devoid of record support." But the Supreme Court made it clear in the benzene case that the agency is "free to use conservative assumptions" in interpreting carcinogenic data, "risking error on the side of overprotection rather than underprotection." In the EtO

---

[283]*Id.* at 656, 8 OSHC at 1603.

[284]*Supra* note 278, at 1486, 12 OSHC at 1910.

[285]*Id.* at 1495, 12 OSHC at 1918.

[286]In its risk assessment, OSHA used a linear, no-threshold model, as did EPA's Carcinogen Assessment Group. This model assumes that EtO exposure and biological response vary proportionately, and that there is no threshold level below which EtO exposure produces no adverse health effects. *Id.* at 1497–1498, 12 OSHC at 1919–1920.

proceeding, the court of appeals said, OSHA had gone to "great lengths, within the bounds of available scientific data," in deciding that there was a significant risk. Thus, OSHA's actions here are in "stark contrast" to its actions in the benzene proceeding.[287]

The court of appeals went on to affirm the EtO 8-hour level, finding substantial evidence supporting the no-threshold model and OSHA's conclusion that the 1-ppm level was reasonably necessary and appropriate to remedy a significant risk at the 50-ppm level. In sum, the court concluded that OSHA had "completed the difficult task" of establishing the 8-hour exposure limit "with a thorough and professional approach."[288]

The agency's conclusion that there is no need for a STEL did not win the same praise from the court.[289] The court's underlying premise was that OSHA *must* set a level that is feasible if it would reduce a significant health risk.[290] In light of OSHA's finding that there would still be a significant risk at a 1-ppm 8-hour level,[291] OSHA would be required to set a STEL, if the STEL would "further" reduce that risk and if it were feasible. Since OSHA never argued nonfeasibility for a STEL, the issue before the court was the effect of a STEL on the significant risk at the 1-ppm level.

---

[287]*Id.* at 1499, 12 OSHC at 1920–1921. In the lead proceeding, the District of Columbia Circuit affirmed the permissible exposure limit, similarly finding that the agency had met the requirements of the benzene case by not relying on assumptions but rather by "amass[ing] voluminous evidence of the specific harmful effects of lead at particular blood-lead levels, and correlat[ing] these blood-lead levels with air-lead levels." *United Steelworkers v. Marshall, supra* note 281, at 1248, 8 OSHC at 1850. In its cotton dust decision, the Supreme Court also affirmed the permissible exposure limit, saying that it was "difficult to imagine what else the agency could do to comply" with the significant risk requirements of the benzene decision. *American Textile Mfrs. Inst. v. Donovan,* 452 U.S. 490, 506 n.25, 9 OSHC 1913, 1919 (1981). In *ASARCO Inc. v. OSHA,* 746 F.2d 483, 11 OSHC 2213 (9th Cir. 1984), the court of appeals upheld OSHA's arsenic standard, rejecting an industry challenge to OSHA's significant risk finding.

[288]*Supra* note 278, at 1503, 12 OSHC at 1924.

[289]As noted below, the court did not decide the legality of OMB influence. See discussion *infra* at notes 294 & 295 and accompanying text. The extent to which the concerns of the court over the procedural deficiencies in the rulemaking may have influenced its substantive determination must remain in the realm of speculation.

[290]The court relied on the statement in §6(b)(5) that OSHA "shall" set the standard which most adequately assumes that no employee will suffer material impairment of health. See also *American Textile Mfrs. Inst. v. Donovan, supra* note 287, at 507 n.27, 9 OSHC at 1920, where the Supreme Court said that in order to satisfy the Act, OSHA must select *"the* standard *** that is most protective."

[291]OSHA expressly found that exposure at the 1-ppm level would cause 12–23 excess deaths. See note 165, *supra.* This finding was unchallenged on appeal. *Supra* note 278.

OSHA's conclusion that the STEL would not reduce the significant risk still existing with a 1-ppm 8-hour limit was based on the inference that in reaching a 1-ppm level, employers will "substantially reduce the magnitude of short-term exposures." The record does not support that conclusion, the court concluded. Thus, there was evidence of short-term exposures as high as 480 ppm for 1 minute, or 96 ppm for 5 minutes each day; yet in these instances, the average daily exposure was within the 1-ppm 8-hour limit. If the agency regulated the short-term exposure in these circumstances, which would be feasible, the court observed, the 8-hour level would be below 1 ppm and thus the significant risk would be reduced.

Since the agency "entirely failed to consider an important aspect of the problem,"[292] the court remanded the case to OSHA for further consideration of the STEL issue, telling the agency to "ventilate the issues on this point thoroughly" and either adopt a STEL or "explain why."[293]

The court finally noted that the parties "vigorously dispute" the legality of OMB's participation in the rulemaking. After noting that one court has decided the issue,[294] and that the issue had been debated in the legal literature, the court pointed out that in light of its disposition on the substantial evidence question, it had "no occasion to reach the difficult constitutional questions involved."[295]

D. The Court of Appeals, Once Again—1987

The court of appeals decision remanding the EtO proceeding to OSHA "to ventilate the issues on [the STEL] point thoroughly" was handed down on July 25, 1986. When OSHA failed to take action on the remand for more than 10 months,

---

[292]The Court here cited the "air bags" case, *Motor Vehicle Mfrs. Ass'n v. State Farm Mut. Automobile Ins. Co.*, 463 U.S. 29, 43 (1983), where the Supreme Court set aside the Department of Transportation rescission of the air bag regulations. The issue of court review of agency deregulatory actions has brought forth an increasing body of legal writings. See, for example, M. Garland, *Deregulation and Judicial Review*, 98 HARV. L. REV. 507 (1985).

[293]*Supra* note 278, at 1507, 12 OSHC at 1927. The court did not discuss explicitly whether OSHA would be required to undertake renewed rulemaking or whether it could modify (or reaffirm) its original standard on the basis of the original record, in light of the court decision.

[294]*Environmental Defense Fund v. Thomas*, 627 F. Supp. 566 (D.D.C. 1986).

[295]The court characterized these "difficult" issues as the "executive's proper role in administrative proceedings" and the "appropriate scope of delegated power from Congress to certain executive agencies."

Public Citizen, on April 5, 1987, asked the court of appeals to hold OSHA in contempt for failing to act.

The Court of Appeals for the District of Columbia Circuit, in a *per curiam* opinion issued on July 21, 1987—four days short of one year from its original decision—ordered OSHA to complete its rulemaking on the STEL issue by March 1988; the court, however, refused to hold the agency in contempt.[296] The court prefaced its decision with a statement of the "delicate position" in which it had been placed. The court said:

> "Although the courts must never forget that our constitutional system gives the Executive Branch a certain degree of breathing space in its implementation of the law, we cannot countenance maneuvering that merely maintains a facade of good faith compliance with the law while actually achieving a result forbidden by court order. We understand that technical questions of health regulation are not easily untangled. We understand that an agency's limited resources may make impossible the rapid development of regulation on several fronts at once. And we understand that the agency before us has far greater medical and public health knowledge than do the lawyers who comprise this tribunal. But we also understand, because we have seen it happen time and time again, that action Congress has ordered for the protection of the public health all too easily becomes hostage to bureaucratic recalcitrance, factional infighting, and special interest politics. At some point, we must lean forward from the bench to let an agency know, in no uncertain terms, that enough is enough."

The court first set forth the sequence of events in the EtO proceeding, noting that OSHA's initial proposal was published in 1982 and that OSHA now asserted that the STEL decision would not be handed down until March 1988. "With lives hanging in the balance, six years is a very long time," the court said. At the same time, the court concluded that OSHA had not acted contemptuously of its remand order. In particular, the court concluded that OSHA's decision to go forth with additional rulemaking on the STEL issue did not exhibit bad faith since the court's 1986 decision did not bar "record supplementation."[297]

At the same time, the court of appeals expressed great concern over the agency's failure to issue a proposal at the

---

[296]*Public Citizen Health Research Group v. Brock*, __ F.2d __, 13 OSHC 1362 (D.C. Cir. 1987).

[297]Cf. *United Steelworkers of America v. Pendergrass*, 819 F.2d 1263, 13 OSHC 1305, 1308–1310 (3d Cir. 1987). In that case, the Court of Appeals for the Third Circuit issued a judgment remanding to OSHA its hazard communication standard for reconsideration of its nonapplication to employer sectors other than manufactur-

same time that it had engaged a consultant to supplement the record on the STEL issue. The court said that it had "hoped" that OSHA could act with "greater alacrity"; the fact that the supplementation of the record requires a more extended time-table does not mean, said the court, that *"any* timetable" is reasonable. Recognizing that deciding whether a delay was unreasonable is a "trickly proposition," the court stated that "fortunately" it had been presented with an agency timetable which represented that the final rule would issue by March 1988. Although "disappointed" by the target date, the court could not find any specific part of the schedule which was "impermissibly" slow. However, the court of appeals, noting that OSHA timetables had "suffered over the years from a persistent excess of optimism," emphasized that OSHA was "treading at the very lip of the abyss of unreasonable delay," and found that any delay beyond the proposed schedule would be unreasonable.

The court ordered OSHA to adhere to its schedule and to submit a progress report every 90 days until the final rule was in place.

On January 21, 1988, OSHA published a notice proposing to add a 5 ppm excursion limit in the EtO standard, averaged over a 15-minute period. OSHA gave the public 30 days to comment and indicated that a public hearing, if convened, would take place on March 3, 1988.[298]

---

ing and ordering its application to those employer sectors. In November 1985, four months later, OSHA issued an Advanced Notice of Proposed Rulemaking, soliciting comment on the issues in the remand. Several unions and Public Citizen then moved to enforce the court's original judgment, and the court of appeals, in a decision on May 29, 1987, ordered OSHA to publish, within 60 days, a hazard communication standard applicable to all workers covered by the Act, or a statement of reasons, based on the existing record, why a hazard communication standard was not feasible. The court's conclusion that further rulemaking would not be allowed was based in part on the language of its original decision, on OSHA's failure to demonstrate that amplification of the record was necessary to comply with the court's remand, and on the additional fact that OSHA failed to explain its further considerable delay in publishing a proposal after the comment period on the Advanced Notice ended on February 25, 1986. The court of appeals also noted that under existing law, an agency may issue a rule with a "substantially different scope" than contained in the proposed rule without engaging in additional rulemaking. The court cited among other cases, *United Steelworkers of America v. Marshall supra,* 647 F.2d 1189, 1221, 8 OSHC 1810, 1828, where the Court of Appeals for the District of Columbia Circuit upheld OSHA's 50 microgram permissible exposure level for lead over objections based on lack of notice. On the question whether OSHA can amend a standard without rule-making on remand from a court, see *Interim Final Rule, Hazard Communication,* 50 FED. REG. 48,750, 48,754–756 (1985) ("good cause exists for OSHA's issuance of the final rule without notice and comment").

[298]53 FED. REG. 1724 (1988).

20

Judicial Review of Standards

OSHA has issued more than 50 health and safety standards after rulemaking.[1] Section 6(f) of the Act[2] allows aggrieved parties to challenge standards issued by OSHA. Most health standards and some safety standards have been challenged, and two challenges have reached the U.S. Supreme Court.[3] Decisions in these standards cases have mostly been favorable to OSHA, although a number of significant standards have been vacated by the courts in whole or in part. These court decisions have been important in influencing the direction of OSHA standards development, and several of these have had considerable impact beyond the OSHA program.[4]

---

[1]These totals are approximate based on the calculations of the Office of Technology Assessment (OTA) in its 1984 report, PREVENTING ILLNESS AND INJURY IN THE WORKPLACE 228–229 (Washington, D.C.: U.S. Congress, Office of Technology Assessment, OTA-H-256, April 1985) (hereafter OTA Report). Subsequent to the OTA Report, OSHA issued a number of final standards. Among the most important of these in the health area are: asbestos (revised) 51 FED. REG. 22,723 (1986); field sanitation, 52 FED. REG. 16,050 (1987); benzene (revised) 52 FED. REG. 34,460 (1987); formaldehyde, 52 FED. REG. 46,168 (1987). There are different ways to count the number of OSHA standards; thus, OTA considers OSHA's 14 identical standards on carcinogens, issued in 1974, as one standard. OTA Report, 228–229. However, these could be considered as 14 separate standards. Further, the distinction between "safety" and "health" standards has also been problematic. See, for example, the statement of Assistant Secretary Bingham quoted in B. Mintz, OSHA: HISTORY, LAW, AND POLICY 38 (1984).

[2]29 U.S.C. §655(f) (1982).

[3]*Industrial Union Dep't v. American Petroleum Inst.*, 448 U.S. 607, 8 OSHC 1586 (1980) (benzene); *American Textile Mfrs. Inst. v. Donovan*, 452 U.S. 490, 9 OSHC 1913 (1981) (cotton dust).

[4]For example, *Industrial Union Dep't, supra* note 3, and *Industrial Union Dep't v. Hodgson*, 499 F.2d 467, 1 OSHC 1631 (D.C. Cir. 1974) (asbestos).

This chapter surveys decisions in cases involving review of OSHA standards, and in suits to compel OSHA action in standards cases. (A chronologically ordered list of the major decisions is provided as an appendix to the chapter text.)

Section 6(f) provides:

> "Any person who may be adversely affected by a standard issued under this section may at any time prior to the sixtieth day after such standard is promulgated file a petition challenging the validity of such standard with the United States court of appeals for the circuit wherein such person resides or has his principal place of business, for judicial review of such standard. A copy of the petition shall be forthwith transmitted by the clerk of the court to the Secretary. The filing of such petition shall not, unless otherwise ordered by the court, operate as a stay of the standard. The determinations of the Secretary shall be conclusive if supported by substantial evidence in the record considered as a whole."[5]

I. Threshold Issues

A. Who May Challenge an OSHA Standard?

Section 6(f) permits court challenges to be brought by "any person who may be adversely affected by a standard."[6] In *Fire Equipment Manufacturers Association v. Marshall*,[7] the OSHA revision of its fire protection standard[8] was challenged on a variety of procedural and substantive grounds by an association of manufacturers of firefighting equipment and individual manufacturers. The Seventh Circuit dismissed the petition for review for lack of standing, finding that the manufacturers were not "adversely affected" by the standard under Section 6(f). Emphasizing that the purpose of the Act was the protection of workers, the court concluded that the manufacturers' interest in profits and in avoiding competitive disadvantage

---

[5]29 U.S.C. §655(f). As will appear, the judicial review provisions of the Administrative Procedure Act, 5 U.S.C. §701 *et seq.*, may also serve as a basis of court review of OSHA standards.

[6]29 U.S.C. §655(f).

[7]679 F.2d 679, 10 OSHC 1649 (7th Cir. 1982), *cert. denied*, 459 U.S. 1105 (1983). The original fire protection standard was issued by OSHA in 1971 under §6(a), 29 U.S.C. §655(a), as a "start-up" standard.

[8]45 FED. REG. 60,656 (1980), codified at 29 C.F.R. §§1910.155–.165 (1986).

was not within the protected "zone of interests" of the Act.[9] In *Calumet Industries v. Brock*,[10] a Notice of Interpretation of the hazard communication standard[11] issued by OSHA was challenged by manufacturers of various types of oils. Although the interpretation did not pertain to products they make, the manufacturers claimed it was invalid and would lead to loss of profits. Citing among other cases *Fire Equipment Manufacturers Association*, the District of Columbia Circuit ruled that petitioners did not meet nonconstitutional, prudential standing requirements for bringing the suit since they came before the court not as "protectors of worker safety" but as "entrepreneurs seeking to protect their competitive interests," and therefore did not come within the "zone of interest" protected by the statute and regulation.[12]

B. What Is a "Standard" Under Section 6(f)?

Section 6(f) authorizes challenges to "a standard issued under this section." Although the matter has not been litigated directly, the routine practice of the courts, supported by the statutory language, has been to entertain challenges to all three types of standards issued under Section 6(f): start-up (or "early") standards (§6(a)), standards issued after rule-making (§6(b)), and emergency temporary standards (§6(c)). However, no direct challenges to start-up standards were filed within 60 days of their issuance in 1971;[13] these challenges

---

[9]The "zone of interest" principle was stated by the Supreme Court in *Association of Data Processing Serv. Org. v. Camp*, 397 U.S. 150 (1970). The court said that a party, even though meeting constitutional standing requirements, must also be "arguably within the zone of interests to be protected or regulated by the statute *** in question." *Id.* at 153.

[10]807 F.2d 225, 13 OSHC 1001 (D.C. Cir. 1986).

[11]29 C.F.R. §1910.1200 (1986).

[12]See also *R.T. Vanderbilt Co. v. OSHRC*, 708 F.2d 570, 11 OSHC 1545 (11th Cir. 1983) (manufacturer of talc may not challenge OSHA's asbestos standard); *Simplex Time Recorder Co. v. Secretary*, 766 F.2d 575, 12 OSHC 1401 (D.C. Cir. 1985), discussed *infra* nn. 39–42 and accompanying text. In the recently decided *National Cottonseed Prods. Ass'n v. Brock*, 825 F.2d 482, 13 OSHC 1353 (D.C. Cir. 1987), the court upheld the standing of the 3M Corp. to challenge the respirator portion of OSHA's revised cotton dust standard, 825 F.2d at 488–492, 13 OSHC at 1357-1360. The court, distinguishing *Fire Equip. Mfrs. Assoc.*, concluded that 3M's interest in selling respirators, and the cotton-processing plant operators interest in buying, were "two sides of the same coin," 825 F.2d at 491–492, 13 OSHC at 1359–1360.

[13]OSHA's start-up standards were promulgated on May 29, 1971, 36 FED. REG. 10,466 (1971), with a delayed effective date of 90 days. Approximately 600 of these

have arisen in the context of administrative proceedings to enforce citations and penalties. In those cases, it was assumed that Section 6(a) standards may be challenged; the threshold issue was whether the challenge may be commenced more than 60 days after the standard was promulgated.[14]

The remaining question has concerned the difference between a "standard," covered by Section 6(f), and a "regulation," which would be reviewable in a district court under Section 706 of the Administrative Procedure Act.[15] In *Louisiana Chemical Association v. Bingham*,[16] the issue was the characterization of OSHA's "rule" on Access to Employee Exposure and Medical Records.[17] The Fifth Circuit refused to accept OSHA's determination that it had issued a "standard." The court said that the most important feature of a "standard" is its function of "correcting or ameliorating a particular hazard." The records access rule, on the other hand, was "aimed primarily at possible detection over a long period of time of significant risks not yet covered by standards."[18] Accordingly, the court held that the rule was "among the more general class of enforcement and detection regulations contemplated by Congress in Section 8."[19] The court also rejected OSHA's argument that the records access rule, even if not a "standard," should be reviewable in a court of appeals because it is closely related to substance-specific standards, which must be reviewed in the courts of appeals.[20] The court said that accepting

---

standards were deleted by OSHA after rulemaking in 1978, on the ground that they did not improve safety and health. 43 FED. REG. 49,726 (1978).

[14]This issue is discussed in Section I.C, *infra*.

[15]5 U.S.C. §706.

[16]657 F.2d 777, 10 OSHC 1017 (5th Cir. 1981). The district court had earlier found that the "rule" was a "standard." *Louisiana Chem. Ass'n v. Bingham*, 496 F. Supp. 1188, 8 OSHC 1950 (W.D. La. 1980).

[17]The records access rule appears at 29 C.F.R. §1910.20. The rule provided for access by OSHA and employees to existing employer monitoring and employee medical records. OSHA subsequently proposed changes to the rule, 47 FED. REG. 30,420 (1983), but no final action has been taken on this proposal.

[18]The "significant risk" requirement was stated by the Supreme Court in the benzene case. *Industrial Union Dep't v. American Petroleum Inst.*, 448 U.S. 607, 8 OSHC 1586 (1980). This case is discussed *infra* Section II.

[19]The court stated that the "closest analogies" to the records access rule are the regulations under 29 C.F.R. pt. 1904 requiring employers to maintain records of employee injuries and illnesses and providing employee access to these records; see 29 C.F.R. §1904.7. These regulations were issued under the authority of §8, 29 U.S.C. §657 (1986).

[20]OSHA substance-specific standards typically provide for access by employees and their representatives to employee records required to be maintained by employers. See, e.g., asbestos standard, "Recordkeeping," 51 FED. REG. 22,733, 22,738 (1986),

this argument would allow OSHA to "bootstrap" virtually every recordkeeping regulation into the accelerated court of appeals review procedure reserved by the Act for standards.[21]

In *United Steelworkers v. Auchter*,[22] the *Louisiana Chemical* ruling was distinguished by the Third Circuit in holding that the hazard communications rule was a standard and not a regulation. The hazard communications "rule" required chemical manufacturers and importers to assess the hazards of the chemicals they produce and import, affix appropriate warnings labels on containers, and provide workers with material data safety sheets containing information on safe use of the products.[23] Petitioners—a union, a public interest group, and states—claimed that the "rule" was a regulation and accordingly was reviewable in a district court and did not preempt state disclosure regulations.[24] The court rejected the argument, concluding that, unlike the records access regulation, hazard communications were "aimed at eliminating the specific hazard that employees handling hazardous substances will be more likely to suffer *** if they are ignorant of the contents of those substances."[25] The court did not accept the state argument that a standard must reduce risk through "improved protection or reduced exposure," which the hazard communications rule did not do. The court instead concluded that "risk of harm can be greatly be reduced by direct warning to employees."[26]

---

codified at 29 C.F.R. §1910.1001(m)(5)(ii)–(iii) (1986). Indeed, in issuing the records access rule, OSHA made conforming changes in parallel provisions in the various substance-specific standards. 45 FED. REG. 35,212, 35,280–284 (1980).

[21]The court observed that Congress wished to avoid the delay occasioned by review proceedings in various district courts by assigning the "review of measures responding to specific dangers already identified" in the courts of appeals. *Louisiana Chem. Ass'n v. Bingham, supra* note 16, 657 F.2d at 785 n.12, 10 OSHC at 1023. This policy consideration, in the view of the court, was not applicable to review of recordkeeping regulations.

[22]763 F.2d 728, 12 OSHC 1337 (3d Cir. 1985). Other aspects of this case are discussed *infra* Section V.C.

[23]The hazard communication "rule" was published at 48 FED. REG. 53,280 (1983), 29 C.F.R. §1910.1200.

[24]Under §18(a), 29 U.S.C. §667(a), a state is preempted from "asserting jurisdiction under state law over any occupational safety or health issue with respect to which [a federal OSHA] standard is in effect under section 6." Cases determining the preemptive effect of the hazard communication "rule" are cited *infra* note 289.

[25]*United Steelworkers v. Auchter, supra* note 22, 763 F.2d at 735.

[26]*Id.* The issue of whether OSHA's cancer policy, 29 C.F.R. pt. 1990 (1986), should be reviewed in a district court or a court of appeals was raised but never decided, since the policy has never been implemented. See *American Petroleum Inst. v. OSHA,* 8 OSHC 2025 (5th Cir. 1980). This case also involved the issue of venue. See discussion at note 57, *infra.*

C. Are Preenforcement Challenges to Standards Exclusive?

Section 6(f) provides that standards challenges may be brought "at any time prior to the sixtieth day after such standard is promulgated." Direct challenges to a standard filed immediately after it is issued have typically been referred to as "preenforcement" challenges.[27] The issue is whether the preenforcement review is exclusive, or whether standards may be challenged after the 60-day period in enforcement proceedings. Unlike the Federal Mine Safety and Health Act of 1977,[28] the OSH Act does not contain an exclusivity provision. The legislative history, as will be discussed, tends to negate the application of exclusivity in the OSHA context. There are competing policy considerations, and the courts of appeals are widely split on the issue.

In an early case, *Atlantic & Gulf Stevedores v. OSHRC*,[29] involving a citation for violation of the start-up hard-hat standard,[30] the employer challenged the validity of the underlying standard on the grounds that its enforcement would be economically infeasible in the maritime industry since it would lead to a costly strike. Relying on legislative history and policy considerations, the Third Circuit entertained the challenge, although rejecting it on the merits. The court noted that the House OSH bill explicitly provided for review in enforcement proceedings of the "validity of any standard," and concluded that this "understanding" was carried forward "if only implicitly" into the statute as enacted.[31] Moreover, the court said,

---

[27]The leading Supreme Court case on preenforcement review is *Abbott Laboratories v. Gardner*, 387 U.S. 136 (1967). There, the government argued that a regulation could be challenged only after it was enforced; the Court held that preenforcement review was generally authorized by the Administrative Procedure Act and not "prohibited" under the federal Food, Drug, and Cosmetic Act, the statute involved in that case. More recent statutes, such as the OSH Act in §6(f), expressly authorize preenforcement review. Indeed, under statutes such as the OSH Act, the opposite question is raised: whether review in enforcement proceedings is permitted.

[28]30 U.S.C. §811(d). That section provides for preenforcement review of agency standards, and states that the procedures in that subsection "shall be the exclusive means of challenging" those standards.

[29]534 F.2d 541, 4 OSHC 1061 (3d Cir. 1976).

[30]29 C.F.R. §1918.105(a), adopted by OSHA in 1971 from the standard issued under the Longshoremen's & Harbor Workers' Compensation Act, 33 U.S.C. §941.

[31]*Supra* note 29, at 549, 4 OSHC at 1066. The court did not mention the more persuasive explicit language of the Senate Labor Committee report. See *Deering Milliken Inc. v. OSHRC*, 630 F.2d 1094, 1099, 9 OSHC 1001 (5th Cir. 1980).

the economic or technological infeasibility of an OSHA standard may not "manifest themselves" until well after the 60-day period, and therefore enforcement proceedings would be "the only available forum for testing the validity of the standard."[32]

In *National Industrial Constructors v. OSHRC*,[33] decided in 1978, the Eighth Circuit distinguished between procedural and substantive challenges in determining whether standards may be reviewed in enforcement proceedings. The court said: "While the unreasonableness of a regulation may only become apparent after a period during which an employer has made a good-faith effort to comply, procedural irregularities need not await the test of time and can be raised immediately."[34] Since the challenge in that case was procedural, the court held that it could not be raised in an enforcement proceeding. The procedural-substantive distinction was rejected by the Fifth Circuit in *Deering Milliken Inc. v. OSHRC*.[35] The case involved a claim that OSHA's revision in 1971 of its start-up cotton dust standard, "in the interest of greater intelligibility and accuracy,"[36] was invalid because it was done without notice and comment. Although rejecting the challenge on the merits, the court disagreed with the view of the Eighth Circuit

---

[32]*Supra* note 29, at 550, 4 OSHC at 1067. *Atlantic & Gulf* involved a substantive feasibility challenge to the hard-hat standard. The decision was ambiguous as to whether it also applied to procedural challenges. The court first said that it would not enforce a standard issued in violation of the Act's "substantive or procedural requirements." *Id.* However, the court also asserted that an employer cannot "defend" itself against a citation "solely on the grounds" that the procedural requirements established by the Third Circuit for the issuance of standards—see *Synthetic Organic Chem. Mfrs. Ass'n v. Brennan* (*SOCMA I*), 503 F.2d 1155, 2 OSHC 1159 (3d Cir. 1974)—"have been ignored." 534 F.2d at 551. The Fifth Circuit later asserted that the Third Circuit in *Atlantic & Gulf* "appears to permit claims of substantive or procedural invalidity" to be raised in enforcement proceedings. *RSR Corp. v. Donovan*, 747 F.2d 294, 300 n.24, 12 OSHC 1073, 1077 (5th Cir. 1984). The Fifth Circuit, on the other hand, had earlier stated that *Atlantic & Gulf* distinguished between the two types of challenges. *Deering Milliken Inc. v. OSHRC*, 630 F.2d 1094, 1098 n.6, 9 OSHC 1001, 1004 (5th Cir. 1980).

[33]583 F.2d 1048, 6 OSHC 1914 (8th Cir. 1978).

[34]*Id.* at 1052, 6 OSHC at 1917. The employer there challenged OSHA's adoption in 1971, under §6(a), of the construction safety standards, which just prior to the adoption had been issued by the Department of Labor under the Construction Safety Act, 40 U.S.C. §333 (1970). The employer argued that OSHA made the adoption before the 30-day delayed effective date, mandated by the Administrative Procedure Act, 5 U.S.C. §553 (1986). The court refused to consider the challenge. The same argument was rejected, on different grounds, by the Fourth Circuit in *Daniel Int'l Corp. v. OSHRC*, 656 F.2d 925, 9 OSHC 2102 (4th Cir. 1981), discussed *infra* at note 39 and accompanying text.

[35]*Supra* note 32.

[36]The revision was published at 35 FED. REG. 15,101 (1971).

that procedural challenges should not be considered, saying that "industry would have found it quite burdensome to comb through every 6(a) regulation and object to inappropriate promulgations within sixty days";[37] it also referred to the "potential number and technical complexity of summarily promulgated regulations."[38]

In a 1981 case, the Fourth Circuit accepted *arguendo* the *Deering Milliken* view, yet refused to consider the challenge on the grounds that the petitioning employer failed to show any prejudice from the OSHA action that was being attacked.[39] In *Simplex Time Recorder Co. v. Secretary*,[40] the employer challenged an OSHA start-up standard because it differed somewhat from the underlying national consensus standards from which it was adopted. The District of Columbia Circuit first refused to allow Simplex to object to a standard "which does not affect it in this case" because the company believed that the procedure by which the standard was adopted was "irregular."[41] On the merits, the court further found that Simplex failed to show that OSHA had brought about "a substantial or meaningful modification of the thrust of the regulation or the meaning."[42]

A different analysis was pursued by the Fifth Circuit in 1984 in *RSR v. Donovan*.[43] RSR, a secondary lead refiner challenged, among other things, the respirator provisions in OSHA's lead standard, which was promulgated in 1978, on

---

[37]*Supra* note 32, at 1099, 9 OSHC at 1004. The employer in *Deering Milliken* was challenging not the initial adoption of the cotton dust standard but rather a revision of the original start-up standard published shortly thereafter. The court treated both OSHA actions, coming within a brief time period of one another, as the same.

[38]*Id.* In *Deering Milliken* the court rejected the merits challenge on the grounds that OSHA had later not "materially altered" the standard and therefore notice and comment was not required. *Id.* at 1103, 9 OSHC at 1008. The Ninth Circuit also rejected the procedural-substantive distinction in *Marshall v. Union Oil Co.*, 616 F.2d 1113, 8 OSHC 1169 (9th Cir. 1980).

[39]*Daniel Int'l Corp. v. OSHRC, supra* note 34. The court said that the purpose of the 30-day delayed effective date (see note 34, *supra*) was to give the public a chance to prepare for enforcement. Since the challenge was mounted 6 years after the 30-day period would have expired, the employer can "hardly claim" it was denied a reasonable time to prepare. *Id.* at 930–931, 9 OSHC at 2106.

[40]766 F.2d 575, 12 OSHC 1401 (D.C. Cir. 1985). The standard at issue in that case was OSHA's spray booth standard. 29 C.F.R. §1910.107(a).

[41]766 F.2d at 583, 12 OSHC at 1406.

[42]*Id.* at 584, 12 OSHC at 1406–1407. In making this finding, the court relied on *Deering Milliken, supra* note 32.

[43]747 F.2d 294, 12 OSHC 1073 (5th Cir. 1984).

the grounds that OSHA unreasonably refused to consider actual protection factors or wearer acceptance. The court distinguished *Deering Milliken* and related cases on the grounds that they involved challenges to a large corpus of technical and summarily promulgated start-up standards. In that context, the court said, industry may well have been "lulled" into a "false sense of security inducing pre-enforcement inaction." However, OSHA's lead standard was issued after extensive rulemaking, and was subject to massive preenforcement judicial challenge, in which RSR actively participated. Since RSR raised no issue in 1984 that did not exist at the earlier time, the court concluded that the agency should not be compelled to defend itself against "identical attacks in successive enforcement actions" and the courts should not be "vexed with a multiplicity" of petitions that raise issues that could have been raised earlier. Despite the "strong" presumption of the availability of judicial review,[44] the court rejected the challenge on the grounds that RSR could have raised the identical challenge in the preenforcement review stage but, without excuse, did not do so.[45]

In sum, although some courts have concluded that standards challenges are available in enforcement proceedings as an abstract legal proposition, as a practical matter courts have, for one reason or another, rejected the challenges.[46]

D. Which Court of Appeals Should Hear the Standards Challenge?

Section 6(f) provides that the standards challenge may be filed by an aggrieved person in a court of appeals "for the

---

[44]The court referred to the leading case of *Abbott Laboratories v. Gardner*, 387 U.S. 136 (1967). See discussion *supra* note 27.

[45]*Supra* note 43, at 302, 12 OSHC at 1078–1079. The court expressly noted that RSR had participated in the lead rulemaking and preenforcement review of the lead standard. *Id.* at 301, 12 OSHC at 1079. It left open the question whether a nonparticipant in the earlier proceedings may later raise an issue in challenging the standard, if the issue had not been raised by parties to the proceeding. The Fifth Circuit recently considered and rejected in an enforcement proceeding a procedural challenge to OSHA's lead standard's medical removal protection provision, *United Steelworkers of Am., Local 8394 v. Schuylkill Metals Corp.*, 828 F.2d 314, 13 OSHC 1393, 1395–1396 (5th Cir. 1987).

[46]The issue was addressed by the Administrative Conference of the United States in Recommendation 82-7, 1 C.F.R. §305.82-7 (1987), which is based in part on Verkuil, *Congressional Limitation on Judicial Review of Rules*, 57 TULANE L. REV. 733 (1983). Prof. Verkuil discusses some of the OSHA cases, saying that they offer a "good framework" for analyzing the problem generally. *Id.* at 760–761.

circuit wherein such person resides or has his principal place of business." Since there are typically numerous petitioners in standards review cases, in effect the standard may properly be challenged in *any* circuit. The marked difference in the attitudes of some courts of appeal to the review of OSHA standards has resulted in major efforts by aggrieved parties to locate review proceedings in a favorable court. This has resulted in the well-known "race to the courthouse."

The first OSHA case involving a "race" occurred in connection with promulgation of OSHA's emergency temporary standard for benzene in April 1977.[47] Since the relevant statute requires OSHA to file the administrative record in the court in which the *first* petition for review was filed,[48] the crucial question in this factually complex case was: When was the challenged standard promulgated? The three judges were unable to agree on that issue;[49] however, Judges Leventhal and Wilkie agreed that the case should be transferred for decision from the District of Columbia Circuit to the Fifth Circuit. Thus, Judge Wilkie believed that the first petition was filed in the Fifth Circuit and Judge Leventhal, although of the view that the Fifth Circuit was not the court of first filing, concluded that the case should be transferred there "for the convenience of the parties in the interest of justice."[50]

All judges in the case agreed with Judge Leventhal's statement that OSHA should promulgate a regulation making clear what constitutes the time of issuance of a standard. Such

---

[47]*Industrial Union Dep't v. Bingham*, 570 F.2d 965, 6 OSHC 1107 (D.C. Cir. 1977). The benzene ETS was published at 42 FED. REG. 22,516 (1977).

[48]28 U.S.C. §2112 provides in pertinent part:
"If proceedings have been instituted in two or more courts of appeals with respect to the same order, the agency, board, commission, or officer concerned shall file the record in that one of such courts in which a proceeding with respect to such order was first instituted. The other courts in which such proceedings are pending shall thereupon transfer them to the court of appeals in which the record has been filed. For the convenience of the parties in the interest of justice such court may thereafter transfer all the proceedings with respect to such order to any other court of appeals."

[49]Judges Leventhal and Fahy believed that the standard was promulgated for purposes of 28 U.S.C. §2112(a) when it was signed by Assistant Secretary Bingham. Accordingly, the Industrial Union Department petition filed shortly thereafter was the first filed in the District of Columbia Circuit. Judge Wilkie, however, was of the view that the standard was not promulgated until its substance was communicated to the public, which was *after* the union petition was filed, and therefore the American Petroleum Institute petition in the Fifth Circuit was the first filed. Because of delay in litigation, the ETS never took effect. See *infra* note 59.

[50]*Supra* note 47, at 972, 6 OSHC at 1108. See quotation in note 48, *supra*.

action, he said, "would permit more expeditious review of OSHA actions without the lengthy and complex wrangling over jurisdiction and venue."[51] OSHA did so, promulgating a regulation providing that a standard is deemed issued for purposes of judicial review when "officially filed in the Office of the Federal Register."[52] To be fair to all parties, OSHA also adopted the practice of announcing to the public the time of filing at the Federal Register office. This, predictably, led to "simultaneous" filings, the situation involved in *United Steelworkers v. Marshall*,[53] raising the question of which court of appeals should decide the challenge to OSHA's lead standard.

The Third Circuit first rejected the Lead Industries Association (LIA) argument that its petition in the Fifth Circuit was "some ten seconds earlier" than the Steelworkers' petition in the Third Circuit. "Unlike race tracks," the court said, "we are not equipped with photoelectric timers" and it declined to "speculate which nose would show as first in a photo finish."[54] The court also rejected LIA's suggestion that it adopt a "relative aggrievement" test and that the employers were "more" aggrieved. To do so, the court said, would lead to an excursion into the merits, which could delay resolution of the preliminary issue of venue. The court also rejected the arguments based on the union's "motivation" to avoid the Fifth Circuit, saying that, likewise, LIA was motivated in seeking a "favorable" forum to its position. Nor did the court find institutional considerations—relative expertise of the courts or size of dockets—to be determinative. The case was decided, however, on the grounds that a closely related EPA case, involving a lead standard issued under the Clean Air Act, was pending for review in the District of Columbia Circuit (in which *no* petition to review the OSHA lead standard was filed)

---

[51]*Id.* at 970–971, 6 OSHC at 1110–1111.

[52]29 C.F.R. §1911.18 (1986). A later proceeding raised a question as to the meaning of the phrase "officially filed." In an unpublished opinion involving the cotton dust standard, the Fourth Circuit held that a standard is "officially filed" when it is date-stamped by the staff of the Federal Register and made available to the public, and *not* when handed to the Federal Register staff and reviewed by them. *American Textile Mfrs. Inst. v. Bingham*, No. 78-1378 (4th Cir. 1978). This case is discussed in Mintz, *supra* note 1, at 212.

[53]592 F.2d 693 (3d Cir. 1979).

[54]*Id.* at 695. The LIA argument was based on factual assertions contained in affidavits describing the mechanics of a typical "race" to the courthouse. See Mintz, *supra* note 1, at 218–220.

and that there were "strong institutional interests" in having the two proceedings decided in the same court.[55]

Although there have been numerous later challenges to OSHA standards, some involving petitions in different circuits,[56] with limited exceptions[57] none have resulted in published decisions on the venue question.[58]

E. Stays Pending Review

Under Section 6(f), the filing of a petition for review does not automatically stay the standard, absent an order of the court. A number of OSHA standards have been stayed pending review, including the benzene standard,[59] and OSHA's second

---

[55]*United Steelworkers v. Marshall, supra* note 53, at 697–698. Under the Clean Air Act, 42 U.S.C. §7607(b)(1), petitions to review EPA ambient air standards *must* be brought in the District of Columbia Circuit. The court noted that its determination would not result in "inconvenience" to the parties, who are not personally involved in appellate review. 592 F.2d at 697.

[56]An unusual factual situation involving venue arose after OSHA issued the hearing conservation amendments to its noise standard. A challenge on the merits of the amendments was filed by the employer association in the Fourth Circuit. OSHA then twice suspended the effective date of the amendments to give the new administration an opportunity to reconsider them. Because the suspension was effected without notice and comment, the AFL-CIO challenged the suspension in the District of Columbia Circuit. OSHA moved that the AFL-CIO suit be transferred to the Fourth Circuit on the grounds that it was the same proceeding as the merits challenge in which the earlier petition was filed in the Fourth Circuit. The AFL-CIO claimed that the two proceedings were different—one involving the merits of the amendment, the other the proceeding for modification. The District of Columbia Circuit agreed with OSHA in an unpublished opinion and transferred the case. The decisions of the Fourth Circuit are discussed *infra* Section V.D.

[57]*American Petroleum Inst. v. OSHA*, 8 OSHC 2025 (5th Cir. 1980), involved review of OSHA's cancer policy. Review petitions were filed in the District of Columbia, Fifth, and Third Circuits. The case was further complicated by the question of whether the cancer policy should be reviewed in a court of appeals or a district court; see discussion *supra* note 26. (Petitions had also been filed in the district court in the jurisdiction of the Fifth Circuit.) The Fifth Circuit ruled that venue was established in that court "in order to save judicial time" since that court would have to pass on the question of the jurisdiction of the district court, and that issue could not be transferred to any other circuit. 8 OSHC at 2026. See 9 OSHR 757 (Jan. 24, 1980); 10 OSHR 284 (Aug. 14, 1980). The merits of that proceeding were never decided. See *supra* note 26.

[58]The Administrative Conference (see note 46, *supra*) also addressed itself to this issue. 1 C.F.R. §305-80-5 (1986). This recommendation was based on a report published as McGarity, *Multi-Party Forum Shopping for Appellate Review of Administrative Action*, 129 U. PA. L. REV. 302 (1980). The Administrative Conference recommendation that venue be determined in some circumstances by a random lottery was included in proposed regulatory reform legislation, which was never passed. See Mintz, *supra* note 1, at 220–223. On January 8, 1988, the President signed legislation, P.L. 100–236, modifying the procedures for selection of a court of appeals where there have been multiple appeals from a single order.

[59]See *Industrial Union Dep't v. Bingham*, 570 F.2d 965, 974 nn.4 & 6, 6 OSHC 1107, 1108 (D.C. Cir. 1977) (benzene ETS). Since the decision in the case was not handed down until December 7, 1977, at the end of the 6-month period during which

asbestos emergency standard.[60] These stays have not been accompanied by published opinions, except in the case of the emergency temporary standard on diving. Thus, in *Taylor Diving & Salvage Co. v. Department of Labor*,[61] the Fifth Circuit applied the four criteria for granting stays set out in *Virginia Petroleum Jobbers Association v. Federal Power Commission*[62] and concluded that the employer association's "prospects of prevailing on the merits are good," that the employers "show danger of irreparable harm from certain requirements" of the emergency standard, and that OSHA's suggestions of ways to mitigate the harm—variances or contests of citations—were "too uncertain to justify the denial of the stay." The court also rejected as "impractical" the prospects of a stay tailored to specific provisions of the standard entailing "the most significant hardships." The court accordingly granted an indefinite stay. From the language of the brief *per curiam* opinion, it was clear that the court was broadly dissatisfied with OSHA's emergency procedures, which it termed "inherently unsatisfactory."[63]

The Sixth Circuit, on the other hand, rejected a stay of the OSHA emergency standard on acrylonitrile in *Vistron Corp. v. OSHA*.[64] The court order noted in particular that the issuance of the emergency standard was based on two studies, one of workers and another of rodents, with the human study showing an excess of cancer cases among workers exposed to

---

an emergency standard may remain in effect, OSHA decided not to litigate the issue further. As a result, the benzene ETS never became effective. On the stay of the final benzene standard, see 1 OSHR 1955 (1978) and 2 OSHR 13 (1978).

[60]*Asbestos Information Ass'n v. OSHA*, 727 F.2d 415, 11 OSHC 1817 (5th Cir. 1984).

[61]537 F.2d 819, 4 OSHC 1511 (5th Cir. 1976). OSHA subsequently withdrew the diving ETS and issued a "permanent" diving standard after rulemaking.

[62]259 F.2d 921, 925 (D.C. Cir. 1958). These criteria are (1) a substantial likelihood of success on the merits; (2) danger of irreparable harm if relief is denied; (3) other parties will not be harmed substantially if relief is granted; and (4) interim relief will not harm the public interest. The court in *Asbestos Information Ass'n v. OSHA, supra* note 60, referred to these four criteria in explaining, as background to its decision to vacate the asbestos ETS, that a stay of OSHA's asbestos ETS had earlier been granted. *Supra* note 60, at 418 and n.4, 11 OSHC at 1818. The court there expedited the hearing on the merits in light of the stay. *Id.* See also *Public Citizen Health Research Group v. Auchter*, 702 F.2d 1150, 1158–1159, 11 OSHC 1209 (D.C. Cir. 1983), discussed *infra* Section VII, where the court, in another context, stayed the district court order and expedited review proceedings.

[63]*Taylor Diving & Salvage Co., supra* note 61, at 821 n.7, 4 OSHC at 1512. The view of the Fifth Circuit on emergency standards is discussed *infra* in text accompanying notes 309–322.

[64]6 OSHC 1483 (6th Cir. 1978).

acrylonitrile at a du Pont textile fibers plant. The court therefore concluded that the issuance of the stay was "not in the public interest or would substantially harm other parties in the litigation."[65] The court indicated that it was prepared to grant a motion for accelerated appeal, but this became unnecessary when the petition for review was withdrawn.[66] In *Public Citizen Health Research Group v. Rowland*,[67] involving the challenge to OSHA's final ethylene oxide standard by an employer association, the District of Columbia Circuit also rejected a motion for a stay pending appeal. Applying four criteria of *Virginia Petroleum*,[68] the court found that the association had not shown a likelihood of success on the merits and a stay "would not be in the public interest as it would continue to expose workers to Ethylene Oxide at levels that both OSHA and this court have determined constitute a 'significant risk.' "[69] The court finally relied on the fact that the association had concluded that "most" employers had already reached the new level, and therefore no irreparable harm was established.[70]

II. Standard of Review

Perhaps the most significant factor in court review of administrative agency standards is the attitude of the court regarding its proper role in overseeing the agency and the scope of its reviewing authority. This is often referred to as the "standard of review." Indeed, in cases involving court review, opinions typically contain a section entitled "Standard of Review" describing the general approach the court intends to employ in undertaking its statutory review responsibilities.[71] The standard of review to be applied implicates "intri-

---

[65]*Id.* at 1484.

[66]The final acrylonitrile standard was issued on Sept. 29, 1978, 43 FED. REG. 45,762 (1978).

[67]12 OSHC 1183 (D.C. Cir. 1986).

[68]See note 62, *supra.*

[69]12 OSHC at 1184. The court's finding on "significant risk" was made in *Public Citizen Health Research Group v. Auchter, supra* note 62, 702 F.2d at 1158.

[70]12 OSHC at 1184. This case was later decided on the merits. *Public Citizen Health Research Group v. Tyson*, 796 F.2d 1479, 12 OSHC 1905 (D.C. Cir. 1986). See discussion *infra* Section III.A.

[71]See, among many other cases, *American Iron & Steel Inst. v. OSHA*, 577 F.2d 825, 830, 6 OSHC 1451, 1454–1455 (3d Cir. 1978), *cert. dismissed*, 448 U.S. 917 (1980) (coke oven emissions).

cate questions pertaining to fact-finding, policy making and statutory construction,"[72] and typically varies depending on the type of issue before the court.[73] In the review of OSHA standards, there have also been significant differences in approach among the courts of appeals.[74]

The OSH Act provides in Section 6(f) that agency determinations "shall be conclusive if supported by substantial evidence in the record as a whole." Traditionally, courts have applied this more searching "substantial evidence" test to decisions made by agencies after adjudication or formal rulemaking.[75] The more deferential "arbitrary and capricious" test has been applied by courts in reviewing agency "policy" rules.[76] The anomaly of applying the substantial evidence test in the review of policy rules such as OSHA standards was confronted by the Second Circuit and District of Columbia Circuit in the first two standards review cases.

The earlier of the cases was *Associated Industries of New York State v. Department of Labor*,[77] involving OSHA's lavatory standard.[78] Pointing to the "anomaly" of subjecting notice-and-comment rulemaking, which is "essentially legislative in nature," to a substantial evidence test, OSHA argued for a "less severe" standard of review than that stated in Section 6(f). The court, relying particularly on the legislative history, rejected the argument and sustained, at least formally, the applicability of the substantial evidence test. Significantly, however, the court went on to say that the controversy was "semantic in some degree" in the context of informal rulemaking, lacking dispositional importance. Ultimately, the court of appeals said, in the review of rules of general applicability

---

[72]*Id.* at 831, 6 OSHC at 1455.

[73]Thus, in the OSHA context, courts have applied different standards of review in considering standards issued after rulemaking and emergency temporary standards. See discussion *infra* Section VI.

[74]The Fifth Circuit has tended to apply more stringent review standards in OSHA cases. See discussion *infra*, this section. This circumstance has brought about the "race to the courthouse," discussed *supra* Section I.D.

[75]Administrative Procedure Act, 5 U.S.C. §706(2)(E); see *Universal Camera Corp. v. National Labor Relations Bd.*, 340 U.S. 474 (1951), for a classic discussion of the meaning of the "substantial evidence" test.

[76]5 U.S.C. §706(2)(A). See *Citizens to Preserve Overton Park v. Volpe*, 401 U.S. 402 (1971).

[77]487 F.2d 342, 1 OSHC 1340 (2d Cir. 1973).

[78]36 FED. REG. 10,593 (1971), 29 C.F.R. §1910.141. The court vacated the standard in part on both procedural and substantive grounds. See *infra* Section IV.B.

made after rulemaking, the substantial evidence and arbitrary and capricious tests "do tend to converge."[79] Thus, the court determined that the realities of rulemaking result in the court applying a less than strict substantial evidence test to OSHA rulemaking.[80]

The same approach was adopted, more explicitly, by the District of Columbia Circuit in *Industrial Union Department v. Hodgson*.[81] In affirming the permissible exposure limit for asbestos of 2 fibers per cubic centimeter of air, the court emphasized that some of the questions involved in OSHA standards development are "on the frontiers of scientific knowledge." For that reason, the court said, the agency's "policy choices" are not "susceptible to the same type of verification or refutation by reference to the record" as, typically, adjudicatory decisions. The main function of the court in reviewing the "formulation of rules for general application in the future is to negate the dangers of arbitrariness and irrationality."[82] Ultimately the touchstone of review is whether the agency's decision is in accord with the statutory policy; in asbestos, the choice of a low permissible level was "doubtless sound" since "the protection of the health of employees is the overriding concern of OSHA."[83]

Subsequent decisions of the District of Columbia Circuit followed this deferential standard for court review. For example, in affirming for the most part OSHA's cotton dust standard, the court emphasized that Congress had created an "uneasy partnership" between the agency and the reviewing court in order "to check extravagant exercises of the agency's authority to regulate risk." The court's role is to "ensure that

---

[79]*Supra* note 77, 487 F.2d at 350, 1 OSHC at 1344.

[80]This and other OSHA standards review cases are discussed in Note: *Convergence of the Substantial Evidence and Arbitrary and Capricious Standards of Review During Informal Rulemaking*, 54 GEO. WASH. L. REV. 541, 555–559 (1986).

[81]499 F.2d 467, 1 OSHC 1631 (D.C. Cir. 1974) (asbestos).

[82]*Id.* at 478, 1 OSHC at 1637, quoting *Automotive Parts & Accessories Ass'n v. Boyd*, 407 F.2d 330, 338 (1968). In his opinion for the court, Judge McGowan, after referring to the court's opinion in *Associated Industries*, stated that judicial review of legislative-type standards will prove feasible if the court will show "wisdom," "restraint," and "flexibility," and will be "mindful" that "at least some legislative judgments cannot be anchored securely, and solely in demonstrable fact." *Industrial Union Dep't v. Hodgson, supra* note 81, at 476, 1 OSHC at 1638.

[83]*Industrial Union Dep't v. Hodgson, supra* note 81, at 477, 1 OSHC at 1637. The court refused, however, to affirm OSHA's delaying the effective date for the 2-fiber level. See discussion *infra* Section III.B.

the regulations resulted from a process of reasoned decision making consistent with the agency's mandate from Congress."[84] Similarly, in affirming for the most part OSHA's lead standard, the court stated that its task was to "ensure public accountability" by "requiring the agency to identify relevant factual evidence to explain the logic and the policies underlying any legislative choice, to state candidly any assumptions on which it relies, and to present its reasons for rejecting significant contrary evidence and argument."[85]

The Third Circuit has followed substantially the same approach in construing the substantial evidence test with flexibility. In reviewing OSHA's ethyleneimine standard, the court relied extensively on *Associated Industries* and *Industrial Union Department* and emphasized the "difficulty of attempting to measure a legislative policy decision against a factual yardstick."[86] It said that the agency had made a "legal rather than factual determination" in extrapolating evidence of carcinogenicity from animals to humans. This, the court said, is in the nature of "prudential legislative action."[87] Judicial review of OSHA standards should encompass at least five elements: most relevantly, if OSHA's determination is based "on factual matters subject to evidentiary development," these determinations must be supported by substantial evidence.[88] The Second Circuit adopted a similar view in affirming OSHA's vinyl chloride standard,[89] saying that the "the factual finger points

---

[84]*AFL-CIO v. Marshall*, 617 F.2d 636, 650 (D.C. Cir. 1979), *aff'd sub nom. American Textile Mfrs. Inst. v. Donovan*, 452 U.S. 490, 9 OSHC 1913 (1981). The opinion in *AFL-CIO* was by Judge Bazelon who emphasized the importance of agency adherence to procedural requirements in order to provide the framework for "reasoned decision-making." The Bazelon approach was largely rejected by the Supreme Court in *Vermont Yankee Nuclear Power Corp. v. Natural Resources Defense Council*, 435 U.S. 519 (1978). See McGarity, *Substantive and Procedural Discretion in Administrative Resolution of Science Policy Questions: Regulating Carcinogens in EPA and OSHA*, 67 GEO. L. J. 729, 796–808 (1979).

[85]*United Steelworkers v. Marshall*, 647 F.2d 1189, 1207, 8 OSHC 1810, 1817 (D.C. Cir. 1980), *cert. denied*, 453 U.S. 913 (1981).

[86]*Synthetic Organic Chem. Mfrs. Ass'n v. Brennan (SOCMA I)*, 503 F.2d 1155, 1158, 2 OSHC 1159, 1161 (3d Cir. 1974), *cert. denied*, 420 U.S. 973 (1975).

[87]*Id.*

[88]503 F.2d at 1160, 2 OSHC at 1163. The other four elements for review as stated by the court are adequate notice, adequate statement of reasons, consideration of statutorily relevant factors, and consideration of alternatives. *SOCMA I* has been followed in other decisions by the Third Circuit: *American Iron & Steel Inst. v. OSHA*, 577 F.2d, 825, 830–831, 6 OSHC 1451 (3d Cir. 1978) (coke oven); *AFL-CIO v. Brennan*, 530 F.2d 109, 114, 3 OSHC 1820, 1823 (3d Cir. 1975) (no-hands-in-dies); *United Steelworkers v. Auchter*, 763 F.2d 728, 736 (3d Cir. 1985) (hazard communication).

[89]*Society of Plastics Indus. v. OSHA*, 509 F.2d 1301, 2 OSHC 1496 (2d Cir.), *cert. denied*, 421 U.S. 992 (1975).

[but] does not conclude. Under the command of OSHA, it remains the duty of the Secretary to protect the workingman *** "90

A considerably stricter standard of review has been applied by the Fifth Circuit. In vacating OSHA's benzene standard, the court insisted that OSHA "regulate on the basis of knowledge rather than on the unknown" and that OSHA may not issue a health standard until it can provide substantial evidence that the benefits to be achieved by reducing the permissible level "bear a reasonable relationship to the costs imposed by the reduction."[91] The Supreme Court in *Industrial Union Department v. American Petroleum Institute*[92] agreed that the benzene standard should be vacated; however, in so doing, it affirmed the courts' traditional deference to the agency. Thus, while the Supreme Court established certain substantive legal requirements for OSHA standards, notably, that the hazard presents a "significant risk" in the workplace, it emphasized that this requirement was not a "mathematical straightjacket," and that the agency was not required to support the significant risk finding "with anything approaching scientific certainty." Moreover, the "reviewing court [must] give OSHA some leeway where its finding must be made on the frontiers of scientific knowledge."[93] In other words, the Supreme Court disagreed with OSHA's legal interpretation of the Act;[94] the Court suggested that it would defer to OSHA's judgments in prescribing a permissible level for benzene, if they were based on a proper legal interpretation of the Act.[95]

---

[90]*Id.* at 1308, 2 OSHC at 1501.

[91]*American Petroleum Inst. v. OSHA*, 581 F.2d 493, 504, 6 OSHC 1959, 1967 (5th Cir. 1978), *aff'd on other grounds sub nom. Industrial Union Dep't v. American Petroleum Inst.*, 448 U.S. 607, 8 OSHC 1586 (1980). The same stringent review criteria were applied by the Fifth Circuit in vacating OSHA's cotton ginning standard in a case decided after the Supreme Court benzene decision. *Texas Indep. Ginners Ass'n v. Marshall*, 630 F.2d 398, 8 OSHC 2205 (5th Cir. 1980).

[92]*Supra* note 91. The case is discussed *infra* Section III.A.

[93]448 U.S. at 653–656, 8 OSHC at 1603. The District of Columbia Circuit in *Public Citizen Health Research Group v. Tyson* (ethylene oxide) construed the Supreme Court decision in this manner as not impinging on the traditional deference due OSHA policy-type decisions, 796 F.2d 1479, 1497, 12 OSHC 1905, 1910 (D.C. Cir. 1986).

[94]The Supreme Court declined to defer to OSHA's interpretation of the Act because of the "inconsistencies in OSHA's position and the legislative history of the Act." 448 U.S. at 651 n.58, 8 OSHC at 1602.

[95]The Supreme Court decision on benzene was similarly interpreted by the Ninth Circuit in affirming OSHA's arsenic standard. *ASARCO Inc. v. OSHA*, 746 F.2d 483, 11 OSHC 2217 (9th Cir. 1984).

In affirming the cotton dust standard a year later, the Supreme Court applied a similarly deferential standard in testing OSHA's feasibility findings. Thus, the Supreme Court concurred in the view of the court of appeals that "[t]he very nature of economic analysis frequently imposes practical limits on the precision [of cost estimates] which reasonably can be required of the agency."[96]

The deferential approach has been rejected, however, by the courts in reviewing OSHA emergency standards which are promulgated without public participation.[97] This view was first stated by the Fifth Circuit in vacating OSHA's pesticide standard.[98] Emphasizing that "extraordinary power" was given to OSHA under Section 6(c), the court insisted that the power be "delicately exercised," and only in "emergency situations" that require it.[99] Similarly, the Third Circuit asserted that emergency standards are an "unusual response to exceptional circumstances" and should not be used to avoid the "ordinary process of rulemaking."[100] In 1984, the Fifth Circuit permanently stayed OSHA's second asbestos standard, reiterating these principles and saying that although OSHA had supported its grave danger finding with several risk assessments,[101] these risk assessments were "precisely" the type of data that should undergo public comment before they are accepted by an agency as a basis for regulation.[102]

In reviewing the adequacy of the procedures used by OSHA in rulemaking, the courts have generally been inclined to judge more strictly the agency's adherence to statutory and regulatory requirements, particularly pertaining to notice and statement of reasons.[103] This theme was articulated by the

---

[96]*American Textile Mfrs. Inst. v. Donovan*, 452 U.S. 490, 529 n.54, 9 OSHC 1913, 1928 (1981), quoting *AFL-CIO v. Marshall*, 617 F.2d 636, 661–662, 7 OSHC 1775, 1791 (D.C. Cir. 1979).

[97]Section 6(c), 29 U.S.C. §655(c). All but one of the OSHA emergency standards that were challenged in court were either vacated or stayed.

[98]*Florida Peach Growers Ass'n v. Department of Labor*, 489 F.2d 120, 1 OSHC 1472 (5th Cir. 1974).

[99]489 F.2d at 129, 1 OSHC at 1480.

[100]*Dry Color Mfrs. Ass'n v. Department of Labor*, 486 F.2d 98, 104 n.9a, 1 OSHC 1331, 1335 (3d Cir. 1973). The court vacated a portion of OSHA's standard for 14 carcinogens because of the inadequacy of the statement of reasons.

[101]Risk assessments are discussed *infra* Section III.A.

[102]*Asbestos Information Ass'n v. OSHA*, 727 F.2d 415, 426, 11 OSHC 1817, 1824 (5th Cir. 1984).

[103]Cases involving procedural objections to OSHA standards are discussed *infra* Section IV.

courts in early decisions. Thus, in *Industrial Union Department v. Hodgson*, the court, after acknowledging the limited nature of its review functions, said that "what we are entitled to in all events is a careful identification by the Secretary *** of the reasons why he chooses to follow one course rather than another."[104] Finally, the question of court review of OSHA legal determinations raises significantly different issues from those arising during review of policy or factual decisions. As a general matter, courts have not been reluctant to overturn agency legal interpretations.[105]

III. Substantive Issues in Court Review of OSHA Standards

The key issue in OSHA's regulation of toxic substances is the permissible exposure limit (PEL) that will be set.[106] Determination of a PEL involves a decision whether the substance is carcinogenic or otherwise toxic, assessment of the risk at various levels of exposure, and the feasibility of the PEL.[107] OSHA has typically separated these determinations into two steps: (1) the health need for the PEL, and (2) feasibility constraints.

A. Health Need for the Permissible Exposure Limit: Significant Risk

In regulating asbestos in 1972, OSHA noted in the preamble to its final standard that there was clear evidence that

---

[104]499 F.2d 467, 475, 1 OSHC 1631, 1637 (D.C. Cir. 1974). See also *SOCMA I*, 503 F.2d 1155, 2 OSHC 1159 (3d Cir. 1974), where at least two of the elements of review are procedural.

[105]See, e.g., *National Petroleum Refiners Ass'n v. Federal Trade Comm'n*, 482 F.2d 672, 692 (D.C. Cir. 1973). A more deferential approach was taken in *Chevron, USA v. National Resources Defense Council*, 467 U.S. 837, 842–845 (1984).

[106]The question of whether a short-term exposure limit (STEL) should be set also has arisen. See discussion *infra* Section III.A.

[107]These determinations are mandated by §6(b)(5), 29 U.S.C. §655(b)(5), which relevantly provides: "The Secretary, in promulgating standards dealing with toxic materials or harmful physical agents under this subsection, shall set the standard which most adequately assures, to the extent feasible, on the basis of the best available evidence, that no employee will suffer material impairment of health or func-

asbestos causes asbestosis, mesothelioma, and lung cancer, but there was generally no "accurate measure" 20 or 30 years ago of the levels giving rise to these diseases. OSHA concluded that because of the "grave consequences" of the substance, the "conflict in the medical evidence is resolved in favor of the health of employees" even though the evidence may not be "as good as scientifically desirable."[108] As discussed in Section II above, the court of appeals upheld the 2-fiber level. It said that OSHA was justified in adopting, "over strong employer objection, a relatively low limit," explaining that "inasmuch as the protection of the health of employees is the over-riding concern of OSHA, this choice is doubtless sound."[109] In *Synthetic Organic Chemical Mfrs. Ass'n v. Brennan*, the Third Circuit affirmed OSHA's ethyleneimine (EI) standard, concluding that it would treat as carcinogenic in humans a chemical such as EI which had been found experimentally carcinogenic only in animals.[110] The court agreed with OSHA that the "responsible and correct approach" would be to regulate the substance as a carcinogen even though evidence of carcinogenicity in humans was not available.[111]

In 1975 the Second Circuit affirmed OSHA's PEL for vinyl chloride of 1 part per million.[112] The employer association had argued on the basis of medical evidence that "no one can say" whether exposure to the chemical at low levels "was safe or

tional capacity even if such employee has regular exposure to the hazard dealt with by the standard for the period of his working life." This section, by its terms, relates to health standards; safety standards do not normally include a permissible exposure limit although feasibility issues are present. The Supreme Court's benzene decision, *Industrial Union Dep't v. American Petroleum Inst.*, 448 U.S. 607, 8 OSHC 1586 (1980), established the importance of the statutory definition of "standard," §2(3), 29 U.S.C. §652(3), in determining the appropriate PEL.

[108]37 FED. REG. 11,318 (1972).

[109]*Industrial Union Dep't v. Hodgson*, 499 F.2d 467, 475 (D.C. Cir. 1974).

[110]*SOCMA I, supra* note 104.

[111]The court also said that the extrapolation of carcinogenicity evidence from rodents to humans was "justified" by the Report of the Ad Hoc Committee on the Evaluation of Low Levels of Environmental Carcinogens to the Surgeon General, which stated that a substance shown "conclusively" to cause tumors in animals should be considered carcinogenic and therefore a "potential cancer hazard for man." *SOCMA I, supra* note 104, at 1159, 2 OSHC at 1161. The courts have followed the view of the court in *SOCMA I*. In 1986, the District of Columbia Circuit affirmed OSHA's 8-hour time-weighted average for ethylene oxide, rejecting, among other arguments, the arguments of the employer association that OSHA improperly relied on an animal study as a data base for its quantitative risk assessment in that proceeding. *Public Citizen Health Research Group v. Tyson*, 796 F.2d 1479, 1497, 12 OSHC 1905, 1918–1920 (D.C. Cir. 1986).

[112]*Society of Plastics Indus. v. OSHA*, 509 F.2d 1301, 2 OSHC 1496 (2d Cir.), *cert. denied*, 421 U.S. 992 (1975).

unsafe." The court rejected the contention, saying the Act commands the agency "to act even in circumstances where existing methodology or research is deficient."[113] It held that OSHA properly extrapolated the carcinogenicity data "from mouse to man" and reduced the permissible level to "the lowest detectable one."[114] Three years later, the Third Circuit affirmed OSHA's coke oven emissions standard, relying, in agreement with the agency, on epidemiological studies showing a significantly higher rate of mortality among coke oven workers than in the general population. The court further relied on expert testimony that "there is no scientific data to show there is a safe level of exposure to carcinogens."[115]

Relying on this line of authority, OSHA in 1978 reduced the PEL for benzene from 10 parts per million parts of air (ppm) to 1 ppm, with a 5-ppm ceiling during any 15-minute period.[116] The evidence showing the carcinogenicity of benzene was limited to employee exposures above 10 ppm, and no animal data existed. OSHA imposed the stringent 1-ppm level, utilizing what has come to be known as the "lowest feasible level" policy. Since a determination of a precise level of benzene that presents no hazard cannot be made, OSHA said, "prudent health policy" requires limiting exposure to the maximum extent feasible.[117]

The Fifth Circuit vacated the standard, and the Supreme Court affirmed.[118] The Supreme Court rejected outright OSHA's "lowest feasible level" policy on the grounds that it permitted OSHA to shift the burden to employers to demonstrate that

---

[113]*Id.* at 1308, 2 OSHC at 1502.

[114]*Id.*

[115]*American Iron & Steel Inst. v. OSHA*, 577 F.2d 825, 832, 6 OSHC 1451, 1455–1456 (3d Cir. 1978). The Supreme Court in the benzene decision expressly referred to the coke oven emission proceeding, indicating that OSHA's calculation, based on epidemiological studies, that among 21,000 coke oven workers, there was an annual excess mortality of over 200, constituted a "rational judgment" about the "relative significance of the risks associated with a particular carcinogen." *Industrial Union Dep't v. American Petroleum Inst.*, 448 U.S. 607, 657 n.64, 8 OSHC 1586, 1604 (1980). The Supreme Court at the same time approved OSHA's calculation in establishing the significance of risk to workers based on animal studies in the vinyl chloride proceedings and in the proceeding to regulate 14 carcinogens (including EI). *Id.*

[116]43 FED. REG. 5918 (1978).

[117]*Id.*

[118]*American Petroleum Inst. v. OSHA*, 581 F.2d 493 (5th Cir. 1978), *aff'd on other grounds sub nom. Industrial Union Dep't v. American Petroleum Inst., supra* note 115. The Supreme Court decision was by a 5-4 vote. There was no majority opinion, and the plurality opinion was written by Justice Stevens.

a substance was safe.[119] Rather, the Supreme Court held, OSHA has the burden of showing through substantial evidence that it is addressing a "significant risk" of harm in the workplace and that the standard would eliminate or reduce that significant risk.[120] Although the Supreme Court did not expressly state how OSHA could show that a significant risk exists, the plurality opinion has generally been understood as requiring a quantitative risk assessment, showing by statistical projections the degree of risk to workers at various levels of exposure.[121] Thus, the plurality noted that OSHA had "a fair amount" of epidemiological evidence on the carcinogenicity of benzene and, while it might not be possible to construct a "precise correlation" between exposure levels and cancer risks, it would "at least be helpful" in determining whether a significant risk existed at the 10-ppm level.[122]

After the Supreme Court's benzene decision, the crucial issue in court review of OSHA standards was whether the agency established a significant risk within the meaning of the benzene decision. Three cases were decided shortly after the benzene decision on this issue, involving OSHA standards issued prior to the Supreme Court decision and therefore not explicitly addressing the significant risk issue. In the first of these, the District of Columbia Circuit upheld for the most part OSHA's lead standard.[123] The court held that OSHA had clearly met the Supreme Court benzene requirements. Nowhere, the court said, did OSHA rely on "categorical assump-

---

[119]*Industrial Union Dep't, supra* note 115, at 653, 8 OSHC at 1605.

[120]In reaching this conclusion, the Court relied on the statutory definition of "standard," §3(8), 29 U.S.C. §652(8), particularly the phrase "reasonably necessary and appropriate" to provide safe and healthful conditions.

[121]See *Industrial Dep't, supra* note 115, at 656–657 & n.64, 8 OSHC at 1604. A quantitative risk assessment typically consists of a dose-response curve reflecting the experimental data, either animal or human, and showing the risk to employees at various levels of exposure. In the benzene proceeding, OSHA had refused to utilize a quantitative risk assessment, claiming that the assessment could yield "markedly different results" depending on the assumptions used. See *Reply Brief for the Federal Parties in Industrial Union Department v. American Petroleum Institute,* quoted in Mintz, *supra* note 1, at 267–268. The Supreme Court rejected OSHA's view, saying, as noted in the text, that risk assessment would "at least be helpful in deciding whether there was a significant risk." There is a considerable body of literature on risk assessments. See, e.g., National Research Council Commission on Life Sciences, RISK ASSESSMENT IN THE FEDERAL GOVERNMENT: MANAGING THE PROCESS (1983).

[122]*Industrial Union Dep't, supra* note 115, at 654–655 & n.64, 8 OSHC at 1604. Since the Supreme Court's decision was largely based on the general definition of "standard," it applied both to safety and health standards.

[123]*United Steelworkers v. Marshall,* 647 F.2d 1189, 8 OSHC 1810 (D.C. Cir. 1980), *cert. denied,* 453 U.S. 913 (1981).

tions" about the effects of lead; rather, the agency "amassed voluminous evidence of the specific harmful effects at particular blood-lead levels and correlated these blood-lead levels with air-lead levels." OSHA was thus able to show significant risk from lead by describing the "actual harmful effects of lead on a worker population at both the current PEL and the new PEL."[124]

Several months later, the Fifth Circuit vacated OSHA's cotton ginning standard. In view of the significantly different ginning conditions in the United States and other countries where some of the ginning studies were conducted, and in view of the lack of evidence of byssinosis in the ginning study conducted in the United States, the court found that these studies did not constitute substantial evidence supporting OSHA's significant risk finding.[125]

The Supreme Court itself applied the significant risk test in upholding the PEL for cotton dust in the textile industry in *American Textile Manufacturers Institute v. Donovan*. The Supreme Court said, "It is difficult to imagine what else the agency could do to comply with this court's decisions in *Industrial Union Department v. American Petroleum Institute*."[126] The court noted that a dose-response curve had been constructed showing the byssinosis risk to workers in the textile industry at various levels of exposure. These studies demonstrated that at the current levels of exposure, cotton dust caused significant health effects in 25 percent of exposed employees and that the prevalence of byssinosis could be "significantly reduced" by lowering the PEL from 500 micrograms per cubic meter to 200 micrograms.[127]

---

[124]647 F.2d at 1248, 8 OSHC at 1850. The court also upheld OSHA's conclusion that the record supported OSHA's finding that lead caused "subclinical effects" in workers and that OSHA was empowered to set a PEL to protect against these effects. *Id.* at 1263, 8 OSHC at 1853–1858. The court further held that OSHA properly required employers to reduce air lead levels in order to assure safe worker blood levels; employers had argued for direct biological monitoring of worker blood levels. *Id.* at 1308. The latter issue had been similarly decided early in the history of OSHA by the Eighth Circuit in an enforcement proceeding, *American Smelting & Ref. Co. v. OSHRC*, 501 F.2d 504, 2 OSHC 1041 (8th Cir. 1974).

[125]*Texas Indep. Ginners Ass'n v. Marshall*, 630 F.2d 398, 8 OSHC 2205 (5th Cir. 1980). OSHA has not issued another cotton ginning standard.

[126]*American Textile Mfrs. Inst. v. Donovan*, 452 U.S. 490, 505 n.25, 9 OSHC 1913, 1919 (1981).

[127]*Id.* The Supreme Court also noted, *id.*, that after its benzene decision (*Industrial Union Dep't, supra* note 115) and prior to the present (cotton dust) decision, OSHA had amended its carcinogens policy to delete its reliance on the "lowest feasible level policy," which the Supreme Court had rejected in its benzene decision. 46 FED. REG. 4889 (1981).

The courts next confronted the significant risk issue in reviewing OSHA's arsenic standard.[128] In this proceeding OSHA had reopened the record to obtain additional data and to make findings on the significant risk issue.[129] Based on several epidemiological studies, OSHA conducted risk assessments showing that at the current 500-microgram per cubic meter level there would be 400 excess deaths per 1,000 employees from arsenic exposure, which would be reduced to 8 excess deaths at the 10-microgram level. On this basis, OSHA concluded the new PEL would substantially reduce a significant risk.[130] Employers challenging the standard argued primarily that OSHA's risk assessment, contrary to the Supreme Court benzene decision, relied on a linear no-threshold dosage model and thus assumed impermissibly that there was no safe level of exposure to arsenic. The Ninth Circuit upheld the no-threshold model on the grounds that OSHA did not rely on "administrative fiat," as the agency did in the benzene proceedings, but rather on scientific reasoning and expert testimony showing that the linear no-threshold model "fits" the epidemiological data "far better" than a model showing a safe threshold.[131]

The District of Columbia Circuit followed closely the reasoning in the arsenic case in upholding the 1 part per million parts of air (1 ppm) time-weighted average in OSHA's final standard for ethylene oxide (EtO).[132] In first affirming OSHA's conclusion that EtO had carcinogenic, mutagenic, and cytogenetic effects and that it posed reproductive hazards, the court emphasized that the employer "fundamentally misconstrues" the court's role in rulemaking; the court's function is not to seek "proof positive," or to "reweigh the evidence and

[128]*ASARCO Inc. v. OSHA*, 746 F.2d 483, 11 OSHC 2217 (9th Cir. 1984).

[129]OSHA had issued the arsenic standard in 1978 prior to the Supreme Court benzene decision. 43 FED. REG. 19,584 (1978). In 1981, the Ninth Circuit remanded on OSHA's request the record to OSHA to perform a risk assessment consistent with the benzene decision. *ASARCO Inc. v. OSHA*, 647 F.2d 1 (1981). On the basis of the new evidence, OSHA reaffirmed the modified PEL and published an additional statement of reasons for lowering the PEL. 48 FED. REG. 1869 (1983).

[130]*ASARCO Inc., supra* note 128, at 488, 11 OSHC at 2221.

[131]*Id.* at 492–493, 11 OSHC at 2222–2223. The court also upheld the "cumulative dosage" aspect of OSHA's model on the grounds that it is not an "unsubstantiated assumption" but was supported by direct evidence and expert opinion. *Id.* at 493.

[132]*Public Citizen Health Research Group v. Tyson*, 796 F.2d 1479, 12 OSHC 1905 (D.C. Cir. 1986).

come to its own conclusions," but rather to "assess the reasonableness of OSHA's conclusions."[133]

In the rulemaking proceeding, OSHA had performed several risk assessments based on animal studies which established, according to OSHA, that the lowered PEL would substantially reduce a significant risk.[134] The main attack on OSHA's finding was that the no-threshold model used in the risk assessment violated the benzene decision because it assumed that EtO is harmful at low doses. Like the court in the arsenic case,[135] the court in the EtO case rejected the argument. In its benzene decision, the Supreme Court "chastised" OSHA for not basing its finding of no safe level on scientific evidence. In the instant case, however, OSHA "assuredly rectified that failure," having gone to "great lengths to calculate within the bounds of available scientific data" the significance of the risk.[136] The court then concluded that substantial evidence supported OSHA's no-threshold model and upheld OSHA's finding based on the risk assessments that the risk found met the significance test set in the benzene case.[137]

The second portion of the court of appeals decision in the EtO case also applied the significant risk requirements, but in a different context; it reversed the agency determination and remanded to OSHA the short-term exposure limit issue. OSHA had rejected the arguments of various parties and refused to establish a 10-ppm short-term exposure limit (STEL).[138] OSHA argued in the court proceedings brought by the Public Citizen Health Research Group that the STEL was unnecessary because the 1-ppm 8-hour PEL would both reduce a significant health risk and simultaneously adequately control short-term exposures. The OSHA argument assumes, the court said, that in "every case" employers, in meeting the 1-ppm 8-hour limit, would reduce short-term exposures below the proposed 10-ppm STEL. But this fact was not established by the record, the court concluded, since employers with a low background level of EtO could allow short-term exposures above

---

[133]*Id.* at 1485–1486, 12 OSHC at 1917–1918.

[134]*Id.* at 1496.

[135]*ASARCO, supra* note 128.

[136]The challenge to OSHA's use of animal data as a basis for its risk assessments is discussed *supra* note 111.

[137]*Public Citizen, supra* note 132, at 1501, 12 OSHC at 1924.

[138]*Id.* at 1506, 12 OSHC at 1924–1926.

10 ppm and still not exceed the 8-hour permitted level. Since OSHA conceded that a 10-ppm STEL was feasible and that there was still a significant risk at the 1-ppm 8-hour average level set in the standard,[139] the court concluded that a 10-ppm STEL, feasibly, could possibly help reduce a significant risk and therefore OSHA's refusal to adopt a STEL was not supported by substantial evidence.[140]

This holding of the District of Columbia Circuit has important implications. The Act as construed by the Supreme Court's benzene decision thus not only requires OSHA to address only significant risks but also requires OSHA, where feasible, to require the elimination or reduction of significant risks.[141]

When, after approximately a year, no action had been taken by OSHA on the remand, HRG returned to the court of appeals, asking that OSHA be found in contempt of the court's remand order. The court in a decision on July 21, 1987, found itself in a "delicate position." Recognizing, on the one hand, that agencies must have "a certain breathing space in their implementation of the law," the court nonetheless emphasized that actions to protect the public health must not "too easily become[] hostage to bureaucratic recalcitrance, factional infighting, and special interest politics." The court agreed that its mandate in the original decision did not preclude OSHA's conclusion that it would undertake further rulemaking before

---

[139]OSHA had found that 634–1,093 excess deaths per 10,000 workers at the then current 50-ppm exposure level is a significant risk and that the new 1-ppm PEL would significantly reduce the risk (12–23 estimated deaths). While OSHA found that at the lower level the risk was still significant, it set the level at 1 ppm because of feasibility factors. The employer association did not directly challenge these central findings, although it argued that OSHA inflated the excess deaths from the current standard.

[140]*Public Citizen, supra* note 132, at 1506, 12 OSHC at 1926–1927. The procedural issues raised respecting the Office of Management and Budget's involvement in OSHA's reopening the record on the STEL issue were not decided by the court of appeals. *Id.* at 1507, 12 OSHC at 1927.

[141]In *National Cottonseed Prods. Ass'n v. Brock, supra* n.12, the Court of Appeals for the District of Columbia Circuit upheld OSHA's imposition of "backup" medical surveillance requirements on the cottonseed industry, despite its failure to find a "significant risk in that industry." 825 F.2d at 484–487, 13 OSHC at 1354–1356.

In issuing health standards, OSHA typically follows this procedure: it (a) makes a determination of the toxic effects of the chemical, whether carcinogenic or otherwise, on the basis of the available scientific data; (b) prepares several risk assessments, estimating the excess risk to workers at various levels of exposure; and (c) determines if the risk is significant and whether a reduction of the PEL will eliminate or reduce the risk. See, e.g., the preamble to OSHA's recently revised asbestos standard, 51 FED. REG. 22,612, 22,615–622 (1986). OSHA then makes its feasibility findings. See discussion *infra* Section III.B.

deciding the STEL issue.[142] However, the court asserted that OSHA's failure to issue a proposal simultaneously with its hiring a contractor to gather information to supplement the record was "more difficult to defend." The court of appeals, while not finding OSHA in contempt, concluded that OSHA was treading "at the very lip of the abyss of unreasonable delay" and held that any delay beyond the agency's own March 1988 deadline for decision of the STEL issue would be "unreasonable."[143]

B. Feasibility of the Standard

Section 6(b)(5) of the Act, in mandating that OSHA assure that "no employee will suffer material impairment of health" even if exposed to a toxic substance for an entire working life, also constrains OSHA in insisting that the protection be achieved "to the extent feasible."

The feasibility issue first arose in *Industrial Union Department v. Hodgson*[144] involving review of OSHA's first asbestos standard. OSHA first determined that a PEL of 2 fibers per cubic centimeter of air was necessary to protect workers.[145] At the same time, OSHA provided a 4-year delay in the 2-fiber level, providing a 5-fiber standard in the interim, because of feasibility considerations. Referring to the need for "extensive redesign and relocation of equipment," OSHA said that a 4-year delay would give employers "reasonable time to comply with the lower limit."[146] The International Union Department in the AFL-CIO and a public interest group challenged the delayed effective date. The District of Columbia Circuit upheld OSHA's view of the meaning of feasibility in the Act,

---

[142]Compare *United Steelworkers v. Pendergrass*, 819 F.2d 1263, 13 OSHC 1305 (3d Cir. 1987), discussed *infra* Section VI, involving a remand of the hazard communication standard, where the court concluded that it could not be "seriously contended" that OSHA should do anything other than decide the remanded issue on the basis of the existing rulemaking record. 819 F.2d at 1267, 13 OSHC at 1308.

[143]*Public Citizen Health Research Group v. Brock*, __ F.2d __, 13 OSHC 1362 (D.C. Cir. 1987). The issue of court suits to compel OSHA to take action in standards proceedings is discussed *infra* Section VI.

[144]499 F.2d 467, 1 OSHC 1631 (D.C. Cir. 1974).

[145]See discussion of this aspect of the standard in text accompanying notes 81–83, *supra*.

[146]37 FED. REG. 11,318–319 (1972). The statement of reasons was not clear as to whether economic feasibility as well technological feasibility was involved in the decision. As will appear, the court of appeals dealt with both issues.

but remanded to the agency for further development of the record on the application of the feasibility provision in this case.

In this major decision, the court agreed with OSHA that feasibility considerations, both technological and economic, may be considered in setting a PEL. The court said that "practical considerations can temper protective requirements" and that a standard was not feasible if it required "protective devices unavailable under existing technology" or if it made "financial viability generally impossible."[147] At the same time, the court said standards would be economically feasible even if they are "financially burdensome and affect profit margins adversely."[148] The court added that the Act does not "necessarily guarantee the continued existence of individual employers" who "lagged" behind the rest of the industry in providing protection to workers. However, the court concluded that OSHA's decision to delay the effective date for all industries was not supported by the record. In light of the serious health hazards to workers occasioned by the delay, the court returned the proceeding to OSHA for further development of the record on the ability of different industries to comply with a lower PEL more quickly.[149]

The Third Circuit in 1975 agreed with the asbestos decision, saying that "[a]n economically impossible standard would in all likelihood prove unenforceable, inducing employers faced with going out of business to evade rather than comply with the regulation."[150] The Third Circuit case also involved a union challenge to an OSHA standards action; OSHA had significantly revised its no-hands-in-dies machine guarding standard,[151] based in part on the economic infeasibility of employers' complying with the stringent requirements of the 1971 national consensus standard. The court agreed with OSHA that cost of compliance was a proper consideration in modifying the standard.[152] The initial approach of the District of Colum-

---

[147]*Industrial Union Dep't, supra* note 143, at 447–478, 1 OSHC at 1639.

[148]*Id.* at 478, 1 OSHC at 1639.

[149]*Id.* at 480–481, 1 OSHC at 1641–1642.

[150]*AFL-CIO v. Brennan*, 530 F.2d 109, 3 OSHC 1820 (3d Cir. 1975).

[151]29 C.F.R. §1910.217(b)(1) (1986).

[152]*AFL-CIO, supra* note 150, at 123, 3 OSHC at 1829–1830. The court rejected the AFL-CIO argument that a court should be less deferential when it reviews OSHA's modifications of a previously adopted national consensus standard than in reviewing, as in the asbestos decision, a standard issued after rulemaking. *Id.* at 115, 3 OSHC

bia Circuit to feasibility issues was followed by that circuit upholding OSHA's lead standard.[153]

In affirming OSHA's cotton dust standard, the Supreme Court approved a somewhat modified articulation of the economic feasibility test earlier stated in the asbestos and lead cases.[154] OSHA asserted, and the Supreme Court agreed, that to meet the requirements of feasibility, "[a]t bottom the Secretary must [and did] determine, that the industry will maintain long term profitability and competitiveness."[155]

The Supreme Court's decision in the cotton dust case also disposed of a long-standing and major issue in OSHA standards development: the use of cost-benefit analysis.[156] Relying on the "language, structure and legislative history of the Act," the Court held that the Act does not require OSHA to perform a cost-benefit analysis in developing health standards under Section 6(b)(5).[157] The Court noted that the Act requires that health standards be "feasible," meaning, in the dictionary def-

---

at 1825. However, at the same time, the court of appeals remanded the standard to OSHA for further explication of its decision under §6(b)(8), 29 U.S.C. §655(b)(8). That section requires OSHA, when modifying a national consensus standard, to publish a statement of reasons why the rule as modified will "better effectuate the purposes of the act than the national consensus standard."

[153]*United Steelworkers v. Marshall*, 647 F.2d 1189, 1264, 8 OSHC 1810, 1864–1865 (D.C. Cir. 1980), *cert. denied*, 453 U.S. 913 (1981). The courts' feasibility findings are discussed *infra* in text accompanying notes 173–182.

[154]*Industrial Union Dep't, supra* note 143 (asbestos standard); *United Steelworkers, supra* note 153 (lead standard).

[155]*American Textile Mfrs. Inst. v. Donovan*, 452 U.S. 490, 530 n.55, 9 OSHC 1913, 1928 (1981). The employer association had argued that under OSHA's view, a standard would be economically feasible "merely" if it "will not put the affected industry out of business." *Id.* This characterization of its view was denied by OSHA. The same definition of economic and technological feasibility was followed by the Ninth Circuit in the arsenic case. *ASARCO Inc. v. OSHA*, 746 F.2d 483, 11 OSHC 2217 (9th Cir. 1984).

[156]There has been a great deal of dispute as to the nature of cost-benefit analysis, particularly in the context of health regulation such as is involved under the OSH Act. In the cotton dust case, the Supreme Court referred to the sharply differing characterizations by the parties of the cost-benefit exercise, but did not itself resolve the dispute except to find that cost-benefit analysis was not required under the OSH Act. *American Textile Mfrs. Inst. v. Donovan, supra* note 155, at 506–507 n.26, 9 OSHC at 1920.

[157]Although the Supreme Court did not explicitly hold that cost-benefit analysis was prohibited, its decision has been universally so construed. See statement of Assistant Secretary of OSHA Auchter in 1981 congressional oversight hearings, quoted in Mintz, *supra* note 1, at 322. In deciding this cost-benefit issue, the Supreme Court resolved a conflict between the District of Columbia Circuit, *AFL-CIO v. Marshall*, 617 F.2d 636, 7 OSHC 1775 (D.C. Cir. 1979) (upholding cotton dust standard and rejecting cost-benefit analysis), and the Fifth Circuit, *American Petroleum Inst. v. OSHA*, 581 F.2d 493, 6 OSHC 1959 (5th Cir. 1978), *aff'd on other grounds sub nom. Industrial Union Dep't v. American Petroleum Inst.*, 448 U.S. 607, 8 OSHC 1586 (1980) (vacating benzene standard and requiring cost-benefit analysis).

inition, "capable of being done," and that when Congress wanted an agency to engage in cost-benefit analysis, "it has clearly indicated such intent on the face of the statute."[158] As to the legislative history, the Court conceded that it was not "crystal clear"; however, "general support" for the rejection of cost-benefit analysis was provided by such statements as that of Senator Yarborough: "We are talking of people's lives, not the indifference of some cost accountants."[159] In sum, the Court concluded that OSHA should not balance costs and benefits because Congress had already done so in establishing the criteria for health standards in Section 6(b)(5).[160]

Applying these broad principles of feasibility to specific factual situations in OSHA standards has produced a range of court opinions, some upholding and others reversing OSHA determinations. As discussed, in *Industrial Union Department v. Hodgson*,[161] the District of Columbia Circuit refused to approve an industrywide delayed effective date for asbestos without further evidence by OSHA on the feasibility need for the delay in particular industries. In a major decision involving technological feasibility, the Second Circuit in *Society of the Plastics Industry v. OSHA*[162] rejected industry's argument that it "will never be able to reduce levels of exposure to 1 ppm through engineering means."[163] The court held that OSHA "may raise standards which require improvements in existing technologies or which require the development of new technology, and [the Secretary] is not limited to issuing standards

---

[158]*American Textile Mfrs. Inst. v. Donovan, supra* note 155, at 508–509, 9 OSHC at 1921.

[159]*Id.* at 521, 9 OSHC at 1925. The court conceded, however (*id.* n.32), that the "reasonably necessary and appropriate" language in §2(3), 29 U.S.C. §652(3), relied on by the Supreme Court in *Industrial Union Dep't, supra* note 153 (benzene), would require OSHA to adopt the least costly standard so long as it adequately addresses the significant risk of material impairment found by OSHA. This exercise has generally been called "cost-effectiveness" analysis and has been adopted by OSHA as one of the elements in its development of standards. REPORT OF THE PRESIDENT TO THE CONGRESS ON OCCUPATIONAL SAFETY AND HEALTH FOR CALENDAR YEAR 1985 at 7–8.

[160]*American Textile Mfrs. Inst., supra* note 155, at 511–512, 9 OSHC at 1922. The Supreme Court agreed with OSHA that §6(b)(5) establishes a two-step process for setting PELs: the health level, constrained only by feasibility. Cost-benefit analysis would undermine that statutory process, the Court said, if the PEL resulting from the cost-benefit balancing was higher than that mandated by the statutory criteria. The cotton dust decision has been the subject of considerable commentary. See, e.g., Fisher, *Controlling Government Regulation: Cost-Benefit Analysis Before and After the Cotton-Dust Case*, 36 AD. L. REV. 179 (1984).

[161]499 F.2d 467, 481, 1 OSHC 1631, 1641–1642 (D.C. Cir. 1974).

[162]509 F.2d 1301, 2 OSHC 1496 (2d Cir.), *cert. denied*, 421 U.S. 992 (1975).

[163]*Id.* at 1308, 2 OSHC at 1502.

based on devices already fully developed."[164] This principle, generally referred to as "technology forcing," has been reaffirmed by other courts of appeals.[165] The court in the vinyl chloride case noted in particular the "vast improvements" in protection made by industry "in a matter of weeks" after OSHA issued an emergency standard on vinyl chloride.[166] In addition, the court, in finding the standard feasible, relied on the fact that OSHA would permit respirators if engineering controls could not feasibly bring the level down to permissible limits.[167]

In *American Iron & Steel Institute v. OSHA*, the Third Circuit upheld OSHA's coke oven standard, finding the 0.15-milligram per cubic meter standard technologically and economically feasible.[168] Relying particularly on the vinyl chloride decision,[169] the court based its decision on the experience at the Fairfeld and Bethlehem steel batteries where "motivated" innovative programs were successful in reaching levels contained in the new PEL in every job classification monitored on a least some of the monitoring days.[170] While the court said

---

[164]*Id.* at 1309, 2 OSHC at 1502.

[165]See, e.g., *United Steelworkers v. Marshall*, 647 F.2d 1189, 1264, 8 OSHC 1810 (D.C. Cir. 1980), *cert. denied*, 453 U.S. 913 (1981) (lead standard); *ASARCO Inc. v. OSHA*, 746 F.2d 483, 495, 11 OSHC 2217, 2225 (9th Cir. 1984) (arsenic standard).

[166]*Society of Plastics Indus. v. OSHA, supra* note 162, at 1309, 2 OSHC at 1503. The emergency standard was not challenged in court.

[167]*Id.* at 1310, 2 OSHC at 1503. The District of Columbia Circuit in the lead case later criticized this analysis of the Second Circuit as "circular." If a standard's feasibility may be established through the use of respirators, the District of Columbia Circuit said, "[h]ow can such a standard ever be infeasible?" *United Steelworkers v. Marshall, supra* note 165, at 1269, 8 OSHC at 1867. The court attempted to resolve this circularity by distinguishing between the requirements for establishing feasibility in the direct challenge to a standard at the preenforcement stage, and the feasibility requirements when the standard is being implemented at a particular establishment. (OSHA health standards typically require the employer to meet the PEL by engineering controls, "except to the extent that such controls are not feasible." See, e.g., the asbestos standard, 29 C.F.R. §1910.1001(f)(1) (1986).) In a direct challenge, the court said, OSHA must show that "the typical firm will be able to develop and install work practice controls that can meet the PEL in most of its operations." The effect of OSHA prevailing on the issue of feasibility at the preenforcement stages would be to establish a presumption that the standard can be met without the use of respirators, so that in an enforcement proceeding a firm would have the burden of overcoming that presumption in showing infeasibility. *United Steelworkers*, 647 F.2d at 1269, 8 OSHC at 1867–1868. A case involving the feasibility of enforcing the noise standard, 29 C.F.R. §1910.95 (1986), at a particular plant is *Donovan v. Castle & Cooke Foods*, 692 F.2d 641 (9th Cir. 1982). In this chapter, we are concerned primarily with feasibility in preenforcement standards challenges.

[168]577 F.2d 825, 836–837, 6 OSHC 1451, 1460 (3d Cir. 1978), *cert. dismissed*, 448 U.S. 917 (1980).

[169]*Society of Plastics Indus. v. OSHA, supra* note 162.

[170]*Supra* note 168, at 834, 6 OSHC at 1458.

that it was "sensitive" to the financial implications of the standard, it concluded that it would not "precipitate anything approaching the 'massive dislocation' " that would characterize an economically infeasible standard.[171] The court also relied on the fact that the union, whose members would suffer as much as, if not more than, any party if the standard were infeasible, supported the standard.[172]

The District of Columbia Circuit in 1980 found substantial evidence supporting the feasibility of OSHA's lead standard in 10 industries, including the major primary and secondary lead smelting and battery industries.[173] In deciding the feasibility issue, the court cited the principles stated in numerous standards decisions,[174] and emphasized that it must make its determinations according to the "best *available* evidence"[175] and that the agency and the courts "cannot let workers suffer while it awaits the Godot of scientific certainty."[176] OSHA's duty in determining technological feasibility is to show that "modern technology has at least conceived some industrial strategies or devices which are likely to be capable of meeting the PEL and which industries are generally capable of adopting."[177] As for economic feasibility, the court said, OSHA must "construct a reasonable estimate of compliance costs and demonstrate a reasonable likelihood that these costs will not threaten the existence or competitive structure of an industry."[178] The court further insisted that OSHA meet

---

[171]The term was first stated by the Third Circuit in the no-hands-in-dies case, *AFL-CIO v. Brennan*, 530 F.2d 109, 123, 3 OSHC 1820, 1830 (3d Cir. 1975).

[172]*Supra* note 168, at 837, 6 OSHC at 1460. The court also upheld OSHA's authority to combine a performance standard, embodying a PEL, and mandated engineering controls, and found that substantial evidence supported the mandated controls. *Id.* at 837–838. The Supreme Court's decision in *Industrial Union Dep't v. American Petroleum Inst.*, 448 U.S. 607, 8 OSHC 1586 (1980) (benzene) did not reach any feasibility issues since the standard was vacated on the grounds that OSHA had not established a significant risk as required by the Act.

[173]*United Steelworkers v. Marshall*, *supra* note 165, at 1311, 8 OSHC at 1901–1902.

[174]*Industrial Union Dep't v. Hodgson*, 499 F.2d 467, 1 OSHC 1631 (D.C. Cir. 1974) (asbestos); *AFL-CIO v. Brennan*, *supra* note 171 (no-hands-in-dies); *American Iron & Steel Inst. v. OSHA*, *supra* note 168 (coke oven emissions); *Society of Plastics Indus. v. OSHA*, *supra* note 162 (vinyl chloride); *AFL-CIO v. Marshall*, 617 F.2d 636, 7 OSHC 1775 (D.C. Cir. 1979) (cotton dust); *Industrial Union Dep't v. American Petroleum Inst.*, *supra* note 172 (benzene).

[175]§6(b)(5), 29 U.S.C. §655(b)(5).

[176]*United Steelworkers v. Marshall*, *supra* note 165, at 1266, 8 OSHC at 1864 (emphasis in original).

[177]*Id.*

[178]*Id.* at 1272, 8 OSHC at 1870.

these feasibility requirements in each covered industry individually, and, for some major industries, with respect to several important operations within the industry.[179] Based on
these general principles and after a detailed analysis of the
"strongest evidence" OSHA presented for each industry and
the "most important" counterevidence and counter analysis,[180]
the court found the standard economically and technologically
feasible in 10 industries.[181] However, as to 39 other industries,
the court found that the agency had not met its feasibility
burden. Thus, as illustrative, the court remanded to OSHA
the feasibility finding for the pigment manufacture industry,
saying that even under the court's generous view of "technology forcing" and its "very flexible meaning of feasibility,"
OSHA had failed to offer "convincing evidence" and had failed
in "sustained logical analysis" to justify the standard.[182]

The Supreme Court in the cotton dust case affirmed the
decision of the District of Columbia Circuit that the standard
was economically and technologically feasible.[183] Applying a
familiar rule that it would not disturb a lower court's substantial evidence findings unless the court "misapprehended
or grossly misapplied" the standard of review, the Supreme
Court first accepted OSHA's cost estimates for the textile industry, even though it agreed with the court of appeals that
these were "not free from imprecision."[184] This imprecision

[179]*Id.* at 1277, 8 OSHC at 1874.

[180]*Id.* n.134.

[181]In finding feasibility in doubtful situations, the court relied in particular on
the delayed effective dates for engineering controls. The court said: "The extremely
remote deadline at which the primary smelters are to meet the final PEL is perhaps
the single most important factor supporting the feasibility of the standard. OSHA's
technology-forcing strategy, and its reliance on embryonic schemes for compliance,
are particularly reasonable in light of such a generous phase-in period." *Id.* at 1278,
8 OSHC at 1874.

[182]*Id.* at 1296, 8 OSHC at 1889. In its cotton dust decision, the District of Columbia Circuit also made an industry-by-industry analysis of feasibility, and rejected
OSHA's feasibility finding for the cottonseed oil industry. *AFL-CIO v. Marshall,*
supra note 174, at 669, 7 OSHC at 1796. The court said: "If the constraint of economic
feasibility is to have any effect on the agency's rulemaking, it demands more serious
consideration than it was given here." *Supra* note 174, at 672, 7 OSHC at 1799. On
remand, OSHA required medical surveillance but did not impose a PEL in the cottonseed industry. 50 FED. REG. 51,120 (1985). The District of Columbia Circuit upheld
the revised standard in *National Cottonseed Prods. Ass'n v. Brock, supra* nn.12 and
141, finding the medical surveillance requirements feasible, 825 F.2d at 487–488.

[183]*American Textile Mfrs. Inst. v. Donovan,* 452 U.S. 490, 536, 9 OSHC 1913,
1930–1931 (1981). The court relied primarily on a leading case involving substantial
evidence review of an NLRB decision, *Universal Camera Corp. v. NLRB,* 340 U.S.
474, 477 (1951).

[184]*Supra* note 183, 452 U.S. at 529, n.54, 9 OSHC at 1928.

resulted from the difficulty in obtaining accurate data, partly because of industry's lack of cooperation and the "inherent crudeness of estimation tools." Though based on "assumptions" of the consultant, the Court said that OSHA was obligated only to "subject such assumptions to careful scrutiny, and to decide how they might affect the correctness of the proferred estimates."[185] Based on this estimate, the Court affirmed OSHA's conclusion that the standard would not seriously threaten the existence of the textile industry; even if OSHA's estimate was "overstated," the court said that it was "fortified" in its conclusion by the fact that OSHA's consultant concluded that a cotton dust standard four times as expensive was feasible.[186]

OSHA's determinations of economic and technological feasibility, mandated by the Act and court opinions, were reiterated by President Reagan in Executive Order 12,291, issued in early 1981.[187] The Order required, for each "major" agency action, a preliminary and final regulatory analysis, describing the cost and benefits of the rule and alternative approaches that could achieve the same regulatory goal at lower cost.[188] As a result, in the preamble to its proposed and final standards, OSHA now summarizes the contents of the regulatory impact analysis (RIA) required under the Executive Order. Thus, a standards preamble will typically discuss technological feasibility, by industry, and economic feasibility, by industry, and present analyses under the Regulatory Flexibility Act[189] and the National Environmental Policy Act.[190]

---

[185]*Id.* There were two cost estimates in the record, one prepared by Research Triangle Institute, an OSHA-contracted group, and one by Hocutt-Thomas, an industry group. OSHA built on the conclusions of each in forming a composite estimate. *Id.* at 523, 9 OSHC at 1926.

[186]*Id.* at 536, 9 OSHC at 1931.

[187]46 FED. REG. 13,193 (1981), codified at 3 C.F.R. §127 (1986).

[188]The requirement of the order that the agency conduct a cost-benefit analysis "to the extent permitted by law" was determined to be inapplicable to OSHA in view of the decision *American Textile Mfrs. Inst. v. Donovan, supra* note 183. The Office of Management and Budget explained the order's requirements in "Interim Regulations Impact Analysis Guidance," reprinted in *Hearings on the Role of OMB in Regulation Before the Subcomm. on Oversight and Investigation of the House Comm. on Energy and Commerce,* 97th Cong., 1st. Sess. 360 (1981).

[189]5 U.S.C. §601 *et seq.*, dealing with the economic burden of the standard on small business.

[190]42 U.S.C. §4325 *et seq.*, requiring the preparation of an environmental impact assessment.

This framework was recently followed, for example, in OSHA's revised asbestos standard, issued in 1986.[191]

OSHA's arsenic standard, issued with a supplemental statement of reasons after the benzene decisions and the promulgation of Executive Order 12,291,[192] was upheld by the Ninth Circuit,[193] which rejected the infeasibility arguments advanced by industry. The court first held that the "limited" respirator use that would be necessary under the standard to meet the PEL does not make the standard infeasible.[194] The court also emphasized that the economic feasibility issue, most serious for ASARCO's Tacoma, Washington, plant, had became moot because the company was discontinuing copper smelting there, thus eliminating arsenic hazards.[195] Noting that OSHA was not "required to write an economic treatise on the smelting industry,"[196] the court concluded that OSHA's economic feasibility findings were supported by substantial evidence.

IV. Procedural Issues

The rulemaking procedures for OSHA standards are specified in Section 6 of the OSH Act,[197] the Administrative Procedure Act,[198] and OSHA regulations on rulemaking procedures.[199] The importance of OSHA adherence to these procedures has been repeatedly emphasized in court decisions. Thus, for example, the District of Columbia Circuit, in affirming OSHA's cotton dust standard, made clear that its role was to "ensure that the regulations resulted from a process of reasoned decisionmaking." This process must include, the court said, "notice to interested parties of issues presented in the

---

[191]51 FED. REG. 22,612, 22,650–675 (1986).

[192]The history of OSHA's arsenic standard is discussed *supra* note 129.

[193]*ASARCO Inc. v. OSHA*, 746 F.2d 483, 11 OSHC 2217 (9th Cir. 1984).

[194]*Id.* at 496–497, 11 OSHC at 2226–2227.

[195]*Id.* at 500, 11 OSHC at 2229.

[196]*Id.* at 501, 11 OSHC at 2230. The recent court decision in the ethylene oxide proceeding did not involve feasibility issues, since the employers conceded the feasibility of the 8-hour PEL and OSHA did not claim that the proposed 10-ppm STEL was infeasible. See discussion *supra* Section III.A.

[197]29 U.S.C. §655.

[198]5 U.S.C. §553.

[199]29 C.F.R. pt. 1911 (1986).

proposed rule" and "opportunities for these parties to offer contrary evidence and arguments." In addition, the court asserted that its responsibility included making certain that the agency has followed statutory procedures, including those procedures prescribed by its own regulations, and that the agency has "explicated the basis for its decision."[200]

A. Notice

Under Section 6(b)(2), OSHA, if it seeks to promulgate, modify, or revoke a standard, "shall publish a proposed rule *** in the Federal Register" and it "shall afford interested persons thirty days after publication to submit written data or comments."[201] The adequacy of OSHA's proposal was a major issue in a number of cases involving the validity of OSHA standards.[202]

In an early case, the Third Circuit set aside the laboratory provisions of OSHA's ethyleneimine (EI) standard because the notice of proposal did not advise the public that "the agency planned to make special provisions regarding EI use in laboratories."[203] In the coke oven case,[204] the Third Circuit set

---

[200]*AFL-CIO v. Marshall*, 617 F.2d 636, 649–650, 7 OSHC 1775, 1782–1783 (D.C. Cir. 1979).

[201]29 U.S.C. §655(b)(2). The Administrative Procedure Act, 5 U.S.C. §553, requires the agency to publish "either the terms or substance of the proposed rule or a description of the subjects and issues involved." OSHA typically publishes the text of the proposed rule and a preamble; it rarely issues only a "description of the subjects or issues involved."

[202]There is a separate issue as to whether OSHA is exempt from the procedural requirements of rulemaking under one of the exceptions stated in 5 U.S.C. §553(b). In *Chlorine Inst. v. OSHA*, 613 F.2d 120, 8 OSHC 1031 (5th Cir.), *cert. denied*, 449 U.S. 826 (1980), the court upheld OSHA's modification, without rulemaking, of its chlorine standard. OSHA concluded that the 8-hour time-weighted average in the start-up standard was a clerical error, and substituted a ceiling limit. The court, with obvious disbelief, accepted OSHA's claim that the time-weighted average was an error and that it took 7 years to "uncover" the error. In 1981, the action by the new administration in suspending the effective dates of the noise and lead standards was challenged by unions on the grounds that no opportunity for public comment had been afforded by the agency. In light of subsequent events, the issue became moot, and no court decision was handed down. These proceedings are discussed in Mintz, *supra* note 1, at 241–242. See also *Chamber of Commerce v. OSHC*, 636 F.2d 464, 8 OSHC 1648 (D.C. Cir. 1980), where the court vacated OSHA's walkaround pay regulation because the agency did not follow notice-and-comment rulemaking; OSHA's argument that its rule was an "interpretation" exempt from notice-and-comment requirements was rejected by the court.

[203]*Synthetic Organic Chem. Mfrs. Ass'n v. OSHA (SOCMA I)*, 503 F.2d 1155, 1160, 2 OSHC 1159, 1163 (3d Cir. 1974).

[204]*American Iron & Steel Inst. v. OSHA*, 577 F.2d 825, 6 OSHC 1451 (3d Cir. 1978), *cert. dismissed*, 448 U.S. 917 (1980).

aside the provisions of the coke oven emissions standard insofar as they applied to construction workers. Because of the unique features of the exposure of construction workers to coke oven emissions, the court determined that specific notice to the construction industry was required on such a "matter of vital importance."[205] On the other hand, in 1978, the District of Columbia Circuit rejected an argument that the pulmonary function tables in OSHA's cotton dust standard should be vacated.[206] Even though the specific tables ultimately adopted and the studies upon which they were based were not part of the record and therefore unavailable for public comment, the court found the public notice adequate because similar tables were contained in the proposal and the "general concept" of these tables was "thoroughly discussed" during the rulemaking. Thus, the court held, the final tables were "a logical outgrowth of the testimony taken during the proceeding."[207]

In OSHA's lead proceeding, the notice issue was sharply contested. OSHA's original notice, published in 1975, proposed reducing the lead standard PEL from 200 micrograms per cubic meter to 100 micrograms. At the completion of the rulemaking in 1978, OSHA reduced the PEL to 50 micrograms, and the employer association vigorously argued that the significant difference between the level proposed and the promulgated level deprived it and its members of an opportunity to comment effectively. Recognizing that the "logical outgrowth" principle is a "verbal formula" which only takes one "so far," the District of Columbia Circuit decided the issue on the basis of a careful comparison of the language of the proposed and final standard "in light of the evidence adduced at the hearing."[208] The court concluded that the language of the proposal "contains enough suggestions of the possibility" of a PEL lower than 100 micrograms to meet the notice requirements. It relied on the language of the "published explanation

---

[205]The unique features of construction industry exposure were also an issue in other contexts involving standards review. See, e.g., lead standard, *infra* note 371.

[206]*AFL-CIO v. Marshall, supra* note 200, at 675, 7 OSHC at 1801–1802.

[207]The "logical outgrowth" test was stated originally in *South Terminal Corp. v. EPA,* 504 F.2d 646, 659 (1st Cir. 1974). In *Taylor Diving & Salvage Co. v. Department of Labor,* 599 F.2d 622, 626, 7 OSHC 1507, 1509–1510 (5th Cir. 1979), the court held that the provision in the final diving standard on employee access to records was a "logical outgrowth" of the proposal.

[208]*United Steelworkers v. Marshall,* 647 F.2d 1189, 1221, 8 OSHC 1810, 1828 (D.C. Cir. 1980), *cert. denied,* 453 U.S. 913 (1981).

accompanying the proposed rule," the language of a supplemental notice[209] announcing an additional hearing which stated that the 100 microgram PEL was "inadequate" to protect exposed pregnant women, and the fact that evidence was presented at the hearing on the subclinical effects of lead, which could necessitate a "lower" PEL.[210] Although the court concluded that OSHA would have "served the parties far better" if it had proposed alternative levels, it nonetheless found that the final level was a "logical outgrowth" of the proposal and, accordingly, met legal requirements.[211]

In rejecting a notice argument and affirming OSHA amendments to its temporary flooring standard,[212] the Fourth Circuit relied particularly on the circumstance that the changes from the proposal—which the employer claimed were "substantial"—were made "in response to comments."[213] The court said that "the fact that numerous comments were submitted on the issue suggests that notice was adequate. To hold otherwise would penalize the agency for benefiting from comments received and further bureaucratize this process."[214] Similarly, in *Louisiana Chemical Association v. Bingham,*[215]

---

[209]The additional hearing was scheduled on the issue of medical removal protection. *Id.* at 1222 n.41, 8 OSHC at 1829. See discussion of medical removal protection in text accompanying notes 265–270, *infra.*

[210]*Id.* at 1222, 8 OSHC at 1829.

[211]The court of appeals also rejected arguments that there was inadequate notice respecting four other issues: the use of respirators as a means of compliance; the differences in effective dates for various industries; the manner of calculating the time-weighted average; and access to employee medical records. The court said that the issue is whether the agency did a "legally adequate job," and not whether it did the "best possible job." The court held that as to all issues, the notice was legally adequate. *Id.* at 1225. Early in its opinion, the court noted that OSHA was "occasionally careless or inefficient in its procedures" and that "procedural purists" will never place this standard in the "Pantheon of administrative proceedings." *Id.* at 1207.

[212]29 C.F.R. §1926.750(b)(2)(1) (1986). The proposal dealt only with changes in the maximum distance below a tier of beams on which certain work was being done requiring a substantial floor. The final standard changed not only the distance requirements but also the description of the work on the beams triggering the requirement, an "issue" not raised in the proposal.

[213]*Daniel Int'l Corp. v. OSHRC*, 656 F.2d 925, 932, 9 OSHC 2102, 2106–2107 (4th Cir. 1981). This case also involved the issue whether an OSHA standard may be challenged in an enforcement proceeding. See discussion in notes 34 and 39 and accompanying text, *supra.*

[214]*Id.* In the recently decided *United Steelworkers Local 8394 v. Schuylkill Metals Corp.*, 828 F.2d 314, 13 OSHC 1393 (5th Cir. 1987), the court of appeals, in an enforcement proceeding, rejected a notice argument involving the meaning of medical removal protection in OSHA's lead standard.

[215]550 F. Supp. 1136, 10 OSHC 2113 (W.D. La. 1982), *aff'd without opinion*, 731 F.2d 280 (5th Cir. 1984).

the employer association claimed that the definition of "toxic substance and harmful physical agent" in the final records access rule was such a "radical expansion" from the proposal that it was effectively denied the right to comment. The district court rejected the claim, saying that there was "adequate notice that the topic would be addressed" and there was "much discussion of the matter during rulemaking and comment proceeding," thus satisfying the "logical outgrowth" test.[216]

B. Statement of Reasons

Under the Act, OSHA when promulgating a standard must "include a statement of reasons for such action."[217] Two early OSHA standards, one a safety standard, the other a health standard, were set aside, at least in part,[218] because of the inadequacy of the statement of reasons. *Associated Industries of New York State v. Department of Labor*,[219] decided in 1973, involved OSHA's lavatory standard. OSHA reduced the lavatory requirements for office employees, after rulemaking, but, rejecting employer arguments, retained the original stringent requirements for industrial establishments. The court vacated the industrial provision because of the lack of substantial evidence supporting the action and because of the inadequacy of the statement of reasons. "In a case where the proposed standard under OSHA has been opposed on grounds as substantial as those presented here," the court said, the Department "has the burden of offering some reasoned explanation."[220]

---

[216]550 F. Supp. at 1148, 10 OSHC at 2122–2123.

[217]§6(e), 29 U.S.C. §655(e). Under §6(b)(8), 29 U.S.C. §655(b)(8), there is a separate requirement for a statement of reasons where a standard modifying a national consensus standard is adopted. See discussion *supra* note 152. See also 29 C.F.R. §1911.18(b) (1986). These requirements reflect the provision in the Administrative Procedure Act requiring the agency to incorporate in adopted rules "a concise general statement of their basis and purpose." 5 U.S.C. §553(c).

[218]It is not always clear whether the court reversal is based on the inadequacy of the statement of reasons or lack of substantial evidence. As demonstrated by the two cases discussed, the two factors are closely connected. For example, in *AFL-CIO v. Marshall*, 617 F.2d 636, 7 OSHC 1775 (D.C. Cir. 1979), the court of appeals found no substantial evidence supporting the economic feasibility findings for the cottonseed oil industry, saying also that "the agency's position is too unclear to permit us to complete our reviewing function." *Id.* at 669.

[219]487 F.2d 342, 1 OSHC 1340 (2d Cir. 1973).

[220]*Id.* at 354, 1 OSHC at 1348.

Shortly thereafter, the Third Circuit in *Dry Color Manufacturers Association v. Department of Labor*[221] set aside OSHA's emergency standard applicable to two carcinogenic substances (3,3'-dichlorobenzidine (DCB) and ethyleneimine (EI)), because of the insufficiency of the statement of reasons. Referring to OSHA's statement, which said only that the substances were carcinogenic and an ETS was necessary, as "conclusory," the court asserted that the statement placed "too great a burden" on the public interested in challenging the standard, and invited the use in subsequent judicial review of *post hoc* "rationalizations" not reflecting the views of the agency.[222] In particular, the court said, the standard did not set forth the "basis" for the agency determination that the substances were carcinogenic[223] and why the standard was "necessary" to protect employees.[224] In its cotton dust decision in 1981, the Supreme Court rejected as legally insufficient OSHA's preamble justification for the income protection provisions of the standard.[225] Although the Supreme Court suggested that the OSHA brief presented legally adequate reasons for the provision, it refused to accept these *post hoc* arguments of attorneys as a substitute for the agency's own statement of the reasons for its action.[226]

C. "Informal" Rulemaking

In 1971, in promulgating procedures for standards rulemaking, OSHA determined on the basis of the statutory structure and the legislative history that its rulemaking was "informal" but that "more than the bare essentials of informal rulemaking should be provided."[227] Specifically, the regula-

---

[221]486 F.2d 98, 1 OSHC 1331 (3d Cir. 1973).

[222]*Id.* at 106, 1 OSHC at 1336.

[223]This could be achieved, the court said, by a "brief statement" that certain scientific data showing carcinogenicity in rodents support the conclusion that the substances were carcinogens in humans as well. *Id.* at 106–107, 1 OSHC at 1337.

[224]*Id.* at 107, 1 OSHC at 1337.

[225]*American Textile Mfrs. Inst. v. Donovan*, 452 U.S. 490, 9 OSHC 1913 (1981). OSHA's authority to provide income protection, or medical removal protection, is discussed in text accompanying notes 265–270, *infra*.

[226]*Id.* at 539, 9 OSHC at 1931–1932.

[227]29 C.F.R. §1911.15(a) (1986). The differences between "informal" rulemaking and "formal" rulemaking and adjudication are discussed in numerous texts. See, e.g., W. Fox, UNDERSTANDING ADMINISTRATIVE LAW 123–137 (1987).

tions required that a hearing examiner (now an administrative law judge) preside at the hearing; that there be "an opportunity for cross-examination on crucial issues"; and that a verbatim transcript be kept of the proceedings.[228] In *Industrial Union Department v. Hodgson*,[229] the District of Columbia Circuit agreed with OSHA's conclusion as to the essentially informal nature of its rulemaking, and at the same time commended OSHA for going beyond the bare procedural requirements of the OSH Act in providing for cross-examination of witnesses.[230] Subsequent court decisions reaffirmed the informal nature of OSHA rulemaking proceedings. In particular, in its review of OSHA's lead standard in 1980, the District of Columbia Circuit predicated its rejection of a number of procedural objections to the rulemaking on the fact that OSHA rulemaking, although involving more elaborate procedures than "pure" informal rulemaking,[231] was nonetheless "legislative in its essential nature, and therefore not subject to the stringent procedures of formal rulemaking or adjudication."[232]

The first issue discussed by the court in the lead case was the alleged bias of Dr. Eula Bingham, the Assistant Secretary who made the ultimate decision on the lead standard. According to the employers, she prejudged the issue of medical removal protection and, under the *Cinderella* rule,[233] the provision should be vacated. Under the *Cinderella* rule, the decision maker would be disqualified if "a disinterested observer may conclude that [he or she] has in some measure adjudged the facts as well as the law of a particular case in advance of hearing it." The court distinguished *Cinderella* because it was applied in an adjudicative context; it is inapplicable to rulemaking, such as in the lead proceeding, where "the factual component of the policy decision is not easily assessed in terms of empirically verifiable conditions." The court concluded: "The

---

[228]29 C.F.R. §1911(b).

[229]499 F.2d 467, 1 OSHC 1631 (D.C. Cir. 1974).

[230]*Id.* at 476, 1 OSHC at 1638.

[231]This type of rulemaking is often referred to as "hybrid" rulemaking. See Kestenbaum, *Rulemaking Beyond APA: Criteria for Trial-Type Procedure and the FTC Improvement Act*, 44 GEO. WASH. L. REV. 679, 682–685 (1976). See Chapter 19, IV.A Hybrid Rulemaking.

[232]*United Steelworkers v. Marshall*, 647 F.2d 1189, 8 OSHC 1810 (D.C. Cir. 1980), cert. denied, 453 U.S. 913 (1981).

[233]*Cinderella Career & Finishing Schools v. FTC*, 425 F.2d 583, 591 (D.C. Cir. 1970).

Cinderella view of a neutral and detached adjudicator is simply an inapposite role model for an administrator who must translate broad statutory commands into concrete social policies."[234]

The next procedural attack in the lead case was directed to the fact that OSHA attorneys, who participated actively in the rulemaking and allegedly were advocates for a strict standard, consulted with the Assistant Secretary in her decision-making activity. This, it was argued, constituted illegal *ex parte* contacts between an adverse side in the rulemaking and the decision maker.[235] The court rejected the contention, making the following basic points: (a) in rulemaking, it makes "little sense" to speak of agency staff advocating for one "side" over the other; (b) while the Administrative Procedure Act bars *ex parte* contacts in adjudications,[236] this prohibition does not apply to OSHA informal rulemaking; (c) even though some cases, such as *Home Box Office v. FCC*,[237] raised questions as to the legality of *ex parte* contacts in informal rulemaking, these involved "massive evidence" of communication from outside interested parties. "Influence from within an agency poses no such threat," the court said. In sum, the court held that "rulemaking is essentially an institutional, not an individual process," and it is "unrealistic to expect an official facing a massive, almost inchoate, record to isolate herself from the people with whom she worked in generating the record."[238]

For many of the same reasons, the court rejected the employer argument that OSHA improperly relied on outside consultants in promulgating the lead standard. The court first

---

[234]In deciding the lead case, *supra* note 232, the court relied particularly on its decision in *Association of Nat'l Advertisers v. FTC*, 627 F.2d 1151, 1168 (D.C. Cir. 1979). The Administrative Conference of the United States adopted a recommendation on this issue entitled "Decisional Proceedings," 1 C.F.R. §305.80-4 (1983), based on Strauss, *Disqualification of Decisional Officers in Rulemaking*, 80 COLUM. L. REV. 990 (1980).

[235]*United Steelworkers v. Marshall, supra* note 232, at 1210, 8 OSHC at 1820.

[236]5 U.S.C. §557(d)(1).

[237]567 F.2d 9, 51–59 (D.C. Cir.), *cert. denied*, 434 U.S. 829 (1977).

[238]*Supra* note 232, at 1216, 8 OSHC at 1824. There have been no court decisions on OSHA standards dealing with *ex parte* contacts with interested parties. In the ethylene oxide case, *Public Citizen Health Research Group v. Tyson*, 796 F.2d 1479, 12 OSHC 1905 (D.C. Cir. 1986), the court reversed on the STEL issue on the merits and did not reach the question whether Office of Management and Budget communications with OSHA were unlawful as *ex parte* contacts. 796 F.2d at 1507, 12 OSHC at 1927. See Asimow, *When the Curtain Falls: Separation of Functions in Federal Administrative Agencies*, 81 COLUM. L. REV. 759 (1981).

found such reliance not only unobjectionable but commendable, showing that "the agency has taken its responsibilities seriously."[239] Nor were the communications between the agency and the consultants prohibited as *ex parte*; these consultants were "functionally" agency staff, the court said, and therefore these communications "were simply part of the deliberative process of drawing conclusions from the public record."[240] Finally, the court rejected the argument of the American Iron and Steel Institute that it was denied the right to cross-examine the authors of the consultants' report on the feasibility issue. The court found no statutory basis for the right to cross-examination and, while the letter of OSHA's regulation was "uncertain," it would "violate the spirit of the rule" to hold that a party has a right to examine not only persons appearing at the hearing but also those who only placed evidence in the record.[241]

D. Advisory Committees

Under Section 6 of the Act,[242] OSHA has discretion to establish standards advisory committees which are expected to make recommendations on the standards development. Section 7[243] defines the composition of standards advisory committees as comprising equal numbers of employer and employee representatives, state representatives, and other experts in safety and health. OSHA's regulations provide that in the formulation of construction standards OSHA "shall" consult with a standing advisory committee, the Construction Safety and Health Advisory Committee.[244] In challenging OSHA's standard to protect workers on sloping roofs of buildings, the National Roofing Contractors Association argued, among other things, that since no roofing industry member was appointed

---

[239]*Supra* note 232, at 1217, 8 OSHC at 1825.

[240]*Id.* at 1218, 8 OSHC at 1827. The court relied heavily on a just-issued decision of the Second Circuit in *Lead Indus. Ass'n v. OSHA*, 610 F.2d 70 (2d Cir. 1979). There, the court concluded that consultants' reports in the lead proceeding were not available under the Freedom of Information Act on the grounds that they were part of the agency's deliberative process and therefore exempt from disclosure.

[241]*Supra* note 232, at 1227–1228, 8 OSHC at 1833–1834.

[242]29 U.S.C. §655.

[243]§656.

[244]29 C.F.R. §1911.10(a) (1986). This requirement was based on the legislative history of the OSH Act. *Id.*

to the advisory committee, it was inadequately represented and the standard was procedurally defective. The Seventh Circuit rejected the argument, saying that there was no evidence in the record demonstrating that general contractors, who were members of the advisory committee, prejudiced the position of roofing employers or were not competent to judge suitable safety standards for the industry.[245] Absent a showing of "specific" prejudice, the court concluded, there was no requirement that OSHA appoint a roofing representative to the committee.[246]

Under the statutory scheme in Section 6(b), the OSHA advisory committee makes its recommendation, and this is followed by the publication of a proposal. In the case of the carcinogen MOCA, OSHA first issued an emergency standard, which, as required by the Act,[247] was immediately followed by a proposed "permanent" standard, and only then did OSHA convene an advisory committee. Because OSHA had breached the statutory sequence, the Third Circuit invalidated the standard, even though special circumstances were involved because of the earlier ETS.[248] In *National Constructors Association v. Marshall*[249] the court remanded to OSHA a construction standard requiring ground fault circuit interrupters because OSHA did not consult with its construction advisory committee a second time after the standard was significantly modified from the version that was originally submitted to the committee.[250] After further OSHA consultation with the advisory committee, the standard was upheld by the court in an unpublished opinion.

E. Time Frames for Rulemaking

Section 6 contains a number of specific deadlines for agency action on rulemaking. Thus, for example, OSHA is required

---

[245]*National Roofing Contractors Ass'n v. Brennan*, 495 F.2d 1294, 1 OSHC 1667 (7th Cir. 1974).

[246]*Id.* at 1296, 1 OSHC at 1668.

[247]29 U.S.C. §655(c)(3).

[248]*Synthetic Organic Chem. Mfrs. Ass'n v. Brennan* (*SOCMA II*), 506 F.2d 385, 2 OSHC 1402 (3d Cir. 1974) *cert. denied*, 423 U.S. 830 (1975). OSHA had argued that it was mandated by the Act to publish the proposal immediately after the ETS without waiting for the recommendations of an advisory committee.

[249]581 F.2d 962, 6 OSHC 1721 (D.C. Cir. 1978).

[250]The court rejected the argument that the revised standard was a "logical outgrowth" of the original version and no further consultation was needed. See *supra* note 207 on the "logical outgrowth" theory in the context of adequacy of notice.

to issue a rule, or decision not to issue a rule, within 60 days after the completion of the hearing or the completion of the comment period.[251] In one of the lengthiest and most controversial OSHA standards proceedings, to be discussed more fully below,[252] the issue was whether the court should order OSHA to issue a field sanitation standard. In the first of its decisions in this proceeding, the District of Columbia Circuit held that the 60-day deadline was not mandatory, and OSHA may "rationally order priorities and reallocate its resources *** at any rulemaking stage" so long as its discretion is "honestly and fairly exercised."[253]

V. Scope of OSHA Authority

OSHA has construed the scope of its authority expansively and its broad interpretations have frequently been challenged in court. This section discusses court decisions on OSHA authority in the context of standards development.[254]

A. Medical Surveillance

Section 6(b)(7)[255] of the Act requires OSHA, "where appropriate," to prescribe in health standards "the type and frequency of medical examinations or other tests which shall be made available by the employer, or at his cost, to employees exposed to such hazards." The section also states that the

---

[251]29 U.S.C. §655(b)(4).

[252]See Section VII, *infra*. The field sanitation standard was finally issued on May 1, 1987. See text accompanying notes 330–347, *infra*.

[253]*National Congress of Hispanic Am. Citizens v. Usery*, 554 F.2d 1196, 1200, 5 OSHC 1255, 1257 (D.C. Cir. 1977). See Tomlinson, *Report on the Experience of Various Agencies With Statutory Time Limits Applicable to Licensing or Clearance Functions and to Rulemaking, Administrative Conference of the United States. Recommendations and Reports*, 119, 195–211 (1978). See also Administrative Conference of the United States, "Time Limits on Agency Action," 1 C.F.R. §305.78-3 (1983).

[254]The issues raised are legal issues, involving primarily the interpretation of the OSH Act. On such issues, the courts would typically give only limited deference to the agency view. See *supra* note 105. However, factual, legal, policy, and other kinds of consideration are not easily separated. Thus, as will be discussed, in treating the medical removal protection issues the court first considered the question of OSHA authority to deal with the issue, and then whether substantial evidence supported its particular determination. See text accompanying notes 265–270, *infra*.

[255]29 U.S.C. §655(b)(7).

"result of such examination or tests" shall be furnished only to OSHA and NIOSH and "at the request of the employee, to his physician."

In reviewing OSHA's first asbestos standard, the court affirmed OSHA's determination that the employer select the physician who conducts the physical examination.[256] The union argued that an employer-selected physician would disclose the results of the examination to the employer, thus breaching physician-patient confidentiality and possibly leading to adverse job action against the employee where the examination showed illness. The court rejected the argument, stating that it was more practical to use the employer's own "medical programs and expertise" and that it did not "make sense" to provide medical examination and yet not permit the employer to know or use the results.[257] OSHA has continued to provide for employer-selected physicians in its standards, but in the diving standard added a multiphysician review procedure. The diving standard, issued in 1977,[258] provided that an employee dissatisfied with the opinion of the employer-selected physician could select his own physician, and if the two opinions differed, the two physicians would select a third who would give a binding opinion. The purpose, OSHA said, was to avoid a situation that would result in any employee being dismissed or refused entry to a job "for a cause which is less than substantial."[259] The Fifth Circuit found the provision beyond OSHA's authority, which is limited to protecting workplace safety and health. The multiphysician review, the court said, was a "mandatory job security provision," and the court held that OSHA "may not impose a ceiling on the medical fitness standard used by employers in hiring divers."[260]

However, in its later review of OSHA's lead standard, the District of Columbia Circuit upheld OSHA's authority to issue

---

[256]*Industrial Union Dep't v. Hodgson,* 499 F.2d 467, 1 OSHC 1631, 1645 (D.C. Cir. 1974).

[257]*Id.*

[258]42 FED. REG. 37,650 (1977).

[259]*Id.* at 37,658, 37,669–670.

[260]*Taylor Diving & Salvage Co. v. Department of Labor,* 599 F.2d 622, 625, 7 OSHC 1507, 1509 (5th Cir. 1979). In vacating the income protection provision in the cotton dust standard, the Supreme Court found OSHA's stated reason for the clause beyond its authority because it was based on economic rather than safety and health protection for employees. *American Textile Mfrs. Inst. v. Donovan,* 452 U.S. 490, 540, 9 OSHC 1913, 1931–1932 (1981).

a similar multiphysician review provision whose purpose, OSHA said, was to "strengthen and broaden the basis for medical determinations" and to assure "employee confidence" in their soundness.[261] In the context of the medical removal provision, which was also included in the lead standard,[262] the court distinguished the diving standard and held that the provision "directly enhances worker health," and is therefore within the ambit of OSHA's authority."[263] The court emphasized that the clause would not operate as a job protection provision since, in light of the wage protection available, there was "little practical incentive" for an employee to abuse the multiphysician review to avoid removal.[264]

The court in the lead proceeding also upheld OSHA's authority to provide medical removal protection (MRP). Under this provision, OSHA required the removal of employees who are determined in employer-provided medical examinations to be at excess risk from lead exposure, and the payment by the employer of equivalent wages and benefits during specified periods of removal.[265] OSHA justified the provision as necessary to encourage "voluntary and meaningful worker participation" in the medical surveillance program. Thus, if employees could lose their jobs because of the results of the medical examination, they would be reluctant to participate in or cooperate with the examinations.[266] OSHA's authority to issue the MRP provision was vigorously challenged by the lead employers association; it argued that OSHA's authority was limited to improving workplace safety and health, not assuring the payment of wages and benefits. The court of appeals agreed with OSHA that MRP was necessary for an effective medical surveillance program, and ruled that MRP

---

[261]43 FED. REG. 52,952, 52,998 (1978).

[262]Under medical removal protection, employees removed from their jobs because of their excess risk from lead exposure are guaranteed wages and other benefits for a period of time. See discussion accompanying notes 265–270, *infra.*

[263]*United Steelworkers v. Marshall,* 647 F.2d 1189, 1239, 8 OSHC 1810, 1842–1843 (D.C. Cir. 1980), *cert. denied,* 453 U.S. 913 (1981).

[264]*Id.* at 1239 n.76, 8 OSHC at 1843. The court also concluded that unlike the clause in the diving standard, the lead provision did not place a "ceiling" on the right of an employer to protect health. *Id.*

[265]43 FED. REG. 52,952 (1978). The background of MRP is discussed in Mintz, *Medical Surveillance of Employees Under the Occupational Safety and Health Administration,* 28 J. OCCUPATIONAL MED. 913, 917–920 (1986).

[266]*Supra* note 263, 647 F.2d at 1237. An earlier form of earnings protection was approved by the court of appeals in OSHA's first asbestos standard. *Industrial Union Dep't, supra* note 256, 499 F.2d at 485.

lies "well within the general range" of OSHA authority and that OSHA could "charge to employers the cost of any new means it devises to protect workers."[267] The court considered as more "serious" the employers' arguments based on Section 4(b)(4),[268] which bars OSHA from "in any manner affect[ing] any workmen's compensation law." According to the employers, MRP would "affect," indeed "replace" in some circumstances, the worker compensation laws. After a detailed analysis of the relationship between MRP and workers' compensation, the court concluded that while MRP would have "a great practical effect," it would "leave state workers' compensation schemes wholly intact as a *legal* matter" and thus not breach Section 4(b)(4).[269] Finally, the court upheld MRP as a "reasonable" exercise of statutory power and "well grounded in the evidence."[270]

B. Access to Records

The first asbestos standard required the retention of employee medical records for a period of at least 20 years, with access to the records provided to OSHA, NIOSH, the employer, and the employees' physician. The union challenged the 20-year period as insufficient, but the court held that OSHA could reasonably conclude that this would "ordinarily encompass the period during which disease could be detected and studied."[271]

---

[267]*Supra* note 263, at 1230–1231, 8 OSHC at 1836–1837. The decision of the Fourth Circuit, *en banc*, upholding the hearing conservation standard, affirmed OSHA's authority to require employers to pay for personal protective equipment, relying on *United Steelworkers v. Marshall. Forging Indus. Ass'n v. Donovan*, 773 F.2d 1436, 1451–1452, 12 OSHC 1472, 1484 (4th Cir. 1985).

[268]29 U.S.C. §653(b)(4).

[269]*Supra* note 263, at 1236, 8 OSHC at 1840 (emphasis in original).

[270]*Id.* at 1237, 8 OSHC at 1841–1842. OSHA's recently amended cotton dust standard contains a limited wage rate protection provision. Under the standard, an employee working in an area where the dust level exceeds the PEL who is not able to wear "any form" of respirator is entitled to a transfer with "no reduction in current wage rate or other benefits." 29 C.F.R. §1043(f)(2)(iv) (1986). OSHA's recently issued revised asbestos standard reaffirmed existing modified wage protection provisions. 51 FED. REG. 22,735, 22,739 (1986), 29 C.F.R. §1910.1001(g)(3)(iv), 29 C.F.R. §1926.58(g)(3)(iv) (1987).

[271]*Industrial Union Dep't v. Hodgson*, 449 F.2d 467, 486, 1 OSHC 1631, 1646 (D.C. Cir. 1974). The court found the 3-year retention period for exposure monitoring to be "surprisingly short," however, and remanded the issue for further consideration. OSHA's second asbestos standard, issued in 1986, provided for retention of monitoring records for at least 30 years. 29 C.F.R. §1910.1001(m)(1) (1986). Compare the requirements of OSHA's regulation on access to records, 29 C.F.R. §1910.20.

As noted, the first asbestos standard provided access to employee medical records to the employer and the employees' designated physician.[272] In its benzene standard, OSHA provided access to medical records not only to the employee's physician but also to the employee himself or to any individual designated by the employee.[273] More detailed access provisions were contained in the lead standard. The medical records were made available to OSHA and NIOSH and "to a physician or any other individual" designated by the employee. In addition, "biological monitoring and medical removal records" were made available to employees "or their authorized representatives."[274] The provision was challenged by the lead employer association on various legal grounds.

The court of appeals had "little difficulty" in upholding OSHA and NIOSH access as "a reasonable exercise of governmental responsibility over public welfare" and as "reasonably related to the general goal of preventing occupational lead disease."[275] The court was more concerned about the access to biological monitoring and removal records, construing the standard as giving union officials an "independent right of access" without employee consent.[276] The court was concerned particularly about violation of the confidential relationship. However, the court emphasized that access was limited to biological monitoring and removal records, records which do not "address actual symptoms of illness" as do records of physical examinations. The court made it clear, however, that there may well be a consitutional violation if unions, without consent, were permitted to examine records containing "the intimate results of physician examinations."[277]

In 1980, OSHA promulgated its records access regulation, providing for OSHA and employee access to medical and exposure records that an employer had previously and volun-

---

[272]The court affirmed the provision. See discussion *supra* Section V.A.

[273]43 FED. REG. 5918, 5967 (1978). The standard was vacated by the Supreme Court on other grounds.

[274]43 FED. REG. 52,952 (1958), codified at 29 C.F.R. §1910 (1986).

[275]*United Steelworkers v. Marshall*, 647 F.2d 1189, 1241, 8 OSHC 1810, 1844 (D.C. Cir. 1980), *cert. denied*, 453 U.S. 913 (1981). The court relied particularly on *Whalen v. Roe*, 429 U.S. 589 (1977), where the Supreme Court upheld New York State's system for recording the identities of persons taking prescribed addictive drugs.

[276]647 F.2d at 1242, 8 OSHC at 1845.

[277]*Id.* at 1243, 8 OSHC at 1846.

tarily created.[278] A broad-ranging challenge to the regulation[279] was mounted by an employer association, but its arguments were rejected by a district court whose opinion was affirmed, without opinion, by the court of appeals.[280]

The district court first stated that even a "cursory examination" of the Act makes it plain that the regulation "bears at least a reasonable relation" to the Act's purpose. In particular, the court found OSHA authority for the regulation in its general rulemaking authority under Section 6(g),[281] and under the authority in Section 8(c)(1),[282] which specifically grants OSHA authority to require recordkeeping and reporting by employers.[283] Noting that access to employee medical records could be obtained only with employee consent, the court upheld OSHA access to these records, relying, as did the court of appeals in the lead case, on the Supreme Court decision in *Whalen v. Roe*.[284] The district court similarly rejected the association argument that the regulation requires disclosure of trade secrets, explaining that the Trade Secrets Act[285] allows the disclosure of trade secrets where "authorized by law" and OSHA's access regulation issued after public participation constitutes such legal authorization.[286] Finally, the court rejected the arguments that the regulation infringes on the jurisdiction of the National Labor Relations Board and was impermissibly overbroad in its definition of "toxic substance."[287]

C. Hazard Communication

In November 1983, OSHA issued a hazard communication standard requiring chemical manufacturers and importers to

---

[278]45 FED. REG. 35,212 (1980), codified at 29 C.F.R. §1910.20 (1986).

[279]The Fifth Circuit had earlier held that the access rule was a regulation, reviewable initially in a district court, and not a standard, which is reviewable in the court of appeals. See discussion *supra* Section I.B.

[280]*Louisiana Chem. Ass'n v. Bingham*, 550 F. Supp. 1136, 10 OSHC 2113 (W.D. La. 1982).

[281]29 U.S.C. §655(g).

[282]§657(c)(1).

[283]*Supra* note 280, at 1140, 10 OSHC at 2115.

[284]*Supra* note 275.

[285]18 U.S.C. §1955.

[286]*Supra* note 280, at 1143–1144, 10 OSHC at 2118–2119. The issue of the meaning of the phrase "authorized by law" in the Trade Secrets Act was discussed by the Supreme Court in *Chrysler Corp. v. Brown*, 441 U.S. 281 (1979).

[287]*Supra* note 280, at 1145, 10 OSHC at 2119–2120.

assess the hazards of chemicals they produce or import and to provide information to their workers and customers about the hazards of these chemicals by means of labels, material data safety sheets, training, and access to records.[288] The standard was challenged as not being sufficiently stringent by unions, states, and a public interest group.[289] In a major decision, the Third Circuit upheld the application of the standard[290] to the manufacturing sector, but in other respects remanded it to OSHA for broadening to other sectors and for other changes.[291] The court initially held that the standard provided for unnecessarily broad trade secret protection for information which the employer makes available to workers. Suggesting that a rule protecting only "formula and process" information but requiring disclosure of hazardous ingredients would afford increased health protection and, at the same time, adequate trade secret protection, the court of appeals remanded to OSHA for reconsideration of the issue.[292] The court at the same time upheld the requirement in the standard that the request for information allegedly involving trade secrets be in writing, with supporting documentation of need, and that, except in a medical emergency, a manufacturer may request a confidentiality agreement containing a liquidated

[288]48 FED. REG. 53,279 (1983), codified at 29 C.F.R. §1910.1200 (1986).

[289]Employer groups did not challenge the hazard communication standard itself. However, there has been extensive subsequent litigation on the extent to which the OSHA hazard communication standard preempts state and city right-to-know statutes, ordinances, and regulations. Under §18(a), 29 U.S.C. §667(a), the states and localities may assert jurisdiction "over any occupational safety or health issue" only if there is "no [Federal OSHA] standard *** in effect under section 6." The major cases dealing with the preemption issue are *New Jersey State Chamber of Commerce v. Hughey*, 774 F.2d 587, 12 OSHC 1489 (3d Cir. 1985); *Manufacturers Ass'n v. Knepper*, 801 F.2d 130, 12 OSHC 1553 (3d Cir. 1986); *Ohio Mfrs. Ass'n v. City of Akron*, 801 F.2d 824, 12 OSHC 2089 (6th Cir. 1986). The issue of the preemption of state regulations by federal OSH standards was also involved, obliquely, in the recent court of appeals decision in the field sanitation case, *Farmworker Justice Fund v. Brock*, 811 F.2d 613, 13 OSHC 1049 (D.C. Cir. 1987), *vac. as moot*, 13 OSHC 1288 (May 7, 1987). See text accompanying notes 338–347.

[290]The court of appeals previously decided the threshold issue that the rule was a standard subject to court of appeals review. See *supra* note 22 and accompanying text.

[291]*United Steelworkers v. Auchter*, 763 F.2d 728 (3d Cir. 1985). The Fifth Circuit had previously ruled that OSHA had authority to require employers to warn not only their own employees of hazards but also to warn customers of the hazards of the products, so that the downstream employers could warn their employees. *American Petroleum Inst. v. OSHA*, 581 F.2d 493, 510, 6 OSHC 1959, 1972 (5th Cir. 1978).

[292]*United Steelworkers, supra* note 291, at 741–741. OSHA subsequently amended the standard, limiting the trade secret protection by conforming the standard's definition of trade secrets to that in the Restatement of Torts, as suggested by the court. 51 FED. REG. 590 (1986).

damages clause.[293] However, the court found the standard's restriction of access to trade secret information to health professionals unsupported by substantial evidence.[294]

D. Hearing Conservation

In November 1984, the Fourth Circuit in an unexpected and, to many, surprising decision, vacated OSHA's revised hearing conservation standard.[295] The standard continued OSHA's 90-decibel PEL for noise but added that where an employee, based on audiometric tests, showed a specified level of hearing loss, hearing protection measures would be required including personal protective equipment for that employee.[296] The court held the provision was beyond OSHA's authority because it failed to distinguish between hearing losses caused by workplace noise and those caused by nonworkplace noise. Regulating nonworkplace hazards, the court said, "was not a problem that Congress delegated to OSHA to remedy."[297]

OSHA and the AFL-CIO requested reconsideration and an *en banc* hearing, arguing that the court's holding would pervasively undermine OSHA regulation of health hazards since it is frequently impossible to distinguish between workplace-caused illness and illness resulting from causes outside the workplace.[298] The full court reconsidered the case and in October 1985 reversed the original decision, thus upholding the hearing conservation standard.[299] The court recognized that some hearing loss may be due to nonworkplace conditions,[300] but insisted that this fact was "scant reason to char-

---

[293]*United Steelworkers, supra* note 291, at 742.

[294]*Id.* at 743. OSHA also amended the regulation to meet the court's objections on this issue. 51 FED. REG. 34,590 (1986). The court also remanded to OSHA the issue of the scope of the standard, concluding that OSHA's decision to limit the coverage of the standard to the manufacturing sector was unsupported. This issue is discussed *infra* Section VII.

[295]*Forging Indus. Ass'n v. Donovan*, 748 F.2d 210, 12 OSHC 1041 (4th Cir. 1984).

[296]48 FED. REG. 9738 (1983); 29 C.F.R. §1910.95.

[297]*Supra* note 295, at 215, 12 OSHC at 1043.

[298]Briefs for OSHA and AFL-CIO, *Forging Indus. Ass'n v. Donovan*, 773 F.2d 1436, 12 OSHC 1472 (4th Cir. 1985).

[299]*Forging Indus. Ass'n v. Donovan, supra* note 298.

[300]*Id.* at 1444, 12 OSHC at 1478. The court emphasized, however, that it was "hard to imagine" nonworkplace noise that would be of the same intensity as workplace noise over a period of 8 hours a day.

acterize the primary risk factor as non-occupational."[301] In any event, OSHA in its standard was regulating occupational noise, which it has authority to do. The analogy was to toxic substances, according to the court: "The presence of unhealthy lungs in the workplace, however, hardly justifies failure to regulate noxious workplace fumes."[302]

VI. Emergency Temporary Standards

OSHA has issued nine emergency temporary standards (ETSs) since the Act took effect. Five of these were either stayed or vacated by a court of appeals: the standards covering the carcinogens dichlorobenzidine (DCB) and ethyleneimine (EI), pesticides, benzene, and diving, and the second asbestos ETS in 1983. One challenged ETS—for acrylonitrile—remained in effect when a stay was refused by the Sixth Circuit.[303] The other emergency standards were not challenged.

The first ETS decision was handed down by the Third Circuit vacating OSHA's standard for EI and DCB.[304] After an extensive discussion of OSHA's evidence,[305] the court stated that it had "doubts—more serious as to EI then to DCB" as to whether the substances were carcinogenic in man and

---

[301]*Id.*

[302]*Id.* The court further found that substantial evidence in the record supported OSHA's findings that the preamendment standard exposed employees to a significant risk and that the amendment would reduce that risk, *id.* at 1445–1446, and that the standard was technologically and economically feasible, *id.* at 1452–1453. Finally, the court upheld OSHA's determination that the individual requirements of the standard, such as audiometric testing, monitoring, and employee information requirements, were "reasonably related" to the purposes of the Act. *Id.* at 1451.

[303]*Vistron Corp. v. OSHA*, 6 OSHC 1483 (6th Cir. 1978), discussed in text accompanying note 64, *supra.*

[304]*Dry Color Mfrs. Ass'n v. Department of Labor*, 486 F.2d 98, 1 OSHC 1331 (3d Cir. 1973). OSHA had issued an ETS covering 14 carcinogens. The ETS was challenged, however, only with respect to these two substances.

[305]In the context of review of an ETS, where no rulemaking proceeding has been conducted, there is often dispute over what constitutes the "record." In *Dry Color*, the court refused to consider the relevant Report of the Ad Hoc Committee on the Evaluation of Environmental Chemicals to the Surgeon General because it was not part of the "record," there being no evidence in the published notice that the report was relied on by OSHA in issuing the ETS. *Supra* note 304, at 104 and n.8, 1 OSHC at 1334. The court said that it would be "better practice" for OSHA to specifically designate those items in the record being certified to the court which had not been read by it prior to the publication of the standard. *Supra* note 304, at 108. The court in *Florida Peach Growers Ass'n, infra* note 309, at 129, 1 OSHC at 1479, also expressed concern about the state of the record before it. See also *Asbestos Information Ass'n v. OSHA*, 727 F.2d 415, 420 n.12, 11 OSHC 1817, 1820 (5th Cir. 1984).

therefore constituted a "grave danger" to employees.[306] The court found it unnecessary to reach the substantive question, however, because it decided that the preamble to the ETS failed to satisfy the requirements of Section 6(e) requiring OSHA to give a "statement of reasons" for its action.[307] The standard as to the two substances was therefore vacated.[308]

OSHA's emergency standard regulating reentry times for farmworkers in fields treated with pesticides was vacated by the Fifth Circuit in *Florida Peach Growers Association v. Department of Labor*.[309] The court initially stated its duty was to determine if OSHA carried out "its essentially legislative task in a manner reasonable under the state of the record" before it and concluded that there was no substantial evidence to support the OSHA determination that a "grave danger" existed. The inherent toxicity of the pesticides involved was conceded, but did not establish a grave danger to exposed farmworkers, the court said. The "ultimate picture" according to the court was that "only a few farmworkers" became ill from exposure "relative to the mass of agricultural workers in contact with treated foliage."[310] The distinction between "incurable, permanent, or fatal consequences" and, as was the case here, "easily curable and fleeting effects" in health, was critical, the court emphasized, in deciding whether emergency measures were necessary.[311]

In August 1976, the Fifth Circuit issued an indefinite stay of OSHA's diving ETS, finding that there was a danger of irreparable harm to diving contractors, who were likely to prevail on the merits.[312] The Fifth Circuit again struck down

---

[306]*Supra* note 304, at 105, 1 OSHC at 1335. The studies showing the carcinogenicity of EI and DCB were all conducted with rodents. See discussion *supra* Section III.A on the issue of extrapolating from "mouse to man."

[307]That portion of the case is discussed *supra* Section IV.B.

[308]The court separately held that the requirements for an environmental impact statement under the Environmental Protection Act, 42 U.S.C. §4321 *et seq.*, were not applicable in the circumstances of this ETS, since the delay involved in preparing the statement would "impair" the prompt protection intended for employees. *Supra* note 304, at 107–108.

[309]489 F.2d 120, 1 OSHC 1472 (5th Cir. 1974).

[310]*Id.* at 132, 1 OSHC at 1482.

[311]*Id.* The court also noted that the investigative committees convened by OSHA on the issue of regulating pesticides did not believe an ETS was necessary. Excerpts from the letter from the chairperson of OSHA's pesticide subcommittee criticizing the ETS are reprinted in Mintz, *supra* note 1, at 101–102.

[312]*Taylor Diving & Salvage Co. v. Department of Labor*, 573 F.2d 819, 4 OSHC 1511 (5th Cir. 1976). This decision is discussed *supra* Section I.E.

an OSHA emergency standard in 1983. OSHA had issued a second ETS on asbestos[313] lowering the PEL from 2 fibers per cubic centimeter of air, the level which became effective in 1976 under the original standard,[314] to 0.5 fibers. The standard was quickly challenged by various asbestos employers and the Fifth Circuit found the standard invalid.[315] In deciding that OSHA had not met either requirement of Section 6(c),[316] i.e., that there exist a "grave danger" and that the standard be "necessary" to protect workers from the danger,[317] the court reached the following major conclusions:

(a) In determining the harm that would accrue from the existing PEL, the Act limits consideration to the 6-month period which is the maximum life of the standard.[318]

(b) OSHA calculated that the ETS would save 80 lives for the 6-month period. On the basis of the record, the court concluded that the number of lives that would be saved was "uncertain" and likely to be "substantially less" than 80. The "gravity of the risk *** and the necessity of an ETS *** are therefore questionable."[319]

(c) OSHA also justified the ETS on the grounds that it "often" takes several years to complete rulemaking.[320] The court answered that the statute did not contemplate that OSHA

---

[313]48 FED. REG. 51,086 (1983).

[314]Although the court in *Industrial Union Dep't v. Hodgson*, 499 F.2d 467, 1 OSHC 1631 (D.C. Cir. 1974), remanded the delayed effective date to OSHA for further consideration, OSHA took no action on the remand, and the lower PEL went into effect in 1976 by operation of the standard.

[315]*Asbestos Information Ass'n v. OSHA*, 727 F.2d 415, 11 OSHC 1817 (5th Cir. 1984).

[316]29 U.S.C. §655(c).

[317]The court of appeals rejected the employer argument that the ETS was not based on new information and therefore no "emergency" existed. The court insisted, however, that the agency give an explanation on the timing of the promulgation since "it had known for years" of the serious health risks involved. *Supra* note 315, at 423, 11 OSHC at 1822.

[318]*Id.* at 422, 11 OSHC at 1821. Under §§6(c)(2) and (3), 29 U.S.C. §§655(c)(2) and (3), the ETS is "superseded" when a "permanent" standard is issued after rulemaking, which "shall" take place no later than 6 months after publication of the ETS.

[319]*Supra* note 315, at 425, 11 OSHC at 1824.

[320]The thrust of OSHA's argument was that rulemaking normally takes years, but because of the "urgency generated by OSHA's finding of grave danger" and the 6-month statutory deadline, OSHA is able to complete a rulemaking much more quickly after issuance of an ETS. See preamble to ETS, 48 FED. REG. 51,086, 51,098 (1983). The court's rejection was based on the view that OSHA was seeking to use the emergency authority for the ulterior purpose of speeding up its rulemaking process.

use its emergency power as a "stop-gap measure" rather than to deal with grave dangers.[321]

(d) Under the existing standard, OSHA could require the use of respirators, which would reduce employee exposure and thus eliminate the grave danger without requiring "resort to the most dramatic weapon in OSHA's enforcement arsenal." The court concluded, therefore, that an ETS was not "necessary" to meet the grave danger.[322]

OSHA subsequently issued, after rulemaking, a "permanent" asbestos standard, lowering the PEL to 0.2 fibers per cubic centimeter.[323]

VII. Court Suits to Compel OSHA to Take Action in Standards Proceedings

The cases discussed earlier have related to suits challenging actions by OSHA in promulgating standards. Thus, the question typically confronting the court in these cases was whether to uphold a standard issued by OSHA as meeting the appropriate substantive and procedural requirements of the Act.[324] In this section we discuss whether, and under what circumstances, a court will order OSHA to initiate or complete rulemaking;[325] and whether a court will order OSHA to issue a standard after it has completed rulemaking and decided that

---

[321]*Supra* note 315, at 422, 11 OSHC at 1822.

[322]*Id.* at 426, 11 OSHC at 1824–1825.

[323]51 FED. REG. 22,612 (1986). Challenges to the new asbestos standard are as of this writing pending in the District of Columbia Circuit.

[324]An illustration of this type of case is *Industrial Union Dep't v. American Petroleum Inst.*, 448 U.S. 607, 8 OSHC 1586 (1980), where OSHA's "permanent" standard on benzene was vacated. This challenge was based on the claim that the standard was unduly "stringent." There are also, of course, challenges based on claims that the standard is insufficiently protective, for example, the challenge to the delay in the effective date for the more stringent level, *Industrial Union Dep't v. Hodgson*, 499 F.2d 467, 1 OSHC 1631 (D.C. Cir. 1974) (asbestos); a challenge to OSHA's amending a standard to delete protective requirements, *AFL-CIO v. Brennan*, 530 F.2d 109, 3 OSHC 1820 (3d Cir. 1975) (no-hand-in-dies); and a challenge to OSHA's issuing a standard with an 8-hour time-weighted average but no short-term limit, *Public Citizen Health Research Group v. Tyson*, 796 F.2d 1479, 12 OSHC 1905 (D.C. Cir. 1986) (ethylene oxide).

[325]The first ethylene oxide decision of the District of Columbia Circuit, *Public Citizen Health Research Group v. Auchter*, 702 F.2d 1150, 11 OSHC 1209 (D.C. Cir. 1983), discussed *infra* this section, involved the issue of OSHA's initiating rulemaking.

no standard should be issued.[326] Review in these situations would be predicated either on Section 6(f)[327] of the Act or the Administrative Procedure Act.[328] The basic principle which the courts have applied in cases of agency failure to act is that any review of agency action is "extremely narrow" because these types of decisions are committed to agency discretion.[329] However, as will appear from the OSHA cases to be considered, the courts of appeals have afforded relief in some circumstances, despite their stated limited review function.

These issues were extensively litigated in the controversy over OSHA's field sanitation standard. The suit was originally brought in 1974 by a migrant action group asking that OSHA be compelled to issue a standard requiring toilets and drinking water for field workers.[330] The case came before the District of Columbia Circuit three times, most recently in February 1987. In its first decision,[331] the court held that the time deadlines in Section 6 were not mandatory and that the agency could reorder its priorities, if its discretion was being "honestly and fairly exercised."[332] OSHA prepared a timetable which indicated that because of more pressing priorities with other standards, a field sanitation standard would not be issued within the next 18 months. In its second opinion,[333] the court of appeals reversed the district court, which had rejected the

---

[326]The 1987 decision of the District of Columbia Circuit on the field sanitation standard, *Farmworker Justice Fund v. Brock*, 811 F.2d 613, 13 OSHC 1049 (D.C. Cir. 1987), *vac. as moot*, 13 OSHC 1288 (May 7, 1987), discussed in text accompanying notes 338–343, *infra*, involved this issue. Similar issues are raised where OSHA, after rulemaking, decides to exempt an industry or industries from the scope of coverage. An example is *United Steelworkers v. Auchter*, 763 F.2d 728 (3d Cir. 1985), discussed as to this issue *infra*, where OSHA exempted nonmanufacturing industries from coverage under the hazard communications standard.

[327]29 U.S.C. §655(f). The explicit language of this section—referring to "a standard issued under this section"—would seem not to cover cases of agency failure to act.

[328]5 U.S.C. §701 *et seq.* In *Public Citizen Health Research Group v. Auchter*, 702 F.2d 1150, 11 OSHC 1209 (D.C. Cir. 1983), the court of appeals ordered OSHA to issue an ethylene oxide proposal under §706(1) of the Administrative Procedure Act ("The reviewing court shall—(1) Compel agency action unlawfully withheld or unreasonably delayed").

[329]See e.g., *WWHT, Inc. v. FCC*, 656 F.2d 807, 819 (D.C. Cir. 1981).

[330]A similar standard had been adopted by OSHA in 1971 for nonagricultural workers.

[331]*National Congress of Hispanic Am. Citizens v. Usery*, 554 F.2d 1196, 5 OSHC 1255 (D.C. Cir. 1977).

[332]*Id.* at 1200, 5 OSHC at 1257. This issue is discussed *supra* Section IV.E.

[333]*National Congress of Hispanic Am. Citizens v. Marshall*, 626 F.2d 882, 7 OSHC 2029 (D.C. Cir. 1979).

timetable and directed OSHA to issue a standard "as soon as possible"; the court of appeals said that "greater respect is due the Secretary's judgment" that other projects warrant higher priority. The court insisted, however, that OSHA prepare another timetable showing *when* a field sanitation standard would be issued, considering a longer time horizon than 18 months. Eventually, the litigation was settled in 1982 on the basis of OSHA's making a good faith effort to issue a standard by February 1985.[334]

OSHA completed the rulemaking hearings on field sanitation pursuant to the settlement, and in 1985 it decided that no field sanitation standard would be issued.[335] Following massive protests, this decision was reversed by a new Secretary of Labor, who decided that a federal field sanitation standard would be issued after two years if, at the end of 18 months, OSHA determined that in the interim states had not responded adequately in issuing their own state field sanitation standards.[336] The validity of that decision came before the court in 1987.[337]

The court of appeals in *Farmworker Justice Fund v. Brock* first confronted the issue of scope of its review where agency *inaction* is involved. This question was particularly important in light of the 1985 Supreme Court decision in *Heckler v. Chaney*.[338] The Court there held that "an agency's decision not to take enforcement action should be presumed immune from judicial review *** [although] the presumption may be rebutted where the substantive statute has provided guidelines for the agency to follow in exercising its enforcement powers."[339] The court of appeals concluded that *Chaney* in no way limits court review of agency decisions that are "contrary to law," since an agency has no "discretion" to violate the law. In addition, the court of appeals ruled, *Chaney* does not preclude judicial review for agency abuse of discretion where the

---

[334]Relevant excerpts from the settlement agreement appear in Mintz, *supra* note 1, at 195–196.

[335]50 FED. REG. 15,086 (1985). The decision was based on a preference for state regulation and enforcement priorities.

[336]50 FED. REG. 42,660 (1985).

[337]*Farmworker Justice Fund v. Brock*, 811 F.2d 613, 13 OSHC 1049 (D.C. Cir. 1987), *vac. as moot* 13 OSHC 1288 (May 7, 1987).

[338]470 U.S. 821 (1985).

[339]*Id.* at 832–833, quoted in *Farmworker Justice Fund v. Brock, supra* note 337, 13 OSHC at 1054.

agency decision is "based on factors which the court is competent to evaluate."[340] These principles of review, the court of appeals said, apply to agency *inaction* as well as *action*, even after the Supreme Court decision in *Chaney*.

Applying the principles to the proceedings in field sanitation, the court of appeals first held that OSHA acted contrary to the OSH Act in delaying a standard which it otherwise thought necessary on the ground that state regulation of field sanitation was preferable to federal regulation.[341] The court also held that in light of extensive prior delays, OSHA abused its discretion by delaying the standard, which it acknowledged was needed, for an additional two years while awaiting state action. Finally, the court decided that there was a "total inadequacy of the evidence in the record" to support this additional two-year delay.[342]

The court therefore ordered OSHA to issue a field sanitation standard within 30 days, hoping "to bring to an end this disgraceful chapter of legal neglect."[343] OSHA issued a final rule on field sanitation on May 1, 1987.[344] Although OSHA had complied with the court directive, the preamble to the standard asserted OSHA's disagreement with the court of appeals reasoning, and noted that OSHA had petitioned for a rehearing and rehearing *en banc* of the court decision.[345] On May 7, 1987, in light of OSHA's having issued the standard, the court of appeals panel "vacated as moot" its opinion and judgment in the case.[346] At the same time, the panel dismissed the petition for rehearing as moot and the full court dismissed the suggestion for rehearing *en banc* as moot.[347]

The ethylene oxide proceeding was begun in 1981, after the commencement of the field sanitation litigation. In Sep-

---

[340]*Farmworker Justice Fund, supra* note 337, at 620–621, 13 OSHC at 1055–1056. The Supreme Court decision in *Chaney*, which involved agency refusal to take enforcement action, left open the question of the scope of judicial review of agency refusal to initiate rulemaking. *Supra* note 338, at 825 n.2.

[341]*Farmworker Justice Fund, supra* note 337, at 627, 13 OSHC at 1060.

[342]*Id.* at 632–633, 13 OSHC at 1564.

[343]*Id.* at 614, 13 OSHC at 1050.

[344]52 FED. REG. 16,050 (1987).

[345]*Id.* at 16,053. OSHA characterized the majority opinion as "judicial overreaching." *Id.*

[346]*Farmworker Justice Fund v. Brock*, No. 85-1824 (D.C. Cir. May 7, 1987) (panel decision).

[347]*Id.* (court of appeals sitting *en banc*).

tember 1981, OSHA denied the petition of the Health Research Group (HRG) for an ETS reducing the PEL for ethylene oxide despite the new evidence showing the carcinogenicity of ethylene oxide. HRG took the case to court, and the district court ordered OSHA to issue an ETS within 30 days, concluding that OSHA had abused its discretion, in light of record evidence showing a "solid and certain foundation" that workers were exposed to grave danger when exposed within the current PEL.[348] The court of appeals considered the case on an expedited basis, reversed the district court order on the ETS, but ordered OSHA to issue a proposal within 30 days.[349] On the issue of the ETS, the court stated that it was "hesitant to *compel* the Assistant Secretary to grant extraordinary relief," particularly since there was no record evidence contradicting OSHA's "estimate" that workers were in fact exposed at a level of 10 parts per million and there was no "clear scientific evidence" on the degree of harm from exposure at 10 ppm or below.[350] However, the court also concluded that workers at the current 50-ppm level were exposed to a "potential grave danger" and that the agency's "unaccounted for delay" in issuing a proposed permanent standard constituted agency action "unreasonably delayed" within the meaning of the Administrative Procedure Act. While the court said it would "hesitate" to order OSHA action if it would "seriously disrupt rulemaking of higher or competing priority," the court made it clear, after examining in detail other OSHA standard projects, that "we do not confront such a case."[351] "In the context of the OSHA Act designed to protect workers' health," the court concluded, "the Assistant Secretary's protracted course

---

[348]*Public Citizen Health Research Group v. Auchter,* 554 F. Supp. 242, 251 (D.D.C. 1983).

[349]*Public Citizen Health Research Group v. Auchter,* 702 F.2d 1150, 11 OSHC 1209 (D.C. Cir. 1983). The court also directed OSHA to issue a permanent standard on a "priority, expedited basis," well in advance of the projected estimate of late 1984. The considerations against judicial review are arguably more compelling where the agency refuses to initiate rulemaking than when the record has been developed and it decides not to issue the rule, the situation in the 1987 field sanitation case. See *Natural Resources Defense Council v. SEC,* 606 F.2d 1031, 1045–1047 (D.C. Cir. 1979).

[350]*Supra* note 349, 702 F.2d at 1156–1157, 11 OSHC at 1213–1214. The issue of whether a court should order OSHA to issue an ETS was also discussed in *United Automobile, Aerospace, & Agricultural Implement Workers v. Donovan, infra* note 355 at 2022 (formaldehyde).

[351]*Supra* note 349, 702 F.2d at 1158, 11 OSHC at 1214.

in the face of potentially grave health risks cannot be characterized as reasonable."[352]

There have been a number of other suits filed to compel OSHA to take action on standards under Section 6. Several of these, filed in the early years of the agency, did not result in court decision because OSHA took the action requested and the suit became moot.[353] Similarly, in December 1975, the Textile Workers Union brought suit to compel OSHA to commence Section 6(b) proceedings to revise the 1971 cotton dust standard. The proposed revisions were published and the suit was revived in 1978 when publication of the final standard was delayed by intervention of the White House. However the final standard was soon issued, and no court decision on the merits was handed down.[354]

More recently, the United Automobile, Aerospace, and Agricultural Implement Workers of America brought suit to compel OSHA to issue an ETS, or to initiate rulemaking, to lower the PEL for formaldehyde. OSHA denied the request for an ETS in 1982; and by July 1984, when the district court issued its ruling,[355] it had not commenced rulemaking. Emphasizing the theme of "deference and restraint" and relying particularly on the court of appeal's decision in *Public Citizen*,[356] the district court decided that in the particular posture of the case, "reappraisal by the agency [of its determination] is the better course."[357] Since OSHA had denied the ETS two years before, the court remanded to OSHA to "review all cur-

---

[352]*Id.* n.30. OSHA initiated rulemaking and issued a "permanent" standard on ethylene oxide in June 1984. 49 FED. REG. 25,734 (1984). The standard was challenged and the District of Columbia Circuit issued a decision partially upholding and partially remanding the standard for consideration of the short-term exposure limit issue. In July 1987, the court of appeals ordered OSHA to act on the remand by March 1988. See discussion *supra* Section III.A.

[353]The suits were to compel issuance of an ETS on pesticides, see 2 OSHR 1221 (Mar. 22, 1973), and to compel issuance of an ETS on 14 carcinogenic substances, see 2 OSHR 1341 (April 19, 1973).

[354]See *Textile Workers v. Marshall*, [1977–1978] OSHD ¶21,914 (D.D.C. 1977). This proceeding is discussed in Mintz, *supra* note 1, at 309 n.47.

[355]*United Automobile, Aerospace, & Agricultural Implement Workers v. Donovan*, 590 F. Supp. 747, 11 OSHC 2017 (D.D.C. 1984).

[356]*Supra* note 349.

[357]*Supra* note 355, at 751, 11 OSHC at 2021. On October 9, 1987, the Court of Appeals for the District of Columbia Circuit denied a union request that OSHA be directed to issue an ETS regulating cadmium, saying that the determination of when to proceed on the basis of complete data is a decision "largely entrusted to the expertise of the agency." *In re Chemical Workers Union*, 830 F.2d 369, 13 OSHC 1402 (D.C. Cir. 1987).

rent scientific data" and "all of its regulatory options" and decide on its course of action. The court emphasized that it was *not* ordering OSHA to issue an ETS or even to commence rulemaking.[358]

Pursuant to a further court order, OSHA developed a timetable for administrative proceedings to consider regulation of formaldehyde. In October 1984, the district court approved the timetable, saying that it would not substitute its judgment for that of the agency.[359] However, the court emphasized that it expected OSHA "to strictly comply with the schedule," and to make "periodic progress reports." It concluded by saying that it "will look with extreme displeasure on any variance from the schedule and will not hesitate to set a date certain for completion of the administrative proceedings if OSHA unreasonably delay[s]" its consideration of the petition for action.[360] The proceeding was then transferred to the District of Columbia Circuit,[361] which accepted the district court timetable. "Mindful of past [agency] delays," the court of appeals ordered OSHA to inform the court immediately of any proposed actions "which might interfere with the timetable." The court of appeals also agreed with the district court that it would look with "extreme displeasure" at any variance from the schedule and, if necessary, would set a "date certain for completion."[362]

In July 1983, almost three years after the Supreme Court decision vacating OSHA's benzene standard, OSHA announced expedited rulemaking on a new benzene standard, to

---

[358]*Id.* at 753, 11 OSHC at 2022.

[359]*United Automobile, Aerospace, & Agricultural Implement Workers v. Donovan,* 12 OSHC 1001 (D.D.C. 1984). The court relied on *National Congress of Hispanic Am. Citizens v. Marshall,* 626 F.2d 882, 7 OSHC 2029 (D.C. Cir. 1979) (the second court of appeals decision on field sanitation) for the proposition that a court should not "substitute [ ] its judgment for that of the agency" on rulemaking priorities.

[360]*United Automobile Workers, supra* note 359.

[361]*United Automobile, Aerospace, & Agricultural Implement Workers v. Donovan,* 756 F.2d 162, 12 OSHC 1201 (D.C. Cir. 1985). The transfer took place under the decision of the District of Columbia Circuit in *Telecommunications Research & Action Center v. FCC,* 750 F.2d 70 (D.C. Cir. 1984) (*TRAC*) holding that where a statute commits final agency action to review in the court of appeals, the same court has jurisdiction to hear suits—such as those seeking to compel agency action—that might affect its future (statutory review). The court of appeals applied *TRAC* retroactively to this case and held that it would give full effect to the district orders, including the timetable.

[362]*United Automobile Workers, supra* note 361, at 165, 12 OSHC at 1203. OSHA issued a proposed standard on formaldehyde in December 1985. 50 FED. REG. 50,412 (1985), and a final standard on December 14, 1987, 52 FED. REG. 46,168 (1987).

be completed in June 1984.[363] When in December 1984 no proposal had been issued, various unions and a public interest group brought suit to compel OSHA to issue the proposal. The case was eventually set down for oral argument before the District of Columbia Circuit in December 1985. Just before the date of argument, OSHA issued a benzene proposal.[364] At oral argument, OSHA submitted a schedule calling for promulgation of a final standard by February 1987. The unions and other petitioners asked the court to reject the schedule. The court of appeals refused, saying that "judicial imposition of an overly hasty timetable at this stage would ill serve the public interest."[365] The court emphasized that any standard would have to be "constructed carefully and thoroughly if the agency's action is to pass judicial scrutiny this time around."[366]

There have been several court cases involving the question, closely related to those being discussed, whether OSHA properly included, or excluded, industries from the scope of a standard's coverage.[367] In the review of OSHA's hazard communications standard,[368] the major issue was whether OSHA properly limited coverage of the standard to manufacturing industries. OSHA argued that the "focus of this standard should remain on the manufacturing sector since that is where the greatest number of chemical source injuries and illnesses are occurring."[369] OSHA argued further that under Section 6(g)[370] dealing with standards priorities, OSHA has "unreviewable discretion" to determine what industries shall be covered by

---

[363]48 FED. REG. 31,412 (1983).

[364]50 FED. REG. 50,512 (1985).

[365]*In re United Steelworkers*, 783 F.2d 1117, 12 OSHC 1673 (D.C. Cir. 1986).

[366]*Id.* at 1120, 12 OSHC at 1674. This consideration—that the agency will likely be compelled to defend the standard in a court suit, which militates against a court's ordering OSHA to issue a standard hastily—is more compelling where an ETS is involved, because of the close scrutiny given to an ETS on review. See *Public Citizen Health Research Group v. Auchter*, 702 F.2d 1150, 1156–1157, 11 OSHC 1209, 1210 (D.C. Cir. 1983). A final new benzene standard was issued on September 11, 1987, 52 FED. REG. 34,460 (1987).

[367]These cases could plausibly be discussed together with those involving challenges to standards issued by OSHA. However, where an industry or industries are not covered, the issue becomes one of OSHA deciding not to regulate at all rather than a question of its regulating, but not sufficiently stringently. However, see the hazard communications case, discussed in text accompanying notes 368–372, *infra*, where the court distinguished between the two types of situations.

[368]*United Steelworkers v. Auchter*, 763 F.2d 728, 12 OSHC 1337 (3d Cir. 1985), discussed respecting other issues *supra* Section V.C.

[369]*Id.* at 737.

[370]27 U.S.C. §655(g).

a standard. The court rejected the argument.[371] While conceding that OSHA has discretion "to set priorities for the use of the agency's resources and to promulgate standards sequentially," the court insisted that once OSHA promulgates a standard, it may exclude a particular industry only if it shows not only that the industries covered present greater hazards but also "why it is not feasible for the same standard to be applied in other sectors, where workers are exposed to similar hazards." Since OSHA had failed to show why nonmanufacturing coverage could have "seriously impeded the rulemaking process," the court allowed the standard to go into effect in the manufacturing sector, but remanded the proceeding to OSHA to apply the standard to other sectors unless OSHA could "state reasons why such application would not be feasible."[372]

In response to the court remand, OSHA, on November 27, 1985, issued an advance notice of proposed rulemaking on the feasibility of expanding the standard.[373] OSHA took no further action, however, and the Steelworkers Union and Public Citizen Health Research Group moved to enforce the court's 1985 judgment. On May 29, 1987, the Third Circuit ordered OSHA, within 60 days, to publish an expanded hazard communications standard on the basis of the existing administrative record or to state reasons why it is not "feasible" to do so. The court asserted that its earlier judgment "did not contemplate going back to square one" and that some of the questions asked by OSHA in its advance notice "could not have been posed with the serious intention of obtaining meaningful information."[374]

---

[371]The court of appeals in reviewing the OSHA action relied on its earlier decision in the lead case approving OSHA's exemption of the construction industry from the lead standard. *United Steelworkers v. Marshall*, 647 F.2d 1189, 1309–1310, 8 OSHC 1810, 1900–1901 (D.C. Cir. 1980), *cert. denied*, 453 U.S. 913 (1981). The court of appeals there "implicitly rejected the contention that the Secretary's priority-setting authority is unreviewable. We do so explicitly." *United Steelworkers v. Auchter*, 763 F.2d at 783, 12 OSHC at 1344.

[372]*United Steelworkers v. Auchter, supra* note 368, at 739, 12 OSHC at 1345. See also *Forging Indus. v. Donovan*, 773 F.2d 1436, 12 OSHC 1472 (4th Cir. 1985), where the Fourth Circuit held that OSHA had not acted arbitrarily and capriciously in refusing to exempt the tree care industry from the noise standard. 773 F.2d at 1454, 12 OSHC at 1487. There have been many articles discussing the issue of court decisions compelling agency standards actions. See, e.g., *Note, Judicial Review of Agency Inaction*, 87 COLUM. L. REV. 627 (1983).

[373]50 FED. REG. 48,794 (1985).

[374]*United Steelworkers v. Pendergrass*, 819 F.2d 1263, 1268, 13 OSHC 1305, 1308–1310 (3d Cir. 1987). OSHA expanded the scope of the hazard communication standard on August 24, 1987, 52 FED. REG. 31,852 (1987).

* * * * *

Judge Harold Leventhal described the relationship between administrative agencies and reviewing courts as a "partnership."[375] This description was refined by Judge Friendly who, in an OSHA case, observed that the "partnership" was an "uneasy" one.[376] The 17 years of review of OSHA standards by the courts of appeals and the Supreme Court confirms the views of both these eminent jurists. This history also demonstrates that the courts are "senior" in the partnership— that while they sometimes constrain the agency and sometimes press the agency to action, they always have the "last" word on standards matters.

---

[375]*Greater Boston Television Corp. v. FCC*, 444 F.2d 841, 850 (D.C. Cir. 1970).

[376]*Associated Industries v. Department of Labor*, 487 F.2d 342, 354, 1 OSHC 1340, 1348 (2d Cir. 1973).

Appendix to Chapter 20

Court Decisions in the Major OSHA Standards Cases

Associated Industries of New York State v. Department of Labor, 487 F.2d 342, 1 OSHC 1340 (2d Cir. 1973) (vacating lavatory standard)

Dry Color Manufacturers Ass'n v. Department of Labor, 486 F.2d 98, 1 OSHC 1331 (3d Cir. 1973) (vacating lavatory standard)

Florida Peach Growers Ass'n v. Department of Labor, 489 F.2d 120, 1 OSHC 1472 (5th Cir. 1974) (vacating pesticide emergency temporary standard)

Industrial Union Department v. Hodgson, 499 F.2d 467, 1 OSHC 1631 (D.C. Cir. 1974) (affirming asbestos standard)

Synthetic Organic Chemical Manufacturers Ass'n v. Brennan (SOCMA I), 503 F.2d 1155, 2 OSHC 1159 (3d Cir. 1974), *cert. denied*, 420 U.S. 973 (1975) (affirming standard for ethyleneimine)

Synthetic Organic Chemical Manufacturers Association v. Brennan (SOCMA II), 506 F.2d 385, 2 OSHC 1402 (3d Cir. 1974), *cert. denied*, 423 U.S. 830 (1975) (vacating MOCA standard)

Society of the Plastics Industry v. OSHA, 509 F.2d 1301, 2 OSHC 1496 (2d Cir.), *cert. denied*, 421 U.S. 992 (1975) (affirming vinyl chloride standard)

AFL-CIO v. Brennan, 530 F.2d 109, 3 OSHC 1820 (3d Cir. 1975) (remanding to OSHA an amendment to no-hands-in-die standard)

American Iron & Steel Institute v. OSHA, 577 F.2d 825, 6 OSHC 1451 (3d Cir. 1978), *cert. dismissed*, 448 U.S. 917 (1980) (affirming coke oven standard)

Industrial Union Department v. American Petroleum Institute, 448 U.S. 607, 8 OSHC 1586 (1980), *affirming on other grounds, American Petroleum Institute v. OSHA*, 581 F.2d 493 (5th Cir. 1978) (vacating benzene standard)

Texas Independent Ginners Ass'n v. Marshall, 630 F.2d 398, 8 OSHC 2205 (5th Cir. 1980) (vacating cotton ginning standard)

United Steelworkers v. Marshall, 647 F.2d 1189, 8 OSHC 1810 (D.C. Cir. 1980), *cert. denied*, 453 U.S. 913 (1981) (lead standard affirmed in part, remanded in part)

American Textile Manufacturers Institute v. Donovan, 452 U.S. 490, 9 OSHC 1913 (1981), *affirming AFL-CIO v. Marshall*, 617 F.2d 636, 7 OSHC 1775 (D.C. Cir. 1979) (upholding cotton dust standard)

Public Citizen Health Research Group v. Auchter, 702 F.2d 1150 (D.C. Cir. 1983) (requiring OSHA to issue ethylene oxide proposal)

ASARCO Inc. v. OSHA, 746 F.2d 489, 11 OSHC 2217 (9th Cir. 1984) (affirming arsenic standard)

Asbestos Information Ass'n v. OSHA, 727 F.2d 415, 11 OSHC 1817 (5th Cir. 1984) (vacating OSHA's 1983 asbestos emergency standard)

United Steelworkers v. Auchter, 763 F.2d 728, 12 OSHC 1337 (3d Cir. 1985) (affirming in part and vacating in part hazard communications standard)

Public Citizen Health Research Group v. Tyson, 796 F.2d 1479, 12 OSHC 1905 (D.C. Cir. 1986) (affirming in part and vacating in part ethylene oxide standard)

Farmworker Justice Fund v. Brock, 811 F.2d 613, 13 OSHC 1049 (D.C. Cir. 1987) (ordering OSHA to issue field sanitation standard)

United Steelworkers v. Pendergrass, 819 F.2d 1263, 13 OSHC 1356 (3d Cir. 1987) (ordering OSHA to expand the scope of the hazard communication standard)

Public Citizen Health Research Group v. Brock, 823 F.2d 626, 13 OSHC 1362 (D.C. Cir. 1987) (setting a deadline for OSHA's acting on the ethylene oxide standard)

National Cottonseed Producers Ass'n v. Brock, 825 F.2d 482, 13 OSHC 1353 (D.C. Cir. 1987) (upholding the medical surveillance provision in cottonseed processing industry in amended cotton dust standard)

Part VI

Other Issues Arising Under and Related to the Occupational Safety and Health Act

21

Other Issues Related to Standards

I. Variances

A. Types of Variances

The OSH Act provides the Secretary of Labor discretion to grant an employer an exemption with respect to a standard where the employer cannot comply in a timely fashion, where there are reasons of national defense, or where the employer can demonstrate that alternatives will provide as safe and healthful a work environment as would exist through compliance with the standard.[1] This exemption is in the form of a variance which can be granted to defer or excuse compliance. Variances are granted in three forms: temporary, permanent, and for reasons of national defense.

1. Temporary Variances

A temporary variance permits an employer to operate in violation of a standard for a limited, definite period.[2] It does not excuse the employer's compliance, but rather permits an employer to continue its operation under specific conditions until such time as it is able to achieve full compliance.[3] Tem-

---

[1]29 U.S.C. §§655(b)(6)(A), 655(d), and 665 (1982).
[2]29 U.S.C. §665(b)(6)(A).
[3]*Id.*

637

porary variances are available only during the period between the time the standard is promulgated and its effective date. After the effective date of a standard, the employer must come into compliance, apply for a permanent variance, or be subject to citation. However, an employer that, having been cited, cannot comply within the abatement period set by the OSHA Area Director, may file a petition for modification of abatement with the Area Director who has the authority to extend the abatement period. Such petition must contain many of the same elements as a temporary variance application.[4]

An employer can apply for such a variance at any time after the promulgation of the standard. The application must show that its inability timely to comply with the standard is truly beyond its control for one of the following reasons: (1) there is a lack of available professional or technical personnel; (2) the necessary material or equipment needed to comply is unavailable; or (3) the necessary construction or alterations to its facility cannot be completed by the effective date of the standard.[5] The employer must have an effective and acceptable program to bring its operation into compliance as quickly as possible. As with all variances, the employer must also satisfy OSHA that until it can comply, it is using every possible means to protect its employees from the hazards that the standard was designed to address.

The duration of temporary variances is limited to the period of necessity shown by the employer, or for one year, whichever period is shorter. However, these variances can be renewed twice for up to 180 days per renewal period.

Temporary variances are also available to permit an employer to participate in research and experiments that will develop or validate new safety and health techniques.[6] The Secretary of Labor or the Secretary of Health and Human Services must approve the research or experiment. Such a variance is limited to that amount of time necessary for the research or experiment to be completed.

---

[4]See OSHA Instruction STD 6.1 (Oct. 30, 1978) at ¶4 (e). Authority to modify abatement periods was placed in the Area Director to avoid the need for employers to petition the Commission to extend abatement periods when compliance within the designated period becomes impossible. Telephone conversation with James J. Concannon, Director, Office of Variance Determination, OSHA, U.S. Dep't of Labor (Aug. 12, 1986).

[5]29 U.S.C. §655(b)(6)(A)(i)–(iii).

[6]29 U.S.C. §655(b)(6)(C).

2. Permanent Variances

Under certain circumstances, an employer may obtain a permanent variance. To qualify, the employer must prove that alternative practices or equipment will provide "employment and places of employment to his employees which are as safe and healthful as those which would prevail if he complied with the standard."[7] The variance, if granted, is narrow in scope and will specify the conditions the employer must meet and/or the alternative practices it must adopt to receive such protection.

Permanent variances, unlike temporary variances, are granted for an indefinite period. While they do not expire, they may be modified or revoked by the Secretary on his own motion or upon application by the employer or affected employees. Such application is made by following the same procedure as for obtaining the variance.[8] No application for modification or revocation of a permanent variance may be made until the variance has been in effect for 6 months.[9]

3. National Defense Variances

The Secretary is given broad latitude to "allow reasonable variations, tolerances and exemptions to and from any or all provisions of the Act" to "avoid serious impairment of the national defense."[10] The Secretary must grant such relief on the record and only after notice and an opportunity for a hearing.

B. Procedure for Obtaining Variances

1. Application

(a) General requirements. An employer initiates the process to obtain all three types of variances by filing an application with the Secretary at the U.S. Department of Labor, Washington, D.C. 20210.[11] However, variance applications

---

[7] 29 U.S.C. §655(d).

[8] See Section I.E, *infra.*

[9] 29 U.S.C. §655(d)

[10] 29 U.S.C. §665.

[11] Temporary—29 C.F.R. §1905.10(a) (1985); permanent—§1905.11(a); national defense—§1905.12(a)

should be made to the federal agency only (1) where the state in which the employment or place of employment is located has no approved state plan with provisions covering the subject area of the variance sought; or (2) where the variance sought will affect employment or places of employment in more than one state, e.g., industrywide variances or variances sought by multistate employers. (See discussion in Subsections (b) and (c) below.)

The application need not follow any particular form, but it must contain certain requisite information and representations. All applications must contain the name and address of the employer; the address of the place or places of employment involved in the application; the specific portions of the standard from which the variance is sought; a request for a hearing, if desired; and a statement describing how affected employees have been informed of the application and their right to petition for a hearing.[12]

In addition, an application for a temporary variance must contain representations that the employer is unable to comply with the standard or portion in question by the effective date, and the reasons for this inability. These representations must be supported by statements from qualified persons with first-hand knowledge of the facts contained in the application.[13] The employer must also outline the steps it is taking to protect employees from the hazards addressed by the standard until it can comply fully.[14] The application should also set up a timetable showing the steps the employer is taking to comply, the dates on which these steps have been or will be taken, and the ultimate date of compliance.[15]

An application for a permanent variance must contain the requisite information outlined above and a statement describing the practices and procedures to be used, with an explanation of how these practices and procedures will meet the requirement of providing as safe or healthful an employment or place of employment as would "prevail if [the employer] complied with the standard."[16]

---

[12]§§1905.10(b), 1905.11(b), and 1905.12(b).
[13]§1905.10(b)(4).
[14]§1905.10(b)(5).
[15]§1905.10(b)(6).
[16]§§1905.11(b)(3) and (4).

(b) State plans: Multistate employment. When the variance being applied for would affect employment or places of employment in more than one state, an employer's application should consider the state plans, if any, in the different states affected. A variance granted by the federal agency is a binding interpretation of any approved state occupational safety and health standard which is identical to the federal standard.[17]

Section 18 of the Act[18] allows a state to promulgate and enforce its own safety and health plan which provides for the development and enforcement of standards "at least as effective as" the federal standards. When a state demonstrates that its plan (a) meets the minimum criteria for certification under the Act;[19] (b) is formally approved by the Secretary; and (c) proves itself through minimum performance for at least 3 years, the federal standards promulgated under the Act will no longer apply with respect to any of the occupational safety and health issues covered by the state plan.[20] Though the Secretary no longer has jurisdiction to enforce the federal standards, he may retain jurisdiction in proceedings commenced prior to the effective date of the state plan.[21] Once a state plan has been approved, the state has an affirmative obligation to maintain its standards "at least as effective as" the comparable federal standard.[22] Thus, if a federal standard is changed

---

[17]§1905.13(c).

[18]29 U.S.C. §667.

[19]The criteria for approval of a state plan require a plan to (1) designate a state agency to administer and enforce the plan; (2) provide for development and enforcement of standards at least as effective as the federal standards; (3) provide for the right of entry and inspection of workplaces; (4) assure that the state agency has sufficient authority and personnel for enforcement; (5) assure the adequacy of funding; (6) assure the development of comparable standards applicable to all state public agencies; (7) require comprehensive reporting from employers to the state, and from the state to the Secretary. 29 U.S.C. §667(c) (1980). Having proposed a plan, a state must then act on its proposal, sharing authority for enforcement with the federal agency for a minimum of 3 years. If, at that time, the Secretary determines that the state plan is an effective one, final approval is given to the plan and the Secretary's enforcement responsibilities cease prospectively in all areas covered by the state plan. 29 U.S.C. §667(e); 29 C.F.R. §§1952.1 *et seq.* (1985). For further discussion of procedures relating to approval of state plans, see Chapter 23 (Role of States in Occupational Safety and Health).

[20]29 U.S.C. §667(e). In a move to strengthen the interrelationship between the federal and state occupational safety and health agencies, revisions were made in 1975 to the regulations governing procedures under the Act. These revisions require the Secretary to communicate with the state agencies regarding action that will, or potentially may, affect employment or places of employment within their jurisdiction. See 40 FED. REG. 25,449–450 (1975).

[21]29 U.S.C. §667(e).

[22]29 C.F.R. §§1953.20 *et seq.*

in any way, the state must make any changes necessary to remain "at least as effective."

Where a variance, if granted, would have an effect on employment or places of employment in more than one state, at least one of which has an approved state plan, the regulations require an employer to identify any state standard identical to the federal standard from which the variance is sought.[23] An employer then makes a side-by-side comparison of the federal and state provisions.[24] The term "identical" does not require that the two standards read exactly the same. Editorial or nonsubstantive variations from the federal standard are not sufficient to render a state standard not identical. Rather it is sufficient that a standard is "identical in substance and requirements."[25]

To avoid conflicting rulings on the same variance application, an employer must certify that no state agency has issued it a citation with respect to the standard in question, and that it has not applied to the state agency for a variance from the "identical" standard.[26] When the application is filed, the state agency is notified and given an opportunity to participate in the variance process by commenting and/or becoming a party.[27] By allowing full participation by the state agency and full disclosure of the identical nature of the provisions, variance proceedings affecting both state and federal standards can be consolidated into one action, the outcome of which is binding upon both.[28]

(c) State plans: Single-state employment. As discussed above, where a state has an approved plan, the federal agency surrenders enforcement jurisdiction in all areas covered by the plan. Thus if an employer is applying for a variance affecting only employment or places of employment within the state, it should apply to the state agency.

---

[23] 29 C.F.R. §§1905.10(b)(11) and 1905.11(b)(8).

[24] §§1905.10(b)(11)(i) and 1905.11(b)(8)(i).

[25] *Id.*

[26] §§1905.10(b)(11)(ii) and (iii); §§1905.11(b)(8)(ii) and (iii).

[27] §1905.14(b)(3).

[28] 40 FED. REG. 25,448 (1975). If a variance is granted with multistate applicability and a state with an "identical" standard was not given notice of an opportunity to comment, the federal variance is nonetheless an authoritative interpretation of the state standard once the employer has filed the required information with the state agency and the state does not object to it. 29 C.F.R. §1905.13(c).

2. *Interim Orders*

A variance application alone does not protect an employer from enforcement of a standard before the variance is granted. Therefore, to protect itself, an employer applying for any of the three types of variances must concurrently or subsequently apply for an interim order allowing it to operate in violation of the standard while its application is being acted upon.[29] By definition, the interim order only remains in effect from the time it is granted until it is revoked or a decision is rendered on the variance application.[30] An application for an interim order need not follow any particular form but must show why the order is justified. This application may be supported by relevant statements and arguments explaining why the interim order should be granted.[31]

Interim orders do not require a hearing and may be granted *ex parte* by the Secretary.[32] If the interim order is denied, the employer is given notice of the fact with a brief explanation of the reasons for the denial.[33] If the interim order is granted, notice of the terms of the interim order is served on the parties to the application and published in the *Federal Register*. It is a condition of the granting of all interim orders that affected employees be notified of the order in the same manner in which they were informed of the variance application.[34] A failure to comply with the terms and conditions of an interim order will result in a citation for violation of the underlying standard.[35]

---

[29]29 C.F.R. §§1905.10(c), 1905.11(c), and 1905.12(c). While the regulations provide that interim orders may be granted pending decisions on temporary, permanent, or national defense variances, the Act specifically mentions interim orders only pending temporary variance determinations. Compare 29 U.S.C. §§655(b)(6)(A), 655(d), and 665.

[30]29 C.F.R. §§1905.10(c)(1), 1905.11(c)(1), and 1905.12(c)(1).

[31]*Id.*

[32]*Id.*

[33]§§1905.10(c)(2), 1905.11(c)(2), and 1905.12(c)(2).

[34]§§1905.10(c)(3), 1905.11(c)(3), and 1905.12(c)(3).

[35]In *B.W. Drilling, Inc.*, [1975–1976] OSHD ¶20,551 (Rev. Comm'n, 1976), the Secretary granted an industrywide interim order allowing employers to violate oil derrick ladder requirements until action could be taken on a pending variance application. The Secretary's interim order specifically required employers to use certain "ladder safety devices" if they were in violation of the standard. B.W. Drilling neither complied with the standard nor used the safety devices required. The administrative law judge upheld a citation and penalties for violating the standard. Failure to meet the conditions of the interim order made it inapplicable to the employer who was thus subject to the requirements of the standard.

Although interim orders can be granted *ex parte*, without state agency participation, they are nonetheless binding on a state agency as a limited exception to any "identical" state standard for the term of the order.[36]

3. Department of Labor Action on Variance Applications

Variance applications that do not contain the necessary information discussed above are considered defective and will be denied by the Secretary. Where the application is defective, notice of denial is sent to the employer with a statement outlining the deficiencies. The employer is then free to file a second application.[37]

If a variance application contains the minimum information required, it is then processed through informal rulemaking procedures.[38] The Secretary first publishes notice of the application in the *Federal Register*; the notice includes details of the application and an invitation to interested persons to submit written comments during a specified period of time.[39] The notice also informs employers, employees, and any affected state agencies that they may request a hearing on the variance application.[40] In addition to publication in the *Federal Register*, the Secretary also sends a copy of the variance application to the affected state agency, if any, and extends an invitation to that agency to comment or participate as a party in the proceedings.[41] The employer must give spe-

---

[36]29 C.F.R. §1905.31(c).

[37]29 C.F.R. §§1905.14(a)(2)–(4).

[38]During passage of the Act, much congressional debate centered around whether action under the Act should be taken by informal or formal rulemaking procedures set forth in the Administrative Procedure Act. Under the formal rulemaking procedures, a hearing is required before any standard can be adopted, modified, or revoked. Congress ultimately decided that actions of the Secretary would be governed by the informal rulemaking procedures. Under these procedures, a hearing is necessary only if requested. H.R. REP. NO. 1765 (Conference Report), 91st Cong., 2d Sess. 34 (1970), *reprinted in* Subcommittee on Labor of the Senate Comm. on Labor and Public Welfare, 92d Cong., 1st Sess., LEGISLATIVE HISTORY OF THE OCCUPATIONAL SAFETY AND HEALTH ACT OF 1970, at 1187 (Comm. Print 1971).

[39]29 C.F.R. §§1905.14(b)(1) and (2).

[40]§1905.14(b)(2). An affected employer, employee, or state agency requests a hearing by following the procedures specified in §1905.15. For further information regarding hearing and comment procedures, see Chapter 19 (The Development of Occupational Safety and Health Standards).

[41]§1905.14(b)(3). The regulations also provide that the Secretary must promptly notify affected state agencies of the final dispositions of all variance applications affecting that agency. §1905.14(b)(4).

cific notice of the application to affected employees and details of the notice given should be included in the variance application.[42]

Before granting any variances, a preannounced inspection may be conducted by an OSHA representative or a representative of an affected state agency. The inspection is limited to those areas involved in the variance request.[43]

The OSHA representative cannot issue citations during this inspection. However, following the inspection, the representative will inform the employer or employees and the Area Director of any observed violations and the need to abate them. The Area Director then has the discretion to investigate these violations further.[44] If the variance is denied, the Area Director will conduct a full compliance inspection within 30 days of the denial.[45] Thus, while the inspection itself will not result in any citations, information about any violations observed will be communicated to enforcement authorities for possible action.

Variance inspections are discretionary in most circumstances. They are required for temporary and experimental variances, for variances where employees have objected, or where a firsthand look at the facility is necessary to make a ruling on the application. Variance inspections are also required for variance requests involving flammable and combustible liquids, toxic and carcinogenic substances, explosives, and electrical equipment.[46]

The time to process a variance application fully is estimated by OSHA to be 120 days: 30 days from receipt to publication, 30 days for public comment, and either 45 days to issue the final order for publication if no comments are received or 60 days to do so if there are comments.[47]

---

[42]§§1905.10(b), 1905.11(b), and 1905.12(b).

[43]OSHA Instruction STD 6.1 (Oct. 30, 1978) at ¶4(f). This instruction refers to the federal representative as an OSHA Variance Representative to distinguish his investigatory role in a variance inspection from a similar role performed in a traditional compliance inspection.

[44]*Id.*

[45]*Id.*

[46]*Id.* at ¶4(g).

[47]*Id.* at ¶4(a).

4. Hearings

Any affected employer, employee, or state agency can request a hearing during the comment period.[48] The Secretary can also hold a hearing on his own motion.[49] A hearing request must be filed with the Secretary in quadruplicate and must contain (1) a brief description of how the variance will affect the employer or employee requesting the hearing; (2) a brief summary of any statements in the variance application that are denied by the person requesting the hearing; (3) the evidence that could be produced at a hearing which would support that denial; and (4) a brief synopsis of views or arguments, either pro or con, that the person requesting the hearing would make on the issues or facts presented at the hearing.[50]

If no hearing request is filed, the Secretary will rule on the application and issue an order. If a hearing is requested, or if the Secretary holds a hearing on his own motion, he will serve notice upon all interested parties, including the applying employer(s), affected employees, all persons commenting on the application, and all affected state agencies.[51] This notice will contain the time, place, and nature of the hearing; the legal authority for holding the hearing; the issues to be considered; and the identity of the hearing examiner.[52]

The hearing is an adversarial proceeding at which the applicant has the burden of going first and proving by a "preponderance of reliable and probative evidence" the facts and circumstances contained in its application.[53] The hearing examiner is empowered with all authority "necessary or appropriate to conduct a fair, full and impartial hearing."[54] His authority is that given by the Act, the regulations, and the Administrative Procedure Act[55] to make all appropriate rul-

---

[48]29 C.F.R. §1905.15(a).

[49]§1905.20(a).

[50]§1905.15(b). By requiring the requesting party to outline the precise issues, arguments, and facts in dispute, the Secretary can better consider the merits of the hearing request and consolidate similar issues for more efficient consideration. See §1905.16.

[51]§1905.20.

[52]*Id.* As a practical matter, hearings are rarely held on variance applications.

[53]§§1905.26(a) and (b), §1905.27(b).

[54]§1905.22(a). Hearing examiners are administrative law judges from the same pool of judges who hear cases for other areas of the Department of Labor, including variance applications and standards review proceedings.

[55]5 U.S.C. §§551 *et seq.* (1980).

ings on evidence, discovery, and procedure.[56] Evidence presented at a hearing can be documentary or by oral or deposition testimony, with all parties having the right to cross-examine witnesses and object to evidence. Discovery, including oral depositions and written interrogatories, may be allowed upon application to the hearing examiner.[57] Objections to evidence are ruled upon by the hearing examiner on the record.[58]

Prior to the actual hearing, the hearing examiner holds a prehearing conference to simplify the proceedings where possible. At the conference, the hearing examiner encourages the parties to enter into stipulations, discuss the documentary evidence, and simplify the issues. He may also request the parties to consider limiting the number of parties or expert witnesses.[59] A record of the conference is made and the agreements reached govern the conduct of the hearing.[60]

At any time, the parties may, at the discretion of the hearing examiner, stop the proceedings to discuss settlement of all or part of the action. A consent settlement requires approval of the hearing examiner. In approving a settlement, the examiner will issue the terms of the agreement as his decision. It has the same effect as a decision after a full hearing. However, by consenting to settlement, the parties expressly waive any objections to the proceedings and any right to challenge or appeal the decision.[61]

If the parties do not reach a settlement, the hearing continues until all relevant evidence is received. Each party can offer proposed findings of fact and conclusions of law. Any party to the hearing has an option to submit a posthearing brief. The hearing examiner will issue his decision within a reasonable time.[62] The decision of the hearing examiner is final unless exceptions are filed, in which case the entire record is sent to the Secretary for determination.

---

[56]29 C.F.R. §§1905.22(a)(8) and (9).

[57]A party must make an application to the hearing examiner designating the time, place, and subject matter of a deposition. Each witness deposed must testify under oath, and all interested parties are given an opportunity to cross-examine the witness. Subject to objections made at the time the deposition is taken, any deposition may be read into the record or offered as evidence at the hearing. §1905.25.

[58]§1905.26(c)(3).

[59]§1905.23(a).

[60]§1905.23(b).

[61]§1905.24.

[62]§1905.27.

A losing party can file written exceptions with the hearing examiner within 20 days after receiving the decision.[63] If objections are filed, the hearing examiner will designate a period within which the prevailing party can file an opposition to the objections.[64] The entire record, including the transcript, briefs, objections, and oppositions, is then sent to the Secretary for review.[65]

In ruling on the exceptions, the Secretary has authority to adopt, modify, or set aside, in whole or in part, the findings and conclusions of the hearing examiner.[66]

The regulations provide for summary decision in variance application procedures. A party requesting summary decision should file a motion to that effect with the hearing examiner at least 20 days before the date set for the hearing. A party may support his motion with appropriate affidavits, although they are not required. Oppositions to the motion for summary decision are due within 10 days after service of the moving papers.[67]

A summary decision is appropriate where there exists no genuine issue of material fact which would require that the matter proceed to a hearing.[68] If the hearing examiner grants summary decision, his disposition becomes final in 20 days.[69] As with decisions made after a full hearing, any party may file exceptions to the summary decision within 20 days. The same procedure is followed for exceptions to summary decisions as for exceptions to decisions after a full hearing.[70]

There is no interlocutory appeal from a denial of a motion for summary decision. However, an exception to this rule permits a hearing examiner to certify that a case involves an important and controversial question of law which would justify an immediate appeal. With such certification, an interlocutory appeal can be taken directly to the Secretary for decision.[71] However, an immediate appeal to the Secretary

---

[63] §1905.28.
[64] *Id.*
[65] §1905.29.
[66] §1905.30.
[67] §1905.40(a).
[68] §1905.40(c).
[69] §1905.41(a)(i).
[70] *Id.* See §§1905.28 *et seq.*
[71] §1905.40(f).

does not automatically stay the underlying action and it will proceed unless the Secretary grants a request to the contrary.[72]

5. *Judicial Review; Renewals and Extensions*

Once a variance application has been processed to completion, the terms of the dispositions are published in the *Federal Register*.[73] Copies of all decisions are sent to affected state agencies.[74]

Neither the regulations nor the Act provides for judicial review of the Secretary's determination. However, such review is available under the Administrative Procedure Act as a final agency decision.[75]

Temporary variances can be renewed up to two times for 180-day periods. Variances for reasons of national defense may likewise be renewed after notice and an opportunity for hearing.[76] Applications for extensions or renewals are made to the Secretary. The procedure for granting an extension or renewal is the same as that for issuing the type of variance being extended or renewed.[77]

C. **Disposition of Variances**

As of June 30, 1986, there were 2,007 variance applications received by the Secretary. Of these, 2,002 were processed to completion and 5 were still in process. Of the applications processed, 10 percent of the requests for variances were granted, 10 percent denied, and 4 percent withdrawn. Sixty-four percent of the applications were treated as requests for clarification for which no variance was necessary. Three percent of the applications were treated as requests for modification or revocation of the standard. Seven percent of the applications were dismissed because the applicant failed to prosecute them to completion. Slightly over 1 percent were disposed of for miscellaneous reasons.[78]

---

[72]*Id.*

[73]29 U.S.C. §655(e); 29 C.F.R. §1905.6.

[74]29 C.F.R. §1905.14(b)(4). See Section I.B.1(b)–(c), *supra*, and note 27, *supra*.

[75]5 U.S.C. §704; 29 C.F.R. §1905.51.

[76]29 U.S.C. §665; 29 C.F.R. §1905.13(b).

[77]29 C.F.R. §1905.13(b).

[78]J. Concannon, QUARTERLY REPORT ON STATUS OF VARIANCE APPLICATIONS, OSHA, U.S. Dep't of Labor (July 1, 1986).

D. Effects of Variances

A variance applies only prospectively.[79] Thus, a variance does not affect pending proceedings concerning the standard. Rather, it defines new criteria to be applied specifically to the employer, group of employees, or industry to whom the variance has been granted. A variance does not alter the existing standard nor does it apply to any employers other than the parties whose application was granted.

Once granted, the variance, rather than the affected standard, prescribes the employer's conduct. However, if the employer deviates from the terms of the variance, the variance will not protect the employer from citation for violation of the standard. Thus, violation of the terms of a variance can result in the usual citations, penalties, and other discipline provided for by the Act.[80]

E. Modifying or Revoking Variances

Variances may be modified or revoked by application of the employer or an affected employee using the same procedure as that for applying for the variance. An application for modification or revocation of a variance must contain the name and address of the applicant, the type of relief sought, and the basis for seeking relief.[81] If the applicant is an employer, it must notify affected employees of the changes being proposed.[82] If the applicant is an employee, a copy of his application must be sent to his employer.[83] As with variance applications, an applicant may include a hearing request which should generally state how the applicant would be affected by the proposed action and what can be shown at a hearing.[84]

---

[79]29 C.F.R. §1905.5.

[80]See, e.g., *M.W. Kellogg Co., Chimney Div.*, 3 OSHC 1471 (Rev. Comm'n 1975) (employer cited and fined for willful violation of general duty clause because of failure to meet terms of permanent variance; penalty vacated owing to lack of employer knowledge of unsafe conditions).

[81]29 C.F.R. §1905.13.

[82]§1905.13(a)(1)(iv).

[83]§1905.13(a)(1)(v).

[84]§1905.13(a)(1)(vi).

The Secretary may also decide to modify or revoke a variance on his own motion.[85] To do so, the Secretary publishes notice of his intent to change or revoke it in the *Federal Register*. This notice invites interested parties to comment or request a hearing during a specified period. The Secretary also gives specific notice of his proposed changes to any affected employer(s) and employees.[86]

The procedures for modification or revocation of a variance require that notice be given to any affected states in the same manner as for variance applications.[87] Just as a variance, when granted, affects an identical state standard as a binding interpretation of that standard, a change in the variance similarly binds the state agency to the modified terms of the variance.[88]

F. Relationship Between Variances and Citations

Since variances, once granted, only excuse compliance with a standard prospectively, a variance subsequently granted will not excuse a citation already issued for violation of a standard.

The adjudication of citations and the granting of variances, depending as they do upon separate decision makers, could lead to potential differences in the separate decisions of the Review Commission and the Secretary. To avoid such conflicts, the regulations provide that the Secretary may refuse to entertain a variance request when a citation has been issued and a proceeding is pending before the Commission or a state agency relating to the same issues.[89] However, as a practical matter, the Secretary has issued variances where appeals proceedings were pending before the Commission.[90] Moreover,

---

[85]§1905.13(a)(2).

[86]*Id.*

[87]§1905.13(c). See Section I.B.1(b)–(c), *supra.*

[88]§1905.13(c).

[89]29 C.F.R. §1905.5.

[90]In the early days of the Act, the Secretary issued variances where appeals proceedings were pending before the Commission. Almost uniformly this was done where the standard being enforced was new and untested and the Secretary felt the employer's alternative procedures were at least as effective as the standard. Today, however, the standards are more clearly defined and tested. Thus, the Secretary is more likely to withhold a decision on a variance pending the outcome of the appeals proceedings. Telephone conversation with James J. Concannon, Director, Office of Variance Determination, OSHA, U.S. Dep't of Labor (Aug. 12, 1986).

the Commission is often willing to permit the variance process to resolve citations where abatement is not feasible. In such circumstances, the assessment of a penalty may not serve the purposes of the Act.

Since the Secretary serves a prosecuting role in the adjudication of citations, he may grant a variance to an employer faced with a citation, and then withdraw his opposition to the employer's citation contest. In *Star Textile & Research*,[91] while two citation contests were pending, the Secretary granted a variance request which effectively rendered the issues moot. The Secretary then moved to vacate the contested citations. Because the variance operated only prospectively, the Commission was not obligated to vacate the citations. However, the Commission treated the Secretary's action as a motion to withdraw and granted the motion.[92]

The Commission has extended the period for abatement where a variance application was pending before the Secretary. Thus, in *Deemer Steel Casting Co.*,[93] the Secretary admitted to the Commission that compliance with the standard was impossible. The Commission approved the administrative law judge's ruling affirming the fact of a violation, but postponed abatement until the employer could file for a variance. The Secretary argued that such a ruling violated the Secretary's variance powers, which should be separate from the Commission's adjudicatory function. However, the Commission stated:

"The Commission must issue an appropriate order following its adjudication. In doing so, it should not act so single-mindedly as to ignore a relevant variance proceeding pending before the Secretary. Under the circumstances of this case we are hard pressed to act concerning the abatement requirement because the Secretary has stipulated as to the extreme difficulty of abatement and because action in the variance proceeding may possibly change future methods of compliance by the employer. Accordingly, we will refrain from setting an abatement require-

---

[91]2 OSHC 1697 (Rev. Comm'n 1975).

[92]In *Lock Joint Pipe Co.*, [1975–1976] OSHD ¶19,898 (Rev. Comm'n J. 1975), the Secretary agreed to withdraw his opposition in a citation contest as part of a settlement that would obligate the employer to comply with interim orders until the Secretary could grant a variance.

[93]2 OSHC 1577 (Rev. Comm'n 1975).

ment until the Secretary acts upon the pending variance application."[94]

Other appeals have been settled by the employer's agreeing to apply for a variance. In *Midland Empire Packing Co.*,[95] the employer argued that it would violate Department of Agriculture standards if it complied with an OSHA standard to provide safety rails. As part of a settlement approved by the administrative law judge, the employer agreed to request a variance from the Secretary.[96]

While the Commission has approved of deferring to the variance process in certain cases, it has refused to reconsider a ruling in light of a subsequent variance application. In *George A. Hormel & Co.*,[97] the employer petitioned the Commission for reconsideration of its adverse decision. Meanwhile, the employer applied for a variance which, if granted, would render the workplace safer than would compliance with the standard. The employer argued that, pending determination of its variance application, it was in the precarious position that it must abate or face penalties; yet if the variance were granted, it would not have to comply with the standard at all. Rejecting the employer's arguments, the Commission denied reconsideration, finding the employer's position exaggerated. Anticipating that the Secretary would act expeditiously, the Commission reasoned that if the employer were cited for failing to timely abate, the totality of circumstances would be considered.

II. Revision of Standards

A. Rulemaking Procedures

The Secretary has power under 29 U.S.C. §655(b) to revise a standard. The procedure outlined in the Act for the revision

---

[94]*Id.* at 1578. See also *Ensign Elec. Div., Harvey Hubbell, Inc.*, 10 OSAHRC 711 (Rev. Comm'n J., No. 7638, Aug. 12, 1974), in which the administrative law judge ruled that an employer should not have to abate a violation while a variance application was pending. The administrative law judge vacated the citation and proposed penalties, and deferred consideration of the issues until 30 days after a ruling from the Secretary on a pending variance application.

[95]1 OSHC 3092 (Rev. Comm'n J. 1972).

[96]See also *Lock Joint Pipe Co., supra* note 92.

[97]2 OSHC 1190 (Rev. Comm'n 1974).

of standards is the same as that followed for their promulgation.[98]

The usual procedure for promulgation, modification, and revocation of standards is found at 29 C.F.R. §§1911.1–1911.18. The Secretary may, on his own motion, or in response to suggestions or requests submitted to him, propose a modification or revision of any standard.[99] However, as is discussed in Sections II.A.3 and II.A.4 below, the Secretary must follow special procedures for revising or modifying certain standards found in 29 C.F.R. Part 1910.[100] As is the case with the promulgation of a standard, the Secretary may appoint and consult with an advisory committee with respect to proposed revisions.

1. Notice and Comment

If the Secretary, on the basis of information provided by interested parties or an advisory committee, determines that a change should be proposed, notice of this proposed change is published in the *Federal Register* and a period of 30 days allowed for comments.[101] During this period, interested persons may express opinions or submit written objections to the Secretary and/or request a hearing. If objections are received and a hearing is requested, the Secretary must schedule a hearing.[102] Notices of hearings are published in the *Federal Register* stating the nature of the comments received and the time and place of the hearing. This notice is given within 30 days after the close of the comment period.[103] Where a hearing is held, the Secretary publishes notice of the revision in the *Federal Register* within 60 days from the date of the hearing.[104] If there is no hearing, the Secretary publishes notice

---

[98]See Chapter 19 (The Development of Occupational Safety and Health Standards).

[99]29 C.F.R. §1911.3 (1985) provides that any individual may submit a written petition for the promulgation, modification, or revocation of a standard. Such petitions should be addressed to Assistant Secretary of Labor, Occupational Safety and Health Administration, U.S. Department of Labor, Washington, D.C. 20210.

[100]These procedures, promulgated in 1976, require that the Secretary consult the public and certain advisory committees *before* proposing changes. 41 FED. REG. 17,100 (1976).

[101]29 U.S.C. §655(b)(2); 29 C.F.R. §1911.11(b).

[102]29 U.S.C. §655(b)(3); 29 C.F.R. §1911.11(d).

[103]29 C.F.R. §1911.11(d).

[104]29 U.S.C. §655(b)(4); 29 C.F.R. §1911.18.

of the revision within 60 days from the close of the comment period.

If the Secretary determines not to issue the proposed revision, notice to that effect must also be published in the *Federal Register* within the required time period. Where the Secretary intends not to change the standard, he may, in his notice, invite further data, views, or comments and postpone a final decision for up to 60 days.[105]

2. *Hearings*

Hearings to consider revising a standard are conducted in the same manner as those for the promulgation of a standard. Unlike hearings concerning the granting of a variance, hearings on modifications and revision are more informal in nature. These hearings are investigative hearings, presided over by a hearing examiner as provided for in the Administrative Procedure Act. Cross-examination is permitted where appropriate, and a complete record is kept.[106] At the conclusion of the hearing, the hearing examiner certifies the complete record with exhibits, transcript, written comments, and post-hearing comments to the Secretary for a final determination.[107]

Rules modifying or revising a standard are considered issued at the time they are filed in the office of the *Federal Register*.[108] Any person adversely affected may petition any U.S. court of appeals that has jurisdiction over the person or his place of business for review of the standard up to 60 days after its promulgation, modification, or revision.[109]

Determination of the Secretary will not be disturbed by a court of appeals if supported by substantial evidence on the record as a whole.[110]

---

[105]29 C.F.R. §1911.18(a)(2).

[106]5 U.S.C. §3105; 29 C.F.R. §1911.15.

[107]29 C.F.R. §1911.17.

[108]29 C.F.R. §1911.18(d); 42 FED. REG. 65,166 (1977). The date of issue becomes relevant for purposes of judicial review under 29 U.S.C. §655(f).

[109]29 U.S.C. §655(f).

[110]29 U.S.C. §655(f); 29 C.F.R. §1911.15(a)(2). See Chapter 20 (Judicial Review of Standards).

3. *Revision of Construction Industry Standards*

As noted, the Secretary must follow additional procedures when modifying or revising certain standards under 29 C.F.R. Part 1910.[111] (The standards which are prescribed in 29 C.F.R. Part 1926 are adopted or incorporated by Section 1910.12.) First, the regulations provide that when the Secretary seeks to modify or revise the Construction Industry Standards, 29 C.F.R. §1910.12, he must consult the Advisory Committee on Construction Safety and Health, established pursuant to Section 107 of the Contract Work Hours and Safety Standards Act. The Secretary is required to submit his proposals, together with all pertinent facts, research, experiments, and data, to the Advisory Committee for its recommendation. The Committee then submits its recommendations within a period prescribed by the Secretary, not to exceed 270 days. The Secretary can then offer his proposed changes by following the procedures outlined above.[112]

4. *Revision of Start-Up Standards*

Standards adopted as national consensus standards or established federal standards under Section 6(a) of the Act[113] were adopted as OSHA standards without following established rulemaking procedures.[114] Many of these standards proved to be inappropriate or insufficient to serve the purposes of the Act. In 1976, in response to congressional concern, the Secretary announced a procedure for considering revisions and modifications of problem standards. Under this procedure, the Secretary publishes a notice in the *Federal Register* that certain subparts of Part 1910 will be under consideration for possible revision or modification. This notice includes the portions of the standards under consideration and the nature of the comments received to date. The Secretary then schedules

---

[111]29 C.F.R. §1911.10(a).

[112]§§1911.10(a) and (b).

[113]29 U.S.C. §655(a).

[114]Under 29 U.S.C. §655(a), the Secretary was empowered to promulgate as a standard any national consensus standard or established federal standard without regard to the rulemaking procedures of the Administrative Procedure Act. The purpose was to establish a broad base of occupational safety and health standards as quickly as possible. See Chapter 3 (Specific Occupational Safety and Health Standards).

public hearings throughout the country to solicit opinions from interested persons on proposals to modify these standards. Written comments are also solicited. After conducting these hearings and considering all data and suggestions, the Secretary then proposes changes to the sections in question by publishing notice of the changes and allowing for comment or hearings according to normal procedures for the promulgation of modifications or revisions to standards.[115]

B. Revision Without Notice and Comment

The regulations also provide for the modification or revision of standards without notice and comment where the change is nonsubstantive in nature. These changes are limited to minor clarifications "in which the public is not particularly interested." To make these changes, the Secretary merely publishes the rule, notes the changes, and incorporates a statement indicating why notice and comment procedures were not followed.[116]

Whether minor changes are truly nonsubstantive in nature has been the subject of dispute. In *Island Steel & Welding*,[117] the employer objected to the Secretary's reissuance of 29 C.F.R. §1926.28 using procedures for nonsubstantive modifications. As the standard previously read, protective equipment was required where there was exposure to hazards "and" if other standards contained in Part 1926 indicated a need for the equipment. When the standard was reissued, OSHA changed the "and" to "or," thus arguably broadening the requirements of the standard. Island Steel objected that the standard was invalid because the Secretary had not followed rulemaking procedures. The Review Commission ruled that the standard was the same before and after the change.

In *Chlorine Institute v. OSHA*,[118] the Fifth Circuit upheld the Secretary's change in the chlorine exposure standard without notice and comment. The standard had been adopted in 1971 as a national consensus standard. Difficulty arose because there existed two standards from equally reliable sources

---

[115]41 FED. REG. 17,100 (1976).

[116]29 C.F.R. §1911.5.

[117]3 OSHC 1101 (Rev. Comm'n 1975).

[118]613 F.2d 120, 8 OSHC 1031 (5th Cir. 1980).

that could be considered national consensus standards. One standard required a one part per million (1 ppm) exposure as a time-weighted average (TWA). The other standard required a 1-ppm ceiling limit for exposure. As originally published, the standard adopted was the 1-ppm TWA. Seven years later, in 1978, the Secretary republished the standard with a 1-ppm ceiling limit exposure. The Chlorine Institute appealed to the court of appeals. The court chastised the agency for delay in publishing the modification, but upheld its action. Because the Secretary's guidelines indicated he intended all along to adopt the national standard that would give to employees the most protection, the court considered the initial promulgation of the less restrictive standard to be, as suggested by the secretary, a clerical error.

III. Administrative Interpretation of Standards

OSHA promulgates directives regarding internal administration and the interpretation and enforcement of standards. These directives are intended to provide guidance to agency personnel (see OSHA's Field Operations Manual).

These directives often substantively interpret standards by establishing methods of compliance. For example, OSHA Instruction STD 3-10.1B (August 14, 1979)[119] provides that laminated planking can be used on scaffolds even though the standard specifies "scaffold grade planking."

These directives also prescribe compliance procedures for OSHA personnel. For example, OSHA Instruction CPL 2.28 required that inspections involving exposure to lead be conducted under the guidance of a specially trained industrial hygienist.[120] And in OSHA Instruction CPL 2.73, OSHA administrators were instructed to emphasize inspections of fireworks manufacturers.[121]

OSHA also will issue definitive interpretations with respect to a particular standard in letters sent in response to inquiries from employers. Although these directives and let-

---

[119]21 OSHR 9111.

[120]This instruction has been canceled.

[121]21 OSHR 9444.

ters interpret standards or means of enforcement, they cannot be relied upon by employers. The Review Commission has consistently rejected attempts by employers to hold the agency to its own directives. The Commission has reasoned that these directives deal with the manner of enforcement and do not substantially affect an employer's liability.

Thus, in *GAF Corp.*,[122] a manufacturer of asbestos siding was found in violation of a standard requiring employers to provide medical examinations to workers exposed to airborne asbestos fibers even where exposure was only to concentrations below permissible limits. In its defense, the employer submitted letters from the Secretary informing two other firms that OSHA would not issue citations for exposure levels of less than 0.10 fibers per cubic centimeter—above the amount its employees were exposed to. The employer argued that the letters were binding on the agency as they defined levels of enforcement, and predated the alleged violation. The administrative law judge held that the standard could not be amended under the guise of interpretation, but may only be revised or modified through the rulemaking procedures set forth in the Act.

OSHA has also established internal enforcement procedures which, while they may result in broader application of a standard, are not subject to rulemaking procedures. Such are contained in the Field Operations Manual. In *Limbach Co.*,[123] the employer challenged OSHA's policy of citing all employers on a multiemployer site whose employees were exposed to any violation, although the employers cited may not be responsible for the violation or have any control over it. The employer argued the procedure was invalid because it had not been published as a proposed rule in the *Federal Register*, nor had it been the subject of public comment. The Commission rejected the employer's argument, stating that the Field Operations Manual "does not purport to, nor does it in fact or law, create liability on an employer" and thus does not contain standards or substantive rules subject to notice and comment procedures.

Just as OSHA creates its own enforcement guidelines without following rulemaking procedures, it is free to modify

---

[122]6 OSHC 1206 (Rev. Comm'n J. 1977).
[123]6 OSHC 1244 (Rev. Comm'n 1977).

or ignore these guidelines. Thus, in *FMC Corp.*,[124] the employer was cited for repeat violations of the housekeeping standard found at 29 C.F.R. §1916.51(a) (now 29 C.F.R. §1915.91). The Area Director did not follow the procedures found in the Field Operations Manual in that he did not consult the Assistant Regional Director for permission to issue a citation for a repeat violation. Thus, the employer argued, the Area Director exceeded his authority in issuing the citation. The Commission rejected this argument, holding that the guidelines in the Manual were for internal application to promote efficiency and "not to create an administrative straightjacket." Accordingly, the Commission held that the Area Director had not exceeded his authority by not following the procedures in the Manual.

IV. Stays of Standards

A person "adversely affected by a standard" may seek review of a standard or determination issued by the Secretary. First, he may petition the court of appeals within 60 days from the date the rule or determination is officially filed in the office of the *Federal Register*.[125] Second, most circuits recognize that an interested party or employer may contest a citation before the Review Commission and appeal to the courts, raising any argument, either substantive or procedural, even after the recognized 60-day review period expires.[126]

In conjunction with this review, or as part of the promulgation, modification, or revision procedure, a party may seek a stay of the standard pending a final determination by the court or the Secretary. A stay will postpone the effective date of the standard or will render it unenforceable during a limited period. A stay differs from a variance in that it affects the standard as applied to all employers, while a variance excuses compliance for a specific employer or group of employers.

The Secretary may issue an interim or permanent stay on his own motion in conjunction with a review of a standard.

---

[124]5 OSHC 1707 (Rev. Comm'n 1977).
[125]29 U.S.C. §655(f). See Chapter 20 (Judicial Review of Standards).
[126]See Chapter 20 (Judicial Review of Standards).

For example, the Secretary may issue a stay where comments received indicate that further investigation is necessary before a standard or modification of a standard is ultimately enforced. In one such case, the Secretary's proposed amendments to the occupational noise exposure standard were administratively stayed pending consideration of numerous comments and objections received by the Secretary. Before expiration of the stay, the Secretary lifted some of its aspects while allowing portions of the modification to go into effect. A one-month interim stay was issued as to other portions.[127]

A party seeking review before the court of appeals can request a stay pending the outcome of the court's consideration. A petition for a stay from a court must demonstrate:

(1) There is a substantial likelihood that the petitioner will prevail on the merits;

(2) The petitioner will suffer irreparable harm if the stay is denied;

(3) The issuance of the stay will not substantially harm other parties, if any, to the proceedings; and

(4) The issuance of the stay will not interfere with the public interest.[128]

If a stay is granted, it only delays enforcement of the standard and is not a decision or comment on the merits of the appeal. A stay will remain in effect as long as the court deems necessary.[129] It can be lifted or denied, in whole or in part, at any time and will not affect review of the standard.[130]

---

[127]46 FED. REG. 42,622 (1981). The final rule was published at 48 FED. REG. 9,738 (1983) (codified at 29 C.F.R. §§1910.95(c)–(p) and Appendices A–I).

[128]*Taylor Diving & Salvage Co. v. Department of Labor*, 537 F.2d 819, 821, 4 OSHC 1511, 1529 (5th Cir. 1976).

[129]*American Iron & Steel Inst. v. OSHA*, 577 F.2d 825, 6 OSHC 1451 (3d Cir. 1977).

[130]*American Iron & Steel Inst. v. OSHA*, 577 F.2d 825, 830, 6 OSHC 1451, 1454 (3d Cir. 1977). In *American Iron*, the Third Circuit issued an interim stay of four sections of the coke oven emissions standard pending consideration of arguments whether a permanent stay should issue. Although the court had not completed review of the standard, it later lifted the interim stay and denied a permanent stay, thus permitting the standard to become fully effective. See also *AFL-CIO v. Marshall*, 617 F.2d 636, 7 OSHC 1775 (D.C. Cir. 1979) (stay of cotton dust standard lifted, all petitions for rehearing denied, and standard to become effective 20 days from date of order).

The grant of a stay is discretionary with the court, which has the power to stay all or part of a standard or to delay the effective date of all or part of a standard, even after the standard has been held valid.[131]

---

[131]See *AFL-CIO v. Marshall, supra* note 130 at 677, 7 OSHC at 1802 (effective date of standard delayed 20 days from order upholding cotton dust standard); and *United Steelworkers v. Marshall,* 592 F.2d 693, 7 OSHC 1001 (D.C. Cir. 1979) (parts of standard requiring employer to incur expense of engineering controls partially stayed, but portions of standard providing substantive protections to employees during judicial review not stayed). See also *American Iron & Steel Inst. v. OSHA, supra* note 130 (only four sections of coke oven emission standard stayed).

22

Discrimination Against Employees Exercising Their Rights Under the Act

The OSH Act not only requires employers to provide their workers with a safe and hazard-free workplace, but also affords employees a broad range of rights. In particular, Section 11(c) of the Act prohibits retaliation against employees who exercise their rights under the Act.[1] This chapter outlines the safeguards provided to employees by Section 11(c); it also briefly discusses Sections 7 and 502 of the National Labor Relations Act[2] (NLRA) and the extent to which those provisions are consistent with—and to some degree overlap with—the protections provided by Section 11(c).

I. Coverage

The prohibitions contained in Section 11(c) are extremely broad, applying to "persons," not just employers. Section 3(4)

---

[1]Section 11(c)(1) provides that:
"No person shall discharge or in any manner discriminate against any employee because such employee has filed any complaint or instituted or caused to be instituted any proceeding under or related to this Act or has testified or is about to testify in any such proceeding or because of the exercise by such employee on behalf of himself or others of any right afforded by this Act."
29 U.S.C. §660(c) (1) (1973).
[2]29 U.S.C. §§158, 143 (1973).

of the Act defines "person" as "one or more individuals, partnerships, associations, corporations, business trusts, legal representatives, or any group of persons." As OSHA regulations make clear, unions, employment agencies, and "any other person in a position to discriminate against an employee" are covered by the prohibitions of Section 11(c). Furthermore, "the prohibitions of Section 11(c) are not limited to actions taken by employers against their own employees" since a person "may be chargeable with discriminatory action against an employee of another person."[3]

The class of persons protected by Section 11(c) is no less broad. Section 11(c) prohibits discrimination against any "employee." The term "employee" has been interpreted to include applicants for employment, former employees, supervisors, and employees of an employer other than the discriminator.[4] Moreover, the determination of whether someone is an employee "is to be based upon economic realities rather than upon common law doctrines and concepts."[5] However, consistent with the scope of the Act's coverage in general, Section 11(c) does not cover employees of state or local governments.[6]

A significant but yet unresolved issue is whether Section 11(c) protects employees who are employed in industries whose safety and health practices are regulated by federal agencies other than OSHA. Section 4(b)(1) of the Act provides that "[n]othing in this Act shall apply to working conditions of employees with respect to which other Federal agencies *** exercise statutory authority to prescribe or enforce standards or regulations affecting occupational safety or health." In *Marshall v. American Atomics*,[7] the court ruled that the language of Section 4(b)(1) does not preclude the Secretary from pursuing a Section 11(c) case on behalf of an employee in an industry in which OSHA is precluded from enforcing its safety and health regulations. The basis for the court's decision was that the limitation of Section 4(b)(1) applied only to the regulation of "working conditions," and not

---

[3]29 C.F.R. §1977.4 (1985).

[4]See *Mangus Firearms*, 3 OSHC 1214 (Rev. Comm'n J. 1975) (partner is employee); *Hayden Elec. Servs.*, 2 OSHC 3069 (Rev. Comm'n J. 1974) (supervisors are employees); 29 C.F.R. §1977.5(b).

[5]29 C.F.R. §1977.5(a).

[6]§1977.5(c).

[7]8 OSHC 1243 (D. Ariz. 1980).

to the protection of employees from retaliatory discrimination. However, in *Donovan v. Texaco*,[8] a federal district court in Texas reached the opposite conclusion, finding that Section 4(b)(1) precluded the Secretary from bringing suit for a retaliatory discharge on behalf of an employee who worked in an industry regulated by the Coast Guard. The court reasoned that the preemption of OSHA by Section 4(b)(1) meant that the employee's safety and health complaints were not related to the Act and therefore were not protected by Section 11(c).

II. Protected Conduct

A wide variety of employee conduct is protected by the Act. On its face, Section 11(c) expressly prohibits discrimination against employees who have filed complaints with OSHA, who have instituted or caused to be instituted proceedings under the Act, or who are about to testify or have testified in an OSHA proceeding.[9] More generally, Section 11(c) has been interpreted to protect employees who have taken virtually any action directly related to the safety of their working conditions. A wide variety of employee activities has been determined to fall within the ambit of the Act. These activities include participating in an OSHA inspection, either by acting as the employee walkaround representative or by providing information of any kind to the OSHA inspector; filing a notice of contest with regard to an abatement date under Section 10(c); participating in a variance proceeding under Section 6(d) or in a judicial challenge to a standard under Section 6(f); and filing complaints with governmental agencies or organizations other than OSHA.[10] In addition, numerous cases have held that employees are also protected when they complain

---

[8]535 F. Supp. 641, 10 OSHC 1532 (E.D. Tex. 1982).

[9]29 U.S.C. §660(c)(1). See *Marshall v. Montgomery Ward & Co.*, 7 OSHC 1049 (M.D. Fla. 1978); *Usery v. Granite Groves, A Joint Venture*, 5 OSHC 1935 (D.D.C. 1977); *Marshall v. Kennedy Tubular Prods.*, 5 OSHC 1467 (W.D. Pa. 1977).

[10]29 C.F.R. §§1977.9–.11. See *Donovan v. Peter Zimmer Am., Inc.*, 10 OSHC 1775 (D.S.C. 1982) (complaint filed with state department of labor); *Dunlop v. Hanover Shoe Farms*, 441 F. Supp. 385, 4 OSHC 1241 (E.D. Pa. 1976) (complaint filed with legal aid society).

to their employer about safety or health conditions.[11] To be protected, such employee complaints must be made in good faith; but employees are protected even if their concerns prove to be unwarranted.[12]

In *Donovan v. R.D. Andersen Construction Co.*,[13] a federal district court in Kansas further expanded the scope of the protections afforded employees by Section 11(c). The court ruled that an employer cannot retaliate against an employee for communicating concerns about workplace health hazards to a newspaper reporter. The employee was discharged after he admitted to his employer that he was the source of the quote that appeared in a newspaper article about asbestos dust at the employer's workplace. The court acknowledged that the situation in the case before it was unusual but it nonetheless rejected the defendant's argument that the employee's communication with the reporter was not a protected activity. Noting that employees are protected when they institute proceedings under the Act, the court reasoned that an employee is similarly protected as long as he engages in activity which could result in the institution of proceedings under the Act.[14]

Finally, it should be noted that Section 11(c) prohibits all forms of employment-related discrimination. It is illegal to reprimand, suspend, reduce the pay of, discharge, or refuse to hire anyone because he has engaged in activity protected by Section 11(c).[15] Perhaps the only major controversy with respect to this aspect of the law under Section 11(c) arose when the Secretary promulgated a regulation stating that an employer's failure to pay employees for time spent participating in an OSHA inspection would be considered illegal discrimination under Section 11(c). The Secretary's regulation was invalidated on procedural grounds in *Chamber of Commerce*

---

[11]E.g., *Marshall v. Klug & Smith Co.*, 7 OSHC 1162 (D.N.D. 1979); *Marshall v. Power City Elec.*, [1979] OSHD ¶23,947 (E.D. Wash. 1979); *Marshall v. P & Z Co.*, 6 OSHC 1587 (D.D.C. 1978), *aff'd*, 600 F.2d 280, 7 OSHC 1633 (D.C. Cir. 1979); *Marshall v. Wallace Bros. Mfg. Co.*, 7 OSHC 1022 (M.D. Pa. 1978); *Marshall v. Springville Poultry*, 445 F. Supp. 2, 5 OSHC 1761 (M.D. Pa. 1977); *Usery v. Granite Groves, A Joint Venture, supra* note 9.

[12]*Marshall v. Klug & Smith Co., supra* note 11.

[13]10 OSHC 2025 (D. Kan. 1982).

[14]Under certain limited circumstances, an employee's refusal to perform hazardous work may also be protected. See discussion in Section III, *infra*.

[15]*Marshall v. P & Z Co., supra* note 11; *Marshall v. Wallace Bros. Mfg. Co., supra* note 11.

v. OSHA.[16] In that case, the court concluded that the walk-around pay regulation was invalid because the agency had failed to comply with the notice and comment procedures for legislative rulemaking established under the Administrative Procedure Act.[17] Subsequent to the court's decision, the Secretary rescinded the walkaround pay regulation and has not yet instituted a rulemaking proceeding on any proposed regulation requiring that employees be paid for time spent on walkaround inspections.[18]

III. Employee Work Refusals Under Section 11(c)

A. OSHA Regulation

The Act does not expressly entitle employees to refuse to perform work because of a potentially hazardous condition at a workplace. However, in 1973 the Secretary promulgated a regulation providing that an employee has a right to refuse to work in certain situations.[19] Noting that the Act did not specifically provide employees with the right to refuse to perform hazardous work, the regulation observed that the overall enforcement scheme of the Act contemplated that in most situations employees would be able to correct hazardous conditions by bringing them to the attention of their employers or, if this fails, by requesting an inspection by OSHA pursuant to Section 8(f) of the Act.

While an employer generally will not violate Section 11(c) if it disciplines an employee who refuses to perform his normal job assignments because of alleged safety or health hazards, the regulation provides that an employee may be protected against discipline or discharge if he is "confronted with a choice between not performing assigned tasks or subjecting himself to serious injury or death arising from the hazardous condition

---

[16]636 F.2d 464, 8 OSHC 1648 (D.C. Cir. 1980).

[17]5 U.S.C. §553 (1977).

[18]Nothing in the Act prohibits an employer from agreeing to pay its employees for time spent assisting in an OSHA inspection, and a number of collective bargaining agreements do in fact provide for walkaround pay.

[19]29 C.F.R. §§1977.12(b)(1)–(2).

at the work place."[20] The regulation sets forth four criteria to be satisfied if the employee's work refusal is to be protected by Section 11(c): First, the employee's refusal to work must be made in good faith. Second, the condition which the employee believes to be hazardous must be of such a nature that a reasonable person, under the circumstances then confronting the employee, would conclude that there is a real danger of death or serious injury. Third, there must be insufficient time to eliminate the hazard through use of the Act's regular enforcement mechanisms. Fianlly, the employee, where possible, must have sought from his employer, and been unable to obtain, a correction of the dangerous condition.[21]

B. Supreme Court Test

In *Whirlpool Corp. v. Marshall,*[22] the Supreme Court unanimously rejected an employer's challenge to the validity of this regulation. In that case, two employees refused to work on an elevated wire mesh screen two weeks after another employee had fallen through the screen to his death. This refusal took place after the employees had filed an OSHA complaint and unsuccessfully voiced their concern over the safety of the elevated screen to members of management. In upholding the validity of the Secretary's regulation, the Court noted that the Act's basic orientation is remedial. It is "prophylactic in nature" and "does not wait for an employee to die or become injured," but authorizes the promulgation of standards and establishes an enforcement mechanism in order "to prevent deaths or injuries from ever occurring."[23] With this statutory purpose in mind, the Court concluded that it would be "anomalous to construe an Act so directed and constructed as prohibiting an employee, with no other reasonable alternative, the freedom to withdraw from a work place environment that he reasonably believes is highly dangerous."[24] In addition, the Court viewed a limited right of refusal to work as "an appropriate aid to the full effectuation of the Act's

---

[20]*Id.*
[21]*Id.*
[22]445 U.S. 1, 8 OSHC 1001 (1980).
[23]*Id.* at 12, 8 OSHC at 1005.
[24]*Id.*

'general duty' clause," which supplements specific OSHA standards by imposing on all employers a general duty to maintain the health and safety of employees.[25] The Court reasoned that because "OSHA inspectors cannot be present around the clock in every work place, the Secretary's regulation insures that employees will in all circumstances enjoy the rights afforded them by the general duty clause."[26]

Although the Court upheld the Secretary's regulation, it did emphasize that the refusal to perform assigned work was an extraordinary remedy. It also stated that Section 11(c) does not require an employee to be paid for any time that he is not on the job owing to his refusal to work. According to the Court, the Act's legislative history establishes that the drafters of the legislation did not intend the Act to be used to protect "strikes with pay."[27] The employer has a right to require the employee to perform alternative work if he refuses an assignment. However, an employee is entitled to back pay and/or reinstatement if, as a result of his refusal to work, he is suspended or discharged rather than given the opportunity to perform safe alternative work.[28]

Cases involving employee work refusals suggest that the courts have interpreted the requirement that the employee's fears be reasonable as calling for some objective evidence that the employee was in danger of death or serious injury.[29] Establishing that similar conditions have caused death or serious injury in the recent past is one way of providing such objective evidence.[30] In accordance with the criteria set forth in the Secretary's regulation, there must also be some proof of urgency indicating that there was insufficient time to eliminate the danger through the regular enforcement channels, although employees do not have to establish that they actually

---

[25]*Id.* at 12–13, 8 OSHC at 1005.

[26]*Id.* at 13, 8 OSHC at 1005.

[27]*Id.* at 17–19, 8 OSHC at 1006–1007.

[28]*Marshall v. Whirlpool Corp.*, 9 OSHC 1038 (N.D. Ohio 1980) (on remand from the Supreme Court).

[29]Compare *Marshall v. National Indus. Constructors*, 8 OSHC 1117 (D. Neb. 1980) (refusal to work unprotected since no objective evidence of danger and walkout motivated by desire for more money), with *Marshall v. Firestone Tire & Rubber Co.*, 8 OSHC 1637 (C.D. Ill. 1980) (objective evidence that 40-foot-high catwalk was ice covered and poorly lighted).

[30]See *Marshall v. N.L. Indus.*, 618 F.2d 1220, 8 OSHC 1166 (7th Cir. 1980) (employee stopped work since similar conditions had resulted in serious explosion one week earlier).

attempted to contact an OSHA office or file an OSHA complaint before refusing to work.[31]

IV. National Labor Relations Act

The protections provided employees by Section 11(c) of the Act parallel those provided by Sections 7 and 502 of the NLRA.

A. Section 7

Section 7 of the NLRA guarantees employees[32] "[t]he right *** to engage in *** concerted activities for the purpose of *** mutual aid or protection."[33] In exploring the scope of Section 7 in a safety work refusal context, the Supreme Court held, in *NLRB v. Washington Aluminum*,[34] that employees have a protected right to refuse to work when faced with unhealthy or unsafe working conditions, and that an employer violates the Act when it disciplines employees who engage in such work stoppages.

To be protected from discipline or discharge under the NLRA, employees do not have to establish through objective evidence that their belief that they were faced with unsafe working conditions was a reasonable one. Rather, they are protected as long as they had a good faith belief that their work assignments were dangerous. For instance, in *NLRB v. Tamara Foods*,[35] the court concluded that a walkout by 11 employees protesting exposure to ammonia fumes was protected activity even though the employer had complied with

---

[31]*Marshall v. Firestone Tire & Rubber Co., supra* note 29; *Marshall v. Seaward Constr. Co.*, 7 OSHC 1244 (D.N.H. 1979).

[32]Under the NLRA, supervisors, independent contractors, and agricultural and domestic workers are not considered to be "employees" and therefore are not entitled to the protections provided by the NLRA. 29 U.S.C. §152(3). Because of the differences between the OSH Act and the NLRA regarding the definition of "employee," Section 11(c) provides protection to many more workers than does the NLRA. See *Marshall v. Whirlpool Corp.*, 592 F.2d 715, 726 n.23, 7 OSHC 1075, 1082 (6th Cir. 1979), *aff'd*, 445 U.S. 1 (1980) (estimating that OSH Act covers 20 million more workers than does NLRA).

[33]29 U.S.C. §147.

[34]370 U.S. 9 (1962). In this case employees walked off the job after their employer repeatedly failed to repair a heating system.

[35]692 F.2d 1171 (8th Cir. 1982).

OSHA standards. The court observed that "[t]he rights guaranteed to employees under the National Labor Relations Act are distinct from and are not subordinate to the provisions of the Occupational Safety and Health Act."[36]

However, employee activity is protected under the NLRA only if it is concerted. There can be little question that when two or more employees join together and refuse to accept job assignments they believe are unsafe, their activity is likely to be deemed "concerted" for purposes of the NLRA. However, if only one employee refuses to perform hazardous work, it is, at best, unclear that his activity is "concerted." The National Labor Relations Board (NLRB) has adopted a broad interpretation of "concerted": where a single employee engages in safety-related activity, he is engaged in concerted activity unless his fellow employees have actually disavowed his actions.[37]

This rule, known as the *Interboro*[38] doctrine, was adopted by the Supreme Court in *NLRB v. City Disposal Systems.*[39] The Court held that an employee's reasonable and honest assertion of a right contained in a collective bargaining agreement is an extension of the concerted activity that produced the agreement and that the assertion of such a right affects the rights of all employees covered by the agreement.

While the protection provided by Section 7 of the NLRA is broader than that provided by Section 11(c) of the OSH Act in that the former does not require the employee to prove that his fears were reasonable, it is narrower in the sense that it only protects concerted activity.

B. Section 502

Employees can, of course, waive their right to engage in work stoppages. This is commonly done when, through a union, they enter into a collective bargaining agreement that contains a no-strike clause. In general, the presence of a no-strike clause in a contract affords employers the right to discipline or discharge employees who engage in work stoppages, for

[36]*Id.* at 1177.
[37]See, e.g., *Transport Serv. Co.*, 263 NLRB 910, 111 LRRM 1107 (Sept. 3, 1982); *Alleluia Cushion*, 221 NLRB 999 (1975).
[38]*Interboro Contractors*, 157 NLRB 1295, 61 LRRM 1537 (1966).
[39]465 U.S. 822, 115 LRRM 3193 (1984).

safety-related reasons or otherwise, without violating the NLRA.[40] However, Section 502 of the NLRA creates a limited exception to the employer's rights under a no-strike clause. That section provides that the "quitting of labor by an employee or employees in good faith because of abnormally dangerous conditions for work at a place of employment [shall not] be deemed a strike."[41] Thus, employees who refuse to engage in "abnormally dangerous" work do not breach a no-strike clause, and may not be disciplined or discharged for their actions. In *Gateway Coal Co. v. United Mine Workers*,[42] the Court decided that for an employee to enjoy the protections of Section 502, he must present "ascertainable, objective evidence supporting [his] conclusion that an abnormally dangerous condition *** exists."[43] As with Section 11(c), therefore, an employee who wishes to invoke Section 502 must support his refusal to work with some objective evidence that the assigned work was hazardous, However, in contrast to Section 11(c), Section 502 also requires a showing that the dangerous conditions were "abnormal." In interpreting this requirement, the NLRB has indicated that "work which is recognized and accepted *** as inherently dangerous does not become 'abnormally dangerous' merely because employee patience with prevailing conditions wears thin."[44]

C. OSHA and NLRB Memorandum of Understanding

To resolve some of the problems of overlap caused by the parallel statutory provisions, including the possibility of duplicative litigation, OSHA and the NLRB have entered into a memorandum of understanding. The memorandum provides

---

[40]The exact scope of a no-strike clause is, of course, a question of contract interpretation. The NLRB has, in a number of decisions, stated tht it will not readily infer a waiver of the right to strike over a particular subject matter from a broad, general no-strike clause. E.g., *Operating Eng'rs, Local 18 (Davis-McKee, Inc.)*, 238 NLRB 652, 99 LRRM 1307 (1978); *Gary Hobart Water Corp.*, 210 NLRB 742 (1974), *enforced*, 511 F.2d 284 (7th Cir.), *cert. denied*, 423 U.S. 925 (1975). Some courts of appeals have rejected this principle of contract interpretation, concluding instead that a general no-strike clause does operate as an absolute waiver of the right to strike. E.g., *Pacemaker Yacht Co. v. NLRB*, 663 F.2d 455, 108 LRRM 2817 (3d Cir. 1981).

[41]29 U.S.C. §143.

[42]414 U.S. 368, 85 LRRM 2049 (1974).

[43]*Id.* at 387, 85 LRRM at 2056.

[44]*Anaconda Aluminum Co.*, 197 NLRB 336, 344 (1972).

that when there is a charge filed with the NLRB that raises issues covered under Section 11(c) of the Act and a complaint involving the same factual matter has been filed with OSHA, the NLRB will defer or dismiss the charge it has received.[45] Where there has been no complaint filed with OSHA, the NLRB will inform the employee of his right to proceed under Section 11(c) of the Act.[46] When the charge falls within the exclusive jurisdiction of the NLRB, but nonetheless involves a question of possible discrimination because of safety-related activities, the memorandum calls for consultations between the two agencies.[47]

In addition, the Secretary has promulgated regulations which provide that if a complainant is pursuing his remedies before another agency, such as the NLRB, the Secretary, in his discretion, may postpone any determination of whether discrimination has occurred under the Act and/or defer to the outcome of the other agency's proceedings.[48] At least one court has ruled that OSHA could not hold a complaint in abeyance pending the outcome of a parallel NLRB proceeding. In *Newport News Shipbuilding & Dry Dock Co. v. Marshall*,[49] the court granted the employer's motion for summary judgment seeking to compel the Secretary to bring suit or dismiss the complaint. The court reasoned that nothing in the Act provides the Secretary with the authority to defer to another agency's proceedings.

V. Arbitration

In addition to seeking remedies under the NLRA and the OSH Act, employees covered by collective bargaining agreements may be able to avoid employer retaliation for refusal to work by pursuing relief under the grievance and arbitration provisions of such agreements. In the context of safety work

---

[45]Memorandum of Understanding Between OSHA and NLRB. 40 FED. REG. 26,083–26,084 (1975).

[46]*Id.* at 26,084.

[47]*Id.*

[48]29 C.F.R. §1977.18.

[49]8 OSHC 1393 (E.D. Va. 1980).

refusals, most arbitrators have recognized a limited exception to the general rule that employees should "work now, grieve later."[50] Thus, employees need not undertake assignments that will clearly endanger their health or safety. However, arbitrators are divided on what an employee must show in order to gain protection under the contract. Many arbitrators will only require an employee to show that he had a good faith belief that the work has hazardous,[51] while some arbitrators require an employee to present objective evidence that a safety hazard did in fact exist.[52]

The possibility of pursuing relief through a collective bargaining agreement raises the issue of the relationship between arbitration proceedings and proceedings under the OSH Act. As already indicated, the Secretary has promulgated regulations which provide that if an employee is pursuing his remedies before another government agency, the Secretary may postpone determination of whether to proceed with the complaint and/or defer to the outcome of the proceedings before the other agency. The same regulations provide that the Secretary may defer to an arbitration proceeding.[53] At least one court has placed the validity of these regulations in doubt, finding that the Secretary has no authority to defer to other proceedings.[54] Whether or not the Secretary can defer to an arbitration proceeding, if he so chooses, several courts have decided that the Secretary is in any case not bound by the results of a prior arbitration proceeding. For example, in *Marshall v. N.L. Industries*,[55] the Seventh Circuit reasoned that since "the OSHA legislation was intended to create a separate and general right of broad social importance existing beyond the perimeters of an individual labor agreement *** giving preclusive effect or even requiring total deference to an arbitrator's decision in this context would be inconsistent with

---

[50]In addition, some collective bargaining agreements expressly provide employees with the right to refuse work that endangers their health or safety. See, e.g., the collective bargaining agreement between The Gillette Co. and Oil, Chemical and Atomic Workers. (See COLLECTIVE BARGAINING, NEGOTIATIONS, AND CONTRACTS (BNA).

[51]E.g., *FMC Corp.*, 45 Lab. Arb. 293 (1965) (McCoy, Arb.).

[52]E.g., *Quaker Oats*, 69 Lab. Arb. 727 (1977) (Hunter, Arb.); *Morgan Eng'g Co.*, 77-1 Lab. Arb. Awards ¶8021 (1976) (Gibson, Arb.).

[53]29 C.F.R. §1977.18.

[54]*Newport News Shipbuilding & Dry Dock Co. v. Marshall, supra* note 49.

[55]618 F.2d 1220, 8 OSHC 1166 (7th Cir. 1980).

the statutory purpose."[56] The court concluded that the Secretary could proceed with the suit even though the employee had accepted the partial relief (reinstatement without back pay) provided to him by the arbitration award.

VI. Procedure

A. Litigation

The Act provides that complaints of alleged discrimination must be filed with the Secretary within 30 days of the alleged discriminatory act.[57] However, the Secretary has, by regulation, provided that this 30-day period may be tolled if the employer has concealed from the employee the grounds for the adverse action against the employee; if the employee has, within the 30-day period, filed a grievance under a collective bargaining agreement or a charge or complaint with another government agency; or for any other equitable reason.[58] In *Donovan v. Peter Zimmer America, Inc.*,[59] the court upheld the validity of this regulation, finding that the statutory time limit was tolled for employees who had filed complaints with a state agency within 30 days. In so doing, the court noted that the 30-day filing requirement was extremely short and concluded that it should not be regarded as a jurisdictional time limit, but rather as a condition precedent subject to equitable tolling. Similarly, in *Donovan v. Hahner, Foreman & Harness, Inc.*,[60] a federal district court concluded that the 30-day limit was tolled when an employer misled an employee into thinking he had been laid off rather than fired.

The Act also provides that the Secretary is to make a determination as to whether to proceed with the case within 90 days of the filing of the complaint.[61] This time period has

---

[56]*Id.* at 1222–1223, 8 OSHC at 1167.

[57]29 U.S.C. §660(c)(2). See *Usery v. Certified Welding Corp.*, 7 OSHC 1069 (10th Cir. 1980); *Powell v. Globe Indus.*, 431 F. Supp. 1096, 5 OSHC 1250 (W.D. Ohio 1977); *Usery v. Northern Tank Line*, 4 OSHC 1964 (D. Mont. 1976).

[58]29 C.F.R. §1977.15(d)(3).

[59]10 OSHC 1775 (D.S.C. 1982).

[60]11 OSHC 1081 (D. Kan. 1982), *aff'd*, 736 F.2d 1421, 11 OSHC 1977 (10th Cir. 1984).

[61]29 U.S.C. §660(c)(3).

been held to be directory, not mandatory.[62] Accordingly, the Secretary's failure to make a determination within the 90-day period is not grounds for dismissing a case which eventually may be filed in the matter. Furthermore, the statutory reference to the 90-day period does not require the Secretary to make a formal finding of merit within the 90-day period before filing suit in a U.S. district court.[63]

If the Secretary elects to proceed with a suit, it must be brought in federal district court. Only the Secretary is authorized to file suit as a plaintiff under Section 11(c); there is no private right of action.[64] Section 11(c) contains no express time limitation on the Secretary's right to file suit, nor have the courts implied any.[65] In addition, at least one court has ruled that the Secretary is not barred by the doctrine of laches since a Section 11(c) suit is designed to enforce a public right.[66] Since Section 11(c) cases are equitable in nature, there is no right to a jury trial.[67]

B. Burden of Proof

The Secretary has the burden of establishing a Section 11(c) violation. In order to meet his burden of proof, the Secretary must establish that the alleged discriminatee engaged in protected activity,[68] that the alleged discriminator had knowledge of the protected activity at the time the discriminatory action was initiated,[69] and that the discriminatory action was motivated by a desire to retaliate for the employee's protected activity. With respect to this last element, various courts have applied different tests for assessing the employer's

---

[62]*Donovan v. Freeway Constr. Co.*, 551 F. Supp. 869, 878 (D.R.I. 1982); *Dunlop v. Bechtel Power Co.*, 6 OSHC 1605 (M.D. La. 1977).

[63]*Marshall v. S.K. Williams Co.*, 6 OSHC 2193 (E.D. Wis. 1978); *Dunlop v. Hanover Shoe Farms*, 441 F. Supp. 385, 4 OSHC 1241 (E.D. Pa. 1976).

[64]*Taylor v. Brighton Corp.*, 616 F.2d 256, 8 OSHC 1010 (6th Cir. 1980); *Powell v. Globe Indus., supra* note 57.

[65]*Marshall v. Intermountain Elec.*, 614 F.2d 260, 7 OSHC 2149 (10th Cir. 1980); *Marshall v. American Atomics*, 8 OSHC 1243 (D. Ariz. 1980).

[66]*Dunlop v. Bechtel Power Co., supra* note 62.

[67]*Dunlop v. Hanover Shoe Farms, supra* note 63.

[68]*Marshall v. P & Z Co.*, 6 OSHC 1587 (D.D.C. 1978); *Dunlop v. Bechtel Power Co., supra* note 62; *Marshall v. Springville Poultry*, 445 F. Supp. 2, 5 OSHC 1761 (M.D. Pa. 1977).

[69]*Marshall v. Montgomery Ward & Co.*, 7 OSHC 1049 (M.D. Fla. 1978); *Marshall v. Dairyman's Creamery Ass'n*, 6 OSHC 2186 (D. Idaho 1978).

motive. The courts have held that the Secretary meets his burden of proof if he has shown that the employee's activity was "a substantial reason" for,[70] "an immediate cause" of,[71] or a "factor" in[72] the adverse action taken against him, or that "but for" the employee's protected activity he would not have been discharged or disciplined.[73] One court has decided that once the Secretary has established that the employee has engaged in protected activity with the knowledge of the employer, it is then up to the employer to establish by a preponderance of the evidence that it would have disciplined or discharged the employee in the absence of the protected conduct.[74] In view of the lengthy controversies that have surrounded burden of proof issues under antidiscrimination provisions in other statutes,[75] this area of Section 11(c) law is likely to remain unsettled for some time.

C. Remedies

There is a wide range of remedies available to the employee who has succeeded in showing that he has suffered unlawful discrimination. Where appropriate, employees may be granted reinstatement, back pay with interest, purging of disciplinary records, reimbursement of expenses incurred in seeking new work, and/or restoration of lost benefits and seniority rights.[76] In computing interest on back pay, the courts have generally adopted the formula utilized by the NLRB, which is tied to the Internal Revenue Service's "adjusted prime

---

[70]*Marshall v. P & Z Co., supra* note 68.

[71]*Dunlop v. Trumbull Asphalt Co.*, 4 OSHC 1847 (E.D. Mo. 1976).

[72]*Donovan v. Peter Zimmer Am. Inc.*, 10 OSHC 1775 (D.S.C. 1982).

[73]*Marshall v. Commonwealth Aquarium*, 469 F. Supp. 690, 7 OSHC 1387 (D. Mass.), *aff'd*, 611 F.2d 1, 7 OSHC 1970 (1st Cir. 1979).

[74]*Donovan v. Freeway Constr. Co.*, 551 F. Supp. 869, 878–879 (D.R.I. 1982).

[75]In *NLRB v. Transportation Mgt. Corp.*, 462 U.S. 393, 113 LRRM 3673 (1983), the Supreme Court adopted the Board's *Wright Line* test for allocating the General Counsel's burden of proof under §8(a)(3) of the NLRA. (See *Wright Line*, 251 NLRB 1083 (1980), *enforced*, 662 F.2d 899 (1st Cir. 1981), *cert. denied*, 455 U.S. 989 (1982).) The Court held that once the General Counsel makes a *prima facie* showing that protected activity was a motivating factor in the decision to discharge, the burden of proof is then properly placed on the employer to prove by a preponderance of the evidence that it would have fired the employee for permissible reasons even in the absence of the employee's protected union activities.

[76]*Donovan v. Peter Zimmer Am. Inc., supra* note 72.

rate."[77] Interim earnings are deductible from back pay awards, and employees are required to attempt to mitigate damages; but the employer bears the burden of proving that the employee unjustifiably failed to mitigate damages by finding interim employment.[78] The courts are divided on the question of whether unemployment compensation payments should be deducted from a back pay award.[79]

---

[77]E.g., *Donovan v. Hahner, Foreman & Harness, Inc.*, 11 OSHC 1081 (D. Kan. 1982), *aff'd*, 736 F.2d 1421, 11 OSHC 1977 (10th Cir. 1984); *Donovan v. Peter Zimmer Am. Inc., supra* note 72. For a description of the "adjusted prime rate," and the NLRB's rationale for using it to calculate interest, see *Olympic Medical Corp.*, 250 NLRB 146 (1980), *aff'd mem.*, 108 LRRM 3059 (9th Cir. 1981).

[78]E.g., *Donovan v. Peter Zimmer Am., Inc., supra* note 72.

[79]Compare *Donovan v. Commercial Sewing*, [1982] OSHD ¶26,268 (D. Conn. 1982) (unemployment compensation deductible) with *Donovan v. Peter Zimmer Am. Inc., supra* note 72 (unemployment compensation not deductible).

23

Role of States in Occupational Safety and Health

I. State Regulation Under the OSH Act

A. Legislative Background

In 1970, few states had comprehensive laws dealing with job safety and health, and fewer still had adequate programs to enforce them. The judgment of Congress that existing state laws were unable to provide meaningful protection to the nation's working men and women is clearly evident in the legislative history of the OSH Act. With some exceptions, state safety and health standards typically were incomplete and outdated; and such enforcement efforts as then existed were faulted as lax, intermittent, and ineffective.[1] Above all, sponsors of federal legislation criticized the lack of uniformity among state safety and health programs, which created competitive

---

[1]See, e.g., S. REP. No. 1282, 91st Cong., 2d Sess. 4, 21 (1970), *reprinted in* Subcommittee on Labor of the Senate Comm. on Labor and Public Welfare, 92d Cong., 1st Sess., LEGISLATIVE HISTORY OF THE OCCUPATIONAL SAFETY AND HEALTH ACT OF 1970, at 144, 161 (Comm. Print 1971) (hereinafter cited as LEGIS. HIST.); Senate debate on OSH Act of 1970 (Oct. 13, 1970)(statement of Sen. Saxbe), *reprinted in* LEGIS. HIST. at 343; Senate debate on OSH Act of 1970 (Nov. 16, 1970)(statements of Sen. Williams, Sen. Yarborough), *reprinted in* LEGIS. HIST. at 413, 444; Senate debate (Nov. 17, 1980)(statements of Sen. Muskie, Sen. Cranston), *reprinted in* LEGIS. HIST. at 513, 519; H.R. REP. No. 1291, 91st Cong., 2d Sess. 15, 31–32 (1970), *reprinted in* LEGIS. HIST. at 845, 861–862.

disadvantages and discouraged nationwide progress toward safer working conditions.[2] These congressional concerns were undeniably an important factor leading to the passage of the OSH Act.

Because, in the words of the Senate report, "the inadequacy of anything less than a comprehensive, nationwide approach has been exemplified by experience,"[3] Congress decided to make maximum use of its power to regulate commerce. All employers engaged in a business affecting commerce were covered, and state authority to regulate safety and health issues addressed by the OSH Act was preempted.[4]

But Congress did not intend, and the Act does not provide for, the complete federalization of occupational safety and heatlh enforcement. Although Congress recognized the shortcomings of contemporary state safety and health programs, the history and text of the OSH Act repeatedly indicate that the states should have an opportunity to particiapte in achieving the Act's far-reaching, ambitious goals.[5] Specifically, the stated objective of assuring safe and healthful conditions for working men and women was to be reached, in part, by "encouraging the States to assume the fullest responsibility for the administration and enforcement of State occupational safety and health laws" by means of federal grants and approved state plans.[6]

Congress had practical reasons for encouraging state participation. Congress understood that it would take time to promulgate new federal standards and create a program for their enforcement. Permitting the state safety and health agencies to continue to function would provide workers at least

---

[2]*Brennan v. OSHRC (John J. Gordon Co.)*, 492 F.2d 1027, 1 OSHC 1580 (2d Cir. 1974), and legislative history cited therein at 1030.

[3]S. REP. NO. 1282, *supra* note 1, at 4, LEGIS. HIST. at 144.

[4]Section 18(a) of the Act, which provides that "[n]othing in this Act shall prevent any State agency or court from asserting jurisdiction under State law over any occupational safety or health issue with respect to which no standard is in effect under section 6," has been read as a statement of congressional intent to prohibit state regulation of any issue addressed by an OSHA standard. See 29 C.F.R. §1901.2; *Five Migrant Farmworkers v. Hoffman*, 345 A.2d 378, 380 (N.J. Super. Ct. 1975); *Columbus Coated Fabrics v. Industrial Comm'n*, 1 OSHC 1361 (S.D. Ohio 1973). On the other hand, state laws concerning *public* safety and health are not preempted. See *Township of Greenwich v. Mobil Oil Corp.*, 504 F. Supp. 1275, 9 OSHC 1337 (D.N.J. 1981).

[5]See, e.g., Senate debate on the OSH Act (Oct. 13, 1970), *reprinted in* LEGIS. HIST. at 339–341.

[6]Section 2(b)(11).

some protection during the interim period. Moreover, there were some states that had administered effective programs prior to passage of the Act. Concern was expressed during the legislative debates that these services not be lost to the overall safety and health effort.[7]

B. Section 18 of the OSH Act[8]

The new federal law incorporated two mechanisms for utilizing state resources. First, to avoid the immediate displacement of state enforcement programs during the transition from state to federal jurisdiction, the Secretary was authorized by Section 18(h) to enter into temporary agreements permitting state enforcement under pre-OSHA laws and procedures. Authority for these agreements expired two years after the effective date of the Act.[9]

Far more significant and enduring, however, were the provisions for state plans contained in Section 18(b) through (g) of the Act. The state plan concept set forth in the Act is, in essence, a scheme of administratively controlled preemption. While the Act generally preempts state enforcement once the federal government regulates, Section 18(b) provides that states desiring to regain responsibility for the development and enforcement of safety and health standards under state law may do so by submitting, and obtaining federal approval of, a state plan which meets the stringent requirements set forth in Section 18(c) (see discussion of these requirements in Section III below). Among the most important conditions that the plan must satisfy are those spelled out in Sections 18(c)(4) and (5):[10]

> "(4) [The plan must] contain[ ] satisfactory assurances that [the state] agency or agencies have or will have the legal authority and qualified personnel necessary for the enforcement of *** standards, [and]

---

[7]E.g., Senate debate, note 5, *supra*; House debate (Nov. 24, 1970)(statement of Rep. Hathaway), *reprinted in* LEGIS. HIST. at 1067. See *AFL-CIO v. Marshall*, 570 F.2d 1030, 1037, 6 OSHC 1257, 1260 (D.C. Cir. 1978).

[8]29 U.S.C. §667 (1970).

[9]See *Industrial Union Dep't, AFL-CIO v. Hodgson*, 499 F.2d 467, 1 OSHC 1631 (D.D.C. 1973).

[10]See discussion in Section III.C below.

"(5) give[ ] satisfactory assurances that [the] State will devote adequate funds to the administration and enforcement of *** standards[.]"

Approval of a state plan by OSHA operates to permit the state to reenter the field of occupational safety and health regulation.

In practical terms, Section 18 is a charter for federal-state cooperation enabling OSHA to assist the states whose plans meet the statutory criteria and, in addition, to be assisted by them. Thus, OSHA provides grants to states under Section 23(g) of the Act[11] as well as technical assistance, training, and other services. Congress's premise was that OSHA's continuing review of state programs to assure that they measure up to the requirements set forth in the Act would induce agencies in most states to develop far more comprehensive and up-to-date programs than would otherwise have been possible. In turn, the state's participation would assist OSHA in, for example, carrying out its enforcement responsibilities under the Act.

Approved state plans play a major role in today's occupational safety and health enforcement efforts. Of the 56 jurisdictions eligible to submit plans, 25 presently operate approved occupational safety and health programs.[12] Forty percent of the workers covered by OSHA are covered by these approved plans, and well over half of the nation's safety and health compliance officers are currently employed by these state agencies.[13] State plans also play an essential role in providing safety and health protection for state and local government employees who, by operation of Section 3(5) of the Act, are excluded from federal OSHA coverage.[14]

---

[11]The federal grants to state programs must be matched by funds from the states.

[12]The definition of "state" set forth in Section 3(7) of the OSH Act includes the 50 states of the United States, the District of Columbia, Puerto Rico, the Virgin Islands, American Samoa, Guam, and the Trust Territories of the Pacific. Approved plans are listed and generally described in 29 C.F.R. pt. 1952. The state plans of Connecticut and New York differ from the remaining 23 (as of July 1, 1986) in that they cover only employees of the state and its political subdivisions (see note 22, *infra*); private sector standards and enforcement are provided by OSHA.

[13]Estimate of covered workers based on Bureau of Labor Statistics, U.S. Dep't of Labor, STATE AND METROPOLITAN AREA EMPLOYMENT AND UNEMPLOYMENT, No. 84–66 (1983).

[14]States wishing to enforce standards in the private sector must maintain an effective safety and health enforcement program for employees of the state and its political subdivisions as a condition of obtaining state plan approval. 29 C.F.R. pt. 1956.

II. Legal Overview of State Plans

In Section 18 Congress adopted an innovative regulatory approach authorizing the joint exercise of authority by state and federal agencies in the safety and health area. Unlike many other federal laws, the OSH Act does not afford the states complete freedom to enact supplementary or complementary requirements in areas addressed by federal standards.[15] On the other hand, the Act does not completely exclude state regulation by totally "occupying the field" of occupational safety and health. As discussed previously, Section 18(b) permits states to remain active in safety and health enforcement through federally approved state plans. The state plans concept set forth in the Act has a number of legally significant features, which warrant brief discussion.

A. Enforcement Authority

First, federal enforcement authority is not delegated to the states. Instead, Section 18 permits states with federally approved plans to enforce state standards under the authority of state law, using state administrative and judicial procedures.[16] For this reason, judicial and Review Commission decisions in cases arising under the OSH Act are not binding upon state courts in cases arising under an approved plan. Nor are state safety and health agencies which administer approved plans necessarily limited by statutory restrictions imposed upon OSHA.[17] However, because state safety and health laws which were enacted to provide the basis for a state plan are generally *in pari materia* with the OSH Act (and

---

[15]Congress rejected a proposal that would have preempted only those state regulations that were "in conflict" with the federal legislation. S. 2788 and H.R. 13373, 91st Cong., 1st Sess. §14(b)(1) (1969), *reprinted in* LEGIS. HIST. at 58, 706.

[16]Section 18(c)(4) requires that a state submitting a plan for approval must provide "the legal authority *** necessary for the enforcement of *** standards," and Section 2(b)(11) encourages states, through the state plan process, to "assume the fullest responsibility for the administration and enforcement of their occupational safety and health laws."

[17]See, e.g., *Green Mountain Power Corp. v. Commissioner of Labor & Indus.*, 383 A.2d 1046, 1051, 6 OSHC 1499, 1502 (Vt. 1978) (federal precedent persuasive but not conclusive); 42 FED. REG. 5356 (1977) (Department of Labor interpretation of appropriations riders limiting OSHA penalties and coverage as nonbinding on states); *United Air Lines v. Occupational Safety & Health Appeals Bd.*, 654 P.2d 157 (1982) (states not bound by §4(b)(1) of OSH Act).

indeed in many cases the two are nearly identical), comparisons are hard to avoid. Accordingly, federal precedents are commonly relied upon in state proceedings. Nevertheless, state courts are free to develop their own interpretations of state law, and significant departures from federal precedent sometimes result.[18]

B. Plan Approval

Next, federal approval of a plan is not a single regulatory event; rather, it occurs in stages.[19] Initial approval of a plan under Section 18(b) is granted if, upon submission of the plan, OSHA finds that the plan meets or will meet the criteria set forth in Section 18(c) and its implementing regulations. During the period of initial approval, OSHA may exercise its enforcement authority concurrently with the state, with employers subject to inspection and citation by either authority.[20] Initial approval is the most significant step in the plan approval process, as it removes the preemption barrier and permits the state both to commence enforcement operations and to receive federal matching grants.

Because the Act expressly permits approval of plans that "meet or will meet" the statutory criteria, OSHA can and does approve plans which are not yet completely developed.[21] These so-called "developmental plans" are assigned a three-year

---

[18]From time to time state court decisions affecting state standards or enforcement procedures may have the effect of rendering the state's plan less effective than the federal law, in which event the state must correct the deficiency by further appeal or by seeking regulatory or legislative changes. 29 C.F.R. §§1902.37(b)(5) and (14).

[19]See note 22, *infra*.

[20]The exercise of this concurrent federal authority is discretionary, because the Act stipulates that the Secretary "may, but shall not be required to" conduct enforcement in an initially approved plan state. However, pursuant to 29 C.F.R. §1954.3, OSHA may decide during this period to enter into a procedural agreement, with a state whose plan meets certain criteria, to the effect that federal enforcement will not be conducted in the state with respect to issues covered by the state. These agreements do not, as a matter of law, deprive OSHA of jurisdiction to inspect worksites or issue citations, since by statute that jurisdiction exists until OSHA issues a final approval determination. But as a practical matter OSHA would be estopped from issuing citations contrary to the terms of its agreements; valid operational status agreements may be raised as an affirmative defense by a cited employer. See 47 FED. REG. 25,323 (1982) (agreements with 10 state plan states); *General Motors Corp., Central Foundry Div.*, 8 OSHC 1298 (Rev. Comm'n 1980); *Willamette Iron & Steel Co.*, 9 OSHC 1900 (Rev. Comm'n 1981); *General Motors Corp., Chevrolet Motor Div.*, 10 OSHC 1293 (Rev. Comm'n 1982).

[21]See *Robinson Pipe Cleaning Co. v. Department of Labor & Indus.*, 2 OSHC 1114 (D.N.J. 1974).

timetable for completing various specified steps. The completion of the developmental process is marked by OSHA certification of the plan.

After a state plan is initially approved or, in the case of a developmental plan, certified, the Secretary is required to evaluate actual state operations under the plan to determine whether the criteria in Section 18(c) are being met (see discussion in Section IV below); if they are, final approval under Section 18(e) is granted and OSHA's concurrent enforcement authority in the state is withdrawn.[22]

OSHA retains authority to monitor an approved plan at all times. If after final approval a state fails to meet its commitments under the plan, final approval can be reconsidered and concurrent federal enforcement reinstated, in effect relegating the state to initial approval status. If the failure is substantial, federal approval of the plan must be withdrawn in accordance with the procedures set forth in Section 18(f). The consequence will be to reinstate OSHA's exclusive enforcement authority.[23]

III. Criteria for State Plans

A. Standards

A state plan must provide for the development of state standards "at least as effective" as corresponding federal standards in providing safe and healthful employment and places of employment. The adoption of standards identical to those

---

[22]Procedures and detailed criteria for initial approval, certification, and final approval of state plans are set forth in 29 C.F.R. pt. 1902. 29 C.F.R. pt. 1952 contains subparts for each state with an approved plan setting forth the approval status of the plan. Of 25 existing state plans which have received initial approval (including New York's and Connecticut's, which cover only state employees), 11 received final approval as of July 1, 1986. See, e.g., Virgin Islands (49 FED. REG. 16,766 (1984)); Alaska (49 FED. REG. 38,252 (1984)); Kentucky (50 FED. REG. 24,884 (1985)). The state plans of Arizona, Iowa, Maryland, Minnesota, Tennessee, Utah, Wyoming, and Hawaii also have been fully approved.

[23]Procedures for reconsideration of final approval determinations are found at 29 C.F.R. §§1902.47–.53; procedures for withdrawal of plan approval are set forth in 29 C.F.R. pt. 1955. Following final approval OSHA retains its authority over safety and health issues not covered by the approved plan, such as shipyards or marine terminals in states that have elected not to cover maritime activities. OSHA also retains its authority to carry out provisions of the Act not listed in §18(e), notably §11(c).

of OSHA will, of course, satisfy this criterion. In fact, the great majority of state standards under approved plans are identical to their federal counterparts; usually this is a matter of state choice, but sometimes it occurs because state legislation facilitates or requires incorporation by reference.[24] Where states elect to develop their own standards, OSHA must evaluate the effectiveness of the state's rulemaking procedure as well as the substantive content of the resulting standards.[25]

State standards that differ from corresponding federal standards must provide equal or better protection than their federal counterparts; moreover, the state must assure that subsequent judicial and administrative interpretations of a state standard do not compromise its effectiveness.[26] For example, in ruling on requests for variances from state standards, states must provide notice to interested parties and an opportunity to participate in hearings; the showing necessary to obtain a state variance must be equivalent to that required federally.[27]

The "at least as effective" provision clearly implies that a state that has an approved plan may adopt standards that are more effective than corresponding federal standards dealing with the same issues.[28] The principal limitation on the standards-setting authority of states with an approved plan is found in the so-called "product standard limitation" or "product clause" of Section 18(c)(2). This provides that state standards, when applicable to products distributed or used in interstate commerce, must be required by compelling conditions and must not unduly burden interstate commerce. This

---

[24]See e.g., NEV. REV. STAT. §618.295.8 (1957) (OSHA standard "shall be deemed Nevada occupational safety and health standards").

[25]29 C.F.R. §1902.4(b) (criteria for state standards adoption procedures).

[26]§§1902.37(b)(4)–(7).

[27]§1902.4(b)(2)(iv) and §§1902.37(b)(6) and (7). Employers who maintain workplaces in several states, including at least one in a state plan state, should consult the multistate variance and variance reciprocity provisions outlined in 29 C.F.R. pt. 1905 and 29 C.F.R. §1952.9. See discussion in Chapter 21 (Other Issues Related to Standards), Section I.

[28]References in the legislative history of the OSH Act also support this conclusion. One Senate report characterized federal safety and health standards as "providing a floor *** which the States can build upon." LEGIS. HIST. at 297. It should be noted that Section 18(b) of the Act permits states to adopt more effective standards only through the vehicle of an approved state plan. Early proposals which would have continued in effect any state standard affording "significantly greater protection" than the federal standard were rejected by Congress. S. 2788 and H.R. 13373, *supra* note 15, §14(b)(2).

two-part test echoes the language of many judicial decisions involving preemption of "burdensome" state regulation under the Commerce Clause of the U.S. Constitution.[29] However, review of a state standard under the product clause is a matter over which the Secretary, not the courts, has primary jurisdiction. At least one court has declined to order declaratory relief from a "more stringent" state standard on exhaustion grounds, i.e., that OSHA review of the state standard had not yet been completed.[30] Moreover, objections that a standard is unduly burdensome or unjustified by local conditions can be raised only in narrow circumstances, i.e., where the more stringent standard is "applicable to products."

The scope of the product standard limitation is difficult to define with precision. As ordinarily used, "products" is an inclusive term which potentially applies to a wide variety of state safety and health requirements. On the other hand, statements in the legislative history of Section 18(c)(2) suggest a more restrictive application.[31]

Finally, each state is required to supplement or revise its standards in response to changes in the federal program and regulations so that its program remains at least as effective as OSHA's. State action in response to required federal program changes must generally be taken within 6 months.[32]

B. Enforcement

Section 18(c)(2) requires state plans to provide enforcement of state standards at least as effective as federal enforcement. The remaining subparagraphs of Section 18(c) contain criteria which emphasize the importance of state pro-

---

[29]See, e.g., *Raymond Motor Transport v. Rice*, 434 U.S. 429 (1978); *Pike v. Bruce Church, Inc.*, 397 U.S. 137 (1970).

[30]*Florida Citrus Packers v. California*, 549 F. Supp. 213, 214, 10 OSHC 2048, 2049 (N.D. Cal. 1982).OSHA has interpreted the product clause as generally not applicable to "optional parts or additions to products." 29 C.F.R §1902.3(c)(2). The agency has usually determined whether state standards are applicable to products on a standard-by-standard basis. 48 FED. REG. 8610 (1983) (California EDB standard applicable to products); 48 FED. REG. 53,280 (1983) (hazard communication requirements applicable to products).

[31]Senate debate (Nov. 17, 1970), in LEGIS. HIST. at 500–501; House debate (Nov. 23, 1970), in LEGIS. HIST. at 1041–1042.

[32]29 C.F.R. §§1953. 20–23. Response time for emergency temporary standards is 30 days (§1953.22(a)(1)).

visions that adequately provide for right of entry and inspection, funding and staffing, and state reporting requirements.[33]

Enforcement procedures generally parallel those used in the federal system. Adoption of a procedural manual resembling the OSHA Field Operations Manual has been a required developmental step in virtually all states, as has been the promulgation by the states of administrative regulations comparable to federal regulations governing such matters as inspections, citations, penalties, review procedures, and variances. Approved plans must, *inter alia*, assure response by the state to employee complaints of unsafe working conditions, and provide opportunities equivalent to those afforded by OSHA for employee participation during inspections and in proceedings involving contested citations, imminent danger, and variances.[34] OSHA regulations also require states to guarantee employees protection from discrimination equivalent to that provided in Section 11(c) of the OSH Act, which is designed to protect employees' exercise of rights under the Act.[35]

The right of state safety and health agencies to enter and inspect all workplaces covered by the plan is explicitly addressed in Section 18(c)(3). It requires state inspection rights to be at least as effective as those available to OSHA under Section 8 of the Act. The Fourth Amendment limitations on OSHA inspections enunciated by the U.S. Supreme Court in *Marshall v. Barlow's, Inc.*[36] apply equally to state inspection. Numerous state courts have issued decisions consistent with *Barlow's*, albeit based upon state as well as federal constitutional requirements.[37] But Section 18(c)(3) places the responsibility upon states to assure that state law limitations on entry are no greater than those imposed upon OSHA by the

---

[33]Implementing regulations in 29 C.F.R. §§1902.3 and 1902.4 set forth more detailed criteria for state plan enforcement.

[34]29 C.F.R. §1902.4(c)(2).

[35]§1902.4(c)(2)(v). As a matter of discretion, OSHA will usually refer complaints of discrimination in state plan states to the agency which administers the plan. 29 C.F.R. §1977.23. However, OSHA retains statutory authority to enforce Section 11(c) in state plan states even after final approval. OSH Act, §18(e); 29 CFR §1954.3(e).

[36]436 U.S. 307, 6 OSHC 1571 (1978).

[37]State court decisions predating *Barlow's* included *Yocum v. Burnette Tractor Co.*, 6 OSHC 1638 (Ky. 1978); *Oregon v. Keith Mfg. Co.*, 6 OSHC 1043 (Ore. Ct. App. 1977) (applying *Camara*, 387 U.S. 523 (1967), and *See*, 387 U.S. 541 (1967), to workplace inspections by states). State court decisions following *Barlow's* include *Indiana v. Komomo Tube Co.*, 10 OSHC 1159 (Ind. Ct. App. 1981), and cases cited therein at 10 OSHC 1164.

federal courts. For this reason, state judicial or administrative decisions on such matters as the showing of "probable cause" sufficient for issuance of a warrant, the availability of *ex parte* or anticipatory warrants, and application of the exclusionary rule all have a bearing on continuing federal approval of the plan.

C. State Staffing—"Benchmarks"

As noted, the OSH Act requires state plans to provide an enforcement program which is "at least as effective as" that of OSHA, a phrase which Congress repeated several times in Section 18(c). The "at least as effective as" standard is not, however, contained in Sections 18(c)(4) and (5), which declare that the Secretary is to approve a plan only if it provides in absolute terms "necessary" qualified personnel and "adequate" funds to carry out the all-important enforcement functions.

Prior to 1978, OSHA had developed numerical staffing requirements or "benchmarks" for each state plan. These benchmarks, in effect, required the state only to have an enforcement staff equivalent to the number of compliance officers OSHA would deploy in that state if no plan had been approved. These early, "at least as effective as" benchmarks were challenged by the AFL-CIO.[38] The AFL-CIO's position was that the benchmarks employed by the Secretary were predicated on federal enforcement levels that were artificially low because the Secretary did not mount a meaningful federal enforcement effort after the OSH Act was passed and, instead, deliberately withheld commitment of adequate resources until he knew the full extent of likely state participation. According to the AFL-CIO, the Secretary's benchmark system produced a result that is directly at odds with the core purpose of the Act. Congress had found that in 1970, 1,600 state and fewer than 100 federal inspectors employed by a handful of agencies constituted a "critically short supply."[39] Yet, using the Secretary's benchmarks, the total number of enforcement personnel for the nation as a whole, including all the states and territories covered by the Act, would be 1,111—a *decrease*

[38]*AFL-CIO v. Marshall*, 570 F.2d 1030, 6 OSHC 1257 (D.C. Cir. 1978).
[39]S. REP. No. 1282, *supra* note 1, at 21, *reprinted in* LEGIS. HIST. at 161.

from the number employed in 1970. The Secretary maintained that given congressional concern over possible state dislocation produced by precipitous federal intervention, he properly refrained from requesting large appropriations that would have been necessary to finance a significant increase in the federal enforcement effort.

The suit culminated in a decision by the District of Columbia Circuit, *AFL-CIO v. Marshall*,[40] reversing the district court's grant of summary judgment in favor of the Secretary. After considering the text, legislative history, and purpose of the OSH Act, the court of appeals concluded (1) that in referring to personnel and funding levels in terms of adequacy and sufficiency, Congress clearly intended that the states assure effective enforcement programs, i.e., that the states would have the resources "necessary to do the job"; and (2) that in granting initial approval to state plans, the Secretary can consider—as interim federal benchmarks—whether the state is willing to provide personnel and funding "at least as effective as" that provided by the federal government; but that "such interim federal benchmarks must be part of an articulated, coherent program calculated to achieve a fully effective program at some point in the forseeable future."[41] Judge Leventhal, speaking for a unanimous panel, went on to explain that the Secretary's regulations and program directives establishing the benchmarks failed to meet that test. The case was remanded with instructions that the Secretary establish criteria (through supplements to the then existing state plan approval process) to be incorporated into a plan that would achieve a fully effective enforcement effort over a reasonable time period.

The district court, on remand, issued an order on December 5, 1978, directing the Secretary to develop a 5-year schedule for each state to meet the "fully effective" enforcement staffing levels, taking into account certain specified factors in each state such as the number of employers and employees; the number of hazardous industries; the number of general schedule inspections that should be conducted; the anticipated number of accident, complaint, and follow-up inspections required; and the number of inspections a compliance officer can

---

[40]*Supra* note 38.
[41]570 F.2d at 1036, 6 OSHC at 1259.

perform.[42] The order also provided that adherence by states to the benchmark levels of the 5-year schedule was a prerequisite of final plan approval under Section 18(c) of the Act. Finally, the order directed the Secretary to develop procedures for revising established benchmarks in light of new data and information.

Implementing the court's order generated as much controversy as did the Secretary's initial regulations and benchmarks. In response to the order, on April 25, 1980, the Secretary submitted a comprehensive "Report to the Court" which, among other things, set forth revised benchmark levels for safety inspectors and industrial hygienists for each state seeking final approval of its plan and a 5-year schedule for achieving those levels (contingent upon the available funds). Those benchmarks would have required the states to more than double their existing staff levels.[43] For that reasons the AFL-CIO concurred in the report. But Congress did not appropriate any additional funds for the federal share of the required increase in state staffing, and those benchmark levels were not put into effect. Instead, in August 1983, OSHA staff personnel, together with representatives of various state plans, launched another review of the benchmark formula which culminated in a major revision of the 1980 benchmarks, substantially reducing the number of safety inspectors and industrial hygienists who would have to be in place in each state before that state would qualify for "final approval." And, on the basis of these revised benchmarks, the Secretary has granted final approval to numerous state plans over the past few years.[44]

IV. OSHA Evaluation of State Plans

OSHA evaluation of state plans begins during the early stages of the plan approval process. During this phase, eval-

---

[42]*AFL-CIO v. Marshall*, 6 OSHC 2128 (D.D.C. 1978).

[43]*Supra* note 38, 570 F.2d at 1033, 6 OSHC at 1257.

[44]One state (with *amicus* support from two others) unsuccessfully brought suit for declaratory and injunctive relief from the 1980 benchmarks. *Baliles v. Donovan*, 549 F. Supp. 661, 10 OSHC 2043 (W.D. Va. 1982). In a 1982 appropriations measure, Congress briefly authorized the Secretary to reformulate benchmarks to provide for state staffing equivalent to federal staffing. Pub. L. 97-257 (1982); see 47 FED. REG. 50,307 (1982). This language has not been included in subsequent appropriations measures. Accordingly, the "fully effective" test announced by the District of Columbia Circuit and set forth in the district court's 1978 remand order remains in effect.

uation consists, in large measure, of comparing state and federal legislation, standards, regulations, and written procedures. This review continues throughout the developmental period and, to some extent, throughout the life of the plan as states develop and submit new standards or enforcement provisions, either in response to changes by OSHA or on their own initiative.[45] The focus of evaluation shifts, however, after a state plan receives initial approval (or after certification, in the case of a developmental plan). At this point, the Secretary's attention is directed by Section 18(e) to evaluating actual operations by the state under its plan.[46]

Regulations outlining OSHA's monitoring responsibilities are found in 29 C.F.R. Part 1954. A more comprehensive and detailed description of OSHA's monitoring procedures has been incorporated in OSHA's State Plans Policies and Procedures Manual.[47] The key concept in evaluating state program performance, like state standards, is the "at least as effective as" test required by Section 18(c) of the Act and 29 C.F.R. Part 1902.[48]

The present OSHA monitoring system (see note 48) does not eliminate the need for informed professional judgments

---

[45]See generally 29 C.F.R. pt. 1953, dealing with changes to state plans.

[46]See also Section 18(f), which provides that "[t]he Secretary shall, on the basis of reports submitted by the State agency and his own inspections make a continuing evaluation of the manner in which each State *** is carrying out [its] plan."

[47]Chapters I, VII, VIII, IX, X, and XI of the Manual were published by the Office of State Programs as OSHA Instruction STP 2.22 (Aug. 26, 1983). These procedures for measuring and reporting state performance are to be supplemented by chapters dealing with initial approval, certification, technical assistance to states, plan changes, and other topics.

[48]OSHA's present monitoring system, as developed in 1981 and 1982 by task groups of federal and state representatives and incorporated in the monitoring chapters of the new State Plans Manual (SPM), is based on over 100 "activities measures," each of which measures a particular aspect of state performance and relates that performance to comparable federal data. For each measured activity there is a "further review level," which sets a statistical range of performance in relation to current federal experience which, if met, will generally result in no further OSHA review of that particular aspect of state performance. The further review levels are not "pass/fail" levels but indicators by OSHA of the state performance levels above or below federal experience (the SPM refers to these statistical variances as "outliers"), which may indicate a need for improvement in the program area being measured. On the other hand, detailed inquiry may show that the outlier reflects only a difference in the design of the state's program which can be adequately explained and does not detract from the overall effectiveness of the plan; indeed, in many cases outliers may reflect superior performance by the state during the monitoring period. Review and evaluation of state-submitted management information may, when necessary, be supplemented by on-site monitoring such as OSHA review of specific state case files, or by monitoring visits to worksites either in conjunction with the state or as a spot-check inspection following a state compliance inspection.

by OSHA personnel in rendering an overall evaluation of each state, but is intended to allow these judgments to be made in a more objective and consistent manner. The results of OSHA monitoring are set forth in periodic evaluation reports for each approved state plan, which can be obtained from the OSHA Office of State Programs or from the appropriate OSHA regional office.

V. Public Participation in State Plans

The existence of state plans has created an opportunity in many states for groups of employees, employers, or other interested parties to comment upon or participate in the development of state standards, regulations, and procedures addressing local occupational safety and health concerns.[49] Moreover, every state plan must guarantee, as a condition of federal approval, rights of employer and employee participation in enforcement proceedings which are equivalent to those available under the OSH Act.[50] In addition, on the federal level there are opportunities for public comments or participation in the administrative decisions OSHA must make in approving a state plan.[51] Similar opportunities are available in federal proceedings to reject a submitted plan;[52] to reconsider a final approval decision under Section 18(e);[53] or to withdraw approval of a state plan under Section 18(f).[54]

While federal approval of a particular state standard is ordinarily a routine matter, state standards (or other plan changes) that generate substantial public interest or controversy are sometimes accompanied by a request from OSHA for public comment prior to a decision on whether to federally approve or reject the standard.[55]

---

[49]29 C.F.R. §§1902.4(b)(2)(iii) and (iv).

[50]§1902.4(c)(2)(x).

[51]29 C.F.R. §§1902.11(d) and (f). These provisions afford an opportunity to submit data, views, arguments, or requests for hearings concerning initial and final approval of specific state plans.

[52]§§1902.17–.19.

[53]§1902.49(c).

[54]29 C.F.R. pt. 1955 (1986).

[55]E.g., 47 FED. REG. 36,449 (1982) (requesting comments on federal approval of a state-initiated voluntary compliance program for small employers); 46 FED. REG. 57,060 (1981) (request for comments on state EDB standard); 42 FED. REG. 56,812 (1977) (proposed rejection of state scaffolding standard).

Finally, any interested person or group may submit a complaint concerning the operation or administration of any aspect of a state plan.[56] Such a complaint—designated a "CASPA"—should be filed with the appropriate OSHA Regional Administrator. He is empowered to investigate the complaint using information supplied by the complainant as well as other sources of information, such as worksite inspections; review of state files; and interviews with members of the public, employers, employees, or state personnel. The state must be notified of the filing of the complaint, and both the complainant and the state must be notified of the results. The regulations provide that the complainant's name, and the names of any other complainants mentioned, shall not appear in any record published or released.

---

[56]29 C.F.R. §§1954.20–.22.

24

National Institute for Occupational Safety and Health

Commonly characterized as OSHA's "research arm,"[1] the National Institute for Occupational Safety and Health (NIOSH) plays an integral part in the congressionally conceived plan to improve conditions affecting the safety and health of America's workers. This chapter reviews NIOSH's statutory authority, the programs carried out under that authority, and the courts' oversight of those programs.[2]

I. Statutory Authority

A. The OSH Act

NIOSH was established under Section 22 of the OSH Act[3] and assigned the task of carrying out the responsibilities of the Secretary of Health, Education, and Welfare (HEW, now

---

[1]See, e.g., *Industrial Union Dep't, AFL-CIO v. American Petroleum Inst.*, 448 U.S. 607, 619, 8 OSHC 1586, 1589 (1980); *United States v. McGee Indus.*, 439 F. Supp. 296, 297, 5 OSHC 1562, 1563 (E.D. Pa. 1977), *aff'd mem.*, 568 F.2d 771, 6 OSHC 1540 (3d Cir. 1978).

[2]NIOSH is also discussed in Chapter 5 (Employer Obligations to Obtain, Maintain, and Disseminate Information), Section VII.

[3]29 U.S.C. §671 (1976).

Health and Human Services or HHS) under the Act.[4] The legislative history of the OSH Act demonstrates Congress's conviction that a competent and objective research program was necessary to support the new regulatory scheme.[5] Congress also recognized the need to assign research responsibilities to an agency separate from the regulatory agency[6] and to one with demonstrated expertise in occupational health research. The obvious candidate was the Department of Health, Education, and Welfare, which had had a modest occupational safety and health research program directed by the Bureau of Occupational Safety and Health (BOSH). BOSH later became the core of the newly created institute.

The idea of a new institute was introduced by the Republican senator from New York, Jacob Javits, as an amendment to S. 2193, the bill that eventually became law. The purpose of Javits' amendment was to

> "elevate the status of occupational health and safety research to place it on an equal footing with the research conducted by HEW into other matters of vital social concern ***. Such an institute will be able to attract the qualified personnel necessary to engage in occupational health and safety research if we are to make any real progress in reducing job-related injury and disease under the Act, and will much more easily attract the substantial increase in funding which will be necessary to achieve the purposes of this act."[7]

Congress recognized the inadequacy of past and then existing research to provide solutions to the problems it sought to address with its new legislation. It was concerned that there was insufficient information regarding known hazards; but it was even more concerned about the unknown potential hazards of "complex, often synergistic, interactions of numerous physical and chemical agents" which were being introduced into the workplace at a rapid pace.[8] As a result, Congress gave

---

[4]29 U.S.C. §§671(a) and (c). "Secretary of HEW" and "NIOSH" are used interchangeably throughout this chapter since NIOSH is designated to carry out most of the Secretary's functions under the Act.

[5]See S. REP. NO. 1282, 91st Cong., 2d Sess. at 19 (1970), and H.R. REP. NO. 1291, 91st Cong., 2d Sess. at 27 (1970).

[6]See, e.g., H.R. REP. NO. 1291, *supra* note 5, at 27 ("The public interest orientation of governmental research efforts should be aided by the statutory definition of both Secretaries' [Labor and HEW] responsibilities. This should also help to assure freedom from even a theoretical conflict of responsibilities").

[7]S. REP. NO. 1282, *supra* note 5, at 57.

[8]*Id.* at 19.

NIOSH a broad range of discretionary authority to conduct research relating to occupational safety and health, including "studies of psychological factors" and exploration of "innovative methods, techniques, and approaches" for solving safety and health problems;[9] investigation of "new problems, including those created by new technology in occupational safety and health" and of "motivational and behavioral factors;"[10] and "industrywide studies on the effect of chronic or low-level exposure to industrial materials, processes, and stresses" on aging adults.[11]

The statute also gives NIOSH specific responsibilities that relate to its support role for OSHA standard-setting activities. These include

- production and annual publication of criteria to enable the Secretary of Labor to formulate standards;[12]
- development and establishment of recommended occupational safety and health standards;[13]
- publication and maintenance of a list of all known toxic substances and the concentration at which toxicity is known to occur. This is commonly known as the RTECS or Registry of Toxic Effects of Chemical Substances;[14]
- investigations to determine toxicity of materials at concentrations used or found in the workplace. These investigations are called Health Hazard Evaluations (HHEs).[15]

---

[9]29 U.S.C. §669(a)(1).

[10]§669(a)(4).

[11]§669(a)(7).

[12]§§669(a)(2) and (3). The term "criteria" is not defined in the Act, although the Act's legislative history suggests what Congress had in mind. The Senate report describes "criteria" as "scientifically determined conclusions, describing medically acceptable tolerance levels of exposure to harmful substances or conditions over a period of time, and may include medical judgments on methods and devices used to control exposure or its effects." S. REP. No. 1282, *supra* note 5, at 19. The House similarly described "criteria" as "scientifically determined conclusions based on the best available medical evidence." H.R. REP. 1291, *supra* note 5, at 27.

[13]§671(c)(1). This function is the only exclusive function of NIOSH, i.e., one that is not derivative from the Secretary of HEW. It is assigned to NIOSH in this section along with one other function—to carry out all of the enumerated functions of the Secretary in §§669 and 670.

[14]§669(a)(6).

[15]HHEs are performed at the request of an employer or authorized representative of employees. If an HHE identifies an unregulated toxic material, NIOSH is required to inform the Secretary of Labor and submit all pertinent criteria to him, *id.*, presumably so the Secretary can then consider development of a new standard.

Under the Act, NIOSH shares responsibility for training and education with the Secretary of Labor. NIOSH is directed to conduct education programs to help provide an adequate supply of qualified personnel to carry out the purposes of the Act.[16] NIOSH is also charged with conducting "informational programs on the importance of and proper use of adequate safety and health equipment."[17]

B. Mine Safety and Health Legislation

NIOSH has another large body of statutory responsibility under the nation's laws protecting the health and safety of miners. Under general public health research authority, BOSH and its predecessors in the U.S. Public Health Service had investigated mine-related health problems since 1914. However, no direct legislative basis for Public Health Service research on the health of miners existed until 1969, when Congress passed the Federal Coal Mine Health and Safety Act.[18] All of the HEW Secretary's responsibilities under this law (except for the Black Lung Benefits program) were administratively delegated to NIOSH in 1971. When this law was amended in 1977,[19] the HEW Secretary's responsibility to conduct, jointly with the Secretary of the Interior, research on health and safety in coal, metal, and nonmetal mines was legislatively delegated to NIOSH.[20] The Secretary of the Interior, through the Bureau of Mines, is charged with performing "safety" research, while NIOSH is given the job of "health" research.[21]

---

[16]§670(a)(1). Of primary concern to the law's drafters was the inadequate number of trained safety specialists and industrial hygienists who would perform the investigations and inspections under the Act. See H.R. REP. NO. 1291, *supra* note 5, at 31, and 116 CONG. REC. 38,366 (1970).

[17]§670(a)(2).

[18]Pub. L. No. 91-173, 83 Stat. 742 (1969) (codified as amended at 30 U.S.C. §§801–962 (1976 & Supp. V 1981).

[19]Federal Mine Safety & Health Amendments Act of 1977, Pub. L. No. 95-164, 91 Stat. 1290 (1977). This law also repealed the Metal & Nonmetallic Mine Safety Act (Pub. L. No. 89-577, formerly codified at 30 U.S.C. §§721 *et seq.*) and incorporated safety and health protection for coal, metal, and nonmetal mines into a single statute.

[20]30 U.S.C. §951(b).

[21]*Id.*

In addition to research authority, the 1977 Amendments Act gives NIOSH responsibility for

- administering a medical surveillance program which includes (a) chest X-rays for detection of pneumoconiosis in underground coal miners, which, in turn, serves as a basis for giving miners the option to transfer to less hazardous jobs in the mine, and (b) autopsies of deceased miners;[22]
- operating a mining health hazard evaluation program similar to the one for general industry;[23]
- operating a testing and certification program for personal protective equipment and hazard measuring instruments;[24]
- appointing and supporting a federal mine health research advisory committee;[25]
- developing criteria and recommending health standards to the Mine Safety and Health Administration of the Department of Labor.[26]

C. Other Statutory Sources of Authority

Statutory authority for NIOSH activities also derives from the Public Health Service Act[27] and from various other laws enacted since 1970. The latter primarily involve participation in interagency committees and scientific studies on occupational health and safety subjects.[28] One such statute, the Comprehensive Environmental Response, Compensation, and Liability Act of 1980, also known as Superfund, assigns NIOSH a major responsibility for protection of workers involved in cleanup of toxic waste sites.[29]

---

[22]30 U.S.C. §843.

[23]§951(a)(11).

[24]§§842(a), (e), and (h), and §844.

[25]§812(b).

[26]§§811(a)(1), 811(a)(6)(B), 842(d), 845, and 846. The Mine Safety & Health Administration is the successor to the Department of Interior's Mine Enforcement & Safety Administration. It was transferred to the Department of Labor and renamed by the 1977 amendments.

[27]Pub. L. No. 78-410, as amended and codified at 42 U.S.C. §§201 *et seq.* (1982). This statute gives direct authority to conduct health research to the Secretary of Health and Human Services, who then delegates it to subordinate agencies.

[28]For example, Pub. L. No. 95-239, §17, 92 Stat. 95, 105 (1978), required a study of occupationally related pulmonary and respiratory diseases.

[29]Pub. L. No. 96-510 (1980).

II. Legal Actions

NIOSH activities have generated a remarkably small amount of legal action. All of the reported cases to which NIOSH has been a party involve NIOSH's right of entry to a private workplace or access to records and documents. Other legal action has involved NIOSH's role in the standard-setting function of the OSH Act (see Section B below) and NIOSH's program of testing and certification of personal protective equipment (see Section C below).

A. Subpoena and Right of Entry Cases

The largest number of cases involving NIOSH fall into two major categories. The majority involve attempts by NIOSH to invoke the subpoena power of Section 8(b) of the Act to obtain records maintained by an employer or a third party. The remainder are adjudications of NIOSH's right of entry under an *ex parte* administrative search warrant. The latter cases have resulted from (1) epidemiologic and industrial hygiene studies, (2) surveillance activities to assist in identifying health problems, or (3) workplace Health Hazard Evaluations. These three types of workplace investigations involve taking samples of air or other media to determine the presence of toxic substances; interviewing management and workers; giving medical examinations; and reviewing medical, personnel, production, and exposure records maintained by the employer. NIOSH has issued regulations for conducting these investigations.[30]

1. Subpoena Power

The NIOSH subpoena cases have established conclusively that NIOSH has authority to issue administrative subpoenas *duces tecum*;[31] that this authority is coextensive with that of

---

[30]42 C.F.R. pt. 85 (1986) sets forth regulations for conducting HHEs, and 42 C.F.R. pt. 85a applies to all other types of investigations. See also notes 49, 50, and 58, *infra*, and accompanying text.

[31]*United States v. Westinghouse Elec. Corp.*, 638 F.2d 570, 575, 8 OSHC 2131, 2134 (3d Cir. 1980); *General Motors Corp. v. Director, NIOSH*, 636 F.2d 163, 165, 9 OSHC 1139, 1140 (6th Cir. 1980), *cert. denied*, 454 U.S. 877 (1981); *United States v. Allis Chalmers Corp.*, 498 F. Supp. 1027, 1028, 9 OSHC 1165, 1167 (E.D. Wis. 1980); *E.I. du Pont de Nemours v. Finklea*, 442 F. Supp. 821, 824, 6 OSHC 1167, 1168 (S.D. W. Va. 1977).

OSHA's;[32] and that it must be exercised in accordance with the general standards for administrative subpoenas set forth in the Supreme Court's decision in *United States v. Morton Salt Co.*[33] It has also been held that this power may be exercised against third-party holders of records, such as insurance companies, as well as against employers.[34]

(a) Privacy issues. An issue not conclusively settled is the extent to which employee medical and personnel records are protected from disclosure by a constitutional claim of privacy.[35] In the first case decided on this issue, *E.I. du Pont de Nemours v. Finklea,* the district court decided that the employees whose medical records were sought had a constitutional right of privacy,[36] but that disclosure to NIOSH, under the terms of the subpoena, would not abridge that right. Relying on *Whalen v. Roe,*[37] the court held that the right to privacy was not violated because NIOSH assured the court that the information obtained would be treated as confidential[38] and there was no countervailing evidence to show that it would be used improperly. To provide an extra measure of assurance, the court ordered NIOSH to maintain the records as confidential and to comply with all applicable statutory and regulatory protections regarding confidentiality.

The courts that have considered this issue since *du Pont* have all agreed that employees have a constitutional right of privacy regarding their medical or personnel records. They have not, however, agreed on what degree of protection the

---

[32]See cases cited *supra* note 31.

[33]*United States v. Westinghouse Elec. Corp.*, *supra* note 31, at 574, 8 OSHC at 2134; *United States v. Allis Chalmers Corp.*, *supra* note 31, at 1029, 9 OSHC at 1167; *United States v. McGee Indus.*, 439 F. Supp. 296, 298, 5 OSHC 1562, 1563 (E.D. Pa. 1977), *aff'd mem.*, 568 F.2d 771, 6 OSHC 1540 (3d Cir. 1978). *United States v. Morton Salt Co.*, 338 U.S. 632, 652 (1950), holds that the inquiry must be within the authority of the agency, the demand for production must not be too indefinite, and the information sought must be reasonably relevant to the authorized inquiry.

[34]*United States v. Amalgamated Life Ins. Co.*, 534 F. Supp. 676, 10 OSHC 1447 (S.D.N.Y. 1982).

[35]See *Griswold v. Connecticut*, 381 U.S. 479, 484 (1965), which held that the Constitution creates "zones of privacy."

[36]*Supra* note 31, at 824, 6 OSHC at 1169. The district court found these medical records to be within the zones of privacy as enunciated in *Griswold*, *supra* note 35, because they contain intimate facts of a personal nature.

[37]429 U.S. 589 (1977).

[38]The court was satisfied that NIOSH assurances of confidentiality, along with departmental regulations prohibiting disclosure and exemptions from disclosure under the Freedom of Information Act, would provide adequate protection. *Supra* note 31 at 825, 6 OSHC at 1169.

courts should insist upon to assure that this right is not abridged. The Sixth Circuit, in *General Motors Corp. v. Director, NIOSH*, held that the constitutional right of employees would not be jeopardized by disclosure to NIOSH because the district court could fashion appropriate protective orders to assure the security of the subpoenaed documents.[39] The circuit court reversed the district court's condition[40] that release of the subpoenaed records was contingent upon deletion of individuals' names and addresses from the files. The protective order device was also used in *United States v. Allis Chalmers Corp.*[41]

A different result was reached by the Third Circuit in *United States v. Westinghouse Electric Corp.*[42] In this case and a subsequent district court case which followed its reasoning,[43] courts, for the first time, incorporated into the analysis of the constitutional issue the "delicate task of weighing competing interests."[44] It imposed the requirement that the societal interest in disclosure must outweigh the competing interest in privacy. In balancing the interests presented by the facts of the case, the court found that "the strong public interest in facilitating the research and investigations of NIOSH justify this minimal intrusion into the privacy which surrounds the employees' medical records."[45] It recognized, however, that there may be highly sensitive information in an employee's medical or personnel records and that an employee's claim of constitutional privacy may not always be outweighed by NIOSH's need for the records. Therefore, the court went one step further than previous courts and ordered NIOSH to give notice to affected employees that it would review the company's medical records and that employees would have an opportunity to object to the release of sensitive medical infor-

---

[39]*Supra* note 31, at 166, 9 OSHC at 1141. See also *United States v. McGee Indus., supra* note 33, at 298, 5 OSHC at 1564. This was a NIOSH subpoena case in which a subpoena issued in furtherance of a national survey of chemicals used in the workplace was enforced only after a protective order was issued. At stake was the proprietary interest of the company in the formulation of certain products. This remedy was not based on constitutional protections, but on the inherent power of the district courts to administer the enforcement of subpoenas.

[40]459 F. Supp. 235, 6 OSHC 1976 (S.D. Ohio 1978).

[41]*Supra* note 31.

[42]638 F.2d 570, 8 OSHC 2131 (3d Cir. 1980).

[43]*United States v. Lasco Indus.*, 531 F. Supp. 256, 10 OSHC 1356 (N.D. Tex. 1981).

[44]*Supra* note 42, at 578, 8 OSHC at 2137.

[45]*Id.* at 580, 8 OSHC at 2138.

mation. If employees did not respond to the notice, their consent to release was implied and the company was required to release the records. The court rejected Westinghouse's contention that written consent from each employee was needed to provide minimal constitutional protection.

(b) Standing. Several of the NIOSH subpoena cases present the issue of whether an employer has standing to raise the constitutional claims of employees who are not parties to the action. Best stated by the Third Circuit, the consensus is that the employer

> "has the necessary concrete adverseness to present the issues. The subpoena at issue is directed to [the company] and requires production of documents in its possession. Failure to comply with the subpoena will subject it to the penalty of a contempt citation. Furthermore, it has an ongoing relationship with its employees and it asserts that an adverse decision on the merits of the constitutional claim regarding employee privacy may adversely affect the flow of medical information which it needs from them.[Footnote omitted.] As a practical matter, the absence of any notice to the employees of the subpoena means that no person other than [the company] would be likely to raise the privacy claim."[46]

One court, however, has suggested in *dictum* that there is a danger in recognizing standing for a party whose interests may in fact be adverse to the employee's. For example, an employer or insurer's interest may be in avoiding liability, and the court considered it unlikely that an employee, given the choice, would press a privacy claim.[47]

(c) Privilege as a defense. The issue of whether a company can assert the confidentiality of the physician-patient relationship as a defense to a NIOSH subpoena for medical records has arisen in several cases. This defense exists in state common law or in certain state statutes. The federal courts that have considered this question have held unanimously that in cases involving a federal statute such as the OSH Act, state law is not controlling authority on the issue and therefore a privilege based on state law may not be asserted. Since there

---

[46]*United States v. Westinghouse Elec. Corp., supra* note 42, at 574, 8 OSHC at 2133. See also *United States v. Allis-Chalmers Corp.,* 498 F. Supp. 1027, 1029, 9 OSHC 1165, 1167 (E.D. Wis. 1980).

[47]*United States v. Amalgamated Life Ins. Co.,* 534 F. Supp. 676, 679 n.5, 10 OSHC 1447, 1449 (S.D.N.Y. 1982).

is no physician-patient privilege in federal statutory or federal common law (and these courts declined to establish new federal common law), the confidential physician-patient relationship cannot be raised as a defense to a properly executed subpoena.[48]

2. Right of Entry

The second group of NIOSH cases involves NIOSH's right of entry to a workplace to perform an investigation or inspection. NIOSH's authority to enter a private workplace to obtain information to carry out its statutory responsibilities is contained in the OSH Act for general industry investigations[49] and the Federal Mine Safety and Health Act for mining-related investigations.[50] All of the reported NIOSH cases have arisen under the OSH Act.

The OSH Act states that for the purpose of carrying out his research responsibilities, "[t]he Secretary of Health, Education, and Welfare is authorized to make inspections and question employers and employees as provided in section 8 of this Act [29 U.S.C.§657]."[51] This latter section details the powers of the Secretary of Labor to make inspections, conduct investigations, and require the keeping of records.

These cases are outgrowths of the Supreme Court's landmark decision in *Marshall v. Barlow's, Inc.*,[52] in which the

---

[48]*General Motors Corp. v. Director, NIOSH*, 636 F.2d 163, 165, 9 OSHC 1139, 1140 (6th Cir. 1980), *cert. denied*, 454 U.S. 877 (1981); *United States v. Amalgamated Life Ins. Co.*, *supra* note 47, at 679–680, 10 OSHC at 1450; *United States v. Allis-Chalmers Corp.*, *supra* note 46, at 1029, 9 OSHC at 1167.

[49]29 U.S.C. §669(b). NIOSH regulations for exercising this authority are in 42 C.F.R. pt. 85 for HHEs and in 42 C.F.R. pt. 85a for all other types of inspections and investigations.

[50]30 U.S.C. §813(a). Section 813(a) of Title 30 requires both the Secretary of Labor and the Secretary of HEW to

"make frequent inspections and investigations in coal or other mines each year for the purpose of (1) obtaining, utilizing, and disseminating information relating to health and safety conditions, the causes of accidents, and the causes of disease and physical impairments originating in such mines, (2) gathering information with respect to mandatory health and safety standards ***. For the purpose of making any inspection or investigation under this Act, the Secretary [of Labor], or the Secretary of Health, Education, and Welfare, with respect to fulfilling his responsibilities under the Act *** shall have a right of entry to, upon, or through any coal or other mine."

NIOSH has implementing regulations for this authority in 42 C.F.R. pt. 85 for HHEs and in 42 C.F.R. pt. 85a for all other inspections and investigations.

[51]29 U.S.C. §669(b).

[52]436 U.S. 307, 6 OSHC 1571 (1978).

Court held that if OSHA first obtains an administrative search warrant based on probable cause, OSHA's exercise of its right of entry under Section 8(a) of the OSH Act[53] does not violate the Fourth Amendment's protection from unreasonable searches and seizures. Section 8(a) has been the subject of extensive litigation; and the case law developed under this section, in addition to the litigation to which NIOSH has been a party, is the controlling precedent for NIOSH activities.[54]

The cases involving NIOSH all upheld NIOSH warrants. Together they stand for the principle that where consent is denied, NIOSH is entitled to enter the premises of a private workplace under the authority of an *ex parte* administrative search warrant based on probable cause.

(a) Probable cause. The *Barlow's* standard for administrative probable cause applies to NIOSH inspections and investigations;[55] but since NIOSH does not have enforcement authority, evidence of a violation of the OSH Act is not a relevant criterion for the initiation of authorized NIOSH investigations. The courts have held, however, that the *Barlow's* criterion of "reasonable legislative *or* administrative standards" is met for investigations conducted pursuant to HHEs and other types of field studies.[56] Legislative standards are found in the OSH Act,[57] and administrative standards are found in the regulations implementing these provisions.[58]

For an HHE investigation, which is initiated only after a request by an employer or authorized representative of em-

---

[53]29 U.S.C. §657(a).

[54]See Chapter 7 (Inspections) for a full explication of the right of entry in OSHA cases.

[55]*In re Inland Steel Co.*, 492 F. Supp. 1310, 1312, 8 OSHC 1725, 1725–1726 (N.D. Ind. 1980).

[56]*In re Establishment Inspection of Keokuk Steel*, 493 F. Supp. 842, 844–845, 8 OSHC 1730, 1731–1732 (S.D. Iowa 1980), *aff'd*, 638 F.2d 42, 9 OSHC 1195 (8th Cir. 1981); *In re Inland Steel Co.*, *supra* note 55, at 1312–1313, 8 OSHC at 1726; *In re Pfister & Vogel Tanning Co.*, 493 F. Supp. 351, 354 (E.D. Wis. 1980).

[57]29 U.S.C. §669(a)(6) refers to HHEs. It states that the Secretary of HEW "shall determine following a written request by any employer or authorized representative of employees, specifying with reasonable particularity the grounds on which the request is made, whether any substance normally found in the place of employment has potentially toxic effects in such concentrations as used or found; and shall submit such determination both to employers and affected employees as soon as possible."
Reasonable legislative standards would appear to exist for HHEs carried out under the authority of the Federal Mine Safety & Health Act as well, since the analogous language in this law is almost identical. See 30 U.S.C. §951(a)(11).

[58]42 C.F.R. pt. 85 and pt. 85a.

ployees, the courts will determine whether these standards have been met by reviewing the application for the warrant and the employee request. (An employer request would not result in the need for a warrant.) Sufficient facts must be alleged, either by sworn employee complaints or by a sworn affidavit from the requesting official, to allow the court to make an independent assessment as to whether an inspection is justified.[59] To be valid, the request must be made by (1) a single employee, if there are three or fewer employees in the particular workplace, (2) an employee who has been authorized by two or more other employees, or (3) an authorized representative of the employees' local or parent international union; and it must include pertinent general and specific information specifying with reasonable particularity the nature of the conditions on which the request is made. Specific employees who have complained or who have personal knowledge of the conditions need not be named. NIOSH must then have reasonable grounds to conclude that the requested investigation is justified.[60]

Other types of investigations (non-HHEs) are not usually initiated on the basis of an external request, but are generally part of a planned program of research. Occasionally, NIOSH is asked to provide technical assistance to OSHA or other public agencies in the investigation of a particular problem or emergency situation. For these investigations, probable cause is satisfied if the warrant demonstrates that a "specific business was chosen for the search on the basis of 'a general administrative plan for the enforcement of the Act derived from neutral sources.' "[61] In issuing the warrant, the "magistrate must determine that (1) there is a reasonable legislative or administrative inspection program, and (2) the desired inspection fits within that program."[62] This standard would appear to be met whenever the particular inspection meets the criteria established in a planned research protocol or in the request for technical assistance.

---

[59]*In re Inland Steel Co.*, *supra* note 55, at 1312, 8 OSHC at 1726; *In re Pfister & Vogel Tanning Co.*, *supra* note 56, at 354.

[60]*In re Establishment Inspection of Keokuk Steel*, *supra* note 56, 493 F. Supp. at 845, 9 OSHC at 1197.

[61]*In re Pfister & Vogel Tanning Co.*, *supra* note 56, at 354.

[62]*Id.*

(b) Scope of the warrant. A second issue arising in NIOSH warrant cases involves the appropriate scope of the warrant. Citing a 1979 decision by the Seventh Circuit, the district court in *In re Pfister & Vogel Tanning Co.* stated the general principle that the scope of a warrant "must be as broad as the subject matter regulated by the statute and restricted only by the limitations imposed by Congress and the reasonableness requirement of the Fourth Amendment."[63] With one exception, the courts have upheld NIOSH's interpretation of what constitutes the necessary scope of a search. The court in *Pfister & Vogel Tanning Co.* recognized that an HHE or research investigation by its very nature requires the magistrate to defer to a great extent to the expertise of the agency.[64]

Physical examination of the workplace and private interviews of employers and employees during regular working hours are specifically authorized by statute.[65] The use of personal monitoring devices by employees to measure exposure to airborne contaminants satisfies the further statutory requirement that inspections are to be made within reasonable limits and in a reasonable manner.[66]

The statute gives no further guidance on the permissibility of other related activities, but NIOSH has interpreted Section 8(a) of the Act to include physical and medical examinations of employees and examination of documents and records.[67] Medical examinations have been held to fit within a reasonable construction of Section 8(a) so long as they are carried out under the limitations established in the regulations, i.e., employee consent is obtained, they are performed at NIOSH's expense, and they are conducted to preclude unreasonable disruption of the employer's operations.[68] This construction is especially true for HHEs, since a determination

---

[63]*Id.* at 355.

[64]*Id.*

[65]29 U.S.C. §657(a)(2).

[66]*In re Establishment Inspection of Keokuk Steel, supra* note 56, 638 F.2d at 46, 9 OSHC at 1197. But see *Plum Creek Lumber Co. v. Hutton,* 608 F.2d 1283, 7 OSHC 1940 (9th Cir. 1979).

[67]29 C.F.R. §§85.5 and 85a.3.

[68]*In re Inland Steel Co.,* 492 F. Supp. 1310, 1313, 8 OSHC 1725, 1727 (N.D. Ind. 1980).

of toxicity is dependent on a medical assessment of exposed persons.[69]

With respect to documents and records, the two cases on point, both district court decisions, are in disagreement. *Pfister & Vogel Tanning Co.* held that a warrant that permitted inspection of medical and personnel records was not overbroad.[70] *Inland Steel*, decided two weeks later, held that Section 8(a) does not authorize medical or personnel records to be examined under the authority of a warrant. Distinguishing the Supreme Court's *Barlow's* decision and three Seventh Circuit cases which "could be read to support the [opposite] view," the *Inland Steel* court concluded that Congress intended that records be obtained with a subpoena under the authority of Section 8(b). The court reasoned that the subpoena offers "superior procedural protection" than a warrant which could be obtained *ex parte*, and courts should not "lightly strip such protection away in the name of administrative convenience."[71]

(c) Ex parte warrants. A third issue in these cases is whether NIOSH may obtain *ex parte* warrants. The *Barlow's* case originally decided that OSHA did not have authority to obtain *ex parte* warrants.[72] Citing *dictum* in *Barlow's*, two courts which have addressed the issue have concluded that NIOSH may constitutionally obtain *ex parte* warrants, despite having no procedural regulations as guidelines for obtaining them.[73]

(d) Fifth Amendment claims. One final issue with respect to warrants deserves mention. In *Southern Indiana Gas & Electric Co. v. Director, NIOSH*, the employer sought an injunction to halt a planned NIOSH HHE inspection. In attempting to demonstrate irreparable injury, the employer alleged that the regulation for HHEs requiring the employer to provide NIOSH with suitable space in the workplace to conduct medical examinations and employee interviews[74] (and

---

[69]*Southern Ind. Gas & Elec. Co. v. Director, NIOSH*, 522 F. Supp. 850, 853, 10 OSHC 1171, 1173–1174 (S.D. Ind. 1981).

[70]*In re Pfister & Vogel Tanning Co.*, 493 F. Supp. 351, 356 (E.D. Wis. 1980).

[71]*In re Inland Steel Co.*, *supra* note 68, at 1314–1316, 8 OSHC at 1728–1729.

[72]The issue is discussed in detail in Chapter 7 (Inspections).

[73]*In re Establishment Inspection of Keokuk Steel*, 493 F. Supp. 842, 845, 8 OSHC 1730, 1732 (S.D. Iowa 1980), *aff'd*, 638 F.2d 42, 44–45, 9 OSHC 1195, 1196 (8th Cir. 1981); *In re Pfister & Vogel Tanning Co.*, *supra* note 70, at 353–354.

[74]29 C.F.R. §85.8.

the associated use of the employer's water and electricity for these purposes) violates the employer's Fifth Amendment right to just compensation for the "taking of property."[75] The court held that the regulation in question is not unconstitutional because it "sets a standard of 'reasonableness' in light of the congressional purpose of promoting worker health."[76]

B. Recommended Occupational Safety and Health Standards

NIOSH's statutory responsibility to produce criteria and recommended standards has been accomplished through the publication of "criteria documents" and similar publications, such as occupational hazard assessments, special hazard reviews, and current intelligence bulletins, which provide analyses, assessments, and recommendations regarding occupational hazards. NIOSH has published over 100 criteria documents recommending standards to OSHA. Many of the other types of documents have also been published.

NIOSH criteria documents and recommendations have figured prominently in most of OSHA's standard-setting activities and have been instrumental in providing evidence to support standards against court challenges.[77] However, in setting standards, OSHA is not legally constrained by NIOSH'S recommendations. For example, in reviewing the OSHA standard for exposure to asbestos fibers, the District of Columbia Circuit held that Congress undoubtedly saw NIOSH recommendations as an important aid to OSHA, but not more than that.[78]

On the other hand, Congress has given NIOSH recommendations for mine standards more weight. When the Mine Safety and Health Administration (MSHA) receives a recommendation from NIOSH to issue a standard or receives criteria for toxic substances or harmful physical agents found

---

[75]*Southern Ind. Gas & Elec. Co. v. Director, NIOSH, supra* note 69, at 854, 10 OSHC at 1173.

[76]*Id.*

[77]See Chapter 20 (Judicial Review of Standards). NIOSH recommendations have also been used as evidence in enforcement proceedings.

[78]*Industrial Union Dep't, AFL-CIO v. Hodgson,* 499 F.2d 467, 477, 1 OSHC 1631, 1639 (D.C. Cir. 1974).

in mines, mandatory action is required. MSHA must (a) within 60 days, commence rulemaking, either by publishing the recommended standard in the *Federal Register* as a proposed rule or convening an advisory committee; or (b) decide not to commence rulemaking and publish reasons for the decision in the *Federal Register*.[79] There is no legal authority, however, that MSHA is any more constrained than OSHA by a NIOSH recommendation when formulating the final rule.

C. Testing and Certification of Respiratory Protection

NIOSH administers a controversial program of testing and certification of respiratory devices (gas masks, dust and fume respirators, etc.). MSHA is a joint sponsor of the program under a formal agreement between the agencies.[80] In a workplace regulated by OSHA or MSHA, use of an uncertified respirator is a violation of applicable OSHA standards[81] or MSHA standards.[82]

Testing is carried out under laboratory conditions. NIOSH regulations detail the procedures for testing and certifying respiratory devices.[83] Outside the laboratory, however, a particular model may not perform in a satisfactory manner. This has occurred in a few instances; but there is no clear authority for NIOSH to invoke mandatory sanctions, such as revocation of certification, when certified equipment fails to perform as expected. NIOSH investigates performance failures and in appropriate circumstances requests manufacturers and distributors to voluntarily stop sales of the product, recall the product, or initiate a retrofit program to repair defective parts. Voluntary action has occurred in most cases where requests have been made, but NIOSH appears to have no legal authority to enforce these requests.

[79]29 U.S.C. §§811(a)(1) and (a)(6)(B).

[80]Memorandum of Understanding Between MSHA and NIOSH, May 4, 1978, MSHR 21:3101.

[81]29 C.F.R. §1910.134(b)(11).

[82]30 C.F.R. pt. 70, subpart D (1985), for coal mines, and 30 C.F.R. §56.5005 for metal and nonmetal mines.

[83]30 C.F.R. pt. 11 (1985).

25

Impact of the OSH Act on Private Litigation and Workers' Compensation Laws

I. Impact on Civil Litigation

A. In General

The OSH Act was designed to create an additional approach to improving industrial safety and health without upsetting preexisting law.[1] The objective of neutrality with respect to preexisting rights arising in the employment context is clearly expressed in Section 4(b)(4):

> "Nothing in this Act shall be construed to supersede or in any manner affect any workmen's compensation law or to enlarge or diminish or affect in any other manner the common law or statutory rights, duties, or liabilities of employers and employees under any law with respect to injuries, diseases, or death of employees arising out of, or in the course of, employment."[2]

---

[1]See S. REP. NO. 1282, 91st Cong., 2d Sess. (1970), *reprinted in* 1970 U.S. CODE CONG. & AD. NEWS at 5178, 5199. Resort to the legislative history of the OSH Act to clarify or expand on the statutory language is often, as in this case, of little help, generally because the numerous compromises in the legislative process did not produce agreement on the present language until after the conference report.

[2]29 U.S.C. §653(b)(4) (1982).

Congress's intention to create a system of workplace safety and health regulation and enforcement without disturbing previously established employer-employee rights and obligations has, for the most part, been accomplished. However, there has been considerable litigation in this area despite the seemingly unambiguous declaration of Section 4(b)(4).

In accordance with Section 4(b)(4), the Act has been held not to preempt any common law or other statutory duties.[3] Instead of focusing on preemption, however, most litigation has concerned whether the Act creates rights in addition to those that already exist or whether, with respect to preexisting rights, the Act or the safety standards promulgated under it impact those rights in any way.

B. Personal Injury and Wrongful Death

Courts uniformly have held that the Act does not expressly or impliedly create a private cause of action for personal injury or wrongful death. Employees injured as a result of a violation of the Act have no cause of action against their own private employer[4] or their employer's executive officers.[5] This has been followed on the state level as well.[6] Similarly, it has been held that the Act does not create any private cause of action for federal employees against their federal agency employers.[7] The underpinning of these cases is Section 4(b)(4).

---

[3]*Shimp v. New Jersey Bell Tel. Co.*, 145 N.J. Super. 516, 368 A.2d 408 (1976); *P & Z Co. v. District of Columbia*, 408 A.2d 1249, 8 OSHC 1078 (D.C. 1979); *Berardi v. Getty Ref. & Mfg. Co.*, 107 Misc.2d 451, 435 N.Y.S.2d 212 (Sup. Ct. Richmond County 1980).

[4]*Byrd v. Fieldcrest Mills, Inc.*, 496 F.2d 1323, 1 OSHC 1743 (4th Cir. 1974) (wrongful death); *National Marine Serv. v. Gulf Oil Co.*, 608 F.2d 522 (5th Cir. 1979), *aff'g mem.* 433 F. Supp. 913 (E.D. La. 1977); *Russell v. Bartley*, 494 F.2d 334, 1 OSHC 1589 (6th Cir. 1974); *Buhler v. Marriott Hotels*, 390 F. Supp. 999, 3 OSHC 1199 (E.D. La. 1974).

[5]*Skidmore v. Travelers Ins. Co.*, 483 F.2d 67, 1 OSHC 1294 (5th Cir. 1973), *cert. denied*, 415 U.S. 949 (1974). But cf. *Davis v. Crook*, 261 N.W.2d 500 (Iowa 1978).

[6]*Aras v. Feather's Jewelers*, 92 N.M. 89, 582 P.2d 1302 (1978). But cf. *Davis v. Crook, supra* note 5.

[7]*Federal Employees for Non-Smokers' Rights v. United States*, 446 F. Supp. 181, 6 OSHC 1407 (D.D.C. 1978), *aff'd mem.*, 598 F.2d 310, 7 OSHC 1634 (D.C. Cir. 1979), *cert. denied*, 444 U.S. 926 (1979); *John Howard Pavilion Defense Fund v. Harris*, [1980] OSHD ¶24,486 (D.D.C. 1980).

Although Section 4(b)(4) appears on its face to leave room for causes of action by employees against others who are not their employers where violations of the Act cause injuries or death, it must be remembered that the Act only imposes obligations on employers vis-à-vis their employees. Accordingly, no basis for a private cause of action for injuries or death has been found in favor of employees against owners,[8] general contractors,[9] employees of general contractors,[10] or consulting engineers.[11] However, in one case where architects specifically undertook by contract to obtain compliance with OSHA standards, an employee injured as a result of the violation of a standard by one of the employers on the job site was permitted to maintain a personal injury action in state court against the architects for their failure to perform in accordance with their agreement.[12] Similarly, another court refused to dismiss an employee's personal injury complaint against the employer's safety inspector for negligent performance of his job responsibilities, holding that a cause of action was stated under the state's Good Samaritan rule.[13]

In contrast to the unanimity of the courts' holdings that no private cause of action is created by OSHA, there is a sharp division among the courts concerning the admissibility of OSHA standards and violations as evidence of negligence in private causes of action.[14] State courts in Connecticut and Indiana have rejected use of OSHA violations to establish negligence *per se*, holding that such use affects the common law rights,

---

[8]*Jeter v. St. Regis Paper Co.*, 507 F.2d 973, 2 OSHC 1591 (5th Cir. 1975); *Cochran v. International Harvester Co.*, 408 F. Supp. 598, 4 OSHC 1385 (W.D. Ky. 1975); *Otto v. Specialties, Inc.*, 386 F. Supp. 1240, 2 OSHC 1424 (N.D. Miss. 1974); *Pruette v. Precision Plumbing*, 27 Ariz. App. 288, 554 P.2d 655 (1976); *Taira v. Oahu Sugar Co.*, 616 P.2d 1026 (Hawaii Ct. App. 1980).

[9]*Horn v. C.L. Osborn Contracting Co.*, 591 F.2d 318, 7 OSHC 1256, *reh'g denied*, 595 F.2d 1221 (5th Cir. 1979); *Knight v. Burns, Kirkley & Williams Constr. Co.*, 331 So.2d 651, 4 OSHC 1271 (Ala. 1976); *Pruette v. Precision Plumbing*, *supra* note 8; *Koll v. Manatt's Transp. Co.*, 253 N.W.2d 265, 5 OSHC 1398 (Iowa 1977); *Kelley v. Howard S. Wright Constr. Co.*, 90 Wash.2d 323, 582 P.2d 500, 6 OSHC 1934 (1978).

[10]*Koll v. Manatt's Transp. Co.*, *supra* note 9.

[11]*Russell v. Bartley*, *supra* note 4.

[12]*Duncan v. Pennington County Hous. Auth.*, 283 N.W.2d 546 (S.D. 1979). However, the case was later dismissed for failure to prosecute. 382 N.W.2d 425 (S.D. 1986).

[13]*Santillo v. Chambersburg Eng'g Co.*, 603 F. Supp. 211 (E.D. Pa. 1985).

[14]Compare *Industrial Tile v. Stewart*, 388 So.2d 171 (Ala. 1980), *cert. denied*, 449 U.S. 1081 (1981) (OSHA citation as evidence), with *McKinnon v. Skil Corp.*, 638 F.2d 270 (1st Cir. 1981) (OSHA standard inadmissible).

duties, and liabilities of employers and employees and is, therefore, clearly proscribed by Section 4(b)(4).[15] On the other hand, courts in two states have thus far held that violation of an OSHA safety regulation is negligence *per se.* In *Bachner & Jones v. Rich & Rich*[16] and *Kelley v. Howard S. Wright Construction Co.,*[17] employees of subcontractors who sued their respective general contractors were permitted recovery on the basis of negligence *per se.* In each case, however, the courts also held that determination of liability on a negligence *per se* theory did not deprive the defendant of the right to an evaluation of the plaintiff's comparative negligence on the damage issue. Similarly, in *Bertholf v. Burlington Northern Railroad,*[18] where an employee was suing his own employer, the federal district court held that violation of an OSHA regulation is negligence *per se* under the Federal Employers' Liability Act,[19] but that the employee's contributory negligence could be shown to reduce liability. However, more recently, the First Circuit held in *Pratico v. Portland Terminal Co.*[20] that violation of an OSHA regulation is not only negligence *per se* but also eliminates contributory negligence as a defense under the Federal Employers' Liability Act.

Longshoremen and seamen injured on the job have attempted to use OSHA regulations and standards in actions against shipowners and employers. It appears that while injured longshoremen covered by the Longshoremen's and Harbor Workers' Compensation Act (LHWCA) may use OSHA standards as evidence,[21] seamen covered by the Jones Act

---

[15]*Ridgefield Constr. Servs. v. Wendland,* 184 Conn. 173, 439 A.2d 954, 10 OSHC 1727 (1981); *Hebel v. Conrail,* 475 N.E.2d 652 (Ind. 1985).

[16]554 P.2d 430 (Alaska 1976). But cf. *Macey v. United States,* 454 F. Supp. 684 (D. Alaska 1978), in which the district court refused to extend *Bachner* to impose negligence *per se* in the wrongful death action of a nonemployee.

[17]*Supra* note 9.

[18]402 F. Supp. 171 (E.D. Wash. 1975).

[19]45 U.S.C. §§51 *et seq.* (1981).

[20]783 F.2d 255, 12 OSHC 1567 (1st Cir. 1985).

[21]33 U.S.C. §§901 *et seq.* (1977). *Arthur v. Flota Mercante Gran Centro Americana, S.A.,* 487 F.2d 561, 1 OSHC 1434 (1973) (negligence *per se* if statute intended to protect class of persons to which plaintiff belongs against risk that actually occurred), *reh'g denied,* 488 F.2d 552 (5th Cir. 1974); *Croshaw v. Koninklijke Nedloyd,* 398 F. Supp. 1224 (D. Ore. 1975) (not negligence *per se* when no explicit agreement by employer to comply with standards); *Davis v. Partenrederei M.S. Normannia,* 657 F.2d 1048 (8th Cir. 1981) (jury properly considered evidence of OSHA standards).

when injured on the high seas may not.[22] The negligence standard under the LHWCA is the same as for "land-based" torts. A different standard is required under the Jones Act. Thus, in applying the Jones Act, the court in *Clary v. Ocean Drilling & Exploration Co.*[23] held that a vessel on the high seas is not a workplace covered by the OSH Act. Although violations of OSHA standards have usually been admissible only against employers,[24] in *Bachtel v. Mammoth Bulk Carriers*[25] the court held an OSHA standard was properly admitted in a suit against a shipowner.

In addition, the evidentiary use of OSHA standards has been permitted in a third party claim for indemnification by a defendant manufacturer against an employer in *Rabon v. Automatic Fasteners*,[26] and in favor of a nonemployee plaintiff in *Vagle v. Picklands Mather & Co.*[27] This view has been followed in several states.[28] The only case ever to hold that a

---

[22]46 U.S.C. §688 (1977). *Barger v. Mayor of Baltimore*, 616 F.2d 730, 8 OSHC 1114 (4th Cir. 1980), *cert. denied*, 449 U.S. 834 (1980). But see *Hicks v. Crowley Marine Corp.*, 538 F. Supp. 285, (S.D. Tex. 1982), *aff'd mem.*, 707 F.2d 514 (5th Cir. 1983), where the court allowed the use of OSHA standards although the injury occurred on the high seas.

[23]429 F. Supp. 905, 5 OSHC 1278 (W.D. La. 1977), *aff'd*, 609 F.2d 1120, 7 OSHC 2209 (5th Cir. 1980).

[24]See *Chavis v. Finnlines Ltd. O/Y*, 576 F.2d 1072 (4th Cir. 1978); *Brown v. Mitsubishi Shintaku Ginko & Kaninichi Kaiun Kaish, Ltd.*, 550 F.2d 331 (5th Cir. 1977).

[25]605 F.2d 438 (9th Cir. 1979), *vacated and remanded on other grounds*, 451 U.S. 978 (1981).

[26]672 F.2d 1231 (5th Cir. 1982). In *dictum*, the Fifth Circuit stated that violation of OSHA would constitute negligence *per se* where the plaintiff could establish that (a) he is within the class protected by the Act, (b) he has suffered the harm the Act intended to prevent, and (c) the violation was the proximate cause of his injury. *Id.* at 1238. This is the standard that has been applied in a long line of cases in the Fifth Circuit under the Longshoremen's and Harbor Workers' Compensation Act, 33 U.S.C. §§901 *et seq.* (1977). *Arthur v. Flota Mercante Gran Centro Americana, S.A.*, *supra* note 21.

[27]611 F.2d 1212 (8th Cir. 1979). See *Kraus v. Alamo Nat'l Bank*, 586 S.W.2d 202 (Tex. Civ. App. 1979), *aff'd*, 616 S.W.2d 908 (1981) (car driver and passenger against owner of building which collapsed). But see *Trowell v. Brunswick Pulp & Paper Co.*, 522 F. Supp. 782, 10 OSHC 1028 (D.S.C. 1981) (nonemployee touring paper mill may not use OSHA standards).

[28]*Scrimager v. Cabot Corp*, 23 Ill. App.3d 193, 318 N.E.2d 521 (1974); *Knight v. Burns, Kirkley & Williams Constr. Co.*, 331 So.2d 651, 4 OSHC 1271 (Ala. 1976); *Dan Dunn Roofing Co. v. Brimer*, 259 Ark. 855, 537 S.W.2d 164, 4 OSHC 1501 (1976); *DiSabatino Bros. v. Baio*, 366 A.2d 508, 4 OSHC 1855 (Del. 1976); *Koll v. Manatt's Transp. Co.*, 253 N.W.2d 265, 5 OSHC 1398 (Iowa 1977); *Michel v. Valdastri, Ltd.*, 59 Hawaii 53, 575 P.2d 1299 (1978); *Parker v. South La. Contractors*, 370 So.2d 1310, *cert. denied*, 374 So.2d 662 (La. 1979); *Ceco Corp. v. Maloney*, 404 A.2d 935 (D.C. 1979); *Kraus v. Alamo Nat'l Bank*, *supra* note 27. See *Trowell v. Brunswick Pulp & Paper Co.*, *supra* note 27.

violation or standard is not admissible for any purpose,[29] decided within the Fifth Circuit, now virtually stands alone.

On the other hand, evidence of compliance with OSHA has been held not to be a dispositive defense in a personal injury action.[30] However, in *Spangler v. Kranco, Inc.*,[31] the Fourth Circuit held that where a crane had been built without a warning device in accordance with the customer's specifications and where no such device was required by OSHA regulations, the defendant manufacturer was relieved from liability.

C. Product Liability

An employee who is injured in the workplace as a result of defective machinery or tools can usually recover workers' compensation without having to prove the negligence of the employer. The amount of the recovery, however, is limited and excludes damages for pain and suffering and permanent injuries. Consequently, employees have sought additional compensation through negligence and product liability actions against the manufacturer of the allegedly defective machinery.[32] The employee may proceed on theories of negligence,[33] breach of warranty,[34] and strict product liability.[35] Actions

---

[29]*Otto v. Specialties, Inc.*, 386 F. Supp. 1240, 2 OSHC 1424 (N.D. Miss. 1974).

[30]*Jackson v. New Jersey Mfr.'s Ins. Co.*, 166 N.J. Super. 448, 400 A.2d 81, *cert. denied*, 81 N.J. 330, 407 A.2d 1204 (1979).

[31]481 F.2d 373, 375 (4th Cir. 1973). See also *Jordan v. Kelly Springfield Tire & Rubber Co.*, 624 F.2d 674 (5th Cir. 1980).

[32]See *Porter v. American Optical Corp.*, 641 F.2d 1128 (5th Cir.), *cert. denied*, 454 U.S. 1109 (1981), *reh'g denied*, 455 U.S. 1009 (1982) (defectively designed respirator used by employee who contracted asbestosis).

[33]See RESTATEMENT (SECOND) OF TORTS §395 (1965) (negligent manufacture of chattel dangerous unless carefully made).

[34]See U.C.C. §2–314(1) (implied warranty of merchantability); *id.* §2–318 (third party beneficiary of express or implied warranties—no privity required).

[35]See generally RESTATEMENT (SECOND) OF TORTS §402A (1965):
"Special Liability of Seller of Product for Physical Harm to User or Consumer
"(1) One who sells any product in a defective condition unreasonably dangerous to the user or consumer or to his product is subject to liability for physical harm thereby caused to the ultimate user or consumer, or to his property, if
    "(a) the seller is engaged in the business of selling such a product, and
    "(b) it is expected to and does reach the user or consumer without substantial change in the condition in which it is sold.
"(2) The rule stated in Subsection (1) applies although
    "(a) the seller has exercised all possible care in the preparation and sale of his product, and
    "(b) the user or consumer has not bought the product from or entered into any contractual relation with the seller."

have also been asserted against the distributors and installers of the industrial machinery.[36]

In a strict product liability or breach of warranty action, OSHA standards may be used as the basis of expert testimony to buttress opinions on the defects of the equipment and to establish a yardstick against which to determine the dangerous nature of the equipment.[37] OSHA regulations may also be introduced as evidence of design standards or industry standards of care in negligence actions.[38]

In *Jackson v. New Jersey Manufacturer's Insurance Co.*,[39] the court refused to admit either an OSHA regulation or an American National Standards Institute standard incorporated under the regulation into evidence in a strict product liability suit. The reason for the refusal to admit them, however, was that they were not in existence at the time the equipment in question was installed in the employer's plant. The clear implication in this case is that such regulations and standards will be admissible with respect to equipment provided to manufacturers while they are in effect. Notably, the court also suggested that a manufacturer's proof of OSHA compliance would not necessarily preclude a finding of negligence, since it might be shown that a reasonable manufacturer would have taken additional precautions.

In *Davis v. Niagara Machine Co.*,[40] an employee sued the manufacturer of his employer's equipment for the manufacturer's failure to have point of operation guards on the machine in accordance with the requirements of OSHA. The manufacturer in turn sought indemnification from the employee's employer on the theory that it was the employer's obligation to its employee under OSHA to ensure the existence of point of operation guards. The court held that the manufacturer could not seek indemnification from the employer,

---

[36]See *Ladwig v. Ermanco, Inc.*, 504 F. Supp. 1229 (E.D. Wis. 1981).

[37]See generally Note, *The Use of OSHA in Products Liability Suits Against the Manufacturers of Industrial Machinery*, 11 VAL. U.L. REV. 37 (1976).

[38]See, e.g., *Scott v. Dreis & Krump Mfg. Co.*, 26 Ill. App.3d 971, 326 N.E.2d 74 (1975).

[39]*Supra* note 30.

[40]90 Wash.2d 342, 581 P.2d 1344 (1978). See *Scott v. Dreis & Krump Mfg. Co.*, *supra* note 38 (manufacturer cannot introduce OSHA standards to show duty on employer to incorporate safety devices because manufacturer has nondelegable duty to product reasonably safe products). Cf. *Bowlus v. North-South Supply Co.*, [1975–1976] OSHD ¶20,409 (Ky. 1976).

since OSHA did not create any obligations to the manufacturer on the part of the employer. On the other hand, in *Rabon v. Automatic Fasteners*,[41] the Fifth Circuit, applying Florida law in a diversity case, affirmed the lower court's holding that an employer was liable to indemnify an equipment manufacturer against whom the employer's employee had recovered on a theory of strict liability. However, it appears that a significant distinction in this case was evidence that the employer had undertaken to share with the manufacturer the obligation of warning the users of the equipment of its dangers. Evidence that the employer had failed to perform as agreed, and that such failure was the proximate cause of the injury, enabled the manufacturer to seek indemnification.

D. Wrongful Discharge

Traditionally, employers have been permitted to discharge "at will" employees who had no contractual right to a term of employment.[42] In the past decade, however, the employment-at-will doctrine has been partially eroded in several states.[43] Courts have generally declined, however, to infer a private right of action based on a statutory prohibition of retaliatory discharge.[44]

Similarly, when an employee who was discriminatorily discharged in retaliation for reporting OSHA violations attempted to maintain an independent cause of action against his employer, the Sixth Circuit held in *Taylor v. Brighton*

---

[41]672 F.2d 1231 (5th Cir. 1982).

[42]See H. Wood, A TREATISE ON THE LAW OF MASTER AND SERVANT §134 (1981).

[43]A cause of action for abusive discharge has been recognized in some states. See, e.g., *Frampton v. Central Ind. Gas Co.*, 260 Ind. 249, 297 N.E.2d 425 (1973). Where the public policy of the state has been implicated or compromised by the discharge of an at-will employee, a cause of action has been recognized. See, e.g., *Molush v. Orkin Exterminating Co.*, 547 F. Supp. 54 (E.D. Pa. 1982); *Savodnick v. Korvettes, Inc.*, 488 F. Supp. 822 (E.D.N.Y. 1980). But see *Murphy v. American Home Prods. Corp.*, 58 N.Y.2d 293, 461 N.Y.S.2d 232 (1983) (no tort cause of action for abusive or wrongful discharge in New York).

[44]Federal courts have held there is no private implied right of action under the Consumer Credit Protection Act for employees who were fired because their wages were garnished. See *McCabe v. City of Eureka*, 664 F.2d 680 (8th Cir. 1981). See also *Loucks v. Star City Glass Co.*, 551 F.2d 745 (7th Cir. 1977) (no cause of action under Illinois law for retaliatory discharge after employee filed a workers' compensation claim).

Corp.[45] that Section 11(c) of the Act[46] is the exclusive remedy in such circumstances. In deciding there was no implied private cause of action under the Act, the Sixth Circuit used the Supreme Court's *Cort v. Ash*[47] analysis. Under this analytical framework, passage of a federal statute is found to give rise to a private cause of action only where (1) the plaintiff is a member of the class benefited by the statute, (2) there is an explicit or implicit legislative intent to grant such a cause of action, (3) a private cause of action is consistent with the legislative scheme, and (4) inference of a private cause of action under federal law would not conflict with a matter traditionally of concern to the state. In this case, the plaintiff was found to be a member of the class benefited by the statute, and there would have been no conflict with state law if a private cause of action were implied. However, as in earlier cases in other contexts, the Sixth Circuit found the explicit legislative intent as expressed in Section 4(b)(4) militated against the inference of a private cause of action and that such a right would not be consistent with the exhaustive remedial scheme of the Act. In *George v. Aztec Rental Center*,[48] the Fifth Circuit adopted the Sixth Circuit's rationale.

E. Union Liability to Members

The Act imposes an obligation of compliance on employers only.[49] Attempts have been made, however, to hold unions liable for the injuries sustained by their members, based on theories of common law negligence and breach of duty of fair

---

[45]616 F.2d 256, 8 OSHC 1010 (6th Cir. 1980). But cf. *McCarthy v. Bark Peking*, 676 F.2d 42, 46 (2d Cir. 1982) (since employee did not file timely complaint and thus failed to exhaust administrative remedies, even if implied right of action existed, it could only be used after filing complaint "proved fruitless"), *vacated on other grounds*, 459 U.S. 1166 (1983), *on remand*, 716 F.2d 130 (2d Cir. 1983), *cert. denied*, 465 U.S. 1078, *reh'g denied*, 466 U.S. 994, *reh'g denied*, 468 U.S. 1250 (1984).

[46]29 U.S.C. §660(c) prohibits discrimination against employees because of OSHA-related activities and provides a procedure for the enforcement of the right of an employee to be free from such discrimination.

[47]422 U.S. 66, 78 (1965).

[48]763 F.2d 184, 12 OSHC 1381 (5th Cir. 1985).

[49]29 U.S.C. §654(a) (1976). Even though §5(b), 29 U.S.C. §654(b), literally obligates employees to comply with the Act and all the rules and regulations promulgated thereunder, such obligation is unenforceable under the Act. See *Atlantic & Gulf Stevedores v. OSHRC*, 534 F.2d 541, 4 OSHC 1061 (3d Cir. 1976); *I.T.O. Corp. v. OSHRC*, 540 F.2d 543, 4 OSHC 1574 (1st Cir. 1976).

representation emanating from the unions' failure to seek compliance with safety rules and standards.[50] Such negligence actions instituted in state court have frequently been subject to the invocation of the preemption doctrine.[51] Unions have asserted, but have not prevailed, on the argument that the jurisdiction of the National Labor Relations Board preempts the jurisdiction of the state court.[52] In *Bryant v. United Mine Workers*, the Sixth Circuit's characterization of the employee's action as a breach of duty of fair representation suit, which is governed by federal labor law, may have effectively precluded recovery on the theory of negligence.[53] Relying on long-standing, well-established precedent emanating more recently from the Supreme Court's decision in *Vaca v. Sipes*,[54] the court held that union liability for a breach of its duty of fair representation is established only where it is shown that the union treated its members arbitrarily, discriminatorily, or in bad faith and not where mere negligence is proved.

Similarly, where a defendant manufacturer in a product liability suit or employer in a negligence or workers' compensation action has impleaded the union, alleging a duty on the union's part to ensure a safe working place, courts have applied the same reasoning and dismissed the action.[55] In *House v. Mine Safety Appliance Co.*,[56] for example, employers sued by miners as a result of a mine fire unsuccessfully attempted

[50]See generally Segall, *The Wrong Pocket: Union Liability for Health and Safety Hazards*, 4 INDUS. REL. L.J. 390 (1981); Note, *Responsibility for Safe Working Conditions: Expanding the Limits of Union Liability*, 32 SYRACUSE L. REV. 681 (1981).

[51]See, e.g., *Condon v. United Steelworkers*, 683 F.2d 590, 110 LRRM 3244 (1st Cir. 1982); *Helton v. Hake*, 564 S.W.2d 313, 98 LRRM 1290 (Mo. Ct. App.), *cert. denied*, 439 U.S. 959 (1978); *Brooks v. New Jersey Mfr.'s Ins. Co.*, 170 N.J. Super. 20, 405 A.2d 466, 103 LRRM 2136 (1979); *Carollo v. Forty-Eight Insulation*, 252 Pa. Super. 422, 381 A.2d 990 (1977); *Farmer v. General Refractories Co.*, 271 Pa. Super. 349, 413 A.2d 701 (1979); *Higley v. Disston*, 92 LRRM 2443 (Wash. 1976).

[52]*Dunbar v. United Steelworkers*, 100 Idaho 523, 602 P.2d 21, 103 LRRM 2434 (1979), *cert. denied*, 446 U.S. 983 (1980). See Note, *supra* note 50, at 684. Although the Steelworkers eventually prevailed at trial on the facts, the Supreme Court of Idaho specifically declined to reconsider the preemption issue in upholding the trial court's dismissal based upon the evidence at trial. *Rawson v. United Steelworkers*, 111 Idaho 630, 726 P.2d 742 (1986).

[53]467 F.2d 1, 81 LRRM 2401 (6th Cir.), *cert. denied*, 410 U.S. 930 (1972). See *House v. Mine Safety Appliance Co.*, 417 F. Supp. 939, 92 LRRM 1033 (D. Idaho 1976). But see *Dunbar v. United Steelworkers*, *supra* note 52.

[54]386 U.S. 171 (1967).

[55]See *Gerace v. Johns-Manville Corp.*, 95 LRRM 3282 (Pa. Ct. Common Pleas Phila. 1977) (employer); *House v. Mine Safety Appliance Co.*, *supra* note 53; *Globig v. Johns-Manville Sales Co.*, 486 F. Supp. 735 (E.D. Wis. 1980) (manufacturer).

[56]*Supra* note 53.

to implead the employees' union on the theory that it had breached the obligation assumed under the safety and health provisions of the collective bargaining agreement, including participation in a joint committee to review safety matters.

Injured employees have prevailed in suits against their union on a theory other than fair representation in cases where the union had affirmatively assumed in a collective bargaining agreement the duty to inspect worksites, inform management of problems, and warn its members.[57] For example, in *Helton v. Hake*,[58] the court held the union liable for negligently failing to enforce a safety rule where the collective bargaining agreement provided that no work should be performed near high tension lines until the union steward saw that the rules were being complied with. A similar result was reached in *Dunbar v. United Steelworkers*.[59] At least one commentator has suggested that these two cases were incorrectly decided, reasoning that federal preemption should operate to bar the imposition of tort liability on unions and that a union's promises of safety enforcement ran to the employer, not to the members of the union.[60] *Michigan Mutual Insurance Co. v. United Steelworkers*[61] provides some substance to this criticism, holding that an insurance company's attempt to obtain a contribution from a union toward settlement of a bona fide claim on the ground that the union negligently performed duties voluntarily assumed under its collective bargaining agreement was preempted by Section 301 of the Labor Management Relations Act.[62]

Various states have enacted statutes that exempt unions from liability for employee injuries covered by workers' compensation.[63]

---

[57]*Dunbar v. United Steelworkers, supra* note 52; *Helton v. Hake, supra* note 51 (contract provided steward *shall* ensure compliance with work conditions).

[58]*Supra* note 51.

[59]*Supra* note 52.

[60]Segall, *supra* note 50, at 406–417.

[61]774 F.2d 104 (6th Cir. 1985).

[62]*Id.* at 106.

[63]See HAWAII REV. STAT. §386–8.5 (Supp. 1982); MICH. COMP. LAWS ANN. §418-827(8) (West Supp. 1981); ORE. REV. STAT. §654.720 (1979); *Gonzales v. R.J. Novick Constr. Co.*, 70 Cal. App.3d 131, 139 Cal. Rptr. 113 (1977) (based on California statute prohibiting use of OSHA in personal injury action), *vacated on other grounds*, 20 Cal.3d 798, 575 P.2d 1190, 144 Cal. Rptr. 408 (1978).

F. Government Liability for Failure to Enforce the OSH Act

Private suits against the federal government under the Federal Tort Claims Act[64] for its failure to enforce the OSH Act have not been successful. Liability requires, at a minimum, that the plaintiff establish that government employees acting within the scope of their employment engaged in negligent acts or omissions causing injuries or death "under circumstances where the United States, if a private person, would be liable to the claimant in accordance with the law of the place where the act or omission occurred."[65] In *Davis v. United States*,[66] a wrongful death action charged that an employee's death in a trench cave-in was due to an OSHA inspector's failure to follow up and determine whether a serious violation that had been cited was ever abated. The accident having occurred in Nebraska, the Eighth Circuit affirmed the district court's dismissal of the case on the ground that there was no counterpart under Nebraska law for the obligations imposed on federal OSHA inspectors. A similar result was reached in *Caldwell v. United States*,[67] but the district court went one step further. After reviewing the legislative history of the OSH Act and finding that a House amendment to the bill which specifically provided for a cause of action in the Court of Claims had been stricken from the Act as passed, the court held that Section 13(d) of the Act[68] does not give employees a cause of action against the United States for alleged failures by the Secretary to seek relief under the Act. Plaintiffs were likewise unsuccessful in *Caamano v. United States*,[69] where it was held that neither the government's performance of investigatory duties nor its discretionary decision to investigate provided a basis for recovery; further, the court noted, evidence of a negligent inspection would not provide a basis for relief, since the agency was not sufficiently involved in su-

[64]28 U.S.C. §1346 (1977).
[65]*Id.*
[66]536 F.2d 758, 4 OSHC 1417 (8th Cir. 1976).
[67]6 OSHC 1410 (D.D.C. 1978).
[68]29 U.S.C. §662(d) gives employees the right in imminent danger situations to seek a writ of mandamus to compel the Secretary to seek an order enjoining such imminent danger and other appropriate relief whenever the Secretary "arbitrarily or capriciously fails to seek relief" under Section 13.
[69][1980] OSHD ¶24,891 (S.D.N.Y. 1980).

pervising the work. In *Mudlo v. United States*,[70] a claim was dismissed for procedural defects with respect to the filing of the claim. Similarly, in *Leftridge v. United States*,[71] a claim was dismissed because it had not been filed within the two-year statute of limitations applicable under the Federal Tort Claims Act.

In similar suits against state governments, the results have for the most part been similarly unsuccessful.[72] However, in at least one state there is a possibility for recovery by plaintiffs against the state for the negligence of state safety inspectors in administering the state-run plan. Unlike those decisions that addressed the discretionary function exemption under the Federal Tort Claims Act and similar state law provisions, *Wallace v. Alaska*[73] denied state immunity on this ground, holding that while making the decision to inspect is a discretionary act, once it is made, performance of the inspection is a ministerial function and negligence is actionable against the state. In this case for wrongful death, the inspector had warned the employer two days before a fatal accident that he was operating a boom impermissibly close to electrical lines, taking no further action with respect to the violation.

II. The OSH Act and Workers' Compensation Laws

A. Background

Despite early legislation to improve workplace safety and health,[74] workers and, in the event of death, their families, were often destitute because of inadequate compensation for injury or death. Many lived in poverty, dependent upon pri-

---

[70]423 F. Supp. 1373 (W.D. Pa. 1976).

[71]612 F. Supp. 631, 12 OSHC 1429 (W.D. Mo. 1985).

[72]*Estate of Klee v. New York*, [1978] OSHD ¶22,887 (N.Y. Ct. Cl. 1976); *Brock v. California*, [1978] OSHD ¶22,834 (Cal. Ct. App. 1978); *White v. Utah*, [1978] OSHD ¶22,905 (Utah 1978).

[73]557 P.2d 1120 (Alaska 1976). Cf. *Blessing v. United States*, 447 F. Supp. 1160 (E.D. Pa. 1978).

[74]See Chapter 1 (Safety and Health Law Prior to the Occupational Safety and Health Act) for a comprehensive discussion of the early development of safety legislation.

vate or public welfare.[75] Something more than the safety and health laws was needed to compensate adequately the injured worker and make accidents the financial responsibility of the employer and not the worker.

State commissions were appointed to study the problem and recommend solutions,[76] and in 1910 a conference of nine state commissions and the U.S. government met for that purpose.[77] The 1910 conference agreed that existing systems were inadequate and turned to the workers' compensation concept as the best method of solving the problem. The principal appeal of this solution was twofold. First, the human toll was to be treated as a cost of production; second, the financial burden of accidents was moved from the worker to the employer, to be paid for by the consumer as part of the cost of the product.[78] Significant support for such a system was provided when both the National Association of Manufacturers and the American Federation of Labor endorsed the principle.[79]

Major obstacles to laws providing for workers' compensation, particularly if the laws were compulsory, were the Fourteenth Amendment of the U.S. Constitution and the guarantees of jury trial in both federal and state constitutions; for the laws in question provided for liability without fault, exempted certain classes of employees or employers, and created administrative tribunals to determine the rights of the parties. The first state statute, adopted by New York in 1910, was a compulsory statute. It was declared unconstitutional by the New York Court of Appeals[80] as violating the due process clauses of both the state and federal constitutions. The New York decision made state legislatures cautious about adopting workers' compensation statutes. Some enacted laws that were

---

[75]REPORT OF NATIONAL COMMISSION ON STATE WORKMEN'S COMPENSATION LAWS 34 (1971) (hereinafter cited as REPORT).

[76]W. Dodd, ADMINISTRATION OF WORKMEN'S COMPENSATION LAWS 27 n.89 (1936).

[77]*Id.* at 27. Earlier, in 1884 Germany had adopted a law which compensated employees as part of a social insurance program, financed in part by employee contributions. Prosser, THE LAW OF TORTS 530 (4th ed. 1971). England in 1897 had enacted a wage replacement law for occupational injuries and illnesses based on liability without fault of either the employer or employee, the costs to be borne by the employer. In 1908, Congress had enacted a law covering certain federal employees, Prosser, *supra*, subsequently amended to provide broad coverage for all federal employees, which included employees of the government of the District of Columbia.

[78]Prosser, *supra* note 77, at 531.

[79]REPORT at 34.

[80]*Ives v. South B. Ry.*, 201 N.Y. 271, 94 N.E. 431 (1911).

elective for both employer and employee, with penalties against the employer who rejected the act, such as imposing liability for damages at common law with the defenses removed. Others amended their state constitution to allow enactment of such statutes.[81]

In rapid succession state after state acted,[82] and the dilemma over whether to adopt compulsory or elective statutes was finally resolved when the New York Court of Appeals and the U.S. Supreme Court, in cases decided on the same day, declared both New York's compulsory law[83] and Iowa's elective law[84] to be constitutional as a valid exercise of the state police power, at the same time approving the exclusion of certain classes of employments as not violating the Equal Protection Clause of the Fourteenth Amendment. By 1920 all but six of the states then in existence had adopted some type of workers' compensation law.[85] Today every state has such a statute.

At the federal level the Congress, having previously legislated as to federal employees, in 1927 enacted the Longshoremen's and Harbor Workers' Compensation Act,[86] applicable to employees other than seamen engaged in maritime employment. Only railroad employees and seamen are not now covered by a workers' compensation law.

B. National Commission

Section 27 of the OSH Act declared that an equitable system of workers' compensation was a necessary adjunct to an effective program of occupational safety and health regulation since it provided the basic economic security for workers suffering disabling injuries or for their families in the event of death.[87] It also noted that serious questions concerning the efficacy of state workers' compensation laws increasingly were being raised in light of economic growth, a changing labor force, increased medical knowledge, new risks attributable to

---

[81]Dodd, *supra* note 76, at 33.

[82]*Id.* at 28.

[83]*White v. New York Cent. R.R.*, 216 N.Y. 653, 110 N.E. 105 (1915).

[84]*Arizona Employers Liability Cases*, 250 U.S. 400 (1915).

[85]Dodd, *supra* note 76, at 28.

[86]33 U.S.C. §§901 *et seq.* (1977).

[87]29 U.S.C. §676(a)(1)(A) (1977).

advancing technology, and a rising level of wages and the cost of living.[88] Section 27 of the OSH Act therefore established the National Commission on State Workmen's Compensation Laws for the purpose of making "an effective study and objective evaluation of State workmen's compensation laws in order to determine if such laws provide an adequate, prompt, and equitable system of compensation for injury or death arising out of or in the course of employment."[89]

In 1972, the Commission issued its report.[90] Generally, the report recommended against a federal statute, but made 84 recommendations for change, of which 19 were denoted as essential if state laws were to be adequate. These changes were set forth within the framework of five broad objectives for a modern workers' compensation program: (1) broad coverage of employees and of work-related injuries and diseases, (2) substantial protection against interruption of income, (3) provision of sufficient medical care and rehabilitation services, (4) encouragement of safety, and (5) an effective delivery system for benefits and services.[91]

Proposals to accomplish broad coverage included making all systems compulsory; eliminating exemptions from coverage based on size; extending coverage to farm, household, and casual workers on the same basis as other employees; making coverage of government employees mandatory; permitting an employee to file for benefits either in the state where he was hired, where he principally worked, or where he was injured; eliminating the "accident" prerequisite for compensability; and providing full coverage of all work-related diseases.[92] It suggested that income protection be effected by reducing waiting periods; increasing the maximum weekly wage benefit; increasing cash and death benefits to 80 percent of an employee's spendable weekly earnings for temporary total disability and death; and permitting beneficiaries of death benefits to receive them for longer periods.[93] It also recommended that for ful-

---

[88] §676(a)(1)(B).

[89] §676(a)(2).

[90] REPORT, *supra* note 75.

[91] *Id.* at 15.

[92] *Id.* at 3–52. In 1972, only 31 states had compulsory coverage. *Id.* at 45. In 1970, approximately 83.4% of the work force was covered by workers' compensation. *Id.* at 44.

[93] *Id.* at 53–75. In 1972, only 31 states provided compensation for life or the period of disability with respect to permanent total disabilities. *Id.* at 65. Only 15 states at

fillment of adequate medical and rehabilitative services, employees be permitted to select physicians, that limits on duration and amount of benefit receipts be eliminated, and that administrative supervision of medical and rehabilitative services be provided.[94] With respect to safety encouragement, the recommendations were that experience rating be used and that insurance carriers be required to provide loss prevention services.[95] Lastly, a number of administrative recommendations concerning efficient and effective delivery of services were made.[96]

The impact which the report of the Commission has had on workers' compensation statutes has been dramatic. The vast majority of states have improved their statutes wherever necessary in accordance with the Commission's recommendations.[97] Most obviously, the Act's authorization of the Commission's study made meaningful change a real possibility. As one authority in the field notes, the Commission's recommendations equaled or exceeded the highest standards of earlier, similar reports by other groups.[98] More importantly, tangible results have been realized with the widespread implementation of the Commission's recommendations. Furthermore, some of the changes recommended will ensure continuing improvements in the years to come. The recommendation concerning work-related injuries and diseases, for example, uses as a test "arising out of and in the course of the employment," thereby eliminating the need for future statutory change as new technologies both create new hazards and make it possible to detect existing ones. Where this recommendation has been adopted, it will no longer be necessary to amend existing statutes to add, say, byssinosis to the list of

---

the time provided for death benefits to an employee's widow until her death or remarriage and to dependent children under the age of 18. *Id.* at 72.

[94]*Id.* at 77–85.

[95]*Id.* at 87–98.

[96]*Id.* at 99–114.

[97]In January of 1981, the Department of Labor, which had been monitoring state compliance, issued a report analyzing state progress for the years 1972 to 1980 on numerous criteria such as full compliance with the essential recommendations by state and full compliance by all states. In accordance with those criteria, in 1972 there was 36.3% total compliance, whereas by 1980 total compliance had increased to 63.6%. Compliance from state to state varies considerably. U.S. Dep't of Labor, REPORT ON STATE COMPLIANCE WITH THE 19 RECOMMENDATIONS OF THE NATIONAL COMMISSION ON STATE WORKMEN'S COMPENSATION LAWS, 1972–1980 (1981).

[98]1 A. Larson, LAW OF WORKMEN'S COMPENSATION §5.30 (1978 & Supp. 1981).

covered occupational diseases. The broad, all-encompassing definition will accomplish coverage much more simply and efficiently.

C. Workers' Compensation Litigation

As expected in light of Section 4(b)(4), states have held that OSHA has no impact on their workers' compensation laws.[99]

However, the relationship between OSHA and workers' compensation arose in an interesting context during the challenges to the lead standard in *United Steelworkers v. Marshall*.[100] Among the employers' challenges was a contention that the medical removal protection provisions violated Section 4(b)(4) of the Act by supplanting state workers' compensation statutes insofar as lead exposure was concerned. The District of Columbia Circuit conceded that the medical removal protection provisions might have an impact, as a practical matter, on the number of claims under state workers' compensation laws, as would any successful health regulation; but it held that there was no violation of Section 4(b)(4) since the medical removal provisions did not alter the laws themselves.

The Act, pursuant to Section 4(b)(4), should also leave undisturbed those causes of action for work-related injuries or diseases previously not precluded by workers' compensation laws. For example, intentional acts by an employer that cause injury to an employee are not within the scope of a compensation law, since such acts are held to be not in the course of the employment. "Accident" and "intentional act" are mutually exclusive words.[101]

The general and majority rule is that an intentional act that will give rise to a suit for damages is one committed by the employer with *actual* intent to injure.[102] Examples of such acts are those constituting fraud, deceit, deliberate intent to injure, false imprisonment, and slander. They give rise to an

[99]*Childers v. International Harvester Co.*, 569 S.W.2d 675 (Ky. Ct. App. 1977); *Green Mountain Power Corp. v. Commissioner of Labor & Indus.*, 136 Vt. 15, 383 A.2d 1046, 6 OSHC 1499 (1978).

[100]647 F.2d 1189, 8 OSHC 1810 (D.C. Cir. 1980), *cert. denied*, 453 U.S. 913 (1981).

[101]*Pagan v. Kaufman*, 25 N.J. Super. 425, 92 A.2d 134 (1952).

[102]See, e.g., *Shearer v. Homestake Mining Co.*, 727 F.2d 707 (8th Cir. 1984).

action for damages at common law.[103] If the intentional act is committed by one the court considers an alter ego of the employer, the latter is subject to suit.[104] Gross negligence, recklessness, culpable or malicious negligence, violation of a safety regulation, or other misconduct do not rise to the level of acts committed with actual intent to injure.[105] Even removal of a machine guard at the direction of the employer has been held not such an intentional act as to permit suit for injury.[106]

However, a West Virginia court, in a decision one commentator calls "a distinctly out of line holding,"[107] decided that allowing employees to work under "known" hazardous conditions was an intentional act and permitted suit for damages.[108] More recently, an Ohio court found a cause of action in the mere allegation that employees were exposed to fumes of certain chemicals and that the employer, knowing of such conditions, failed to correct, to warn of, or to report conditions to state and federal authorities.[109] A California case with a contrary but anomalous result held that similar allegations were insufficient, but that a complaint alleging concealment of the existence of an occupational disease from the plaintiff and examining physicians, thus preventing plaintiff from getting treatment, stated a cause of action for aggravation of the disease.[110]

Misrepresentation of the result of a physical examination, by reason of which the employee did not timely file a workers' compensation claim, has been held to be actionable,[111] as has the failure of an employer to advise employees that they had

[103]Larson, supra note 98, at §78.13. But cf. Houston v. Bechtel Assocs., 522 F. Supp. 1094 (D.D.C. 1981) (recovery under Longshoremen's & Harbor Workers' Compensation Act exclusive remedy).

[104]Larson, supra note 98, at §78.13.

[105]Larson, supra note 98, at §68.13.

[106]Santiago v. Brill Monfort Co., 11 A.D.2d 1041, 205 N.Y.S.2d 919 (1960), aff'd, 10 N.Y.2d 718, 219 N.Y.S.2d 266 (1961).

[107]Larson, supra note 98, at 65 (1982 Supp.).

[108]Mandolidis v. Elkins Indus., 161 W. Va. 695, 246 S.E.2d 907 (1978). In 1983, the West Virginia legislature severely limited the scope of Mandolidis in H.B. 1201 amending §23-4-2 of the West Virginia Code.

[109]Blankenship v. Cincinnati Milacron Chems., 69 Ohio St.2d 608, 433 N.E.2d 572 (1982), cert. denied, 459 U.S. 857 (1983).

[110]Johns-Manville Prods. Corp. v. Contra Costa Superior Court, 27 Cal.2d 465, 612 P.2d 948, 165 Cal. Rptr. 858 (1980).

[111]Woodward v. Standard Forging, 112 F.2d 271 (7th Cir. 1940).

an occupational disease allegedly as the result of workplace exposure.[112]

An interesting development of comparatively recent origin is the dual capacity doctrine, which holds that an employer who has either assumed a second role with respect to the employee or breached a duty owed to the public which led to the injury of an employee may be sued. With respect to the assumption of a second role, for example, a member of a joint venture who leased a defective truck was held liable for damages to an employee of the joint venture who was injured because of the defect.[113] Illustrative of the breach of duty owed to the public is a case in which a tire manufacturer was held liable to an employee truck driver when a defective tire manufactured by the employer caused injury to the employee while driving a company-owned truck.[114] The court held that the manufacturer's duty to make a safe product reached to the employee when he was exposed to the same hazard he might have been exposed to as a member of the public. Similarly, an employee of a hospital who became ill at work was sent to the emergency room for treatment and was injured when the footstand of an X-ray table came loose, causing her to fall. The court held that as a patient she was entitled to the same duty of care as a paying patient.[115] Whether these decisions signal a general trend or are only minor deviations will be a matter for close attention in the coming years.

---

[112]*In re Johns-Manville/Asbestosis Cases*, 511 F. Supp. 1235 (N.D. Ill. 1981).

[113]*Smith v. Metropolitan Sanitary Dist.*, 61 Ill. App.3d 103, 377 N.E.2d 1088 (1978), *aff'd*, 396 N.E.2d 524 (1979).

[114]*Mercer v. Uniroyal, Inc.*, 49 Ohio App.2d 279, 361 N.E.2d 492 (1976).

[115]*Tatrai v. Presbyterial Univ. Hosp.*, 497 Pa. 247, 439 A.2d 1162 (1982).

26

The Federal Mine Safety and Health Act of 1977

All mines are subject to the jurisdiction of the Mine Safety and Health Administration (MSHA) rather than OSHA. Although there are a number of similarities between MSHA and OSHA, the two legislative and regulatory schemes are quite different. This chapter will deal with mine safety procedures and requirements.

I. Legislative History

On March 3, 1891, Congress passed "an Act for the protection of the lives of miners in the Territories."[1] This relatively brief piece of legislation was the first federal statute governing mine safety. It provided for inspection of coal mines only in the territories. The 1891 Act required mines to have at least two shafts, slopes, or other outlets; required minimum ventilation of 3,300 cubic feet of air per minute per 50 men; required safety catches on man hoists; required speaking tubes; and prohibited use of furnace shafts as escape shafts. Mine operators also were prohibited from employing any children under 12 years of age.

---

[1]Ch. 564, 26 Stat. 1104 (1891).

Notices of unsafe conditions could be issued to mine operators. Failure to correct the condition within the time period of the notice was a misdemeanor which could lead to a fine of up to $500. Section 16 of the 1891 Act provided that "as a cumulative remedy" courts could issue an injunction restraining mining operations until problems in notices were corrected.

On May 16, 1910, Congress established a Bureau of Mines in the Department of the Interior.[2] Section 2 of this Act provided as follows:

> "[I]t shall be the province and duty of said bureau and its director, under the direction of the Secretary of the Interior to make diligent investigation of the methods of mining, especially in relation to the safety of miners, and the appliances best adapted to prevent accidents, the possible improvement of conditions under which mining operations are carried on, the treatment of ores and other mineral substances, the use of explosives and electricity, the prevention of accidents, and other inquiries and technologic investigations pertinent to said industries, and from time to time to make such public reports of the work, investigations, and information obtained as the Secretary of said department may direct, with the recommendations of such bureau."[3]

This initial legislation establishing the Bureau of Mines specifically denied all Bureau officers or employees "any right or authority in connection with the inspection or supervision of mines or metallurgical plants."[4] The lack of any inspection authority was a shortcoming. The Bureau was not given authority to make inspections until May 1941.[5] The 1941 legislation provided for authority to make "annual or necessary inspections."[6] It had no provision for promulgating health or safety standards for coal mines or for achieving compliance with the standards or recommendations of the Bureau or the Secretary of the Interior. With no enforcement tools, and providing only a generalized verbal commitment in favor of health and safety, the force of the legislation remained unclear.

With the passage of the Federal Coal Mine Safety Act of 1952,[7] the current legislation first began to take shape. The

---

[2] Ch. 240, 36 Stat. 369 (1910).
[3] *Id.* at 370.
[4] *Id.*
[5] Ch. 87, 55 Stat. 177 (1941).
[6] *Id.*
[7] Pub. L. No. 82-552, 66 Stat. 692 (1952).

1952 Act provided for the issuance of violation notices and imminent danger withdrawal orders. There were no provisions for regulations, and safety requirements were thus limited to those outlined in Section 209. There were penalty provisions providing fines for noncompliance with withdrawal orders (up to $2,000) or failure to give inspectors access to mine property (up to $500), but no penalties for noncompliance with the safety provisions.

The first federal statute directly regulating noncoal mines did not appear until the passage of the Federal Metal and Nonmetallic Mine Safety Act of 1966.[8] This 1966 Act provided for the promulgation of standards and for inspections and investigations,[9] yet its enforcement scheme was minimal. It provided that employees could be barred from entering areas of mines where imminent dangers existed[10] and provided for the issuance of citations for violations of mandatory standards.[11] If the violation of a standard was not corrected, orders could be issued barring personnel from the area until the violation was corrected.[12] If an operator failed to comply with any of these orders, the Secretary of the Interior could pursue injunctive relief in federal district court.[13] Additionally, a criminal penalty for failure to comply could be pursued.[14] In practice, these remedies seldom were used.

In 1969, Congress passed the Federal Coal Mine Health and Safety Act of 1969 (1969 Coal Act).[15] This legislation was more comprehensive than any previous federal legislation regulating the mining industry. In addition to the types of remedies encompassed in the 1966 Metal and Nonmetallic Act, the 1969 Coal Act also provided for civil and criminal penalties for violations of health and safety standards.[16] Titles II and

---

[8]Pub. L. No. 89-577, 80 Stat. 772 (1966), codified at 30 U.S.C. §§721 *et seq.* (1976) (repealed 1977).

[9]30 U.S.C. §725 (1976).

[10]§727(a).

[11]§727(b).

[12]*Id.*

[13]§733(a).

[14]§733(b).

[15]Pub. L. No. 91-173, 83 Stat. 742 (1969), codified at 30 U.S.C. §§801 *et seq.* (1976).

[16]30 U.S.C. §819 (1976).

III included a number of interim health and safety standards.[17] Title IV of the legislation provided benefits for employees who develop pneumoconiosis (black lung) as a result
of their employment in underground coal mines.[18]

In 1977, Congress passed the Federal Mine Safety and
Health Act of 1977 (Mine Act), the legislation that is the focus
of this chapter.[19] The Mine Act was an amendments act which
amended the 1969 Coal Act in a number of significant ways.
The Mine Act consolidated all federal regulation of coal mining as well as noncoal mining under a single statutory scheme.
Thus, the 1966 Metal and Nonmetallic Act was repealed. The
Mine Act also established an independent regulatory commission, the Federal Mine Safety and Health Review Commission, to review citations and orders.[20] The 1966 Metal and
Nonmetallic Act had established an independent review board.[21]
However, because of the nature of the enforcement proceedings under that Act, there had been little activity on the part
of the board. The 1969 Coal Act provided for review by the
Secretary of the Interior, who had established the Board of
Mine Operations Appeals under his jurisdiction.[22] The enforcement mechanisms under the Mine Act are generally similar to the 1969 Coal Act, although some new remedies were
added. Miners' rights also were broadened by the Mine Act.

A primary factor motivating Congress in the passage of
the various pieces of legislation was the hazards in the mining
industry and number of mining disasters that occurred over
the years. The Sunshine Silver Mine disaster of May 1972,
the Buffalo Creek Coal Mine impoundment dam break in February 1972, and the Scotia Coal Mine disaster in March 1976
played a particularly significant role in the passage of the
Mine Act.[23]

---

[17]§§841 *et seq.*
[18]§§901 *et seq.*
[19]Pub. L. No. 95-164, 91 Stat. 1290 (1977).
[20]30 U.S.C. §823 (Supp. II 1978).
[21]*Supra* note 8, at §729.
[22]*Supra* note 15, at §815.
[23]H.R. REP. NO. 312, 95th Cong., 1st Sess. 3–7, *reprinted in* 1977 U.S. CODE CONG.
& AD. NEWS 3401; S. REP. NO. 181, 95th Cong., 1st Sess. 4, *reprinted in* 1977 U.S.
CODE CONG. & AD. NEWS 3404.

II. The 1977 Act

A. Overview

As detailed in the balance of this chapter, the Mine Act is enforced by the Secretary of Labor through the Mine Safety and Health Administration.[24] The Mine Act provides for the promulgation of regulations by the Secretary of Labor. Inspections, during which inspectors may issue a variety of citations and orders, are authorized.

Citations or orders that have been issued may be accepted by the mine operator, with any accompanying fine being paid, or challenged through administrative litigation. If a citation or order is challenged, the matter is initially heard by an administrative law judge. The decision of the administrative law judge may be reviewed at the discretion of the Federal Mine Safety and Health Review Commission (Mine Commission). Following Mine Commission action, a matter may be appealed to an appropriate court of appeals.

B. Jurisdiction—Definition of "Mine"

The Mine Act applies only to a "coal or other mine, the products of which enter commerce, or the operations or products of which affect commerce."[25] "Coal or other mine" is defined as follows:

"(A) An area of land from which minerals are extracted in non-liquid form or, if in liquid form, are extracted with workers underground,

"(B) private ways and roads appurtenant to such area, and

"(C) lands, excavations, underground passageways, shafts, slopes, tunnels and workings, structures, facilities, equipment, machines, tools, or other property including impoundments, retention dams, and tailing ponds, on the surface or underground, used in, or to be used in, or resulting from, the work of extracting such minerals from their natural deposits in nonliquid form, or if in liquid form, with workers underground, or used in, or to be used in, the milling of such minerals, or the work of preparing coal or other minerals, and includes custom coal preparation facilities."[26]

---

[24]§302, Pub. L. No. 95-164, 91 Stat. 1290 (1977).

[25]30 U.S.C. §803 (Supp. II 1978).

[26]§802(h)(1).

The definition of "coal or other mine" also provides that in determining what constitutes "mineral milling," the Secretary of Labor is to give due consideration to the "convenience of administration" that can result from having all authority with respect to the health and safety of miners at one physical establishment delegated to one assistant secretary.[27]

On first reading, it might seem that the definition of "mine" would be understood by all, with little controversy. Probably because of the differences between OSHA and MSHA regulations, however, a number of questions have been raised regarding the definition of "mine" and the scope of MSHA jurisdiction.

The definitional issue initially was raised in the context of search warrants. The Supreme Court ruled in *Marshall v. Barlow's, Inc.*[28] that an employer may require OSHA to obtain a search warrant before entering its premises for an inspection or investigation. One of the early decisions regarding search warrants also involved an issue regarding the definition of "mine." In *Marshall v. Stoudt's Ferry Preparation Co.,*[29] the court concluded that a company which purchased material dredged by the Commonwealth of Pennsylvania from the Schuylkill River was a mine. The company purchased the material and by use of a front-end loader and conveyer belts transported the dredged material to its plant. There, through a sink and float process, the material was separated into sand and gravel and burnable materials, which were used for fuel. Noting that Congress had made clear that the definition of "mine" was to be given the broadest possible interpretation, the court of appeals concluded that Stoudt's Ferry was subject to MSHA jurisdiction.

The legislative history of the Mine Act, relied upon in *Stoudt's Ferry*, directs that "mine" be interpreted broadly:

> "The Committee notes that there may be a need to resolve jurisdictional conflicts but it is the Committee's intention that what is considered to be a mine and to be regulated under this Act be given the broadest possibly [sic] interpretation, and it is the intent of this Committee that doubts be resolved in favor of inclusion of a facility within the coverage of the Act."[30]

---

[27]*Id.*

[28]436 U.S. 307, 6 OSHC 1571 (1978).

[29]602 F.2d 589, 1 MSHC 2097 (3d Cir. 1979), *cert. denied*, 444 U.S. 1015 (1980).

[30]S. REP. NO. 181, *supra* note 23, at 14, 1977 U.S. CODE CONG. & AD. NEWS at 3414.

It is apparent from the legislative history that "mine" is intended to be construed broadly in determining MSHA's jurisdiction. Further, the reference in the definition itself to convenience of administration in avoiding duplicative jurisdiction for a single facility suggests some discretion in enforcement. It should be added that since the OSH Act provides that it will apply to working conditions not covered by some other federal agency,[31] the issue is who will regulate, not whether regulation will occur.

The Mine Commission has issued several decisions on this issue. In *Secretary of Labor v. Cyprus Industrial Mineral Co.*,[32] the issue was whether exploratory work is included in the definition of "mine." There the administrative law judge had found that an employee who was fatally injured in an accident had been engaged in blasting, drilling, cutting, and other mining-type activities. It was conceded that there was an existing ore body. The individuals were engaged in establishing a portal and driving an exploration drift. It was concluded that these were mining activities and thus jurisdiction was found. This decision was affirmed by the Ninth Circuit.[33]

This decision is not particularly surprising. While it did not involve actual extraction of minerals, it did include traditional mining activities preparatory to the actual extraction of minerals. Where the activities involved are performed underground, it is difficult to hypothesize a situation where MSHA jurisdiction would not be asserted and upheld. Surface activities, on the other hand, may in some instances be similar, or identical, to activities traditionally covered by OSHA. In that situation, there may be more dispute as to jurisdiction. Cases to date have not focused upon this sort of distinction.

The Mine Commission also has held that a company reclaiming coal from an above-ground refuse pile was a mine.[34] The reclamation work was being done after the mine itself had been sealed. The work involved using an end-loader to remove material from the refuse pile and deposit it in trucks. The trucks took the material to a screening plant where the coal was separated from other materials and stockpiled for

---

[31] 29 U.S.C. §653(b)(1) (1976).

[32] 2 MSHC 1128 (FMSHRC No. DENV 78-558-M, 1981).

[33] 664 F.2d 1116, 2 MSHC 1554 (9th Cir. 1981).

[34] *Secretary of Labor v. Alexander Bros.*, 2 MSHC 1670 (FMSHRC No. HOPE 79-221-P, 1982).

transportation to a cleaning plant. Following cleaning and some additional processing, the coal was then sold. In finding that the operation was a mine, the Mine Commission focused upon the definition of "coal or other mine" and the inclusion in that definition of "the work of preparing coal."[35] "Work of preparing coal" includes the following: "[T]he breaking, crushing, sizing, cleaning, washing, drying, mixing, storing, and loading of bituminous coal, lignite, or anthracite, and such other work of preparing such coal as is usually done by the operator of the coal mine."[36] The administrative law judge had found, and the Mine Commission agreed, that the company's processes included "breaking, crushing, sizing, cleaning, washing, drying, mixing, storing, and loading" of coal and thus the operation was held to be a coal mine and subject to MSHA jurisdiction.

Certain facilities have been held not to be mines. One such facility was a commercial loading dock on the Ohio River. In *Secretary of Labor v. Oliver M. Elam, Jr., Co.*,[37] Elam operated a dock facility which dealt with a variety of commodities. Forty to sixty percent of the tonnage loaded at the dock was coal. Among the activities performed in addition to the unloading and stockpiling of coal and loading of coal on barges was crushing of coal to a single size. Although noting that Elam did engage in several of the functions included in the definition of "work of preparing coal" (storing, breaking, crushing, and loading), the Mine Commission concluded that these functions were performed "solely to facilitate its loading business and not to meet customers' specifications nor to render the coal fit for any particular use."[38]

MSHA jurisdiction was rejected by the Mine Commission in *Secretary of Labor v. Carolina Stalite Co.*[39] Carolina Stalite produced a lightweight construction material ("Stalite") from slate gravel. The gravel was purchased from an adjacent quarry which was an entity independent of Carolina Stalite. The gravel was delivered to Carolina Stalite by conveyer belts. Carolina

---

[35]30 U.S.C. §802(h)(1) (Supp. II 1978).

[36]§802(i).

[37]2 MSHC 1572 (1982). *Cf. Secretary of Labor v. Mineral Coal Sales*, 3 MSHC 1755 (FMSHRC No. VA 83-26, 1985). See also *Donovan v. Inland Terminals*, 3 MSHC 1893 (S.D. Ind. 1985) (loading facility not a "mine").

[38]2 MSHC at 1574.

[39]2 MSHC 1665 (FMSHRC No. BARB 79-319-M, 1982).

Stalite heated the slate gravel in rotary kilns and subsequently crushed and sized the material. The material was sold primarily for production of lightweight masonry blocks. The Mine Commission concluded that the work done by Carolina Stalite was "a manufacturing process that results in a product, rather than 'milling' process under the Mine Act."[40] The Mine Commission thus held that Carolina Stalite was not subject to MSHA jurisdiction. Commissioner Lawson dissented, focusing upon the definition of "mine" and the legislative history stating that the scope of the Mine Act should be construed broadly. The court of appeals reversed, holding that the broad definition of "mine" and the legislative history of the Mine Act give the Secretary of Labor relatively wide discretion in determining what constitutes "mineral milling." The court ruled that the Secretary acted within his power in asserting Mine Act jurisdiction.[41]

One would expect litigation to continue in this area. The likelihood is that the focus will continue to be on surface operations which are on the fringes of the mining process. Where work is actually being performed underground, it is difficult to hypothesize a situation where the Mine Commission would deny jurisdiction. Because the statute gives the Secretary of Labor some discretion in determining jurisdiction between OSHA and MSHA regarding mineral milling facilities, and because of the significant differences in regulatory schemes, further refinements may well be pursued in litigation.[42]

C. Promulgations of Regulations

Many regulations are in force under the Mine Act covering a wide variety of health and safety issues. These substantive regulations appear in the *Code of Federal Regulations*.[43] The substantive regulations promulgated by MSHA are beyond the scope of this treatise. Procedures for promulgation of regulations will be reviewed, however.

---

[40]*Id.* at 1666.

[41]734 F.2d 1547, 3 MSHC 1337 (D.C. Cir. 1984).

[42]Office facilities where engineering personnel are working on the design for construction of a shaft and underground storage of nuclear waste are not a "mine." *Paul v. P.B.-K.B.B., Inc.*, 3 MSHC 2006, *aff'd*, 812 F.2d 717 (D.C. Cir. 1987).

[43]Substantive regulations are in 30 C.F.R. pts. 11–100 (1985). Mine Commission procedural regulations appear at 29 C.F.R. pt. 2700 (1985).

The Mine Act provides that mandatory standards that had been issued by the Secretary of the Interior under the 1966 Metal and Nonmetallic Act and standards and regulations that had been issued under the 1969 Coal Act were to be continued in effect until the Secretary of Labor issued new or revised standards.[44] Procedures for the promulgation of mandatory safety and health standards are set forth in Section 101 of the Mine Act.[45]

Section 102 provides for the appointment of advisory committees to conduct research on various health and safety matters.[46] Additionally, the Secretary of Labor may receive information from organizations or individual representatives of employers or employees, the National Institute for Occupational Safety and Health (NIOSH), state or political subdivisions, or any other interested person.[47] The Secretary also may develop information himself pursuant to inspections or investigations conducted under the Mine Act.[48] When this safety or health information is made available to the Secretary, he may either request the recommendation of one of his advisory committees or act on the information himself. The statute provides, however, that if the recommendation with appropriate criteria is received from NIOSH, within 60 days after receipt of the recommendation, (1) the recommendation must be referred to an advisory committee, (2) a proposed rule must be published in the *Federal Register*, or (3) a determination not to accept the recommendation must be published in the *Federal Register*, accompanied by the reasons for the determination.[49]

Proposed rules promulgating, modifying, or revoking mandatory health or safety standards must be published in the *Federal Register*. Interested persons are to be given 30 days after publication to submit written data or comments on the rule, subject to extensions by the Secretary of Labor. A request may also be made for a public hearing. Within 60 days after the filing of comments, the Secretary is obligated to

---

[44]§301(b)(1), Pub. L. No. 95-164, 91 Stat. 1290 (1977).
[45]30 U.S.C. §811 (Supp. II 1978).
[46]§812.
[47]§811(a)(1).
[48]§813(a).
[49]§811(a)(1).

publish in the *Federal Register* notification of the standards commented upon and propose a schedule for a hearing, if requested. The hearing must commence no later than 60 days after publication of the hearing notice. Interested parties are to be allowed to present oral or written comments and the Secretary may utilize subpoena power to compel attendance of witnesses.[50]

Within 90 days after certification of the record of a hearing, or within 90 days after the close of the comment period if no hearing has been requested, the Secretary by rule is to promulgate, modify, or revoke the health or safety standard and publish his reasons for the action. If a proposed standard is not to be promulgated within the 90-day time period, then the reasons for delay are also to be published in the *Federal Register*. New standards are effective upon publication in the *Federal Register* unless otherwise provided.[51]

The statute has additional provisions regarding regulations relating to toxic materials or harmful substances. The Secretary is directed to "set standards which most adequately assure on the basis of the best available evidence that no miner will suffer material impairment of health or functional capacity even if such miner has regular exposure to the hazards dealt with by such standard for the period of his working life."[52] In addition to attaining "the highest degree of health and safety protection for the miner," other considerations for the Secretary are to evaluate the latest available scientific data, feasibility of the standard, and experience under health and safety laws.[53]

Standards are to provide for labels or other appropriate warnings where necessary, as well as suitable protective equipment to maximize the protection of miners.[54] Additionally, where appropriate, mandatory standards are to provide that if "a determination is made that a miner may suffer material impairment of health or functional capacity by reason of exposure to the hazard covered by [a] mandatory stan-

---

[50]§811(a)(2) and (3).
[51]§811(a)(4) and (5).
[52]§811(a)(6)(A).
[53]*Id.*
[54]§811(a)(7).

dard," then the miner is to be removed from the exposure and reassigned to a different position "at no less pay than the regular rate of pay for miners in the classification such miner held immediately prior to his transfer."[55]

The Secretary is directed, to the extent practicable, to promulgate separate standards for mine surface construction activity.[56] Additionally, the Mine Act prohibits the promulgation of any standard that would "reduce the protection afforded miners by an existing mandatory health or safety standard" at the time of the passage of the Mine Act.[57]

The Mine Act also provides that if the Secretary determines that "miners are exposed to grave danger," an emergency temporary standard may be promulgated. Where that occurs, a mandatory health or safety standard must be promulgated under the procedures of Section 101(a) just outlined within 9 months after publication of the emergency temporary standard.[58] This provision has not been utilized by the Secretary of Labor.

Any person adversely affected by a mandatory health or safety standard promulgated under Section 101 may file a petition challenging the validity of the standard within 60 days after the promulgation of the standard. That challenge may be filed in either the District of Columbia Circuit or the circuit in which the aggrieved person resides or has his principal place of business. No objection that has not been raised with the Secretary can be considered by the court unless good cause is shown for failure to raise the objection.[59]

The Mine Act also provides that the Secretary is to send a copy of every proposed health or safety standard or regulation at the time of publication in the *Federal Register* to each mine operator and representative of miners for posting on the mine bulletin board. Failure to receive a notice does not relieve anyone of the obligation to comply with the standard or regulation, however.[60]

---

[55]*Id.*
[56]§811(a)(8).
[57]§811(a)(9).
[58]§811(b).
[59]§811(d).
[60]§811(e).

D. Inspections and Investigations

Inspections and investigations are covered in Section 103 of the Mine Act.[61] There are to be "frequent" inspections and investigations for the following purposes:

"(1) [O]btaining, utilizing, and disseminating information relating to health and safety conditions, the causes of accidents, and the causes of diseases and physical impairments originating in such mines;

"(2) gathering information with respect to mandatory health or safety standards;

"(3) determining whether an imminent danger exists; and

"(4) determining whether there is compliance with the mandatory health or safety standards or with any citation, order, or decision issued under this subchapter or other requirements of this chapter."[62]

Giving advance notice of inspections is prohibited, except for inspections conducted by the Secretary of Health, Education, and Welfare (Health and Human Services) under clauses 1 and 2 noted above.[63]

The Secretary is to make a complete inspection of all underground mines at least four times a year and of all surface mines at least twice a year.[64] Special provision is also made for additional inspections of mines that liberate excessive quantities of methane or other explosive gases or have had a death or serious injury from a methane or other gas ignition in the prior 5 years.[65] The Secretary of Labor may develop guidelines for additional inspections based on criteria such as a mine's hazards and experience under the Mine Act and other health and safety laws.[66] Depending upon the volume of methane or other explosive gases liberated, there may be a spot inspection as frequently as once every 5 working days.

Although the provision is seldom used, the Secretary is empowered to conduct public hearings, after notice, in connection with investigation of any accident or other occurrence relating to health and safety.[67] The Mine Act also provides

---

[61]30 U.S.C. §813 (Supp. II 1978).

[62]§813(a).

[63]*Id.*

[64]*Id.*

[65]§813(i).

[66]§813(a).

[67]§813(b).

that accidents are to be investigated by the operator or his agent to determine the accident's cause and means of preventing the accident, and that operator records regarding the accident and investigation are to be made available to the Secretary.[68] Detailed regulations have been promulgated by the Secretary of Labor regarding not only accident investigations, but also recordkeeping in connection with accidents.[69]

1. Search Warrants

Companies regulated by OSHA can require a search warrant before an inspector is allowed on the property to inspect the plant.[70] The Supreme Court has held, however, that mines cannot insist upon a search warrant before an inspection.[71] A Mine Commission administrative law judge held, following the Supreme Court, that a mine operator's refusal to allow an inspector to inspect without a search warrant was a violation of Section 103(a) of the Mine Act.[72]

An open question is whether a search warrant may be required before an MSHA inspector reviews mine records. This involves two issues which may not necessarily lead to the same result. The first is whether a search warrant can be insisted upon in order to review records the mine operator is required to keep under the Mine Act or its regulations. The second issue is whether a warrant can be insisted upon before the agency is allowed to review mine records that the mine operator is *not* required by law to keep, but keeps as a part of the ongoing management of its operations. It should go without saying, too, that a number of practical, as well as legal, issues are involved in determining whether to insist on a warrant before allowing review of documents by MSHA.

A Mine Commission administrative law judge fined Peabody Coal Co. $500 for insisting on a warrant before disclosing records required to be kept by the Mine Act. The Mine Commission affirmed the judge's holding that there is no expec-

---

[68]§813(d).

[69]30 C.F.R. pt. 50 (1985).

[70]*Marshall v. Barlow's, Inc.*, 436 U.S. 307, 6 OSHC 1571 (1978). See text accompanying note 28, *supra.*

[71]*Donovan v. Dewey*, 452 U.S. 594, 2 MSHC 1321 (1981).

[72]*Secretary of Labor v. John Cullen Rock Crushing*, 2 MSHC 1989 (FMSHRC J. No. WEST 82-33-M, 1982).

tation of privacy in records a mine operator is required to keep by the Mine Act or regulations.[73]

In *United States v. Consolidation Coal Co.*,[74] a case that arose under the 1969 Coal Act, the Sixth Circuit held that a search warrant could be insisted upon even if the government were seeking to inspect records required to be kept by statute. The court held that the proper standard for such a warrant was administrative, not criminal, probable cause, even though the litigation involved alleged criminal violations.

Even if a search warrant is required, the impact of that requirement is uncertain. The Sixth Circuit has indicated that the scope of the exclusionary rule may be limited in this context. In *United States v. Blue Diamond Coal Co.*,[75] which also arose under the 1969 Coal Act, the court concluded that an inspector's seizure of records required to be kept under the Coal Act, without a warrant or operator consent, was improper. The court refused to exclude the records from evidence in this criminal proceeding, however, saying that the mine operator's expectation of privacy was insufficient to justify exclusion.

2. Accident Investigations

In addition to the general provision of Section 103(a) for the investigation of accidents, specific powers are also granted to the Secretary in connection with those investigations. As already noted, the Secretary may set a public hearing as a part of an accident investigation, although that is seldom done. Additionally, the Mine Act provides that the Secretary or his authorized representative (inspector) shall be allowed to review company accident investigation records.[76] Should an accident occur, the Secretary may, as a part of the investigation of the accident, issue orders to ensure the safety of personnel and also to provide for rescue or recovery of personnel and returning affected areas to normal.[77]

---

[73]*Peabody Coal Co.*, 3 MSHC 1234 (FMSHRC No. KENT 80-318-R, 1984).

[74]560 F.2d 214, 1 MSHC 1549 (6th Cir. 1977), *vacated and remanded*, 436 U.S. 942 (for further consideration in light of *Marshall v. Barlow's, Inc., supra* note 70), *judgment reinstated*, 579 F.2d 1011, 1 MSHC 1664 (6th Cir. 1978), *cert. denied*, 439 U.S. 1069 (1979).

[75]667 F.2d 510, 2 MSHC 1521 (6th Cir. 1981), *cert. denied*, 456 U.S. 1007 (1982).

[76]30 U.S.C. §813(d) (Supp. II 1978).

[77]§813(k).

The Secretary has promulgated extensive regulations regarding accident notification and investigation records.[78] These regulations require immediate notification to MSHA if an accident occurs.[79] They also require that mine operators preserve the accident scene until all investigations have been completed, "except to the extent necessary to rescue or recover an individual, prevent or eliminate an imminent danger, or prevent destruction of mining equipment."[80] The failure to immediately notify the Interior Department's Mine Enforcement and Safety Administration, MSHA's predecessor, of a methane ignition, and to preserve the scene, led to two $7500 penalties for the mine operator.[81]

A key point to note in these regulations is that there are separate definitions for "accident," "occupational injury," and "occupational illness." If the incident which occurs is an accident, it must be immediately reported to MSHA.[82] After such immediate notification, MSHA will then determine whether it wishes to conduct an immediate investigation of the accident.[83] In any event, the mine operator must complete the reporting requirements spelled out in the regulations.

If, however, the incident is merely an occupational injury or occupational illness, then the only reporting obligation is completion of the forms described in the regulations. These forms must be mailed to MSHA within 10 working days after the accident or occupational injury occurs or the occupational illness is diagnosed.[84] The mine operator is also required to maintain copies of these forms as well as its own investigation reports for a period of at least 5 years.[85]

A mine operator is to report an accident or injury even if there is no causal connection to the miner's work. Thus a mine operator was found to have violated the reporting requirements by failing to report the hospitalization of a miner who

[78] 30 C.F.R. pt. 50 (1985).
[79] 30 C.F.R. §50.10.
[80] §50.12.
[81] *Secretary of Labor v. Itmann Coal Co.*, 2 MSHC 1786 (FMSHRC No. HOPE 76-197-P, 1982).
[82] §50.10.
[83] §50.11.
[84] §50.20(a).
[85] §50.40.

had a history of back trouble and suffered severe pain while putting on his work boots before the start of a shift.[86]

The key question to note in applying these regulations is whether the incident that has occurred is an "accident." The definitional criteria are fairly straightforward, with one exception. An accident includes an injury "which has a reasonable potential to cause death."[87] The application of this definition is unclear. To date, the Mine Commission has not spoken to this issue. Where an injury, but no fatality, has occurred, careful evaluation of the situation would be appropriate.

E. Citations and Orders

1. Section 104(a) Citation

The most frequently issued MSHA enforcement action is a citation under Section 104(a).[88] This citation may be used for violations of the Mine Act, violations of mandatory health or safety standards, or violations of other rules and regulations promulgated under the Mine Act. The citation is to describe "with particularity" the nature of the violation, including a reference to the provision of the Mine Act or regulations alleged to have been violated. Additionally, the citation is to set a "reasonable time" for the abatement (correction) of the violation.

2. "Significant and Substantial" Violations

A key issue for a mine operator is whether a citation it receives is designated "significant and substantial." If a citation is not deemed "significant and substantial," then under the current penalty assessment regulations a mine operator may receive a minimum penalty of $20.[89] Additionally, whether a particular citation is deemed significant and substantial is

---

[86]*Secretary of Labor v. Freeman United Coal Mining*, 3 MSHC 1447 (FMSHRC No. LAKE 82-89, 1984).

[87]§50.20(h)(2).

[88]30 U.S.C. §814(a) (Supp. II 1978).

[89]30 C.F.R. §100.4 (1985). Penalties are discussed in greater detail *infra* in text accompanying notes 136–161.

critical on issues regarding unwarrantable failure citations and pattern of violation notices, discussed below.

In examining the $20 assessment provision of the Secretary's regulation, the Mine Commission has concluded that it would look at all the statutory criteria in determining whether a $20 penalty was proper.[90] The Mine Commission previously had held that it is *not* bound by the Secretary's penalty assessment regulation in contested penalty cases.[91]

The legislative history of the Mine Act provides some guidance on the issue of the "significant and substantial" designation for violations. The Senate report approved a construction of that term by the Board of Mine Operations Appeals under the 1969 Coal Act in *Alabama By-Products Corp.*[92] There the Board held that a violation would *not* be significant and substantial in two kinds of situations:

> "[1] [V]iolations posing no risk of injury at all, that is to say, purely technical violations, and [2] violations posing a source of *any* injury which has only a remote or speculative chance of coming to fruition. A corollary of this proposition is that a notice of violation may be issued under Section 104(c)(1) without regard for the seriousness or gravity of the injury likely to result from the hazard posed by the violation, that is, an inspector need not find a risk of serious bodily harm, let alone of death."[93]

The key decision interpreting "significant and substantial" under the Mine Act is *Secretary of Labor v. National Gypsum Co.*[94] At issue for the Mine Commission in *National Gypsum* was how inclusive the concept should be. The Mine Commission rejected an interpretation that would include virtually all violations on the basis that such an interpretation would render the statutory language superfluous. It also indicated that a broad, virtually all-encompassing definition of "significant and substantial" would have an "untenable effect" on the application of the pattern of violation provisions of the Mine Act. The standard set forth by the Mine Commission is that a violation will be deemed significant and substantial "if,

[90]*Secretary of Labor v. U.S. Steel Mining Co.*, 3 MSHC 1363 (FMSHRC No. PENN 82-328, 1984).

[91]*Secretary of Labor v. Sellersburg Stone*, 2 MSHC 2010 (FMSHRC No. LAKE 80-363-M, 1983), *aff'd*, 736 F.2d 1147, 3 MSHC 1385 (7th Cir. 1984).

[92]S. Rep. No. 181, *supra* note 23, at 31, 1977 U.S. Code Cong. & Ad. News at 3431.

[93]83 Interior Dec. 574, 578, 1 MSHC 1484, 1487 (IBMA No. 76-29, 1976).

[94]2 MSHC 1201 (FMSHRC No. VINC 79-154-PM, 1981).

based upon the particular facts surrounding that violation, there exists a reasonable likelihood that the hazard contributed to will result in an injury or illness of a reasonably serious nature."[95]

Subsequent to the issuance of the *National Gypsum* decision, which was not appealed, MSHA issued a memorandum outlining its interpretation of "significant and substantial."[96] Because the designation is important, both in penalty assessment and in other areas, it will probably continue as a subject of litigation.

3. Failure-to-Abate Closure Orders

As already noted, Section 104(a) requires that a citation issued under that subsection must "fix a reasonable time for the abatement of the violation." That abatement time begins to run at the time of the issuance of the citation, not at the time the citation becomes a final order.[97] If an inspector determines in a follow-up inspection that a Section 104(a) citation has not been totally abated within the time period set in the original citation or in any subsequent amendments, and that the time for abatement should not be extended, then the inspector is to issue a closure order. The closure order is to determine "the area affected by the violation" and to prohibit persons from entering the affected area until the violation has been abated.[98]

Section 104(c) of the Mine Act does allow certain categories of persons to enter areas under a failure-to-abate closure order. These include individuals needed to eliminate the condition described in the order, public officials whose duties require them to enter the area, representatives of miners, and consultants to any of the foregoing.[99]

---

[95]*Id.* at 1203 (1981). See also *Secretary of Labor v. Mathies Coal Co.*, 3 MSHC 1184, 1186 (FMSHRC No. PENN 82-3-R, 1984) (restating the test). There is a rebuttable presumption that a respirable coal dust violation is "significant and substantial." *Consolidation Coal Co. v. Secretary of Labor*, 4 MSHC 1001 (FMSHRC No. WEVA 82-245, 1986), *aff'd*, 824 F.2d 1071 (D.C. Cir. 1987).

[96]MINE SAFETY & HEALTH REP. [2 Current Rep. 540–541] (May 20, 1981).

[97]Cf. 29 U.S.C. §659(b) (1976).

[98]30 U.S.C. §814(b) (Supp. II 1978).

[99]§814(c).

4. Unwarrantable Failure Citations and Orders

Section 104(d) provides for the issuance of "unwarrantable failure" citations and closure orders.[100] Conditions for issuance of an unwarrantable failure citation are (1) violation of a mandatory health or safety standard; (2) a situation that does *not* cause an imminent danger; (3) a significant and substantial violation; and (4) an "unwarrantable failure" by the mine operator to comply with the mandatory health or safety standard.[101] The two key areas of focus are the "significant and substantial" finding and the "unwarrantable failure" determination.

The Mine Commission has held that "unwarrantable failure" "means aggravated conduct, constituting more than ordinary negligence. . . ."[102] The legislative history also treats the question. The Senate report specifically approved *Zeigler Coal Co.,* a prior decision of the Board of Mine Operation Appeals defining "unwarrantable failure."[103] The Board held

> "that an inspector should find that a violation of any mandatory standard was caused by an unwarrantable failure to comply with such standard if he determines that the operator involved has failed to abate the conditions or practices constituting such violation, conditions or practices the operator knew or should have known existed or which it failed to abate because of a lack of due diligence, or because of indifference or lack of reasonable care. The inspector's judgment in this regard must be based upon a thorough investigation and must be reasonable."[104]

A citation may contain "unwarrantable failure" and "significant and substantial" findings even if the violative condition no longer exists.[105]

---

[100]30 U.S.C. §814(d) (Supp. II 1978).

[101]§814(d)(1).

[102]*Secretary of Labor v. Emery Mining Corp.,* 4 MSHC 1585 (FMSHRC No. WEST 86-35-R, 1987); *Secretary of Labor v. Youghiogheny & Ohio Coal Co.,* 4 MSHC 1590 (FMSHRC No. LAKE 86-21-R, 1987).

[103]S. REP. NO. 181, *supra* note 23, at 32, 1977 U.S. CODE CONG. & AD. NEWS at 3432, *citing with approval Zeigler Coal Co.,* 84 Interior Dec. 127, 1 MSHC 1518 (IBMA No. 74-37, 1977).

[104]*Id.* at 135, 1 MSHC at 1524.

[105]*Secretary of Labor v. Nacco Mining Co.,* 4 MSHC 1505 (FMSHRC No. LAKE 85-87-R, 1987), *appeal filed,* No. 88-1053 (D.C. Cir. Jan. 27, 1988); *Secretary of Labor v. Emerald Mines Corp.,* 4 MSHC 1535 (FMSHRC No. PENN 85-298-R, 1987), *appeal filed,* No. 87-1819 (D.C. Cir. Dec. 23, 1987); *Secretary of Labor v. Pennsylvania Mines Corp.,* 4 MSHC 1528 (FMSHRC No. PENN 85-188-R, 1987); *Secretary of Labor v. White County Coal Corp.,* 4 MSHC 1540 (FMSHRC No. LAKE 86-58-R).

After an unwarrantable failure citation has been issued, closure orders may follow if the criteria of the statute are met. If, during the same inspection in which the unwarrantable failure citation was issued, or during any subsequent inspection within 90 days after the citation is issued, an inspector finds another violation of a health or safety standard and finds that the violation was caused by an unwarrantable failure to comply, the inspector is then to issue an unwarrantable failure closure order.[106] Until the inspector determines that the violation has been abated, the unwarrantable failure closure order will prohibit all persons from entering the area affected by the violation other than those persons allowed to enter the area under Section 104(c) of the Act.[107]

An interpretative question in the context of the closure order is whether the second violation leading to the closure order must be a significant and substantial violation. It should be noted that identical language under the 1969 Coal Act providing for unwarrantable failure closure orders has been interpreted by one court and the Mine Commission to allow the issuance of a closure order, even though the violation leading to the closure order was not a significant and substantial violation.[108]

Once an unwarrantable failure citation and subsequent closure order have been issued, the mine operator will continue to receive withdrawal orders for subsequent unwarrantable failure violations issued anywhere in the mine until there has been a full mine inspection with no unwarrantable failure violations.[109] The key statutory language in this continuing issuance of unwarrantable failure closure orders is that the violations involved are to be "similar to those that resulted in the issuance of the withdrawal order." The Mine Commission has not spoken directly to the question. The law under the 1969 Coal Act, which contained identical language,

---

[106]30 U.S.C. §814(d)(1) (Supp. II 1978).

[107]*Id.*

[108]*UMWA v. Kleppe*, 532 F.2d 1403, 1 MSHC 1421 (D.C. Cir. 1976), *cert. denied*, 429 U.S. 858 (1976). Accord *Secretary of Labor v. Old Ben Coal Co.*, 1 MSHC 2241 (FMSHRC No. VINC 74-11, 1979).

[109]30 U.S.C. §814(d)(2) (Supp. II 1978). A full inspection may be either a regular full mine inspection or a series of inspections covering the full mine. *Kitt Energy Corp.*, 3 MSHC 1463 (FMSHRC No. WEVA 83-65-R, 1984), *aff'd*, 768 F.2d 1477, 3 MSHC 1868 (D.C. Cir. 1985).

was that this "similarity" provision requires that subsequent withdrawal orders can be issued only for unwarrantable failure violations as opposed to non-unwarrantable failure violations. There need not be any substantive similarity in the standards violated between the initial unwarrantable failure closure order and subsequent unwarrantable failure closure orders.[110]

A second issue which has been raised involves the interpretation of the language "until such time as an inspection of such mine discloses no similar violations."[111] The issue here is whether a partial mine inspection with no unwarrantable failure violations would discontinue the issuance of unwarrantable failure orders or whether the entire mine must be inspected. The Mine Commission has answered this question, stating that there must be a " 'clean' inspection of the entire mine" in order to stop the chain of issuance of unwarrantable failure closure orders under Section 104(d)(2).[112] In each of the cited cases, the Mine Commission held that the Secretary of Labor, as a part of presenting a *prima facie* case, must show that there was no intervening "clean" inspection of the entire mine in the period between the issuance of the initial unwarrantable failure withdrawal order and subsequent unwarrantable failure withdrawal orders.

5. *Pattern of Violations*

Section 104(e) provides for pattern of violations notices and closure orders.[113] If the Secretary of Labor determines that a mine operator has a "pattern of violations of mandatory health or safety standards," and that the violations are "significant and substantial," then a written notice of the existence of a pattern is to be issued to the mine operator. If, during any inspection within 90 days after the issuance of the

[110]*Eastern Associated Coal Corp.*, 81 Interior Dec. 567, 1 MSHC 1179 (IBMA No. 74-18, 1974).

[111]30 U.S.C. §814(d)(2) (Supp. II 1978).

[112]*Kitt Energy Corp., supra* n. 109; *Secretary of Labor v. CF&I Steel Corp.*, 2 MSHC 1057 (FMSHRC No. DENV 76-46, 1980) (1969 Coal Act); *Secretary of Labor v. U.S. Steel Corp.*, 2 MSHC 1100 (FMSHRC No. HOPE 75-708, 1981) (1969 Coal Act); *Secretary of Labor v. Old Ben Coal Corp.*, 2 MSHC 1305 (FMSHRC No. VINC 74-157, 1981) (1969 Coal Act).

[113]30 U.S.C. §814(e) (Supp. II 1978).

notice, the inspector finds a violation of a mandatory health or safety standard which is "significant and substantial," then the inspector is to issue a withdrawal order prohibiting all persons from entering the affected area other than those persons referred to in Section 104(c) of the Mine Act. If such a withdrawal order is issued within 90 days after the issuance of a pattern of violations notice, then withdrawal orders will continue to be issued for all significant and substantial violations of health or safety standards until such time as there has been inspection of the entire mine with no finding of a significant substantial health or safety violation. When such an inspection occurs, the pattern of violations notice then no longer applies and pattern of violation closure orders cannot be issued until such time as any new pattern of violations notice is issued.

The statute provides that the Secretary of Labor is to promulgate rules establishing criteria for when a pattern of violations exists.[114] To date, no regulations have been promulgated. Proposed rules were published in the *Federal Register* in August 1980.[115] The proposed rules, which have now been withdrawn, spelled out a variety of criteria which look to mine size, the number and types of citations and orders issued at the mine, the accident/injury/illness/fatality incidence rate at the mine, and inspectors' statements regarding citations and orders. It is difficult to predict the extent, if any, to which the final regulations, yet to be promulgated, will track the earlier proposed rules.

The Senate committee in its report on the 1977 Mine Act references the Scotia Coal Mine disaster in discussing the pattern of violations provisions. The committee report states that Scotia and other mines investigated "had an inspection history of recurrent violations ***. The Committee's intention is to provide an effective enforcement tool to protect miners when the operator demonstrates his disregard for the health and safety of miners through an established pattern of violations."[116] The Senate report goes on to state the intention

---

[114]§814(e)(4).

[115]Proposed 30 C.F.R. pt. 104, 45 FED. REG. 54,656 (1980). These proposed rules have been withdrawn with an indication that the Secretary is considering whether to resume rulemaking proceedings. 50 FED. REG. 5470 (1985).

[116]S. REP. NO. 181, *supra* note 23, at 32, 1977 U.S. CODE CONG. & AD. NEWS at 3432.

that the Secretary have "broad discretion in establishing criteria for determining when a pattern of violations exists."[117] The committee states that "a pattern is more than an isolated violation," but it "does not necessarily mean a prescribed number of violations of predetermined standards nor does it presuppose any element of intent or state of mind of the operator."[118] Both the Senate report and the conference report note that a "pattern" can be established by either an accumulation of violations of one standard or a series of violations of several standards.[119]

6. Respirable Dust Withdrawal Orders

If, as a result of analysis of coal dust samples taken under Section 202(a),[120] or of samples for respirable dust taken during the course of an inspection, the Secretary determines that violations have occurred, a citation will be issued under Section 104(f).[121] If a citation is issued, samples will be taken in the affected area during each production shift; if dust levels remain high, closure orders may be issued under Section 104(f). If a closure order is issued, persons knowledgeable in the methods and means of controlling and reducing respirable dust are then to analyze the situation to determine what steps should be taken to ensure the health of persons in the mine.

7. Untrained Miners Withdrawal Orders

As discussed below, Section 115 provides for extensive safety training for miners. If, during an inspection, it is determined that a miner has not received the training required by the Mine Act, then the inspector may issue an order declaring the miner to be a hazard to himself and others and requiring that the miner be withdrawn from the mine and prohibited from reentering until such time as he has received the required training.[122] A miner who is directed to be re-

---

[117]*Id.* at 33, 1977 U.S. CODE CONG. & AD. NEWS at 3433.

[118]*Id.*

[119]*Id.*, and H.R. REP. NO. 655 (Conference Report), 95th Cong., 1st Sess. 48–49, *reprinted in* 1977 U.S. CODE CONG. & AD. NEWS 3401, 3496–3497.

[120]30 U.S.C. §842(a) (1976).

[121]30 U.S.C. §814(f) (Supp. II 1978).

[122]30 U.S.C. §814(g)(1) (Supp. II 1978).

moved from the work area for training cannot be discharged or discriminated against because of the removal, or denied compensation during the period necessary for training.[123]

8. *Imminent Danger Withdrawal Orders*

Section 107(a) provides that whenever an inspector finds that an "imminent danger" exists, the inspector is to determine the area of the mine affected by the danger and to order that all miners be withdrawn, except for those persons specified in Section 104(c) of the Act.[124] "Imminent danger" is defined as "the existence of any condition or practice in a coal or other mine which could reasonably be expected to cause death or serious physical harm before such condition or practice can be abated."[125] The 1969 Coal Act contained a similar provision for issuance of imminent danger closure orders. The definition of imminent danger is identical under both Acts. The 1969 Coal Act imminent danger provisions were construed as follows:

> "The question in every case is essentially the proximity of the peril to life and limb. Put another way: would a reasonable man, given a qualified inspector's education and experience, conclude that the facts indicate an impending accident or disaster, threatening to kill or cause serious physical harm, likely to occur at any moment, but not necessarily immediately? The uncertainty must be of a nature that would induce a reasonable man to estimate that, if normal operations designed to extract coal in the disputed area proceeded, it is just as probable as not that the feared accident or disaster would occur before elimination of the danger."[126]

The potential of the feared disaster under the 1969 Coal Act had to be "greater than a mere possibility" and of the type "which a reasonable man would expect to occur at any moment."[127] It should also be noted that under the 1969 Coal Act, a mine operator had the burden of proving that an imminent danger did *not* exist if a challenge was made to the

[123]§814(g)(2).
[124]30 U.S.C. §817(a) (Supp. II 1978).
[125]30 U.S.C. §802(j) (Supp. II 1978).
[126]*Freeman Coal Mining Corp.*, 80 Interior Dec. 610, 616, 1 MSHC 1073, 1077 (IBMA No. 73-15, 1973), *aff'd*, 504 F.2d 741, 1 MSHC 1192 (7th Cir. 1974).
[127]*Rochester & Pittsburgh Coal Co.*, 82 Interior Dec. 368, 372, 1 MSHC 1318, 1320 (IBMA No. 75-13, 1975).

order.[128] To date, Mine Commission rules and decisions have enunciated no position on this issue.[129]

The Mine Commission has indicated that under the 1977 Act, it "will examine anew the question of what conditions or practices constitute an imminent danger."[130] The Mine Commission has specifically reserved the issue of whether or not the " 'as probable as not' gloss upon the language of section 3(j)" should be utilized.[131] While the Mine Commission has indicated it will not feel bound by 1969 Coal Act interpretations of "imminent danger," it is as yet unclear how the Mine Commission will interpret that language differently, if at all.

Portions of the legislative history of the Mine Act are also relevant. The Senate committee in its report

> "disavow[ed] any notion that imminent danger can be defined in terms of a percentage of probability that an accident will happen; rather, the concept of imminent danger requires an examination of the potential of the risk to cause serious physical harm at any time. It is the Committee's view that the authority under this section is essential to the protection of miners and should be construed expansively by inspectors and the Commission."[132]

The Mine Act also provides for the issuance of closure orders for potential imminent dangers. Under Section 107(b), if an inspector determines that (1) conditions exist that have not resulted in an imminent danger, (2) the condition cannot be effectively corrected through existing technology, and (3) reasonable assurance cannot be provided that continued mining will not result in an imminent danger, then the inspector is to determine the affected area and issue a notice to the mine operator.[133] Following an investigation, including an opportunity for a public hearing if requested by an interested

---

[128]*Old Ben Coal Co. v. IBMA*, 523 F.2d 25, 34–36, 1 MSHC 1290, 1296–1298 (7th Cir. 1975).

[129]Compare Interim Procedural Rules, 29 C.F.R. §2700.48, 43 FED. REG. 10,320, 10,326 (1978), in which the position was taken that a mine operator had the burden of disproving the existence of an imminent danger, with the final Rules of Procedure, 29 C.F.R. pt. 2700, which take no position on the issue.

[130]*Pittsburgh & Midway Coal Mining Co. v. Secretary of Labor*, 1 MSHC 2354, 2355 (FMSHRC No. 74-666, 1980).

[131]*Id.*

[132]S. REP. NO. 181, *supra* note 23, at 38, 1977 U.S. CODE CONG. & AD. NEWS 3401, 3438.

[133]30 U.S.C. §817(b)(1) (Supp. II 1978).

party, the Secretary of Labor is to make findings and either cancel the notice or issue a closure order covering the affected area. As with other closure orders, the only persons who can enter the area are those listed in Section 104(c).[134] To date, this provision of the Act has apparently not been used by the Secretary of Labor.[135]

F. Civil and Criminal Penalties Under the Act

Section 110 contains the Mine Act's penalty provisions.[136] Civil and criminal penalties are provided for.

The primary penalty provision is Section 110(a), which provides for a mandatory civil penalty of up to $10,000 for *each* violation of a mandatory health or safety standard or of a provision of the Act itself.[137] Each occurrence of a violation may constitute a separate offense. Thus, if the same violation occurs on 15 pieces of equipment, that could be written as either one or 15 violations.

As discussed above, if a mine operator fails to correct a citation which it has been issued under Section 104(a) within the period prescribed in the citation or in any amendments to the citation, a closure order may be issued for the area affected by the violation. In addition to that closure order, under Section 110(b) a civil penalty of up to $1000 per day for each day during which the violation continues uncorrected may be issued.[138]

MSHA regulations contain a variety of provisions relating to smoking or the carrying of smoking materials, matches, or lighters. Any miner who willfully violates these regulations may be subject to a civil penalty of up to $250 for each occurrence of a violation.[139]

In addition to civil penalties, the Act also provides for a variety of criminal penalties. Any mine operator who willfully

---

[134]30 U.S.C. §107(b)(2) (Supp. II 1978).

[135]The Interior Board of Mine Operations Appeals discussed similar provisions under the 1969 Coal Act in *Buffalo Mining Co.*, 80 Interior Dec. 630, 643–645, 1 MSHC 1086, 1094–1095 (IBMA No. 73-18, 1973). No potential danger notice had been issued there.

[136]30 U.S.C. §820 (Supp. II 1978).

[137]§820(a).

[138]§820(b).

[139]§820(g).

violates a health or safety standard or knowingly violates, or fails or refuses to comply with, an order issued under Section 107(a) of the Act or any final decision issued under the Act may, upon conviction, be fined $25,000 or imprisoned for up to one year, or both. A second conviction may lead to a fine of up to $50,000 or 5 years' imprisonment, or both.[140]

Anyone who knowingly makes a false statement, representation, or certification in any application, record, report, plan, or other document filed or required to be kept under the Mine Act may be fined up to $10,000 or imprisoned up to 5 years, or both.[141] Anyone who knowingly distributes, sells, offers to sell, introduces, or delivers in commerce any mining equipment which is represented as complying with Mine Act provisions or regulations, but does not comply with those provisions, may be subject to the same fine or imprisonment.[142]

The Act also provides that directors, officers, or agents of corporate operators may be subject to the civil or criminal penalties of Subsections (a) and (d) of Section 110. If a corporate operator has violated the health or safety standards or knowingly violates, or fails or refuses to comply with, an order issued under the Act, any director, officer, or agent of the corporation who "knowingly authorized, ordered, or carried out" the violation, failure, or refusal may be subject either to civil penalties or fines and imprisonment.[143]

This provision has been challenged on the basis that it is a violation of equal protection of the laws in that it imposes liability only on corporate agents. In *Secretary of Labor v. Richardson*,[144] the Mine Commission upheld comparable language under the 1969 Coal Act providing solely for corporate agent liability. Commissioner Backley dissented in the 3-to-1 decision, stating that he could find no rational basis for singling out agents of corporate operators and excusing other agents of noncorporate operators. The Mine Commission's decision was affirmed by the court of appeals.[145]

---

[140]§820(d).

[141]§820(f).

[142]§820(h).

[143]§820(c).

[144]2 MSHC 1114 (FMSHRC No. BARB 78-600-P, 1981).

[145]689 F.2d 632, 2 MSHC 1865 (6th Cir. 1982), *cert. denied*, 461 U.S. 928 (1983). The corporate agent must know or have reason to know of the violation. Where a supervisor did not know or have reason to know his subordinates would subsequently

The Mine Act also provides criminal penalties for any person who gives advance notice of any inspection to be conducted under the Mine Act. The penalty is a fine of up to $1000 or imprisonment of up to 6 months, or both.[146] Section 103(a) specifically prohibits giving advance notice of inspections, except for certain inspections conducted by the Secretary of Health, Education and Welfare (Health and Human Services).[147]

In determining civil penalties, the Mine Commission is to consider (1) the mine operator's history of prior violations; (2) the size of the business of the mine operator; (3) whether the mine operator was negligent; (4) the effect of the penalty on the mine operator's ability to continue in operation; (5) the gravity of the violation; and (6) the demonstrated good faith in pursuing rapid compliance after notification of the violation.[148] OSHA criteria are similar, but do not include operator negligence or ability to continue business.[149]

MSHA has promulgated extensive regulations which set forth a fairly mechanical approach to determining a proposed penalty after a citation has been issued.[150] This approach does not encompass all citations that are issued, but does provide a basis for proposing a penalty in the overwhelming percentage of instances.

Two exceptions should be noted. The first of these is that if the violation is abated within the time set by the inspector and involves a situation "not reasonably likely to result in a reasonably serious injury or illness," then an assessment of $20 may be imposed.[151] This proposal for a minimum penalty is a new provision which was promulgated on May 21, 1982.[152]

---

perform an assigned task in violation of regulatory standards, the Mine Commission has held that the supervisor was not liable under §110(c). *Secretary of Labor v. Glenn*, 3 MSHC 1457 (FMSHRC No. WEST 80-158-M, 1984).

[146]30 U.S.C. §820(e) (Supp. II 1978).

[147]30 U.S.C. §813(a) (Supp. II 1978).

[148]30 U.S.C. §820(i) (Supp. II 1978). The Mine Commission has ruled that all these factors must be considered even if the violation poses little probable harm. *Secretary of Labor v. U.S. Steel Mining Co.*, 3 MSHC 1362 (FMSHRC No. PENN 82-328, 1984).

[149]Cf. 29 U.S.C. §666(i) (1976).

[150]30 C.F.R. pt. 100 (1985).

[151]30 C.F.R. §100.4 (1985). If a matter is litigated, the Mine Commission will make its determination of the appropriate penalty. See text *supra* at notes 89–91 and 148.

[152]47 FED. REG. 22,294, 22,296 (1982).

The second exception is that certain kinds of situations may be subject to a special assessment rather than the regular assessment formula. These include matters such as violations in connection with fatalities or serious injuries, unwarrantable failure violations, operation of a mine in the face of a closure order, refusal to allow an inspector to perform an inspection or investigation, situations involving corporate agent liability, violations involving an imminent danger order, violations involving discrimination under Section 105(c), and violations involving a high degree of negligence, gravity, or other unique circumstances.[153]

After notice by MSHA, parties initially have 10 days within which to submit additional information to the appropriate MSHA district manager or his designee, or to request a safety and health conference with the district manager or designee. MSHA has discretion as to whether or not to grant the conference. If a conference is conducted, parties may submit any additional relevant information related to the alleged violation. If the review of the matter indicates that no violation occurred, then the citation or order will be vacated. Otherwise, the district manager will refer the citation or order to MSHA's Office of Assessments for proposal of a penalty.[154]

A notice of proposed penalty is then issued by the Office of Assessments and served on the mine operator by certified mail, and by regular mail to the representative of miners. It is important to emphasize that the Office of Assessments is a separate office of MSHA which is concerned only with reviewing the documentation of MSHA inspectors or investigators and proposing a penalty. There is thus a second, independent review which may lead to variations. Assessment personnel may not alter citations or orders, however.

After issuance of a proposed penalty, the mine operator has 30 days either to pay the proposed penalty or to notify MSHA of an intention to contest the proposed penalty.[155] If a proposed penalty is not contested within 30 days, it is deemed a final order of the Mine Commission and is not subject to review either by the Mine Commission or a court or agency.[156]

---

[153] 30 C.F.R. §100.5 (1985).

[154] §100.6.

[155] §100.7.

[156] 30 U.S.C. §815(d) (Supp. II 1978); 30 C.F.R. §100.7(c) (1982). Particular issues regarding the timing of litigation and litigation procedures are discussed in greater detail *infra*, Section III.D.

If a civil penalty is contested, any settlement of the penalty must be approved by the Mine Commission.[157] In exercising this authority, the Mine Commission has required its judges to articulate reasons for approving a settlement, and to do so on the record so that the settlement can be fairly evaluated.[158] The Mine Commission has rejected a settlement agreement containing exculpatory language that could be used in future litigation to avoid a pattern-of-violations finding or limit the prior history of violations determinations in assessing a penalty.[159] The Mine Commission has approved a settlement agreement with a nonadmission clause, however.[160] Commissioner Lawson dissented, objecting that a mine operator cannot deny the existence of a violation and yet agree to pay a civil penalty.

It should be emphasized also that the Mine Commission reviews penalty issues independently of the Secretary of Labor's determination. The Mine Commission is not bound by the proposed penalty.[161]

In evaluating penalties, the Mine Commission has held that while negligence of a mine operator's supervisory personnel may be imputed to the operator in determining a penalty, negligence of rank-and-file employees may not be so imputed.[162] The Mine Commission held that the operator's supervision, discipline, training, and other actions taken to prevent rank-and-file miners' misconduct may be considered to determine if the operator had taken reasonable steps to prevent the rank-and-file miner's violative conduct. Commissioner Lawson filed a strong dissent to the reduction of two $10,000 penalties to $5,000 each. He questioned both the pro-

---

[157]30 U.S.C. §820(k) (Supp. II 1978).

[158]E.g., *Secretary of Labor v. Republic Steel Corp.*, 1 MSHC 1709 (FMSHRC No. PITT 78-156-P, 1978); *Secretary of Labor v. Davis Coal Co.*, 1 MSHC 2305 (FMSHRC No. HOPE 78-627-P, 1980).

[159]*Secretary of Labor v. AMAX Lead Co.*, 2 MSHC 1716 (FMSHRC No. CENT 81-63-M, 1982).

[160]*Secretary of Labor v. Central Ohio Coal Co.*, 2 MSHC 1766 (FMSHRC No. LAKE 81-78, 1982).

[161]*Secretary of Labor v. Sellersburg Stone Co.*, 2 MSHC 2010 (FMSHRC No. LAKE 80-363-M, 1983), *aff'd*, 736 F.2d 1147, 3 MSHC 1385 (7th Cir. 1984). However, if an operator objects before a hearing that MSHA violated its penalty proposal regulations when it proposed a penalty, the Mine Commission may order MSHA to re-propose a penalty. *Secretary of Labor v. Youghiogheny & Ohio Coal Co.*, 4 MSHC 1329, 1332-4 (FMSHRC No. LAKE 84-98, 1987).

[162]*Secretary of Labor v. Southern Ohio Coal Co.*, 2 MSHC 1825 (FMSHRC No. VINC 79-227-P, 1982).

priety of Commission review of the penalty amount, and the "artificial" allocation of penalty dollars which might lead operators to avoid supervisory responsibilities.

G. Injunctions

The Mine Act also authorizes the Secretary of Labor to pursue a civil action seeking relief, including temporary or permanent injunctive relief or a restraining order.[163] The Secretary may pursue a civil action if a mine operator or its agent refuses to comply with an order or decision issued under the Mine Act; "interferes with, hinders, or delays" the Secretary of Labor or of Health, Education, and Welfare (Health and Human Services) from carrying out provisions under the Mine Act; or refuses to allow an inspection or investigation, or to provide other information or access to the mine.[164]

The Secretary of Labor also may file a civil action for relief if the Secretary believes that an operator has engaged in a pattern of violation of mandatory health or safety standards where the Secretary believes that the pattern "constitutes a continuing hazard to the health or safety of miners."[165] The interaction of this provision with the pattern-of-violation provisions of Section 104(e)[166] has yet to be explored in any cases. To date, the Secretary of Labor has not cited any mine operator for a pattern of violations.

H. Posting of Orders and Decisions

Each mine is required to maintain an office with a conspicuous sign designating it as the mine office.[167] At that office, or in a conspicuous place near the entrance of the mine, there is to be a bulletin board where all orders, citations, notices, or decisions required to be posted shall be posted.[168] All orders, citations, notices, or decisions addressed to the operator are to be posted on the bulletin board.[169] The mine

---

[163]30 U.S.C. §818 (Supp. II 1978).
[164]§818(a)(1) (A)–(F).
[165]§818(a)(2).
[166]§814(e).
[167]30 U.S.C. §819(a) (Supp. II 1978).
[168]*Id.*
[169]*Id.*

operator is to mail copies of these materials to a representative of miners if there is one at the mine, and the appropriate state official or agency, if any, responsible for administering state health or safety laws at the mine.[170]

I. Identification of Mine Operators and Representatives of Miners

Each operator of a mine is to file an identification notice with the Secretary of Labor.[171] Regulations provide for the procedures to be followed in accomplishing this.[172]

If miners designate a representative, they are also to file an identification notice with the Secretary of Labor.[173] Failure to file such a notice does not preclude a representative of miners from participating in an inspection or exercising other rights under the Mine Act.[174]

The definition of "operator" in the Mine Act includes not only owners or operators of mines, but also independent contractors performing services or construction at a mine.[175] Independent contractors also are required to file identification notices with the Secretary of Labor.[176]

Now that regulations have been promulgated, the primary question relating to independent contractors is whether a citation or order should properly be issued to a mine operator or an independent contractor.[177] In its enforcement guidelines published with the independent contractor identification regulations, MSHA indicates that both independent contractors and mine operators are responsible for compliance with the Mine Act. MSHA takes the position that there may be circumstances in which it will be appropriate to issue a citation

---

[170]§819(b).

[171]30 U.S.C. §819(d) (Supp. II 1978).

[172]30 C.F.R. pt. 41 (1982).

[173]30 C.F.R. pt. 40 (1982).

[174]*Consolidation Coal Co. v. Secretary of Labor*, 2 MSHC 1185 (FMSHRC No. WEVA 80-333-R, 1981).

[175]30 U.S.C. §802(d) (Supp. II 1978).

[176]30 C.F.R. pt. 45 (1985).

[177]This issue was an especially difficult one prior to the promulgation of regulations. In *Secretary of Labor v. Phillips Uranium Corp.*, 2 MSHC 1697 (FMSHRC No. CENT 79-281-M, 1982), the history of this problem as well as issues of statutory construction and policy are thoroughly analyzed in the decision of the Commission and dissent.

or order to both the independent contractor and the mine operator. Action taken by MSHA will depend upon whether the mine operator has contributed to the violation, the comparative culpability of the mine operator and the independent contractor, the hazard to the mine operator's versus the independent contractor's employees, and the control which the mine operator has over correction of the situation.[178] Two courts of appeals have stated that a citation may be issued to a mine operator, independent contractor, or both, depending upon the circumstances.[179] It is likely that issues as to operator or contractor liability will continue to be raised in Mine Commission and court litigation, since the issue of comparative responsibility is one that must be assessed according to the facts of each case.[180]

J. Petitions for Modification

Section 101(c) provides that either a mine operator or a representative of miners may petition the Secretary of Labor to modify the application of any mandatory safety standard. Before granting a modification, the Secretary must determine either that the alternative method which has been proposed for achieving the result sought by the standard will at all times guarantee no less than the same measure of protection afforded miners under the standard, or that application of the standard in the particular mine will result in a diminution of safety to miners.[181]

---

[178] 30 C.F.R. pt. 45 app. A. Appendix appears at 45 FED. REG. 44,497–44,498 (1980).

[179] *Harman Mining Corp. v. FMSHRC*, 671 F.2d 794, 2 MSHC 1551 (4th Cir. 1981); *Cyprus Indus. Minerals Co. v. FMSHRC*, 664 F.2d 1116, 2 MSHC 1554 (9th Cir. 1981).

[180] The Fourth Circuit recently reversed the Mine Commission on an independent contractor determination. The court held that an electric utility having a substation on mine property for the purpose of supplying electrical power to the mine was not an independent contractor under the Mine Act. Rather, OSHA jurisdiction should attach. *Old Dominion Power Co. v. Donovan*, 772 F.2d 92, 3 MSHC 1913 (4th Cir. 1985), *rev'g* 3 MSHC 1505 (FMSHRC No. VA 81-40-R, 1984). The Mine Commission has held that the Secretary of Labor improperly cited a mine operator in addition to an independent contractor where the hazard insufficiently impacted the mine operator's employees and the operator had no control over the situation. *Cathedral Bluffs Shale Oil Co.*, 3 MSHC 1519 (FMSHRC No. WEST 81-186-M, 1984), *rev'd*, 796 F.2d 533, 4 MSHC 1033 (D.C. Cir. 1986). See also *Calvin Black Enters.*, 3 MSHC 1897 (FMSHRC No. WEST 80-6-M, 1985) (owner operator properly cited for independent contractor violation).

[181] 30 U.S.C. §811(c) (Supp. II 1978). See generally *Mine Workers v. MSHA*, 830 F.2d 289 (D.C. Cir. 1987).

Procedures for filing a petition for modification are set forth in MSHA regulations.[182] A petition is to be filed with the Assistant Secretary of Labor for Mine Safety and Health.[183] Petitions are often investigated by local MSHA personnel. MSHA District Managers usually are willing to meet to review issues of interpretation prior to the filing of a petition, as well as during the petition review process.

The petition should outline the nature of modification sought and the factual basis for the modification, among other things.[184] Notice of the petition is published in the *Federal Register*, allowing 30 days for comment.[185] After review, a decision is made on the petition, following which a hearing may be requested.[186] It should be noted that these matters are handled by the Department of Labor, with hearings in the event of appeals being held by administrative law judges of the Department of Labor, not the Mine Commission.[187] Appeals from the administrative law judge are to the Assistant Secretary of Labor for Mine Safety and Health.[188] A decision by the Assistant Secretary is final agency action for purposes of judicial review under the Administrative Procedure Act.[189]

The Mine Commission has ruled on several occasions that the so-called "diminution of safety" defense is to be pursued and resolved in the context of modification proceedings. Such a defense is not allowed in enforcement proceedings, although more recent cases suggest some possible narrowing of this broad prohibition.[190]

III. Mine Commission Procedures

A. Mine Commission Organization

Section 113 of the 1977 Mine Act establishes the Federal Mine Safety and Health Review Commission. It consists of

---

[182]30 C.F.R. pt. 44 (1985).

[183]30 C.F.R. §44.10.

[184]§44.11.

[185]§44.12.

[186]§§44.13, 44.14 (1985). Hearing Procedures at §§44.20–.41.

[187]§§44.20, 44.2(d).

[188]§44.33.

[189]§44.51. In general, MSHA may not grant interim relief pending disposition of a modification petition. *Mine Workers v. MSHA*, 823 F.2d 608 (D.C. Cir. 1987).

[190]Compare *Secretary of Labor v. Florence Mining Co.*, 2 MSHC 1993 (FMSHRC No. PITT 77-15, 1983), and *Secretary of Labor v. Penn Allegh. Coal Co.*, 2 MSHC

five members appointed with the advice and consent of the Senate for terms of 6 years.[191] The Mine Commission may delegate its powers to any group of three or more members.[192] To date, all decisions have been made by the full complement of Commissioners sitting at the time of the decision. The Mine Commission appoints administrative law judges to hear matters under the Mine Act, and those judges are charged with writing a decision which may then be appealed to the Mine Commission.[193]

B. Mechanics of Challenging a Citation or Order

After a citation or order has been issued, a mine operator may file a challenge to the citation or order. That challenge may be filed after the citation or order is issued, or it may be deferred until after a civil penalty is proposed. Such a challenge must be filed within 30 days of the receipt of the citation or order, or the proposed penalty.[194] Initially, the matter will be heard by an administrative law judge, who will hold a hearing and then issue a written decision disposing of the matter.[195] Under Mine Commission rules, matters may be resolved by summary decision.[196] Any person adversely affected or aggrieved by a decision of an administrative law judge may file a petition for discretionary review with the Mine Commission within 30 days after the issuance of the decision.[197] Petitions must be *received* by the Mine Commission within this 30-day petitioning period.[198] Petitions for discretionary review may be filed only upon the basis that a finding or conclusion of material fact is not supported by substantial evidence; a necessary legal conclusion is erroneous; the decision is contrary to law or promulgated rules or decisions of

---

1353 (FMSHRC No. PITT 78-97-P, 1981), with *Secretary of Labor v. Westmoreland Coal Co.*, 3 MSHC 1939 (FMSHRC No. WEVA 82-152-R, 1985), and *Secretary of Labor v. Sewell Coal Co.*, 3 MSHC 1132 (FMSHRC No. WEVA 79-31, 1983).

[191]30 U.S.C. §823(a) and (b) (Supp. II 1978).

[192]§823(c).

[193]§823(d).

[194]30 U.S.C. §815(d) (Supp. II 1978).

[195]30 U.S.C. §823(d)(1) (Supp. II 1978); 29 C.F.R. §2700.65 (1985).

[196]29 C.F.R. §2700.64.

[197]30 U.S.C. §823(d)(A)(i).

[198]29 C.F.R. §2700.70(a).

the Mine Commission; a substantial question of law, policy, or discretion is involved; or a prejudicial error of procedure was committed.[199] Issues in a petition for discretionary review are to be numbered separately and concisely stated with detailed citations to the record or statutes, regulations, or authorities to be relied upon. Except for good cause shown, no issue can be reviewed by the Mine Commission unless it was raised with the administrative law judge.[200]

Additionally, the Mine Commission, at its discretion, may, within 30 days after the issuance of a decision, order that a case be reviewed. That review may only be ordered on the ground that the decision may be contrary to law or Mine Commission policy or that a novel question of policy has been presented.[201]

If the Mine Commission denies a petition for discretionary review, or after a petition has been granted and a decision issued by the Mine Commission, any person adversely affected or aggrieved by the order of the Mine Commission may seek review in a U.S. court of appeals. That review must be sought in either the court of appeals where the violation is alleged to have occurred or the District of Columbia Circuit. Review must be sought within 30 days after the issuance of the Mine Commission order. Additionally, no objection that was not raised before the Mine Commission can be considered by the court unless the failure or neglect to raise the objection is excused by "extraordinary circumstances."[202] The Secretary of Labor may also seek review and enforcement of a Mine Commission order in the court of appeals where the violation is alleged to have occurred or in the District of Columbia Circuit.[203]

C. Commission Procedural Rules

The Mine Commission has promulgated detailed rules of procedure.[204] These rules are generally similar to the Federal

---

[199]30 U.S.C. §823(d)(2)(A)(ii).
[200]§823(d)(2)(A)(iii).
[201]§823(d)(2)(B).
[202]30 U.S.C. §816(a) (Supp. II 1978).
[203]§816(b).
[204]29 C.F.R. pt. 2700 (1985).

Rules of Civil Procedure and provide that "so far as practicable" pertinent provisions of the Federal Rules of Civil Procedure can be looked to, as appropriate, on procedural rules themselves, or the Administrative Procedure Act.[205] In practice, it appears that administrative law judges vary widely in their approaches to both procedural rules and rules of evidence.

Certain unique procedural provisions should be noted. While the rules provide for discovery, those rules also provide that discovery procedures should be initiated within 20 days after the litigation has been commenced and completed within 60 days after the litigation has been commenced.[206] These provisions provide that the time periods may be extended for good cause, but extensions should not be presumed. Additionally, while the rules provide for the taking of depositions and submission of written interrogatories or requests for admission, they state that production of documents or inspection and photographing of documents or objects is to occur only on order of the judge for good cause shown.[207]

As discussed in Section II.J above, the Mine Commission has rejected a "diminution of safety" defense. It has held also that a defense of impossibility of compliance will have limited application, if any. In *Secretary of Labor v. Sewell Coal Co.*,[208] the Mine Commission ruled that impossibility due to a reduced work force during a strike was not a valid defense. The violative condition was corrected quickly after the citation was issued. The Mine Commission thus stated compliance was difficult but not impossible. These factors can be properly considered in assessing a penalty, however.

D. Timing of a Challenge

A question that generated considerable disagreement during the early months after the passage of the Mine Act was the timing for a challenge to citations or orders. It is clear,

---

[205]29 C.F.R. §2700.1(b). The F.R. Evid. are not applicable, however. See *Secretary of Labor v. Mid-Continental Resources, Inc.*, 3 MSHC 1353 (FMSHRC WEST 82-174, 1984) (admitting and relying on hearsay).

[206]29 C.F.R. §2700.55(a)(b).

[207]§§2700.56, 2700.57.

[208]2 MSHC 1333 (FMSHRC No. HOPE 78-744-P, 1981), *aff'd*, 686 F.2d 1060, 2 MSHC 1819 (4th Cir. 1982).

under Section 107(e)(1),[209] that challenges to imminent danger orders issued under Section 107 must be filed with the Mine Commission within 30 days after issuance of the order. The uncertain issue, however, is the timing for the challenge of citations and special findings issued under Section 104 of the Act. That matter is dealt with in Section 105 of the Act.

Section 105(a) provides that, within a reasonable time after termination of an inspection or investigation, the mine operator is to be notified by certified mail of a proposed civil penalty and that the operator has 30 days within which to contest the citation or proposed civil penalty. If no notice is filed, then the citation and civil penalty become final orders of the Mine Commission and are not subject to further review by a court or any agency.[210] Section 105(d) provides in pertinent part as follows:

> "If within 30 days of receipt thereof, an operator of a coal or other mine notifies the Secretary that he intends to contest the issuance or modification of an order issued under section 104, or citation or a notification of proposed assessment of penalty issued under subsection (a) or (b) of this section, or the reasonableness of the length of abatement time fixed in a citation or modification thereof issued under section 104, or any miner or representative of miners notifies the Secretary of an intention to contest the issuance, modification, or termination of any order issued under section 104, or the reasonableness of the length of time set for abatement by a citation or modification thereof issued under section 104, the Secretary shall immediately advise the Commission of such notification, and the Commission shall afford an opportunity for a hearing ***."[211]

The issue that was raised under this statutory language was whether the allegation of violation in a citation, an unwarrantable failure finding, or a significant and substantial finding issued under Section 104 could be challenged in litigation *before* the issuance of a civil penalty. The focus was the language in Section 105(d), referring to a challenge to a "citation or a notification of proposed assessment of penalty issued under subsection (a) or (b)." The argument accepted by some administrative law judges was that, since Section 105(a) spoke to issuance of proposed penalties, no challenge could be made to the citation until a penalty was proposed.

---

[209] 30 U.S.C. §817(e)(1) (Supp. II 1978).
[210] 30 U.S.C. §815(a) (Supp. II 1978).
[211] §815(d).

The question takes on practical significance for mine operators because MSHA has taken a position through its penalty assessment regulations that no penalty will be issued until a citation has been abated.[212] The basis for this interpretation presumably is the penalty assessment criterion relating to demonstrated good faith efforts by the mine operator to achieve rapid compliance.[213] Thus, in more complex matters such as noise or dust control, a penalty assessment may not be issued for some time, and a challenge to the underlying citation would effectively be precluded. It is clear under Section 105(d) that a failure to abate or other withdrawal order issued under Section 104 may be litigated without delay. However, a mine operator sometimes may wish to challenge a citation, unwarrantable failure finding, or significant and substantial finding before a civil penalty or withdrawal order has been issued.

The Mine Commission initially spoke to this issue in *Energy Fuels Corp. v. Secretary of Labor*.[214] The Mine Commission in a 4-to-1 decision ruled that it had jurisdiction to review special findings of an unwarrantable failure where a violation was abated but no penalty had been assessed. This conclusion was based upon the Mine Commission's reading of admittedly unclear statutory language and the legislative history, placing primary reliance upon the construction which "would best implement the 1977 Act."[215] The Mine Commission noted that litigation of an unwarrantable failure finding would often be in the interest of the mine operator because of the impact which subsequent unwarrantable failure closure orders might have on the mining operation. Additionally, the Mine Commission stated that immediate contestability did not appear to be an approach that would adversely affect the interest of the Secretary of Labor. The dissenting opinion relied essentially upon a different reading of the legislative history and statutory language and a concern that early review might be detrimental to miners' interests. A primary concern was a fear that "early review" might give a mine operator "two bites at the apple," thus increasing litigation.

---

[212]30 C.F.R. §100.6 (1985).

[213]30 U.S.C. §820(i) (Supp. II 1978).

[214]1 MSHC 2013 (FMSHRC No. DENV 78-410, 1979).

[215]*Id.* at 2018. Cf. *Consolidation Coal Co. v. FMSHRC*, 824 F.2d 1071 (D.C. Cir. 1987) (Article III court may review "significant and substantial" findings of citation).

The Mine Commission has reached a similar result in holding that abated Section 104(a) citations involving no special findings may be litigated, even though a penalty has not been assessed.[216] The Mine Commission has not dealt with the question of whether citations issued under Section 104 can properly be litigated before they have been abated.[217]

It should be added that these decisions do not *mandate* prepenalty assessment litigation of citations. Mine Commission procedural rules specifically allow an operator to defer litigation of a Section 104 citation until after a penalty has been assessed.[218]

IV. Miners' Rights

The Mine Act provides a wide variety of rights to miners. These include protections against discrimination, transfer rights, and rights of participation in matters under the Mine Act. It should be pointed out that the definition of "miner" encompasses supervisory as well as nonsupervisory employees.[219] The supervisor-nonsupervisor distinction under the La-

---

[216]*Helvetia Coal Co. v. Secretary of Labor*, 1 MSHC 2024 (FMSHRC No. PITT 78-322, 1979); *Peter White Coal Mining Corp. v. Secretary of Labor*, 1 MSHC 2026 (FMSHRC No. HOPE 78-374-382, 1979). The Mine Commission has also held that an operator who does not contest an unwarrantable failure order within 30 days after its issuance may contest the unwarrantable failure finding in a civil penalty proceeding. *Secretary of Labor v. Quinland Coals, Inc.*, 4 MSHC 1562, 1566-1569 (FMSHRC No. WEVA 85-169, 1987).

[217]Two administrative law judges have ruled in complex cases involving noise and dust control that preabatement litigation may be pursued under the language of §105(d) providing for litigation of the "reasonableness of the length of time set for abatement." One judge ruled that since feasibility of potential engineering dust controls is both a necessary element of a violation and a necessary element for determining the requirements and timing for abatement, the merits of the alleged violations could be litigated. *Climax Molybdenum Co. v. Secretary of Labor*, No. DENV 79-83-M (FMSHRC J. Jan. 9, 1979). Another judge, in dealing with noise control issues, ruled more broadly that reasonableness of the length of abatement time always includes a review of the elements of the alleged violation. *Climax Molybdenum Co. v. Secretary of Labor*, 1 FMSHRC Dec. 213 (April 9, 1979). Neither of these decisions was reviewed by the Mine Commission.

[218]29 C.F.R. §2700.22 (1985). If, however, an operator contests a citation and then deliberately pays a proposed penalty, the citation contest will be dismissed. *Secretary of Labor v. Old Ben Coal Co.*, 3 MSHC 1686 (FMSHRC No. LAKE 85-50-R, 1985).

[219]30 U.S.C. §802(g) (Supp. II 1978). An employee need not work directly in the mining process to be protected. A bookkeeper is also a miner. *Donovan v. Stafford Constr. Co.*, 732 F.2d 954, 3 MSHC 1321 (D.C. Cir. 1984), *rev'g* 2 MSHC 2081 (FMSHRC No. WEST 80-71-DM, 1983).

bor Management Relations Act is thus of no moment in evaluating miners' rights under the Mine Act.

A. Section 105(c)—Protection Against Discrimination

1. Summary of Protections

Probably the most significant right of miners under the Mine Act is Section 105(c)'s protections against discrimination.[220] Section 105(c) specifically protects not only miners, but also a representative of miners and applicants for employment who have engaged in protected activity. Discharge and any other type of discrimination are prohibited. Section 105(c) specifically provides for

(a) Protection against interference with the exercise of statutory rights;

(b) Protection for filing or making complaints with the Secretary of Labor, a representative of miners, or the mine operator regarding an alleged danger or safety or health violation in the mine;

(c) Protection for initiating, or causing initiations of, any proceeding under the Mine Act;

(d) Protection against retaliation for testifying, or planning to testify, in any proceeding under the Mine Act; and

(e) Protection if the individual is subject to medical evaluations and potential transfers under standards published pursuant to Section 101 of the Mine Act.

As discussed below, Section 105(c) also has been construed to include protections against discharge or other discrimination for refusals to work in unsafe conditions.

Section 105(c) provides that discrimination complaints are to be filed within 60 days after the alleged discrimination occurs. The Mine Commission, relying upon legislative history, has indicated that this time period may be extended under "justifiable circumstances."[221]

---

[220]30 U.S.C. §815(c) (Supp. II 1978).

[221]*Herman v. IMCO Servs.*, 2 MSHC 1929 (FMSHRC No. WEST 81-109-DM, 1982). See additional cases note 259 *infra*.

2. *Other State and Federal Limitations*

It should be emphasized that a given situation under Section 105(c) also may raise issues under a variety of additional federal and state statutes. The Labor Management Relations Act (LMRA) prevents employers from retaliating against their employees for engaging in protected concerted activity.[222] That Act also prohibits retaliation against employees engaged in union activities[223] and requires employers to bargain collectively with duly elected representatives of their employees regarding a variety of matters, including health and safety.[224] It is important to keep in mind that particular safety activity may be protected under the LMRA although unprotected under the Mine Act, and vice versa. Additionally, if the mine has a collective bargaining agreement under the LMRA, that agreement may impose additional contractual requirements on the mine operator and confer added contractual rights upon miners.

Title VII of the Civil Rights Act of 1964 prohibits discrimination on the basis of race, color, religion, sex, or national origin.[225] The Age Discrimination in Employment Act prohibits employers from taking adverse action against employees who are at least 40 years of age but less than 70 years of age on the basis of the employee's or job applicant's age.[226] If the mining operation is a contractor with the federal government or a subcontractor for a government contractor, the operation might be subject to handicap discrimination requirements of Sections 503 and 504 of the Rehabilitation Act of 1973.[227] State antidiscrimination laws also may impact employee relations requirements.

---

[222]29 U.S.C. §158(a)(1) (1976). See e.g., *Alleluia Cushion Co.*, 221 NLRB 999, 91 LRRM 1131 (1975) (filing Cal-OSHA complaint protected activity). *Alleluia Cushion* was reversed by the NLRB in *Meyers Indus.*, 268 NLRB 493, 115 LRRM 1025 (1984), *rev'd and rem'd*, 755 F.2d 941, 118 LRRM 2649 (D.C. Cir. 1985), *reaffirmed on remand*, 281 NLRB No. 118, 123 LRRM 1137 (1986). See also *NLRB v. City Disposal Sys.*, 465 U.S. 822, 115 LRRM 3193 (1984) (individual activity to pursue rights under collective bargaining agreement protected).

[223]29 U.S.C. §158(a)(3) (1976).

[224]§158(a)(5). See, e.g., *Gulf Power Co.*, 156 NLRB 622, 61 LRRM 1073 (1966), *enforced*, 384 F.2d 822, 66 LRRM 2501 (5th Cir. 1967) (safety mandatory subject of collective bargaining and government standards establish starting point).

[225]42 U.S.C. §§2000e *et seq.* (1976).

[226]29 U.S.C. §§621 *et seq.* (1976).

[227]29 U.S.C. §§793 and 794 (1976).

The presence of these other requirements may lead to multiple statutory violations for a single incident. These requirements are mutually exclusive and should be kept in mind in considering issues under Section 105(c).

3. Criteria for a Section 105(c) Violation

The first key decision of the Mine Commission in the area of evaluating discrimination matters was *Pasula v. Consolidation Coal Co.*[228] *Pasula* outlined the basic criteria which the Mine Commission will look to in this area and thus it warrants fairly thorough discussion.

Pasula refused to operate a continuous mining machine which he believed was too noisy. He shut down the machine and explained that the noise had given him an "extreme headache," made his ears hurt, and made him nervous. The union safety committeeman listened to the machine and agreed with company personnel that the machine was not too loud to operate. Pasula still refused to operate the machine, and when his helper was asked to operate the machine, Pasula stated that "nobody's going to run it."[229] No sound level readings were taken at the time of this incident, but readings taken later that day revealed that gear noise was 93 decibels when the machine was not cutting coal and 103 decibels when it was cutting coal.[230] Under the coal mine noise standard, a miner can operate a mining machine which has a noise level of 103 decibels for approximately 1½ hours without hearing protection.[231] The decision does not reflect how long Pasula had been operating the continuous mining machine. Additionally, there is no suggestion in the decision that Pasula was ever offered any personal hearing protection equipment (earmuffs or earplugs). Pasula subsequently was discharged, and an arbitrator upheld the discharge. An administrative law judge held that the discharge violated Section 105(c), and the Mine Commission affirmed.

---

[228]2 MSHC 1001 (FMSHRC No. PITT 78-458, 1980), *rev'd*, 663 F.2d 1211, 2 MSHC 1465 (3d Cir. 1981).

[229]2 MSHC at 1002 (FMSHRC No. PITT 78-458, 1980).

[230]*Id.*

[231]30 C.F.R. §70.510 (1985).

Initially, the Mine Commission held that there is a right to refuse to work in unsafe or unhealthy conditions.[232] While this result is not clearly mandated by the language of the Mine Act itself, the notion that an employee has a right to refuse to work under certain unsafe or unhealthy conditions is clearly articulated in the legislative history discussed by the Mine Commission.[233] The right to refuse has not really been an issue. Rather, the question has been when, and how, that right can be exercised.

In stating the test for determining whether or not discrimination has occurred, the Mine Commission rejected both the "in any part" test of the discrimination and the "but for" test.[234] Briefly summarized, the "in any part" test would hold that if the action against an employee was in any way motivated by the employee's protected health or safety activity, then the action was unlawful. The "but for" test would find no violation unless it could be shown that the adverse action would not have been taken but for the protected activity. The Mine Commission chose instead to follow a balancing approach which had recently been articulated by the Supreme Court in a First Amendment case. That approach involves varying allocations of burden of persuasion rather than focusing solely on causation.[235]

In outlining the order of proof for discrimination cases, the Mine Commission in *Pasula* held that the complaining party must initially establish that the miner engaged in some sort of protected health or safety activity and the discipline, discharge, or other adverse action taken against the miner was motivated in some part by that protected activity. Complainant has the ultimate burden of persuasion on those issues. If the complaining party makes this initial showing, the mine operator must then go forward with evidence. The mine operator has the ultimate burden of persuasion that the discipline, discharge, or other adverse action against the miner was taken because he engaged in some unprotected activity,

---

[232]2 MSHC at 1003–1004.

[233]*Id.* at 1004–1006.

[234]*Id.* at 1008–1009.

[235]*Id.* at 1009–1010, relying upon *Mt. Healthy City Bd. of Educ. v. Doyle*, 429 U.S. 274 (1977). But compare *Texas Dep't of Community Affairs v. Burdine*, 450 U.S. 248, 25 FEP 113 (1981) (Title VII of Civil Rights Act of 1964), with *NLRB v. Transportation Mgt.*, 462 U.S. 393, 113 LRRM 2857 (1983).

such as insubordination or some other job misconduct. The
mine operator must prove that the misconduct alone would
have led to the same discipline in any event.[236] The Mine
Commission found that Pasula's discharge was unlawful. The
Third Circuit reversed that finding, holding that Pasula's misconduct in shutting down equipment and refusing to permit
others to operate it was unprotected and sufficient to justify
the termination. The Third Circuit did not specifically deal
with the analytical framework set forth by the Mine Commission, focusing rather upon the factual findings in the case.[237]

The Sixth Circuit initially rejected *Pasula's* shifting burden approach, holding that a mine operator need only produce
"evidence of a legitimate business purpose sufficient to create
a genuine issue of fact."[238] This decision was vacated following
the Supreme Court's ruling, taking a contrary approach under
the Labor Management Relations Act.[239]

Another important aspect of the *Pasula* decision is its
discussion of issues regarding deferral to arbitration. Pasula's
termination had been arbitrated and the termination had been
upheld. In this instance, relying on an analogy to Title VII of
the Civil Rights Act of 1964, the Mine Commission held that
the decision of the arbitrator was not controlling.[240] The Mine
Commission stated that an administrative law judge might
accord weight to the findings of an arbitrator if he chose to
do so, but emphasized that proceedings before the judge are
"still de novo and it is the responsibility of the judge to render
a decision in accordance with his own view of the facts, not
that of the arbitrator."[241] Arbitral findings, even those addressing issues perfectly congruent with those before the judge,
are not controlling upon the judge. The Mine Commission
concluded that its administrative law judge was correct in

---

[236]2 MSHC at 1010.

[237]*Supra* note 228.

[238]*W.B. Coal Co. v. FMSHRC*, 704 F.2d 275, 2 MSHC 2041, 2047 (6th Cir. 1983),
vacated, 719 F.2d 194, 3 MSHC 1065 (6th Cir. 1983).

[239]*NLRB v. Transportation Mgt.*, *supra* note 235. The Mine Commission's *Pasula*
test for dual motive has been reiterated by the Commission and applied by several
courts. See *Robinette v. United Castle Coal Co.*, 2 MSHC 1213 (FMSHRC No. VA 79-
141-D, 1981); *Eastern Assoc. Coal Corp. v. FMSHRC*, 813 F.2d 639, 642 (4th Cir.
1987); *Donovan v. Stafford Constr. Co.*, 732 F.2d 954, 958–959 (D.C. Cir. 1984); *Boich
v. FMSHRC*, 719 F.2d 194, 195–196 (6th Cir. 1983).

[240]2 MSHC at 1007, relying on *Alexander v. Gardner-Denver Co.*, 415 U.S. 36
(1974).

[241]2 MSHC at 1007.

giving little or no weight to the arbitral findings in *Pasula*, because the gravity element involved under the collective bargaining agreement was "far narrower" than the gravity element under the Mine Act.[242] The Third Circuit specifically held that the decision of the arbitrator was not binding upon either the administrative law judge or the Mine Commission.[243]

In *Robinette v. United Castle Coal Co.*,[244] the Mine Commission ruled that the employee's refusal to work must have been based upon a reasonable good faith belief that the hazard existed. The Mine Commission held that employee good faith in refusing to work "implies an accompanying rule requiring validation of the reasonable belief."[245] The Mine Commission rejected the "stringent" test of requiring " 'objective, ascertainable' evidence" which is the test "usually satisfied only by the introduction of physicial evidence, 'disinterested' corroborative testimony, and—not infrequently—expert testimony."[246] The Mine Commission stated that reasonableness may "be established at the minimum through the miner's own testimony *** [which is to be] evaluated for its detailed, inherent logic, and overall credibility."[247] Physical testimony or other expert evidence can be presented by all parties, and the ultimate decision as to an employee's reasonable good faith will be made "on the basis of all the evidence."[248] This standard would apply both to a refusal to work or to any affirmative "self-help" action, such as adjusting or shutting off equipment to eliminate or protect against a hazard.[249]

In *Bush v. Union Carbide Corp.*,[250] the Mine Commission held that the complaining miner has the burden of proving the good faith and reasonableness of his belief that a hazard existed. Commissioner Lawson dissented on the basis that the

---

[242]*Id.*

[243]*Supra* note 228, 663 F.2d at 1219, 2 MSHC at 1471.

[244]2 MSHC 1213 (FMSHRC No. VA 79-141-D, 1981).

[245]*Id.* at 1219.

[246]*Id* at 1219–1220.

[247]*Id.* at 1220.

[248]*Id.* The Mine Commission has ruled that a good faith belief of a hazard to another employee may also support a refusal to work. *Cameron v. Consolidation Coal Co.*, 3 MSHC 1721 (FMSHRC No. WEVA 82-190-D, 1985), *aff'd*, 795 F.2d 364 (4th Cir. 1986).

[249]*Supra* note 244, at 1217, 1220.

[250]2 MSHC 2152 (FMSHRC No. WEST 81-115-DM, 1983).

Commission's conclusion that the work refusal was not based on a reasonable belief was not supported.

In *Chacon v. Phelps-Dodge Corp.*,[251] the Mine Commission stated, in effect, that it would not second-guess the "business and personal judgment" of an operator, even if the judgment may in some abstract sense be "unjust." Chacon was an employee who had raised a number of safety complaints and was given a 3-day suspension following his second train derailment. Commissioner Lawson filed a strong dissent. That dissent seems to focus more on the application of this principle to the facts in *Chacon* than the principle articulated by the majority opinion. The principle involved is not a new one. The issue is essentially one of whether or not to discipline and, if the discipline is given, the severity of discipline for particular types of conduct: "Courts are generally less competent than employers to restructure business practices, and unless mandated to do so by Congress they should not attempt it."[252] Certainly, one would expect unanimity from the Mine Commission that a defense of appropriate discipline will not be accepted if a majority finds either that the alleged misconduct did not occur or that the discipline was inconsistent with prior discipline of the complaining miner or other miners. The court of appeals reversed the Mine Commission, holding that the administrative law judge's findings were supported by substantial evidence and the Commission is statutorily bound to uphold those findings.[253]

Another case that defines the parameters of a protected work refusal is *Dunmire v. Northern Coal Co.*[254] In *Dunmire*, the Mine Commission stated that a miner "should ordinarily communicate, or at least attempt to communicate to some representative of the operator his belief in the safety or health hazard at issue."[255]

Bradley v. Belva Coal Co.[256] involved issues of deferral to state proceedings on discrimination matters. Bradley, a sec-

---

[251] 2 MSHC 1505 (FMSHRC No. WEST 79-349-DM, 1981), *rev'd*, 709 F.2d 51, 2 MSHC 2145 (D.C. Cir. 1983).

[252] *Furnco Constr. Corp. v. Waters*, 438 U.S. 567, 578, 17 FEP 1062, 1066 (1978) (Title VII).

[253] *Chacon v. Phelps-Dodge Corp.*, *supra* note 251.

[254] 2 MSHC 1585 (FMSHRC No. WEST 80-313-D, 1982).

[255] *Id.* at 1590. Accord *Miller v. FMSHRC*, 687 F.2d 194, 2 MSHC 1817 (7th Cir. 1982) (refusing to work without prompt reporting to management unprotected).

[256] 2 MSHC 1729 (FMSHRC No. WEVA 80-708-D, 1982).

tion foreman, had filed discrimination charges under West Virginia law. West Virginia authorities concluded that safety issues were not involved in Bradley's termination and upheld the termination. The Mine Commission refused to defer to the West Virginia finding. The decision does not rule out the possibility of applying *res judicata* or collateral estoppel doctrines under Section 105(c) with respect to prior state proceedings. It does apply deferral doctrines narrowly, however. The Mine Commission distinguished the Supreme Court's decision involving the same issue under Title VII.[257]

The Mine Commission has also held that it is improper to interrogate an employee coercively regarding his exercise of Section 105(c) rights. The same decision found it improper to discharge an employee suspected of protected activity, even though no protected activity had occurred.[258]

The Mine Act requires that discrimination claims be filed within 60 days after the discrimination occurs. The Mine Commission has held that this time limit may be excused for justifiable reasons. Circumstances are to be evaluated on a case-by-case basis.[259]

4. Temporary Reinstatement

Section 105(c)(2) provides that the Secretary of Labor may seek "immediate reinstatement" of the miner if the Secretary determines the complaint was "not frivolously brought."[260] Initial Mine Commission rules provided that a petition for temporary reinstatement was to be granted unless it was found that the Secretary's conclusion was "arbitrarily or capriciously" made.[261] The Mine Commission subsequently con-

---

[257]*Kremer v. Chemical Constr. Corp.*, 456 U.S. 461, 28 FEP Cases 1412 (1982).

[258]*Moses v. Whitley Dev. Corp.*, 2 MSHC 1835 (FMSHRC No. KENT 79-366-D, 1982).

[259]Compare *Hollis v. Consolidation Coal Co.*, 3 MSHC 1169 (FMSHRC No. WEVA 81-480-D, 1984) (miner knew of rights and failed to act, therefore delay *not* excused) with *Schulte v. Lizza Indus.*, 3 MSHC 1176 (FMSHRC No. 81-53-DM, 1984) (miner's lack of knowledge excused delay).

[260]30 U.S.C. §815(c)(2) (Supp. II 1978). See *Price v. Jim Walter Resources, Inc.*, 4 MSHC 1475 (FMSHRC No. SE 87-87-D, 1987), *appeal filed*, No. 87-7484 (11th Cir. Aug. 8, 1987).

[261]44 FED. REG. 38,227, 38,230–231 (1979) (codified at 29 C.F.R. §2700.44 (1981)).

cluded that this narrow standard violated due process of law because it did not provide a mine operator with sufficient opportunity to show that temporary reinstatement was improper.[262] The new Mine Commission rule eliminates the "arbitrary or capricious" standard.[263]

5. Individual Miner Discrimination Actions

If the Secretary of Labor chooses not to pursue a discrimination action in a miner's behalf, the miner may pursue it himself.[264] If a miner ultimately prevails in such an action, he may receive costs and expenses of litigation, including his attorney's fees.[265]

B. Transfer Rights

The Mine Act contains two provisions relating to transfers. The first of these is Section 101(a)(7).[266] The second is Section 203(b).[267] The language in Section 203(b) is carried over from the 1969 Coal Act.

Section 101(a)(7) deals with the promulgation of mandatory health or safety standards relating to toxic substances and warning of hazards and symptoms. If, in promulgating regulations, it is determined that "a miner may suffer material impairment of health or functional capacity by reason of exposure to the hazard covered by the mandatory standard,"

---

[262]*Gooslin v. Kentucky Carbon Corp.*, 2 MSHC 1385 (FMSHRC No. KENT 80-145-D, 1981).

[263]29 C.F.R. §2700.44(a) (1985), promulgated at 46 FED. REG. 39,137 (1981).

[264]30 U.S.C. §815(c)(3) (Supp. II 1978). The miner may not, however, file his action until the Secretary determines that Section 105(c) was not violated. *Gilbert v. Sandy Fork Mining Co.*, 4 MSHC 1464 (FMSHRC No. KENT 86-76-D, 1987).

[265]*Id.* E.g., *Bradley v. Belva Coal Co.*, 2 MSHC 1729 (FMSHRC No. WEVA 80-78-D, 1982); *Dunmire v. Northern Coal Co.*, 2 MSHC 1585 (FMSHRC No. WEST 80-313-D, 1982) (recovery of hearing expenses allowed); *Bailey v. Arkansas-Carbona Co.*, 3 MSHC 1145 (FMSHRC No. 81-13-D, 1983) (interest computation; fringe benefits not paid by company, not recoverable). Attorney's fees may be awarded for private counsel where the action is presented by the Secretary of Labor. Private attorney's fees will be reduced to the extent that the private attorney's efforts duplicate the Secretary's. *Ribel v. Eastern Associated Coal Corp.*, 3 MSHC 2017 (FMSHRC No. WEVA 84-33-D, 1985). A miner's remedies also may be limited by postdischarge misconduct. *Cruz v. Puerto Rican Cement Co.*, 3 MSHC 1751 (FMSHRC No. SE 83-62-M, 1985) (alleged life threat, if proved, may preclude reinstatement and toll accrual of back pay).

[266]30 U.S.C. §811(a)(7) (Supp. II 1978).

[267]30 U.S.C. §843(b) (Supp. II 1978).

then the miner is to be "removed from such exposure and reassigned."[268] To date, MSHA has not promulgated any regulations under this statutory provision.

Key language requiring interpretation relates to the pay rate which the transferred miner is to receive:

> "Any miner transferred as a result of such exposure shall continue to receive compensation for such work at no less than the regular rate of pay for miners in the classification such miner held immediately prior to his transfer. In the event of the transfer of a miner pursuant to the preceding sentence, increases in wages of the transferred miner shall be based upon the new work classification."[269]

The question under this statutory language is how to compute the future rate of pay.

To illustrate the situation: Suppose that a miner is making $10 per hour and is transferred to a $9 per hour job. Section 101(a)(7) is clear that the miner's pay could never be reduced below $10 per hour. It is also clear that if the next wage increase in the $9 classification is 10 cents, although the wage increase in the $10 classification is 15 cents, the transferred miner's wage increase will be no more than 10 cents per hour. The question is whether the language that states that the transferred miner's wage increases "shall be based upon the new work classification" means that the transferred miner would continue at $10 per hour until the $9 per hour classification had reached $10 or above, or that the transferred miner would receive wage increases given personnel in the $9 per hour classification so that he continued to paid $1 per hour more than the miners working in the $9 per hour classification who had not received medical transfers. This issue presumably will be addressed when regulations are promulgated under Section 101(a)(7) and could well be the subject of litigation.

Although the issue of pay rights has not been addressed to date under Section 101(a)(7), it has been addressed under Section 203.[270] Section 203 provides for medical examinations of employees working in coal mines. If a miner shows evidence of the development of pneumoconiosis, the miner is to be given the option of transferring to a position in another area of the

---

[268]30 U.S.C. §811(a)(7) (Supp. II 1978).
[269]*Id.*
[270]30 U.S.C. §843 (1976).

mine where the concentration of respirable dust is lower. A miner transferred under that provision "shall receive compensation for [work in his new position] at not less than the regular rate of pay received by him immediately prior to his transfer."[271] This language varies significantly from the language in Section 101(a)(7). Notwithstanding that fact, interpretations of the cases under that provision may be instructive.

In *Higgins v. Marshall*,[272] the District of Columbia Circuit ruled that the language of Section 203(b)(3) required that a transferred miner continue to receive the pay rate he had received in his pretransfer classification. The court held, however, that a miner transferred under Section 203(b) was not entitled to pay increases which he would have received had he remained in the old classification. The court stated: "[A]t the very least, the legislative history surrounding the enactment of Section 811(a)(7) is consonant with our holding."[273] The Fourth Circuit has reached a similar result.[274]

Two additional decisions are of note in this area. A court of appeals has ruled that where a miner was spending a substantial portion of his working time performing tasks in a job classification that had a higher rate of pay than his formal classification, the transferred miner was entitled to have that higher pay rate considered in computing the rate of pay to be given to him after a transfer.[275] The Mine Commission also has ruled that a miner discharged because he was the subject of medical evaluation and potential transfer may recover under the antidiscrimination provision in Section 105(c) of the 1977 Mine Act, despite the provision of similar protections in the Black Lung Benefits Act.[276]

C. Employee Walkaround Pay Rights

The Mine Act provides that mine operators and representatives of miners have the right to accompany inspectors

---

[271] §843(b)(3).

[272] 584 F.2d 1035, 1 MSHC 1656 (D.C. Cir. 1978), *cert. denied*, 441 U.S. 931 (1979).

[273] *Id.* at 1039, 1 MSHC at 1659–1660.

[274] *Matala v. Consolidation Coal Co.*, 647 F.2d 427, 2 MSHC 1265 (4th Cir. 1981).

[275] *Mullins v. Andrus*, 664 F.2d 297, 2 MSHC 1089 (D.C. Cir. 1980).

[276] *Goff v. Youghiogheny & Ohio Coal Co.*, 3 MSHC 2002 (FMSHRC No. LAKE 84-86-D, 1985); see also *Mullins v. Beth-Elkhorn Coal Corp.*, 4 MSHC 1361 (FMSHRC No. KENT 83-268-D, 1987). The Mine Commission had reached a different conclusion under the 1969 Coal Act. See *Matala v. Consolidation Coal Co.*, 1 MSHC 2001 (FMSHRC No. MORG 76-53, 1979).

during inspections or investigations at a mine.[277] Although
the statute provides that the Secretary of Labor may promulgate regulations regarding rights to accompany inspectors, that has not been done. MSHA has issued a policy
memorandum on the subject, however.[278]

The central issue that has been raised in this area is not
the right to accompany inspectors, but rather the nature of
the pay rights of employee representatives accompanying the
inspectors. Section 103(f) provides that if a representative of
miners is also an employee of the operator, the representative
"shall suffer no loss of pay during the period of his participation in the inspection made under this subsection."[279]

The Mine Commission issued three decisions dealing with
the issue of whether a "spot inspection" was included in the
pay provision. Spot inspections were defined by the Mine Commission to include both inspections required under Section
103(i)[280] of the Mine Act for mines which liberate excessive
quantities of methane,[281] and specialized inspections such as
electrical inspections.[282] The Mine Commission held, by a 3-
to-2 vote, that employee walkaround representatives on the
regular full mine inspections contemplated by Section 103(a)
were to be paid, but that miners accompanying inspectors on
the spot inspections were not entitled to walkaround pay. The
District of Columbia Circuit reversed these three decisions
and held that miners were entitled to walkaround pay on spot
inspections, as well as on regular inspections.[283]

Another walkaround pay question dealt with the issue of
multiple inspection teams. The Mine Commission ruled that
if there are two or more inspection parties simultaneously

---

[277]30 U.S.C. §813(f) (Supp. II 1978).

[278]43 FED. REG. 17,546 (1978).

[279]30 U.S.C. §813(f) (Supp. II 1978).

[280]§813(i).

[281]*Secretary of Labor v. Helen Mining*, 1 MSHC 2193 (FMSHRC No. PITT 79-11-P, 1979); *Sparks v. Allied Chem. Corp.*, 1 MSHC 2232 (FMSHRC No. WEVA 79-148-O, 1979).

[282]*Kentland-Elkhorn Coal Corp. v. Secretary of Labor*, 1 MSHC 2230 (FMSHRC No. PIKE 78-399, 1979).

[283]*UMWA v. FMSHRC*, 671 F.2d 615, 2 MSHC 1601 (D.C. Cir. 1982), *cert. denied*, 459 U.S. 927 (1982). Accord *Consolidation Coal Co. v. FMSHRC*, 740 F.2d 271, 3 MSHC 1489 (3rd Cir. 1984); *Monterey Coal Co. v. FMSHRC*, 743 F.2d 589, 3 MSHC 1561 (7th Cir. 1984). Cf. *Truex v. Consolidation Coal Co.*, 4 MSHC 1130 (FMSHRC No. WEVA 85-151-D, 1986).

involved in a regular inspection at a time, there is to be one paid miners' representative in each inspection party.[284] This decision was affirmed by the Ninth Circuit.[285]

D. Pay Entitlement of Miners Idled by Closure Orders

Section 111 of the Mine Act provides that miners are to continue to receive full pay for a certain period of time if they are idled by withdrawal orders.[286] If a mine or a portion of a mine is closed by an order issued under Sections 103, 104, or 107 of the Mine Act, all idled miners who were working on that shift are entitled to full compensation at the regular rate of pay for the time period they were idled or for the balance of the shift, whichever is less. If the order has not been lifted before the next working shift, all miners who were idled by the order are entitled to compensation at the regular rate of pay for the time they are idled for up to 4 hours of the shift. Additionally, if the mine is closed by an order issued under Sections 104 or 107 for failure to comply with a mandatory health or safety standard, miners are to be compensated at their regular rate of pay for the time they were idled or for one week, whichever is less. If an operator fails to comply with a closure order issued under Sections 103, 104, or 107 of the Mine Act, all miners who would have been idled by the order, but instead were directed to continue to work, are entitled to full compensation by the operator at their regular rate of pay in addition to the pay they received for the work they performed between the time the order was issued and complied with, vacated, or terminated. A mine operator is entitled to request a hearing for a determination of these pay entitlement issues.

It should be noted that these pay entitlements accrue to idled miners regardless of the outcome of any litigation challenging the withdrawal order.[287] This provision, which was

---

[284]*Magma Copper Co. v. Secretary of Labor*, 1 MSHC 2227 (FMSHRC No. DENV 78-533-M, 1979). The Commission has also held that walkaround representatives must be paid for the time they spent in post-inspection conferences with MSHA inspectors. *Secretary of Labor v. Southern Ohio Coal Co.*, 3 MSHC 2113 (FMSHRC No. LAKE 82-93-R, 1986), *aff'd without opinion*, 4 MSHC 1288 (6th Cir. 1987).

[285]645 F.2d 694, 2 MSHC 1289 (9th Cir. 1981), *cert. denied*, 454 U.S. 940 (1981).

[286]30 U.S.C. §821 (Supp. II 1978).

[287]*Id.* In addition, compensation for miners idled by an imminent danger withdrawal order must be paid even if the necessary allegation of violation of a standard

also in the 1969 Coal Act, was upheld under that Act.[288] The mining company's objection that this provision violates due process was denied by the court.

Providing alternative work for part, but not all, of a shift will not relieve a mine operator of a compensation obligation. Thus, the Mine Commission has ruled, under the identical 1969 Coal Act language, that where miners are given 4 hours of work on the shift following issuance of an order, but are then sent home, they are entitled to compensation for the last 4 hours of the shift.[289] Four hours' compensation has also been directed where miners on the shift after issuance of an order were offered 4 hours of alternative work, but not a full shift of work.[290]

Miners have been denied compensation where the idling resulted from the oral directions of an inspector and no order was issued.[291] Likewise, compensation has been denied where at the time of the issuance of the withdrawal order the miners had already withdrawn under collective bargaining agreement provisions providing for a noncompensated "memorial period."[292]

E. Miner Participation in Proceedings

The Mine Act makes specific provision for miner participation. The legislative history strongly emphasizes the need for miner participation. In addition to the right to participate in inspections already noted, miners may request an inspection or investigation by MSHA if they believe there is a vio-

---

is alleged not in the order but in a citation or a subsequent modification of the order. *UMWA, Local 1889 v. Westmoreland Coal Co.*, 4 OSHC 1121, 1127–1129 (FMSHRC No. WEVA 81-256-C, 1986).

[288]*Rushton Mining Co. v. Morton*, 520 F.2d 716, 1 MSHC 1307 (3d Cir. 1975).

[289]*UMWA, Local 5869 v. Youngstown Mines Corp.*, 1 MSHC 2114 (FMSHRC No. HOPE 76-231, 1979) (1969 Coal Act).

[290]*UMWA, Local 3453 v. Kanawha Coal Co.*, 1 MSHC 2156 (FMSHRC No. HOPE 77-193, 1979) (1969 Coal Act); *reh'g denied*, 1 MSHC 2157 (FMSHRC No. HOPE 77-193, 1979).

[291]*UMWA, Local 6843 v. Williamson Shaft Contracting Co.*, 2 MSHC 1130 (FMSHRC No. VA 80-17-C, 1981).

[292]*UMWA, Local 781 v. Eastern Associated Coal Corp.*, 2 MSHC 1296 (FMSHRC No. WEVA 80-473-C, 1981). But see *UMWA, Local 1889 v. Westmoreland Coal Co.*, 4 MSHC 1121, 1125–1127 (FMSHRC No. WEVA 81-256-C, 1986) (miners idled by §103 control order to be compensated if imminent danger withdrawal order is also issued).

lation of the Act or of a mandatory health or safety standard, or that an imminent danger exists.[293] Miners are entitled to participate in the litigation of citations, penalty assessments, and orders issued under the Act.[294] They also may seek and receive party status in litigation.[295]

Section 105(d) of the Mine Act clearly provides that any miner or representative of miners may contest the issuance, modification, or termination of any order issued under Section 104 or the reasonableness of the length of abatement time set for abatement of a citation or a modification of a citation.[296] Similarly, Section 107(e)(1) provides that any representative of miners may apply for reinstatement, modification, or vacation of an imminent danger order.[297]

A question has been raised concerning the language of Section 105(d) that addresses the scope of miner participation in proceedings on citations. In *UMWA v. Secretary of Labor*,[298] the Mine Commission held that miners and their representatives do not have the statutory authority to initiate review of citations. Commissioner Lawson dissented, arguing that the legislative history and health and safety policy warrant allowing miner action rather than leaving matters to the "prosecutorial discretion" of the Secretary of Labor. It is interesting to note that miners' representatives as well as two coal companies joined in the position that miners and their representatives should be allowed to initiate proceedings to challenge a citation or to present evidence that a citation should have been categorized as an unwarrantable failure or imminent danger.

F. Training of Miners

Section 115 of the Mine Act requires that miners be given extensive training. New miners who have no prior under-

---

[293]30 U.S.C. §813(g) (Supp. II 1978).
[294]§815(d).
[295]29 C.F.R. §2700.4(b) (1985).
[296]30 U.S.C. §815(d) (Supp. II 1978).
[297]§817(e).
[298]2 MSHC 2097 (FMSHRC No. CENT 81-223-R, 1983), *aff'd*, 725 F.2d 126, 3 MSHC 1137 (D.C. Cir. 1983). See also *UMWA v. Secretary of Labor*, 3 MSHC 1049 (FMSHRC No. LAKE 82-70-R, 1983) (miners cannot challenge Secretary's vacation of withdrawal order).

ground experience must receive at least 40 hours of training before they work underground. Training is to include a variety of subjects, such as the statutory rights of miners and their representatives under the Mine Act, use of self-rescue and respiratory devices, hazard recognition, and the health and safety aspects of the tasks to which they will be assigned.[299] New miners who will work in surface operations must receive at least 24 hours of training covering essentially the same range of subjects required to be covered for new underground miners.[300] Additionally, all miners are to receive at least 8 hours of refresher training no less frequently than once each 12 months.[301] When a miner is assigned to a new task in which he has no prior work experience, he must receive task training in accordance with the training plan approved by MSHA under the regulations.[302]

The Secretary of Labor has promulgated regulations outlining the requirements for training and retraining of underground miners and surface miners.[303] These rules provide for the review and approval of training plans by the Secretary of Labor and outline the procedures for review of proposed plans and appeal procedures. Mine operators, miners, or representatives of miners have appeal rights under the regulations.[304] The regulations also require that mine operators maintain complete records of the miner training that has been conducted.[305]

Several training issues should be noted. The first of these involves the question of whether or not nonemployee representatives of miners are entitled to monitor a mine operator's classes under a training program. A Mine Commission administrative law judge concluded that nonemployee representatives of miners are entitled to attend those sessions and that it was a violation of Section 105(c) for a mine operator to bar the representatives from those sessions. This decision was reversed by the Mine Commission, which held that nonemployee

---

[299] 30 U.S.C. §825(a)(1) (Supp. II 1978).
[300] §825(a)(2).
[301] §825(a)(3).
[302] §825(a)(4).
[303] 30 C.F.R. pt. 48 (1985).
[304] 30 C.F.R. §§48.12, 48.32.
[305] §§48.9, 48.29.

representatives of miners have no right to monitor training classes held for miners on mine property.[306]

A second issue that should be noted is whether mine operators may lawfully require new employees, as a precondition to employment, to obtain training at their expense. An administrative law judge concluded that a requirement that prospective employees obtain 32 hours of training at their expense as a precondition to employment violated not only Section 115(c) of the Act, but also Section 105(c). The Mine Commission affirmed.[307] The Tenth Circuit denied enforcement of the Mine Commission's decision, holding that nothing in the Mine Act, or its legislative history, suggests that an employee must be paid wages and expenses for time spent in a course which the employee took on his own before he was employed.[308]

The Mine Commission has ruled that in recalling employees from layoff, a mine operator can give preference to those miners who kept up their required health and safety training. Training was obtained at miner expense, but recalled employees were reimbursed for their expenses.[309]

Another training issue recently decided by the Mine Commission involves the timing of the 8-hour retraining required by Section 115(a)(3) and the regulations. As noted above, this subsection requires that miners shall receive 8 hours of refresher training "no less frequently than once each 12 months." The issue that arose was whether that requirement mandates that retraining occur once each calendar year or within 12 months after the initial new miner training or refresher training. The administrative law judge vacated citations issued for violation of the training regulations, holding that training is required by the regulations only once each calendar year. Thus, a miner trained in January 1986 need not be trained until December 1987. The Mine Commission reversed the decision

---

[306]*Council of S. Mountains v. Martin County Coal Corp.*, 3 MSHC 1245 (FMSHRC No. KENT 80-222-D, 1984), *aff'd*, 751 F.2d 1418, 3 MSHC 1665 (D.C. Cir. 1985).

[307]*Bennett v. Emery Mining Corp.*, 2 MSHC 2236 (FMSHRC No. WEST 8-489-D, 1983).

[308]*Emery Mining Corp. v. Secretary of Labor*, 783 F.2d 155, 3 MSHC 2073 (10th Cir. 1986).

[309]*Rowe v. Peabody Coal Co.*, 3 MSHC 1929 (FMSHRC No. KENT 82-103-D, 1985), *aff'd*, 822 F.2d 1134 (D.C. Cir. 1987). Accord *Acton v. Jim Walter Resources*, 3 MSHC 1934 (FMSHRC No. SE 84-31-D, 1985) (recall preference to miners who kept up training, no violation; failure to reimburse miners found to be violation), *aff'd*, 822 F.2d 1134 (D.C. Cir. 1987).

and held that retraining is required within 12 months rather than only once each calendar year.[310] The Tenth Circuit affirmed.[311]

G. Black Lung Benefits

Title IV of the Mine Act contains extensive provisions providing for compensation for employees suffering from pneumoconiosis (black lung). The details of this coverage are beyond the scope of this treatise; however, the miners' entitlement should be noted.

Essentially, these statutory provisions provide workers' compensation coverage for employees who contract black lung where state laws do not provide adequate coverage for pneumoconiosis. Under the Mine Act, claims are to be filed under the applicable state workers' compensation law if that coverage is adequate.[312] The Secretary of Labor is to publish a list of those states whose compensation system is deemed adequate for black lung under the criteria set forth in the statute.[313] Detailed regulations have been promulgated for determining the adequacy of state laws.[314] If the statutory and regulatory criteria are not met, an employee is then entitled to seek federal benefits. To date, no state plans have been found adequate by the Secretary of Labor.

Disability benefit payments are set forth in the statute.[315] These payments cover not only a miner who is totally disabled owing to pneumoconiosis, but also the surviving spouse or dependent children of a miner who has died of pneumoconiosis. Regulations have been promulgated covering requirements to be met to show entitlement and for determination and payments of benefits.[316]

---

[310]*Secretary of Labor v. Emery Mining Corp.*, 3 MSHC 1001 (FMSHRC No. WEST 81-400-R, 1983).

[311]744 F.2d 1411, 3 MSHC 1585 (10th Cir. 1984).

[312]30 U.S.C. §931(a) (1976).

[313]§931(b).

[314]20 C.F.R. pt. 722 (1985).

[315]30 U.S.C. §922 (1976).

[316]20 C.F.R. pt. 410 (1985).

V. Key Differences Between the Mine Act and the OSH Act

The general regulatory schemes of the OSH Act and Mine Act are, in many respects, similar. Both statutes provide for the promulgation of regulations. Under each Act, inspectors are authorized to inspect a particular work situation and issue citations that may lead to civil or criminal penalties. Actions taken by inspectors are subject to administrative review by an independent commission and the commission's action is subject to judicial review.

Notwithstanding the similarities, there are a number of significant differences between the two statutes. Enforcement tools available to MSHA are different from OSHA in a number of respects already discussed. Additionally, it should be noted that MSHA and OSHA are separate administrations within the Department of Labor, and interpretations or approaches of one agency cannot be assumed in dealing with the other agency. This final section is designed to highlight briefly some of the key differences that an OSHA practitioner might otherwise overlook in first dealing with MSHA.

A. General Duty Clause

The Mine Act has no general duty clause comparable to Section 5 of the OSH Act.[317] The imminent danger order provisions in Section 107 of the Mine Act are in some sense analogous, however. An imminent danger order may be issued under the Mine Act, even though the situation leading to the closure order does not give rise to a violation of a mandatory health or safety standard.

B. Time for Abating Citations

Under the OSH Act, the time that an employer is given to correct a violation does not begin to run until the entry of a final order by the Review Commission.[318] If a violation is contested in litigation, it could easily be a year or more before an employer is obligated to begin correcting the problem. By contrast, a mine operator is obligated to begin correcting a

---

[317]29 U.S.C. §654 (1976).
[318]29 U.S.C. §659(b) (1976).

violation immediately, even if the matter is challenged in litigation. If the matter is in litigation, that might operate to delay the abatement time if an agreement is reached with MSHA. MSHA is not obligated to delay abatement for litigation and can issue Section 104(b) closure orders for failure to abate if it chooses to do so.

C. Imminent Danger Orders

As detailed above, the Mine Act provides for the issuance of imminent danger orders. Failure to comply with these orders can lead not only to civil penalties, but also to criminal sanctions under Section 110(d) of the Mine Act.[319] By contrast, the OSH Act provides that inspectors are to advise employees and employers of a conclusion that an imminent danger exists and that the inspector will be recommending that judicial relief be sought.[320] There is no requirement in the OSH Act, however, that the employer remove employees from exposure to the imminent danger situation until injunctive relief has been sought in the appropriate U.S. district court.[321] Employees who refuse to work in the dangerous situation after being advised of the inspector's view of the situation presumably would be protected from discipline or discharge under OSHA regulations protecting refusals to work.[322]

D. Unwarrantable Failure and Pattern of Violations Provisions

The OSH Act provides for more serious penalties for "willful" or "repeated" violations than for other violations.[323] While the "willful" and "repeated" concepts find some analogy in the criteria for the issuance of "unwarrantable failure" citations and orders and "pattern of violation" notices and orders, the provisions of the two Acts are radically different. The Mine Act specifically provides for closure orders in the unwarrant-

---

[319]30 U.S.C. §820(d) (Supp. II 1978).
[320]29 U.S.C. §662(c) (1976).
[321]§662(a).
[322]29 C.F.R. §1977.12 (1985).
[323]Compare 29 U.S.C. §666(a) with 29 U.S.C. §§666(b) and (c) (1976).

able failure and pattern-of-violation context, while no such orders may be issued at all under the OSH Act unless an imminent danger is found to exist. OSHA must then go to court, while MSHA can issue penalties for noncompliance with a withdrawal order.

E. Miners' Rights

1. Protection of Certain Nonemployees

The antidiscrimination provisions of Section 105(c) of the Mine Act protect not only employees, but also applicants for employment and representatives of miners.[324] By contrast, the antidiscrimination provisions of the OSH Act apply only to employees.[325] Also, under the OSH Act, only the Secretary of Labor can bring a discrimination proceeding; the Mine Act permits individuals to pursue their own claims.

2. Pay Rights

The Mine Act provides for pay for employee miners' representatives who accompany inspectors.[326] It also provides for pay for miners idled by closure orders.[327] There are no such pay provisions under the OSH Act. In 1977, OSHA issued an interpretation requiring employers to pay employees for time spent accompanying inspectors.[328] That interpretive requirement was rejected by the District of Columbia Circuit.[329] The court held that such an interpretation was legislative and must be promulgated in accord with the requirements of the OSH Act and Administrative Procedure Act. Since that had not occurred, the pay requirement was invalidated. OSHA subsequently revoked its interpretation[330] and proposed a new

---

[324]30 U.S.C. §815(c)(1) (Supp. II 1978).
[325]29 U.S.C. §660(c)(1) (1976).
[326]30 U.S.C. §813(f) (Supp. II 1978).
[327]30 U.S.C. §821 (Supp. II 1978).
[328]42 FED. REG. 47,344 (1977).
[329]636 F.2d 464, 8 OSHC 1648 (D.C. Cir. 1980).
[330]45 FED. REG. 72,118 (1980).

regulation providing for walkaround pay.[331] That proposed regulation has never been promulgated.

F. State Plans

The Mine Act does not allow for state plans similar to the OSHA allowance. A state may enforce its own mine safety laws to the extent that those laws are consistent with or more strict than Mine Act requirements.[332]

---

[331]45 FED. REG. 75,232 (1980).
[332]30 U.S.C. §955 (Supp. II 1978).

27

Other Federal Legislation and Regulation Affecting Workplace Safety and Health

The OSH Act is not, and was not meant to be, the nation's only safety and health legislation. Other federal laws and regulations impact workplace safety and health, either by design or operation. This chapter identifies such federal legislation and regulation, explains its relationship to the OSH Act, and discusses relevant case law.

I. Section 4(b)(1) of the OSH Act—Partial Exemption From Coverage[1]

Existing safety and health regulations of many federal agencies were not preempted by passage of the OSH Act. Congress provided, in Section 4(b)(1) of the Act, that "[n]othing in this chapter shall apply to working conditions of employees with respect to which other Federal agencies *** exercise statutory authority to prescribe or enforce standards or regulations affecting occupational safety or health."[2] The purpose of

---

[1]See also discussion in Chapter 15 (Defenses), Section II.C.
[2]29 U.S.C. §653(b)(1) (1985).

the foregoing exemption was to avoid duplication in the regulation of safety and health standards in the workplace. Recognizing the importance of such a statutory purpose, the Review Commission, in *Mushroom Transportation Co.*,[3] accordingly held that "[o]nce a Federal agency exercises its authority over specific working conditions, [OSHA] cannot enforce its own regulations covering the same conditions."[4]

A four-prong test, evolved from court and Review Commission decisions, must be satisfied, however, before the Section 4(b)(1) exemption may properly be claimed by an employer. First, the employer must be covered by a federal statute. Second, the purpose of, or the policy behind, the statute must be to assure safe and healthful working conditions for employees. Third, the federal agency must have actually *exercised* its statutory authority to prescribe or enforce safety and health standards or regulations. Finally, the particular working conditions involved must specifically be covered by the preempting agency's standards.

Initially, to be exempt under Section 4(b)(1), the federal agency must possess jurisdiction to prescribe regulations affecting the safety of the workplace in question. The Review Commission, in *Northwest Airlines*,[5] held that the federal agency's claim of authority, whether or not explicitly granted by statute, is controlling, provided such authority is reasonably implied from that agency's enabling legislation. Mere statutory authority to regulate the workplace, however, is insufficient to preempt OSHA's regulations. More specifically, as stated previously under the second prong of the test, the actual policy behind such statutory authority must be the safety and health of the employee.

A policy or purpose test was first expressed by the Review Commission in *Fineberg Packing Co.*[6] *Fineberg* involved an employer who was subject to, and therefore had to comply with, specific sanitation standards imposed under the Wholesome Meat Act,[7] by the Department of Agriculture. As it was already subject to such regulation, the employer in *Fineberg*

---

[3]1 OSHC 1390 (Rev. Comm'n 1973).
[4]*Id.* at 1392.
[5]8 OSHC 1982 (Rev. Comm'n 1980).
[6]1 OSHC 1598 (Rev. Comm'n 1974).
[7]21 U.S.C. §§601–695.

claimed that OSHA was precluded from exercising any further authority over its employees' working conditions.

The Review Commission held against the employer, however, as it determined that the real purpose behind the Wholesome Meat Act was to protect consumers, rather than employees. As such, it found that the Section 4(b)(1) exemption was not properly applicable to the employer. More specifically, in setting forth the policy or purpose test, the Review Commission stated that "[t]o be cognizable under section 4(b)(1), we conclude that a different statutory scheme and rules thereunder must have a policy or purpose that is consonant with that of the Occupational Safety and Health Act. That is, there must be a policy or purpose to include employees in the class of persons to be protected thereunder."[8]

The third prong of the Section 4(b)(1) exemption test, as stated, is that the controlling federal agency actually must have exercised its statutory authority to prescribe or enforce the standards or regulations. Thus, in *Southern Pacific Transportation Co. v. Usery*,[9] the mere proposal of future rules for railroad safety was held to be insufficient action to trigger the applicability of the Section 4(b)(1) exemption. The court stated in *Southern Pacific* that existing and operative OSHA regulations could not be displaced by mere statements of intent to do something in the future.[10]

Finally, the fourth prong in the Section 4(b)(1) exemption test is that the specific working conditions involved must be covered by the other agency's standards. In other words, Section 4(b)(1) cannot be interpreted as providing an "industry-wide" exemption when an agency merely has exercised partial authority over certain working conditions of an industry.[11]

---

[8]*Fineberg*, supra note 6, at 1599. See also *Haas & Haynie Corp.*, 4 OSHC 1911 (Rev. Comm'n 1976) (GSA regulations issued under procurement statutes do not preempt OSHA standards); *Gearhart-Owen Indus.*, 2 OSHC 1568 (Rev. Comm'n 1975) (DOD procurement regulations do not preempt OSHA standards). Accord *Ensign-Bickford Co. v. OSHRC*, 717 F.2d 1419 (D.C. Cir. 1983).

[9]539 F.2d 386 (5th Cir. 1976), *cert. denied*, 434 U.S. 874 (1977).

[10]*Id.* at 392–393. See also *Pennsuco Cement & Aggregates*, 8 OSHC 1378 (Rev. Comm'n 1980) (MESA's suspension of regulation over kilns did not affect preemption where MESA exercised jurisdiction before and after suspension); *Northwest Airlines*, supra note 5 (FAA enforcement scheme requiring airlines to comply with airplane manufacturers' maintenance manuals as modified by airline constituted exercise of jurisdiction sufficient to preempt OSHA). Cf. *Conrail* cases, discussed in Chapter 15 in Section II.C.3.

[11]*Supra* note 9, at 390; *Marshall v. Northwest Orient Airlines*, 574 F.2d 119, 122 (2d Cir. 1978).

Therefore, an analysis of the scope of the applicability of an agency's standards must be made in each case.

In defining the scope of "working conditions," the Fifth Circuit stated in *Southern Pacific* that the term encompasses not only a worker's surroundings, but also the hazards incident to his work.[12] The Fourth Circuit, in *Southern Railway v. OSHRC*,[13] more broadly defined the term "working conditions" as "the environmental area in which an employee customarily goes about his daily tasks."[14]

Once the Review Commission has determined that a rule promulgated by another agency has the "force and effect of law," it need not inquire further as to the manner of enforcement. In other words, a regulation or standard, found applicable to an employer under the foregoing four-prong test, need not be precisely the same or equally as stringent to preclude OSHA from enforcing its standards.[15] Thus, although some enforcement must be shown on the other agency's part, Section 4(b)(1) does not permit the Review Commission to oversee the adequacy of such enforcement efforts.[16]

In order to alleviate, in part, the inevitable problems arising from conflicting standards and enforcement responsibility, OSHA has concluded formal and/or informal understandings with a number of agencies, including the Department of the Treasury (Bureau of Alcohol, Tobacco, and Firearms), the Department of Agriculture (Food Safety and Inspection Service and Extension Service), the Consumer Product Safety Commission, and the Food and Drug Administration.[17] (OSHA negotiated, but failed to reach, a memorandum of understanding with the Federal Aviation Administration.[18]) In addition, OSHA and the National Institute for Occupational Safety and Health (NIOSH) have entered into the OSHA-NIOSH Interagency Agreement on Employee Protection;[19] OSHA and the Mine Safety and Health Administration (MSHA) concerted

---

[12]*Supra* note 9, at 390.

[13]539 F.2d 335 (4th Cir. 1976), *cert. denied*, 429 U.S. 999 (1976).

[14]*Id.* at 339. See also *Columbia Gas v. Marshall*, 636 F.2d 913, 916 (3d Cir. 1980) (court joined Fourth Circuit's broad definition of "working conditions").

[15]Northwest Airlines, *supra* note 5, at 1990.

[16]*Pennsuco Cement & Aggregates, supra* note 10.

[17]These memoranda have not be published.

[18]11 OSHR 587 (1981).

[19]44 FED. REG. 22,834 (1979), OSHR [Reference File 21:7001].

their efforts to provide a safe workplace for employees in the OSHA-MSHA Interagency Agreement;[20] and the Environmental Protection Agency (EPA) and OSHA in 1981 agreed upon the Memorandum of Understanding Between the EPA and OSHA.[21] Moreover, the U.S. Coast Guard has entered into several interagency agreements with OSHA, including the OSHA-Coast Guard Memorandum of Understanding on the Outer Continental Shelf,[22] the OSHA-Coast Guard Memorandum of Agreement 2-80 for Workplaces Aboard Inspected Vessels,[23] and the OSHA-Coast Guard Memorandum of Understanding on Enforcing Standards for Seamen Aboard Inspected Vessels.[24]

II. Federal Agencies

A. Department of Transportation

The Department of Transportation (DOT) was established in an effort to provide "fast, safe, efficient, and convenient transportation" for all persons in the United States.[25] To accomplish this purpose, the Secretary of Transportation has been granted broad powers to formulate transportation policies and regulations. Included among his powers is the authority to promote "industrial harmony and stable employment conditions in all modes of transportation."[26]

The Secretary of Transportation regulates the safety and health of employees pursuant to powers granted him by the Federal Aviation Act of 1958, as amended,[27] the Interstate Commerce Act,[28] the Federal Railroad Safety Act of 1970,[29]

---

[20]44 FED. REG. 22,827 (1979), OSHR [Reference File 21:7071].

[21]OSHR [Reference File 21:7191]. See also the February 6, 1986, Memorandum of Understanding Between the U.S. Environmental Protection Agency and the Department of Labor, OSHR [Reference File 21:7201].

[22]45 FED. REG. 9142 (1980), OSHR [Reference File 21:7101].

[23]45 FED. REG. 14,987 (1980), OSHR [Reference File 21:7151].

[24]48 FED. REG. 11,550 (1983), OSHR [Reference File 21:7161].

[25]49 U.S.C. §101(a) (1986).

[26]§301(4).

[27]§§1301 *et seq.*

[28]§§10101 *et seq.*

[29]45 U.S.C. §§421 *et seq.* (1972).

the Federal Safety Appliance Act,[30] the Hazardous Liquid Pipeline Safety Act of 1979,[31] the Independent Safety Board Act of 1974,[32] and various other federal statutes. Additionally, the Secretary of Transportation has issued regulations concerning employee safety in the operation or use of motor carriers and in the transportation of natural or other gas by pipeline.[33] To assist the Secretary of Transportation, the National Transportation Safety Board was set up to investigate civil aircraft, railroad grade-crossing, highway, major marine, and pipeline accidents.[34]

1. Air Commerce

The Department of Transportation, through the Secretary of Transportation, is empowered to develop "[s]uch reasonable rules and regulations *** as the Secretary of Transportation may find necessary to provide adequately for national security and safety in air commerce."[35]

The scope of this authority is far-reaching. The DOT not only has authority to promulgate regulations concerning the safety of the working conditions of airline employees in flight, but its authority extends to working conditions of airline maintenance personnel engaged in ground operations involving aircraft servicing.[36] Thus, the DOT, by requiring airlines to develop maintenance manuals that include safety instructions for maintenance personnel, effectively, through promulgation of its own regulations, precludes OSHA from regulating the working conditions of airline ground personnel.[37] OSHA, however, as stated previously, retains authority to regulate specific working conditions of airline-related employment that are not regulated already.[38]

---

[30]45 U.S.C.§§1 *et seq.* (1986).

[31]49 U.S.C. §§2001 *et seq.* (1986).

[32]49 U.S.C. §§1901 *et seq.* (1976).

[33]See, e.g., 49 C.F.R. §§190.1 *et seq.* (1985).

[34]49 C.F.R. §800.3 (1985).

[35]49 U.S.C. §1421(a)(6)(1986).

[36]§§1421–1432.

[37]*Northwest Airlines*, 8 OSHC 1982 (Rev. Comm'n 1980).

[38]See, e.g., *Allegheny Airlines*, 6 OSHC 1147 (Rev. Comm'n 1977); *Fortec Constructors*, 1 OSHC 3208 (Rev. Comm'n 1973).

2. Motor Carriers

Motor carriers are regulated by standards promulgated by DOT pursuant to authority granted it under the Motor Carrier Act of 1980.[39] Accordingly, DOT has issued regulations in such areas as the handling of hazardous materials in interstate trucking[40] and the reporting of violations and accidents. Truckers and motor vehicle common carriers, however, are subject to OSHA standards where DOT has not regulated the specific working conditions in issue.[41]

3. Pipeline Regulation

Pursuant to authorization granted by the Natural Gas Pipeline Safety Act of 1968,[42] the Secretary of Transportation has prescribed minimum safety requirements for pipeline facilities and the transportation of natural gas. The promulgated regulations pertain to the materials used in pipe construction, the design of pipe, general construction installation and maintenance guidelines for pipelines, and testing requirements for pipeline strength and durability.[43]

Also included in the Natural Gas Pipeline Safety Act's provisions are safety regulations for the operation of pipeline facilities. For example, pipeline facility operators and owners must instigate operating plans for their plants which set out specific standards for the safety of employees and provide for regular inspections.[44] In addition, each owner or operator must formulate an emergency plan that sets out detailed steps to be taken by employees when a plant emergency arises.[45]

Authority to enforce the provisions and regulations of the Natural Gas Pipeline Safety Act is delegated to the Secretary of Transportation.[46] The Secretary is authorized to inspect the pipeline facilities to determine whether the owner or operator is in compliance with the Act.[47] The owner must maintain

---

[39]49 U.S.C. §11101 *et seq.* (1986).

[40]49 C.F.R. §§171.1 *et seq.* (1984).

[41]See, e.g., *Lee Way Motor Freight,* 4 OSHC 1968 (Rev. Comm'n 1977).

[42]49 U.S.C. §§1671 *et seq.* (1986).

[43]49 C.F.R. §§192.1 *et seq.* (1985).

[44]§192.605.

[45]§192.615.

[46]49 U.S.C. §§1671 *et seq.* (1985).

[47]§1681.

and make available to the Secretary of Transportation such reports and records that would enable him to determine whether the federal standards required by the Natural Gas Pipeline Safety Act are being followed.[48]

The exercise of the Secretary of Transportation's authority under the Natural Gas Pipeline Safety Act is sufficient, under certain working conditions, to exempt employers engaged in the transmission, sale, and storage of natural gas from the OSH Act, pursuant to Section 4(b)(1). In *Texas Eastern Transmission Corp.*,[49] for example, the Review Commission determined that DOT's minimal federal safety standards preempted OSHA standards regarding the working conditions giving rise to the alleged violations. A contractor engaged in repairing a pipeline, however, would be subject to OSHA regulations because the regulations issued under the Natural Gas Pipeline Safety Act apply only to those persons engaged in the transportation of gas or operation of pipeline facilities.[50]

4. Federal Railway Safety Regulations

For the purpose of promoting safety in all areas of railroad operations, Congress has enacted legislation that makes railroads subject to federal regulation in many, if not most, areas of operation. The Federal Railroad Safety Act of 1970,[51] the Federal Safety Appliance Act and Boiler Inspection Act,[52] and the Federal Employer's Liability Act,[53] for example, regulate working conditions for railroad employees.

The Federal Railroad Safety Act of 1970 was enacted in an effort "to promote safety in all areas of railroad operations and to reduce railroad-related accidents, and to reduce deaths and injuries to persons and to reduce damage to property caused by accidents involving any carriers of hazardous materials."[54] The Act grants the Secretary of Transportation extensive powers to promulgate rules, regulations, orders, and standards to

---

[48]§1680.

[49]3 OSHC 1601 (Rev. Comm'n 1975).

[50]*Id.* See, e.g., *Northern Border Pipeline Co. v. Jackson County*, 512 F. Supp. 1261 (D. Minn. 1981); *Columbia Gas v. Marshall*, 636 F.2d 913 (3d Cir. 1980).

[51]45 U.S.C. §§421 *et seq.* (1972).

[52]§§22 *et seq.*

[53]§§51 *et seq.*

[54]§421.

implement the purpose of the Act.[55] Such powers include authority to designate hazardous situations, or unsafe conditions of facilities or equipment, and to issue orders prohibiting the use of such facilities or equipment.[56]

Pursuant to authority granted under the Federal Railroad Safety Act, the Secretary of Transportation established the Federal Railroad Administration. The Federal Railroad Administration promulgates rules concerning railroad car and track safety standards; railroad operating rules; hours of service of railroad employees; and railroad installation, maintenance, inspection, and repair regulations. In addition to establishing the Federal Railroad Administration, the Federal Railroad Safety Act provides for protection of railroad employees from discharge or discrimination when they have reported safety violations or refused to work in hazardous conditions.[57] It provides for civil and criminal penalties against any railroad operator or owner who disobeys or fails to adhere to any rule, regulation, order, or standard prescribed pursuant to the Federal Railroad Safety Act.[58]

In addition to the safety regulations promulgated pursuant to the Federal Railroad Safety Act, Congress has enacted laws for the purpose of safeguarding employees involved in transporting property through interstate commerce. Two such laws, the Federal Safety Appliance Act and the Boiler Inspection Act, for example, regulate the safety of railroad appliances and equipment. In addition, both are construed and applied in conjunction with the Federal Employer's Liability Act.

The Federal Safety Appliance Act, enacted for the protection of railroad employees as well as the general public, requires carriers engaged in interstate commerce to equip their cars with various safety devices and appliances and to maintain such appliances in sufficient condition.[59] That Act, moreover, declares it unlawful for a carrier subject to the requirements to haul any car not equipped according to the

---

[55]45 U.S.C. §431 (1986).

[56]45 U.S.C. §432 (1972).

[57]45 U.S.C. §441 (1986).

[58]§438. The standard of liability of a railroad operator or owner is "known or should have known." *Fort Worth & D. Ry. v. Lewis*, 693 F.2d 432 (5th Cir. 1982).

[59]45 U.S.C. §§1–7 (1986).

standards prescribed by the Act.[60] A breach of the Act may constitute negligence under the general liability provision of the Federal Employer's Liability Act. Moreover, an employer may be held strictly liable for injuries to any employee caused by an employer's failure to comply with a specific safety requirement of the Federal Safety Appliance Act, irrespective of a showing that an agent of the employer acted negligently.[61] Liability in such circumstances follows from mere use or maintenance by the employer of defective equipment.[62]

The Boiler Inspection Act provides specific safety regulations to be followed in the operation of any locomotive. The safety regulations issued in regard to locomotives are similar to those promulgated pursuant to the Federal Safety Appliance Act. A violation of the Boiler Inspection Act constitutes negligence *per se* on the part of the railroad subject to the requirement.[63]

As noted above, liability under the Federal Employer's Liability Act (FELA) may be predicated on a carrier's violation of the Federal Safety Appliance Act and the Boiler Inspection Act.[64] Thus, although violation of the latter acts confers no express right of action for damages against a carrier, FELA does provide for such damages. More specifically, FELA states that a railroad engaged in interstate commerce is liable in damages to any person suffering from injury while employed by the carrier, where the injury results, in whole or in part, from the negligence of officers, agents, or employees of the carrier or by reasons of defect or insufficiency due to the negligence of the carrier in its cars, engines, appliances, or other equipment.[65] Furthermore, it provides that in the event of an employee's death as a result of such negligence, the carrier is liable to the deceased's personal representatives for the benefit of designated survivors.[66] FELA also states that contributory negligence on the part of the railroad employee does not bar recovery for his injuries or death in an action against an em-

---

[60]§6.

[61]§13.

[62]*Id.*

[63]45 U.S.C. §14 (1972).

[64]See, e.g., *Green v. River Terminal Ry.*, 763 F.2d 805 (6th Cir. 1985); *Coleman v. Burlington N.*, 681 F.2d 542 (8th Cir. 1982).

[65]45 U.S.C. §51 (1972).

[66]§59.

ployer; it does, however, operate to reduce recoveries on a percentage basis (pure comparative negligence theory).[67]

It should be noted that FELA's absolute liability provisions are inapplicable to violations of safety regulations promulgated under the OSH Act. In *Bertholf v. Burlington Northern Railroad*,[68] for example, the plaintiff sued to recover for injuries sustained while repairing a defective spring on a railroad car. The plaintiff argued that his employer was liable under FELA, regardless of the plaintiff's negligence, because the employer had violated an OSHA standard. The *Bertholf* court disagreed, stating that FELA bars diminution of damages in proportion to an employee's contributory negligence only where the employer has violated the Safety Appliance Act or the Boiler Inspection Act, but is inapplicable to violations of an OSHA standard.[69]

Owing to the fact that railway regulations have been issued with respect to certain aspects of the safety and health of railroad employees, railroads have contended that their entire industry is exempt from OSHA regulations under Section 4(b)(1) of the Act. Both the Review Commission and the courts, however, have held that OSHA does not provide an industrywide exemption.[70] More specifically, railroads are exempt from OSHA only where there is an actual exercise of authority by the Federal Railroad Administration.[71]

The issuance of a policy statement by the Federal Railroad Administration, for example, would not constitute an exercise of authority, within the meaning of Section 4(b)(1), over certain working conditions at an employer's railroad repair facility.[72] Railroads, however, would be exempt from OSHA recordkeeping and certain workplace inspection regulations in the event similar Federal Railroad Administration regulations were promulgated concerning the same specific working conditions or places.[73]

---

[67]§53.

[68]402 F. Supp. 171 (E. D. Wash. 1975).

[69]*Id.* at 173.

[70]See, e.g., *Norfolk & W. Ry.*, 4 OSHC 1842 (Rev. Comm'n 1976); *Southern Pac. Transp. Co. v. Usery*, 539 F.2d 386 (5th Cir. 1976), *cert. denied*, 434 U.S. 874 (1977).

[71]See, e.g., *Union Pac. R.R.*, 5 OSHC 1702 (Rev. Comm'n 1977).

[72]See, e.g., *PBR, Inc. v. Secretary of Labor*, 643 F.2d 890 (1st Cir. 1981).

[73]See, e.g., *Dunlop v. Burlington N.R.R.*, 395 F. Supp. 203 (D. Mont. 1975).

B. Environmental Protection Agency

The Environmental Protection Agency investigates, monitors, and regulates the safety and condition of the environment. Pursuant to authority granted under numerous federal statutes, it regulates such areas as waste disposal, clean air, toxic and hazardous substances, and pesticides. As with other federal agencies, when the EPA issues regulations concerning the working conditions and safety of employees in the workplace, OSHA's regulations in that area may be preempted pursuant to Section 4(b)(1) of OSHA.

The following federal statutes empower the EPA to establish certain health and safety standards for employees in the workplace:

1. Federal Insecticide, Fungicide, and Rodenticide Act

EPA, pursuant to the authority granted it by the Federal Insecticide, Fungicide, and Rodenticide Act (FIFRA),[74] oversees the registration and regulation of pesticides. FIFRA specifically requires the registration of hazardous pesticides, sets out extensive labeling requirements, makes it unlawful to distribute misbranded pesticides, and authorizes the EPA Administrator to inspect industries using or manufacturing hazardous pesticides and to investigate alleged violations of FIFRA.[75]

Included in FIFRA's pesticide regulations are standards governing the use of pesticides in the workplace. Such standards address, for example, the protection of farm workers who perform field operations subsequent to the application of pesticides.[76] More specifically, however, they prohibit the application of pesticides in such a manner so as to expose workers to unsafe conditions.[77] Additionally, specific timing intervals, during which employees may reenter areas where hazardous pesticides have been applied, are set out pursuant to FIFRA.[78] As might be expected, the foregoing exercise of authority by

---

[74]7 U.S.C. §136 *et seq.* (1980).
[75]*Id.*
[76]40 C.F.R. §§170.1 *et seq.* (1985).
[77]*Id.*
[78]40 C.F.R. §170.3(b) (1985).

the EPA, under FIFRA, precludes the Secretary of Labor from promulgating and enforcing OSHA standards for farm workers exposed to the regulated pesticides.

2. *Toxic Substances Control Act*

The Toxic Substances Control Act[79] empowers the EPA to regulate the manufacturing, processing, distribution, use and disposal of chemical substances and mixtures and to collect industry-generated data regarding the production, use, and health effects of such chemicals.[80] If the EPA Administrator concludes that a new substance poses an unreasonable risk to health or the environment, he may prohibit the manufacture or sale of the substance or adopt regulations conditioning its use.[81] Additionally, regulations concerning employee health and safety in the manufacture and use of toxic substances have been promulgated.[82]

3. *Clean Air Act*

The Clean Air Act[83] authorizes EPA to regulate emission control standards. Also, the Clean Air Act allows the EPA Administrator to promulgate rules with regard to the direct monitoring of factories and businesses to ensure that its safety standards for employees are being met.

4. *Public Health Service Act and Federal Water Pollution Control Act*

Congress has authorized EPA, through the EPA Administrator, to regulate the health standards of drinking water under the Public Health Service Act.[84] Minimum and maximum contaminant levels, as well as extensive inspection procedures, are established in order to assure that the regulated standards are being met.[85]

---

[79]15 U.S.C. §§2601 *et seq.* (1982).
[80]*Id.*
[81]*Id.*
[82]40 C.F.R. §§702–799 (1985).
[83]42 U.S.C. §§7401 *et seq.* (1983).
[84]42 U.S.C. §§300f *et seq.* (1982).
[85]40 C.F.R. §§142 *et seq.* (1985).

In an additional effort to establish health standards for water in the United States, the Federal Water Pollution Control Act[86] was enacted. The Water Pollution Control Act establishes water pollution standards, as well as inspection and enforcement procedures, for industries coming within the scope of its jurisdiction.[87]

5. *Solid Waste Disposal Act*

The Solid Waste Disposal Act[88] has created the Office of Solid Waste to assist the EPA in the regulation of the safe disposal of waste. Sanitation criteria and standards are imposed on the operators of disposal facilities in order to protect the health of the facilities' employees, the public, and the environment in general.[89]

6. *National Environmental Policy Act of 1969*

The purpose of the National Environmental Policy Act[90] (NEPA) is to establish a federal policy that will improve the quality of the human environment. The NEPA requires that environmental impact statements be prepared and presented to the Council on Environmental Quality (CEQ) by federal agencies when significant federal action, such as the promulgation of safety or health standards, would significantly affect the quality of the human environment.[91] In consultation with the CEQ, federal agencies must develop methods and procedures on environmental quality, and present them through impact statements, in order to ensure that the present environment will be preserved.[92]

A limited exemption from the foregoing environmental impact statement provision, however, has been held to exist for OSHA. More specifically, in situations where temporary emergency OSHA standards are mandated to provide speedy

---

[86]33 U.S.C. §§1251 *et seq.* (1986).
[87]*Id.*
[88]42 U.S.C. §§6901 *et seq.* (1986).
[89]See 40 C.F.R. §§257.1 *et seq.* (1985).
[90]42 U.S.C. §§4321 *et seq.* (1977).
[91]*Id.*
[92]*Id.*

808 Occupational Safety and Health Law Ch. 27

protection from great dangers to employee health, the environmental impact statement regulations of the NEPA do not apply.[93] As explained by one court, the sacrifice of the policy behind the NEPA that occurs when such statements are not prepared is "mitigated somewhat by the fact that an emergency temporary standard must be replaced within six months by a permanent standard, for which a complete impact statement is required."[94]

C. Nuclear Regulatory Commission

The Atomic Energy Act of 1954[95] confers authority upon the Nuclear Regulatory Commission (NRC) to establish standards and instructions governing the possession and use of nuclear materials and their by-products. Those standards and instructions are meant to protect the health, and to minimize the danger to life or property, of those exposed to such nuclear materials. The NRC thus possesses authority to prescribe standards and restrictions governing the design, location, and operation of facilities maintained pursuant to the Atomic Energy Act of 1954. This includes authority to promulgate regulations concerning the safety of employees exposed to, or working with, nuclear materials.

It has been held that extensive workplace safety regulations, promulgated pursuant to the Atomic Energy Act of 1954, supersede comparable OSHA regulations. In *Newport News Shipbuilding & Drydock Co.*,[96] for example, the employer was engaged in the construction of nuclear-powered warships for the U.S. Navy. After having been cited for a serious OSHA violation, the employer argued that it was not subject to the relevant OSHA regulations owing to their having been preempted by Navy safety standards.

The Review Commission agreed, holding that the Navy standards, promulgated under authority of the Atomic Energy Act of 1954, preempted OSHA's authority. The Commission

---

[93]*Dry Color Mfrs.' Ass'n v. Department of Labor*, 486 F.2d 98, 107–108 (3d Cir. 1973).

[94]*Id.* at 108.

[95]42 U.S.C. §§2011 *et seq.* (1973).

[96]2 OSHC 3262 (Rev. Comm'n 1975).

based its holding on the fact that the primary purpose of the regulations was to provide for the safe handling of nuclear materials. Moreover, it recognized that the Navy affirmatively exercised its statutory authority to prescribe or enforce standards affecting the health and safety of employees with respect to the cited working conditions, so that the cited OSHA standards were inapplicable to such working conditions.[97]

D. Consumer Product Safety Commission

The Consumer Product Safety Commission, under authority granted it by the Federal Hazardous Substances Act,[98] regulates the manufacture, transportation, and sale of hazardous substances. The Commission is granted the power to define hazardous substances, require the specific labeling of substances, regulate the inspection of manufacturers, and control the sale of such hazardous material. Included in the Commission's regulations are specific provisions concerning the health and safety of employees involved in the manufacture, sale, or distribution of such hazardous substances.[99]

E. Food and Drug Administration

The Federal Food, Drug, and Cosmetic Act[100] authorizes the Food and Drug Administration (FDA) to define health and safety standards for food, drugs, and cosmetics. Such standards are related to the labeling, packaging, advertising, and composition of various foods, drugs, and cosmetics. The FDA is authorized to investigate various industries that manufacture the regulated foods, drugs, and cosmetics, in order to determine whether they comply with its federal regulations. The FDA also regulates the sanitation conditions of the workplace and, in turn, regulates employee safety and health conditions in such industries.[101]

---

[97]*Id.* at 3264.
[98]15 U.S.C. §§1261 *et seq.* (1982).
[99]See 16 C.F.R. §§1500 *et seq.* (1985).
[100]21 U.S.C. §§301–392 (1972).
[101]See also 21 C.F.R. §§5.1 *et seq.* (1985).

F. Maritime Employment

The working conditions of seamen are regulated by the U.S. Coast Guard and other maritime agencies under authority granted such agencies through various federal statutes. The maritime regulations involve standards covering not only the safety and health of employees working on marine vessels, but also those that extend to employees working on vessels in drydock, provided they are considered crew members of a ship in navigation.

1. Coast Guard

The U.S. Coast Guard has general maritime jurisdiction to promulgate and enforce regulations for the promotion of the safety of life and property on the high seas and water in all matters not specifically delegated to another executive department.[102] Pursuant to that jurisdiction, it has promulgated safety standards concerning vessel construction, maintenance, and repair; provided inspection proceedings and emergency guidelines; and issued regulations involving the transportation and use of hazardous materials and commercial diving operations.[103]

In addition to issuing such preventive regulations, the Coast Guard has set forth extensive procedures to be followed in the event of marine casualties and accidents. Those procedures include comprehensive reporting and investigative requirements.[104]

2. Outer Continental Shelf Lands Act

The Outer Continental Shelf Lands Act[105] provides that U.S. district courts have original jurisdiction over controversies arising out of any marine operations conducted in the outer Continental Shelf.[106] Additionally, it sets forth an ex-

---

[102]14 U.S.C. §§1 *et seq.* (1956).

[103]46 C.F.R. §§1 *et seq.* (1985).

[104]§4. The Coast Guard, not OSHA, has the authority to investigate casualties on vessels in navigable waters. *Dunlop v. Avondale Shipyards*, 3 OSHC 1950 (Rev. Comm'n 1976).

[105]43 U.S.C. §§1331 *et seq.* (1986).

[106]§1349(b)(1).

tensive schedule for the federal regulation of working conditions on offshore drilling platforms.[107] Pursuant to this statutory authority, both the U.S. Coast Guard and the U.S. Geological Survey have enacted safety rules regarding offshore drilling.[108]

The general policy of the Outer Continental Shelf Act is to provide that offshore drilling operations be conducted in a safe manner in order to minimize dangers to workers. In accordance with this policy, the U.S. Geological Survey has promulgated specific orders for platform procedures to be followed in the event of a blowout, the installation of safety equipment on drilling rigs, and mandatory safety training for personnel on offshore platforms.[109]

Moreover, the Coast Guard has promulgated regulations concerning the safety of persons and property on outer Continental Shelf facilities, vessels, and other units engaged in outer Continental Shelf activities.[110] The Coast Guard also regulates such items as platform design, insofar as they are related to employee safety, and sets out regulations regarding lifesaving equipment, fire fighting devices, and escape procedures in the event of emergencies. Investigative procedures with regard to the foregoing also are established by the Coast Guard.[111]

In addition to maritime employee safety regulations issued by the Geological Survey and the Coast Guard, the Interior Department, pursuant to authority granted it by the Outer Continental Shelf Lands Act, also has prescribed maritime workplace safety standards. More specifically, when the federal government executes a lease of offshore lands to be used for drilling operations, the Interior Department has the power to issue safety regulations concerning production operations.[112] These regulations provide that the lessee must perform all operations in a safe and workmanlike manner for the protection, health, and safety of all persons employed on them. The lessee, for example, must take all necessary precautions to prevent, and shall immediately remove, any haz-

---

[107]§1334.

[108]30 C.F.R. §§250.0 *et seq.* (1985).

[109]*Id.*

[110]See 33 C.F.R. §§140.1 *et seq.* (1985).

[111]*Id.*

[112]30 C.F.R. §250.0 *et seq.* (1985).

ardous oil and gas accumulations or other health, safety, or fire hazards.[113]

3. *Applicability of the Section 4(b)(1) Exemption to Maritime Workplaces*

The Review Commission and the courts consistently recognize that the Coast Guard and other maritime agencies retain authority, to the exclusion of OSHA, to regulate the working conditions of "seamen" or vessels operating on the high seas.[114] In order to determine whether an employee is a "seaman," as that term is defined for OSHA preemption purposes, the nature of the employee's duties, in relation to his vessel, must be examined.[115]

Once an employee is found to be a "seaman" and, therefore, subject to maritime agency regulations, a determination must be made as to whether the Coast Guard, or another maritime agency, affirmatively has exercised its authority over the specific working condition at issue. As previously noted, no industrywide Section 4(b)(1) exemption for seamen exists.[116] Preemption, therefore, requires an affirmative exercise of authority by the maritime agency over the specific working condition at issue. Interestingly, however, as the result of extensive regulation over offshore drilling operations, courts generally hold that OSHA lacks jurisdiction over the working conditions of seamen on offshore drilling platforms on the outer Continental Shelf.[117]

Owing to the fact that the Coast Guard and OSHA regulations overlap at times, those organizations have entered into inspection agreements with regard to workplaces aboard inspected vessels.[118] Under those agreements, although the Coast Guard maintains authority over the vessels, it has agreed to notify and cooperate with OSHA on health hazards not specifically covered by Coast Guard regulations.

---

[113]*Id.*

[114]See, e.g., *Clary v. Ocean Drilling & Exploration Co.*, 609 F.2d 1120 (5th Cir. 1980); *Prudential Lines*, 3 OSHC 1532 (Rev. Comm'n 1975).

[115]*Offshore Co. v. Robison*, 266 F.2d 769 (5th Cir. 1959).

[116]*Petrolane Offshore Constr. Servs.*, 3 OSHC 1156 (Rev. Comm'n 1975).

[117]*Marshall v. Nichols*, 486 F. Supp. 615 (E.D. Tex. 1980).

[118]45 FED. REG. 14,987 (1980).

4. Longshoremen's and Harbor Workers' Compensation Act

While the Coast Guard, the Department of the Interior, and the U.S. Geological Survey regulate the safe working conditions of seamen, the Longshoremen's and Harbor Workers' Compensation Act regulates the working conditions of longshoremen.[119] Although specific safety standards concerning such regulation were promulgated under the Act,[120] they subsequently have been adopted under OSHA as established OSHA standards.

In regulating the safety and health conditions of longshoremen, however, OSHA is not limited to the scope of standards derived from the Longshoremen's and Harbor Workers' Compensation Act.[121] In fact, OSHA may promulgate broader health and safety standards for such workers than the regulations issued pursuant to the Longshoremen's and Harbor Worker's Compensation Act.

III. Superseding Effects of Section 4(b)(2) of the OSH Act

When Congress enacted the OSH Act, it incorporated, by reference, several existing federal occupational safety and health statutes. More specifically, Section 4(b)(2) of the Act expressly incorporated the safety and health provisions of the Walsh-Healey Act, the Service Contract Act, the Construction Safety Act, the Longshoremen's and Harbor Workers' Compensation Act, and the National Foundation on Arts and Humanities Act of 1965. Safety and health standards promulgated under these statutes have been deemed to be OSHA safety and health standards.

IV. Federal Employment

In addition to exempting certain workplaces from OSHA regulations under Section 4(b)(1), OSHA specifically excludes

---

[119]33 U.S.C. §§901 *et seq.* (1986).

[120]29 C.F.R. §§1910 *et seq.* (1985).

[121]*Brown & Root, Inc.*, 9 OSHC 1407 (Rev. Comm'n 1981).

federal agencies and their employees from its jurisdiction.[122] Section 19 of the Act, however, requires each federal agency head to establish and maintain an occupational safety and health program for employees which is consistent with standards promulgated by OSHA.[123] Subsequent executive orders are supportive of this requirement.

In 1980, President Carter issued Executive Order 12,196, which introduced greater safety protection to federal government employees.[124] Executive Order 12,196 requires federal agency heads to comply with OSHA standards, provided, of course, the Secretary of Labor has not already approved alternative standards. Federal agencies must, moreover, render workplaces free of recognized hazards that cause, or are likely to cause, death or serious physical harm to federal employees.[125] As such, those agencies have had to establish and maintain their own health and safety standards.

In addition, the Secretary of Labor and/or his representatives are authorized to conduct periodic and, in certain circumstances, unannounced, inspections of the federal workplace.[126] The Executive Order sets forth provisions that protect employees against discrimination for exercising certain rights, such as filing reports about unsafe working conditions.[127] The Order also encourages the formation of joint labor-management occupational safety and health committees to review agency programs, monitor their performances, and consult and advise those agencies on the operation of their programs.[128] Executive Order 12,196 supersedes all prior executive orders dealing with safety and health protection for federal government employees.[129] It applies to all executive branches and agencies, with the exception of military personnel and certain military equipment, systems, and operations.

Executive Order 12,196 also provides for the continued existence of the Federal Advisory Council on Occupational

---

[122]29 U.S.C. §652(5) (1985).

[123]§667.

[124]45 FED. REG. 12,769 (1980). The effective date of Exec. Order 12,196 was amended by Exec. Order 12,223. 45 FED. REG. 45,235 (1980).

[125]Exec. Order No. 12,196.

[126]*Id.*

[127]*Id.*

[128]*Id.*

[129]Exec. Order 12,196 supersedes Exec. Order 11,807.

Safety and Health (FACOSH).[130] FACOSH is today utilized in the public sector as an avenue for submitting material for consideration by federal agencies in setting safety and health standards.

V. Fair Labor Standards Act

In 1938, Congress enacted the Fair Labor Standards Act[131] (FLSA) to regulate the hours and wages of employees engaged in services or the production of goods in commerce, in an effort to eradicate from interstate commerce "labor conditions detrimental to the maintenance of the minimum standard of living necessary for health, efficiency, and general well-being of workers."[132] In addition to establishing wage and hour regulations, the FLSA prohibits use of oppressive child labor and sets forth guidelines for the employment of child labor.[133] "Oppressive child labor" is defined as "employment of a minor in an occupation for which he does not meet the minimum age standards of the Act, [which are set at 16 years of age]."[134] The child working standards that have been promulgated are based on such factors as type of employment and age.[135] Child labor occupations considered particularly hazardous include coal mining, meatpacking, certain agricultural jobs, field labor, motor vehicle and driver positions, and jobs involving the use of or exposure to radioactive materials.[136]

To enforce these regulations, the Administrator of the Wage and Hour Division is authorized to investigate and gather information regarding the working conditions and practices of industries subject to the FLSA.[137] Employers that violate the provisions of the FLSA are subject to specific civil and

---

[130]FACOSH, which was created to aid the Secretary of Labor in evaluating, regulating, and inspecting federal workplaces, should not be confused with NACOSH (National Advisory Committee on Occupational Safety and Health), who are appointed to aid the Secretary of Health and Human Resources on matters relating to administration of the OSH Act.

[131]29 U.S.C. §§201 *et seq.* (1978).

[132]§202(a).

[133]29 U.S.C. §212 (1986).

[134]29 C.F.R. §§570.1(b) and 570.2(a)(1) (1985).

[135]§§570.1 *et seq.*

[136]§§570.50 *et seq.*

[137]29 U.S.C. §211 (1986).

criminal penalties, including fines, imprisonment, and attorney's fees.[138] More importantly, employees who report violative conditions are specifically protected against retaliatory discharge and discrimination by their employer.[139]

More recent developments in child labor FLSA regulations relate to the employment of children in crop harvesting. Section 13(c)(4)(A) of the FLSA, for example, permits the Secretary of Labor to waive restrictions on the employment of children in agriculture harvesting if such employment would not be detrimental to their health.[140] In *National Association of Farm Workers Organizations v. Marshall*,[141] however, the court struck down regulations permitting such waivers for children entering fields sprayed with certain pesticides because objective data ensuring the safety of the children were not presented, so that the effect of the pesticides on such children was unknown. Also, the regulations were struck down because they were promulgated invalidly under the Administrative Procedure Act.[142] A final regulation prohibiting waiver in certain cases involving pesticides with known carcinogenic, mutagenic, or teratogenic properties was promulgated shortly after the *Marshall* decision.

VI. Federal Contractors

The federal government, through various federal agencies and statutes, has attempted to regulate the safety and health of persons who perform services in the accomplishment of public works and the execution of public contracts.

A. Section 4(b)(2) of the OSH Act

Prior to the enactment of the OSH Act, federal safety and health standards concerning government contractors were promulgated under such federal statutes as the Walsh-Healey

---

[138]§216.
[139]§215(a)(3).
[140]§213(c)(4)(A).
[141]628 F.2d 604 (D.C. Cir. 1980).
[142]*Id.*

Act[143] and the Service Contract Act of 1965.[144] When Congress enacted the OSH Act, it provided that some previous federal job safety regulations were to be incorporated into the Act. As Section III above states, pursuant to Section 4(b)(2) of the Act, the safety and health standards of the Walsh-Healey Act and the Service Contract Act were superseded by corresponding OSHA standards. As a result, standards issued under these earlier laws are now considered OSHA standards.

B. Contract Work Hours and Safety Act

Section 4(b)(2) did not, however, incorporate all federal safety standards for government contractors promulgated by previous federal statutes. For example, the Contract Work Hours and Safety Standards Act, a statute prohibiting federal contractors from requiring or permitting employees to work under unsanitary, hazardous, or dangerous conditions,[145] was not incorporated into the Act. Safety standards regulating the workplace of such contractors have been promulgated by the Secretary of Labor pursuant to authority granted him by the Contract Work Hours and Safety Standards Act.[146] Those standards relate to the use and storage of tools and equipment; the level of noise in the workplace; and the degree of exposure to radiation, gases, and fumes. Additionally, regulations requiring the instruction of personnel on safety procedures, the establishment of notification procedures for accidents, and the availability of medical services and first aid care have been promulgated.[147]

C. Section 4(b)(1) Exemption

Government contractors often argue that they fall within the scope of the Section 4(b)(1) exemption provision of the OSH Act. Such contractors, however, are not exempt from OSHA regulations unless another federal agency has actually exer-

---

[143]41 U.S.C. §§35–45 (1986).
[144]§§351 *et seq.*
[145]40 U.S.C. §§327–332 (1986).
[146]29 C.F.R. §§5.1 *et seq.* (1985).
[147]*Id.*

cised its authority to prescribe occupational safety and health standards over them. In *Dela Enterprises*,[148] for example, a defense contractor engaged in the manufacturing of explosives argued that its safety program was supervised by the Defense Contract Administration Service, a Department of Defense agency, and therefore was outside the coverage of OSHA standards. The Review Commission ruled against the employer, however, stating the OSHA's regulations were applicable to the contractor because there had been no showing of the existence of any statutory authority of the Defense Department to regulate safety practices in the contractor's facilities.[149]

The Review Commission must examine the legislative purpose behind the grant of power to an agency in order to ensure that it is meant to provide safe and healthful working conditions to the exclusion of OSHA standards. For example, in *Gearhart-Owen Industries*,[150] OSHA regulations were held to apply to a federal ammunitions contractor despite its assertions that its workplace was exclusively regulated by Department of Defense procurement regulations. More specifically, the Review Commission held that the purpose of the Department of Defense regulations primarily was to assure the efficient and economical production of defense materials and supplies, so that any benefits of employee safety and health arising from the regulations were incidental.[151]

Alternatively, however, the U.S. Navy's standards, promulgated under authority of the Atomic Energy Act, were held to preempt OSHA standards in *Newport News Shipbuilding & Drydock Co.*[152] The Review Commission in *Newport News* ruled that an employer engaged in the construction of a nuclear warship for the U.S. Navy was subject to Navy safety standards and therefore was exempt from OSHA coverage. The Navy regulations were promulgated with the primary intent of protecting the safety and health of those exposed to and using nuclear materials.[153]

[148]2 OSHC 3001 (Rev. Comm'n 1974).
[149]*Id.* at 3002.
[150]2 OSHC 1568 (Rev. Comm'n 1975).
[151]*Id.* at 1570–1571.
[152]2 OSHC 3262 (Rev. Comm'n 1975).
[153]*Id.*

VII. Title VII of the Civil Rights Act of 1964

Employers with policies relating to workplace safety and health have been subject to charges of discrimination by individuals protected under Title VII of the Civil Rights Act of 1964, as amended,[154] and related statutes. Examples of such policies include removing employees from certain positions in the workplace, avoiding the placement of employees in certain working environments, or requiring certain grooming standards. These policies involve such issues as fetal protection, genetic screening, sickle cell anemia, handicap, and the wearing of beards. For example, in recent years, employers have segregated women of childbearing age from employment conditions that involve actual or potential exposure to toxic substances where such exposure may create serious health threats to the woman's reproductive system, unborn fetus, or future offspring. Such issues constitute an emerging area of the law which may result in conflicts between equal employment laws and worker safety and health laws.

The Equal Employment Opportunity Commission (EEOC) and the Office of Federal Contract Compliance Programs (OFCCP), created pursuant to Executive Order 11,246, issued interpretive guidelines on employment discrimination and reproductive hazards.[155] The basic premise of these guidelines was that an exclusionary policy based on reproductive hazards could be justified by employers only after a thorough investigation of all alternatives, and then only when there was specific scientific evidence that a threat to female employees from continued exposure existed and that no equal danger existed to male employees. The EEOC and OFCCP withdrew the guidelines in 1981, however, because they concluded that case-by-case investigation and enforcement was the more appropriate method of eliminating employment discrimination in the workplace where potential exposure to reproductive hazards existed.[156]

In *Wright v. Olin Corp.*,[157] an employee brought suit against Olin Corp. maintaining that Olin's fetal vulnerability policy

---

[154]42 U.S.C. §§2000e *et seq.* (1981). Title VII prohibits discrimination on the basis of sex, race, color, religion or national origin. *Id.* at §2000e-2.

[155]45 FED. REG. 7514 (1980).

[156]46 FED. REG. 3916 (1981).

[157]697 F.2d 1172 (4th Cir. 1982).

discriminated against females and therefore violated Title VII. The policy barred women of childbearing capability from working in areas where there was exposure to known or suspected abortifacient or teratogenic agents. The purpose of the policy was to protect an unborn fetus at a time it was most vulnerable to exposure to harmful chemicals. Nonpregnant females were permitted to work in areas where the nature of the work required only limited contact with such chemicals, provided the employees agreed to notify their supervisors if they became pregnant. The court held that Olin's fetal vulnerability policy did not violate Title VII because the policy was instituted for sound medical and humane reasons, based upon years of research and monitoring of chemical exposure at the plant. The *Olin* court concluded that the policy was not instituted or maintained with the intent to discriminate against women because of their sex.[158]

A similar corporate fetus protection policy was upheld by a federal district court in *Doerr v. B. F. Goodrich Co.*[159] The *Doerr* court denied injunctive relief against the implementation of the policy, because the employee failed to show that the policy would cause irreparable harm to her career development.[160] In *Hayes v. Shelby Hospital*, however, an employer who fired a pregnant X-ray technician was held to have violated Title VII of the Civil Rights Act of 1964.[161] Although the employer in *Hayes* argued that the firing was necessary to protect the unborn fetus from harmful radiation, the court held that the dismissal was an unnecessarily extreme action and discriminated against the employee on the basis of her pregnancy.[162]

A fetus protection policy, however, was held not to be a "hazard" within the meaning of the general duty clause of the OSH Act. In *American Cyanamid Co.*[163] the employer enacted a fetus protection policy that excluded women of childbearing age from production jobs in lead pigmentation departments, unless the women could prove they had been surgically ster-

---

[158]*Id.* at 1180.
[159]484 F. Supp. 320 (N.D. Ohio 1979).
[160]*Id.* at 324.
[161]726 F.2d 1543 (11th Cir. 1984).
[162]*Id.*
[163]9 OSHC 1596 (Rev. Comm'n 1981), *aff'd*, 741 F.2d 444 (D.C. Cir. 1984).

ilized. The Review Commission held that a "hazard," as the term is used in the OSH Act's general duty clause, referred to processes and materials that cause injury or disease to employees during work or work-related activities. The fetus protection policy, it concluded, was not such a process or material.[164]

Another example of the impact of employment discrimination laws on workplace safety and health involves a skin malady known as pseudofolliculitis barbae (PFB), which afflicts black males. Black employees afflicted with PFB are often advised to grow a beard, as shaving worsens the condition. In certain work environments, the OSH Act requires that respirators, often requiring a skintight fit and thus a shaven face, be available to, or actually be used by, employees. Employees required to use such respirators may claim race discrimination because a work rule mandates use of such respirators and thereby effectively precludes their ability to hold such jobs.[165]

VIII. Miscellaneous Safety Regulations

A. Black Lung Legislation

An important aspect of federal mining regulations is the Black Lung benefits program, which provides compensation to miners who are disabled by miners' pneumoconiosis or, as it is commonly called, black lung disease. The Black Lung Benefits Act of 1972[166] entitles miners who are disabled owing to black lung disease, as well as dependents of those who have died from that disease, to compensatory benefits.

---

[164]*Id.* As with other Title VII cases, the defenses of business necessity and bona fide occupational qualification (BFOQ) have been raised in cases involving the firing of women on the basis of their pregnancies. See, e.g., *Hayes, supra* note 161. In order to show that a business necessity justifies a particular employment practice, it must be shown that the practice is necessary to safe and efficient job performance. The business purpose of the policy must be sufficiently compelling to override any discriminatory impact. *Dothard v. Rawlinson*, 433 U.S. 321 (1977). In order to successfully assert the BFOQ defense, the defendant must prove that (1) the class qualification invoked is reasonably necessary to the essence of the employer's business, and (2) the employer has a reasonable factual basis for believing that the excluded class would be unable to safely and efficiently perform the duties of the job. See, e.g. *Harriss v. Pan Am. World Airways*, 649 F.2d 670 (9th Cir. 1980); *Arritt v. Grisell*, 567 F.2d 1267 (4th Cir. 1977).

[165]See *EEOC v. Greyhound Lines*, 635 F.2d 188 (3d Cir. 1980).

[166]30 U.S.C. §§901 *et seq.* (1986).

Pursuant to the authority granted by the Black Lung Benefits Act, the Secretary of Labor has promulgated procedures for the recovery of benefits and has established the Black Lung Disability Trust Fund to assure a source of future benefits for disabled miners.[167] The Social Security Administration was given the responsibility for processing and paying claims up to June 1973, at which time the mining industry assumed such responsibility.

B. District of Columbia Industrial Safety Act

The enactment of the OSH Act did not automatically preclude state occupational safety standards. To the contrary, states, as well as the District of Columbia, maintain authority to promulgate and enforce safety standards and reporting requirements where no federal standard has been enacted. In *P & Z Co. v. District of Columbia*,[168] the defendant argued that the OSH Act preempted employee injury reporting requirements of the District of Columbia Industrial Safety Act.[169] The court rejected the defendant's argument, stating that unless a regulation was expressly superseded by a conflicting OSHA regulation, Congress intended state and federal regulations to coexist.

C. Rehabilitation Act of 1973

Employers that either directly or, in some instances, indirectly receive federal financial assistance on federal contracts may be required to comply with regulations concerning the use and accessibility of their facilities by handicapped individuals, including employees. Authority for these regulations stems from the Rehabilitation Act of 1973, as amended.[170]

For example, any employer or organization receiving financial assistance from the Office of Personnel Management

---

[167]42 C.F.R. §§55a.101 *et seq.* (1985).
[168]408 A.2d 1249 (D.C. 1979).
[169]*Id.*
[170]29 U.S.C. §§701 *et seq.* (1985).

must comply with regulations concerning handicapped individuals issued by that agency.[171] These regulations forbid discrimination against qualified handicapped persons in employment in the operation of programs and activities receiving federal financial assistance. Moreover, the regulations require all recipients of funds to have facilities accessible to, and usable by, handicapped persons. This includes, if necessary, the making of structural changes.[172] Interim guidelines for use in such facilities are, by regulation, available in "American National Standard Specifications for Making Buildings and Facilities Accessible to, and Usable by, the Physically Handicapped."[173] Permanent guidelines will be adopted from the Architectural and Transportation Barriers Compliance Board's "Minimum Guidelines and Requirements for Accessible Design."[174]

Handicap Rules for Federally Assisted Programs have also been promulgated by the Department of Health and Human Services.[175] The coordination of enforcement of these regulations was transferred to the Department of Justice,[176] which implemented Section 1-303 of Executive Order 12,250,[177] thereby requiring the Department of Justice to coordinate implementation of Section 504 of the Rehabilitation Act of 1973, as amended.[178] These regulations require all federal agencies empowered to extend financial assistance to assume responsibility for issuing regulations that prohibit the facilities of a recipient of federal assistance from being inaccessible to, or unusable by, handicapped persons.[179] The regulations require such structural changes as necessary in existing facilities,[180] and new facilities must be designed and constructed to be readily accessible to, and usable by, handicapped persons.[181]

---

[171]5 C.F.R. §§900.701 *et seq.* (1986).
[172]§900.705.
[173]American National Standards Institute (ANSI A117.1-1961 R1971). 5 C.F.R. §900.705(f) (1986).
[174]45 FED. REG. 55,010 (1980).
[175]43 FED. REG. 2132 (1978).
[176]46 FED. REG. 40,686 (1981).
[177]28 C.F.R. §§41.1 *et seq.* (1985).
[178]29 U.S.C. §794 (1985).
[179]28 C.F.R. §§41.1 *et seq.* (1985).
[180]§41.57.
[181]§41.58.

Section 503 of the Rehabilitation Act of 1973[182] mandates that employers with federal contracts in excess of $2500 "take affirmative action to employ and advance in employment qualified handicapped individuals."[183] This includes such reasonable accommodations as structural modifications affecting all employees and individual modifications made for particular workers in particular jobs.[184]

---

[182] 29 U.S.C. §793 (1985).

[183] *Id.*

[184] 41 C.F.R. §§60–741.1 *et seq.* (1985). See "A Study of Accommodations Provided to Handicapped Employees by Federal Contractors," prepared for the U.S. Department of Labor, Employment Standards Administration, by Berkeley Planning Associates, Berkeley, Calif., June 1982. FED. CONT. COMPL. MAN. ¶21,142 (1983).

Appendix A

Occupational Safety and Health Act of 1970*

*P.L. 91–596, 91st Cong., S. 2193, Dec. 29, 1970

An Act

To assure safe and healthful working conditions for working men and women;
by authorizing enforcement of the standards developed under the Act; by
assisting and encouraging the States in their efforts to assure safe and healthful working conditions; by providing for research, information, education, and
training in the field of occupational safety and health; and for other purposes.

*Be it enacted by the Senate and House of Representatives of the
United States of America in Congress assembled,* That this Act may
be cited as the "Occupational Safety and Health Act of 1970".

Occupational
Safety and
Health Aot of
1970.

CONGRESSIONAL FINDINGS AND PURPOSE

SEC. (2) The Congress finds that personal injuries and illnesses arising out of work situations impose a substantial burden upon, and are
a hindrance to, interstate commerce in terms of lost production, wage
loss, medical expenses, and disability compensation payments.

(b) The Congress declares it to be its purpose and policy, through
the exercise of its powers to regulate commerce among the several
States and with foreign nations and to provide for the general welfare,
to assure so far as possible every working man and woman in the
Nation safe and healthful working conditions a.id to preserve our
human resources—

   (1) by encouraging employers and employees in their efforts
to reduce the number of occupational safety and health hazards
at their places of employment, and to stimulate employers and
employees to institute new and to perfect existing programs for
providing safe and healthful working conditions;

   (2) by providing that employers and employees have separate
but dependent responsibilities and rights with respect to achieving safe and healthful working conditions;

   (3) by authorizing the Secretary of Labor to set mandatory
occupational safety and health standards applicable to businesses
affecting interstate commerce, and by creating an Occupational
Safety and Health Review Commission for carrying out adjudicatory functions under the Act;

   (4) by building upon advances already made through employer
and employee initiative for providing safe and healthful working
conditions;

   (5) by providing for research in the field of occupational
safety and health, including the psychological factors involved,
and by developing innovative methods, techniques, and
approaches for dealing with occupational safety and health
problems;

   (6) by exploring ways to discover latent diseases, establishing
causal connections between diseases and work in environmental
conditions, and conducting other research relating to health problems, in recognition of the fact that occupational health standards
present problems often different from those involved in occupational safety;

   (7) by providing medical criteria which will assure insofar as
practicable that no employee will suffer diminished health, functional capacity, or life expectancy as a result of his work
experience;

   (8) by providing for training programs to increase the number and competence of personnel engaged in the field of occupational safety and health;

827

84 STAT. 1591

(9) by providing for the development and promulgation of occupational safety and health standards;

(10) by providing an effective enforcement program which shall include a prohibition against giving advance notice of any inspection and sanctions for any individual violating this prohibition;

(11) by encouraging the States to assume the fullest responsibility for the administration and enforcement of their occupational safety and health laws by providing grants to the States to assist in identifying their needs and responsibilities in the area of occupational safety and health, to develop plans in accordance with the provisions of this Act, to improve the administration and enforcement of State occupational safety and health laws, and to conduct experimental and demonstration projects in connection therewith;

(12) by providing for appropriate reporting procedures with respect to occupational safety and health which procedures will help achieve the objectives of this Act and accurately describe the nature of the occupational safety and health problem;

(13) by encouraging joint labor-management efforts to reduce injuries and disease arising out of employment.

DEFINITIONS

SEC. 3. For the purposes of this Act—

(1) The term "Secretary" mean the Secretary of Labor.

(2) The term "Commission" means the Occupational Safety and Health Review Commission established under this Act.

(3) The term "commerce" means trade, traffic, commerce, transportation, or communication among the several States, or between a State and any place outside thereof, or within the District of Columbia, or a possession of the United States (other than the Trust Territory of the Pacific Islands), or between points in the same State but through a point outside thereof.

(4) The term "person" means one or more individuals, partnerships, associations, corporations, business trusts, legal representatives, or any organized group of persons.

(5) The term "employer" means a person engaged in a business affecting commerce who has employees, but does not include the United States or any State or political subdivision of a State.

(6) The term "employee" means an employee of an employer who is employed in a business of his employer which affects commerce.

(7) The term "State" includes a State of the United States, the District of Columbia, Puerto Rico, the Virgin Islands, American Samoa, Guam, and the Trust Territory of the Pacific Islands.

(8) The term "occupational safety and health standard" means a standard which requires conditions, or the adoption or use of one or more practices, means, methods, operations, or processes, reasonably necessary or appropriate to provide safe or healthful employment and places of employment.

(9) The term "national consensus standard" means any occupational safety and health standard or modification thereof which (1), has been adopted and promulgated by a nationally recognized standards-producing organization under procedures whereby it can be determined by the Secretary that persons interested

and affected by the scope or provisions of the standard have reached substantial agreement on its adoption, (2) was formulated in a manner which afforded an opportunity for diverse views to be considered and (3) has been designated as such a standard by the Secretary, after consultation with other appropriate Federal agencies.

(10) The term "established Federal standard" means any operative occupational safety and health standard established by any agency of the United States and presently in effect, or contained in any Act of Congress in force on the date of enactment of this Act.

(11) The term "Committee" means the National Advisory Committee on Occupational Safety and Health established under this Act.

(12) The term "Director" means the Director of the National Institute for Occupational Safety and Health.

(13) The term "Institute" means the National Institute for Occupational Safety and Health established under this Act.

(14) The term "Workmen's Compensation Commission" means the National Commission on State Workmen's Compensation Laws established under this Act.

APPLICABILITY OF THIS ACT

SEC. 4. (a) This Act shall apply with respect to employment performed in a workplace in a State, the District of Columbia, the Commonwealth of Puerto Rico, the Virgin Islands, American Samoa, Guam, the Trust Territory of the Pacific Islands, Wake Island, Outer Continental Shelf lands defined in the Outer Continental Shelf Lands Act, Johnston Island, and the Canal Zone. The Secretary of the Interior shall, by regulation, provide for judicial enforcement of this Act by the courts established for areas in which there are no United States district courts having jurisdiction.
67 Stat. 462.
43 USC 1331 note.

(b)(1) Nothing in this Act shall apply to working conditions of employees with respect to which other Federal agencies, and State agencies acting under section 274 of the Atomic Energy Act of 1954, as amended (42 U.S.C. 2021), exercise statutory authority to prescribe or enforce standards or regulations affecting occupational safety or health.
73 Stat. 688.

(2) The safety and health standards promulgated under the Act of June 30, 1936, commonly known as the Walsh-Healey Act (41 U.S.C. 35 et seq.), the Service Contract Act of 1965 (41 U.S.C. 351 et seq.), Public Law 91-54, Act of August 9, 1969 (40 U.S.C. 333), Public Law 85-742, Act of August 23, 1958 (33 U.S.C. 941), and the National Foundation on Arts and Humanities Act (20 U.S.C. 951 et seq.) are superseded on the effective date of corresponding standards, promulgated under this Act, which are determined by the Secretary to be more effective. Standards issued under the laws listed in this paragraph and in effect on or after the effective date of this Act shall be deemed to be occupational safety and health standards issued under this Act, as well as under such other Acts.
49 Stat. 2036.
79 Stat. 1034.
83 Stat. 96.
72 Stat. 835.
79 Stat. 845;
Ante, p. 443.

(3) The Secretary shall, within three years after the effective date of this Act, report to the Congress his recommendations for legislation to avoid unnecessary duplication and to achieve coordination between this Act and other Federal laws.
Report to Congress.

84 STAT. 1593

(4) Nothing in this Act shall be construed to supersede or in any manner affect any workmen's compensation law or to enlarge or diminish or affect in any other manner the common law or statutory rights, duties, or liabilities of employers and employees under any law with respect to injuries, diseases, or death of employees arising out of, or in the course of, employment.

DUTIES

SEC. 5. (a) Each employer—
(1) shall furnish to each of his employees employment and a place of employment which are free from recognized hazards that are causing or are likely to cause death or serious physical harm to his employees;
(2) shall comply with occupational safety and health standards promulgated under this Act.
(b) Each employee shall comply with occupational safety and health standards and all rules, regulations, and orders issued pursuant to this Act which are applicable to his own actions and conduct.

OCCUPATIONAL SAFETY AND HEALTH STANDARDS

SEC. 6. (a) Without regard to chapter 5 of title 5, United States Code, or to the other subsections of this section, the Secretary shall, as soon as practicable during the period beginning with the effective date of this Act and ending two years after such date, by rule promulgate as an occupational safety or health standard any national consensus standard, and any established Federal standard, unless he determines that the promulgation of such a standard would not result in improved safety or health for specifically designated employees. In the event of conflict among any such standards, the Secretary shall promulgate the standard which assures the greatest protection of the safety or health of the affected employees.

80 Stat. 381;
81 Stat. 195.
5 USC 500.

(b) The Secretary may by rule promulgate, modify, or revoke any occupational safety or health standard in the following manner:
(1) Whenever the Secretary, upon the basis of information submitted to him in writing by an interested person, a representative of any organization of employers or employees, a nationally recognized standards-producing organization, the Secretary of Health, Education, and Welfare, the National Institute for Occupational Safety and Health, or a State or political subdivision, or on the basis of information developed by the Secretary or otherwise available to him, determines that a rule should be promulgated in order to serve the objectives of this Act, the Secretary may request the recommendations of an advisory committee appointed under section 7 of this Act. The Secretary shall provide such an advisory committee with any proposals of his own or of the Secretary of Health, Education, and Welfare, together with all pertinent factual information developed by the Secretary or the Secretary of Health, Education, and Welfare, or otherwise available, including the results of research, demonstrations, and experiments. An advisory committee shall submit to the Secretary its recommendations regarding the rule to be promulgated within ninety days from the date of its appointment or within such longer or shorter period as may be prescribed by the Secretary, but in no event for a period which is longer than two hundred and seventy days.

Advisory
committee,
recommendations.

84 STAT. 1594

(2) The Secretary shall publish a proposed rule promulgating, modifying, or revoking an occupational safety or health standard in the Federal Register and shall afford interested persons a period of thirty days after publication to submit written data or comments. Where an advisory committee is appointed and the Secretary determines that a rule should be issued, he shall publish the proposed rule within sixty days after the submission of the advisory committee's recommendations or the expiration of the period prescribed by the Secretary for such submission. Publication in Federal Register.

(3) On or before the last day of the period provided for the submission of written data or comments under paragraph (2), any interested person may file with the Secretary written objections to the proposed rule, stating the grounds therefor and requesting a public hearing on such objections. Within thirty days after the last day for filing such objections, the Secretary shall publish in the Federal Register a notice specifying the occupational safety or health standard to which objections have been filed and a hearing requested, and specifying a time and place for such hearing. Hearing, notice. Publication in Federal Register.

(4) Within sixty days after the expiration of the period provided for the submission of written data or comments under paragraph (2), or within sixty days after the completion of any hearing held under paragraph (3), the Secretary shall issue a rule promulgating, modifying, or revoking an occupational safety or health standard or make a determination that a rule should not be issued. Such a rule may contain a provision delaying its effective date for such period (not in excess of ninety days) as the Secretary determines may be necessary to insure that affected employers and employees will be informed of the existence of the standard and of its terms and that employers affected are given an opportunity to familiarize themselves and their employees with the existence of the requirements of the standard.

(5) The Secretary, in promulgating standards dealing with toxic materials or harmful physical agents under this subsection, shall set the standard which most adequately assures, to the extent feasible, on the basis of the best available evidence, that no employee will suffer material impairment of health or functional capacity even if such employee has regular exposure to the hazard dealt with by such standard for the period of his working life. Development of standards under this subsection shall be based upon research, demonstrations, experiments, and such other information as may be appropriate. In addition to the attainment of the highest degree of health and safety protection for the employee, other considerations shall be the latest available scientific data in the field, the feasibility of the standards, and experience gained under this and other health and safety laws. Whenever practicable, the standard promulgated shall be expressed in terms of objective criteria and of the performance desired. Toxic material.

(6)(A) Any employer may apply to the Secretary for a temporary order granting a variance from a standard or any provision thereof promulgated under this section. Such temporary order shall be granted only if the employer files an application which meets the requirements of clause (B) and establishes that (i) he is unable to comply with a standard by its effective date because of unavailability of professional or technical personnel or of materials and equipment needed to come into compliance with the standard or because necessary construction or alteration of facilities cannot be completed by the effective date, (ii) he is taking all available steps to safeguard his employees against the hazards covered by the standard, and (iii) he has an effective program for coming into compliance with the standard as quickly as Temporary variance order.

84 STAT. 1595

Notice,
hearing.

Renewal.

Time limita-
tion.

practicable. Any temporary order issued under this paragraph shall prescribe the practices, means, methods, operations, and processes which the employer must adopt and use while the order is in effect and state in detail his program for coming into compliance with the standard. Such a temporary order may be granted only after notice to employees and an opportunity for a hearing: *Provided,* That the Secretary may issue one interim order to be effective until a decision is made on the basis of the hearing. No temporary order may be in effect for longer than the period needed by the employer to achieve compliance with the standard or one year, whichever is shorter, except that such an order may be renewed not more than twice (I) so long as the requirements of this paragraph are met and (II) if an application for renewal is filed at least 90 days prior to the expiration date of the order. No interim renewal of an order may remain in effect for longer than 180 days.

(B) An application for a temporary order under this paragraph (6) shall contain:

(i) a specification of the standard or portion thereof from which the employer seeks a variance,

(ii) a representation by the employer, supported by representations from qualified persons having firsthand knowledge of the facts represented, that he is unable to comply with the standard or portion thereof and a detailed statement of the reasons therefor,

(iii) a statement of the steps he has taken and will take (with specific dates) to protect employees against the hazard covered by the standard,

(iv) a statement of when he expects to be able to comply with the standard and what steps he has taken and what steps he will take (with dates specified) to come into compliance with the standard, and

(v) a certification that he has informed his employees of the application by giving a copy thereof to their authorized representative, posting a statement giving a summary of the application and specifying where a copy may be examined at the place or places where notices to employees are normally posted, and by other appropriate means.

A description of how employees have been informed shall be contained in the certification. The information to employees shall also inform them of their right to petition the Secretary for a hearing.

(C) The Secretary is authorized to grant a variance from any standard or portion thereof whenever he determines, or the Secretary of Health, Education, and Welfare certifies, that such variance is necessary to permit an employer to participate in an experiment approved by him or the Secretary of Health, Education, and Welfare designed to demonstrate or validate new and improved techniques to safeguard the health or safety of workers.

Labels, etc.

Protective
equipment,
etc.

(7) Any standard promulgated under this subsection shall prescribe the use of labels or other appropriate forms of warning as are necessary to insure that employees are apprised of all hazards to which they are exposed, relevant symptoms and appropriate emergency treatment, and proper conditions and precautions of safe use or exposure. Where appropriate, such standard shall also prescribe suitable protective equipment and control or technological procedures to be used in connection with such hazards and shall provide for monitoring or measuring employee exposure at such locations and intervals, and in such manner as may be necessary for the protection of employees. In

84 STAT. 1596

addition, where appropriate, any such standard shall prescribe the type and frequency of medical examinations or other tests which shall be made available, by the employer or at his cost, to employees exposed to such hazards in order to most effectively determine whether the health of such employees is adversely affected by such exposure. In the event such medical examinations are in the nature of research, as determined by the Secretary of Health, Education, and Welfare, such examinations may be furnished at the expense of the Secretary of Health, Education, and Welfare. The results of such examinations or tests shall be furnished only to the Secretary or the Secretary of Health, Education, and Welfare, and, at the request of the employee, to his physician. The Secretary, in consultation with the Secretary of Health, Education, and Welfare, may by rule promulgated pursuant to section 553 of title 5, United States Code, make appropriate modifications in the foregoing requirements relating to the use of labels or other forms of warning, monitoring or measuring, and medical examinations, as may be warranted by experience, information, or medical or technological developments acquired subsequent to the promulgation of the relevant standard.

Medical examinations.

80 Stat. 383.

(8) Whenever a rule promulgated by the Secretary differs substantially from an existing national consensus standard, the Secretary shall, at the same time, publish in the Federal Register a statement of the reasons why the rule as adopted will better effectuate the purposes of this Act than the national consensus standard.

Publication in Federal Register.

(c) (1) The Secretary shall provide, without regard to the requirements of chapter 5, title 5, United States Code, for an emergency temporary standard to take immediate effect upon publication in the Federal Register if he determines (A) that employees are exposed to grave danger from exposure to substances or agents determined to be toxic or physically harmful or from new hazards, and (B) that such emergency standard is necessary to protect employees from such danger.'

Temporary standard. Publication in Federal Register. 80 Stat. 381; 81 Stat. 195. 5 USC 500.

(2) Such standard shall be effective until superseded by a standard promulgated in accordance with the procedures prescribed in paragraph (3) of this subsection.

Time limitation.

(3) Upon publication of such standard in the Federal Register the Secretary shall commence a proceeding in accordance with section 6(b) of this Act, and the standard as published shall also serve as a proposed rule for the proceeding. The Secretary shall promulgate a standard under this paragraph no later than six months after publication of the emergency standard as provided in paragraph (2) of this subsection.

(d) Any affected employer may apply to the Secretary for a rule or order for a variance from a standard promulgated under this section. Affected employees shall be given notice of each such application and an opportunity to participate in a hearing. The Secretary shall issue such rule or order if he determines on the record, after opportunity for an inspection where appropriate and a hearing, that the proponent of the variance has demonstrated by a preponderance of the evidence that the conditions, practices, means, methods, operations, or processes used or proposed to be used by an employer will provide employment and places of employment to his employees which are as safe and healthful as those which would prevail if he complied with the standard. The rule or order so issued shall prescribe the conditions the employer must maintain, and the practices, means, methods, operations, and processes which he must adopt and utilize to the extent they

Variance rule.

84 STAT. 1597

differ from the standard in question. Such a rule or order may be modified or revoked upon application by an employer, employees, or by the Secretary on his own motion, in the manner prescribed for its issuance under this subsection at any time after six months from its issuance.

Publication in Federal Register.

(e) Whenever the Secretary promulgates any standard, makes any rule, order, or decision, grants any exemption or extension of time, or compromises, mitigates, or settles any penalty assessed under this Act, he shall include a statement of the reasons for such action, which shall be published in the Federal Register.

Petition for judicial review.

(f) Any person who may be adversely affected by a standard issued under this section may at any time prior to the sixtieth day after such standard is promulgated file a petition challenging the validity of such standard with the United States court of appeals for the circuit wherein such person resides or has his principal place of business, for a judicial review of such standard. A copy of the petition shall be forthwith transmitted by the clerk of the court to the Secretary. The filing of such petition shall not, unless otherwise ordered by the court, operate as a stay of the standard. The determinations of the Secretary shall be conclusive if supported by substantial evidence in the record considered as a whole.

(g) In determining the priority for establishing standards under this section, the Secretary shall give due regard to the urgency of the need for mandatory safety and health standards for particular industries, trades, crafts, occupations, businesses, workplaces or work environments. The Secretary shall also give due regard to the recommendations of the Secretary of Health, Education, and Welfare regarding the need for mandatory standards in determining the priority for establishing such standards.

ADVISORY COMMITTEES; ADMINISTRATION

Establishment; membership.

SEC. 7. (a)(1) There is hereby established a National Advisory Committee on Occupational Safety and Health consisting of twelve members appointed by the Secretary, four of whom are to be designated by the Secretary of Health, Education, and Welfare, without

80 Stat. 378. 5 USC 101.

regard to the provisions of title 5, United States Code, governing appointments in the competitive service, and composed of representatives of manangement, labor, occupational safety and occupational health professions. and of the public. The Secretary shall designate one of the public members as Chairman. The members shall be selected upon the basis of their experience and competence in the field of occupational safety and health.

(2) The Committee shall advise, consult with, and make recommendations to the Secretary and the Secretary of Health, Education, and Welfare on matters relating to the administration of the Act. The Committee shall hold no fewer than two meetings during each calendar year. All meetings of the Committee shall be open to the public and a transcript shall be kept and made available for public inspection.

Public transcript.

(3) The members of the Committee shall be compensated in accordance with the provisions of section 3109 of title 5, United States Code.

80 Stat. 416.

(4) The Secretary shall furnish to the Committee an executive secretary and such secretarial, clerical, and other services as are deemed necessary to the conduct of its business.

(b) An advisory committee may be appointed by the Secretary to assist him in his standard-setting functions under section 6 of this Act. Each such committee shall consist of not more than fifteen members

84 STAT. 1598

and shall include as a member one or more designees of the Secretary of Health, Education, and Welfare, and shall include among its members an equal number of persons qualified by experience and affiliation to present the viewpoint of the employers involved, and of persons similarly qualified to present the viewpoint of the workers involved, as well as one or more represent..tives of health and safety agencies of the States. An advisory committee may also include such other persons as the Secretary may appoint who are qualified by knowledge and experience to make a useful contribution to the work of such committee, including one or more representatives of professional organizations of technicians or professionals specializing in occupational safety or health, and one or more representatives of nationally recognized standards-producing organizations, but the number of persons so appointed to any such advisory committee shall not exceed the number appointed to such committee as representatives of Federal and State agencies. Persons appointed to advisory committees from private life shall be compensated in the same manner as consultants or experts under section 3109 of title 5, United States Code. The Secretary shall pay to any State which is the employer of a member of such a committee who is a representative of the health or safety agency of that State, reimbursement sufficient to cover the actual cost to the State resulting from such representative's membership on such committee. Any meeting of such committee shall be open to the public and an accurate record shall be kept and made available to the public. No member of such committee (other than representatives of employers and employees) shall have an economic interest in any proposed rule.

80 Stat. 416.

Recordkeeping.

(c) In carrying out his responsibilities under this Act, the Secretary is authorized to—

(1) use, with the consent of any Federal agency, the services, facilities, and personnel of such agency, with or without reimbursement, and with the consent of any State or political subdivision thereof, accept and use the services, facilities, and personnel of any agency of such State or subdivision with reimbursement; and

(2) employ experts and consultants or organizations thereof as authorized by section 3109 of title 5, United States Code, except that contracts for such employment may be renewed annually; compensate individuals so employed at rates not in excess of the rate specified at the time of service for grade GS–18 under section 5332 of title 5, United States Code, including traveltime, and allow them while away from their homes or regular places of business, travel expenses (including per diem in lieu of subsistence) as authorized by section 5703 of title 5, United States Code, for persons in the Government service employed intermittently, while so employed.

Ante, p. 198-1.

80 Stat. 499;
83 Stat. 190.

INSPECTIONS, INVESTIGATIONS, AND RECORDKEEPING

Sec. 8. (a) In order to carry out the purposes of this Act, the Secretary, upon presenting appropriate credentials to the owner, operator, or agent in charge, is authorized—

(1) to enter without delay and at reasonable times any factory, plant, establishment, construction site, or other area, workplace or environment where work is performed by an employee of an employer; and

84 STAT. 1599

(2) to inspect and investigate during regular working hours and at other reasonable times, and within reasonable limits and in a reasonable manner, any such place of employment and all pertinent conditions, structures, machines, apparatus, devices, equipment, and materials therein, and to question privately any such employer, owner, operator, agent or employee.

Subpoena power.

(b) In making his inspections and investigations under this Act the Secretary may require the attendance and testimony of witnesses and the production of evidence under oath. Witnesses shall be paid the same fees and mileage that are paid witnesses in the courts of the United States. In case of a contumacy, failure, or refusal of any person to obey such an order, any district court of the United States or the United States courts of any territory or possession, within the jurisdiction of which such person is found, or resides or transacts business, upon the application by the Secretary, shall have jurisdiction to issue to such person an order requiring such person to appear to produce evidence if, as, and when so ordered, and to give testimony relating to the matter under investigation or in question, and any failure to obey such order of the court may be punished by said court as a contempt thereof.

Recordkeeping.

(c) (1) Each employer shall make, keep and preserve, and make available to the Secretary or the Secretary of Health, Education, and Welfare, such records regarding his activities relating to this Act as the Secretary, in cooperation with the Secretary of Health, Education, and Welfare, may prescribe by regulation as necessary or appropriate for the enforcement of this Act or for developing information regarding the causes and prevention of occupational accidents and illnesses. In order to carry out the provisions of this paragraph such regulations may include provisions requiring employers to conduct periodic inspections. The Secretary shall also issue regulations requiring that employers, through posting of notices or other appropriate means, keep their employees informed of their protections and obligations under this Act, including the provisions of applicable standards.

Work-related deaths, etc.; reports.

(2) The Secretary, in cooperation with the Secretary of Health, Education, and Welfare, shall prescribe regulations requiring employers to maintain accurate records of, and to make periodic reports on, work-related deaths, injuries and illnesses other than minor injuries requiring only first aid treatment and which do not involve medical treatment, loss of consciousness, restriction of work or motion, or transfer to another job.

(3) The Secretary, in cooperation with the Secretary of Health, Education, and Welfare, shall issue regulations requiring employers to maintain accurate records of employee exposures to potentially toxic materials or harmful physical agents which are required to be monitored or measured under section 6. Such regulations shall provide employees or their representatives with an opportunity to observe such monitoring or measuring, and to have access to the records thereof. Such regulations shall also make appropriate provision for each employee or former employee to have access to such records as will indicate his own exposure to toxic materials or harmful physical agents. Each employer shall promptly notify any employee who has been or is being exposed to toxic materials or harmful physical agents in concentrations or at levels which exceed those prescribed by an applicable occupational safety and health standard promulgated under section 6, and shall inform any employee who is being thus exposed of the corrective action being taken.

84 STAT. 1600

(d) Any information obtained by the Secretary, the Secretary of Health, Education, and Welfare, or a State agency under this Act shall be obtained with a minimum burden upon employers, especially those operating small businesses. Unnecessary duplication of efforts in obtaining information shall be reduced to the maximum extent feasible.

(e) Subject to regulations issued by the Secretary, a representative of the employer and a representative authorized by his employees shall be given an opportunity to accompany the Secretary or his authorized representative during the physical inspection of any workplace under subsection (a) for the purpose of aiding such inspection. Where there is no authorized employee representative, the Secretary or his authorized representative shall consult with a reasonable number of employees concerning matters of health and safety in the workplace.

(f)(1) Any employees or representative of employees who believe that a violation of a safety or health standard exists that threatens physical harm, or that an imminent danger exists, may request an inspection by giving notice to the Secretary or his authorized representative of such violation or danger. Any such notice shall be reduced to writing, shall set forth with reasonable particularity the grounds for the notice, and shall be signed by the employees or representative of employees, and a copy shall be provided the employer or his agent no later than at the time of inspection, except that, upon the request of the person giving such notice, his name and the names of individual employees referred to therein shall not appear in such copy or on any record published, released, or made available pursuant to subsection (g) of this section. If upon receipt of such notification the Secretary determines there are reasonable grounds to believe that such violation or danger exists, he shall make a special inspection in accordance with the provisions of this section as soon as practicable, to determine if such violation or danger exists. If the Secretary determines there are no reasonable grounds to believe that a violation or danger exists he shall notify the employees or representative of the employees in writing of such determination.

(2) Prior to or during any inspection of a workplace, any employees or representative of employees employed in such workplace may notify the Secretary or any representative of the Secretary responsible for conducting the inspection, in writing, of any violation of this Act which they have reason to believe exists in such workplace. The Secretary shall, by regulation, establish procedures for informal review of any refusal by a representative of the Secretary to issue a citation with respect to any such alleged violation and shall furnish the employees or representative of employees requesting such review a written statement of the reasons for the Secretary's final disposition of the case.

(g)(1) The Secretary and Secretary of Health, Education, and Welfare are authorized to compile, analyze, and publish, either in summary or detailed form, all reports or information obtained under this section. *Reports, publication.*

(2) The Secretary and the Secretary of Health, Education, and Welfare shall each prescribe such rules and regulations as he may deem necessary to carry out their responsibilities under this Act, including rules and regulations dealing with the inspection of an employer's establishment. *Rules and regulations.*

84 STAT. 1601

CITATIONS

SEC. 9. (a) If, upon inspection or investigation, the Secretary or his authorized representative believes that an employer has violated a requirement of section 5 of this Act, of any standard, rule or order promulgated pursuant to section 6 of this Act, or of any regulations prescribed pursuant to this Act, he shall with reasonable promptness issue a citation to the employer. Each citation shall be in writing and shall describe with particularity the nature of the violation, including a reference to the provision of the Act, standard, rule, regulation, or order alleged to have been violated. In addition, the citation shall fix a reasonable time for the abatement of the violation. The Secretary may prescribe procedures for the issuance of a notice in lieu of a citation with respect to de minimis violations which have no direct or immediate relationship to safety or health.

(b) Each citation issued under this section, or a copy or copies thereof, shall be prominently posted, as prescribed in regulations issued by the Secretary, at or near each place a violation referred to in the citation occurred.

Limitation.

(c) No citation may be issued under this section after the expiration of six months following the occurrence of any violation.

PROCEDURE FOR ENFORCEMENT

SEC. 10. (a) If, after an inspection or investigation, the Secretary issues a citation under section 9(a), he shall, within a reasonable time after the termination of such inspection or investigation, notify the employer by certified mail of the penalty, if any, proposed to be assessed under section 17 and that the employer has fifteen working days within which to notify the Secretary that he wishes to contest the citation or proposed assessment of penalty. If, within fifteen working days from the receipt of the notice issued by the Secretary the employer fails to notify the Secretary that he intends to contest the citation or proposed assessment of penalty, and no notice is filed by any employee or representative of employees under subsection (c) within such time, the citation and the assessment, as proposed, shall be deemed a final order of the Commission and not subject to review by any court or agency.

(b) If the Secretary has reason to believe that an employer has failed to correct a violation for which a citation has been issued within the period permitted for its correction (which period shall not begin to run until the entry of a final order by the Commission in the case of any review proceedings under this section initiated by the employer in good faith and not solely for delay or avoidance of penalties), the Secretary shall notify the employer by certified mail of such failure and of the penalty proposed to be assessed under section 17 by reason of such failure, and that the employer has fifteen working days within which to notify the Secretary that he wishes to contest the Secretary's notification or the proposed assessment of penalty. If, within fifteen working days from the receipt of notification issued by the Secretary, the employer fails to notify the Secretary that he intends to contest the notification or proposed assessment of penalty, the notification and assessment, as proposed, shall be deemed a final order of the Commission and not subject to review by any court or agency.

(c) If an employer notifies the Secretary that he intends to contest a citation issued under section 9(a) or notification issued under subsection (a) or (b) of this section, or if, within fifteen working days

84 STAT. 1602

of the issuance of a citation under section 9(a), any employee or representative of employees files a notice with the Secretary alleging that the period of time fixed in the citation for the abatement of the violation is unreasonable, the Secretary shall immediately advise the Commission of such notification, and the Commission shall afford an opportunity for a hearing (in accordance with section 554 of title 5, United States Code, but without regard to subsection (a)(3) of such section). The Commission shall thereafter issue an order, based on findings of fact, affirming, modifying, or vacating the Secretary's citation or proposed penalty, or directing other appropriate relief, and such order shall become final thirty days after its issuance. Upon a showing by an employer of a good faith effort to comply with the abatement requirements of a citation, and that abatement has not been completed because of factors beyond his reasonable control, the Secretary, after an opportunity for a hearing as provided in this subsection, shall issue an order affirming or modifying the abatement requirements in such citation. The rules of procedure prescribed by the Commission shall provide affected employees or representatives of affected employees an opportunity to participate as parties to hearings under this subsection.

80 Stat. 384.

JUDICIAL REVIEW

SEC. 11. (a) Any person adversely affected or aggrieved by an order of the Commission issued under subsection (c) of section 10 may obtain a review of such order in any United States court of appeals for the circuit in which the violation is alleged to have occurred or where the employer has its principal office, or in the Court of Appeals for the District of Columbia Circuit, by filing in such court within sixty days following the issuance of such order a written petition praying that the order be modified or set aside. A copy of such petition shall be forthwith transmitted by the clerk of the court to the Commission and to the other parties, and thereupon the Commission shall file in the court the record in the proceeding as provided in section 2112 of title 28, United States Code. Upon such filing, the court shall have jurisdiction of the proceeding and of the question determined therein, and shall have power to grant such temporary relief or restraining order as it deems just and proper, and to make and enter upon the pleadings, testimony, and proceedings set forth in such record a decree affirming, modifying, or setting aside in whole or in part, the order of the Commission and enforcing the same to the extent that such order is affirmed or modified. The commencement of proceedings under this subsection shall not, unless ordered by the court, operate as a stay of the order of the Commission. No objection that has not been urged before the Commission shall be considered by the court, unless the failure or neglect to urge such objection shall be excused because of extraordinary circumstances. The findings of the Commission with respect to questions of fact, if supported by substantial evidence on the record considered as a whole, shall be conclusive. If any party shall apply to the court for leave to adduce additional evidence and shall show to the satisfaction of the court that such additional evidence is material and that there were reasonable grounds for the failure to adduce such evidence in the hearing before the Commission, the court may order such additional evidence to be taken before the Commission and to be made a part of the record. The Commission may modify its findings as to the facts, or make new findings, by reason of additional evidence so taken and filed, and it shall file such modified or new findings, which findings with respect to questions of fact, if supported by substantial evi-

72 Stat. 941;
80 Stat. 1323.

84 STAT. 1603

dence on the record considered as a whole, shall be conclusive, and its recommendations, if any, for the modification or setting aside of its original order. Upon the filing of the record with it, the jurisdiction of the court shall be exclusive and its judgment and decree shall be final, except that the same shall be subject to review by the Supreme Court of the United States, as provided in section 1254 of title 28, United States Code. Petitions filed under this subsection shall be heard expeditiously.

62 Stat. 928.

(b) The Secretary may also obtain review or enforcement of any final order of the Commission by filing a petition for such relief in the United States court of appeals for the circuit in which the alleged violation occurred or in which the employer has its principal office, and the provisions of subsection (a) shall govern such proceedings to the extent applicable. If no petition for review, as provided in subsection (a), is filed within sixty days after service of the Commission's order, the Commission's findings of fact and order shall be conclusive in connection with any petition for enforcement which is filed by the Secretary after the expiration of such sixty-day period. In any such case, as well as in the case of a noncontested citation or notification by the Secretary which has become a final order of the Commission under subsection (a) or (b) of section 10, the clerk of the court, unless otherwise ordered by the court, shall forthwith enter a decree enforcing the order and shall transmit a copy of such decree to the Secretary and the employer named in the petition. In any contempt proceeding brought to enforce a decree of a court of appeals entered pursuant to this subsection or subsection (a), the court of appeals may assess the penalties provided in section 17, in addition to invoking any other available remedies.

(c) (1) No person shall discharge or in any manner discriminate against any employee because such employee has filed any complaint or instituted or caused to be instituted any proceeding under or related to this Act or has testified or is about to testify in any such proceeding or because of the exercise by such employee on behalf of himself or others of any right afforded by this Act.

(2) Any employee who believes that he has been discharged or otherwise discriminated against by any person in violation of this subsection may, within thirty days after such violation occurs, file a complaint with the Secretary alleging such discrimination. Upon receipt of such complaint, the Secretary shall cause such investigation to be made as he deems appropriate. If upon such investigation, the Secretary determines that the provisions of this subsection have been violated, he shall bring an action in any appropriate United States district court against such person. In any such action the United States district courts shall have jurisdiction, for cause shown to restrain violations of paragraph (1) of this subsection and order all appropriate relief including rehiring or reinstatement of the employee to his former position with back pay.

(3) Within 90 days of the receipt of a complaint filed under the subsection the Secretary shall notify the complainant of his determination under paragraph 2 of this subsection.

THE OCCUPATIONAL SAFETY AND HEALTH REVIEW COMMISSION

Establishment; membership.

SEC. 12. (a) The Occupational Safety and Health Review Commission is hereby established. The Commission shall be composed of three members who shall be appointed by the President, by and with the advice and consent of the Senate, from among persons who by reason

84 STAT. 1604

of training, education, or experience are qualified to carry out the functions of the Commission under this Act. The President shall designate one of the members of the Commission to serve as Chairman.

(b) The terms of members of the Commission shall be six years   Terms. except that (1) the members of the Commission first taking office shall serve, as designated by the President at the time of appointment, one for a term of two years, one for a term of four years, and one for a term of six years, and (2) a vacancy caused by the death, resignation, or removal of a member prior to the expiration of the term for which he was appointed shall be filled only for the remainder of such unexpired term. A member of the Commission may be removed by the President for inefficiency, neglect of duty, or malfeasance in office.

(c) (1) Section 5314 of title 5, United States Code, is amended by   80 Stat. 460. adding at the end thereof the following new paragraph:

"(57) Chairman, Occupational Safety and Health Review Commission."

(2) Section 5315 of title 5, United States Code, is amended by add-   Ante, p. 776. ing at the end thereof the following new paragraph:

"(94) Members, Occupational Safety and Health Review Commission."

(d) The principal office of the Commission shall be in the District   Location. of Columbia. Whenever the Commission deems that the convenience of the public or of the parties may be promoted, or delay or expense may be minimized, it may hold hearings or conduct other proceedings at any other place.

(e) The Chairman shall be responsible on behalf of the Commission for the administrative operations of the Commission and shall appoint such hearing examiners and other employees as he deems necessary to assist in the performance of the Commission's functions and to fix their compensation in accordance with the provisions of chapter 51 and subchapter III of chapter 53 of title 5, United States Code,   5 USC 5101, relating to classification and General Schedule pay rates: *Provided*,   5331. That assignment, removal and compensation of hearing examiners   Ante, p. 198-1. shall be in accordance with sections 3105, 3344, 5362, and 7521 of title 5, United States Code.

(f) For the purpose of carrying out its functions under this Act, two   Quorum. members of the Commission shall constitute a quorum and official action can be taken only on the affirmative vote of at least two members.

(g) Every official act of the Commission shall be entered of record,   Public records. and its hearings and records shall be open to the public. The Commission is authorized to make such rules as are necessary for the orderly transaction of its proceedings. Unless the Commission has adopted a different rule, its proceedings shall be in accordance with the Federal Rules of Civil Procedure.   28 USC app.

(h) The Commission may order testimony to be taken by deposition in any proceedings pending before it at any state of such proceeding. Any person may be compelled to appear and depose, and to produce books, papers, or documents, in the same manner as witnesses may be compelled to appear and testify and produce like documentary evidence before the Commission. Witnesses whose depositions are taken under this subsection, and the persons taking such depositions, shall be entitled to the same fees as are paid for like services in the courts of the United States.

(i) For the purpose of any proceeding before the Commission, the provisions of section 11 of the National Labor Relations Act (29 U.S.C. 161) are hereby made applicable to the jurisdiction and powers   61 Stat. 150; of the Commission.   Ante, p. 930.

84 STAT. 1605

Report.

(j) A hearing examiner appointed by the Commission shall hear, and make a determination upon, any proceeding instituted before the Commission and any motion in connection therewith, assigned to such hearing examiner by the Chairman of the Commission, and shall make a report of any such determination which constitutes his final disposition of the proceedings. The report of the hearing examiner shall become the final order of the Commission within thirty days after such report by the hearing examiner, unless within such period any Commission member has directed that such report shall be reviewed by the Commission.

(k) Except as otherwise provided in this Act, the hearing examiners shall be subject to the laws governing employees in the classified civil service, except that appointments shall be made without regard to

80 Stat. 453.

section 5108 of title 5, United States Code. Each hearing examiner shall receive compensation at a rate not less than that prescribed for

Ante, p. 198-1.

GS–16 under section 5332 of title 5, United States Code.

PROCEDURES TO COUNTERACT IMMINENT DANGERS

SEC. 13. (a) The United States district courts shall have jurisdiction, upon petition of the Secretary, to restrain any conditions or practices in any place of employment which are such that a danger exists which could reasonably be expected to cause death or serious physical harm immediately or before the imminence of such danger can be eliminated through the enforcement procedures otherwise provided by this Act. Any order issued under this section may require such steps to be taken as may be necessary to avoid, correct, or remove such imminent danger and prohibit the employment or presence of any individual in locations or under conditions where such imminent danger exists, except individuals whose presence is necessary to avoid, correct, or remove such imminent danger or to maintain the capacity of a continuous-process operation to resume normal operations without a complete cessation of operations, or where a cessation of operations is necessary, to permit such to be accomplished in a safe and orderly manner.

(b) Upon the filing of any such petition the district court shall have jurisdiction to grant such injunctive relief or temporary restraining order pending the outcome of an enforcement proceeding pursuant to this Act. The proceeding shall be as provided by Rule 65 of the Fed-

28 USC app.

eral Rules, Civil Procedure, except that no temporary restraining order issued without notice shall be effective for a period longer than five days.

(c) Whenever and as soon as an inspector concludes that conditions or practices described in subsection (a) exist in any place of employment, he shall inform the affected employees and employers of the danger and that he is recommending to the Secretary that relief be sought.

(d) If the Secretary arbitrarily or capriciously fails to seek relief under this section, any employee who may be injured by reason of such failure, or the representative of such employees, might bring an action against the Secretary in the United States district court for the district in which the imminent danger is alleged to exist or the employer has its principal office, or for the District of Columbia, for a writ of mandamus to compel the Secretary to seek such an order and for such further relief as may be appropriate.

84 STAT. 1606

REPRESENTATION IN CIVIL LITIGATION

SEC. 14. Except as provided in section 518(a) of title 28, United States Code, relating to litigation before the Supreme Court, the Solicitor of Labor may appear for and represent the Secretary in any civil litigation brought under this Act but all such litigation shall be subject to the direction and control of the Attorney General.

80 Stat. 613.

CONFIDENTIALITY OF TRADE SECRETS

SEC. 15. All information reported to or otherwise obtained by the Secretary or his representative in connection with any inspection or proceeding under this Act which contains or which might reveal a trade secret referred to in section 1905 of title 18 of the United States Code shall be considered confidential for the purpose of that section, except that such information may be disclosed to other officers or employees concerned with carrying out this Act or when relevant in any proceeding under this Act. In any such proceeding the Secretary, the Commission, or the court shall issue such orders as may be appropriate to protect the confidentiality of trade secrets.

62 Stat. 791.

VARIATIONS, TOLERANCES, AND EXEMPTIONS

SEC. 16. The Secretary, on the record, after notice and opportunity for a hearing may provide such reasonable limitations and may make such rules and regulations allowing reasonable variations, tolerances, and exemptions to and from any or all provisions of this Act as he may find necessary and proper to avoid serious impairment of the national defense. Such action shall not be in effect for more than six months without notification to affected employees and an opportunity being afforded for a hearing.

PENALTIES

SEC. 17. (a) Any employer who willfully or repeatedly violates the requirements of section 5 of this Act, any standard, rule, or order promulgated pursuant to section 6 of this Act, or regulations prescribed pursuant to this Act, may be assessed a civil penalty of not more than $10,000 for each violation.

(b) Any employer who has received a citation for a serious violation of the requirements of section 5 of this Act, of any standard, rule, or order promulgated pursuant to section 6 of this Act, or of any regulations prescribed pursuant to this Act, shall be assessed a civil penalty of up to $1,000 for each such violation.

(c) Any employer who has received a citation for a violation of the requirements of section 5 of this Act, of any standard, rule, or order promulgated pursuant to section 6 of this Act, or of regulations prescribed pursuant to this Act, and such violation is specifically determined not to be of a serious nature, may be assessed a civil penalty of up to $1,000 for each such violation.

(d) Any employer who fails to correct a violation for which a citation has been issued under section 9(a) within the period permitted for its correction (which period shall not begin to run until the date of the final order of the Commission in the case of any review proceeding under section 10 initiated by the employer in good faith and not solely for delay or avoidance of penalties), may be assessed a civil penalty of not more than $1,000 for each day during which such failure or violation continues.

84 STAT. 1607

(e) Any employer who willfully violates any standard, rule, or order promulgated pursuant to section 6 of this Act, or of any regulations prescribed pursuant to this Act, and that violation caused death to any employee, shall, upon conviction, be punished by a fine of not more than $10,000 or by imprisonment for not more than six months, or by both; except that if the conviction is for a violation committed after a first conviction of such person, punishment shall be by a fine of not more than $20,000 or by imprisonment for not more than one year, or by both.

(f) Any person who gives advance notice of any inspection to be conducted under this Act, without authority from the Secretary or his designees, shall, upon conviction, be punished by a fine of not more than $1,000 or by imprisonment for not more than six months, or by both.

(g) Whoever knowingly makes any false statement, representation, or certification in any application, record, report, plan, or other document filed or required to be maintained pursuant to this Act shall, upon conviction, be punished by a fine of not more than $10,000, or by imprisonment for not more than six months, or by both.

65 Stat. 721;
79 Stat. 234.

(h)(1) Section 1114 of title 18, United States Code, is hereby amended by striking out "designated by the Secretary of Health, Education, and Welfare to conduct investigations, or inspections under the Federal Food, Drug, and Cosmetic Act" and inserting in lieu thereof "or of the Department of Labor assigned to perform investigative, inspection, or law enforcement functions".

62 Stat. 756.

(2) Notwithstanding the provisions of sections 1111 and 1114 of title 18, United States Code, whoever, in violation of the provisions of section 1114 of such title, kills a person while engaged in or on account of the performance of investigative, inspection, or law enforcement functions added to such section 1114 by paragraph (1) of this subsection, and who would otherwise be subject to the penalty provisions of such section 1111, shall be punished by imprisonment for any term of years or for life.

(i) Any employer who violates any of the posting requirements, as prescribed under the provisions of this Act, shall be assessed a civil penalty of up to $1,000 for each violation.

(j) The Commission shall have authority to assess all civil penalties provided in this section, giving due consideration to the appropriateness of the penalty with respect to the size of the business of the employer being charged, the gravity of the violation, the good faith of the employer, and the history of previous violations.

(k) For purposes of this section, a serious violation shall be deemed to exist in a place of employment if there is a substantial probability that death or serious physical harm could result from a condition which exists, or from one or more practices, means, methods, operations, or processes which have been adopted or are in use, in such place of employment unless the employer did not, and could not with the exercise of reasonable diligence, know of the presence of the violation.

(l) Civil penalties owed under this Act shall be paid to the Secretary for deposit into the Treasury of the United States and shall accrue to the United States and may be recovered in a civil action in the name of the United States brought in the United States district court for the district where the violation is alleged to have occurred or where the employer has its principal office.

84 STAT. 1608

STATE JURISDICTION AND STATE PLANS

Sec. 18. (a) Nothing in this Act shall prevent any State agency or court from asserting jurisdiction under State law over any occupational safety or health issue with respect to which no standard is in effect under section 6.

(b) Any State which, at any time, desires to assume responsibility for development and enforcement therein of occupational safety and health standards relating to any occupational safety or health issue with respect to which a Federal standard has been promulgated under section 6 shall submit a State plan for the development of such standards and their enforcement.

(c) The Secretary shall approve the plan submitted by a State under subsection (b), or any modification thereof, if such plan in his judgment—

(1) designates a State agency or agencies as the agency or agencies responsible for administering the plan throughout the State,

(2) provides for the development and enforcement of safety and health standards relating to one or more safety or health issues, which standards (and the enforcement of which standards) are or will be at least as effective in providing safe and healthful employment and places of employment as the standards promulgated under section 6 which relate to the same issues, and which standards, when applicable to products which are distributed or used in interstate commerce, are required by compelling local conditions and do not unduly burden interstate commerce,

(3) provides for a right of entry and inspection of all workplaces subject to the Act which is at least as effective as that provided in section 8, and includes a prohibition on advance notice of inspections,

(4) contains satisfactory assurances that such agency or agencies have or will have the legal authority and qualified personnel necessary for the enforcement of such standards,

(5) gives satisfactory assurances that such State will devote adequate funds to the administration and enforcement of such standards,

(6) contains satisfactory assurances that such State will, to the extent permitted by its law, establish and maintain an effective and comprehensive occupational safety and health program applicable to all employees of public agencies of the State and its political subdivisions, which program is as effective as the standards contained in an approved plan,

(7) requires employers in the State to make reports to the Secretary in the same manner and to the same extent as if the plan were not in effect, and

(8) provides that the State agency will make such reports to the Secretary in such form and containing such information, as the Secretary shall from time to time require.

(d) If the Secretary rejects a plan submitted under subsection (b), he shall afford the State submitting the plan due notice and opportunity for a hearing before so doing. *Notice of hearing.*

(e) After the Secretary approves a State plan submitted under subsection (b), he may, but shall not be required to, exercise his authority under sections 8, 9, 10, 13, and 17 with respect to comparable standards promulgated under section 6, for the period specified in the next sentence. The Secretary may exercise the authority referred to above until he determines, on the basis of actual operations under the

84 STAT. 1609

State plan, that the criteria set forth in subsection (c) are being applied, but he shall not make such determination for at least three years after the plan's approval under subsection (c). Upon making the determination referred to in the preceding sentence, the provisions of sections 5(a) (2), 8 (except for the purpose of carrying out subsection (f) of this section), 9, 10, 13, and 17, and standards promulgated under section 6 of this Act, shall not apply with respect to any occupational safety or health issues covered under the plan, but the Secretary may retain jurisdiction under the above provisions in any proceeding commenced under section 9 or 10 before the date of determination.

Continuing evaluation.

(f) The Secretary shall, on the basis of reports submitted by the State agency and his own inspections make a continuing evaluation of the manner in which each State having a plan approved under this section is carrying out such plan. Whenever the Secretary finds, after affording due notice and opportunity for a hearing, that in the administration of the State plan there is a failure to comply substantially with any provision of the State plan (or any assurance contained therein), he shall notify the State agency of his withdrawal of approval of such plan and upon receipt of such notice such plan shall cease to be in effect, but the State may retain jurisdiction in any case commenced before the withdrawal of the plan in order to enforce standards under the plan whenever the issues involved do not relate to the reasons for the withdrawal of the plan.

Plan rejection, review.

(g) The State may obtain a review of a decision of the Secretary withdrawing approval of or rejecting its plan by the United States court of appeals for the circuit in which the State is located by filing in such court within thirty days following receipt of notice of such decision a petition to modify or set aside in whole or in part the action of the Secretary. A copy of such petition shall forthwith be served upon the Secretary, and thereupon the Secretary shall certify and file in the court the record upon which the decision complained of was issued as provided in section 2112 of title 28, United States Code. Unless the court finds that the Secretary's decision in rejecting a proposed State plan or withdrawing his approval of such a plan is not supported by substantial evidence the court shall affirm the Secretary's decision. The judgment of the court shall be subject to review by the Supreme Court of the United States upon certiorari or certification as provided in section 1254 of title 28, United States Code.

72 Stat. 941;
80 Stat. 1323.

62 Stat. 928.

(h) The Secretary may enter into an agreement with a State under which the State will be permitted to continue to enforce one or more occupational health and safety standards in effect in such State until final action is taken by the Secretary with respect to a plan submitted by a State under subsection (b) of this section, or two years from the date of enactment of this Act, whichever is earlier.

FEDERAL AGENCY SAFETY PROGRAMS AND RESPONSIBILITIES

SEC. 19. (a) It shall be the responsibility of the head of each Federal agency to establish and maintain an effective and comprehensive occupational safety and health program which is consistent with the standards promulgated under section 6. The head of each agency shall (after consultation with representatives of the employees thereof)—

(1) provide safe and healthful places and conditions of employment, consistent with the standards set under section 6;

(2) acquire, maintain, and require the use of safety equipment, personal protective equipment, and devices reasonably necessary to protect employees;

84 STAT. 1610

(3) keep adequate records of all occupational accidents and illnesses for proper evaluation and necessary corrective action;

Recordkeeping.

(4) consult with the Secretary with regard to the adequacy as to form and content of records kept pursuant to subsection (a)(3) of this section; and

(5) make an annual report to the Secretary with respect to occupational accidents and injuries and the agency's program under this section. Such report shall include any report submitted under section 7902(e)(2) of title 5, United States Code.

Annual report.

80 Stat. 530.

(b) The Secretary shall report to the President a summary or digest of reports submitted to him under subsection (a)(5) of this section, together with his evaluations of and recommendations derived from such reports. The President shall transmit annually to the Senate and the House of Representatives a report of the activities of Federal agencies under this section.

Report to President.

Report to Congress.

(c) Section 7902(c)(1) of title 5, United States Code, is amended by inserting after "agencies" the following: "and of labor organizations representing employees".

(d) The Secretary shall have access to records and reports kept and filed by Federal agencies pursuant to subsections (a)(3) and (5) of this section unless those records and reports are specifically required by Executive order to be kept secret in the interest of the national defense or foreign policy, in which case the Secretary shall have access to such information as will not jeopardize national defense or foreign policy.

Records, etc.; availability.

RESEARCH AND RELATED ACTIVITIES

SEC. 20. (a)(1) The Secretary of Health, Education, and Welfare, after consultation with the Secretary and with other appropriate Federal departments or agencies, shall conduct (directly or by grants or contracts) research, experiments, and demonstrations relating to occupational safety and health, including studies of psychological factors involved, and relating to innovative methods, techniques, and approaches for dealing with occupational safety and health problems.

(2) The Secretary of Health, Education, and Welfare shall from time to time consult with the Secretary in order to develop specific plans for such research, demonstrations, and experiments as are necessary to produce criteria, including criteria identifying toxic substances, enabling the Secretary to meet his responsibility for the formulation of safety and health standards under this Act; and the Secretary of Health, Education, and Welfare, on the basis of such research, demonstrations, and experiments and any other information available to him, shall develop and publish at least annually such criteria as will effectuate the purposes of this Act.

(3) The Secretary of Health, Education, and Welfare, on the basis of such research, demonstrations, and experiments, and any other information available to him, shall develop criteria dealing with toxic materials and harmful physical agents and substances which will describe exposure levels that are safe for various periods of employment, including but not limited to the exposure levels at which no employee will suffer impaired health or functional capacities or diminished life expectancy as a result of his work experience.

(4) The Secretary of Health, Education, and Welfare shall also conduct special research, experiments, and demonstrations relating to occupational safety and health as are necessary to explore new problems, including those created by new technology in occupational safety and health, which may require ameliorative action beyond that

84 STAT. 1611

which is otherwise provided for in the operating provisions of this Act. The Secretary of Health, Education, and Welfare shall also conduct research into the motivational and behavioral factors relating to the field of occupational safety and health.

Toxic substances, records.

(5) The Secretary of Health, Education, and Welfare, in order to comply with his responsibilities under paragraph (2), and in order to develop needed information regarding potentially toxic substances or harmful physical agents, may prescribe regulations requiring employers to measure, record, and make reports on the exposure of employees to substances or physical agents which the Secretary of Health, Education, and Welfare reasonably believes may endanger the health or safety of employees. The Secretary of Health, Education, and Welfare also is authorized to establish such programs of medical examinations and tests as may be necessary for determining the incidence of occupational illnesses and the susceptibility of employees to such illnesses. Nothing in this or any other provision of this Act shall be deemed to authorize or require medical examination, immunization, or treatment for those who object thereto on religious grounds, except where such is necessary for the protection of the health or safety of others. Upon the request of any employer who is required to measure and record exposure of employees to substances or physical agents as provided under this subsection, the Secretary of Health, Education, and Welfare shall furnish full financial or other assistance to such employer for the purpose of defraying any additional expense incurred by him in carrying out the measuring and recording as provided in this subsection.

Medical examinations.

Toxic substances, publication.

(6) The Secretary of Health, Education, and Welfare shall publish within six months of enactment of this Act and thereafter as needed but at least annually a list of all known toxic substances by generic family or other useful grouping, and the concentrations at which such toxicity is known to occur. He shall determine following a written request by any employer or authorized representative of employees, specifying with reasonable particularity the grounds on which the request is made, whether any substance normally found in the place of employment has potentially toxic effects in such concentrations as used or found; and shall submit such determination both to employers and affected employees as soon as possible. If the Secretary of Health, Education, and Welfare determines that any substance is potentially toxic at the concentrations in which it is used or found in a place of employment, and such substance is not covered by an occupational safety or health standard promulgated under section 6, the Secretary of Health, Education, and Welfare shall immediately submit such determination to the Secretary, together with all pertinent criteria.

Annual studies.

(7) Within two years of enactment of this Act, and annually thereafter the Secretary of Health, Education, and Welfare shall conduct and publish industrywide studies of the effect of chronic or low-level exposure to industrial materials, processes, and stresses on the potential for illness, disease, or loss of functional capacity in aging adults.

Inspections.

(b) The Secretary of Health, Education, and Welfare is authorized to make inspections and question employers and employees as provided in section 8 of this Act in order to carry out his functions and responsibilities under this section.

Contract authority.

(c) The Secretary is authorized to enter into contracts, agreements, or other arrangements with appropriate public agencies or private organizations for the purpose of conducting studies relating to his responsibilities under this Act. In carrying out his responsibilities

84 STAT, 1612

under this subsection, the Secretary shall cooperate with the Secretary of Health, Education, and Welfare in order to avoid any duplication of efforts under this section.

(d) Information obtained by the Secretary and the Secretary of Health, Education, and Welfare under this section shall be disseminated by the Secretary to employers and employees and organizations thereof.

(e) The functions of the Secretary of Health, Education, and Welfare under this Act shall, to the extent feasible, be delegated to the Director of the National Institute for Occupational Safety and Health established by section 22 of this Act.

Delegation of functions.

TRAINING AND EMPLOYEE EDUCATION

SEC. 21. (a) The Secretary of Health, Education, and Welfare, after consultation with the Secretary and with other appropriate Federal departments and agencies, shall conduct, directly or by grants or contracts (1) education programs to provide an adequate supply of qualified personnel to carry out the purposes of this Act, and (2) informational programs on the importance of and proper use of adequate safety and health equipment.

(b) The Secretary is also authorized to conduct, directly or by grants or contracts, short-term training of personnel engaged in work related to his responsibilities under this Act.

(c) The Secretary, in consultation with the Secretary of Health, Education, and Welfare, shall (1) provide for the establishment and supervision of programs for the education and training of employers and employees in the recognition, avoidance, and prevention of unsafe or unhealthful working conditions in employments covered by this Act, and (2) consult with and advise employers and employees, and organizations representing employers and employees as to effective means of preventing occupational injuries and illnesses.

NATIONAL INSTITUTE FOR OCCUPATIONAL SAFETY AND HEALTH

SEC. 22. (a) It is the purpose of this section to establish a National Institute for Occupational Safety and Health in the Department of Health, Education, and Welfare in order to carry out the policy set forth in section 2 of this Act and to perform the functions of the Secretary of Health, Education, and Welfare under sections 20 and 21 of this Act.

Establishment.

(b) There is hereby established in the Department of Health, Education, and Welfare a National Institute for Occupational Safety and Health. The Institute shall be headed by a Director who shall be appointed by the Secretary of Health, Education, and Welfare, and who shall serve for a term of six years unless previously removed by the Secretary of Health, Education, and Welfare.

Director, appointment, term.

(c) The Institute is authorized to—

(1) develop and establish recommended occupational safety and health standards; and

(2) perform all functions of the Secretary of Health, Education, and Welfare under sections 20 and 21 of this Act.

(d) Upon his own initiative, or upon the request of the Secretary or the Secretary of Health, Education, and Welfare, the Director is authorized (1) to conduct such research and experimental programs as he determines are necessary for the development of criteria for new and improved occupational safety and health standards, and (2) after

84 STAT. 1613

consideration of the results of such research and experimental programs make recommendations concerning new or improved occupational safety and health standards. Any occupational safety and health standard recommended pursuant to this section shall immediately be forwarded to the Secretary of Labor, and to the Secretary of Health, Education, and Welfare.

(e) In addition to any authority vested in the Institute by other provisions of this section, the Director, in carrying out the functions of the Institute, is authorized to—

(1) prescribe such regulations as he deems necessary governing the manner in which its functions shall be carried out;

(2) receive money and other property donated, bequeathed, or devised, without condition or restriction other than that it be used for the purposes of the Institute and to use, sell, or otherwise dispose of such property for the purpose of carrying out its functions;

(3) receive (and use, sell, or otherwise dispose of, in accordance with paragraph (2)), money and other property donated, bequeathed, or devised to the Institute with a condition or restriction, including a condition that the Institute use other funds of the Institute for the purposes of the gift;

(4) in accordance with the civil service laws, appoint and fix the compensation of such personnel as may be necessary to carry out the provisions of this section;

(5) obtain the services of experts and consultants in accordance with the provisions of section 3109 of title 5, United States Code;

80 Stat. 416.

(6) accept and utilize the services of voluntary and noncompensated personnel and reimburse them for travel expenses, including per diem, as authorized by section 5703 of title 5, United States Code;

83 Stat. 190.

(7) enter into contracts, grants or other arrangements, or modifications thereof to carry out the provisions of this section, and such contracts or modifications thereof may be entered into without performance or other bonds, and without regard to section 3709 of the Revised Statutes, as amended (41 U.S.C. 5), or any other provision of law relating to competitive bidding;

(8) make advance, progress, and other payments which the Director deems necessary under this title without regard to the provisions of section 3648 of the Revised Statutes, as amended (31 U.S.C. 529); and

(9) make other necessary expenditures.

Annual report to HEW, President, and Congress.

(f) The Director shall submit to the Secretary of Health, Education, and Welfare, to the President, and to the Congress an annual report of the operations of the Institute under this Act, which shall include a detailed statement of all private and public funds received and expended by it, and such recommendations as he deems appropriate.

GRANTS TO THE STATES

SEC. 23. (a) The Secretary is authorized, during the fiscal year ending June 30, 1971, and the two succeeding fiscal years, to make grants to the States which have designated a State agency under section 18 to assist them—

(1) in identifying their needs and responsibilities in the area of occupational safety and health,

(2) in developing State plans under section 18, or

(3) in developing plans for—
(A) establishing systems for the collection of information concerning the nature and frequency of occupational injuries and diseases;
(B) increasing the expertise and enforcement capabilities of their personnel engaged in occupational safety and health programs; or
(C) otherwise improving the administration and enforcement of State occupational safety and health laws, including standards thereunder, consistent with the objectives of this Act.

(b) The Secretary is authorized, during the fiscal year ending June 30, 1971, and the two succeeding fiscal years, to make grants to the States for experimental and demonstration projects consistent with the objectives set forth in subsection (a) of this section.

(c) The Governor of the State shall designate the appropriate State agency for receipt of any grant made by the Secretary under this section.

(d) Any State agency designated by the Governor of the State desiring a grant under this section shall submit an application therefor to the Secretary.

(e) The Secretary shall review the application, and shall, after consultation with the Secretary of Health, Education, and Welfare, approve or reject such application.

(f) The Federal share for each State grant under subsection (a) or (b) of this section may not exceed 90 per centum of the total cost of the application. In the event the Federal share for all States under either such subsection is not the same, the differences among the States shall be established on the basis of objective criteria.

(g) The Secretary is authorized to make grants to the States to assist them in administering and enforcing programs for occupational safety and health contained in State plans approved by the Secretary pursuant to section 18 of this Act. The Federal share for each State grant under this subsection may not exceed 50 per centum of the total cost to the State of such a program. The last sentence of subsection (f) shall be applicable in determining the Federal share under this subsection.

(h) Prior to June 30, 1973, the Secretary shall, after consultation with the Secretary of Health, Education, and Welfare, transmit a report to the President and to the Congress, describing the experience under the grant programs authorized by this section and making any recommendations he may deem appropriate. Report to President and Congress.

STATISTICS

Sec. 24. (a) In order to further the purposes of this Act, the Secretary, in consultation with the Secretary of Health, Education, and Welfare, shall develop and maintain an effective program of collection, compilation, and analysis of occupational safety and health statistics. Such program may cover all employments whether or not subject to any other provisions of this Act but shall not cover employments excluded by section 4 of the Act. The Secretary shall compile accurate statistics on work injuries and illnesses which shall include all disabling, serious, or significant injuries and illnesses, whether or not involving loss of time from work, other than minor injuries requiring only first aid treatment and which do not involve medical treatment, loss of consciousness, restriction of work or motion, or transfer to another job.

84 STAT. 1615

(b) To carry out his duties under subsection (a) of this section, the Secretary may—

    (1) promote, encourage, or directly engage in programs of studies, information and communication concerning occupational safety and health statistics;

    (2) make grants to States or political subdivisions thereof in order to assist them in developing and administering programs dealing with occupational safety and health statistics; and

    (3) arrange, through grants or contracts, for the conduct of such research and investigations as give promise of furthering the objectives of this section.

(c) The Federal share for each grant under subsection (b) of this section may be up to 50 per centum of the State's total cost.

(d) The Secretary may, with the consent of any State or political subdivision thereof, accept and use the services, facilities, and employees of the agencies of such State or political subdivision, with or without reimbursement, in order to assist him in carrying out his functions under this section.

Reports.

(e) On the basis of the records made and kept pursuant to section 8(c) of this Act, employers shall file such reports with the Secretary as he shall prescribe by regulation, as necessary to carry out his functions under this Act.

(f) Agreements between the Department of Labor and States pertaining to the collection of occupational safety and health statistics already in effect on the effective date of this Act shall remain in effect until superseded by grants or contracts made under this Act.

AUDITS

SEC. 25. (a) Each recipient of a grant under this Act shall keep such records as the Secretary or the Secretary of Health, Education, and Welfare shall prescribe, including records which fully disclose the amount and disposition by such recipient of the proceeds of such grant, the total cost of the project or undertaking in connection with which such grant is made or used, and the amount of that portion of the cost of the project or undertaking supplied by other sources, and such other records as will facilitate an effective audit.

(b) The Secretary or the Secretary of Health, Education, and Welfare, and the Comptroller General of the United States, or any of their duly authorized representatives, shall have access for the purpose of audit and examination to any books, documents, papers, and records of the recipients of any grant under this Act that are pertinent to any such grant.

ANNUAL REPORT

SEC. 26. Within one hundred and twenty days following the convening of each regular session of each Congress, the Secretary and the Secretary of Health, Education, and Welfare shall each prepare and submit to the President for transmittal to the Congress a report upon the subject matter of this Act, the progress toward achievement of the purpose of this Act, the needs and requirements in the field of occupational safety and health, and any other relevant information. Such reports shall include information regarding occupational safety and health standards, and criteria for such standards, developed during the preceding year; evaluation of standards and criteria previously developed under this Act, defining areas of emphasis for new criteria and standards; an evaluation of the degree of observance of applicable occupational safety and health standards, and a summary

84 STAT. 1616

of inspection and enforcement activity undertaken; analysis and evaluation of research activities for which results have been obtained under governmental and nongovernmental sponsorship; an analysis of major occupational diseases; evaluation of available control and measurement technology for hazards for which standards or criteria have been developed during the preceding year; description of cooperative efforts undertaken between Government agencies and other interested parties in the implementation of this Act during the preceding year; a progress report on the development of an adequate supply of trained manpower in the field of occupational safety and health, including estimates of future needs and the efforts being made by Government and others to meet those needs; listing of all toxic substances in industrial usage for which labeling requirements, criteria, or standards have not yet been established; and such recommendations for additional legislation as are deemed necessary to protect the safety and health of the worker and improve the administration of this Act.

NATIONAL COMMISSION ON STATE WORKMEN'S COMPENSATION LAWS

SEC. 27. (a) (1) The Congress hereby finds and declares that—

(A) the vast majority of American workers, and their families, are dependent on workmen's compensation for their basic economic security in the event such workers suffer disabling injury or death in the course of their employment; and that the full protection of American workers from job-related injury or death requires an adequate, prompt, and equitable system of workmen's compensation as well as an effective program of occupational health and safety regulation; and

(B) in recent years serious questions have been raised concerning the fairness and adequacy of present workmen's compensation laws in the light of the growth of the economy, the changing nature of the labor force, increases in medical knowledge, changes in the hazards associated with various types of employment, new technology creating new risks to health and safety, and increases in the general level of wages and the cost of living.

(2) The purpose of this section is to authorize an effective study and objective evaluation of State workmen's compensation laws in order to determine if such laws provide an adequate, prompt, and equitable system of compensation for injury or death arising out of or in the course of employment.

(b) There is hereby established a National Commission on State Workmen's Compensation Laws.   *Establishment.*

(c) (1) The Workmen's Compensation Commission shall be composed of fifteen members to be appointed by the President from among members of State workmen's compensation boards, representatives of insurance carriers, business, labor, members of the medical profession having experience in industrial medicine or in workmen's compensation cases, educators having special expertise in the field of workmen's compensation, and representatives of the general public. The Secretary, the Secretary of Commerce, and the Secretary of Health, Education, and Welfare shall be ex officio members of the Workmen's Compensation Commission: *Membership.*

(2) Any vacancy in the Workmen's Compensation Commission shall not affect its powers.

(3) The President shall designate one of the members to serve as Chairman and one to serve as Vice Chairman of the Workmen's Compensation Commission.

84 STAT. 1617

Quorum.

(4) Eight members of the Workmen's Compensation Commission shall constitute a quorum.

Study.

(d)(1) The Workmen's Compensation Commission shall undertake a comprehensive study and evaluation of State workmen's compensation laws in order to determine if such laws provide an adequate, prompt, and equitable system of compensation. Such study and evaluation shall include, without being limited to, the following subjects: (A) the amount and duration of permanent and temporary disability benefits and the criteria for determining the maximum limitations thereon, (B) the amount and duration of medical benefits and provisions insuring adequate medical care and free choice of physician, (C) the extent of coverage of workers, including exemptions based on numbers or type of employment, (D) standards for determining which injuries or diseases should be deemed compensable, (E) rehabilitation, (F) coverage under second or subsequent injury funds, (G) time limits on filing claims, (H) waiting periods, (I) compulsory or elective coverage, (J) administration, (K) legal expenses, (L) the feasibility and desirability of a uniform system of reporting information concerning job-related injuries and diseases and the operation of workmen's compensation laws, (M) the resolution of conflict of laws, extraterritoriality and similar problems arising from claims with multistate aspects, (N) the extent to which private insurance carriers are excluded from supplying workmen's compensation coverage and the desirability of such exclusionary practices, to the extent they are found to exist, (O) the relationship between workmen's compensation on the one hand, and old-age, disability, and survivors insurance and other types of insurance, public or private, on the other hand, (P) methods of implementing the recommendations of the Commission.

Report to
President
and Congress.

(2) The Workmen's Compensation Commission shall transmit to the President and to the Congress not later than July 31, 1972, a final report containing a detailed statement of the findings and conclusions of the Commission, together with such recommendations as it deems advisable.

Hearings.

(e)(1) The Workmen's Compensation Commission or, on the authorization of the Workmen's Compensation Commission, any subcommittee or members thereof, may, for the purpose of carrying out the provisions of this title, hold such hearings, take such testimony, and sit and act at such times and places as the Workmen's Compensation Commission deems advisable. Any member authorized by the Workmen's Compensation Commission may administer oaths or affirmations to witnesses appearing before the Workmen's Compensation Commission or any subcommittee or members thereof.

(2) Each department, agency, and instrumentality of the executive branch of the Government, including independent agencies, is authorized and directed to furnish to the Workmen's Compensation Commission, upon request made by the Chairman or Vice Chairman, such information as the Workmen's Compensation Commission deems necessary to carry out its functions under this section.

(f) Subject to such rules and regulations as may be adopted by the Workmen's Compensation Commission, the Chairman shall have the power to—

(1) appoint and fix the compensation of an executive director, and such additional staff personnel as he deems necessary, without regard to the provisions of title 5, United States Code, governing appointments in the competitive service, and without regard to the provisions of chapter 51 and subchapter III of chapter 53 of such title relating to classification and General Schedule

80 Stat. 378.
5 USC 101.

5 USC 5101,
5331.

84 STAT. 1618

pay rates, but at rates not in excess of the maximum rate for GS–18 of the General Schedule under section 5332 of such title, and

Ante, p. 198-1.

(2) procure temporary and intermittent services to the same extent as is authorized by section 3109 of title 5, United States Code.

80 Stat. 416.
Contract authorization.

(g) The Workmen's Compensation Commission is authorized to enter into contracts with Federal or State agencies, private firms, institutions, and individuals for the conduct of research or surveys, the preparation of reports, and other activities necessary to the discharge of its duties.

(h) Members of the Workmen's Compensation Commission shall receive compensation for each day they are engaged in the performance of their duties as members of the Workmen's Compensation Commission at the daily rate prescribed for GS–18 under section 5332 of title 5, United States Code, and shall be entitled to reimbursement for travel, subsistence, and other necessary expenses incurred by them in the performance of their duties as members of the Workmen's Compensation Commission.

Compensation; travel expenses.

(i) There are hereby authorized to be appropriated such sums as may be necessary to carry out the provisions of this section.

Appropriation.

(j) On the nineteenth day after the date of submission of its final report to the President, the Workmen's Compensation Commission shall cease to exist.

Termination.

ECONOMIC ASSISTANCE TO SMALL BUSINESSES

SEC. 28. (a) Section 7(b) of the Small Business Act, as amended, is amended—

72 Stat. 387;
83 Stat. 802.
15 USC 636.

(1) by striking out the period at the end of "paragraph (5)" and inserting in lieu thereof "; and"; and

(2) by adding after paragraph (5) a new paragraph as follows:

"(6) to make such loans (either directly or in cooperation with banks or other lending institutions through agreements to participate on an immediate or deferred basis) as the Administration may determine to be necessary or appropriate to assist any small business concern in effecting additions to or alterations in the equipment, facilities, or methods of operation of such business in order to comply with the applicable standards promulgated pursuant to section 6 of the Occupational Safety and Health Act of 1970 or standards adopted by a State pursuant to a plan approved under section 18 of the Occupational Safety and Health Act of 1970, if the Administration determines that such concern is likely to suffer substantial economic injury without assistance under this paragraph."

(b) The third sentence of section 7(b) of he Small Business Act, as amended, is amended by striking out "or (5)" after "paragraph (3)" and inserting a comma followed by "(5) or (6)".

(c) Section 4(c)(1) of the Small Business Act, as amended, is amended by inserting "7(b)(6)," after "7(b)(5),".

80 Stat. 132.
15 USC 633.

(d) Loans may also be made or guaranteed for the purposes set forth in section 7(b)(6) of the Small Business Act, as amended, pursuant to the provisions of section 202 of the Public Works and Economic Development Act of 1965, as amended.

79 Stat. 556.
42 USC 3142.

ADDITIONAL ASSISTANT SECRETARY OF LABOR

SEC. 29. (a) Section 2 of the Act of April 17, 1946 (60 Stat. 91) as amended (29 U.S.C. 553) is amended by—

75 Stat. 338.

84 STAT. 1619

(1) striking out "four" in the first sentence of such section and inserting in lieu thereof "five"; and

(2) adding at the end thereof the following new sentence, "One of such Assistant Secretaries shall be an Assistant Secretary of Labor for Occupational Safety and Health.".

80 Stat. 462.

(b) Paragraph (20) of section 5315 of title 5, United States Code, is amended by striking out "(4)" and inserting in lieu thereof "(5)".

ADDITIONAL POSITIONS

SEC. 30. Section 5108(c) of title 5, United States Code, is amended by—

(1) striking out the word "and" at the end of paragraph (8);

(2) striking out the period at the end of paragraph (9) and inserting in lieu thereof a semicolon and the word "and"; and

(3) by adding immediately after paragraph (9) the following new paragraph:

"(10) (A) the Secretary of Labor, subject to the standards and procedures prescribed by this chapter, may place an additional twenty-five positions in the Department of Labor in GS–16, 17, and 18 for the purposes of carrying out his responsibilities under the Occupational Safety and Health Act of 1970;

"(B) the Occupational Safety and Health Review Commission, subject to the standards and procedures prescribed by this chapter, may place ten positions in GS–16, 17, and 18 in carrying out its functions under the Occupational Safety and Health Act of 1970."

EMERGENCY LOCATOR BEACONS

72 Stat. 775.
49 USC 1421.

SEC. 31. Section 601 of the Federal Aviation Act of 1958 is amended by inserting at the end thereof a new subsection as follows:

"EMERGENCY LOCATOR BEACONS

"(d) (1) Except with respect to aircraft described in paragraph (2) of this subsection, minimum standards pursuant to this section shall include a requirement that emergency locator beacons shall be installed—

"(A) on any fixed-wing, powered aircraft for use in air commerce the manufacture of which is completed, or which is imported into the United States, after one year following the date of enactment of this subsection; and

"(B) on any fixed-wing, powered aircraft used in air commerce after three years following such date.

"(2) The provisions of this subsection shall not apply to jet-powered aircraft; aircraft used in air transportation (other than air taxis and charter aircraft); military aircraft; aircraft used solely for training purposes not involving flights more than twenty miles from its base; and aircraft used for the aerial application of chemicals."

SEPARABILITY

SEC. 32. If any provision of this Act, or the application of such provision to any person or circumstance, shall be held invalid, the remainder of this Act, or the application of such provision to persons or circumstances other than those as to which it is held invalid, shall not be affected thereby.

84 STAT. 1620

APPROPRIATIONS

SEC. 33. There are authorized to be appropriated to carry out this Act for each fiscal year such sums as the Congress shall deem necessary.

EFFECTIVE DATE

SEC. 34. This Act shall take effect one hundred and twenty days after the date of its enactment.

Approved December 29, 1970.

Appendix B

Rules of Procedure of the Occupational Safety and Health Review Commission*

---

*29 C.F.R. pts. 2200 and 2204. Pt. 2200 revised Dec. 8, 1986, corrected April 27, 1987; pt. 2204 revised Feb. 23, 1987.

SUBPART A - GENERAL PROVISIONS

§ 2200.1 Definitions.

As used herein:

(a) "Act" means the Occupational Safety and Health Act of 1970, 29 U.S.C. §§ 651-678.

(b) "Commission," "person," "employer," and "employee" have the meanings set forth in § 3 of the Act.

(c) "Secretary" means the Secretary of Labor or his duly authorized representative.

(d) "Executive Secretary" means the Executive Secretary of the Commission.

(e) "Affected employee" means an employee of a cited employer who is exposed to or has access to the hazard arising out of the allegedly violative circumstances, conditions, practices or operations.

(f) "Judge" means an Administrative Law Judge appointed by the Chairman of the Commission pursuant to § 12(j) of the Act, 29 U.S.C. § 661(j), as amended by Pub. L. 95-251, 92 Stat. 183, 184 (1978).

(g) "Authorized employee representative" means a labor organization that has a collective bargaining relationship with the cited employer and that represents affected employees.

(h) "Representative" means any person, including an authorized employee representative, authorized by a party or intervenor to represent him in a proceeding.

(i) "Citation" means a written communication issued by the Secretary to an employer pursuant to § 9(a) of the Act.

(j) "Notification of proposed penalty" means a written communication issued by the Secretary to an employer pursuant to § 10(a) or (b) of the Act.

(k) "Day" means a calendar day.

(l) "Working day" means all days except Saturdays, Sundays, or Federal holidays.

(m) "Proceeding" means any proceeding before the Commission or before a Judge.

861

(n) "Pleadings" are complaints and answers filed under § 2200.34, statements of reasons and contestants' responses filed under § 2200.38, and petitions for modification of abatement and objecting parties' responses filed under § 2200.37. A motion is not a "pleading" within the meaning of these rules.

§ 2200.2 Scope of rules; Applicability of Federal Rules of Civil Procedure; Construction.

(a) Scope. These rules shall govern all proceedings before the Commission and its Judges.

(b) Applicability of Federal Rules of Civil Procedure. In the absence of a specific provision, procedure shall be in accordance with the Federal Rules of Civil Procedure.

(c) Construction. These rules shall be construed to secure an expeditious, just and inexpensive determination of every case.

§ 2200.3 Use of gender and number.

(a) Number. Words importing the singular number may extend and be applied to the plural and vice versa.

(b) Gender. Words importing the masculine gender may be applied to the feminine gender.

§ 2200.4 Computation of time.

(a) Computation. In computing any period of time prescribed or allowed in these rules, the day from which the designated period begins to run shall not be included. The last day of the period so computed shall be included unless it is a Saturday, Sunday or Federal holiday, in which event the period runs until the end of the next day which is not a Saturday, Sunday, or Federal holiday. When the period of time prescribed or allowed is less than 11 days, intermediate Saturdays, Sundays and Federal holidays shall be excluded from the computation.

(b) Service by mail. Where service of a document, other than a petition for discretionary review, is made by mail pursuant to § 2200.7, three days shall be added to the prescribed period for the filing of a response. The period of time for filing a petition for discretionary review is governed by § 2200.91(b). Service within the meaning of this rule includes issuance of documents by the Commission or Judge.

§ 2200.5 Extensions of time.

Upon motion of a party for good cause shown, the Commission or Judge may enlarge any time prescribed by these rules or prescribed by an order. All such motions shall be in writing but, in exigent circumstances in cases pending before Judges, an oral request may be made and followed by a written motion. A request for an extension of time should be received in advance of the date on which the pleading or document is due to be filed. However, an extension of time may be granted even though the request was filed after the designated time for filing has expired, but in such circumstances, the party requesting the extension must show good cause for his failure to make the request before the time prescribed for the filing had expired. The motion may be acted upon before the time for response has expired.

§ 2200.6 Record address.

Every pleading or document filed by any party or intervenor shall contain the name, current address and telephone number of his representative or, if he has no representative, his own name, current address and telephone number. Any change in such information shall be communicated promptly in writing to the Judge, or the Executive Secretary if no Judge has been assigned, and to all other parties and intervenors. A party or intervenor who fails to furnish such information shall be deemed to have waived his right to notice and service under these rules.

§ 2200.7 Service and notice.

(a) When service is required. At the time of filing pleadings or other documents a copy thereof shall be served by the filing party or intervenor on every other party or intervenor. Every paper relating to discovery required to be served on a party shall be served on all parties.

(b) Service on represented parties or intervenors. Service upon a party or intervenor who has appeared through a representative shall be made only upon such representative.

(c) How accomplished. Unless otherwise ordered, service may be accomplished by postage pre-paid first class mail or by personal delivery. Service is deemed effected at the time of mailing (if by mail) or at the time of personal delivery (if by personal delivery).

(d) Proof of service. Proof of service shall be accomplished by a written statement of the same which sets forth the date and manner of service. Such statement shall be filed with the pleading or document.

(e) <u>Proof of posting</u>. Where service is accomplished by posting, proof of such posting shall be filed not later than the first working day following the posting.

(f) <u>Service on represented employees</u>. Service and notice to employees represented by an authorized employee representative shall be deemed accomplished by serving the representative in the manner prescribed in paragraph (c) of this section.

(g) <u>Service on unrepresented employees</u>. In the event that there are any affected employees who are not represented by an authorized employee representative, the employer shall, immediately upon receipt of notice of the docketing of the notice of contest or petition for modification of the abatement period, post, where the citation is required to be posted, a copy of the notice of contest and a notice informing such affected employees of their right to party status and of the availability of all pleadings for inspection and copying at reasonable times. A notice in the following form shall be deemed to comply with this paragraph:

(Name of employer)

> Your employer has been cited by the Secretary of Labor for violation of the Occupational Safety and Health Act of 1970. The citation has been contested and will be the subject of a hearing before the OCCUPATIONAL SAFETY AND HEALTH REVIEW COMMISSION. Affected employees are entitled to participate in this hearing as parties under terms and conditions established by the OCCUPATIONAL SAFETY AND HEALTH REVIEW COMMISSION in its Rules of Procedure. Notice of intent to participate must be filed no later than 10 days before the hearing. This notice should be sent to:

> > Occupational Safety and Health
> > Review Commission
> > Office of the Executive Secretary
> > 1825 K Street, N. W.
> > Washington, D. C.   20006-1246

All papers relevant to this matter may be inspected at: (Place reasonably convenient to employees, preferably at or near workplace.)

Where appropriate, the second sentence of the above notice will be deleted and the following sentence will be substituted:

> The reasonableness of the period prescribed by the Secretary of Labor for abatement of the violation has been contested and will be the subject of a hearing before the OCCUPATIONAL SAFETY AND HEALTH REVIEW COMMISSION.

(h) <u>Special service requirements; Authorized employee representatives</u>. The authorized employee representative, if any, shall be served with the notice set forth in paragraph (g) of this section and with a copy of the notice of contest.

(i) Notice of hearing to unrepresented employees. Immediately upon receipt, a copy of the notice of the hearing to be held before the Judge shall be served by the employer on affected employees who are not represented by an authorized employee representative by posting a copy of the notice of such hearing at or near the place where the citation is required to be posted.

(j) Notice of hearing to represented employees. Immediately upon receipt, a copy of the notice of the hearing to be held before the Judge shall be served by the employer on the authorized employee representative of affected employees in the manner prescribed in paragraph (c) of this section, if the employer has not been informed that the authorized employee representative has entered an appearance as of the date such notice is received by the employer.

(k) Employee contest; Service on other employees. Where a notice of contest is filed by an affected employee who is not represented by an authorized employee representative and there are other affected employees who are represented by an authorized employee representative, the unrepresented employee shall, upon receipt of the statement filed in conformance with § 2200.38, serve a copy thereof on such authorized employee representative in the manner prescribed in paragraph (c) of this section and shall file proof of such service.

(l) Employee contest; Service on employer. Where a notice of contest is filed by an affected employee or an authorized employee representative, a copy of the notice of contest and response filed in support thereof shall be provided to the employer for posting in the manner prescribed in paragraph (g) of this section.

(m) Employee contest; Service on other authorized employee representatives. An authorized employee representative who files a notice of contest shall be responsible for serving any other authorized employee representative whose members are affected employees.

(n) Duration of posting. Where posting is required by this section, such posting shall be maintained until the commencement of the hearing or until earlier disposition.

§ 2200.8 Filing.

(a) Where to file. Prior to the assignment of a case to a Judge, all papers shall be filed with the Executive Secretary at 1825 K Street, N. W., Washington, D. C. 20006. Subsequent to the assignment of the case to a Judge, all papers shall be filed with the Judge at the address given in the notice informing of such assignment. Subsequent to the docketing of the Judge's report, all papers shall be filed with the Executive Secretary, except as provided in § 2200.90(b)(3).

(b) How to file. Unless otherwise ordered, all filing may be accomplished by postage-prepaid first class mail or by personal delivery.

(c) Number of copies. Unless otherwise ordered or stated in this Part:

(1) If a case is before a Judge or if it has not yet been assigned to a Judge, only the original of a document shall be filed.

(2) If a case is before the Commission for review, the original and four copies of a document shall be filed.

(d) Filing date. Filing is deemed effected at the time of mailing (if by mail) or at the time of personal delivery (if by personal delivery), except that petitions for discretionary review are deemed to be filed at the time of receipt. See § 2200.91.

§ 2200.9 Consolidation.

Cases may be consolidated on the motion of any party, on the Judge's own motion, or on the Commission's own motion, where there exist common parties, common questions of law or fact or in such other circumstances as justice or the administration of the Act require.

§ 2200.10 Severance.

Upon its own motion, or upon motion of any party or intervenor, the Commission or the Judge may, for good cause, order any proceeding severed with respect to some or all issues or parties.

§ 2200.11 Protection of Claims of Privilege.

(a) Scope. This section applies to all claims of privilege, whenever asserted. It applies to privileged information, such as trade secrets and other matter protected by 18 U.S.C. § 1905, and other information the confidentiality of which is protected by law. As it is used in this section, "privileged information" encompasses such confidential information.

(b) Assertion of a privilege. A person claiming that information is privileged shall claim the privilege in writing or, if during a hearing, on the record. The claim shall (1) identify the information that would be disclosed and for which a privilege is claimed, and (2) allege with specificity the facts showing that the information is privileged. The claim

shall be supported by affidavits, depositions or testimony and shall specify the relief sought. The claim may be accompanied by a motion for a
protective order or by a motion that the allegedly privileged information
be received and the claim be ruled upon in camera, that is, with the record
and hearing room closed to the public, or ex parte, that is, without the
participation of parties and their representatives.

(c) Opposition to the claim. A party wishing to make a response opposing
a claim of privilege, or asserting a substantial need for disclosure in
the event a qualified privilege exists, must do so within 15 days but, if
the motion is made during a hearing, the Judge may prescribe a shorter
time or require that the response be made during the hearing. A response
contravening the facts stated by the claimant of the privilege shall be
supported by affidavits, depositions, or testimony.

(d) Examination of claim. In examining a claim of privilege, the Judge
may enter such orders and impose such terms and conditions on his examination as justice may require, including orders designed to assure that
the alleged privileged information not be disclosed until after the examination is completed. The Judge may:

   (1) Receive the allegedly privileged information in camera; he may
temporarily seal the portions of the record containing the allegedly privileged information and may exclude the public from the hearing room.

   (2) Receive the allegedly privileged information ex parte; he may
order that the allegedly privileged information not be heard or served on
all parties and their representatives; he may hear or examine it without
the presence of all parties and their representatives.

      (3) Order the preparation of a summary of the allegedly privileged
information; he may order that a copy of a document be prepared with the
allegedly privileged information excised; he may order that such sum-
maries or documents be served upon other parties or their representatives.

      (4) Enter a protective order. See paragraphs (e) and (f) of this
section.

(e) Upholding of claim. If a claim of privilege is upheld, the Judge may
enter such orders and impose such terms and conditions as justice may
require, including orders that the privileged information not be
disclosed or be disclosed in a specified manner. The Judge may: exclude
the privileged information from the record; enter orders under
§ 2200.52(d), including an order that discovery not be had; revoke or
modify a subpena; and permanently seal that portion of the record or
other files of the Commission containing the privileged information, permitting access only to the Commission and any reviewing court. The Judge
may also permit the information to be disclosed only to persons covered
by protective orders under § 2200.52(d) and paragraph (f) of this section.

(f) <u>Protective Orders</u>. To govern the examination of a claim of privilege or to govern the treatment of privileged information, the Judge may enter protective orders under §2200.52(d). The Judge may decline to permit disclosure to persons against whom the Commission could not enforce the order. The order may require that--

(1) An attorney or other representative not disclose the allegedly privileged information to any person, including his client.

(2) Any person to whom the material will be disclosed sign a written confidentiality agreement that the material will not be disclosed except under stated terms and conditions and that stipulates a reasonable pre-estimate of likely damages.

(3) In the case of an entry upon land, the case be stayed to allow the party seeking entry an opportunity to seek an order of a court or search warrant with protective conditions.

(g) <u>Rejection of claim</u>. If the Judge overrules a claim of privilege, the person claiming the privilege may obtain as of right an order sealing from the public those portions of the record containing the allegedly privileged information pending interlocutory or final review of the ruling, or final disposition of the case, by the Commission. Interlocutory review of such an order shall be given priority consideration by the Commission.

§ 2200.12 <u>References to cases</u>.

(a) <u>Citing decisions by Commission and Judges</u>.

(1) <u>Generally</u>. Parties citing decisions by the Commission should include in the citation the name of the employer, a citation to either the Bureau of National Affairs' Occupational Safety & Health Cases ("BNA OSHC") or Commerce Clearing House's Occupational Safety and Health Decisions ("CCH OSHD"), the OSHRC docket number and the year of the decision. For example, <u>Clement Food Co.</u>, 11 BNA OSHC 2120 (No. 80-607, 1984).

(2) <u>Parenthetical statements</u>. When citing the decision of a Judge, the digest of an opinion, or the opinion of a single Commissioner, a parenthetical statement to that effect should be included. For example, <u>Rust Engineering Co.</u>, 1984 CCH OSHD ¶ 27,023 (No. 79-2090, 1984)(<u>view of Chairman   </u>), <u>vacating direction for review of</u> 1980 CCH OSHD ¶ 24,269 (1980)(ALJ)(digest).

(3) <u>Additional reference to OSAHRC Reports optional</u>. A parallel reference to the Commission's official reporter, OSAHRC Reports, which

prints the full text of all Commission and Judges' decisions in micro-
fiche form, may also be included. For example, Texaco, Inc., 80 OSAHRC
74/B1, 8 BNA OSHC 1758 (No. 77-3040, 1980). See generally 29 C.F.R.
§ 2201.4(c)(on OSAHRC Reports).

    (b) References to court decisions.

       (1) Parallel references to BNA and CCH reporters. When citing a
court decision, a parallel reference to either the Bureau of National
Affairs' Occupational Safety & Health Cases ("BNA OSHC") or Commerce
Clearing House's Occupational Safety and Health Decisions ("CCH OSHD") is
desirable. For example, Simplex Time Recorder Co. v. Secretary of Labor,
766 F.2d 575, 12 BNA OSHC 1401 (D.C. Cir. 1985); Deering Milliken, Inc.
v. OSHRC, 630 F.2d 1094, 1980 CCH OSHD ¶ 24,991 (5th Cir. 1980).

       (2) Name of employer to be indicated. When a court decision is
cited in which the first-listed party on each side is either the
Secretary of Labor (or the name of a particular Secretary of Labor), the
Commission, or a labor union, the citation should include in parenthesis
the name of the employer in the Commission proceeding. For example,
Donovan v. Allied Industrial Workers (Archer Daniels Midland Co.), 760
F.2d 783, 12 BNA OSHC 1310 (7th Cir. 1985); Donovan v. OSHRC (Mobil Oil
Corp.), 713 F.2d 918, 1983 CCH OSHD ¶ 26,627 (2d Cir. 1983).

SUBPART B - PARTIES AND REPRESENTATIVES

§ 2200.20 Party status.

    (a) Affected employees. Affected employees and authorized employee
representatives, by notice of election filed at least ten days before the
hearing, may elect party status concerning any matter in which the Act
confers a right to participate. A notice of election filed less than ten
days prior to the hearing is ineffective unless good cause is shown for
not timely filing the notice. A notice of election shall be served on
all other parties in accordance with § 2200.7.

    (b) Employee contest. Where a notice of contest is filed by an
employee or by an authorized employee representative with respect to the
reasonableness of the period for abatement of a violation, the employer
charged with the responsibility of abating the violation may elect party
status by a notice filed at least ten days before the hearing. A notice
filed less than ten days prior to the hearing is ineffective unless good
cause is shown for not timely filing the notice.

§ 2200.21 Intervention; Appearance by non-parties.

    (a) When allowed. A petition for leave to intervene may be filed at
any time prior to ten days before commencement of the hearing. A petition filed less than ten days prior to the commencement of the hearing

will be denied unless good cause is shown for not timely filing the petition. A petition shall be served on all parties in accordance with § 2200.7.

(b) Requirements of petition. The petition shall set forth the interest of the petitioner in the proceeding and show that the participation of the petitioner will assist in the determination of the issues in question, and that the intervention will not unduly delay the proceeding.

(c) Granting of petition. The Commission or Judge may grant a petition for intervention to such an extent and upon such terms as the Commission or the Judge shall determine.

§ 2200.22 Representation of parties and intervenors.

(a) Representation. Any party or intervenor may appear in person, through an attorney, or through another representative who is not an attorney. A representative must file an appearance in accordance with § 2200.23. In the absence of an appearance by a representative, a party or intervenor will be deemed to appear for himself. A corporation or unincorporated association may be represented by an authorized officer or agent.

(b) Affected employees in collective bargaining unit. Where an authorized employee representative (see § 2200.1(g)) elects to participate as a party, affected employees who are members of the collective bargaining unit may not separately elect party status. If the authorized employee representative does not elect party status, affected employees who are members of the collective bargaining unit may elect party status in the same manner as affected employees who are not members of the collective bargaining unit. See paragraph (c) of this section.

(c) Affected employees not in collective bargaining unit. Affected employees who are not members of a collective bargaining unit may elect party status under § 2200.20(a). If more than one employee so elects, the Judge shall provide for them to be treated as one party.

(d) Control of proceeding. A representative of a party or intervenor shall be deemed to control all matters respecting the interest of such party or intervenor in the proceeding.

§ 2200.23 Appearances and withdrawals.

(a) Entry of appearance.

(1) General. A representative of a party or intervenor shall enter an appearance by signing the first document filed on behalf of the

party or intervenor in accordance with paragraph (a)(2) of this section, or thereafter by filing an entry of appearance in accordance with paragraph (a)(3).

(2) Appearance in first document or pleading. If the first document filed on behalf of a party or intervenor is signed by a representative, he shall be recognized as representing that party. No separate entry of appearance by him is necessary, provided the document contains the information required by § 2200.6.

(3) Subsequent appearance. Where a representative has not previously appeared on behalf of a party or intervenor, he shall file an entry of appearance with the Executive Secretary, or Judge if the case has been assigned. The entry of appearance shall be signed by the representative and contain the information required by § 2200.6.

(b) Withdrawal of counsel. Any counsel or representative of record desiring to withdraw his appearance, or any party desiring to withdraw the appearance of counsel or representative of record for him, must file a motion with the Commission or Judge requesting leave therefor, and showing that prior notice of the motion has been given by him to his client or counsel or representative, as the case may be. The motion of counsel to withdraw may, in the discretion of the Commission or Judge, be denied where it is necessary to avoid undue delay or prejudice to the rights of a party or intervenor.

SUBPART C - PLEADINGS AND MOTIONS

§ 2200.30 General rules.

(a) Format. Pleadings and other documents (other than exhibits) shall be typewritten, double spaced, on letter size opaque paper (approximately 8½ inches by 11 inches). All margins shall be approximately 1½ inches. Pleadings and other documents shall be fastened at the upper left corner.

(b) Clarity. Each allegation or response of a pleading or motion shall be simple, concise and direct.

(c) Separation of claims. Each allegation or response shall be made in separate numbered paragraphs. Each paragraph shall be limited as far as practicable to a statement of a single set of circumstances.

(d) Alternative pleading. A party may set forth two or more statements of a claim or defense alternatively or hypothetically. When two or more statements are made in the alternative and one of them would be sufficient if made independently, the pleading is not made insufficient by the insufficiency of one or more of the alternative statements. A party may state as many separate claims or defenses as he has regardless of consistency or the grounds on which based. All statements shall be made subject to the signature requirements of § 2200.32.

(e) Content of motions and miscellaneous pleadings. A motion shall contain a caption complying with § 2200.31, a signature complying with § 2200.32, and a clear and plain statement of the relief that is sought together with the grounds therefor. These requirements also apply to any pleading not governed by more specific requirements in this Subpart.

(f) Burden of persuasion. The rules of pleading established by this Subpart are not determinative in deciding which party bears the burden of persuasion on an issue. By pleading a matter affirmatively, a party does not waive its right to argue that the burden of persuasion on the matter is on another party.

(g) Enforcement of pleading rules. The Commission or the Judge may refuse for filing any pleading or motion that does not comply with the requirements of this Subpart.

§ 2200.31 Caption; Titles of cases.

(a) Notice of contest cases. Cases initiated by a notice of contest shall be titled:

Secretary of Labor,

Complainant,

v.

(Name of Contestant),

Respondent.

(b) Petitions for modification of abatement period. Cases initiated by a petition for modification of the abatement period shall be titled:

(Name of employer),

Petitioner,

v.

Secretary of Labor,

Respondent.

(c) Location of title. The titles listed in paragraphs (a) and (b) of this section shall appear at the left upper portion of the initial page of any pleading or document (other than exhibits) filed.

(d) Docket number. The initial page of any pleading or document (other than exhibits) shall show, at the upper right of the page, opposite the title, the docket number, if known, assigned by the Commission.

§ 2200.32 Signing of pleadings and motions.

Pleadings and motions shall be signed by the filing party or by the party's representative. The signature of a representative constitutes a representation by him that he is authorized to represent the party or parties on whose behalf the pleading is filed. The signature of a representative or party also constitutes a certificate by him that he has read the pleading, motion, or other paper, that to the best of his knowledge, information, and belief, formed after reasonable inquiry, it is well grounded in fact and is warranted by existing law or a good faith argument for the extension, modification, or reversal of existing law, and that it is not interposed for any improper purpose, such as to harass or to cause unnecessary delay or needless increase in the cost of litigation.

§ 2200.33 Notices of contest.

Within 15 working days after receipt of--

(a) notification that the employer intends to contest a citation or proposed penalty under section 10(a) of the Act, 29 U.S.C. § 659(a); or

(b) notification that the employer wishes to contest a notice of a failure to abate or a proposed penalty under section 10(b) of the Act, 29 U.S.C. § 659(b); or

(c) a notice of contest filed by an employee or representative of employees under section 10(c) of the Act, 29 U.S.C. § 659(c),

the Secretary shall notify the Commission of the receipt in writing and shall promptly furnish to the Executive Secretary of the Commission the original of any documents or records filed by the contesting party and copies of all other documents or records relevant to the contest.

§ 2200.34 Employer contests.

(a) Filing deadline for complaint. The Secretary shall file with the Commission a complaint conforming to the requirements of § 2200.35 no later than 30 days after the filing of the Secretary's notice to the Commission pursuant to § 2200.33.

(b) Motion for more definite statement. Upon a showing by the employer that it cannot frame a responsive answer to the allegations of the complaint, the employer may move for a more definite statement of the Secretary's allegations before filing an answer. The motion shall be filed within twenty days after service of the complaint and shall point out the defects complained of and the details desired. The prompt filing of an amended complaint meeting the objections of the moving party may obviate the necessity for the Judge to rule on the motion.

(c) Order to file amended complaint. In response to a motion for more definite statement, the Secretary may be ordered to file an amended complaint supplying such additional information or further particularization of the complaint's allegations as will enable the employer to frame a responsive answer to the allegations of the complaint.

(d) Time to file answer.

(1) Generally. Except as provided in paragraph (d)(2) of this section, the employer shall file with the Commission an answer conforming to the requirements of § 2200.36 within 30 days after service of the complaint.

(2) Exceptions. If a motion to dismiss or a motion for a more definite statement has been filed, the answer shall be filed within 15 days after the motion is· denied. If a motion to amend the complaint or a motion for a more definite statement has been granted, or if an amended complaint has been filed voluntarily under § 2200.35(f) before an answer is served, the answer shall be filed within 30 days after service of the amended complaint.

§ 2200.35 Complaints.

(a) General requirements. The purpose of this section is to insure the early ascertainment of the issues to be litigated. Attachment of the citation or notification of failure to abate to the complaint and incorporation of its terms by reference do not comply with this section. The complaint shall contain the following allegations in separately designated paragraphs:

(1) The employer is engaged in a business affecting commerce within the meaning of § 3(5) of the Act, 29 U.S.C. § 652(5);

(2) The employer's name, principal place of business and type of business conducted as of the date of the alleged violation or failure to abate; and

(3) The time and place of each alleged violation or failure to abate.

(b) Complaints concerning contested alleged violations. Each alleged violation shall be set out in a separate numbered paragraph, which shall have the subparagraphs described below. All allegations that relate to the same alleged violation shall be placed in one paragraph. A paragraph alleging a violation shall in separate subparagraphs state clearly and concisely--

(1) What provision of the Act, standard, regulation, rule or order was violated and the item and citation number in which the alleged violation is set forth;

(2) The factual basis for each allegation necessary to establish that the standard, regulation or rule applied, and what scope or application provision governs its applicability;

(3) The factual basis for each allegation necessary to establish that the cited circumstances, conditions, practices or operations violated the cited provision of the Act, standard, regulation, rule or order;

(4) Where pertinent, the factual basis for the allegation that employees had access to or were exposed to the cited circumstances, conditions, practices or operations;

(5) That the employer knew or could have known with the exercise of reasonable diligence of the cited circumstances, conditions, practices or operations;

(6) Any allegation that the alleged violation was serious or that the employer willfully committed the alleged violation;

(7) Any allegation that the employer repeatedly committed the alleged violation, each prior citation and item number that serves as the basis for the classification, and the date that each became a final order of the Commission;

(8) That the proposed penalty is appropriate, specifying the amount;

(9) That the proposed abatement date is reasonable, specifying the date.

(c) Additional requirements for complaints alleging violations of the General Duty Clause. With respect to each alleged violation of § 5(a)(1) of the Act, 29 U.S.C. § 654(a)(1), the complaint shall also identify the alleged hazard and specify the feasible means by which the employer could have eliminated or materially reduced the alleged hazard.

(d) Additional requirements for complaints alleging violations of general standards. With respect to each alleged violation of any standard or regulation under which the obligation of the employer is contingent upon the existence of a hazard (e.g., 29 C.F.R. §§ 1910.94(d)(7)(iii), 1910.94(d)(9)(i), 1910.132(a) and 1926.28(a)), the complaint shall also identify the particular hazard created by the circumstances, conditions, practices or operations that are the basis for the alleged violation. With respect to each alleged violation of any standard or regulation that does not specify a means of abatement and does not provide a specific performance criterion, the complaint shall also identify a feasible means by which the employer could have abated the allegedly violative condition.

(e) Complaints alleging failure to abate. With respect to each contested allegation of failure to abate a violation, the complaint shall allege with particularity the failure to abate, specifying its date, location and circumstances. The complaint also shall state the penalty proposed and allege that the penalty is "appropriate" under § 17(j) of the Act, 29 U.S.C. § 666(j). The complaint shall also identify the citation and item number in which the violation was previously cited, the date on which this prior citation became a final order of the Commission, and the date by which abatement was required.

(f) Amendment of the citation and complaint. A contested citation, notification of proposed penalty, or notification of failure to abate may be amended once as a matter of course in the complaint before an answer is served if (1) the amended allegation arises out of the same conduct, occurrence or hazard described in the citation; (2) the amendment does not result in incurable harm to the employer in the preparation or presentation of its case; and (3) the complaint clearly identifies the change that is being made in the allegation. All other amendments of the Secretary's allegations, as well as any amendments of the employer's responses, are governed by Federal Rule of Civil Procedure 15.

§ 2200.36 Content of the answer.

(a) Response to the Secretary's allegations. General denials shall not be accepted. The answer shall contain in short and plain terms a response to each allegation of the complaint. It shall specifically admit or deny each allegation or, if the employer is without knowledge of the facts, the answer shall so state. A statement of lack of knowledge has the effect of a denial. A failure to respond to an allegation shall be treated as an admission that the allegation is true. Amendment of the answer to correct a failure to respond may be permitted when the presentation of the merits of the case will be subserved thereby and the party who obtained the admission fails to satisfy the Commission or Judge that the amendment will prejudice him in presenting his case or defense on the merits.

(b) Affirmative defenses.

(1) The employer shall state in its answer in separate numbered paragraphs any matter that may constitute an avoidance or an affirmative defense and, to the extent they are known or with reasonable diligence could have been known, the facts that are the basis of the defense. Such matters include, but are not limited to, the following: creation of a greater hazard by complying with a cited standard; exemption under § 4(b)(1) of the Act, 29 U.S.C. § 653(b)(1); failure to issue a citation with reasonable promptness; infeasibility of compliance; invalidity of the cited standard; preemption of § 5(a)(1) of the Act, 29 U.S.C. § 654(a)(1), by a specific standard; preemption of a standard by a more specifically applicable standard under 29 CFR § 1910.5(c)(1); res judicata; the six-month limitation period in § 9(c) of the Act, 29 U.S.C. § 658(c); or unpreventable employee conduct.

(2) By pleading an avoidance or affirmative defense, the employer does not waive its right to argue that the Secretary has the burden of persuasion concerning the matter. See § 2200.30(f).

§ 2200.37 Petitions for modification of the abatement period.

(a) Grounds for modifying abatement date. An employer may file a petition for modification of abatement date when such employer has made a good faith effort to comply with the abatement requirements of a citation, but such abatement has not been completed because of factors beyond the employer's reasonable control.

(b) Contents of petition. A petition for modification of abatement date shall be in writing and shall include the following information:

(1) All steps taken by the employer, and the dates of such action, in an effort to achieve compliance during the prescribed abatement period.

(2) The specific additional abatement time necessary in order to achieve compliance.

(3) The reasons such additional time is necessary, including the unavailability of professional or technical personnel or of materials and equipment, or because necessary construction or alteration of facilities cannot be completed by the original abatement date.

(4) All available interim steps being taken to safeguard the employees against the cited hazard during the abatement period.

(c) When and where filed; Posting requirement; Responses to petition. A petition for modification of abatement date shall be filed with the Area Director of the United States Department of Labor who issued the citation no later than the close of the next working day following the date on which abatement was originally required. A later-filed petition shall be accompanied by the employer's statement of exceptional circumstances explaining the delay.

(1) A copy of such petition shall be posted in a conspicuous place where all affected employees will have notice thereof or near each location where the violation occurred. The petition shall remain posted for a period of 10 days.

(2) Affected employees or their representatives may file an objection in writing to such petition with the aforesaid Area Director. Failure to file such objection within 10 working days of the date of posting of such petition shall constitute a waiver of any further right to object to said petition.

(3) The Secretary or his duly authorized agent shall have the authority to approve any uncontested petition for modification of abatement date filed pursuant to paragraphs (b) and (c). Such uncontested petitions shall become final orders pursuant to sections 10(a) and (c) of the Act.

(4) The Secretary or his authorized representative shall not exercise his approval power until the expiration of 15 working days from the date the petition was posted pursuant to paragraphs (c) (1) and (2) by the employer.

(d) Contested petitions. Where any petition is objected to by the Secretary or affected employees, such petition shall be processed as follows:

(1) The Secretary shall forward the petition, citation and any objections to the Commission within 10 working days after the expiration of the 15 working day period set out in paragraph (c)(4).

(2) The Commission shall docket and process such petitions as expedited proceedings as provided for in § 2200.103 of this Part.

(3) An employer petitioning for a modification of the abatement period shall have the burden of proving in accordance with the requirements of section 10(c) of the Act, 29 U.S.C. § 659(c), that such employer has made a good faith effort to comply with the abatement requirements of the citation and that abatement has not been completed because of factors beyond the employer's control.

(4) Within 10 working days after the receipt of notice of the docketing by the Commission of any petition for modification of the abatement date, each objecting party shall file a response setting forth

the reasons for opposing the abatement date requested in the petition.

§ 2200.38 Employee contests.

(a) Secretary's statement of reasons. Where an affected employee or authorized employee representative files a notice of contest with respect to the abatement period, the Secretary shall, within 10 days from his receipt of the notice of contest, file a clear and concise statement of the reasons the abatement period prescribed by him is not unreasonable.

(b) Response to Secretary's statement. Not later than 10 days after receipt of the statement referred to in paragraph (a) of this section, the contestant shall file a response.

(c) Expedited proceedings. All contests under this section shall be handled as expedited proceedings as provided for in § 2200.103 of this Part.

§ 2200.39 Statement of position.

At any time prior to the commencement of the hearing before the Judge, any person entitled to appear as a party, or any person who has been granted leave to intervene, may file a statement of position with respect to any or all issues to be heard. The Judge may order the filing of a statement of position.

§ 2200.40 Motions and requests.

(a) How to make. A request for an order shall be made by motion. Motions shall be in writing or, unless the Judge directs otherwise, may be made orally during a hearing on the record and shall be included in the transcript. In exigent circumstances in cases pending before Judges, a motion may be made telephonically if it is reduced to writing and filed within a short time. A motion shall state with particularity the grounds on which it is based and shall set forth the relief or order sought. A motion shall not be included in another document, such as a brief or a petition for discretionary review, but shall be made in a separate document. Unless a motion is made by all parties, the moving party shall state in the motion any opposition or lack of opposition of which he is aware.

(b) When to make. A motion filed in lieu of an answer pursuant to § 2200.34(b) shall be filed no later than twenty days after the service of the complaint. Any other motion shall be made as soon as the grounds therefor are known.

(c) <u>Responses</u>. Any party or intervenor upon whom a motion is served shall have ten days from service of the motion to file a response. A procedural motion may be ruled upon prior to the expiration of the time for response; a party adversely affected by the ruling may within five days of service of the ruling seek reconsideration.

(d) <u>Postponement not automatic upon filing of motion</u>. The filing of a motion, including a motion for a postponement, does not automatically postpone a hearing. See § 2200.62 with respect to motions for post-ponement.

§ 2200.41 <u>Failure to obey rules</u>.

(a) <u>Sanctions</u>. When any party has failed to plead or otherwise proceed as provided by these rules or as required by the Commission or Judge, he may be declared to be in default either: (1) on the initiative of the Commission or Judge, after having been afforded an opportunity to show cause why he should not be declared to be in default; or (2) on the motion of a party. Thereafter, the Commission or Judge, in their discre-tion, may enter a decision against the defaulting party or strike any pleading or document not filed in accordance with these rules.

(b) <u>Motion to set aside sanctions</u>. For reasons deemed sufficient by the Commission or Judge and upon motion expeditiously made, the Commission or Judge may set aside a sanction imposed under paragraph (a) of this rule. See § 2200.90(b)(3).

(c) <u>Discovery sanctions</u>. This section does not apply to sanctions for failure to comply with orders compelling discovery, which are governed by § 2200.52(e).

SUBPART D - PREHEARING PROCEDURES AND DISCOVERY

§ 2200.51 <u>Prehearing conferences and orders</u>.

Prehearing conferences are encouraged. Prehearing conferences may be conducted by a telephone conference call. In addition to the prehearing and scheduling procedures set forth in Fed.R.Civ.P. 16, the Judge may upon his own initiative or on the motion of a party direct the parties to confer among themselves to consider settlement, stipulation of facts or any other matter that may expedite the hearing. Where a prehearing con-ference is not held, the Judge may in his discretion require the parties or their representatives to prepare and submit an agreed prehearing order setting forth any stipulations among the parties, the disputed issues of fact and law, the names and addresses of witnesses expected to be called, the exhibits expected to be introduced by each party in its case-in-chief, the possibility of settlement, the estimated hearing time, and a proposed hearing date or dates.

§ 2200.52 General provisions governing discovery.

(a) General.

(1) Methods and limitations. In conformity with these rules, any party may, without leave of the Commission or Judge, obtain discovery by one or more of the following methods: (i) production of documents or things or permission to enter upon land or other property for inspection and other purposes (§ 2200.53); (ii) requests for admission to the extent provided in § 2200.54; and (iii) interrogatories to the extent provided in § 2200.55. Discovery is not available under these rules through depositions except to the extent provided in § 2200.56. In the absence of a specific provision, procedure shall be in accordance with the Federal Rules of Civil Procedure.

(2) Time for discovery. A party may initiate all forms of discovery in conformity with these Rules at any time after the filing of the first responsive pleading or motion that delays the filing of an answer, such as a motion to dismiss. Discovery shall be initiated early enough to permit completion of discovery no later than seven days prior to the date set for hearing, unless the Judge orders otherwise.

(3) Service of discovery papers. Every paper relating to discovery required to be served on a party shall be served on all parties.

(b) Scope of discovery. The information or response sought through discovery may concern any matter that is not privileged and that is relevant to the subject matter involved in the pending case. It is not ground for objection that the information or response sought will be inadmissible at the hearing, if the information or response appears reasonably calculated to lead to discovery of admissible evidence, regardless of which party has the burden of proof.

(c) Limitations. The frequency or extent of the discovery methods provided by these rules may be limited by the Commission or Judge if it is determined that: (1) the discovery sought is unreasonably cumulative or duplicative, or is obtainable from some other source that is more convenient, less burdensome, or less expensive; (2) the party seeking discovery has had ample opportunity to obtain the information sought by discovery in the action; or (3) the discovery is unduly burdensome or expensive, taking into account the needs of the case, limitations on the parties' resources, and the importance of the issues in litigation.

(d) <u>Protective orders</u>.   In connection with any discovery procedure, the Commission or Judge may make any order which justice requires to protect a party or person from annoyance, embarrassment, oppression, or undue burden or expense, including one or more of the following:

(1) That the discovery not be had;

(2) That discovery may be had only on specified terms and conditions, including a designation of the time and place, or that the scope of discovery be limited to certain matters;

(3) That discovery be conducted with no one present except persons designated by the Commission or Judge; and

(4) That confidential information not be disclosed or that it be disclosed only in a designated way. See also § 2200.11 on trade secrets.

(e) <u>Failure to cooperate; Sanctions</u>. A party may apply for an order compelling discovery when another party refuses or obstructs discovery. For purposes of this paragraph, an evasive or incomplete answer is to be treated as a failure to answer. If a Judge enters an order compelling discovery and there is a failure to comply with that order, the Judge may make such orders with regard to the failure as are just. The orders may issue upon the initiative of a Judge, after affording an opportunity to show cause why the order should not be entered, or upon the motion of a party. The orders may include any sanction stated in Fed.R.Civ.P. 37, including the following:

(1) An order that designated facts shall be taken to be established for purposes of the case in accordance with the claim of the party obtaining that order;

(2) An order refusing to permit the disobedient party to support or to oppose designated claims or defenses, or prohibiting it from introducing designated matters in evidence;

(3) An order striking out pleadings or parts thereof, or staying further proceedings until the order is obeyed; and

(4) An order dismissing the action or proceeding or any part thereof, or rendering a judgment by default against the disobedient party.

(f) <u>Unreasonable delays</u>. None of the discovery procedures set forth in these rules shall be used in a manner or at a time which shall delay or impede the progress of the case toward hearing status or the hearing of the case on the date for which it is scheduled, unless, in the interests of justice, the Judge shall order otherwise. Unreasonable delays in utilizing discovery procedures may result in termination of the party's right to conduct discovery.

§ 2200.53 Production of documents and things.

(a) Scope. At any time after the filing of the first responsive pleading or motion that delays the filing of an answer, such as a motion to dismiss, any party may serve on any other party a request to:

(1) Produce and permit the party making the request, or a person acting on his or her behalf, to inspect and copy any designated documents, or to inspect and copy, test, or sample any tangible things which are in the possession, custody, or control of the party upon whom the request is served;

(2) Permit entry upon designated land or other property in the possession or control of the party upon whom the request is served for the purpose of inspection and measuring, surveying, photographing, testing or sampling the property or any designated object or operation thereon.

(b) Procedure. The request shall set forth the items to be inspected, either by individual item or by category, and describe each item and category with reasonable particularity. It shall specify a reasonable time, place and manner of making the inspection and performing the related acts. The party upon whom the request is served shall serve a written response within 30 days after service of the request. The Commission or Judge may allow a shorter or longer time. The response shall state, with respect to each item or category, that inspection and related activities will be permitted as requested, unless the request is objected to in whole or in part, in which event the reasons for objection shall be stated. If objection is made to part of an item or category, that part shall be specified. To obtain a ruling on an objection by the responding party, the requesting party shall file a motion with the Judge and shall annex thereto his request, together with the response and objections, if any.

§ 2200.54 Requests for admissions.

(a) Scope. At any time after the filing of the first responsive pleading or motion that delays the filing of an answer, such as a motion to dismiss, any party may serve upon any other party written requests for admissions, for purposes of the pending action only, of the genuineness and authenticity of any document described in or attached to the requests, or of the truth of any specified matter of fact. Each matter of which an admission is requested shall be separately set forth. The number of requested admissions shall not exceed 25, including subparts, without an order of the Commission or Judge. The party seeking to serve more than 25 requested admissions, including subparts, shall have the burden of persuasion to establish that the complexity of the case or the number of citation items necessitates a greater number of requested admissions. The original of the request shall be filed with the Judge.

(b) <u>Response to requests</u>. Each matter is deemed admitted unless, within 30 days after service of the requests or within such shorter or longer time as the Commission or Judge may allow, the party to whom the requests are directed serves upon the requesting party (1) a written answer specifically admitting or denying the matter involved in whole or in part, or asserting that it cannot be truthfully admitted or denied and setting forth in detail the reasons why this is so, or (2) an objection, stating in detail the reasons therefor. The response shall be made under oath or affirmation and signed by the party or his representative. The original shall be filed with the Judge.

(c) <u>Effect of admission</u>. Any matter admitted under this section is conclusively established unless the Judge or Commission on motion permits withdrawal or modification of the admission. Withdrawal or modification may be permitted when the presentation of the merits of the case will be subserved thereby, and the party who obtained the admission fails to satisfy the Commission or Judge that the withdrawal or modification will prejudice him in presenting his case or defense on the merits.

§ 2200.55 <u>Interrogatories</u>.

(a) <u>General</u>. At any time after the filing of the first responsive pleading or motion that delays the filing of an answer, such as a motion to dismiss, any party may serve interrogatories upon any other party. The number of interrogatories shall not exceed 25 questions, including subparts, without an order of the Commission or Judge. The party seeking to serve more than 25 questions, including subparts, shall have the burden of persuasion to establish that the complexity of the case or the number of citation items necessitates a greater number of interrogatories.

(b) <u>Answers</u>. All answers shall be made in good faith and as completely as the answering party's information will permit. The answering party is required to make reasonable inquiry and ascertain readily obtainable information. An answering party may not give lack of information or knowledge as an answer or as a reason for failure to answer, unless he states that he has made reasonable inquiry and that information known or readily obtainable by him is insufficient to enable him to answer the substance of the interrogatory.

(c) <u>Procedure</u>. Each interrogatory shall be answered separately and fully under oath or affirmation. If the interrogatory is objected to, the objection shall be stated in lieu of the answer. The answers are to be signed by the person making them and the objections shall be signed by the party or his counsel. The party on whom the interrogatories have been served shall serve a copy of his answers or objections upon the propounding party within 30 days after the service of the interrogatories. The Judge may allow a shorter or longer time. The burden shall be on the party submitting the interrogatories to move for an order with respect to any objection or other failure to answer an interrogatory.

§ 2200.56 Depositions.

(a) General. Depositions of parties, intervenors, or witnesses shall be allowed only by agreement of all the parties, or on order of the Commission or Judge following the filing of a motion of a party stating good and just reasons. All depositions shall be before an officer authorized to administer oaths and affirmations at the place of examination. The deposition shall be taken in accordance with the Federal Rules of Civil Procedure, particularly Fed.R.Civ.P. 30.

(b) When to file. A motion to take depositions may be filed after the filing of the first responsive pleading or motion that delays the filing of an answer, such as a motion to dismiss.

(c) Notice of taking. Any depositions allowed by the Commission or Judge may be taken after ten days' written notice to the other party or parties. The ten-day notice requirement may be waived by the parties.

(d) Expenses. Expenses for a court reporter and the preparing and serving of depositions shall be borne by the party at whose instance the deposition is taken.

(e) Use of depositions. Depositions taken under this rule may be used for discovery, to contradict or impeach the testimony of a deponent as a witness, or for any other purpose permitted by the Federal Rules of Evidence and the Federal Rules of Civil Procedure, particularly Fed.R.Civ.P. 32.

§ 2200.57 Issuance of subpenas; Petitions to revoke or modify subpenas; Right to inspect or copy data.

(a) Issuance of subpenas. On behalf of the Commission or any member thereof, the Judge shall, on the application of any party, issue to the applying party subpenas requiring the attendance and testimony of witnesses and the production of any evidence, including relevant books, records, correspondence or documents, in his possession or under his control. The party to whom the subpena is issued shall be responsible for its service. Applications for subpenas, if filed prior to the assignment of the case to a Judge, shall be filed with the Executive Secretary at 1825 K Street, N.W., Washington, D.C. 20006. After the case has been assigned to a Judge, applications shall be filed with the Judge. Applications for subpenas shall be made ex parte. The subpena shall show on its face the name and address of the party at whose request the subpena was issued.

(b) Revocation or modification of subpenas. Any person served with a subpena, whether ad testificandum or duces tecum, shall, within 5 days after the date of service of the subpena upon him, move in writing to revoke or modify the subpena if he does not intend to comply. All motions to revoke or modify shall be served on the party at whose request the subpena was issued. The Judge or the Commission shall revoke or modify the subpena if in its opinion the evidence whose production is required does not relate to any matter under investigation or in question in the proceedings or the subpena does not describe with sufficient particularity the evidence whose production is required, or if for any other reason sufficient in law the subpena is otherwise invalid. The Judge or the Commission shall make a simple statement of procedural or other grounds for the ruling on the motion to revoke or modify. The motion to revoke or modify, any answer filed thereto, and any ruling thereon shall become a part of the record.

(c) Rights of persons compelled to submit data. Persons compelled to submit data or evidence at a public proceeding are entitled to retain or, on payment of lawfully prescribed costs, to procure copies of transcripts of the data or evidence submitted by them.

(d) Failure to comply with subpena. Upon the failure of any person to comply with a subpena issued upon the request of a party, the Commission by its counsel shall initiate proceedings in the appropriate district court for the enforcement thereof, if in its judgment the enforcement of such subpena would be consistent with law and with policies of the Act. Neither the Commission nor its counsel shall be deemed thereby to have assumed responsibility for the effective prosecution of the same before the court.

SUBPART E - HEARINGS

§ 2200.60 Notice of hearing; Location.

Except by agreement of the parties, or in an expedited proceeding under § 2200.103, notice of the time, place, and nature of the first setting of a hearing shall be given to the parties and intervenors at least thirty days in advance of the hearing. If a hearing has been previously postponed or if exigent circumstances are present, at least ten days notice shall be given. The Judge will designate a place and time of hearing that involves as little inconvenience and expense to the parties as is practicable.

§ 2200.61 Submission without hearing.

A case may be fully stipulated by the parties and submitted to the Commission or Judge for a decision at any time. The stipulation of facts shall be in writing and signed by the parties or their representatives. The submission of a case under this rule does not alter the burden of proof, the requirements otherwise applicable with respect to adducing proof, or the effect of failure of proof. Motions for summary judgment are covered by Fed.R.Civ.P. 56.

§ 2200.62 Postponement of hearing.

(a) Motion to postpone. A hearing may be postponed by the Judge on his own initiative or for good cause shown upon the motion of a party. A motion for postponement shall state the position of the other parties, either by a joint motion or by a representation of the moving party. The filing of a motion for postponement does not automatically postpone a hearing.

(b) Grounds for postponement. A motion for postponement grounded on conflicting engagements of counsel or employment of new counsel shall be filed promptly after notice is given of the hearing, or as soon as the conflict is learned of or the engagement occurs.

(c) When motion must be received. A motion to postpone a hearing must be received at least seven days prior to the hearing. A motion for postponement received less than seven days prior to the hearing will generally be denied unless good cause is shown for late filing.

(d) Postponement in excess of 60 days. No postponement in excess of 60 days shall be granted without the concurrence of the Chief Administrative Law Judge. The original of any motion seeking a postponement in excess of 60 days shall be filed with the Judge and a copy sent to the Chief Administrative Law Judge.

§ 2200.63 Stay of proceedings.

(a) Motion for stay. Stays are not favored. A party seeking a stay of a case assigned to a Judge shall file a motion for stay with the Judge and send a copy to the Chief Administrative Law Judge. A motion for a stay shall state the position of the other parties, either by a joint motion or by the representation of the moving party. The motion shall set forth the reasons a stay is sought and the length of the stay requested.

(b) Ruling on motion to stay. The Judge, with the concurrence of the Chief Administrative Law Judge, may grant any motion for stay for the period requested or for such period as is deemed appropriate.

(c) Periodic reports required. The parties in a stayed proceeding shall be required to submit periodic reports on such terms and conditions as the Judge may direct.

§ 2200.64 Failure to appear.

(a) Attendance at hearing. The failure of a party to appear at a hearing may result in a decision against that party.

(b) Requests for reinstatement. Requests for reinstatement must be made, in the absence of extraordinary circumstances, within five days after the scheduled hearing date. See § 2200.90(b)(3).

(c) Rescheduling hearing. The Commission or the Judge, upon a showing of good cause, may excuse such failure to appear. In such event, the hearing will be rescheduled.

§ 2200.65 Payment of witness fees and mileage; Fees of persons taking depositions.

Witnesses summoned before the Commission or the Judge shall be paid the same fees and mileage that are paid witnesses in the courts of the United States, and witnesses whose depositions are taken and the persons taking the same shall severally be entitled to the same fees as are paid for like services in the courts of the United States. Witness fees and mileage shall be paid by the party at whose instance the witness appears, and the person taking a deposition shall be paid by the party at whose instance the deposition is taken.

§ 2200.66 Transcript of testimony.

(a) Hearings. Hearings shall be transcribed verbatim. A copy of the transcript of testimony taken at the hearing, duly certified by the reporter, shall be filed with the Judge before whom the matter was heard.

(b) Payment for transcript. The Commission shall bear all expenses for court reporters' fees and for copies of the hearing transcript received by it. Each party is responsible for securing and paying for its copy of the transcript.

(c) Correction of errors. Error in the transcript of the hearing may be corrected by the Judge on his own motion, on joint motion by the parties, or on motion by any party. The motion shall state the error in the transcript and the correction to be made. Corrections will be made by hand with pen and ink and by the appending of an errata sheet.

§ 2200.67 Duties and powers of Judges.

It shall be the duty of the Judge to conduct a fair and impartial hearing, to assure that the facts are fully elicited, to adjudicate all issues and avoid delay. The Judge shall have authority with respect to cases assigned to him, between the time he is designated and the time he issues his decision, subject to the rules and regulations of the Commission, to:

(a) Administer oaths and affirmations;

(b) Issue authorized subpenas;

(c) Rule upon petitions to revoke subpenas;

(d) Rule upon offers of proof and receive relevant evidence;

(e) Take or cause depositions to be taken whenever the needs of justice would be served;

(f) Regulate the course of the hearing and, if appropriate or necessary, exclude persons or counsel from the hearing for contemptuous conduct and strike all related testimony of witnesses refusing to answer any proper questions;

(g) Hold conferences for the settlement or simplification of the issues;

(h) Dispose of procedural requests or similar matters, including motions referred to the Judge by the Commission and motions to amend pleadings; also to dismiss complaints or portions thereof, and to order hearings reopened or, upon motion, consolidated prior to issuance of his decision;

(i) Make decisions in conformity with § 557 of title 5, United States Code;

(j) Call and examine witnesses and to introduce into the record documentary or other evidence;

(k) Request the parties to state their respective positions concerning any issue in the case or theory in support thereof;

(l) Adjourn the hearing as the needs of justice and good administration require;

(m) Take any other action necessary under the foregoing and authorized by the published rules and regulations of the Commission.

§ 2200.68 Disqualification of the Judge.

(a) Discretionary withdrawal. A Judge may withdraw from a proceeding whenever he deems himself disqualified.

(b) <u>Request for withdrawal</u>. Any party may request the Judge, at any time following his designation and before the filing of his decision, to withdraw on ground of personal bias or disqualification, by filing with him promptly upon the discovery of the alleged facts an affidavit setting forth in detail the matters alleged to constitute grounds for disqualification.

(c) <u>Granting request</u>. If, in the opinion of the Judge, the affidavit referred to in paragraph (b) of this section is filed with due diligence and is sufficient on its face, the Judge shall forthwith disqualify himself and withdraw from the proceeding.

(d) <u>Denial of request</u>. If the Judge does not disqualify himself and withdraw from the proceedings, he shall so rule upon the record, stating the grounds for his ruling and shall proceed with the hearing, or, if the hearing has closed, he shall proceed with the issuance of his decision, and the provisions of § 2200.90 shall thereupon apply.

§ 2200.69 <u>Examination of witnesses</u>.

Witnesses shall be examined orally under oath or affirmation.

Opposing parties have the right to cross-examine any witness whose testimony is introduced by an adverse party. All parties shall have the right to cross-examine any witness called by the Judge pursuant to § 2200.67(j).

§ 2200.70 <u>Exhibits</u>.

(a) <u>Marking exhibits</u>. All exhibits offered in evidence by a party shall be marked for identification before or during the hearing. Exhibits shall be marked with the case docket number, with a designation identifying the party or intervenor offering the exhibit, and numbered consecutively.

(b) <u>Removal or substitution of exhibits in evidence</u>. Unless the Judge finds it impractical, a copy of each exhibit shall be given to the other parties and intervenors. A party may remove an exhibit from the official record during the hearing or at the conclusion of the hearing only upon permission of the Judge. The Judge, in his discretion, may permit the substitution of a duplicate for any original document offered into evidence.

(c) <u>Reasons for denial of admitting exhibit</u>. A Judge may, in his discretion, deny the admission of any exhibit because of its excessive size, weight, or other characteristic that prohibits its convenient transportation and storage. A party may offer into evidence photographs, models or other representations of any such exhibit.

(d) <u>Rejected exhibits</u>. All exhibits offered but denied admission into evidence, except exhibits referred to in paragraph (c), shall be placed in a separate file designated for rejected exhibits.

(e) <u>Return of physical exhibits</u>. A party may on motion request the return of a physical exhibit within 30 days after expiration of the time for filing a petition for review of a Commission final order in a United States Court of Appeals under section 11 of the Act, 29 U.S.C. § 660, or within 30 days after completion of any proceedings initiated thereunder. The motion shall be addressed to the Executive Secretary and provide supporting reasons. The exhibit shall be returned if the Executive Secretary determines that it is no longer necessary for use in any Commission proceeding.

(f) <u>Request for custody of physical exhibit</u>. Any person may on motion to the Executive Secretary request custody of a physical exhibit for use in any court or tribunal. The motion shall state the reasons for the request and the duration of custody requested. If the exhibit has been admitted in a pending Commission case, the motion shall be served on all parties to the proceeding. Any person granted custody of an exhibit shall inform the Executive Secretary of the status every six months of his continuing need for the exhibit and return the exhibit after completion of the proceeding.

(g) <u>Disposal of physical exhibit</u>. Any physical exhibit may be disposed of by the Commission's Executive Secretary at any time more than 30 days after expiration of the time for filing a petition for review of a Commission final order in a United States Court of Appeals under section 11 of the Act, 29 U.S.C. § 660, or 30 days after completion of any proceedings initiated thereunder.

§ 2200.71 <u>Rules of evidence</u>.

The Federal Rules of Evidence are applicable.

§ 2200.72 <u>Objections</u>.

(a) <u>Statement of objection</u>. Any objection with respect to the conduct of the hearing, including any objection to the introduction of evidence or a ruling by the Judge, may be stated orally or in writing, accompanied by a short statement of the grounds for the objection, and shall be included in the record. No such objection shall be deemed waived by further participation in the hearing.

(b) <u>Offer of proof</u>. Whenever evidence is excluded from the record, the party offering such evidence may make an offer of proof, which shall be included in the record of the proceeding.

§ 2200.73 Interlocutory review.

(a) General. Interlocutory review of a Judge's ruling is discretionary with the Commission. A petition for interlocutory review may be granted only where the petition asserts and the Commission finds: (1) that the review involves an important question of law or policy about which there is substantial ground for difference of opinion and that immediate review of the ruling may materially expedite the final disposition of the proceedings; or (2) that the ruling will result in a disclosure, before the Commission may review the Judge's report, of information that is alleged to be privileged.

(b) Petition for interlocutory review. Within five days following the receipt of a Judge's ruling from which review is sought, a party may file a petition for interlocutory review with the Commission. Responses to the petition, if any, shall be filed within five days following service of the petition. A copy of the petition and responses shall be filed with the Judge. The petition is denied unless granted within 30 days of the date of receipt by the Commission's Executive Secretary.

(c) Denial without prejudice. The Commission's action in denying a petition for interlocutory review shall not preclude a party from raising an objection to the Judge's interlocutory ruling in a petition for discretionary review.

(d) Stay.

(1) Trade secret matters. The filing of a petition for interlocutory review of a Judge's ruling concerning an alleged trade secret shall stay the effect of the ruling until the Commission denies the petition or rules on the merits.

(2) Other cases. In all other cases, the filing or granting of a petition for interlocutory review shall not stay a proceeding or the effect of a ruling unless otherwise ordered.

(e) Judge's comments. The Judge may be requested to provide the Commission with his written views on whether the petition is meritorious. The Judge shall serve copies of these comments on all parties when he files them with the Commission.

(f) Briefs. Should the Commission desire briefs on the issues raised by an interlocutory review, it shall give notice to the parties. See § 2200.93--Briefs before the Commission.

§ 2200.74 Filing of briefs and proposed findings with the Judge; Oral argument at the hearing.

(a) General. A party is entitled to a reasonable period at the close of the hearing for oral argument, which shall be included in the stenographic report of the hearing. Any party shall be entitled, upon request made before the close of hearing, to file a brief, proposed findings of fact and conclusions of law, or both, with the Judge. In lieu of briefs, the Judge may permit or direct the parties to file memoranda or statements of authorities.

(b) Time. Briefs shall be filed simultaneously on a date established by the Judge. A motion for extension of time for filing any brief shall be made at least three days prior to the due date and shall recite that the moving party has advised the other parties of the motion. Reply briefs shall not be allowed except by order of the Judge.

(c) Untimely briefs. Untimely briefs will not be accepted unless accompanied by a motion setting forth good cause for the delay.

SUBPART F - POSTHEARING PROCEDURES

§ 2200.90 Decisions of Judges.

(a) Contents. The Judge shall prepare a decision that constitutes his final disposition of the proceedings. The decision shall be in writing and shall include findings of fact, conclusions of law, and the reasons or bases for them, on all the material issues of fact, law or discretion presented on the record. The decision shall include an order affirming, modifying or vacating each contested citation item and each proposed penalty, or directing other appropriate relief. A decision finally disposing of a petition for modification of the abatement period shall contain an order affirming or modifying the abatement period.

(b) The Judge's report.

(1) Mailing to parties. The Judge shall mail or otherwise transmit a copy of his decision to each party.

(2) Docketing of Judge's report by Executive Secretary. On the twenty-first day after the transmittal of his decision to the parties, the Judge shall file his report with the Executive Secretary for docketing. The report shall consist of the record, including the Judge's decision, any petitions for discretionary review and statements in opposition to such petitions. Promptly upon receipt of the Judge's report, the Executive Secretary shall docket the report and notify all parties of the docketing date. The date of docketing of the Judge's report is the date that the Judge's report is made for purposes of section 12(j) of the Act, 29 U.S.C. § 661(j).

(3) Correction of errors; Relief from default. Until the Judge's report has been directed for review or, in the absence of a direction for review, until the decision has become a final order, the Judge may correct clerical errors and errors arising through oversight or inadvertence

in decisions, orders or other parts of the record. If a Judge's report
has been directed for review, the decision may be corrected during the
pendency of review with leave of the Commission. Until the Judge's
report has been docketed by the Executive Secretary, the Judge may
relieve a party of default or grant reinstatement under §§ 2200.41(b),
2200.52(e) or 2200.64(b).

(c) Filing documents after the docketing date. Except for papers
filed under paragraph (b)(3) of this section, which shall be filed
with the Judge, on or after the date of the docketing of the Judge's
report all documents shall be filed with the Executive Secretary.

(d) Judge's decision final unless review directed. If no
Commissioner directs review of a report on or before the thirtieth day
following the date of docketing of the Judge's report, the decision of
the Judge shall become a final order of the Commission.

§ 2200.91 Discretionary review; Petitions for discretionary review;
            Statements in opposition to petitions.

(a) Review discretionary. Review by the Commission is not a right.
A Commissioner may, as a matter of discretion, direct review on his own
motion or on the petition of a party.

(b) Petitions for discretionary review. A party adversely affected
or aggrieved by the decision of the Judge may seek review by the
Commission by filing a petition for discretionary review. Discretionary
review by the Commission may be sought by filing with the Judge a petition for discretionary review within the twenty-day period provided by
§ 2200.90(b). Review by the Commission may also be sought by filing
directly with the Executive Secretary a petition for discretionary
review. A petition filed directly with the Executive Secretary shall be
filed within 20 days after the date of docketing of the Judge's report.
The earlier a petition is filed, the more consideration it can be given.
A petition for discretionary review may be conditional, and may state
that review is sought only if a Commissioner were to direct review on the
petition of an opposing party.

(c) Cross-petitions for discretionary review. Where a petition for
discretionary review has been filed by one party, any other party adversely affected or aggrieved by the decision of the Judge may seek
review by the Commission by filing a cross-petition for discretionary
review. The cross-petition may be conditional. See paragraph (b) of
this section. A cross-petition shall be filed with the Judge during the
twenty days provided by § 2200.90(b) or directly with the Executive
Secretary within 27 days after the date of docketing of the Judge's
report. The earlier a cross-petition is filed, the more consideration it can be given.

(d) <u>Contents of the petition</u>. No particular form is required for a petition for discretionary review. A petition should state why review should be directed, including: (whether the Judge's decision raises an important question of law, policy or discretion; whether review by the Commission will resolve a question about which the Commission's Judges have rendered differing opinions; whether the Judge's decision is contrary to law or Commission precedent; whether a finding of material fact is not supported by a preponderance of the evidence; whether a prejudicial error of procedure or an abuse of discretion was committed.) A petition should concisely state the portions of the decision for which review is sought and should refer to the citations and citation items (for example, citation 3, item 4a) for which review is sought. A petition shall not incorporate by reference a brief or legal memorandum. Brevity and the inclusion of precise references to the record and legal authorities will facilitate prompt review of the petition.

(e) <u>When filing effective</u>. A petition for discretionary review is filed when received. If a petition has been filed with the Judge, another petition need not be filed with the Commission.

(f) <u>Failure to file</u>. The failure of a party adversely affected or aggrieved by the Judge's decision to file a petition for discretionary review may foreclose court review of the objections to the Judge's decision. <u>See</u> Keystone Roofing Co. v. Dunlop, 539 F.2d 960 (3d Cir. 1976).

(g) <u>Statements in opposition to petition</u>. Statements in opposition to petitions for discretionary review may be filed in the manner specified in this section for the filing of petitions for discretionary review. Statements in opposition shall concisely state why the Judge's decision should not be reviewed with respect to each portion of the petition to which it is addressed.

(h) <u>Number of copies</u>. An original and three copies of a petition or of a statement in opposition to a petition shall be filed.

§ 2200.92 <u>Review by the Commission</u>.

(a) <u>Jurisdiction of the Commission; Issues on review</u>. Unless the Commission orders otherwise, a direction for review establishes jurisdiction in the Commission to review the entire case. The issues to be decided on review are within the discretion of the Commission but ordinarily will be those stated in the direction for review, those raised in the petitions for discretionary review, or those stated in any later order.

(b) <u>Review on a Commissioner's motion; Issues on review</u>. At any time within 30 days after the docketing date of the Judge's report, a Commissioner may, on his own motion, direct that a Judge's decision be reviewed. In the absence of a petition for discretionary review, a Commissioner will normally not direct review unless the case raises novel questions of law or policy or questions involving conflict in Administrative Law Judges' decisions. When a Commissioner directs review on his own motion, the issues ordinarily will be those specified in the direction for review or any later order.

(c) <u>Issues not raised before Judge</u>. The Commission will ordinarily not review issues that the Judge did not have the opportunity to pass upon. In exercising discretion to review issues that the Judge did not have the opportunity to pass upon, the Commission may consider such factors as whether there was good cause for not raising the issue before the Judge, the degree to which the issue is factual, the degree to which proceedings will be disrupted or delayed by raising the issue on review, whether the ability of an adverse party to press a claim or defense would be impaired, and whether considering the new issue would avoid injustice or ensure that judgment will be rendered in accordance with the law and facts.

§ 2200.93 <u>Briefs before the Commission</u>.

(a) <u>Requests for briefs</u>. The Commission ordinarily will request the parties to file briefs on issues before the Commission. After briefs are requested, a party may, instead of filing a brief, file a letter setting forth its arguments, a letter stating that it will rely on its petition for discretionary review or previous brief, or a letter stating that it wishes the case decided without its brief. The provisions of this section apply to the filing of briefs and letters filed in lieu of briefs.

(b) <u>Filing briefs</u>. Unless the briefing notice states otherwise:

(1) <u>Time for filing briefs</u>. The party required to file the first brief shall do so within 40 days after the date of the briefing notice. All other parties shall file their briefs within 30 days after the first brief is served. Any reply brief permitted by these rules or by order shall be filed within 15 days after the second brief is served.

(2) <u>Sequence of filing</u>. (i) If one petition for discretionary or interlocutory review has been filed, the petitioning party shall file the first brief.

(ii) If more than one petition has been filed but only one was granted, the party whose petition was granted shall file the first brief.

(iii) If more than one petition has been filed, and more than one has been granted or none has been granted, the Secretary shall file the first brief.

(iv) If no petition has been filed, the Secretary shall file the first brief.

(3) Reply briefs. The party who filed the first brief may file a reply brief. Additional briefs are otherwise not allowed except by leave of the Commission.

(c) Motion for extension of time for filing brief. An extension of time to file a brief will ordinarily not be granted except for good cause shown. A motion for extension of time to file a brief shall be filed within the time limit prescribed in paragraph (b) of this section, shall comply with § 2200.40, and shall include the following information: when the brief is due; the number and duration of extensions of time that have been granted to each party; the length of extension being requested; the specific reason for the extension being requested; and an assurance that the brief will be filed within the time extension requested.

(d) Consequences of failure to timely file brief. The Commission may decline to accept a brief that is not timely filed. If a petitioning party fails to respond to a briefing notice or expresses no interest in review, the Commission may vacate the direction for review, or it may decide the case without that party's brief. If the non-petitioning party fails to respond to a briefing notice or expresses no interest in review, the Commission may decide the case without that party's brief. If a case was directed for review upon a Commissioner's own motion, and any party fails to respond to the briefing notice, the Commission may either vacate the direction for review or decide the case without briefs.

(e) Length of brief. Except by permission of the Commission, a main brief, including briefs and legal memorandums it incorporates by reference, shall contain no more than 35 pages of text. A reply brief, including briefs and legal memorandums it incorporates by reference, shall contain no more than 20 pages of text.

(f) Table of contents. A brief in excess of 15 pages shall include a table of contents.

(g) Failure to meet requirements. The Commission may return briefs that do not meet the requirements of paragraphs (e) and (f) of this section.

(h) Number of copies. The original and four copies of a brief shall be filed. See § 2200.8(c)(2).

§ 2200.94 Stay of final order.

(a) Who may file. Any party aggrieved by a final order of the Commission may, while the matter is within the jurisdiction of the Commission, file a motion for a stay.

(b) Contents of motion. Such motion shall set forth the reasons a stay is sought and the length of the stay requested.

(c) Ruling on motion. The Commission may order such stay for the period requested or for such longer or shorter period as it deems appropriate.

§ 2200.95   Oral argument before the Commission.

(a) General policy. Oral argument before the Commission ordinarily will not be allowed.

(b) Notice of oral argument. In the event the Commission desires to hear oral argument with respect to any matter, it will advise all parties to the proceeding of the date, hour, place, time allotted, and scope of such argument at least 10 days prior to the date set.

SUBPART G - MISCELLANEOUS PROVISIONS

§ 2200.100   Settlement.

(a) Policy. Settlement is permitted and encouraged by the Commission at any stage of the proceedings.

(b) Requirements. The Commission does not require that the parties include any particular language in a settlement agreement, but does require that the agreement specify the terms of settlement for each con- tested item, specify any contested item or issue that remains to be decided (if any remain), and state whether any affected employees who have elected party status have raised an objection to the reasonableness of any abatement time. Unless the settlement agreement states otherwise, the withdrawal of a notice of contest, citation, notification of proposed penalty, or petition for modification of abatement period will be with prejudice.

(c) Filing; Service and notice. A settlement submitted for approval after the Judge's report has been directed for review shall be filed with the Executive Secretary. When a settlement agreement is filed with the Judge or the Executive Secretary, proof of service shall be filed with the settlement agreement, showing service upon all parties and authorized employee representatives in the manner prescribed by § 2200.7(c) and the posting of notice to non-party affected employees in the manner pre- scribed by § 2200.7(g). The parties shall also file a final consent order for adoption by the Judge. If the time has not expired under these rules for electing party status, or if party status has been elected, an order terminating the litigation before the Commission because of the settlement shall not be issued until at least ten days after service to consider any affected employee's or authorized employee representative's objection to the reasonableness of any abatement time. The affected employee or authorized employee representative shall file any such objec- tion within this time. If such objection is filed or stated in the

settlement agreement, the Commission or the Judge shall provide an opportunity for the affected employees or authorized employee representative
to be heard and present evidence on the objection, which shall be limited
to the reasonableness of the abatement time.

(d) Form of settlement document. It is preferred that settlement
documents be typewritten in conformance with § 2200.30(a). However, a
settlement document that is hand-written or printed in ink and is legible
shall be acceptable for filing.

§ 2200.101 Settlement Judge procedure.

(a) Appointment of Settlement Judge.

(1) This section applies only to notices of contests by employers and to applications for fees under the Equal Access to Justice
Act and 29 CFR Part 2204.

(2) Upon motion of any party following the filing of the
pleadings (or notice of simplified proceedings), or otherwise with the
consent of the parties at any time in the proceedings, the Chief
Administrative Law Judge or the Chairman may assign a case to a
Settlement Judge for processing under this section whenever it is determined that there is a reasonable prospect of substantial settlement with
the assistance of mediation by a Settlement Judge. In the event either
the Secretary or the employer objects to the use of a Settlement Judge
procedure, such procedure shall not be imposed.

(3) The settlement negotiations under this section shall be for
a period not to exceed 45 days.

(b) Powers and duties of Settlement Judges.

(1) The Judge shall confer with the parties on subjects and
issues of whole or partial settlement of the case.

(2) The Judge may allow or suspend discovery during the time of
assignment.

(3) The Judge may suggest privately to each attorney or other
representative of a party what concessions his or her client should consider, and assess privately with each attorney or other representative
the reasonableness of the party's case or settlement position.

(4) The Judge shall seek resolution of as many of the issues in
the case as is feasible.

(c) Settlement conference and other communication.

(1) Types of conferences. In general it is expected that the
Settlement Judge shall communicate with the parties by a conference

telephone call. The Settlement Judge, however, may schedule a personal conference with the parties under one or more of the following circumstances:

(i) it is possible for the Settlement Judge to schedule in one day three or more cases for conference at or near the same location;

(ii) the offices of the attorneys or other representatives of the parties, as well as that of the Settlement Judge, are located in the same metropolitan area;

(iii) a conference may be scheduled in a place and on a day that the Judge is scheduled to preside in other proceedings under this Part;

(iv) any other suitable circumstances in which, with the concurrence of the Chief Administrative Law Judge, the Settlement Judge determines that a personal meeting is necessary for a resolution of substantial issues in a case and the holding of a conference represents a prudent use of resources.

(2) Participation in conference. The Settlement Judge may recommend that the attorney or other representative who is expected to try the case for each party be present, and, without regard to the scope of the attorney's or other representative's powers, may also recommend that the parties, or agents having full settlement authority, be present. The parties, their representatives, and attorneys are required to be completely candid with the Settlement Judge so that he may properly guide settlement discussions. The failure to be present at a settlement conference or the refusal to cooperate fully within the spirit of this rule may result in the termination of the settlement proceeding under this section. The Settlement Judge may make such other and additional requirements of the parties and persons having an interest in the outcome as to him shall seem proper in order to expedite an amicable resolution of the case. No evidence of statements or conduct in proceedings under this section will be admissible in any subsequent hearing, except by stipulation of the parties. Documents disclosed in the settlement process may not be used in litigation unless obtained through appropriate discovery or subpena.

(d) Report of Settlement Judge.

(1) With the consent of the parties, the Settlement Judge may request from the Chief Administrative Law Judge an enlargement of the time of the settlement period not exceeding 20 days. This request, and any action of the Chief Administrative Law Judge in response thereto, may be written or oral.

(2) Under other circumstances the Settlement Judge, following the expiration of the settlement period or at such earlier date that he

determines further negotiations would be fruitless, shall promptly notify the Chief Administrative Law Judge in writing of the status of the case. If he has not approved a full settlement pursuant to § 2200.100 of these rules, such report shall include written stipulations embodying the terms of such partial settlement as has been achieved during the assignment.

(3) At the termination of the settlement period without a full settlement, the Chief Administrative Law Judge shall promptly assign the case to a different Administrative Law Judge for appropriate action on the remaining issues, unless the parties request otherwise. The Settlement Judge shall not discuss the merits of the case with any Administrative Law Judge or other person, nor be called as a witness in any hearing of the case.

(e) Non-reviewability. Any decision concerning the assignment of a particular Settlement Judge or the decision by any party or Settlement Judge to terminate proceedings under this section is not subject to review by, appeal to, or rehearing by any subsequent presiding officer, the Chief Administrative Law Judge, or the Commission.

§ 2200.102 Withdrawal.

A party may withdraw its notice of contest, citation, notification of proposed penalty, or petition for modification of abatement period at any stage of a proceeding. The notice of withdrawal shall be served in accordance with § 2200.7(c) upon all parties and authorized employee representatives that are eligible to elect, but have not elected, party status. It shall also be posted in the manner prescribed in § 2200.7(g) for the benefit of any affected employees not represented by an authorized employee representative who are eligible to elect, but have not elected, party status. Proof of service shall accompany the notice of withdrawal.

§ 2200.103 Expedited proceeding.

(a) When ordered. Upon application of any party or intervenor or upon its own motion, the Commission may order an expedited proceeding. When an expedited proceeding is ordered by the Commission, the Executive Secretary shall notify all parties and intervenors.

(b) Automatic expedition. Cases initiated by employee contests and petitions for modification of abatement period shall be expedited.

(c) Effect of ordering expedited proceeding. When an expedited proceeding is required by these rules or ordered by the Commission, it shall take precedence on the docket of the Judge to whom it is assigned, or on the Commission's review docket, as applicable, over all other classes of cases, and shall be set for hearing or for the submission of briefs at the earliest practicable date.

(d) Time sequence set by Judge. The assigned Judge shall make rulings with respect to time for filing of pleadings and with respect to all other matters, without reference to times set forth in these rules, may order daily transcripts of the hearing, and shall do all other things appropriate to complete the proceeding in the minimum time consistent with fairness.

§ 2200.104 Standards of conduct.

(a) General. All representatives appearing before the Commission and its Judges shall comply with the letter and spirit of the Model Rules of Professional Conduct of the American Bar Association.

(b) Misbehavior before a Judge.

(1) Exclusion from a proceeding. A Judge may exclude from participation in a proceeding any person, including a party or its representative, who engages in disruptive behavior, refuses to comply with orders or rules of procedure, continuously uses dilatory tactics, refuses to adhere to standards of orderly or ethical conduct, or fails to act in good faith. The cause for the exclusion shall be stated in writing, or may be stated in the record if the exclusion occurs during the course of the hearing. Where the person removed is a party's attorney or other representative, the Judge shall suspend the proceeding for a reasonable time for the purpose of enabling the party to obtain another attorney or other representative.

(2) Appeal rights if excluded. Any attorney or other representative excluded from a proceeding by a Judge may, within five days of the exclusion, appeal to the Commission for reinstatement. No proceeding shall be delayed or suspended pending disposition of the appeal.

(c) Disciplinary action by the Commission. If an attorney or other representative practicing before the Commission engages in unethical or unprofessional conduct or fails to comply with any rule or order of the Commission or its Judges, the Commission may, after reasonable notice and an opportunity to show cause to the contrary, and after hearing, if requested, take any appropriate disciplinary action, including suspension or disbarment from practice before the Commission.

§ 2200.105 Ex parte communication.

(a) General. Except as permitted by § 2200.101 or as otherwise authorized by law, there shall be no ex parte communication with respect to the merits of any case not concluded, between any Commissioner, Judge, employee, or agent of the Commission who is employed in the decisional process and any of the parties or intervenors, representatives or other interested persons.

(b) Disciplinary action. In the event an ex parte communication occurs, the Commission or the Judge may make such orders or take such actions as fairness requires. The exclusion of a person by a Judge from a proceeding shall be governed by § 2200.104(b). Any disciplinary action by the Commission, including suspension or disbarment, shall be governed by § 2200.104(c).

(c) Placement on public record. All ex parte communications in violation of this section shall be placed on the public record of the proceeding.

§ 2200.106 Amendment to rules.

The Commission may at any time upon its own motion or initiative, or upon written suggestion of any interested person setting forth reasonable grounds therefor, amend or revoke any of the rules contained herein. The Commission invites suggestions from interested parties to amend or revoke rules of procedure. Such suggestions should be addressed to the Executive Secretary of the Commission at 1825 K Street, N. W., Washington, D. C. 20006.

§ 2200.107 Special circumstances; Waiver of rules.

In special circumstances not contemplated by the provisions of these rules or for good cause shown, the Commission or Judge may, upon application by any party or intervenor or on their own motion, after three days notice to all parties and intervenors, waive any rule or make such orders as justice or the administration of the Act requires.

§ 2200.108 Official Seal of the Occupational Safety and Health Review Commission.

The seal of the Commission shall consist of: A gold eagle outspread, head facing dexter, a shield with 13 vertical stripes superimposed on its breast, holding an olive branch in its claws, the whole superimposed over a plain solid white Greek cross with a green background, encircled by a white band edged in black and inscribed "Occupational Safety and Health Review Commission" in black letters.

SUBPART M - SIMPLIFIED PROCEEDINGS

§ 2200.200   Purpose.

(a) The purpose of this subpart is to provide simplified procedures for resolving contests under the Occupational Safety and Health Act of 1970, so that parties before the Commission may save time and expense while preserving fundamental procedural fairness. The rules shall be construed and applied to accomplish these ends.

(b) Procedures under this subpart are simplified in a number of ways. The major differences between these procedures and those provided in Subparts A through G of the Commission's rules of procedure are the following: (1) Pleadings generally are not required or permitted. Early discussions among the parties will inform the parties of the legal and factual matters in dispute and narrow the issues to the extent possible. (2) Discovery is generally not permitted. (3) The Federal Rules of Evidence do not apply. (4) Interlocutory appeals are not permitted.

§ 2200.201   Application.

The rules in this subpart shall govern proceedings before an Administrative Law Judge when (a) the case is eligible for simplified proceedings under § 2200.202, (b) any party requests simplified proceedings, (c) no party files an objection to the request, and (d) simplified proceedings are not discontinued pursuant to § 2200.204.

§ 2200.202   Eligibility for simplified proceedings.

A case is eligible for simplified proceedings unless it concerns an alleged violation of section 5(a)(1) of the Act (29 U.S.C. 654(a)(1)) or an alleged failure to comply with a standard listed in table A.

Table A

All standards listed are found in Title 29 of the Code of Federal Regulations.

§ 1910.94	§ 1910.96
§ 1910.95	§ 1910.97

Sections 1910.1000 to 1910.1045, and any occupational health standard that may be added to Subpart Z of Part 1910.

§ 1926.52	§ 1926.55
§ 1926.53	§ 1926.57
§ 1926.54	§ 1926.800(c)

§ 2200.203 Commencing simplified proceedings.

   (a) Requesting simplified proceedings.

      (1) Who may request. Any party may request simplified pro-
ceedings.

      (2) When to request. After the Commission receives an employer's
or employees' notice of contest or petition for modification of abatement,
the Executive Secretary shall issue a notice indicating that the case has
been docketed. A request for simplified proceedings, if any, shall be
filed within 10 days after the notice of docketing is received, unless the
notice of docketing states otherwise.

      (3) How to request. A simple statement is all that is necessary.
For example, "I request simplified proceedings" will suffice. The request
shall be filed with the Executive Secretary and served on all of the following: (i) the employer, (ii) the Secretary of Labor, and, (iii) any
authorized employee representatives. The request also shall be posted for
the benefit of any unrepresented affected employees. (To serve the Secretary
of Labor, the request should be mailed to the regional solicitor named in
the notice of docketing.)

      (4) Effect of the request. For those cases eligible under
§ 2200.202, simplified proceedings are in effect when any party requests
simplified proceedings and no party files a timely objection to the request.

   (b) Objecting to simplified proceedings.

      (1) Who may object. Any party may object to a request for simpli-
fied proceedings.

      (2) When to object. An objection shall be filed within 15 days
after the request for simplified proceedings is served.

      (3) How to object. A simple statement is all that is necessary.
For example, "I object to simplified proceedings" will suffice. An
objection shall be filed with the Executive Secretary and served in the
manner prescribed for requests for simplified proceedings in paragraph (a)
(3) of this section.

      (4) Effect of the objection. The filing of a timely objection
shall preclude the institution of simplified proceedings.

   (c) Notice.

      (1) When the period for objecting to simplified proceedings ex-
pires and no objection has been filed, the Commission shall notify all
parties that simplified proceedings are in effect.

(2) When a party files a timely objection to a request for simplified proceedings, the Commission shall notify all parties that the case shall continue under conventional procedures (Subparts A through G).

(d) Time for filing complaint or answer under § 2200.33. The time for filing a complaint or answer shall not run if a request for simplified proceedings is filed. If the Commission later notifies the parties under § 2200.203(c) that the case is to continue under conventional procedures, the period for filing a complaint or answer shall begin upon receipt of the notice.

§ 2200.204 Discontinuance of simplified proceedings.

At any time prior to the commencement of the hearing, a party may move to discontinue application of the simplified proceeding rules to the case. The motion shall be granted if all parties consent or if sufficient reason is shown for application of the conventional rules to the case. If the motion is granted, the parties shall not file a complaint or answer unless ordered to do so by the Judge.

§ 2200.205 Filing of pleadings.

(a) Complaint and Answer. There shall be no complaint or answer in simplified proceedings. If the Secretary has filed a complaint under § 2200.33, a response to an employee contest under § 2200.35, or a response to a petition under § 2200.34, no response to these documents shall be required.

(b) Motions. A primary purpose of simplified proceedings is to eliminate, as much as possible, motions and similar documents. A motion will not be viewed favorably if the subject of the motion has not been first discussed among the parties prior to the conference/hearing.

§ 2200.206 Discussion among parties.

Within a reasonable time before the conference/hearing, the parties shall meet, or confer by telephone, and discuss the following: Settlement of the case; the narrowing of issues; an agreed statement of issues and facts; defenses; witnesses and exhibits; motions; and any other pertinent matter.

§ 2200.207 Conference/Hearing.

(a) The Judge shall schedule and preside over a conference/hearing, which shall be divided into two segments: a conference and a hearing. The Judge may schedule the hearing to occur one or more days after the conference if, in his discretion, he determines that such a schedule would result in more practical or efficient case disposition.

(b) <u>Conference</u>. At the beginning of the conference, the Judge shall enter into the record all agreements reached by the parties as well as defenses raised during the discussion set forth in § 2200.205. The parties and the Judge then shall attempt to resolve or narrow the remaining issues. At the conclusion of the conference, the Judge shall enter into the record any further agreements reached by the parties.

(c) <u>Hearing</u>. The Judge shall hold a hearing on any issue that remains in dispute at the conclusion of the conference. The hearing shall be in accordance with 5 U.S.C. 554.

(1) <u>Evidence</u>. Oral or documentary evidence shall be received, but the Judge may exclude irrelevant or unduly repetitious evidence. Testimony shall be given under oath. The Federal Rules of Evidence shall not apply.

(2) <u>Oral and written argument</u>. Each party may present oral argument at the close of the hearing. Parties wishing to present written argument shall notify the Judge at the conference/hearing so that the Judge may set a reasonable period for the prompt filing of written argument.

§ 2200.208 <u>Reporter present; transcripts</u>.

A reporter shall be present at the conference/hearing. An official verbatim transcript of the hearing shall be prepared and filed with the Judge. Parties may purchase copies of the transcript from the reporter.

§ 2200.209 <u>Decision of the Judge</u>.

(a) The Judge shall issue a written decision in accordance with § 2200.90.

(b) After the issuance of the Judge's decision, the case shall proceed in the conventional manner (Subparts F and G).

§ 2200.210 <u>Discovery</u>.

Discovery, including requests for admissions, shall not be allowed except by order of the Judge.

§ 2200.211 <u>Interlocutory appeals not permitted</u>.

Appeals to the Commission of a ruling made by a Judge which is not the Judge's final disposition of the case are not permitted.

§ 2200.212 Applicability of Subparts A through G.

Sections 2200.6, 2200.33, 2200.34(d)(4), 2200.35, 2200.36, 2200.38, and 2200.75 shall not apply to simplified proceedings. All other rules contained in Subparts A through G of the Commission's rules of procedure shall apply when consistent with the rules in this subpart governing simplified proceedings.

RULES IMPLEMENTING THE EQUAL ACCESS TO JUSTICE ACT

Subpart A -General Provisions

§ 2204.101 Purpose of these rules.

The Equal Access to Justice Act, 5 U.S.C. 504, provides for the award of attorney or agent fees and other expenses to eligible individuals and entities who are parties to certain administrative proceedings (called "adversary adjudications") before the Occupational Safety and Health Review Commission. An eligible party may receive an award when it prevails over the Secretary of Labor, unless the Secretary's position in the proceeding was substantially justified or special circumstances make an award unjust. The rules in this part describe the parties eligible for awards and the proceedings that are covered. They also explain how to apply for awards and the procedures and standards that the Commission uses to make awards.

§ 2204.102 Definitions.

For the purposes of this part,

(a) The term "agent" means any person other than an attorney who represents a party in a proceeding before the Commission pursuant to § 2200.22;

(b) The term "Commission" means the Occupational Safety and Health Review Commission;

(c) The term "EAJ Act" means the Equal Access to Justice Act, 5 U.S.C. 504;

(d) The term "judge" means an Administrative Law Judge appointed by the Commission under 29 U.S.C. 661(i);

(e) The term "OSH Act" means the Occupational Safety and Health Act of 1970, 29 U.S.C. 651-678;

(f) The term "Secretary" means the Secretary of Labor.

§ 2204.103 When the EAJ Act applies.

The EAJ Act applies to adversary adjudications before the Commission pending or commenced on or after August 5, 1985. The EAJ Act also applies to adversary adjudications commenced on or before October 1, 1984, and finally disposed of before August 5, 1985, if an application for an award of fees and expenses, as described in Subpart B of these rules, has been filed with the Commission within 30 days after August 5, 1985.

§ 2204.104 Proceedings covered.

The EAJ Act applies to adversary adjudications before the Commission. These are adjudications under 5 U.S.C. 554 and 29 U.S.C. 659(c) in which the position of the Secretary is represented by an attorney or other representative. The types of proceedings covered are the following proceedings under section 10(c), 29 U.S.C. 659(c), of the OSH Act:

(a) Contests of citations, notifications, penalties, or abatement periods by an employer;

(b) Contests of abatement periods by an affected employee or authorized employee representative; and

(c) Petitions for modification of the abatement periods by an employer.

§ 2204.105 Eligibility of applicants.

(a) To be eligible for an award of attorney or agent fees and other expenses under the EAJ Act, the applicant must be a party to the adversary adjudication. The term "party" is defined in 5 U.S.C. 551(3). The applicant must show that it satisfies the conditions of eligibility set out in this subpart and subpart B.

(b) The types of eligible applicants are as follows:

(1) The sole owner of an unincorporated business who has a net worth of not more than $7 million, including both personal and business interests, and employs not more than 500 employees;

(2) A charitable or other tax-exempt organization described in section 501(c)(3) of the Internal Revenue Code (26 U.S.C. 501(c)(3)) with not more than 500 employees;

(3) A cooperative association as defined in section 15(a) of the Agricultural Marketing Act (12 U.S.C. 1141j(a)) with not more than 500 employees;

(4) Any other partnership, corporation, association, or public or private organization that has a net worth of not more than $7 million and employs not more than 500 employees; and

(5) An individual with a net worth of not more than $2 million.

(c) For the purpose of eligibility, the net worth and number of employees of an applicant shall be determined as of the date the notice of contest was filed, or, in the case of a petition for modification of abatement period, the date the petition was received by the Commission under § 2200.34(d).

(d) An applicant who owns an unincorporated business shall be considered as an "individual" rather than a "sole owner of an unincorporated business" only if the issues on which the applicant prevails are related primarily to personal interests rather than business interests.

(e) For the purpose of determining eligibility under the EAJ Act, the employees of an applicant include all persons who regularly perform services for remuneration for the applicant under the applicant's direction and control. Part-time employees shall be included on a proportional basis.

§ 2204.106 Standards for awards.

(a) A prevailing applicant may receive an award for fees and expenses in connection with a proceeding, or in a discrete substantive portion of the proceedings, unless the position of the Secretary was substantially justified. The position of the Secretary includes, in addition to the position taken by the Secretary in the adversary adjudication, the action or failure to act by the Secretary upon which the adversary adjudication is based. The burden of persuasion that an award should not be made to an eligible prevailing applicant because the Secretary's position was substantially justified is on the Secretary.

(b) An award shall be reduced or denied if the applicant has unduly or unreasonably protracted the proceeding. An award shall be denied if special circumstances make an award unjust.

§ 2204.107 Allowable fees and expenses.

(a) Awards shall be based on rates customarily charged by persons engaged in the business of acting as attorneys, agents and expert witnesses, even if the services were made available without charge or at a reduced rate to the applicant.

(b) An award for the fee of an attorney or agent under these rules shall not exceed $75 per hour, unless the Commission determines by regulation that an increase in the cost of living or a special factor, such as the limited availability of qualified attorneys or agents for Commission proceedings, justifies a higher fee. An award to compensate an expert witness shall not exceed the highest rate at which the Secretary pays expert witnesses. However, an award may include the reasonable expenses of the attorney, agent or witness as a separate item, if the attorney, agent or witness ordinarily charges clients separately for such expenses.

(c) In determining the reasonableness of the fee sought for an attorney, agent or expert witness, the Commission shall consider the following:

(1) If the attorney, agent, or witness is in private practice, his or her customary fee for similar services, or, if an employee of the applicant, the fully allocated cost of the services;

(2) The prevailing rate for similar services in the community in which the attorney, agent, or witness ordinarily perform services;

(3) The time actually spent in the representation of the applicant;

(4) The time reasonably spent in light of the difficulty or complexity of the issues in the proceeding; and

(5) Such other factors as may bear on the value of the services provided.

(d) The reasonable cost of any study, analysis, engineering report, test, project or similar matter prepared on behalf of a party may be awarded, to the extent that the charge for the service does not exceed the prevailing rate for similar services, and the study or other matter was necessary for preparation of the applicant's case.

§ 2204.108 Delegation of authority.

The Commission delegates to each judge authority to entertain and, subject to § 2204.309, take final action on applications for an award of fees and expenses arising from the OSH Act cases that are assigned to the judge under section 12(j) of the OSH Act, 29 U.S.C. 661(i). However, the Commission retains its right to consider an application for an award of fees and expenses without assignment to a judge or to assign such application to a judge other than the one to whom the underlying OSH Act case is assigned. When entertaining an application for an award of fees and expenses pursuant to this section, each judge is authorized to take any action that the Commission may take under this part, with the exception of actions provided in §§ 2204.309 and 2204.310.

<u>Subpart B - Information Required From Applicants</u>

§ 2204.201 <u>Contents of application.</u>

(a) An application for an award of fees and expenses under the EAJ Act shall identify the applicant and the proceeding for which an award is sought. The application shall show that the applicant has prevailed and identify the position of the Secretary that the applicant alleges was not substantially justified. The application also shall state the number of employees of the applicant and describe briefly the type and purpose of its organization or business.

(b) The application also shall include a statement that the applicant's net worth does not exceed $2 million (if an individual) or $7 million (for all other applicants). However, an applicant may omit this statement if:

(1) It attaches a copy of a ruling by the Internal Revenue Service that it qualifies as an organization described in section 501(c)(3) of the Internal Revenue Code (26 U.S.C 501(c)(3)) or, in the case of a tax-exempt organization not required to obtain a ruling from the Internal Revenue Service on its exempt status, a statement that describes the basis for the applicant's belief that it qualifies under such section; or

(2) It states that it is a cooperative association as defined in section 15(a) of the Agricultural Marketing Act (12 U.S.C. 1141j(a)).

(c) The application shall state the amount of fees and expenses for which an award is sought.

(d) The application also may include any other matters that the applicant wishes the Commission to consider in determining whether and in what amount an award should be made.

(e) The application shall be signed by the applicant or an authorized officer or attorney of the applicant. It also shall contain or be accompanied by a written verification under oath or under penalty of perjury that the information provided in the application is true.

§ 2204.202 <u>Net worth exhibit.</u>

(a) Each applicant except a qualified tax-exempt organization or cooperative association shall provide with its application a detailed exhibit showing the net worth of the applicant as of the date specified by § 2204.105(c). The exhibit may be in any form convenient to the applicant that provides full disclosure of the applicant's assets and liabilities and is sufficient to determine whether the applicant qualifies under the standards in this part. The Commission may require an applicant to file additional information to determine its eligibility for an award.

(b)(1) The new worth exhibit shall be included in the public record of the proceeding except as provided in subsection (b)(2) of this section.

(2) An applicant that objects to public disclosure of information in any portion of the exhibit and believes there are legal grounds for withholding it from disclosure may submit that portion of the exhibit in a sealed envelope labeled "Confidential Information," accompanied by a motion to withhold the information from public disclosure. The motion shall describe the information sought to be withheld and explain, in detail, why it falls within one or more of the specific exemptions from mandatory disclosure under the Freedom of Information Act, 5 U.S.C. 552(b) (1) - (9), why public disclosure of the information would adversely affect the applicant, and why disclosure is not required in the public interest. The material in question shall be served on the Secretary but need not be served on any other party to the proceeding. If the Commission finds that the information should not be withheld from disclosure, it shall be placed in the public record of the proceeding. Otherwise, any request to inspect or copy the exhibit shall be disposed of in accordance with the Commission's procedures under the Freedom of Information Act, Part 2201.

§ 2204.203 Documentation of fees and expenses.

The application shall be accompanied by full documentation of the fees and expenses, including the cost of any study, analysis, engineering report, test, project of similar matter, for which an award is sought. A separate itemized statement shall be submitted for each professional firm or individual whose services are covered by the application, showing the hours spent in connection with the proceeding by each individual, a description of the specific services performed, the rate at which each fee has been computed, any expenses for which reimbursement is sought, the total amount claimed, and the total amount paid or payable by the applicant or by any other person or entity for the services provided. The Commission may require the applicant to provide vouchers, receipts, or other substantiation for any fees or expenses claimed.

Subpart C - Procedures for Considering Applications

§ 2204.301 Filing and service of documents.

An application for an award and any other pleading or document related to an application shall be filed and served on all parties to the proceeding in accordance with §§ 2200.7 and 2200.8, except as provided in § 2204.202(b) for confidential financial information.

§ 2204.302   When an application may be filed.

(a) An application may be filed whenever an applicant has prevailed in a proceeding or in a discrete substantive portion of the proceeding, but in no case later than thirty days after the Commission's final disposition of the proceeding.

(b) If Commission review is sought or directed of a judge's decision as to which an application for a fee award has been filed, proceedings concerning the award of fees shall be stayed until there is a final Commission disposition of the case and the period for seeking review in a court of appeals expires.

(c) If review of a Commission decision, or any item or items contained in that decision, is sought in the court of appeals under section 11 of the OSH Act, 29 U.S.C. 660, an application for an award filed with the Commission with regard to that decision shall be dismissed under 5 U.S.C. 554(c)(1) as to the item or items of which review is sought. If the petition for review in the court of appeals is thereafter withdrawn, the applicant may reinstate its application before the Commission within thirty days of the withdrawal.

(d) For purposes of this section, the date of final disposition is:

(1) The date on which the order of the Judge disposing of the case becomes final under section 12(j) of the OSH Act, 29 U.S.C. 661(i); or

(2) The date on which the order of the Commission affirming, modifying or vacating the Secretary's citation or proposed penalty or directing other appropriate relief becomes final under section 10(c) of the OSH Act, 29 U.S.C. 659(c).

§ 2204.303   Answer to application.

(a) Within 30 days after service of an application, the Secretary shall file an answer to the application.

(b) If the Secretary and the applicant believe that the issues in the fee application can be settled, they may jointly file a statement of their intent to negotiate a settlement. The filing of this statement shall extend the time for filing an answer for an additional 30 days, and further extensions may be granted upon request.

(c) The answer shall explain in detail any objections to the award requested and identify the facts relied on in support of the Secretary's position. If the answer is based on any alleged facts not already in the record of the proceeding, the Secretary shall include with the answer either supporting affidavits or a request for further proceedings under § 2204.307.

§ 2204.304   Reply.

Within 15 days after service of an answer, the applicant may file a
reply. If the reply is based on any alleged facts not already in the
record of the proceeding, the applicant shall include with the reply
either supporting affidavits or a request for further proceedings under
§ 2204.307.

§ 2204.305   Comments by other parties.

Any party to a proceeding other than the applicant and the Secretary
may file comments on an application within 30 days after it is served or
on an answer within 15 days after it is served. A commenting party may
not participate further in proceedings on the application unless the
Commission determines that the public interest requires such participation in order to permit full exploration of matters raised in the comments.

§ 2204.306   Settlement.

The applicant and the Secretary may agree on a proposed settlement
of the award before final action on the application, either in connection
with a settlement of the underlying proceeding, or after the underlying
proceeding has been concluded. If a prevaling party and the Secretary
agree on a proposed settlement of an award before an application has been
filed, the application shall be filed with the proposed settlement.

§ 2204.307   Further proceedings.

(a)(1) The determination of an award shall be made on the basis of
the record made during the proceeding for which fees and expenses are
sought, except as provided in paragraphs (a)(2) and (a)(3) of this section.

(2) On the motion of a party or on the judge's own initiative, the
judge may order further proceedings, including discovery and an evidentiary hearing, as to issues other than substantial justification (such as
the applicant's eligibility or substantiation of fees and expenses).

(3) If the proceeding for which fees and expenses are sought ended
before the Secretary had an opportunity to introduce evidence supporting
the citation or notification of proposed penalty (for example, a citation
was withdrawn or settled before an evidentiary hearing was held), the
Secretary may supplement the record with affidavits or other documentary evidence of substantial justification.

§ 2204.308   Decision.

This preparation and issuance of decision shall be in accordance
with § 2200.90. Additionally, the judge's decision shall include written
findings and conclusions on the applicant's eligibility and status as a
prevailing party and an explanation of the reasons for any difference

between the amount requested and the amount awarded. The decision shall
also include, if at issue, findings on whether the Secretary's position
was substantially justified, whether the applicant unduly protracted the
proceedings, or whether special circumstances make an award unjust.

§ 2204.309 Commission review.

Commission review shall be in accordance with §§ 2200.91 and
2200.92. The applicant, the Secretary, or both may seek review of the
judge's decison on the fee application, and the Commission may grant such
petitions for review or direct review of the decision on the Commission's
own initiative. The Commission delegates to each of its members the
authority to order review of a Judge's decision concerning a fee application. Whether to review a decision is a matter within the discretion
of each member of the Commission. If the Commission does not direct
review, the judge's decision on the application shall become a final
decision of the Commission 30 days after it is received and docketed by
the Executive Secretary of the Commission. If review is directed, the
Commission shall issue a final decision on the application or remand the
application to the judge for further proceedings.

§ 2204.310 Waiver.

After reasonable notice to the parties, the Commission may waive,
for good cause shown, any provision contained in this part as long as
the waiver is consistent with the terms and pupose of the EAJ Act.

§ 2204.311 Payment of award.

An applicant seeking payment of an award shall submit to the officer
designated by the Secretary a copy of the Commission's final decision
granting the award.

Table of Cases

Cases are referenced to chapter and footnote number(s), e.g., 2:21 indicates the case is cited in chapter 2, footnote 21. Names of cases discussed in text are italicized; the corresponding footnote number indicating the location of the discussion in text also is italicized to distinguish the discussion from other footnotes which merely cite the case.

917

Central Meat Co.—*Contd.*
Comm'n 1977)   3: 120
Central Mine Equip. Co., In re Inspection of, 7 OSHC 1185 (E.D. Mo.), *vacated on other grounds,* 608 F.2d 719, 7 OSHC 1907 (8th Cir. 1979)   7: 148, 165, 169, 171, 172, 175, 219, 221; 18: 31
Central Ohio Coal Co.; Secretary of Labor v., 2 MSHC 1766 (FMSHRC No. LAKE 81-78, 1982)   26: 160
Central Soya of Puerto Rico, 8 OSHC 2074 (Rev. Comm'n 1980)   3: 127
Central Tire Co., 1 OSHC 3315 (Rev. Comm'n Nos. 720 & 737, 1974)   5: 48
Cerro Copper Prods., Co., In re Establishment Inspection of, 752 F.2d 280 (7th Cir. 1985)   7: 168
Cerro Metal Products Div., 12 OSHC 1821 (Rev. Comm'n 1986)   4: 138
—v. Marshall, 620 F.2d 964 (3d Cir. 1980)   7: 92, 97
Certified Welding Corp.; Usery v., 7 OSHC 1069 (10th Cir. 1980)   22: 57
Chacon v. Phelps-Dodge Corp., 2 MSHC 1505 (FMSHRC No. WEST 79-349-DM, 1981), *rev'd,* 709 F.2d 51, 2 MSHC 2145 (D.C. Cir. 1983)   26: *251, 253*
Chamber of Commerce v. OSHA, 636 F.2d 464, 8 OSHC 1648 (D.C. Cir. 1980)   20: 202; 22: *16*
Champlin Petroleum Co. v. OSHRC, 593 F.2d 647, 7 OSHC 1241 (5th Cir. 1979)   4: 3, 132; 15: 128
Chanal Plastics Corp., 9 OSHC 1844 (Rev. Comm'n 1981)   13: 48
Chapman & Stephens Co., 5 OSHC 1395 (Rev. Comm'n 1977)   6: 86, 87
Chavis v. Finnlines Ltd. O/Y, 576 F.2d 1072 (4th Cir. 1978)   25: 24
Chemical Workers, In re, 830 F.2d 369, 13 OSHC 1402 (D.C. Cir. 1987)   20: 357
Chesapeake & O. Ry., 3 OSHC 1762 (Rev. Comm'n 1975)   13: *25*
Chevron, USA v. National Resources Defense Council, 467 U.S. 837 (1984)   20: 105
Chicago Bridge & Iron Co.; Brennan v., 514 F.2d 1082, 3 OSHC 1056 (7th Cir. 1975)   8: 14; 18: 24

Chicago, M & St. P. Ry. v. Ross, 112 U.S. 377 (1884)   1: 22, 38, 139
Chicago & N.W. Ry. v. Booten, 57 F.2d 786 (8th Cir. 1932)   1: 124
Childers v. International Harvester Co., 569 S.W.2d 675 (Ky. Ct. App. 1977)   25: 99
Chlorine Inst. v. OSHA, 613 F.2d 120, 8 OSHC 1031 (5th Cir.), *cert. denied,* 449 U.S. 826 (1980)   3: 301; 20: 202; 21: *118*
Chobee Steel Erectors, 8 OSHC 1094 (Rev. Comm'n J. 1979)   5: 116
Christine; United States v., 687 F.2d 749 (3d Cir. 1982)   7: 238
Chromalloy Am. Corp., 8 OSHC 1188 (Rev. Comm'n 1979)   3: 133
Chrysler Corp., 7 OSHC 1578 (Rev. Comm'n 1979)   5: *43*
—v. Brown, 441 U.S. 281 (1979)   20: 286
Church Drilling Inc.; Marshall v., 9 OSHC 1391 (W.D. Mo. 1981)   18: 34
Cinderella Career & Finishing Schools v. FTC, 425 F.2d 583 (D.C. Cir. 1970)   14: 89; 20: 233
Circle T. Drilling Co., 8 OSHC 1681 (Rev. Comm'n 1980)   14: 47
Citizens to Preserve Overton Park v. Volpe, 401 U.S. 402 (1971)   20: 76
City Disposal Sys.; NLRB v., 465 U.S. 822, 115 LRRM 3193 (1984)   22: *39*; 26: 222
City Mills Co., 5 OSHC 1129 (Rev. Comm'n 1977)   13: 75
Clark, Allen, Inc., 1 OSHC 3047 (Rev. Comm'n J. 1973)   11: 105
Clark, John T., & Son, 4 OSHC 1913 (Rev. Comm'n 1976)   6: 76
Clarkson Constr. Co. v. OSHRC, 531 F.2d 451, 3 OSHC 1880 (10th Cir. 1976)   8: 30; 11: 25; 18: 66
Clary v. Ocean Drilling & Exploration Co., 429 F. Supp. 905, 5 OSHC 1278 (W.D. La. 1977), *aff'd,* 609 F.2d 1120, 7 OSHC 2209 (5th Cir. 1980)   6: 65; 15: 45; 25: *23*; 27: 114
Claude Neon Fed. Co., 5 OSHC 1546 (Rev. Comm'n 1977)   6: 28
Cleveland Consol. v. OSHRC, 649 F.2d 1160, 9 OSHC 2043 (5th Cir. 1981)   15: 90, 92, 96
Cleveland Elec. Illuminating Co.; United States v., No. M80-2128

Hudson Fabricating Co., 8 OSHC 1647 (Rev. Comm'n J. 1980) 9: 56; 10: *46*

Huffines Steel Co.; Marshall v., 478 F. Supp. 986 (N.D. Tex.), *complaint dismissed on other grounds*, 488 F. Supp. 995 (N.D. Tex. 1979), *aff'd without opinion*, 645 F.2d 288 (5th Cir. 1981) 7: 97, 106, 108

Hughes, B.J., Inc., 10 OSHC 1545 (Rev. Comm'n 1982) 6: 32, 46; 15: 12, *166*

—7 OSHC 1471 (Rev. Comm'n 1979) 8: 23, *24*; 13: *10, 33, 52, 55*

Hughes Tool Co., 7 OSHC 1666 (Rev. Comm'n Nos. 78-1490 & 78-1932, 1979) 3: 120

Hullenkremer Constr. Co., 6 OSHC 1469 (Rev. Comm'n 1978) 11: 51; 16: *60*

Hurlock Roofing Co., 7 OSHC 1867 (Rev. Comm'n 1979) 14: 145

—7 OSHC 1108 (Rev. Comm'n 1979) 15: 106

Hutchinson v. York, Newcastle & Berwick Ry., 155 Eng. Rep. 150 (1850) 1: *15, 16*

Hydraform Prods. Corp., 7 OSHC 1995 (Rev. Comm'n J. 1979) 11: 66

Hydrate Battery Corp., 2 OSHC 1719 (Rev. Comm'n 1975) 9: 20; 14: 124

Hyman, George, Constr. Co. v. OSHRC, 582 F.2d 834, 6 OSHC 1855 (4th Cir. 1978) 9: 40; 16: 111

I

IBEW, Local 1031 (Stewart-Warner), 7 OSHC 1674 (Rev. Comm'n J. 1979), *vacated*, 8 OSHC 1316 (Rev. Comm'n 1980) 10: 44

IMC Chem. Group, 6 OSHC 2075 (Rev. Comm'n 1980) 17: 25

I.T.O. Corp. v. OSHRC, 540 F.2d 543, 4 OSHC 1574 (1st Cir. 1976) 6: 75; 25: 49

ITT Grinnell Corp., 11 OSHC 1464 (Rev. Comm'n 1983) 10: 44

—v. Donovan, 744 F.2d 344, 11 OSHC 2257 (3d Cir. 1984) 16: 25

Idaho Forest Indus., 2 OSHC 3147 (Rev. Comm'n 1974) 6: 170, 175

Ideal of Idaho, 2 OSHC 3171 (Rev. Comm'n J. 1974) 11: 74

Illinois v. Film Recovery Sys., [1984–1985] OSHD ¶27,356 (Cook County Cir. Ct. 1985) 12: *45*

Illinois v. Gates, 462 U.S. 213 (1983) 7: *143*, 144

Illinois Cent. Gulf R.R., [1977–1978] OSHD ¶22,148 (Rev. Comm'n 1977) 3: 59

Illinois Cent. Ry. Co. v. Gilbert, 157 Ill. 354, 41 N.E. 724 (1895) 1: 126

Imbus Roofing Co., 8 OSHC 2166 (Rev. Comm'n J. 1980) 4: 70

Immigration & Naturalization Serv. v. Chadha, 462 U.S. 919 (1983) 19: 156, *275*

Imperial Lumber Co., 4 OSHC 1908 (Rev. Comm'n 1976) 13: 32, 68

Indiana v. Komomo Tube Co., 10 OSHC 1159 (Ind. Ct. App. 1981) 23: 37

Industrial Steel Erectors, 1 OSHC 1497 (Rev. Comm'n 1974) 15: 108

Industrial Tile v. Stewart, 388 So.2d 171 (Ala. 1980), *cert. denied*, 449 U.S. 1081 (1981) 25: 14

Industrial Union Dep't
—v. *American Petroleum Inst.*, 448 U.S. 607, 8 OSHC 1586 (1980) 2: 54; 3: 31, 309, 311; 19: 77, 82, *87*, *282–283*; 20: 3, 4, 18, *92–94*, 107, 115, *119–122*, 172, 174, 324; 24: 1
—v. *Bingham*, 570 F.2d 965, 6 OSHC 1107 (D.C. Cir. 1977) 19: 26, *47–51*, 59, 210, 218
—v. *Hodgson*, 499 F.2d 467, 1 OSHC 1631 (D.C. Cir. 1974) 3: 347; 19: 40, *68, 74, 89*, 244, 279, 280; 20: 4, *81–83, 104, 109, 143–149, 154, 161*, 174, *229–230, 256–259*, 266, *271*, 314, 324; 23: 9; 24: 78

Ingersoll Rand Co. v. Donovan, 540 F. Supp. 222 (M.D. Pa. 1982) 7: 93, 154, 158, 165, 207

Inland Terminals; Donovan v., 3 MSHC 1893 (S.D. Ind. 1985) 26: 37

Inland Steel Co., 12 OSHC 1968 (Rev. Comm'n 1986) 4: 132, 138

—*In re*, 492 F. Supp. 1310, 8 OSHC 1725 (N.D. Ind. 1980) 7: 190, 191, 192, 194; 24: 55, 56, 59, 68, *71*

Interboro Contractors, 157 NLRB 1295, 61 LRRM 1537 (1966) 22: *38*

Intercounty Constr. Co. v. OSHRC, 522 F.2d 777, 3 OSHC 1337 (4th

Plum Creek Lumber Co.—Contd.
1979) 7: 126, 128, 133, 203, *208*, *209*, 210, *211*, 213; 24: 66
Pohl, R.A., Constr. Co. v. Marshall, 640 F.2d 266, 9 OSHC 1224 (10th Cir. 1981) 8: 40
Poller v. Columbia Broadcasting Sys., 368 U.S. 464 (1962) 14: 26
Port Allen Marine Serv.; United States v., Magis. No. 81-77A U.S. (M.D. La. 1981) 12: 15
Porter v. American Optical Corp., 641 F.2d 1128 (5th Cir.), *cert. denied*, 454 U.S. 1109 (1981), *reh'g denied*, 455 U.S. 1009 (1982) 25: 32
Porter, H.K., Co., 1 OSHC 1600 (Rev. Comm'n 1974) 10: 16; 16: 24
Potlatch Corp., 7 OSHC 1061 (Rev. Comm'n 1979) 9: *32*, 34, *35*, *36*, 37, 47; 11: 120; 16: 105
Poughkeepsie Yacht Club, 7 OSHC 1725 (Rev. Comm'n 1979) 15: 18, 21
Powell v. Globe Indus., 431 F. Supp. 1096, 5 OSHC 1250 (W.D. Ohio 1977) 22: 57, 64
Power City Elec.; Marshall v., [1979] OSHC ¶23,947 (E.D. Wash. 1979) 22: 11
Power Plant Div., Brown & Root, Inc. v. Donovan, 659 F.2d 1291, 10 OSHC 1066 (5th Cir. 1981), *rehearing*, 672 F.2d 111, 10 OSHC 1529 (1982) 18: 43
Power Sys. Div., United Technologies Corp., 9 OSHC 1813 (Rev. Comm'n 1981) 14: 142, 145, *147*
Pratico v. Portland Terminal Co., 783 F.2d 255, 12 OSHC 1567 (1st Cir. 1985) 25: *20*
Pratt & Whitney Aircraft, see United Technologies Corp.
Prestolite Co., Div. of Eltra Co., 1 OSHC 3190 (Rev. Comm'n J. 1973) 10: 39
Prestressed Sys., 9 OSHC 1864 (Rev. Comm'n 1981) 9: 2, 27; 14: 143
Priestly v. Fowler, 150 Eng. Rep. 1030 (1838) 1: 8, 11, *12–13*, *36*
Price v. Jim Walter Resources, Inc., 4 MSHC 1475 (FMSHRC No. SE 87-87-D, 1987), *appeal filed*, No. 87-7484 (11th Cir. Aug. 8, 1987) 26: 260
Price, James L., 4 OSHC 1024 (Rev. Comm'n 1976) 3: 172
Prokosch, A., & Sons Sheet Metal, 8 OSHC 2077 (Rev. Comm'n 1980) 3: 90; 15: 12, 186–187; 16: 78
Prudential Lines, 3 OSHC 1532 (Rev. Comm'n 1975) 27: 114
Pruette v. Precision Plumbing, 27 Ariz. App. 288, 554 P.2d 655 (1976) 25: 8, 9
Psaty Fuhrman, Inc., 2 OSHC 1006 (Rev. Comm'n 1974) 11: 56
Public Citizen v. Steed, 733 F.2d 93 (D.C. Cir. 1984) 19: 245, 253
Public Citizen Health Research Group
—*v. Auchter*, 554 F. Supp. 242, 11 OSHC 1049 (D.D.C.), *rev'd in part*, 702 F.2d 1150, 11 OSHC 1209 (D.C. Cir. 1983) 19: 23, *50–54*; 20: 62, 69, 325, 328, *348*, *349–352*, 366
—*v. Brock*, ___ F.2d ___, 13 OSHC 1362 (D.C. Cir. 1987) 19: *8*, *296*; 20: 143
—*v. Rowland*, 12 OSHC 1183 (D.C. Cir. 1986) 20: *67–70*
——No. 84-1252 (D.C. Cir. 1984) 19: 201, *211*, *243*, 245–251, 253, 254–255, 256–269, 270–277
—*v. Tyson*, 796 F.2d 1479, 12 OSHC 1905 (D.C. Cir. 1986) 3: 316; 19: 7, 214, *278–295*; 20: 93, 111, *132–141*, 238, 324
——No. 81-2343 (D.D.C. 1984) 19: *223–224*
Puffer's Hardware v. Donovan, 742 F.2d 12, 11 OSHC 2197 (1st Cir. 1984) 9: 2
Puget Sound Tug & Barge, 9 OSHC 1764 (Rev. Comm'n 1981), *appeals dismissed*, Nos. 81-7405 & 81-7406 (9th Cir. Mar. 21, 1983), No. 81-4243 (5th Cir. July 5, 1983) 5: *24*, *25*; 15: *46*, 48
Pugh, R.T., Motor Transp., 3 OSHC 1592 (Rev. Comm'n J. 1975) 5: 51
Puterbaugh Enters., 2 OSHC 1030 (Rev. Comm'n 1974) 5: 60; 16: 76

Q

Quaker Oats, 69 LA 727 (1977) 22: 52
Quality Prods., In re Worksite Inspection of, 6 OSHC 1663 (D.R.I.

3,3'-Dichlorobenzidine (DCB)
 standard 99, 160, 509 n.26,
 607, 620–621
4-Dimethylaminoazobenzene
 standard 99, 161
"Diminution of safety" defense 764,
 765, 768
Dip tanks 89, 90
Discipline and discharge (*see*
 Discrimination against
 employees)
Disclosure of information (*see*
 Confidential information;
 Hazard communication
 standards; Records access)
Discovery 363, 366, 367–373, 393,
 473, 647, 768
Discrimination against
 employees 663–678
 arbitration remedy 673–675
 burden of proof 676–677
 employee work refusals 667–670,
 772, 774–778, 791
 federal employees 814
 fetal protection policy 125–128,
 819–812
 FLSA coverage 816
 handicap 773, 822–824
 industries regulated by other federal
 agencies 664–665
 legal procedures 675–676, 792
 miners 760, 771–780, 782, 792
 NLRA coverage 670–673
 order of proof 775
 OSHA coverage 55, 663–664, 792
 private right of action 718–719,
 780, 792
 protected conduct 665–667
 racial 821
 railroad employees 802
 remedies 677–678
 state plans 688
Disqualification of ALJ or Review
 Commission member 379–
 380, 458–459
District of Columbia 682 n.12
District of Columbia Industrial Safety
 Act 822
Diving standard (*see* Commercial
 diving standard)
Dockboards 79
Doctor-patient privilege 394, 703–704
Dominick, Peter 39, 40, 112, 443, 444
Drills 85
Drinking water 806
Drug standards 809

Dual capacity doctrine 730
Duty of fair representation 719–720
Duty owed to public 730

E

Early (start-up) standards 44–45, 50,
 60–62, 506–508, 569–570,
 656–657
Education and Labor Committee,
 House 40, 111, 445
Effective dates 50–51
 ethylene oxide standard 546–547
 final orders of Review
 Commission 462
 start-up standards 569 n.13
Egress 79–80
Electric transmission equipment 170
Electrical protective devices 70–71
Electrical systems 80–81
Emergency temporary standards 45,
 51–52, 64 n.45, 506, 509–511,
 548, 569, 585, 620–623, 742,
 807–808
Emission control standards 806
Employee representatives
 abatement date, right to contest
 employer petition for
 modification of 479
 abatement period, right to
 contest 469–471
 access to illness and injury
 records 147
 citations sent to 212
 discrimination prohibited 664, 772,
 792
 identification requirement 763
 inspection participation 213–214
 liability for employee injury 719–
 721
 mandamus actions in imminent
 danger situations 479–480
 monitoring of training
 programs 787–788
 notice requirements to 156, 157,
 314, 763
 party status at hearings 471–474
 regulatory definition 468 n.1
 request for NIOSH
 investigation 705–706
 Review Commission decisions, right
 to seek judicial review
 of 478–479
 settlement agreements, no right to
 request review of 474–477

P

Paper mills (*see* Pulp, paper, and paperboard mills)
Passageways and exits   77, 79–80, 86
Pattern-of-violations notices and orders   752–754, 762, 791–792
Peak concentrations   98
Peel, Robert   16
Peer reviewers   552
Penalties (*see also* Civil penalties; Criminal penalties)
   for retaliation   55
   for trade secret disclosure   210
Pennsylvania   18, 20
Performance standards   65, 103, 167
Permanent variances   639, 640
Permissible exposure limit (PEL)   98, 101, 504 n.3
   feasibility constraints   594–602
   health need determinations   586–594
"Person," definition   664
Personal injury actions   712–716
Personal protective equipment   64, 86, 93
   as abatement method   94, 98, 279–280, 544–546
   maritime and longshore industries   173, 174, 177
   mine safety   741
   NIOSH certification   699, 700, 710
   standard challenged   428–428, 466
   standards   61, 68–74, 103–104
Personal samplers   240–241, 707
Personnel Management, Office of   822
Personnel referral services   141
Pesticides
   child farm labor   816
   EPA regulation   805–806
   OSHA standards   509 n.26, 510, 512, 585, 620, 621
Peterloo (Manchester) Massacre   14
Petitions
   for exception to BLS recordkeeping requirements   156
   for judicial review   286–287, 483–484, 568, 767
   for modification of abatement date   156, 281–282, 479, 638
   for modification of application of standard   764–765
   for discretionary review   766–767
Phosphorus exposure   21
Photographing of documents   768

Physician-patient privilege   394, 703–704
Physicians   613, 615–616
Pipeline regulation   799, 800
Pittsburgh Survey   19–20
Pittsburgh University   401–402
Plastics industry   83
Platforms   78, 80, 436, 811
Pleadings   359–361
Political subdivisions of states   400, 401
Portable power tools   85
Possibiity test   129, 131, 132
Posting requirement (*see* Notice)
Power presses   82, 84
Power transmission apparatus   82, 84–95
Preambles   517–526, 538, 601, 607, 621, 626
Preemption
   OSHA neutrality   711–712
   Sec. 4(b)(1) jurisdiction preemption   47, 66–67, 96, 141–142, 365, 402–411, 664–665, 794–798, 801, 804, 805, 812, 813, 817–818
   of Sec. 5(a)(1)   437–438
   of state law   571, 720, 822
Pregnant women   125–128, 605, 819–821
Prehearing conferences   363–364, 647
Prehearing orders   363, 383, 384
Preliminary Regulatory Impact Analysis   523–525, 601
Preponderance-of-evidence standard   376, 398, 458
Presumptions   388
Prior statements   389
Privacy of records (*see* Confidential information; Records access)
Privileged information   368, 393–395, 703–704
Probable cause standard   224–232, 324–325
Procedural defenses   431–432
Procedural invalidity of standard   426–427
Product clause   686–687
Product liability   716–718
Programmed inspections   199, 200, 209, 210
Programmed related inspections   199–200
Proof requirements (*see also* Evidence)
   burden of proof   376, 397–398, 425–426, 433, 676–677